Discrete Time Systems and Signal Processing

S. Palani

Discrete Time Systems and Signal Processing

Second Edition

Ane Books Pvt. Ltd.

Springer

S. Palani
National Institute of Technology
Tiruchirapalli, Tamil Nadu, India

ISBN 978-3-031-32420-8 ISBN 978-3-031-32421-5 (eBook)
https://doi.org/10.1007/978-3-031-32421-5

Jointly published with ANE Books India
The print edition is not for sale in South Asia (India, Pakistan, Sri Lanka, Bangladesh, Nepal and Bhutan) and Africa. Customers from South Asia and Africa can please order the print book from: ANE Books Pvt.Ltd.
ISBN of the Co-Publisher's edition: 978-9-383-65627-1

This Springer imprint is published by the registered company Springer Nature Switzerland AG
The registered company address is: Gewerbestrasse 11, 6330 Cham, Switzerland

PINGALA (200BC) in India invented that any number can merely be represented by the binary system using 1s and 0s. A spiritual significance of this is that 1 represented a symbol for GOD while 0 represented nothingness. God created the universe out of nothing. This book is dedicated to the father of the Universe, THE GOD.

Preface to the Second Edition

We have ventured to bring out the second edition of the book titled 'Discrete Time System and Signal Processing' in the new form due to the success and wide patronage extended to the previous edition by the members of teaching faculty and students community. The first edition was published with the title 'Digital Signal Processing' for EEE and E&I branches. The present edition, as in the previous edition, covers strictly the undergraduate syllabus in Discrete Time System and Signal Processing prescribed by Anna University. A large number of numerical problems have been included in the first chapter which describes signals and system representation.

The contents of chapter two include z-transform, Discrete time Fourier transform and linear convolution. The numerical examples worked out on these topics in the previous edition are retained here. The third chapter describes DFT and Computation. A large number of problems worked out in the previous edition on the topics of DFT and FFT are retained. In the fourth chapter on the design of IIR digital filter, the contents in the first edition are retained and a few numerical problems have been worked out and added. The contents of the Design of FIR filter and Digital Signal Processors, which form the fifth and sixth chapters, respectively, are the same as given in the first edition. A few errors which appeared in the first edition have been removed. A few topics which are included in the new syllabus have been given detailed descriptions in the new edition. Solution using MATLAB program and other topics which have no relevance to the prescribed syllabus now have been removed for want of space.

The author takes this opportunity to thank ANE Books Pvt. Ltd for taking up this difficult job. I would also like to express my gratitude to Shri Sunil Sexana, Managing Director, ANE Books Pvt. Ltd., India, and to Shri A. Rathinam, General Manager, for the encouragement given to me for writing a series of books on various titles.

Tiruchirapalli, India

Dr. S. Palani

Preface to the First Edition

Signal processing is all about taking a signal, applying some changes to it and then getting a new signal out. The change might be amplification or filtration or something else, but nearly all electronic circuits can be considered as signal processors. Thus, the signal processor might be composed of discrete components like capacitors and resistors, or it could be a complex integrated circuit, or it could be a digital system that accepts a signal on its input and outputs the changed signal. Digital Signal Processing (DSP) is the processing of signals by digital means. The term 'digital' comes from 'digit', meaning a number and so 'digital' literally means numerical. A signal carries a stream of information representing anything from stock prices to data from a remote sensing satellite. If they are represented in the form of a stream of numbers, they are called digital signals. The processing of a digital signal is done by Digital Signal Processor (DSP) by performing numerical calculations. Digital signal processors require several things to work properly. The processor should be fast enough with enough precision to support the required mathematics it needs to implement. It requires memory to store programming, samples, intermediate results and final results. It also requires A/D and D/A converters to bring real signals into and out of the digital domain. Further, it requires programming to do the job.

The main application of DSP are audio signal processing, audio compression, digital image processing, video compression, speech processing, speech recognition, digital communications, Radar, Sonar, Seismology and biomedicine. Specific examples include speech compression and transmission in digital mobile phones, room matching equalization of sound, analysis and control of industrial processes, seismic data processing, medical imaging such as CAT scans and MRI, MP3 compression, image manipulation, computer-generated animations in movies, high fidelity loudspeaker cross overs and equalization, audio effects, *etc.* Digital signal processing is often implemented using specialized microprocessors such as the DSP 56000, the TMS320 or the SHARC. Multi-core implementations of DSPs have started to emerge from companies including Free scale and stream processors.

The book is divided into seven chapters. Chapter 1 deals with the representations of discrete signals and systems. It motivates the reader as to what signals and systems are and how they are related to other areas such as communication systems, signal processing, computer science and engineering, electrical engineering, information technology, data compression , *etc.* Further, in this chapter, various terminologies related to signals and systems are defined. Mathematical descriptions, representations and classifications of signals and systems are also explained.

In Chapter 2, a comprehensive treatment of discrete time Fourier series of periodic signals and Fourier transform of non-periodic signals are explained with numerical examples.

The discrete time sequence $x[n]$ can be transformed as $X(j\omega)$ by Fourier transform and can be analyzed using a digital computer. However, $X(j\omega)$ is a continuous function of frequency ω and computational difficulties are encountered while analyzing $X(j\omega)$ using DSP. In Chapter 3, we convert $X(j\omega)$ into equally spaced samples. Such a sequence is called discrete Fourier transform (DFT) which is a powerful computational tool for the frequency analysis of discrete time signals. Several methods are available for computing DFT. However, fast Fourier transform (FFT) algorithms eliminate redundant calculations and offer rapid frequency domain analysis. In Chapter 3, the properties of DFT, FFT algorithms, decimation in time and decimation in frequency, linear filtering and correlation are discussed.

Chapter 4, is devoted to the z-transform and its applications to discrete time signals and systems. The properties of z-transform and inverse z-transform are discussed with numerical examples. The use of z-transform to solve difference equation is also explained.

The digital filters are classified as Infinite Impulse Response (IIR) and Finite Impulse Response (FIR) filters. The digital IIR filter design procedure is the extension of analog filter design. In Chapter 5, designs of IIR filters using the impulse invariant method and bilinear transformation are described. IIR filter is also designed using system function $H(s)$. Low pass IIR Butterworth Chebyshev digital filter designs are also described in this chapter.

Design techniques for finite impulse response digital filters are discussed in Chapter 6. These filters are designed using windows such as rectangular window, Hamming window, Kaiser window and Hanning window. FIR filters are designed using the frequency sampling method.

Digital signal processing is the processing of signals by digital means. The digital signal processor which processes the signal should be fast, with enough precision, and should have supporting memory to store programming, samples, intermediate and final results. In Chapter 7, we describe different type of DSP architecture, advanced addressing modes, pipe lining and overview of instructions set of TMS320C5X and C54x.

The notable features of this book include the following:

1. The syllabus content of Digital Signal Processing at the undergraduate level of most of the Indian Universities has been well covered.
2. The organization of the chapters is sequential in nature.
3. Large number of numerical examples have been worked out.
4. Learning objectives and summary are given in each chapter.

5. For the students to practice, short and long questions with answers are given at the end of each chapter.

The author takes this opportunity to thank Shri Sunil Saxena, Managing Director, Ane Books Pvt. Ltd, India, for coming forward to publish this book. I would like to express my sincere thanks to Shri A. Rathinam, General Manager (South), Ane Books Pvt. Ltd., who took the initiative to publish the book in a short span of time. I would like to express our sincere thanks to Mr. V. Ashok, who has done a wonderful job to key the voluminous book like this in a very short time and beautifully too. Suggestions and constructive criticisms are welcome from staff and students.

Tiruchirapalli, India S. Palani

Contents

About the Author

Dr. S. Palani obtained his B.E. degree in Electrical Engineering in the year 1966 from University of Madras, M.Tech. in Control Systems Engineering from IIT Kharagpur in 1968 and Ph.D. in Control Systems Engineering from University of Madras in 1982 with academic excellence. He has wide teaching experience of over four decades. He started his teaching career in the year 1968 at the erstwhile Regional Engineering College (Now National Institute of Technology), Tiruchirapalli in the department of EEE and occupied various positions. As Professor and Head, he took the initiatives to start the Instrumentation and Control Engineering Department. After a meritorious service of more than three decades in REC, Tiruchirapalli, he joined Sudharsan Engineering College, Pudukkottai as the founder Principal.

He has published more than hundred research papers in reputed international journals and has won many cash awards. Under his guidance eighteen scholars were awarded Ph.D. He has carried out several research projects funded by Government of India AICTE and other agencies. As the theme leader of the Indo—UK, REC Project on energy, he has visited many Universities and industries in United Kingdom. He is the author of the books titled Control Systems Engineering, Automatic Control Systems, Digital Signal Processing and Signals and Systems. His areas of research include design of Controllers for Dynamic systems, Digital Signal Processing and Image Processing.

Chapter 1
Representation of Discrete Signals and Systems

Learning Objectives

After completing this chapter, you should be able to:

- ✠ define various terminologies related to signals and systems.
- ✠ classify signals and systems.
- ✠ give mathematical description and representation of signals and systems.
- ✠ perform basic operations on DT signals.
- ✠ classify DT signals as periodic and non-periodic, odd and even and power and energy signals.
- ✠ classify DT systems.
- ✠ obtain the step response, causality and stability of DT systems.
- ✠ describe sampling techniques, quantization, Nyquist rate and aliasing effect.

1.1 Introduction

Most of the signals encountered in science and engineering are analog in nature. That is, the signals are functions of a continuous variable, such as time or space, and usually take on values in a continuous range. Such signals may be processed directly by appropriate analog systems (such as filters or frequency analyzers or frequency multipliers) for the purpose of changing their characteristics or extracting some desired information. In such a case, the signal has been processed directly in its analog form. Both the input signal and the output signal are in analog form and are shown in Fig. 1.1a.

S. Palani, *Discrete Time Systems and Signal Processing*,
https://doi.org/10.1007/978-3-031-32421-5_1

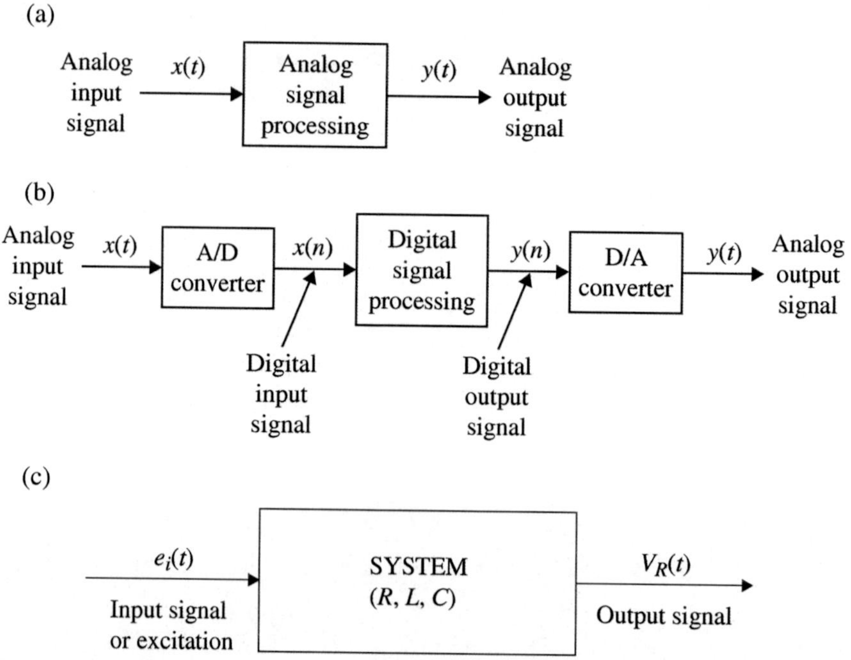

Fig. 1.1 **a** Analog signal processing **b** Digital signal processing **c** Block diagram representation of signals and systems

Digital signal processing provides an alternative method for processing the analog signal and is shown in Fig. 1.1b. To perform processing digitally, there is a need for an interface between the analog signal and the digital processor. This interface is called an analog to digital (A/D) converter. The output of A/D converter is a digital signal that is applied as an input to the digital processor.

The digital signal processor may be a large programmable digital computer or a small microprocessor programmed to perform the desired operations on the input signal. Programmable machines provide the flexibility to change the signal processing operation through a change in the software, whereas hardwired machines are difficult to reconfigure. In an application where the digital output from the digital signal processor is to be given to the user in analog form, we must provide an interface and this is called a digital to analog (D/A) converter. Thus, the signal is provided to the user in analog form.

Digital signal processing has developed very rapidly over the past five decades mainly due to the advances in digital computer technology and very large-scale integrated electronic circuits. These inexpensive smaller but faster and more powerful digital computers are capable of performing very complex signal processing functions which are usually too difficult to perform by analog circuitry. The following are the advantages of digital signal processing (DSP) over analog processing:

1.	Flexibility	:	Digital programmable systems allow flexibility in reconfiguring the DSP operations by simply changing the program.
2.	Accuracy	:	DSP provides better control of accuracy requirements, while tolerance limits have to be met in the analog counterpart.
3.	Easy storage	:	Digital signals can be easily stored in magnetic media without deterioration or loss of signal fidelity. They can also be easily transportable and can be processed off-time in remote laboratories.
4.	Processing	:	DSP allows for the implementation of more sophisticated signal processing than its analog counterpart.
5.	Cost effective	:	With advancement in VLSI technology digital implementation of the signal process system is cheaper.

The limitation of DSP is that the conversion speed of ADC and the processing speed of signal processors should be very high to perform real time processing. Signals of high bandwidth require fast sampling rate ADCs and fast processors.

Some of the applications of the digital signal processors are speech processing, signal transmission on telephone channels, image processing, biomedical, seismology and consumer electronics.

1.	Speech processing	:	Speech compression and decompression for voice storage system and for transmission and reception of voice signals.
2.	Communication	:	Elimination of noise by filtering and echo cancellation by adaptive filtering in transmission channels.
3.	Biomedical	:	Spectrum analysis of ECG signals to identify various disorders in the heart. Spectrum analysis of EEG signals to study the malfunction or disorders in the brain.
4.	Consumer electronics	:	Music synthesis, digital audio and video.
5.	Seismology	:	Spectrum analysis of seismic signals can be used to predict the earthquake, nuclear explosions and earth movement.
6.	Image processing	:	Two dimensional filtering on images for image enhancement, finger print matching, identifying hidden images in the signals received by radars, etc.

The concepts of signals and systems play a very important role in many areas of science and technology. These concepts are very extensively applied in the field of circuit analysis and design, long-distance communication, power system generation and distribution, electron devices, electrical machines, biomedical engineering, aeronautics, process control, speech and image processing to mention a few. **Signals represent some independent variables that contain some information about the behaviour of some natural phenomenon**. Voltages and currents in electrical and electronic circuits, electromagnetic radio waves, human speech and sounds produced by animals are some of the examples of signals. **When these signals are operated on some objects, they give out signals in the same or modified form. These objects are called systems.** A system is, therefore, defined as the interconnection of objects with a definite relationship between objects and attributes. Signals appearing at various stages of the system are attributes. R, L, C components, spring, dash-pots, mass, etc. are the objects. The electrical and electronic circuits comprising of R, L, C components and amplifiers, the transmitter and receiver in a communication system, the petrol and diesel engines in an automobile, chemical plants, nuclear reactor, human beings, animals, a government establishment, etc. are all examples of systems. **In this book, we deal with only discrete signals and systems.**

1.2 Terminologies Related to Signals and Systems

Before we give mathematical descriptions and representations of various terminologies related to signals and systems, the following terminologies which are very frequently used are defined as follows.

1.2.1 Signal

A signal is defined as a physical phenomenon that carries some information or data. The signals are usually functions of independent variable time. There are some cases where the signals are not functions of time. The electrical charge distributed in a body is a signal which is a function of space and not time.

1.2.2 System

A system is defined as a set of interconnected objects with a definite relationship between objects and attributes. The interconnected components provide the desired function.

Objects are parts or components of a system. For example, switches, springs, masses, dash-pots, etc. in a mechanical systems, and inductors, capacitors and resistors in an electrical system are the objects. The displacement of mass, spring and dash pot and the current flow and the voltage across the inductor, capacitor and resistor are the attributes. There is a definite relationship between the objects and attributes. The voltages across R, L, C series components can be expressed as $V_R = iR$; $V_L = L\frac{di}{dt}$ and $V_C = \frac{1}{C}\int i dt$. If this series circuit is excited by the voltage source $e_i(t)$, the $e_i(t)$ is the input attribute or the input signal. If the voltage across any of the objects R, L and C is taken, then such an attribute is called the output signal. The block diagram representation of input and output (voltage across the resistor) signals and the system is shown in Fig. 1.1c.

1.3 Continuous and Discrete Time Signals

Signals are broadly classified as:

1. Continuous time signal (CT signal).
2. Discrete time signal (DT signal).
3. Digital signal.

The signal that is specified for every value of time t is called continuous time signal and is denoted by $x(t)$. On the other hand, the signal that is specified at the discrete value of time is called the discrete time signal. The discrete time signal is represented as a sequence of numbers and is denoted by $x[n]$, where n is an integer. Here time t is divided into n discrete time intervals. The continuous time signal (CT) and discrete time signal (DT) are represented in Figs. 1.2 and 1.3, respectively.

It is to be noted that, in continuous time signal representation, the independent variable t, which has unit as s, is put in the parenthesis $(\cdot)$, and in discrete time signal, the independent variable n, which is an integer, is put inside the square parenthesis $[\cdot]$. Accordingly, the dependent variables of the continuous time signal/system are denoted as $x(t)$, $g(t)$, $u(t)$, etc. Similarly, the dependent variables of discrete time signals/systems are denoted as $x[n]$, $g[n]$, $u[n]$, etc.

A discrete time signal $x[n]$ is represented by the following two methods:

1.

$$x[n] = \begin{cases} \left(\frac{1}{a}\right)^n & n \geq 0 \\ 0 & n < 0 \end{cases} \tag{1.1}$$

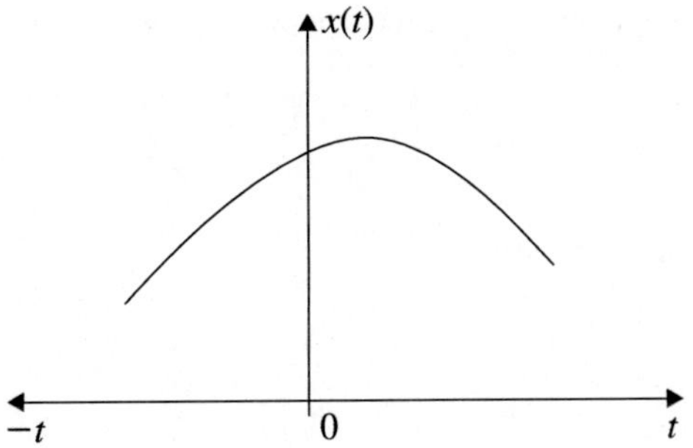

Fig. 1.2 CT signal

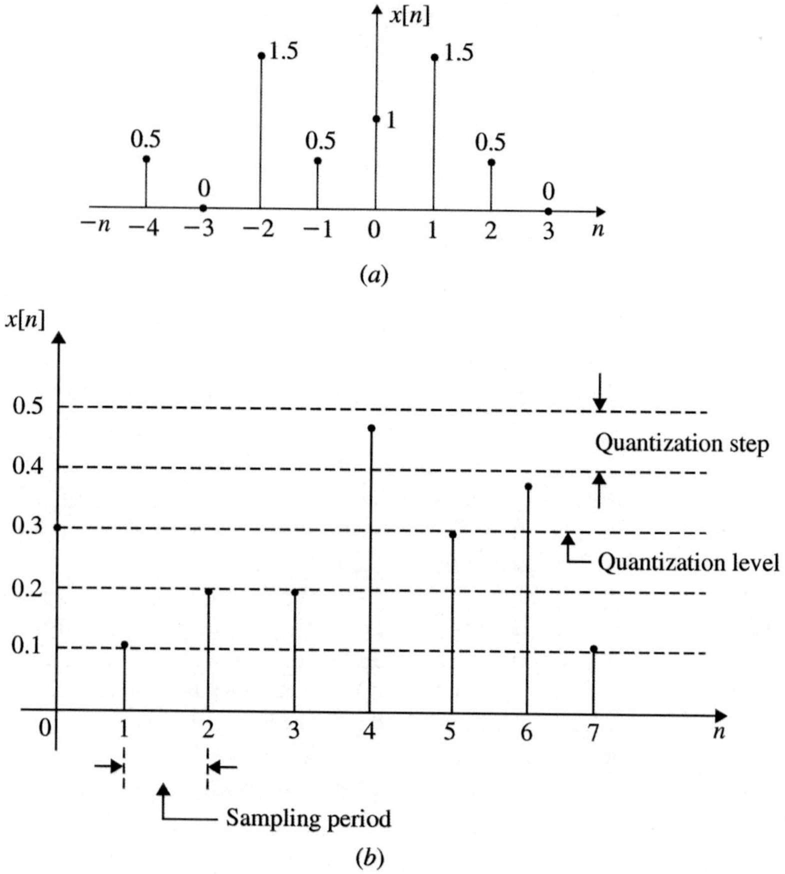

Fig. 1.3 **a** Discrete time (DT) signal **b** Digital signal

Substituting various values of n where $n \geq 0$ in Equation (1.1), the sequence for $x[n]$, which is denoted by $x\{n\}$, is written as follows:

$$x\{n\} = \left\{1, \frac{1}{a}, \frac{1}{a^2}, \ldots, \frac{1}{a^n}\right\}$$

2. The sequence is also represented as

$$x\{n\} = \{3,\ 2,\ \underset{\uparrow}{5},\ 4,\ 6,\ 8,\ 2\}$$

The arrow indicates the value of $x[n]$ at $n = 0$, which is 5 in this case. The numbers to the left of the arrow indicate the negative sequence $n = -1, -2$, etc.

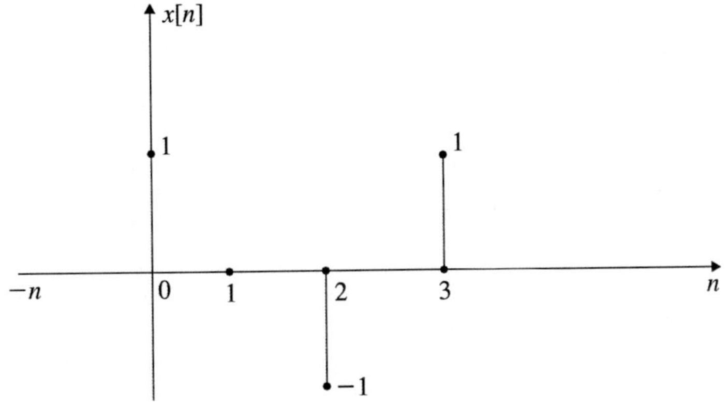

Fig. 1.4 Graphical representation of $x[n]$

The numbers to the right of the arrow correspond to $n = 1, 2, 3, 4$, etc. Thus, for the above sequence $x[-1] = 2$; $x[-2] = 3$; $x[0] = 5$; $x[1] = 4$; $x[2] = 6$; $x[3] = 8$ and $x[4] = 2$. If no arrow is marked for a sequence, the sequence starts from the first term in the extreme left. Consider the sequence

$$x\{n\} = \{5,\ 3,\ 4,\ 2\}$$

Here $x[0] = 5$; $x[1] = 3$; $x[2] = 4$ and $x[3] = 2$. There is no negative sequence here.

A digital signal is not very much different from a discrete signal except that a digital signal amplitude is quantized at a certain specific level. This is because a digital computer can accept only sequences of numbers that are expressed in terms of bits. The representation of the digital signal is shown in Fig. 1.3b. The amplitude at any discrete time interval can be only one of quantization levels. If the amplitude is chosen from a finite set of numbers, the amplitude is said to be discritized. The quantization level is equidistance and is called quantization step. An analog signal can be converted into a digital signal by means of sampling and quantizing. It is then digitized by rounding off its value to the closest permissible level. Thus, error exists when an analog signal is discritized and quantized. This is called quantization error. The digital signal is, therefore, discretized in time and quantized in amplitude.

Example 1.1 Graphically represent the following sequence:

$$x\{n\} = \{1,\ 0,\ -1,\ 1\}$$

Solution The graphical representation $x\{n\} = \{1,\ 0,\ -1,\ 1\}$ is shown in Fig. 1.4.

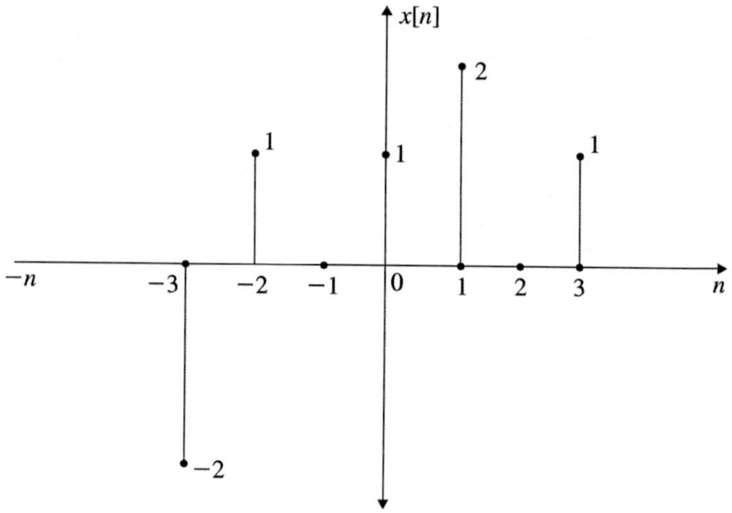

Fig. 1.5 Graphical representation of $x[n]$

Example 1.2 Graphically represent the following sequence:

$$x\{n\} = \{-2,\ 1,\ 0,\ \underset{\uparrow}{1},\ 2,\ 0,\ 1\}$$

Solution The sequence

$$x\{n\} = \{-2,\ 1,\ 0,\ \underset{\uparrow}{1},\ 2,\ 0,\ 1\}$$

is represented in Fig. 1.5.

1.4 Basic Discrete Time Signals

Similar to continuous time signals, basic discrete signals are available. However, these signals are represented at discrete intervals of time 'n' where n is an integer. Representation of basic discrete time signals is discussed below.

1.4.1 The Unit Impulse Sequence

The basic impulse sequence is shown in Fig. 1.6. The unit impulse sequence also called the sample is defined as

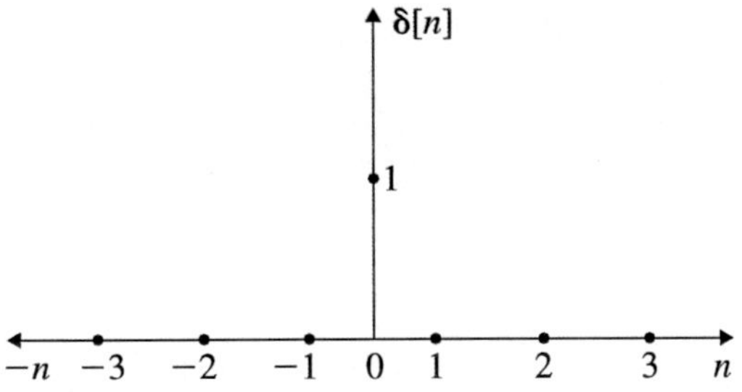

Fig. 1.6 Basic unit impulse sequence

$$\delta[n] = \begin{cases} 1 & n = 0 \\ 0 & n \neq 0 \end{cases} \tag{1.2}$$

1.4.2 The Basic Unit Step Sequence

The basic unit step sequence is represented in Fig. 1.7. It is denoted by $u(n)$. It is defined as

$$u[n] = \begin{cases} 1 & n \geq 0 \\ 0 & n < 0 \end{cases} \tag{1.3}$$

Any discrete sequences $x[n]$ for $n \geq 0$ is expressed as $x[n]u[n]$. For $n < 0$, it is expressed as $x[n]u[-n]$. It is noted that at $n = 0$, the value of $u[n] = 1$.

1.4.3 The Basic Unit Ramp Sequence

The basic unit ramp sequence which is denoted by $r[n]$ is represented in Fig. 1.8. It is defined as

$$r[n] = \begin{cases} n & n \geq 0 \\ 0 & n < 0 \end{cases} \tag{1.4}$$

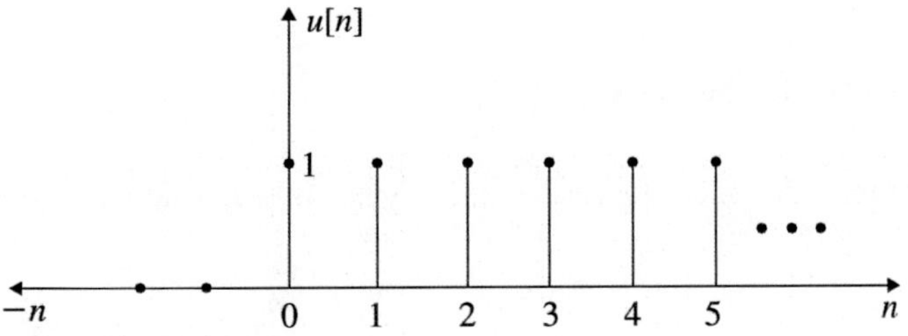

Fig. 1.7 Basic unit step sequence

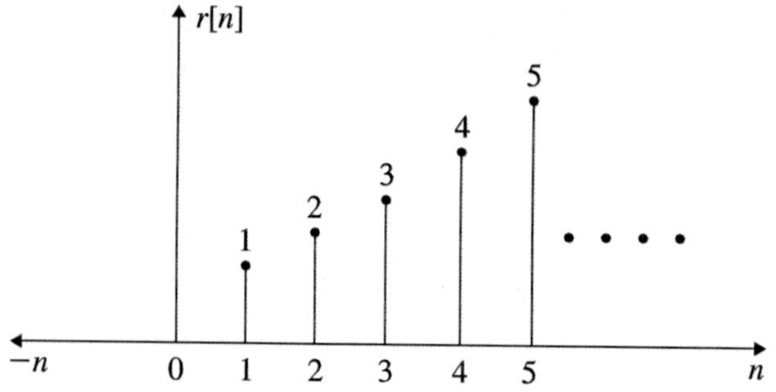

Fig. 1.8 Basic unit ramp sequence

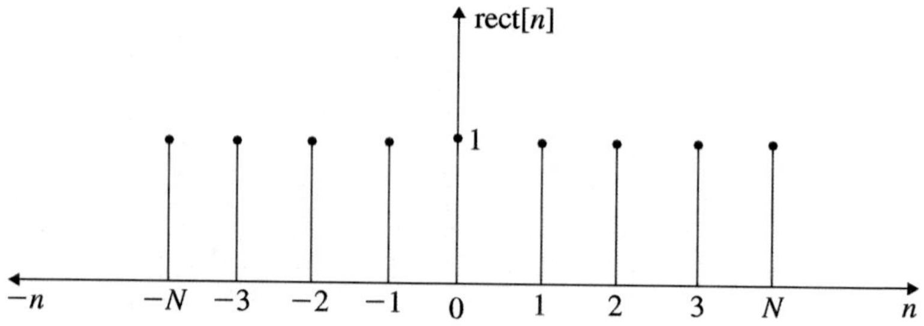

Fig. 1.9 Unit rectangular sequence

1.4.4 Unit Rectangular Sequence

The discrete time unit rectangular sequence is shown in Fig. 1.9. It is defined as

$$\text{rect}[n] = \begin{cases} 1 & |n| \le N \\ 0 & |n| > N \end{cases} \tag{1.5}$$

The above equation can also be expressed as

$$\text{rect}[n] = 1 \qquad -N \le n \le N.$$

1.4.5 Sinusoidal Sequence

The discrete time sinusoidal signal is defined by the following mathematical expression:

$$x[n] = Ae^{-\alpha n}\sin(\omega_0 n + \phi) \tag{1.6}$$

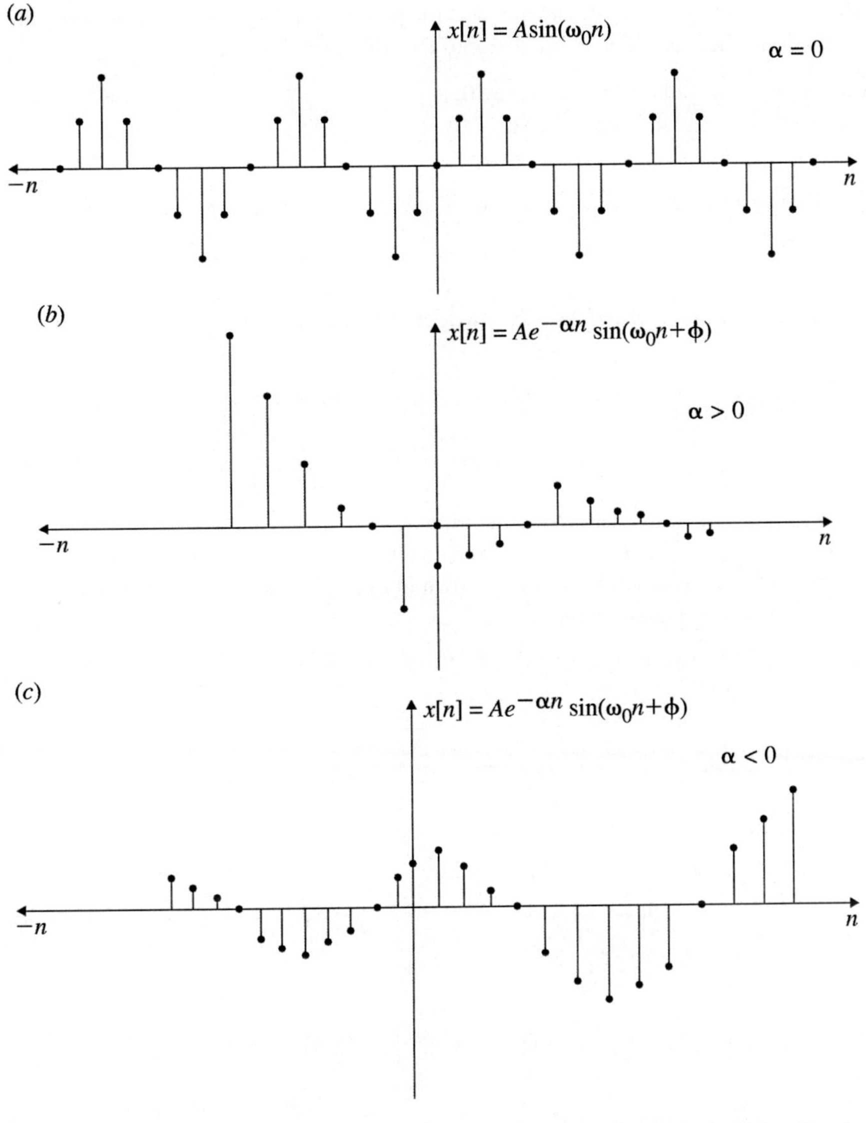

Fig. 1.10 Discrete time sinusoidal signal. **a** Purely sinusoidal; **b** Decaying sinusoidal; **c** Growing sinusoidal

where A and α are real numbers and ϕ is the phase shift. Depending on the value of α, the sinusoidal sequence is divided into the following categories:

- A purely sinusoidal sequence ($\alpha = 0$).
- Decaying sinusoidal sequence ($\alpha > 0$).
- Growing sinusoidal sequence ($\alpha < 0$).

The above sinusoidal sequences are illustrated in Fig. 1.10a, b and c, respectively.

1.4.6 Discrete Time Real Exponential Sequence

The general complex exponential sequence is defined as

$$x[n] = A\alpha^n \tag{1.7}$$

where A and α are in general complex numbers.

In Eq. (1.7), if A and α are real, the sequence is called real exponential. These sequences for various values of α are shown in Fig. 1.11. Depending on the value of α, the sequence is classified as:

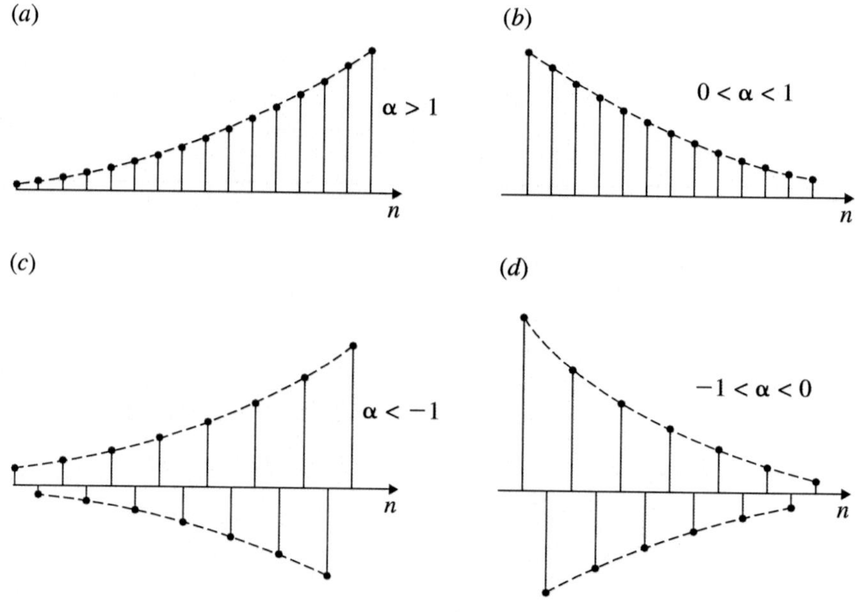

Fig. 1.11 Discrete time real exponential sequences. **a** $\alpha > 1$; **b** $0 < \alpha < 1$; **c** $\alpha < -1$; **d** $-1 < \alpha < 0$

1. Exponentially growing signal ($\alpha > 1$, Fig. 1.11a).
2. Exponentially decaying signal ($0 < \alpha < 1$, Fig. 1.11b).
3. Exponentially growing for the alternate value of n ($\alpha < -1$, Fig. 1.11c).
4. Exponentially decaying for the alternate value of n ($-1 < \alpha < 0$, Fig. 1.11d).

1.5 Basic Operations on Discrete Time Signals

The basic operations that are applied to continuous time signals are also applicable to discrete time signals. The time t in CT signal is replaced by n in DT signals. The basic operations as applied to DT signals are explained below.

1.5.1 Addition of Discrete Time Sequence

Addition of discrete time sequence is done by adding the signals at every instant of time. Consider the signals $x_1[n]$ and $x_2[n]$ shown in Fig. 1.12a and b respectively. The addition of these signals at every n is done and represented as $y[n] = x_1[n] + x_2[n]$. This is shown in Fig. 1.12c.

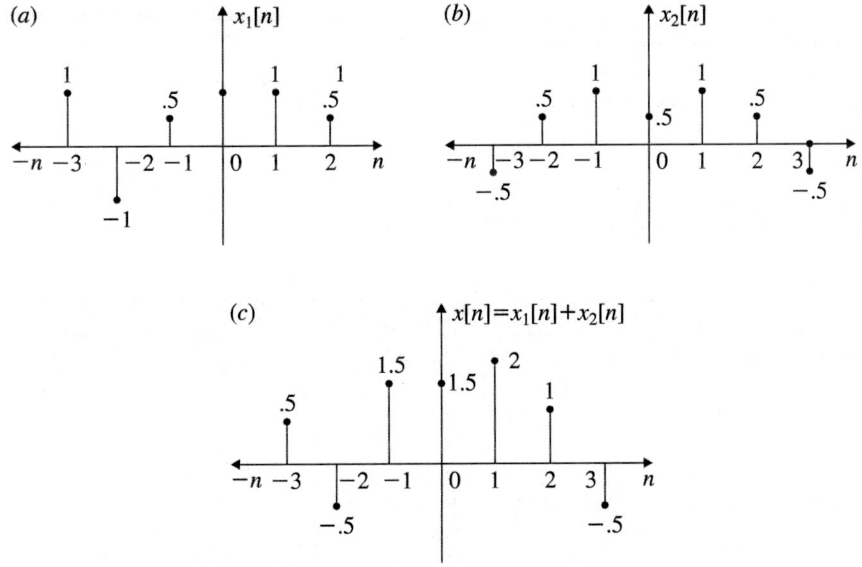

Fig. 1.12 Addition of DT signals

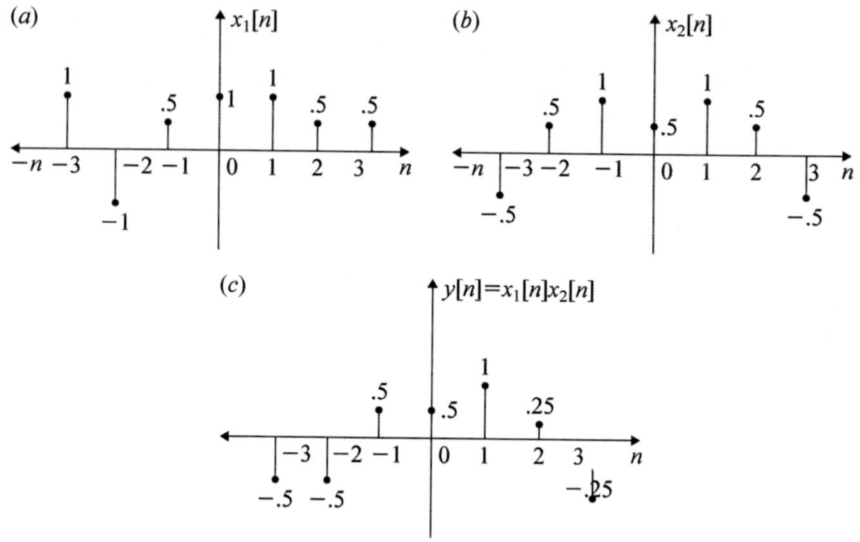

Fig. 1.13 Multiplications of two DT signals

1.5.2 Multiplication of DT Signals

The multiplication of two DT signals $x_1[n]$ and $x_2[n]$ is obtained by multiplying the signal values at each instant of time n. Consider the signal $x_1[n]$ and $x_2[n]$ represented in Fig. 1.13a and b. At each instant of time n, the samples of $x_1[n]$ and $x_2[n]$ are multiplied and represented as shown in Fig. 1.13c.

1.5.3 Amplitude Scaling of DT Signal

Let $x[n]$ be a discrete time signal. The signal $Ax[n]$ is represented by multiplying the amplitude of the sequence by A at each instant of time n. Consider the signal $x[n]$ shown in Fig. 1.14a. The signal $2x[n]$ is represented and shown in Fig. 1.14b.

1.5.4 Time Scaling of DT Signal

The time compression or expansion of a DT signal in time is known as time scaling. Consider the signal $x[n]$ shown in Fig. 1.15a. The time compressed signal $x[2n]$ and time expanded signal $x[\frac{n}{2}]$ are shown in Fig. 1.15b and c, respectively. One should note that while doing compression and expansion of DT signal, **only for integer**

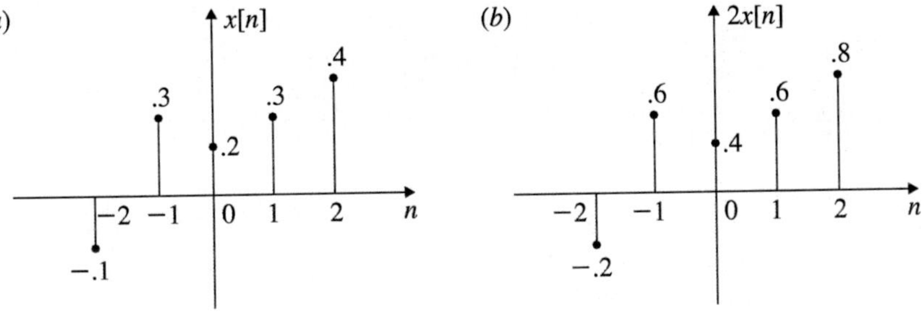

Fig. 1.14 Amplitude scaling of DT signals

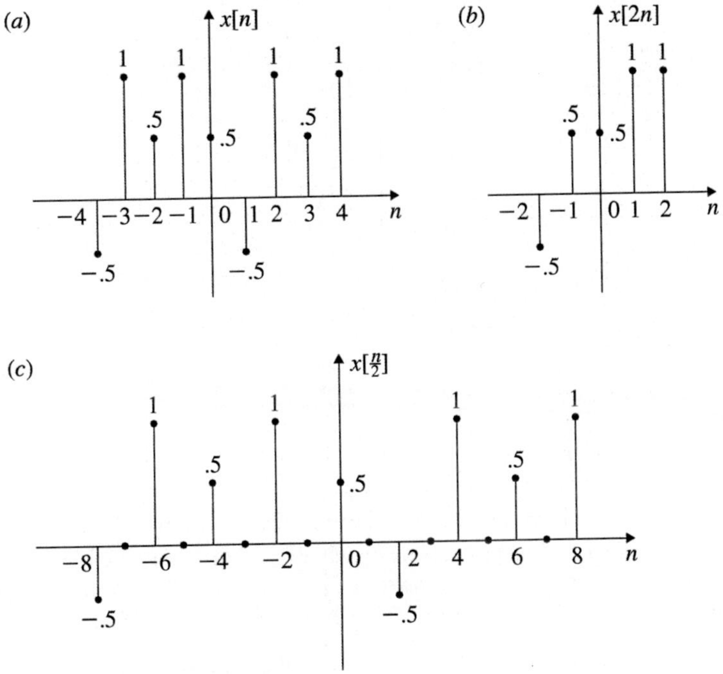

Fig. 1.15 Time scaling of DT signal

value of n the samples exist. For non-integer value of n, the samples do not exist.

Time Compression

Let

$$y[n] = x[2n]$$
$$y[-2] = x[-4] = -0.5$$
$$y[-1] = x[-2] = 0.5$$
$$y[0] = x[0] = 0.5$$

$$y[1] = x[2] = 1$$
$$y[2] = x[4] = 1$$

The plot of $x[2n]$ is shown in Fig. 1.15b.

Time Expansion

Let

$$y[n] = x\left[\frac{n}{2}\right]$$
$$y[-8] = x[-4] = -0.5$$
$$y[-6] = x[-3] = 1$$
$$y[-4] = x[-2] = 0.5$$
$$y[-2] = x[-1] = 1$$
$$y[0] = x[0] = 0.5$$
$$y[2] = x[1] = -0.5$$
$$y[4] = x[2] = 1$$
$$y[6] = x[3] = 0.5$$
$$y[8] = x[4] = 1$$

The plot of $x[\frac{n}{2}]$ is shown in Fig. 1.15c.

1.5.5 Time Shifting of DT Signal

As in the case of CT signal, time shifting property is applied to DT signal also. Let $x[n]$ be the DT signal. Let n_0 be the time by which $x[n]$ is time shifted. Since n is an integer, n_0 is also an integer. The following points are applicable while DT signal is time shifted:

- For the DT signals $x[-n - n_0]$ and $x[n + n_0]$, the signals $x[-n]$ and $x[n]$ are to be left shifted by n_0.
- For the DT signals $x[n - n_0]$ and $x[-n + n_0]$, the signals $x[n]$ and $x[-n]$ are to be right shifted by n_0.

Figure 1.16 shows the time shifting of the DT signal.

In Fig. 1.16a, the sequence $x[n]$ is shown. The sequence $x[n - 2]$ which is right shifted by 2 samples is shown in Fig. 1.16b. $x[-n]$ which is the folded signal is shown in Fig. 1.16c. $x[-n + 2]$ which is left shifted of $x[-n]$ is shown in Fig. 1.16d. $x[n + 2]$ which is right shifted of $x[n]$ is shown in Fig. 1.16e. $x[-n - 2]$ which is left shifted of $x[-n]$ is shown in Fig. 1.16f.

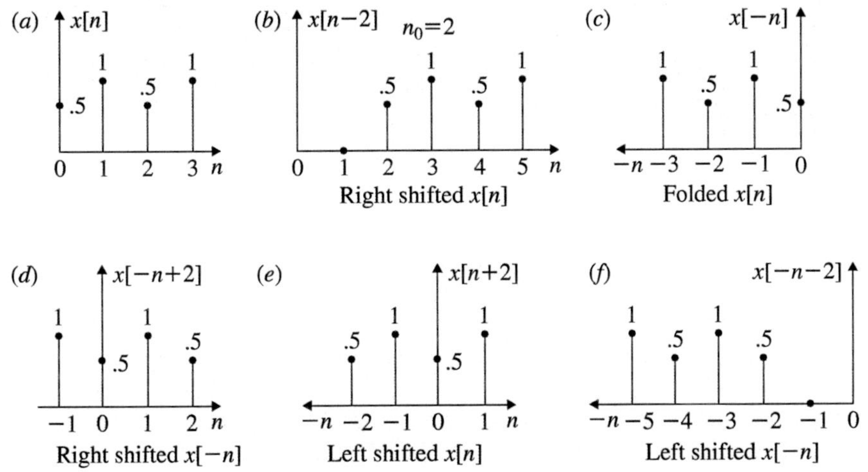

Fig. 1.16 Time shifting of DT signal

1.5.6 Multiple Transformation

The transformations, namely amplitude scaling, time reversal, time shifting, time scaling, etc. are applied to represent DT sequence. The sequence of operation of these transformations is important and followed as described below.

Consider the following DT signal:

$$y[n] = Ax\left[-\frac{n}{a} + n_0\right]$$

1. Plot $x[n]$ sequence and obtain $Ax[n]$ by amplitude scaling.
2. Using time reversal (folding), plot $Ax[-n]$.
3. Using time shifting, plot $Ax[-n + n_0]$ where $n_0 > 0$. The time shift is to be right of $x[-n]$ by n_0 samples.
4. Using time scaling, plot $Ax[-\frac{n}{a} + n_0]$ where a is in integer. In the above case, keeping amplitude constant, time is expanded by a.

The following examples illustrate the above operations:

Example 1.3 Let $x[n]$ and $y[n]$ be as given in Fig. 1.17a and b, respectively. Plot

$(a)\ \ x[2n]$	$(c)\ \ x[n-2] + y[n-2]$
$(b)\ \ x[3n-1]$	$(d)\ \ y[1-n]$

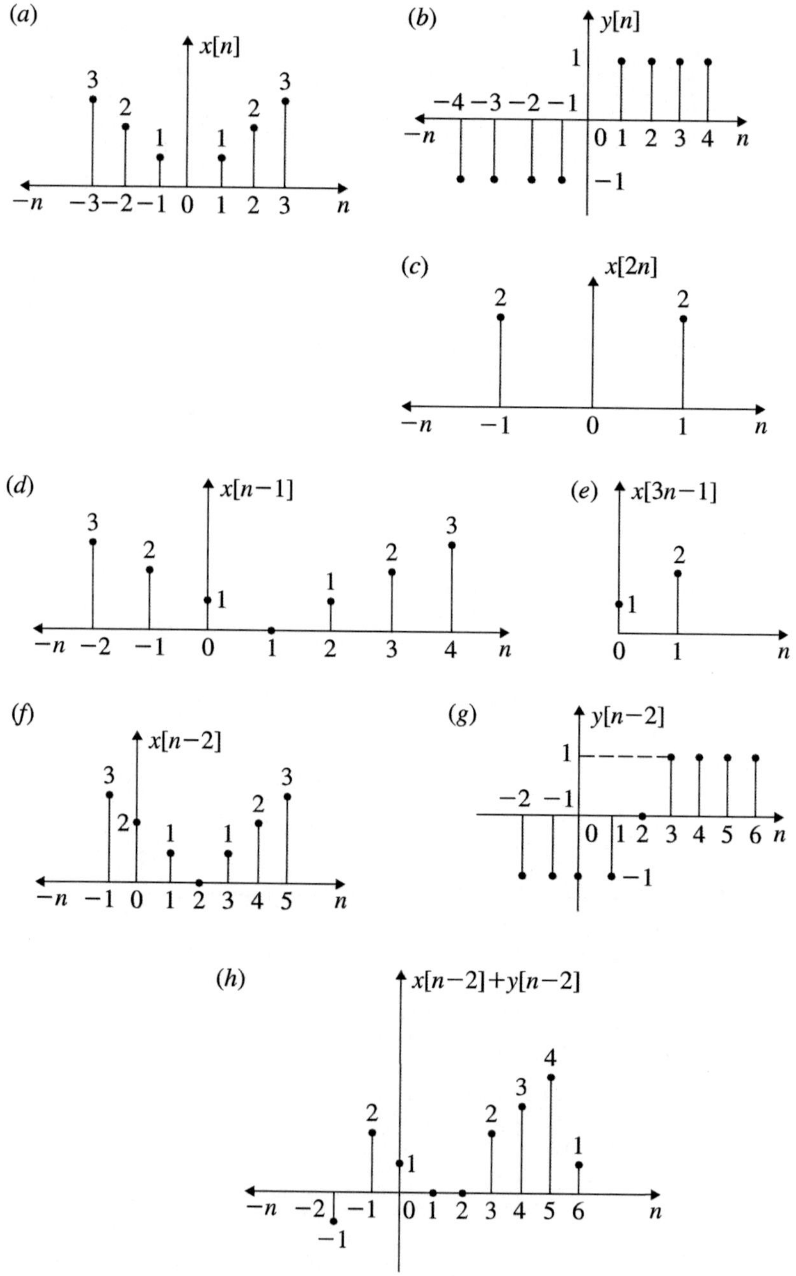

Fig. 1.17 Two discrete sequences

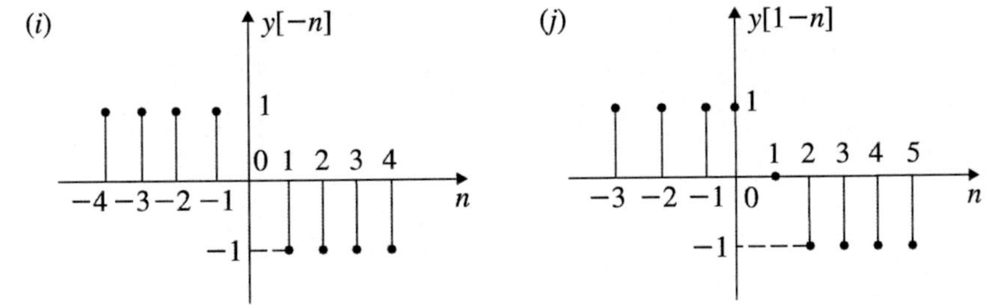

Fig. 1.17 (continued)

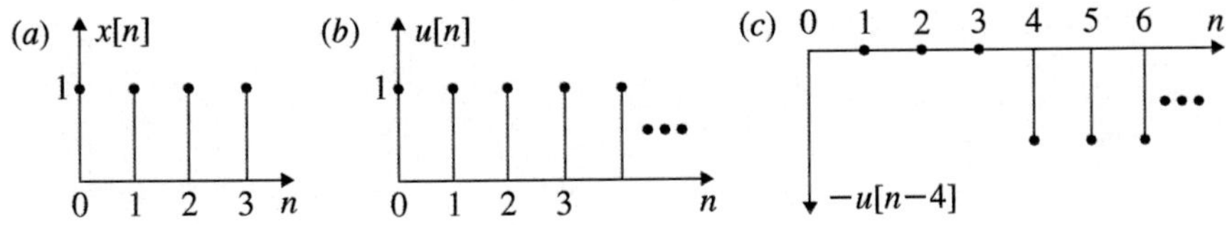

Fig. 1.18 Sequences expressed in terms of step sequences

Solution (*a*) **To plot $x[2n]$**

Here the DT sequence is time compressed by a factor of 2. Hence, the samples only with even numbers are divided by a factor of 2 and the corresponding amplitudes marked and shown in Fig. 1.17c. When odd values of n are divided by the factor 2, it becomes a fraction and they are skipped.

(*b*) **To plot $x[3n-1]$**

The plot of $x[n-1]$ is obtained by right shifting of $x[n]$ by $n_0 = 1$. This is shown in Fig. 1.17d. When $x[n-1]$ is time compressed by a factor 3, $x[3n-1]$ is obtained. Only integers which are divisible by 3 in the sequence $x[n-1]$ are to be taken to plot $x[3n-1]$. Thus, samples for $n = 0$ and $n = 3$, will be plotted as shown in Fig. 1.17e.

(*c*) **To plot $x[n-2] + y[n-2]$**

The sequence $x[n-2]$ is obtained by right shifting of $x[n]$ by 2 and is shown in Fig. 1.17f. Similarly, the sequence $y[n-2]$ is obtained by right shifting of $y[n]$ by 2 and is shown in Fig. 1.17(*g*). The sequence $x[n-2] + y[n-2]$ is obtained by summing up the sequences in Fig. 1.17f and (*g*) for all n and is shown in Fig. 1.17h.

(*d*) **To plot $y[1-n]$**

The sequence $y[-n]$ is obtained by folding $y[n]$ and is shown in Fig. 1.17(*i*). $y[-n]$ is right shifted by 1 sample to get the sequence $y[1-n]$. This is shown in Fig. 1.17(*j*).

Example 1.4 Consider the sequence shown in Fig. 1.18a. Express the sequence in terms of step function.

Solution The unit step sequence $u[n]$ is shown in Fig. 1.18b. The unit negative step sequence with a time delay of $n_0 = 4$ is shown in Fig. 1.18b. It is evident form Fig.

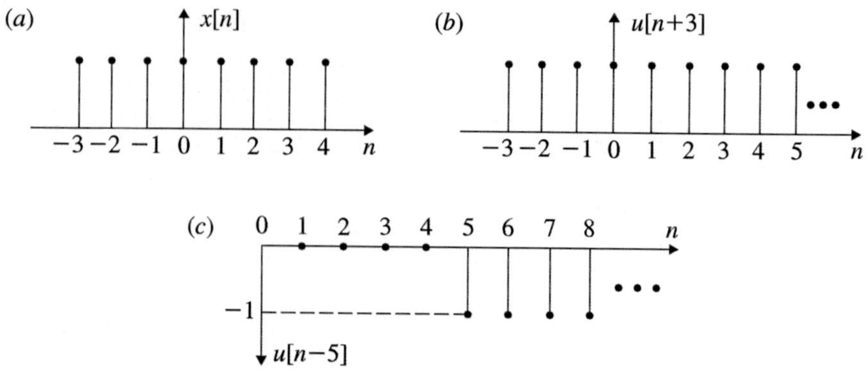

Fig. 1.19 DT sequences expressed in terms of step sequences

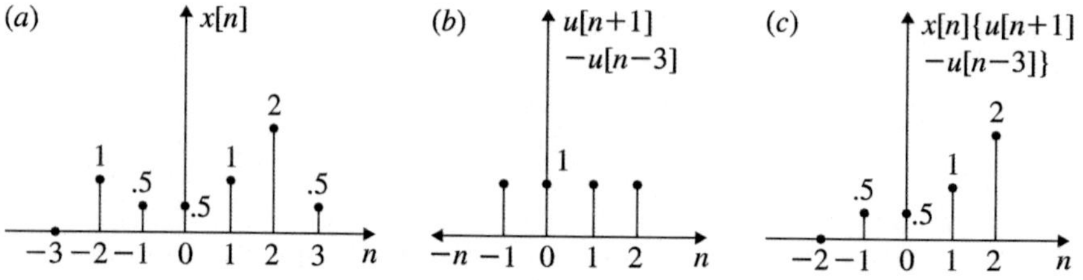

Fig. 1.20 Multiplication of DT sequences

1.18 that $\{u[n] - u[n-4]\}$ gives the required $x[n]$ sequence which is represented in Fig. 1.18a. Thus, $x[n] = \{u[n] - u[n-4]\}$.

Example 1.5 Consider the sequence shown in Fig. 1.19a. Express the sequence in terms of step function.

Solution 1. Figure 1.19a represents the sequence $x[n]$ in the interval $-3 \leq n \leq 4$.
2. Consider $u[n+3]$ which is represented in Fig. 1.19b. The sequence interval is $-3 \leq n < \infty$.
3. Consider the step sequence with a time delay of $n_0 = 5$ and inverted. This can be written as $-u[n-5]$ for the interval $5 \leq n < \infty$. This is represented in Fig. 1.19c.
4. Now consider the sum of the sequences $u[n+3]$ and $-u[n-5]$. This is nothing but $x[n]$. Thus

$$\boxed{x[n] = u[n+3] - u[n-5]}$$

Example 1.6 A discrete time sequence $x[n]$ is shown in Fig. 1.20a. Find

$$x[n]\{u[n+1] - u[n-3]\}$$

Solution 1. $x[n]$ sequence is represented in Fig. 1.20a.

2. $\{u[n+1] - u[n-3]\}$ sequence is nothing but the time delayed unit step sequence with $n_0 = 3$, being subtracted from the time advanced unit step sequence with $n_0 = 1$. This sequence is represented in Fig. 1.20b.
3. Multiplying sample wise of Fig. 1.20a and b, the required sequence $x[n]\{u[n+1] - u[n-3]\}$ is obtained and represented in Fig. 1.20c.

Example 1.7 Sketch $x[n] = a^n$ where $-2 \leq n \leq 2$ for the two cases shown below:

$$(1) \qquad a = \left(-\frac{1}{4}\right)$$

$$(2) \qquad a = -4$$

Solution For $x[n] = (-\frac{1}{4})^n$ and $x[n] = (-4)^n$ where $-2 \leq n \leq 2$, $x[n]$ is found and tabulated below:

n	-2	-1	0	1	2
$x[n] = (-\frac{1}{4})^n$	16	-4	1	$-\frac{1}{4}$	$\frac{1}{16}$
$x[n] = (-4)^n$	$\frac{1}{16}$	$-\frac{1}{4}$	1	-4	16

The samples of $x[n]$ are plotted and shown in Fig. 1.21. $x[n] = (-\frac{1}{4})^n$ is represented in Fig. 1.21a and $x[n] = (-4)^n$ is represented in Fig. 1.21b.

Example 1.8 Express

$$x[n] = (-1)^n \qquad -2 \leq n \leq 2$$

as a sum of scaled and shifted step function.

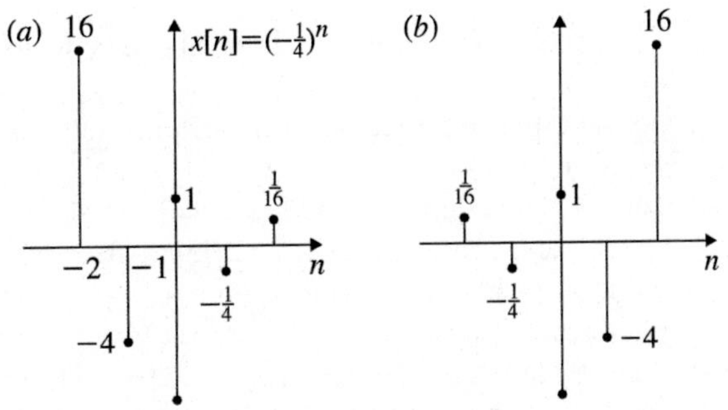

Fig. 1.21 DT sequences of Example 1.7

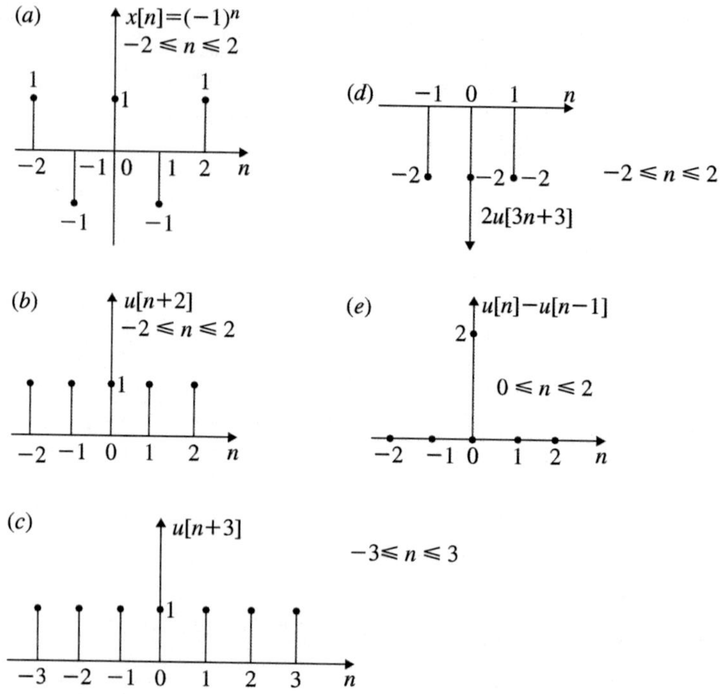

Fig. 1.22 DT sequences of Example 1.8

Solution (1) $x[n] = (-1)^n$ is tabulated for $-2 \le n \le 2$.

n	−2	−1	0	1	2
$x[n]$	1	−1	1	−1	1

The samples corresponding to the above table are sketched and shown as $x[n]$ in Fig. 1.22a.

(2) Consider the step sequence $u[n+2]$ for $-2 \le n \le 2$. The samples are shown in Fig. 1.22b.

(3) Consider unit step sequence $u[n+3]$ for $-3 \le n \le 3$. This is represented in Fig. 1.22c. From Fig. 1.22c, $-2u[3n+3]$ is obtained by amplitude inversion and multiplication and time scaling (compression). This is represented in Fig. 1.22d for $-2 \le n \le 2$.

(4) Consider the step sequence $2\{u[n] - u[n-1]\}$ for $n \ge 0$. This is represented in Fig. 1.22e. This is nothing but the sample of strength 2 at $n = 0$.

(5) Now, by adding the samples in Fig. 1.22b, d and e, it can be easily verified that

$$\boxed{x[n] = u[n+2] - 2u[3n+3] + 2[u[n] - u[n-1]]} \quad -2 \le n \le 2$$

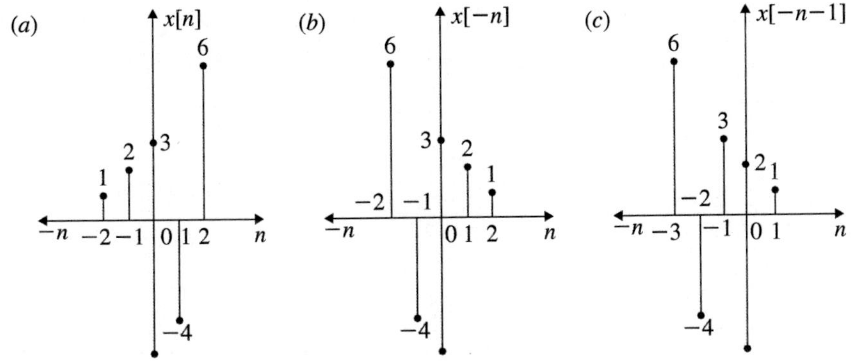

Fig. 1.23 DT sequences of Example 1.9

Example 1.9 Given

$$x[n] = \{1,\ 2,\ \underset{\uparrow}{3},\ -4,\ 6\}$$

Plot the signal $x[-n-1]$.

Solution 1. The sequence $x[n]$ is represented in Fig. 1.23a.
2. By folding $x[n]$, $x[-n]$ is obtained and represented in Fig. 1.23b.
3. $x[-n]$ is shifted to the left by one sample and $x[-n-1]$ is obtained. This is represented in Fig. 1.23c.

1.6 Classification of Discrete Time Signals

Discrete time signals are classified as:

1. Periodic and non-periodic signals.
2. Odd and even signals.
3. Power and energy signals.

They are discussed below with suitable examples.

1.6.1 Periodic and Non-periodic DT Signals

A discrete time signal (sequence) $x[n]$ is said to be periodic with period N which is a positive integer if

$$x[n+N] = x[n] \quad \text{for all } n \tag{1.8}$$

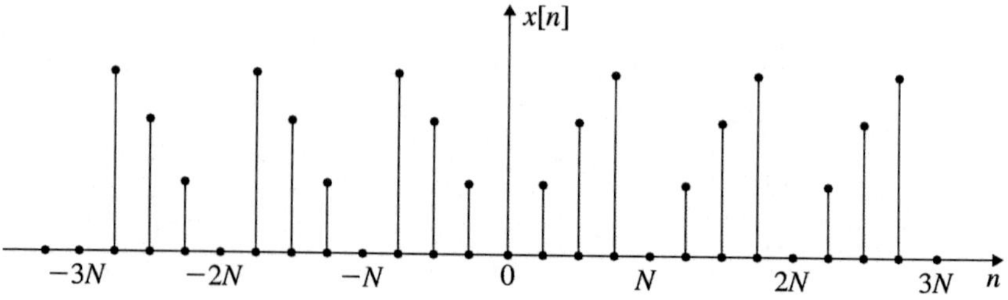

Fig. 1.24 Periodic sequence

Consider the DT sequence shown in Fig. 1.24. The signal gets repeated for every N. For Fig. 1.24, the following equation is written:

$$x[n + mN] = x[n] \quad \text{for all } n \tag{1.9}$$

where m is any integer. The smallest positive integer N in Eq. (1.9) is called the fundamental period N_0. Any sequence which is not periodic is said to be non-periodic or aperiodic.

Example 1.10 Show that complex exponential sequence $x[n] = e^{j\omega_0 n}$ is periodic and find the fundamental period.

Solution

$$\begin{aligned} x[n] &= e^{j\omega_0 n} \\ x[n+N] &= e^{j\omega_0 (n+N)} \\ &= e^{j\omega_0 n} e^{j\omega_0 N} \\ &= e^{j\omega_0 n} \quad \text{if } e^{j\omega_0 N} = 1 \\ \omega_0 N &= m2\pi \quad \text{where } m \text{ is any integer.} \end{aligned}$$

$$\boxed{N = m\frac{2\pi}{\omega_0}}$$

or

$$\frac{\omega_0}{2\pi} = \frac{m}{N} = \text{rational number.}$$

Thus, $e^{j\omega_0 n}$ is periodic if $\frac{m}{N}$ is rational. For $m = 1$, $N = N_0$. The corresponding frequency $F_0 = \frac{1}{N_0}$ is the fundamental frequency. F_0 is expressed in cycles and not Hz. Similarly, ω_0 is expressed in radians and not in radians per second.

Example 1.11 Consider the following DT Signal:

$$x[n] = \sin(\omega_0 n + \phi)$$

Under what condition, the above signal is periodic.

Solution

$$\begin{aligned} x[n] &= \sin(\omega_0 n + \phi) \\ x[n+N] &= \sin(\omega_0(n+N) + \phi) \\ &= \sin(\omega_0 n + \omega_0 N + \phi) \\ &= \sin(\omega_0 n + \phi) \quad \text{if } \omega_0 N = 2\pi m \qquad \text{where } m \text{ is an integer} \\ &= x[n] \end{aligned}$$

$$\boxed{\frac{\omega_0}{2\pi} = \frac{m}{N} = \text{rational}}$$

Example 1.12 If $x_1[n]$ and $x_2[n]$ are periodic, then show that the sum of the composite signal $x[n] = x_1[n] + x_2[n]$ is also periodic with the least common multiple (LCM) of the fundamental period of individual signal.

Solution Let N_1 and N_2 be the fundamental periods of $x_1[n]$ and $x_2[n]$, respectively. Since both $x_1[n]$ and $x_2[n]$ are periodic,

$$\begin{aligned} x_1[n] &= x_1[n + mN_1] \\ x_2[n] &= x_2[n + kN_2] \\ x[n] &= x_1[n] + x_2[n] \\ &= x_1[n + mN_1] + x_2[n + kN_2] \end{aligned}$$

For $x[n]$ to be periodic with period N,

$$\begin{aligned} x[n+N] &= x_1[n+N] + x_2[n+N] \\ x[n] &= x[n+N] \\ x_1[n + mN_1] + x_2[n + kN_2] &= x_1[n+N] + x_2[n+N] \end{aligned}$$

The above equation is satisfied if

$$mN_1 = kN_2 = N$$

m and k which are integers are chosen to satisfy the above equation. It implies that N is the LCM of N_1 and N_2.

On similar line it can be proved that if $x_1[n]$ and $x_2[n]$ are periodic signals with fundamental period N_1 and N_2, respectively, then $x[n] = x_1[n]x_2[n]$ is periodic if

$$mN_1 = kN_2 = N$$

Example 1.13 Find whether the following signals are periodic. If periodic, determine the fundamental period

$$(a) \qquad x[n] = e^{j\pi n}$$
$$(b) \qquad x[n] = \cos\left[\frac{n}{8} - \pi\right]$$
$$(c) \qquad x[n] = \sin^2 \frac{\pi}{4} n$$

Solution (*a*) $\boldsymbol{x[n] = e^{j\pi n}}$

$$\omega_0 = \pi$$
$$N = \frac{2\pi}{\omega_0} m$$

$$\boxed{N = \frac{2\pi}{\pi} = 2 \qquad \text{if } m = 1}$$

$x[n]$ is periodic with fundamental period 2.

(*b*) $\boldsymbol{x[n] = \cos\left[\frac{n}{8} - \pi\right]}$

$$\omega_0 = \frac{1}{8}$$
$$N = \frac{2\pi}{\omega_0} m = 16\pi m$$

For any integer value of m, N is not integer. Hence, $x[n]$ is not periodic.

$$\boxed{x[n] \text{ is not periodic}}$$

(*c*) $\boldsymbol{x[n] = \sin^2 \frac{\pi}{4} n}$

$$\begin{aligned} x[n] &= \sin^2 \frac{\pi}{4} n \\ &= \frac{1}{2} - \frac{1}{2} \cos \frac{2\pi}{4} n \\ &= x_1[n] + x_2[n] \\ x_1[n] &= \frac{1}{2} = \frac{1}{2}(1)^n \text{ is periodic with } N_1 = 1 \\ x_2[n] &= -\frac{1}{2} \cos \frac{\pi}{2} n \\ \omega_0 &= \frac{\pi}{2} \end{aligned}$$

$$N_2 = \frac{2\pi}{\omega_0}m = 4m = 4 \qquad \text{for } m = 1$$
$$\frac{N_1}{N_2} = \frac{1}{4}$$
$$\text{or} \quad 4N_1 = N_2 = N$$

$$\boxed{N = 4}$$

Example 1.14 Find the periodicity of the following DT signal:

$$x[n] = \sin\frac{2\pi}{3}n + \cos\frac{\pi}{2}n$$

Solution

$$\begin{aligned}
x[n] &= \sin\frac{2\pi}{3}n + \cos\frac{\pi}{2}n \\
&= x_1[n] + x_2[n] \\
x_1[n] &= \sin\frac{2}{3}\pi n \\
\omega_1 &= \frac{2}{3}\pi \\
N_1 &= \frac{2\pi}{\omega_1}m_1 = \frac{2\pi}{2\pi}3m_1 = 3 \quad \text{for } m_1 = 1 \\
x_2[n] &= \cos\frac{\pi}{2}n \\
\omega_2 &= \frac{\pi}{2} \\
N_2 &= \frac{2\pi}{\omega_2}m_2 = \frac{2\pi}{\pi}2m_2 = 4 \quad \text{for } m_2 = 1 \\
\frac{N_1}{N_2} &= \frac{3}{4} \text{ or } 4N_1 = 3N_2 = N
\end{aligned}$$

$$\boxed{N = 12}$$

Example 1.15 Determine whether the following signal is periodic. If periodic, find its fundamental period:

$$x[n] = \cos\left(\frac{n\pi}{2}\right)\cos\left(\frac{n\pi}{4}\right)$$

Solution

$$\begin{aligned} x[n] &= \cos\left(\frac{n\pi}{2}\right)\cos\left(\frac{n\pi}{4}\right) \\ &= x_1[n]x_2[n] \\ x_1[n] &= \cos\frac{n\pi}{2} \\ \omega_1 &= \frac{\pi}{2} \\ N_1 &= \frac{2\pi}{\omega_1}m_1 = \frac{2\pi}{\pi}2m_1 = 4 \quad \text{for } m_1 = 1 \end{aligned}$$

$$\begin{aligned} x_2[n] &= \cos\frac{n\pi}{4} \\ \omega_2 &= \frac{\pi}{4} \\ N_2 &= \frac{2\pi}{\omega_2}m_2 = \frac{2\pi}{\pi}4m_2 = 8 \quad \text{for } m_2 = 1 \\ \frac{N_1}{N_2} &= \frac{4}{8} = \frac{1}{2} \quad \text{or} \\ 2N_1 &= N_2 = N \end{aligned}$$

$$\boxed{N = 8}$$

The signal is periodic and the fundamental period $N = 8$.

Example 1.16 Test whether the following signals are periodic or not, and if periodic, calculate the fundamental period.

$$(a) \qquad x[n] = \cos\left(\frac{\pi}{2}n\right) + \sin\left(\frac{\pi}{8}n\right) + 3\cos\left(\frac{\pi}{4}n + \frac{\pi}{3}\right)$$

$$(b) \qquad x[n] = e^{j\frac{2\pi}{3}n} + e^{j\frac{3\pi}{4}n}$$

Solution (*a*)

$$\begin{aligned} x[n] &= \cos\left(\frac{\pi}{2}n\right) + \sin\left(\frac{\pi}{8}n\right) + 3\cos\left(\frac{\pi}{4}n + \frac{\pi}{3}\right) \\ &= x_1[n] + x_2[n] + x_3[n] \\ x_1[n] &= \cos\frac{\pi}{2}n \\ \omega_1 &= \frac{\pi}{2}; \quad N_1 = \frac{2\pi}{\omega_1} = \frac{2\pi\, 2}{\pi} \quad \text{for } m_1 = 1 \\ N_1 &= 4 \end{aligned}$$

$$x_2[n] = \sin\left(\frac{\pi}{8}n\right)$$

$$\omega_2 = \frac{\pi}{8}; \quad N_2 = \frac{2\pi}{\omega_2}m_2 = \frac{2\pi\ 8}{\pi} \quad \text{for } m_2 = 1$$

$$N_2 = 16$$

$$x_3[n] = 3\cos\left(\frac{\pi}{4}n + \frac{\pi}{3}\right)$$

$$\omega_3 = \frac{\pi}{4}; \quad N_3 = \frac{2\pi}{\omega_3}m_3 = \frac{2\pi\ 4}{\pi} \quad \text{for } m_3 = 1$$

$$N_3 = 8$$

To find the LCM of N_1, N_2 and N_3

4	4,	8,	16
2	1,	2,	4
	1,	1,	2

$$\text{LCM} = 4 \times 2 \times 2 = 16$$

$$\boxed{N = 16}$$

The signal is periodic

(*b*)

$$x[n] = e^{j\frac{2\pi}{3}n} + e^{j\frac{3\pi}{4}n}$$

$$= x_1[n] + x_2[n]$$

$$x_1[n] = e^{j\frac{2\pi}{3}n}$$

$$\omega_1 = \frac{2\pi}{3}; \quad N_1 = \frac{2\pi}{\omega_1}m_1 = \frac{2\pi}{2\pi}3 \quad \text{for } m_1 = 1$$

$$N_1 = 3$$

$$x_2[n] = e^{j\frac{3\pi}{4}n}$$

$$\omega_2 = \frac{3\pi}{4}; \quad N_2 = \frac{2\pi}{\omega_2}m_2 = \frac{2\pi}{3\pi}4m_2$$

$$N_2 = 8 \quad \text{for } m_2 = 3$$
$$\frac{N_1}{N_2} = \frac{3}{8}$$
$$8\,N_1 = 3\,N_2 = N = 24$$

$$\boxed{N = 24}$$

The signal is periodic with fundamental period $N = 24$.

1.6.2 *Odd and Even DT Signals*

DT signals are classified as odd and even signals. The relationships are analogous to CT signals.

A discrete time signal $x[n]$ is said to be an even signal if

$$x[-n] = x[n] \tag{1.10}$$

A discrete time signal $x[n]$ is said to be an odd signal if

$$x[-n] = -x[n] \tag{1.11}$$

The signal $x[n]$ can be expressed as the sum of odd and even signals as

$$x[n] = x_e[n] + x_0[n] \tag{1.12}$$

The even and odd components of $x[n]$ can be expressed as

$$x_e[n] = \frac{1}{2}\left[x[n] + x[-n]\right] \tag{1.13}$$
$$x_0[n] = \frac{1}{2}\left[x[n] - x[-n]\right] \tag{1.14}$$

It is to be noted that

- An even function has an odd part which is zero.
- An odd function has an even part which is zero.
- The product of two even signals or of two odd signals is an even signal.
- The product of an odd and an even signal is an odd signal.
- At $n = 0$, the odd signal is zero.

The even and odd signals are represented in Fig. 1.25a and b, respectively.

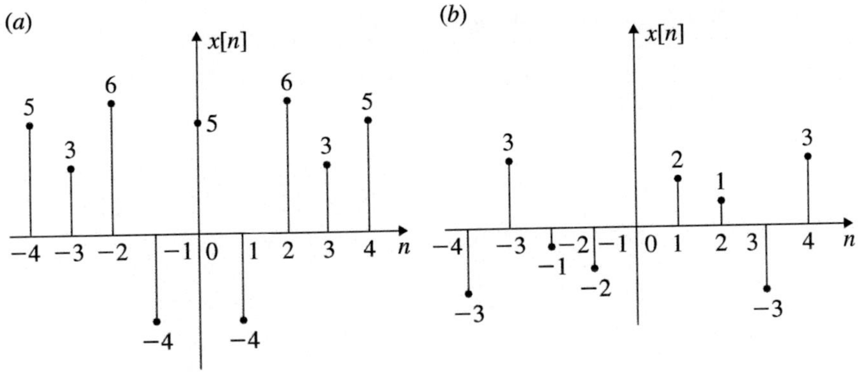

Fig. 1.25 **a** Even function and **b** Odd function

Example 1.17 Determine whether the following functions are odd or even:

$$(a) \qquad x[n] = \sin 2\pi n$$
$$(b) \qquad x[n] = \cos 2\pi n$$

Solution (*a*) $\boldsymbol{x[n] = \sin 2\pi n}$

$$x[-n] = \sin(-2\pi n) = -\sin 2\pi n$$
$$= -x[n]$$

This is an odd signal.

(*b*) $\boldsymbol{x[n] = \cos 2\pi n}$

$$x[-n] = \cos(-2\pi n) = \cos 2\pi n$$
$$= x[n]$$

This is an even signal.

Example 1.18 Find the even and odd components of DT signal given below. Verify the same by graphical method.

$$x[n] = \{-2,\ 1,\ \underset{\uparrow}{3},\ -5,\ 4\}$$

Solution $x[-n]$ is obtained by folding $x[n]$. Thus,

$$x[-n] = \{4,\ -5,\ \underset{\uparrow}{3},\ 1,\ -2\}$$

$$-x[-n] = \{-4,\ 5,\ \underset{\uparrow}{-3},\ -1,\ 2\}$$

$$\begin{aligned} x_e[n] &= \frac{1}{2}\left[x[n] + x[-n]\right] \\ &= \frac{1}{2}\left[\{-2,\ 1,\ \underset{\uparrow}{3},\ -5,\ 4\} + \{4,\ -5,\ \underset{\uparrow}{3},\ 1,\ -2\}\right] \\ &= \frac{1}{2}\left[(-2+4),\ (1-5),\ \underset{\uparrow}{(3+3)},\ (-5+1),\ (4-2)\right] \end{aligned}$$

$$\boxed{x_e[n] = \{1,\ -2,\ 3,\ -2,\ \underset{\uparrow}{1}\}}$$

$$\begin{aligned} x_o[n] &= \frac{1}{2}\left[x[n] - x[-n]\right] \\ &= \frac{1}{2}\left[\{-2,\ 1,\ \underset{\uparrow}{3},\ -5,\ 4\} + \{-4,\ 5,\ \underset{\uparrow}{-3},\ -1,\ 2\}\right] \\ &= \frac{1}{2}\left[(-2-4),\ (1+5),\ \underset{\uparrow}{(3-3)},\ (-5-1),\ (4+2)\right] \end{aligned}$$

$$\boxed{x_o[n] = \{-3,\ 3,\ 0,\ -3,\ \underset{\uparrow}{3}\}}$$

Odd and even components by graphical method.

Solution 1. $x[n]$ is represented in Fig. 1.26a.
2. $x[-n]$ is obtained by folding $x[n]$ which is represented in Fig. 1.26b.
3. $-x[n]$ is obtained by inverting $x[-n]$ of Fig. 1.26b. This is represented in Fig. 1.26c.

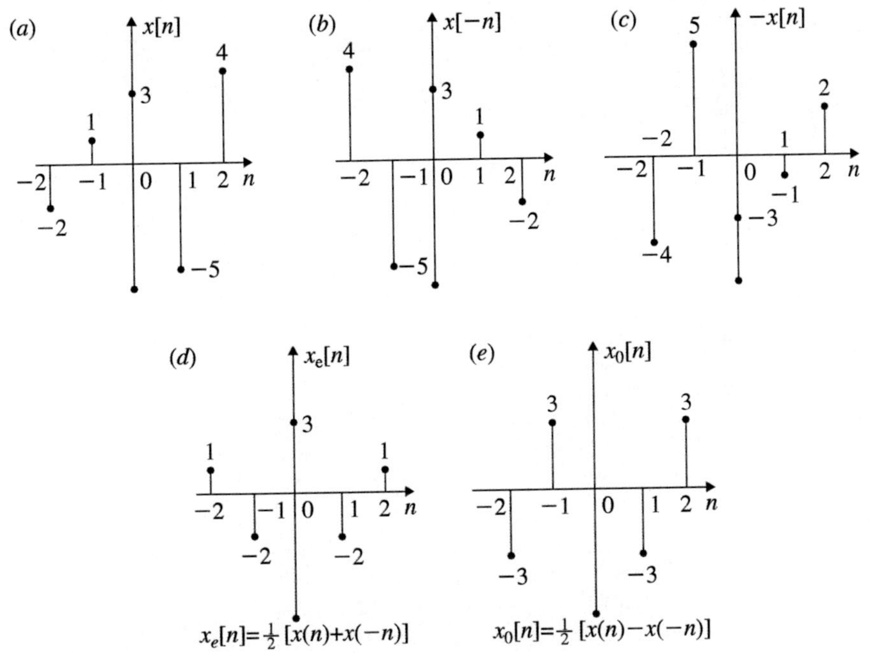

Fig. 1.26 Graphical determination of even and odd components from $x[n]$

4. $x_e[n] = \frac{1}{2}\left[x[n] + x[-n]\right]$. Figure 1.26a and b sample wise are added and their amplitudes are divided by the factor 2. This gives $x_e[n]$ and is represented in Fig. 1.26d.
5. $x_0[n] = \frac{1}{2}\left[x[n] - x[-n]\right]$. Figure 1.26a and c sample wise are added and their amplitudes are divided by a factor 2 to get $x_0[n]$. This is represented in Fig. 1.26e.

Example 1.19 Find the even and odd components of the following DT signal and sketch the same:

$$x[n] = \{-2,\ 1,\ 2,\ -1,\ 3\}$$

Solution

$$\begin{aligned}
x[n] &= \{-2,\ 1,\ 2,\ -1,\ 3\} \\
x[-n] &= \{3,\ -1,\ 2,\ 1,\ -2\} \\
&\qquad\qquad\quad\ \uparrow \\
x_e[n] &= \frac{1}{2}\left\{x[n] + x[-n]\right\} \\
&= \frac{1}{2}\left[\{-2,\ 1,\ 2,\ -1,\ 3\} + \{3,\ -1,\ 2,\ 1,\ -2\}\right]
\end{aligned}$$

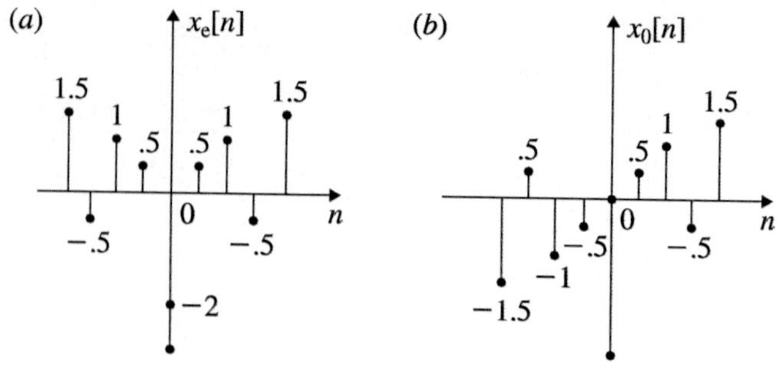

Fig. 1.27 **a** Even components and **b** Odd components

$$\uparrow \qquad\qquad\qquad\qquad\qquad\qquad\qquad\qquad \uparrow$$

$$= \{1.5,\ -.5,\ 1,\ .5,\ -2,\ .5,\ 1,\ -.5,\ 1.5\}$$

$$\uparrow$$

$$x_0[n] = \frac{1}{2}\left[x[n] - x[-n]\right]$$

$$= \frac{1}{2}\left[\{-2,\ 1,\ 2,\ -1,\ 3\} - \{3,\ -1,\ 2,\ 1,\ -2\}\right]$$

$$\uparrow \qquad\qquad\qquad\qquad\qquad\qquad\qquad\qquad \uparrow$$

$$x_0[n] = \{-1.5,\ .5,\ -1,\ -.5,\ 0,\ .5,\ 1,\ -.5,\ 1.5\}$$

$$\uparrow$$

Even and odd components of $x[n]$ are represented in Fig. 1.27a and b, respectively.

1.6.3 Energy and Power of DT Signals

For a discrete time signal $x[n]$, the total energy is defined as

$$E = \sum_{n=-\infty}^{\infty} |x[n]|^2 \tag{1.15}$$

The average power is defined as

$$P = \underset{N\to\infty}{Lt} \frac{1}{(2N+1)} \sum_{n=-N}^{N} |x[n]|^2 \tag{1.16}$$

From the definitions of energy and power, the following inferences are derived:

1. $x[n]$ is an energy sequence iff $0 < E < \infty$. For finite energy signal, the average power $P = 0$.
2. $x[n]$ is a power sequence iff $0 < P < \infty$. For a sequence with average power P being finite, the total energy $E = \infty$.
3. Periodic signal is a power signal and *vice versa* is not true. Here the energy of the signal per period is finite.
4. Signals which do not satisfy the definitions of total energy and average power, neither termed as a power signal nor an energy signal. The following summation formulae are very often used while evaluating the average power and total energy of the DT sequence.

1.

$$\sum_{n=0}^{N-1} a^n = \frac{(1-a^n)}{(1-a)} \qquad a \neq 1 \tag{1.17}$$
$$= N \qquad a = 1$$

2.

$$\sum_{n=0}^{\infty} a^n = \frac{1}{(1-a)} \qquad a < 1 \tag{1.18}$$

3.

$$\sum_{n=m}^{\infty} a^n = \frac{a^m}{(1-a)} \qquad a < 1 \tag{1.19}$$

4.

$$\sum_{n=0}^{\infty} na^n = \frac{a}{(1-a)^2} \qquad a < 1 \tag{1.20}$$

Example 1.20 Determine whether the following signals are energy signals or power signals:

$$\begin{aligned}
&(a) \qquad x[n] = A\delta[n] \\
&(b) \qquad x[n] = u[n] \\
&(c) \qquad x[n] = \text{ramp}\, n \\
&(d) \qquad x[n] = A \\
&(e) \qquad x[n] = 2e^{j(\pi n+\theta)} \\
&(f) \qquad x[n] = \cos \frac{\pi}{2} n
\end{aligned}$$

Solution (*a*) $\boldsymbol{x[n] = A\delta[n]}$

$$
\begin{aligned}
x[n] &= A\delta[n] \\
&= A \qquad n = 0 \\
&= 0 \qquad n \neq 0
\end{aligned}
$$

$$\text{Energy } E = \sum_{n=0}^{0} (A)^2$$

$$\boxed{E = A^2}$$

For unit impulse, $A = 1$ and $E = 1$.

(*b*) $\boldsymbol{x[n] = u[n];\ n \geq 0}$

$$
\begin{aligned}
P &= \underset{N\to\infty}{Lt} \frac{1}{(2N+1)} \sum_{n=0}^{N} |x(n)|^2 \\
&= \underset{N\to\infty}{Lt} \frac{1}{(2N+1)} \sum_{n=0}^{N} 1
\end{aligned}
$$

But $\sum_{n=0}^{N} 1 = (N+1)$

$$
\begin{aligned}
P &= \underset{N\to\infty}{Lt} \frac{(N+1)}{(2N+1)} \\
&= \underset{N\to\infty}{Lt} \frac{N(1+\frac{1}{N})}{N(2+\frac{1}{N})} = \frac{1}{2}
\end{aligned}
$$

$$\boxed{\begin{aligned} P &= \frac{1}{2} \\ E &= \infty \end{aligned}}$$

(*c*) $\boldsymbol{x[n] = \text{ramp}\, n;\ n \geq 0}$

$$P = \underset{N\to\infty}{Lt} \frac{1}{(2N+1)} \sum_{n=0}^{N} |x[n]|^2$$

$$P = \underset{N\to\infty}{Lt} \frac{1}{(2N+1)} \sum_{n=0}^{N} n^2$$

But $\sum_{n=0}^{N} n^2 = \frac{N(N+1)(2N+1)}{6}$

$$P = \underset{N\to\infty}{Lt} \frac{N(N+1)(2N+1)}{(2N+1)6}$$

$$\boxed{P = \infty}$$

$$E = \underset{N\to\infty}{Lt} \sum_{n=0}^{N} n^2$$
$$= \underset{N\to\infty}{Lt} \frac{N(N+1)(2N+1)}{6} = \infty$$

$$\boxed{E = \infty}$$

The signal $x[n] = n$ is neither a power signal nor an energy signal.

(*d*) $\boldsymbol{x[n] = A}$

$$P = \underset{N\to\infty}{Lt} \frac{1}{(2N+1)} \sum_{n=-\infty}^{\infty} A^2$$
$$= \underset{N\to\infty}{Lt} \frac{A^2}{(2N+1)}(2N+1) \qquad \left[\sum_{n=-\infty}^{\infty} 1 = (2N+1)\right]$$

$$\boxed{\begin{array}{c} P = A^2 \\ E = \infty \end{array}}$$

(*e*) $\boldsymbol{x[n] = 2e^{j(\pi n+\theta)}}$

$$P = \underset{N\to\infty}{Lt} \frac{1}{(2N+1)} \sum_{-N}^{N} |2e^{j(n\pi+\theta)}|^2$$
$$P = \underset{N\to\infty}{Lt} \frac{1}{2N+1} 4 \sum_{-N}^{N} |e^{j(n\pi+\theta)}|^2$$

But $|e^{j(n\pi+\theta)}| = 1$ and $\sum_{-N}^{N} 1 = (2N+1)$

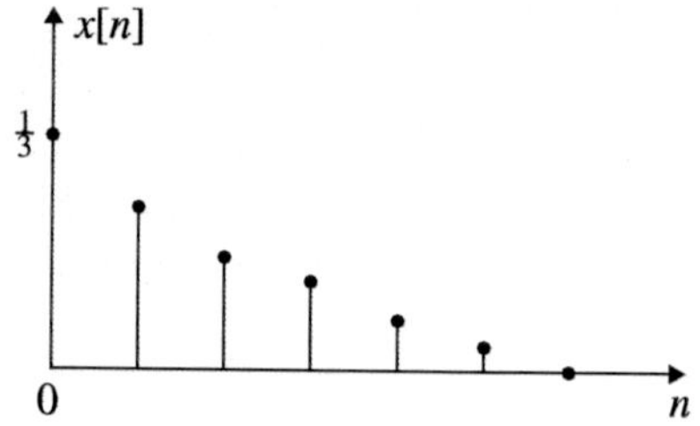

Fig. 1.28 $x[n] = \left(\frac{1}{3}\right)^2 u[n]$

$$P = \underset{N\to\infty}{Lt}\ 4\frac{(2N+1)}{(2N+1)} = 4$$

$$\boxed{\begin{array}{c} P = 4 \\ E = \infty \end{array}}$$

(*f*) $\boldsymbol{x[n] = \cos \frac{\pi}{2} n}$

$$P = \frac{1}{(2N+1)} \sum_{-N}^{N} \cos^2 \frac{\pi}{2} n$$

Since $\sum_{-N}^{N} \cos \pi n = 0$,

$$\begin{aligned} P &= \underset{N\to\infty}{Lt} \frac{1}{(2N+1)} \sum_{-N}^{N} \frac{(1+\cos \pi n)}{2} \\ &= \frac{1}{2} \underset{N\to\infty}{Lt} \frac{(2N+1)}{(2N+1)} \\ &= \frac{1}{2} \end{aligned}$$

$$\boxed{\begin{array}{c} P = \frac{1}{2} \\ E = \infty \end{array}}$$

Example 1.21 Determine the energy of the signal shown in Fig. 1.28 whose

$$x[n] = \left(\frac{1}{3}\right)^n u[n]$$

Solution

$$E = \underset{N\to\infty}{Lt} \sum_{n=0}^{N} \left(\frac{1}{3}\right)^{2n}$$

$$= \underset{N\to\infty}{Lt} \sum_{n=0}^{N} \left(\frac{1}{9}\right)^{n}$$

$$= 1 + \frac{1}{9} + \left(\frac{1}{9}\right)^{2} + \ldots$$

$$= \frac{1}{1-\frac{1}{9}}$$

$$\boxed{\begin{aligned} E &= \frac{9}{8} \\ P &= 0 \end{aligned}}$$

Example 1.22 Find the energy of the following sequence:

$$x[n] = n \qquad 0 \le n \le 4$$

Solution

$$x[n] = n$$

$$= \{0,\ 1,\ 2,\ 3,\ 4\}$$

$$E = \sum_{n=0}^{4} n^2$$

$$= 0 + 1 + 4 + 9 + 16$$

$$\boxed{E = 30}$$

Example 1.23 Determine the average power and the energy per period of the sequence shown in Fig. 1.29.

Solution The fundamental period N of the signal is 5. Hence, the average power per period is

$$P = \frac{1}{5}\sum_{n=0}^{4} n^2 = \frac{1}{5}[0 + 1 + 4 + 9 + 16]$$

$$\boxed{P = 6}$$

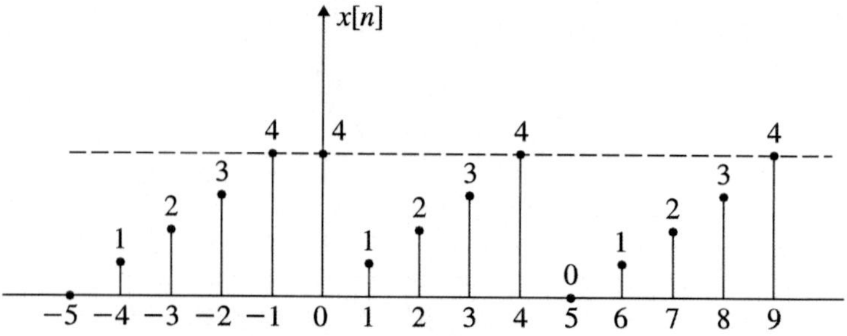

Fig. 1.29 $x[n]$ of Example 1.23

Average energy per period is

$$E = \sum_{n=0}^{4} n^2$$
$$= [0 + 1 + 4 + 9 + 16]$$

$$\boxed{E = 30}$$

Example 1.24 Find the energy and power of the following signal:

$$x[n] = a^n u[n]$$

for the following cases:

$$(a) \quad |a| < 1$$
$$(b) \quad |a| = 1$$
$$(c) \quad |a| > 1$$

Solution (a) $\mathbf{x[n] = a^n u[n]}$ **where** $\mathbf{|a| < 1}$ **and** $\mathbf{n \geq 0}$

$$E = \sum_{n=0}^{\infty} (a^n)^2$$
$$= 1 + a^2 + a^4 + \ldots$$

$$\boxed{E = \frac{1}{1 - |a|^2}}$$

$$P = 0$$

(b) **$x[n] = a^n u[n]$ where $|a| = 1$**

$$E = \underset{N\to\infty}{Lt} \sum_{0}^{N} 1^n = \underset{N\to\infty}{Lt} (N+1)$$

$$\boxed{E = \infty}$$

$$P = \underset{N\to\infty}{Lt} \frac{1}{(2N+1)} \sum_{0}^{N} (1)^n$$

$$P == \underset{N\to\infty}{Lt} \frac{(N+1)}{(2N+1)}$$

$$= \underset{N\to\infty}{Lt} \frac{N(1+\frac{1}{N})}{N(2+\frac{1}{N})}$$

$$\boxed{P = \frac{1}{2}}$$

(c) **$x[n] = a^n u[n]$ where $|a| > 1$**

$$E = \underset{N\to\infty}{Lt} \sum_{0}^{N} a^n$$

$$= 1 + a + a^2 + \cdots + a^N$$

$$\boxed{E = \infty}$$

$$P = \underset{N\to\infty}{Lt} \frac{1}{(2N+1)} \sum_{n=0}^{N} a^n$$

$$= \underset{N\to\infty}{Lt} \frac{1}{(N+1)} \frac{(1-a^{N+1})}{(1-a)}$$

$$\boxed{P = \infty}$$

The signal is neither an energy nor a power signal.

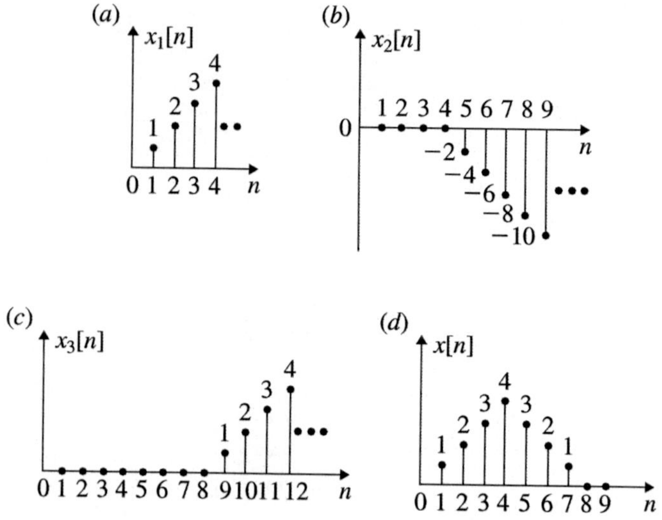

Fig. 1.30 DT energy signal of Example 1.25

Example 1.25 Find the energy of the following signal:

Solution $$x[n] = \text{ramp}[n] - 2\,\text{ramp}[n-4] + \text{ramp}[n-8]$$

$$\begin{aligned} x[n] &= \text{ramp}[n] - 2\,\text{ramp}[n-4] + \text{ramp}[n-8] \\ &= x_1[n] + x_2[n] + x_3[n] \end{aligned}$$

$x_1[n]$, $x_2[n]$ and $x_3[n]$ are shown in Fig. 1.30a, b and c, respectively. Figure 1.30d represents $x[n]$. From Fig. 1.30d, the energy of the signal $x[n]$ is obtained as

$$E = 1^2 + 2^2 + 3^2 + 4^2 + 3^2 + 2^2 + 1^2$$

$$\boxed{E = 44}$$

Example 1.26 Determine the value of power and energy of each of the following signals:

$$(a) \quad x[n] = e^{j\left(\frac{\pi n}{2} + \frac{\pi}{8}\right)}$$

$$(b) \quad x[n] = \left(\frac{1}{2}\right)^n u[n]$$

Solution (*a*) $x[n] = e^{j(\frac{\pi n}{2}+\frac{\pi}{8})}$

$$P = \underset{N\to\infty}{Lt} \frac{1}{2N+1} \sum_{-N}^{N} |e^{j(\frac{\pi n}{2}+\frac{\pi}{8})}|^2$$

$$= \underset{N\to\infty}{Lt} \frac{1}{(2N+1)} \sum_{-N}^{N} 1$$

$$P = \frac{(2N+1)}{(2N+1)} = 1$$

$$\boxed{P = 1 \text{ and } E = \infty}$$

(*b*) $x[n] = \left[\frac{1}{2}\right]^n u[n]$

$$E = \underset{N\to\infty}{Lt} \sum_{0}^{N} \left(\frac{1}{2}\right)^{2n}$$

$$= \underset{N\to\infty}{Lt} \sum_{0}^{N} \left(\frac{1}{4}\right)^{n} = \frac{1}{1-\frac{1}{4}} = \frac{4}{3}$$

$$\boxed{E = \frac{4}{3} \text{ and } P = 0}$$

Example 1.27 Find the energy of the following DT signal:

$$x[n] = \left(\frac{1}{2}\right)^n \qquad n \geq 0$$
$$= 3^n \qquad n < 0$$

Solution

$$E = \left[\sum_{-\infty}^{-1} (3)^{2n} + \sum_{0}^{\infty} \left(\frac{1}{2}\right)^{2n}\right]$$

$$= \left[\sum_{-\infty}^{-1} (9)^{n} + \sum_{0}^{\infty} \left(\frac{1}{4}\right)^{n}\right]$$

$$= \left[\sum_{1}^{\infty}(9)^{-n} + \frac{1}{(1-\frac{1}{4})}\right]$$

$$= \left[\sum_{1}^{\infty}\left(\frac{1}{9}\right)^{n} + \frac{4}{3}\right]$$

$$= \left[\frac{1}{9} + \frac{1}{9^2} + \frac{1}{9^3} + \cdots\right] + \frac{4}{3}$$

$$= \frac{1}{9}\left[1 + \frac{1}{9} + \frac{1}{9^2} + \frac{1}{9^3} + \cdots\right] + \frac{4}{3}$$

$$E = \frac{1}{9}\frac{1}{[1-\frac{1}{9}]} + \frac{4}{3}$$

$$= \frac{1}{8} + \frac{4}{3}$$

$$\boxed{E = \frac{35}{24}}$$

1.7 Discrete Time System

The block diagram of a discrete time system is shown in Fig. 1.31. $x[n]$ is the excitation (input) signal and $y[n]$ is the response (output) signal of the DT system. H represents the functional relationship between the input and output which is described by the difference equation. The input–output signals appear at discrete intervals of time n where $n = 0, 1, 2 \ldots$ which is an integer. n can also take the negative value of an integer.

$$x_1[n] + x_2[n] = y_1[n] + y_2[n]$$

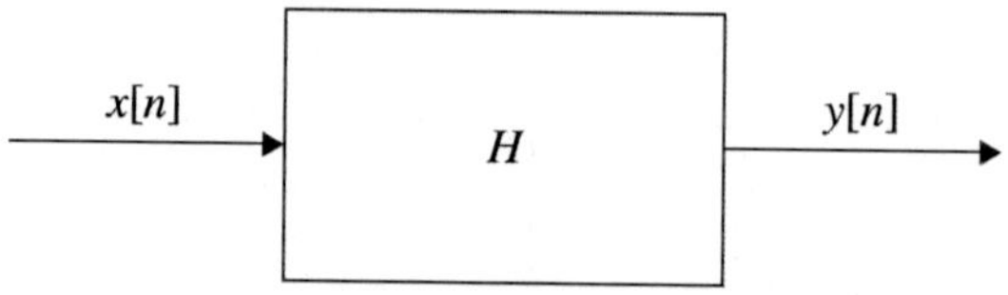

Fig. 1.31 Block diagram representation of discrete time system

1.8 Properties of Discrete Time System

Like CT systems, DT systems also possess similar properties which are given below:

1. Linear and Non-linear Systems.
2. Time Invariant and Time Varying DT Systems.
3. Causal and Non-causal DT Systems.
4. Stable and Unstable Systems.
5. Static and Dynamic Systems.
6. Invertible and Inverse Discrete Time Systems.

1.8.1 Linear and Non-linear Systems

A linear discrete time system obeys the property of superposition. As discussed for the CT system, the superposition property is composed of homogeneity and additivity. Let $x_1[n]$ excitation produce $y_1[n]$ response and $x_2[n]$ produce $y_2[n]$ response. According to additivity property of superposition theorem, if both $x_1[n]$ and $x_2[n]$ are applied simultaneously, then

$$x_1[n] + x_2[n] = y_1[n] + y_2[n]$$

Let $a_1x_1[n]$ and $a_2x_2[n]$ be the inputs. According to the homogeneity (scaling) property, when these signals are separately applied,

$$a_1x_1[n] = a_1y_1[n]$$
$$a_2x_2[n] = a_2y_2[n]$$

If $a_1x_1[n] + a_2x_2[n]$ are simultaneously applied, the output is obtained by applying superposition theorem as

$$a_1x_1[n] + a_2x_2[n] = a_1y_1[n] + a_2y_2[n]$$

In the above equation, $a_1x_1[n] + a_2x_2[n]$ is called the weighted sum of input and $a_1y_1[n] + a_2y_2[n]$ is called the weighted sum of the output. Therefore, the following procedure is followed to test the linearity of a DT system:

1. Express

$$y_1[n] = f(x_1[n])$$
$$y_2[n] = f(x_2[n])$$

2. Find the weighted sum of the output as

$$y_3[n] = a_1 y_1[n] + a_2 y_2[n]$$

3. Find the output $y_4[n]$ due to the weighted sum of input as

$$y_4[n] = f(a_1 x_1[n] + a_2 x_2[n])$$

4. If $y_3[n] = y_4[n]$, then the given DT system is linear. Otherwise, it is non-linear.

The following examples illustrate the method of testing a DT system for its linearity.

Example 1.28 Test whether the following DT systems are linear or not:

$$\begin{aligned} &(a) \quad y[n] = x^2[n] \\ &(b) \quad y[n] = x[4n+1] \\ &(c) \quad y[n] = x[n] + \frac{1}{x[n+1]} \\ &(d) \quad y[n] = x[n^2] \\ &(e) \quad y[n] = x[n] + nx[n+1] \end{aligned}$$

Solution (*a*) $\boldsymbol{y[n] = x^2[n]}$

$$\begin{aligned} y_1[n] &= x_1^2[n] \\ y_2[n] &= x_2^2[n] \end{aligned}$$

1. The weighted sum of the output $y_3[n]$ is

$$\begin{aligned} y_3[n] &= a_1 y_1[n] + a_2 y_2[n] \\ &= a_1 x_1^2[n] + a_2 x_2^2[n] \end{aligned}$$

2. The output due to the weighted sum of the input $y_4[n]$ is

$$\begin{aligned} y_4[n] &= [a_1 x_1[n] + a_2 x_2[n]]^2 \\ &= a_1^2 x_1^2[n] + a_2^2 x_2^2[n] + 2a_1 a_2 x_1[n] x_2[n] \end{aligned}$$

3.

$$y_3[n] \neq y_4[n]$$

$$\boxed{\text{The system is Non-linear.}}$$

(b) $\mathbf{y[n] = x[4n+1]}$

$$\begin{aligned} a_1 y_1[n] &= a_1 x_1[4n+1] \\ a_2 y_2[n] &= a_2 x_2[4n+1] \\ y_3[n] &= a_1 y_1[n] + a_2 y_2[n] \end{aligned}$$

1. The weighted sum of the output is

$$\begin{aligned} y_3[n] &= a_1 y_1[n] + a_2 y_2[n] \\ &= a_1 x_1[4n+1] + a_2 x_2[4n+1] \end{aligned}$$

2. The output due to the weighted sum of the input is

$$y_4[n] = a_1 x_1[4n+1] + a_2 x_2[4n+1]$$

3.

$$y_3[n] = y_4[n]$$

$$\boxed{\text{The system is Linear.}}$$

(c) $\mathbf{y[n] = x[n] + \frac{1}{x(n+1)}}$

$$\begin{aligned} a_1 y_1[n] &= a_1 \left[x_1[n] + \frac{1}{x_1(n+1)} \right] \\ a_2 y_2[n] &= a_2 \left[x_2[n] + \frac{1}{x_2(n+1)} \right] \end{aligned}$$

1. The weighted sum of the output $y_3[n]$ is,

$$\begin{aligned} y_3[n] &= a_1 y_1[n] + a_2 y_2[n] \\ &= a_1 \left[x_1[n] + \frac{1}{x_1(n+1)} \right] + a_2 \left[x_2[n] + \frac{1}{x_2(n+1)} \right] \end{aligned}$$

2. The output due to the weighted sum of the input is

$$\begin{aligned} y_4[n] &= f[a_1 x_1[n] + a_2 x_2[n]] \\ &= a_1[x_1[n] + a_2 x_2[n]] + \left[\frac{1}{a_1 x_1[n+1] + a_2 x_2[n+1]} \right] \end{aligned}$$

3.

$$y_3[n] \neq y_4[n]$$

The system is Non-linear.

(*d*) $\mathbf{y[n] = x[n^2]}$

$$a_1 y_1[n] = a_1 x_1[n^2]$$
$$a_2 y_2[n] = a_2 x_2[n^2]$$

1. The weighted sum of the output $y_3[n]$ is

$$\begin{aligned} y_3[n] &= a_1 y_1[n] + a_2 y_2[n] \\ &= a_1 x_1[n^2] + a_2 x_2[n^2] \end{aligned}$$

2. The output $y_4[n]$ due to the weighted sum of input is,

$$y_4[n] = a_1 x_1[n^2] + a_2 x_2[n^2]$$

3.

$$y_3[n] = y_4[n]$$

The system is Linear.

(*e*) $\mathbf{y[n] = x[n] + nx[n+1]}$

$$a_1 y_1[n] = a_1[x_1[n] + nx_1[n+1]]$$
$$a_2 y_2[n] = a_2[x_2[n] + nx_2[n+1]]$$

1. The weighted sum of the output is,

$$\begin{aligned} y_3[n] &= a_1 y_1[n] + a_2 y_2[n] \\ &= a_1[x_1[n] + nx_1[n+1]] + a_2[x_2[n] + nx_2[n+1]] \end{aligned}$$

2. The output due to the weighted sum of the input is

$$y_4[n] = a_1 x_1[n] + a_2 x_2[n] + a_1 n x_1[n+1] + a_2 n x_2[n+1]$$

3.

$$y_3[n] = y_4[n]$$

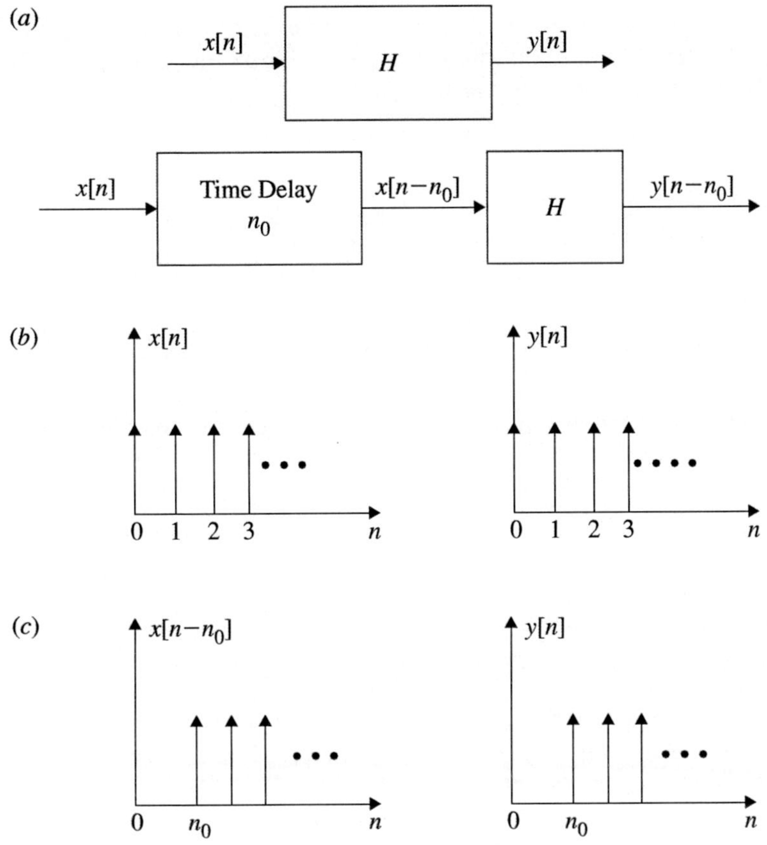

Fig. 1.32 Block diagram and signal representation to illustrate time invariancy of DT system

The system is Linear.

1.8.2 Time Invariant and Time Varying DT Systems

Consider the discrete time system represented in the block diagram of Fig. 1.32a. If the input is $x[n]$, then the output is $y[n]$. If the input is time delayed by n_0, which becomes $x[n - n_0]$, the output becomes $y[n - n_0]$. The signal representation and the delayed signals are shown in Fig. 1.32b and c, respectively. Such systems are called time invariant.

If an arbitrary excitation $x[n]$ of a system causes a response $y[n]$ and the delayed excitation $x[n-n_0]$ where n_0 is any arbitrary integer causes $y[n-n_0]$ then the system is said to be time invariant.

Procedure to Check Time Invariance of DT Systems

1. For the delayed input $x[n-n_0]$ find the output $y[n, n_0]$.
2. Obtain the delayed output $y[n-n_0]$ by substituting $n = n - n_0$ in $y[n]$.
3. If $y[n, n_0] = y[n-n_0]$, the system is time invariant. If not, the system is time variant.

The following examples illustrate the method of testing the time invariancy of DT systems.

Example 1.29 Determine whether the following systems are time invariant or not:

$$(a) \quad y[n] = nx[n]$$
$$(b) \quad y[n] = x[2n]$$
$$(c) \quad y[n] = x[-n]$$
$$(d) \quad y[n] = \sin(x[n])$$
$$(e) \quad y[n] = x[n]x[n-1]$$

Solution (a) $\boldsymbol{y[n] = nx[n]}$

1. The output for the delayed input $x[n-n_0]$ is

$$y[n, n_0] = nx[n-n_0]$$

2. The delayed output for the input $x[n]$ is

$$y[n-n_0] = (n-n_0)x[n-n_0]$$

3.

$$y[n, n_0] \neq y[n-n_0]$$

The system is Time Variant.

(b) $\boldsymbol{y[n] = x[2n]}$

$$y[n, n_0] = x[2n-n_0]$$
$$y[n-n_0] = x[2(n-n_0)]$$
$$= x[2n-2n_0]$$
$$y[n, n_0] \neq y[n-n_0]$$

The system is Time Variant.

(*c*) $\mathbf{y[n] = x[-n]}$

$$\begin{aligned} y[n, n_0] &= x[-n - n_0] \\ y[n - n_0] &= x[-(n - n_0)] \\ &= x[-n + n_0] \\ y[n, n_0] &\neq y[n - n_0] \end{aligned}$$

The system is Time Variant.

(*d*) $\mathbf{y[n] = \sin(x[n])}$

$$\begin{aligned} y[n, n_0] &= \sin(x[n - n_0]) \\ y[n - n_0] &= \sin(x[n - n_0]) \\ y[n, n_0] &= y[n - n_0] \end{aligned}$$

The system is Time Invariant.

(*e*) $\mathbf{y[n] = x[n]x[n - 1]}$

$$\begin{aligned} y[n, n_0] &= x[n - n_0]x[n - n_0 - 1] \\ y[n - n_0] &= x[n - n_0]x[n - n_0 - 1] \\ y[n, n_0] &= y[n - n_0] \end{aligned}$$

The system is Time Invariant.

1.8.3 Causal and Non-causal DT Systems

A discrete time system is said to be causal if the response of the system depends on the present or the past inputs applied. The system is non-causal if the output depends on the future input.

The following examples illustrate the method of identifying causal and non-causal systems.

Example 1.30 Determine whether the following systems are causal or not:

$$
\begin{aligned}
&(a) \quad y[n] = x[n-1] \\
&(b) \quad y[n] = x[n] + x[n-1] \\
&(c) \quad y[n-1] = x[n] \\
&(d) \quad y[n] = \sin(x[n]) \\
&(e) \quad y[n] = \sum_{k=-\infty}^{n+4} x(k) \\
&(f) \quad y[n] = \sum_{k=0}^{-3} x(k)
\end{aligned}
$$

Solution (a) $\boldsymbol{y[n] = x[n-1]}$

$$
\begin{aligned}
y[0] &= x[-1] \\
y[1] &= x[0]
\end{aligned}
$$

The output depends on the past value of $x[n]$. Hence

$$\boxed{\text{The system is Causal.}}$$

(b) $\boldsymbol{y[n] = x[n] + x[n-1]}$

$$
\begin{aligned}
y[0] &= x[0] + x[-1] \\
y[1] &= x[1] + x[0]
\end{aligned}
$$

here $x[1]$ is present value and $x[0]$ is past value. The output depends on the present and past inputs. Hence

$$\boxed{\text{The system is Causal.}}$$

(c) $\boldsymbol{y[n-1] = x[n]}$

$$
\begin{aligned}
y[-1] &= x[0] \quad \text{(Future input)} \\
y[0] &= x[1] \quad \text{(Future input)}
\end{aligned}
$$

The output depends on the future inputs. Hence

$$\boxed{\text{The system is Non-causal.}}$$

(d) $\boldsymbol{y[n] = \sin x[n]}$

$$y[0] = \sin x[0]$$
$$y[-1] = \sin x[-1]$$

The output depends on the present input. Hence

The system is Causal.

(*e*) $\boldsymbol{y[n] = \sum_{k=-\infty}^{n+4} x[k]}$

$$y[0] = \sum_{-\infty}^{4} x[k]$$
$$= x[-\infty] + x[-\infty + 1] + \cdots + x[-1] + x[0] + x[1] + x[2] + x[3] + x[4]$$

$$x[-\infty] + x[-\infty + 1], \ldots, x[-1] = \text{Past input}$$
$$x[0] = \text{Present input}$$
$$x[0],\ x[1],\ x[2],\ x[3] \text{ and } x[4] = \text{Future input}$$

The output depends on past, present and future input. Hence

The system is Non-causal.

(*f*) $\boldsymbol{y[n] = \sum_{k=0}^{n-3} x[k]}$

$$y[0] = \sum_{k=0}^{-3} x[k]$$
$$= x[0] + x[-1] + x[-2] + x[-3]$$
$$x[0] = \text{Present input}$$
$$x[-1] = \text{Past input}$$

The output depends on the present and past input. Hence

The system is Causal.

1.8.4 Stable and Unstable Systems

A discrete time system is said to be stable, if for any bounded input, it produces a bounded output. This implies that the impulse response

$$\boldsymbol{y[n] = \sum_{-\infty}^{\infty} |h[n]| < \infty}$$

is absolutely summable.

For a bounded input

$$|x[n]| \leq M_x < \infty$$

the output

$$|y[n]| \leq M_y < \infty$$

From the above two conditions, it can be obtained that

$$y[n] = \sum_{-\infty}^{\infty} |h[n]| < \infty$$

The following examples illustrate the above procedure.

Example 1.31 Check whether the DT systems described by the following equations are stable or not:

$$(a) \qquad y[n] = \sin x[n]$$

$$(b) \qquad y[n] = \sum_{k=0}^{n+1} x[k]$$

$$(c) \qquad y[n] = e^{x[n]}$$

$$(d) \qquad h[n] = 3^n u[n+3]$$

$$(e) \qquad y[n] = \delta[-n-3]$$

$$(f) \qquad y[n] = \delta[n-1] + \delta[n] + \delta[n+1]$$

$$(g) \qquad h[n] = e^{-|n|}$$

$$(h) \qquad h[n] = n\, u[n]$$

$$(i) \qquad h[n] = 3^n u[n-3]$$

$$(j) \qquad h[n] = 2^n u[-n]$$

Solution (a) $\boldsymbol{y[n] = \sin x[n]}$

If $x[n]$ is bounded, then $\sin x[n]$ is also bounded and so $y[n]$ is also bounded

The system is Stable.

(*b*) $\boldsymbol{y[n] = \sum_{k=0}^{n+1} x[k]}$
Here as $n \to \infty$, $y[n] \to \infty$ and the output is unbounded. For bounded input, n should be a finite number. In that case, $y[n]$ is bounded and the system is stable.

The system is Stable.	for $n =$ finite
The system is Unstable.	for $n = \infty$

(*c*) $\boldsymbol{y[n] = e^{x[n]}}$
For $|x[n]|$ bounded, $e^{|x[n]|}$ is bounded and the system is stable.

The system is Stable.

(*d*) $\boldsymbol{h[n] = 3^n u[n+3]}$

$$\begin{aligned} |y[n]| &= \sum_{n=-3}^{\infty} 3^n \\ &= (3)^{-3} + (3)^{-2} + (3)^{-1} + (3)^0 + (3)^1 + \cdots + (3)^{\infty} \\ &= \infty \end{aligned}$$

The output is unbounded.

The system is Unstable.

(*e*) $\boldsymbol{y[n] = \delta[-n-3]}$

$$\begin{aligned} y[n] &= \delta[-n-3] \\ &= 1 \qquad n = -3 \\ &= 0 \qquad \text{otherwise} \end{aligned}$$

The system is Stable.

(*f*) $\boldsymbol{y[n] = \delta[n-1] + \delta[n] + \delta[n+1]}$

$$\begin{aligned} y[0] &= \delta[-1] + \delta[0] + \delta[1] = 0 + 1 + 0 = 1 \\ y[1] &= \delta[0] + \delta[1] + \delta[2] = 1 + 0 + 0 = 1 \\ y[-1] &= \delta[-2] + \delta[-1] + \delta[0] = 0 + 0 + 1 = 1 \\ y[-2] &= \delta[1] + \delta[2] + \delta[3] = 0 + 0 + 0 = 0 \\ y[2] &= \delta[1] + \delta[2] + \delta[3] = 0 + 0 + 0 = 0 \\ y[n] &= \sum_{-\infty}^{\infty} |h[k]| = 1 + 1 + 1 = 3 < \infty \end{aligned}$$

The system is Stable.

(*g*) $\boldsymbol{h[n] = e^{-|n|}}$

$$
\begin{aligned}
y[n] &= \sum_{-\infty}^{\infty} e^{|n|} = \sum_{-\infty}^{-1} e^{|n|} + \sum_{0}^{\infty} e^{-|n|} \\
&= \sum_{1}^{\infty} e^{-n} + \sum_{0}^{\infty} e^{-n} \\
&= e^{-1} + e^{-2} + \cdots + 1 + e^{-1} + e^{-2} + \cdots \\
&= e^{-1}[1 + e^{-1} + e^{-2} + \cdots] + 1 + e^{-1} + e^{-2} + \cdots \\
&= e^{-1}\frac{1}{[1 - e^{-1}]} + \frac{1}{[1 - e^{-1}]} \\
&= \frac{e^{-1}}{(1 - e^{-1})} + \frac{1}{(1 - e^{-1})} \\
&= \frac{[1 + e^{-1}]}{[1 - e^{-1}]} < \infty
\end{aligned}
$$

The system is Stable.

(*h*) $\boldsymbol{h[n] = n\,u[n]}$

$$
y[n] = \sum_{0}^{\infty} n = 1 + 2 + \cdots + \infty = \infty
$$

The system is Unstable.

(*i*) $\boldsymbol{h[n] = 3^n u[n-3]}$

$$
y[n] = \sum_{3}^{\infty} 3^n = 3^3 + 3^2 + \cdots + \infty = \infty
$$

The system is Unstable.

(*j*) $\boldsymbol{h[n] = 2^n u[-n]}$

$$
y[n] = \sum_{-\infty}^{-1} 2^n = \sum_{1}^{\infty} \left(\frac{1}{2}\right)^n
$$

$$
\begin{aligned}
&= \frac{1}{2} + \left(\frac{1}{2}\right)^2 + \cdots \\
&= \frac{1}{2}\left[1 + \frac{1}{2} + \left(\frac{1}{2}\right)^2 + \cdots\right] \\
&= \frac{1}{2}\left[\frac{1}{1-\frac{1}{2}}\right] = 1 < \infty
\end{aligned}
$$

$$\boxed{\text{The system is Stable.}}$$

1.8.5 *Static and Dynamic Systems*

A discrete time system is said to be static (memoryless or instantaneous) if the output response depends on the present value only and not on the past and future values of excitation. Discrete systems described by difference equations require memory and hence they are dynamic systems.

The following examples illustrate the method of identifying static and dynamic discrete systems.

Example 1.32 Identify whether the following systems are static or dynamic:

$$
\begin{aligned}
&(a) \qquad y[n] = x[3n] \\
&(b) \qquad y[n] = \sin(x[n]) \\
&(c) \qquad y[n-1] + y[n] = x[n] \\
&(d) \qquad y[n] = \text{sgn}|x[n]|
\end{aligned}
$$

Solution (*a*) $\boldsymbol{y[n] = x[3n]}$

$$
\begin{aligned}
y[1] &= x[3] \\
y[-1] &= x[-3]
\end{aligned}
$$

The output $y[1]$ and $y[-1]$ depends on the future value $x[3]$ and the past input $x[-3]$, respectively. Hence,

$$\boxed{\text{The system is Dynamic.}}$$

(*b*) $\boldsymbol{y[n] = \sin(x[n])}$

$$
\begin{aligned}
y[0] &= \sin(x[0]) \\
y[1] &= \sin(x[1])
\end{aligned}
$$

The output depends on the present input at all times. Hence,

The system is Static.

(*c*) $\mathbf{y[n-1]+y[n]=x[n]}$
The system is described by first order difference equation which require memory. Hence,

The system is Dynamic.

(*d*) $\mathbf{y[n]=\mathrm{sgn}|x[n]|}$

$$\begin{aligned}\mathrm{sgn}|x[n]| &= 1 && \text{for } n>0\\ &= -1 && \text{for } n<0\\ y[1] &= x[1]=1\\ y[-1] &= x[-1]=-1\end{aligned}$$

The output depends on the present value of the input. Hence,

The system is Static.

1.8.6 Invertible and Inverse Discrete Time Systems

A discrete time system is said to be invertible if distinct input leads to distinct output. If a system is invertible then an inverse system exists.

Consider the system shown in Fig. 1.33. The input $x[n]$ produces the output $y[n]$. This system is in cascade with its inverse system. The output of this system is nothing but the difference of the two successive inputs $y[n]-y[n-1]$. This is the input to the original system. Thus, by connecting an inverse system in cascade with the original system, the excitation signal $x[n]$ is re-established provided the original system is invertible. The concept of invertibility is very widely used in communications.

Example 1.33 Determine whether the following systems are static, causal, time invariant, linear and stable.

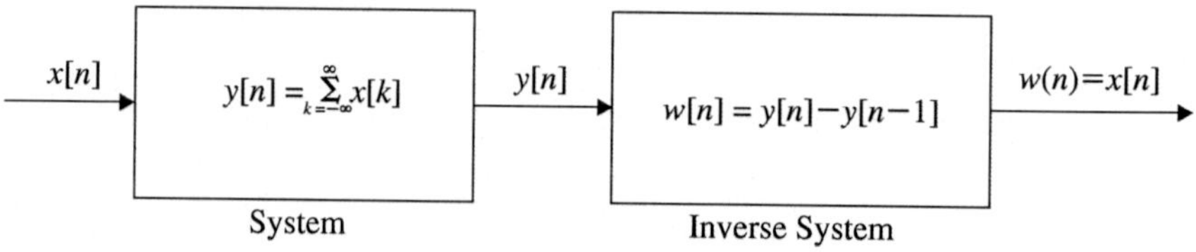

Fig. 1.33 Inverse discrete time system

$$
\begin{array}{ll}
(a) & y[n] = x[4n+1] \\
(b) & y[n] = x[n] + n\,x[n+1] \\
(c) & y[n] = x[n]u[n] \\
(d) & y[n] = \log_{10} x[n] \\
(e) & y[n] = x^2[n]
\end{array}
$$

Solution (*a*) $\boldsymbol{y[n] = x[4n+1]}$

1.

$$y[0] = x[1]$$

The output depends on future input. Hence,

$$\boxed{\text{The system is Dynamic and Non-causal.}}$$

2. The output due to the delayed input is,

$$y[n, n_0] = x[4n - n_0 + 1]$$

The delayed output due to the input is,

$$
\begin{aligned}
y[n - n_0] &= x[4(n - n_0) + 1] \\
&= x[4n - 4n_0 + 1] \\
y[n, n_0] &\neq y[n - n_0]
\end{aligned}
$$

$$\boxed{\text{The system is Time Variant.}}$$

3.

$$
\begin{aligned}
a_1 y_1[n] &= a_1 x_1[4n+1] \\
a_2 y_2[n] &= a_2 x_2[4n+1] \\
y_3[n] &= a_1 y_1[n] + a_2 y_2[n] \\
&= a_1 x_1[4n+1] + a_2 x_2[4n+1] \\
y_4[n] &= a_1 x_1[4n+1] + a_2 x_2[4n+1] \\
y_3[n] &= y_4[n]
\end{aligned}
$$

$$\boxed{\text{The system is Linear.}}$$

4. The input is time shifted and time compressed signal. As long as the input is bounded the output is also bounded.

The system is Stable.

The system is

(1) Dynamic, (2) Non-causal, (3) Time Variant, (4) Linear and (5) Stable.

(*b*) $\mathbf{y[n] = x[n] + n\,x[n+1]}$

1.

$$y[0] = x[0] + 0 \times x[1]$$

The output depends on present and future inputs.

The system is Dynamic and Non-causal.

2. The output due to the delayed input is,

$$y[n, n_0] = x[n - n_0] + nx[n - n_0 + 1]$$

The delayed output due to the input is

$$y[n - n_0] = x[n - n_0] + (n - n_0)x[n - n_0 + 1]$$
$$y[n, n_0] \neq y[n - n_0]$$

The system is Time Variant.

3. The weighted sum of the output due to the input is

$$y_3[n] = a_1 y_1[n] + a_2 y_2[n] = a_1 x_1[n] + a_1 n x_1[n+1] + a_2 x_2[n] + a_2 n x_2[n+1]$$

The output due to the weighted sum of the input is

$$y_4[n] = a_1\{x_1[n] + nx_1[n+1]\} + a_2\{x_2[n] + a_2 n x_2[n+1]\}$$
$$y_3[n] = y_4[n]$$

The system is Linear.

4. As long as $x[n]$ is bounded, $y[n]$ increases as n increases and the system is unstable.

The system is Unstable.

The system is

(1) Dynamic, (2) Non-causal, (3) Time Variant, (4) Linear and (5) Unstable.

(c) $\boldsymbol{y[n] = x[n]u[n]}$

1.

$$y[0] = x[0]u[0]$$
$$y[1] = x[1]u[1]$$

The output depends on the present input only.

The system is Static and Causal.

2. For a causal signal $n \geq 0$. The weighted sum of the output due to input is,

$$\begin{aligned} y_3[n] &= a_1 y_1[n] + a_2 y_2[n] \\ &= \{a_1 x_1[n] + a_2 x_2[n]\}u[n] \end{aligned}$$

The output due to the weighted sum of input is,

$$y_4[n] = \{a_1 x_1[n] + a_2 x_2[n]\}u[n]$$
$$y_3[n] = y_4[n]$$

The system is Linear.

3. The output due to the delayed input is

$$y[n, n_0] = x_1[n - n_0]u[n]$$

The delayed output due to the input is

$$y[n - n_0] = x_1[n - n_0]u[n - n_0]$$
$$y[n, n_0] \neq y[n - n_0]$$

The system is Time Variant.

4. As long as $x[n]$ is bounded, $y[n]$ is also bounded.

The system is Stable.

The system is

(1) Static, (2) Causal, (3) Linear, (4) Time Variant and (5) Stable.

(*d*) $\mathbf{y[n] = \log_{10} x[n]}$

1.

$$y[0] = \log_{10} x[0]$$
$$y[1] = \log_{10} x[1]$$
$$y[-1] = \log_{10} x[-1]$$

The output depends on the present input only.

The system is Static and Causal.

2. The weighted sum of the output due to input is

$$y_3[n] = a_1 y_1[n] + a_2 y_2[n]$$
$$= a_1 \log_{10} x_1[n] + a_2 \log_{10} x_2[n]$$

The output due to the weighted sum of input is

$$y_4[n] = \log_{10}(a_1 x_1[n] + a_2 x_2[n])$$
$$y_3[n] \neq y_4[n]$$

The system is Non-linear.

3. The output due to the delayed input is,

$$y[n, n_0] = \log_{10}[n - n_0]$$

The delayed output due to input is

$$y[n - n_0] = \log_{10}[n - n_0]$$
$$y[n, n_0] = y[n - n_0]$$

The system is Time Invariant.

4. As long as $x[n]$ is bounded, $\log_{10} x[n]$ is bounded and $y[n]$ is also bounded.

The system is Stable.

The system is

(1) Static, (2) Causal, (3) Non-linear, (4) Time invariant and (5) Stable.

(*e*) $\mathbf{y[n] = x^2[n]}$

1.

$$y[0] = x^2[0]$$
$$y[1] = x^2[1]$$

The output depends on the present input only.

The system is Static and Causal.

2. The weighted sum of the output due to input is,

$$\begin{aligned} y_3[n] &= a_1 y_1[n] + a_2 y_2[n] \\ &= a_1 x_1^2[n] + a_2 x_2^2[n] \end{aligned}$$

The output due to weighted sum of input is

$$\begin{aligned} y_4[n] &= \{a_1 x_1[n] + a_2 x_2[n]\}^2 \\ &= a_1^2 x_1^2[n] + a_2^2 x_2^2[n] + 2a_1 a_2 x_1[n] x_2[n] \\ y_3[n] &\neq y_4[n] \end{aligned}$$

The system is Non-linear.

3. The output due to the delayed input is

$$y[n, n_0] = x^2[n - n_0]$$

The delayed output due to the input is

$$y[n - n_0] = x^2[n - n_0]$$
$$y[n, n_0] = y[n - n_0]$$

The system is Time Invariant.

4. If $x[n]$ is bounded, $x^2[n]$ is bounded and $y[n]$ is also bounded.

The system is Stable.

The system is

(1) Static, (2) Causal, (3) Non-linear, (4) Time invariant and (5) Stable.

1.9 Sampling

1.9.1 Sampling Process

Sampling is a process of converting a continuous time signal into discrete time signal. After sampling, the signal is defined at discrete instants of time and the time interval between two subsequent sampling instants is called sampling interval. In this process of sampling, one of the important factors that we have to consider is that the sampling rate must be kept sufficiently high so that the original signal can be reconstructed from its samples.

The sampling operation can be represented by a switch as shown in Fig. 1.34. The switch is closed for a very short interval of time τ, during which the signal is available at the out. Therefore, if the input is $x(t)$, then the sampled output is $x(nT)$, $n = 0, \pm 1, \pm 2, \ldots$ and $x(nT)$ is called the sampled sequence of $x(t)$ where T is the time interval between successive samples and the sampling frequency is given by $f_s = \frac{1}{T}$ Hz. Therefore, the discrete time signal

$$x(nT) = x_s(t) = x(t)\delta_T(t) \tag{1.21}$$

where

$$\delta_T(t) = \sum_{n=-\infty}^{\infty} \delta(t - nT) = \frac{1}{T} \sum_{m=-\infty}^{\infty} e^{jm\Omega_0 t} \tag{1.22}$$

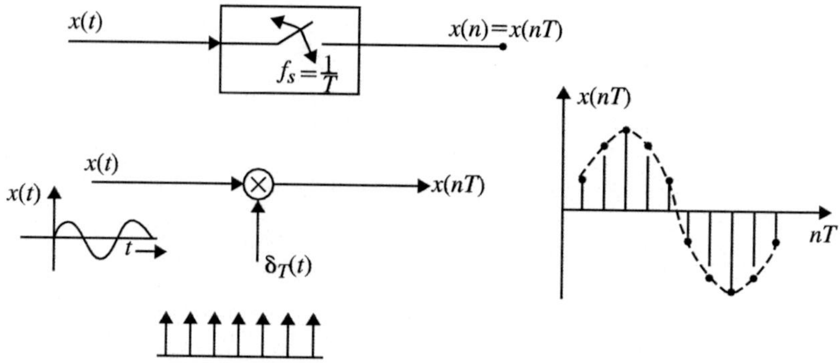

Fig. 1.34 Sampling process

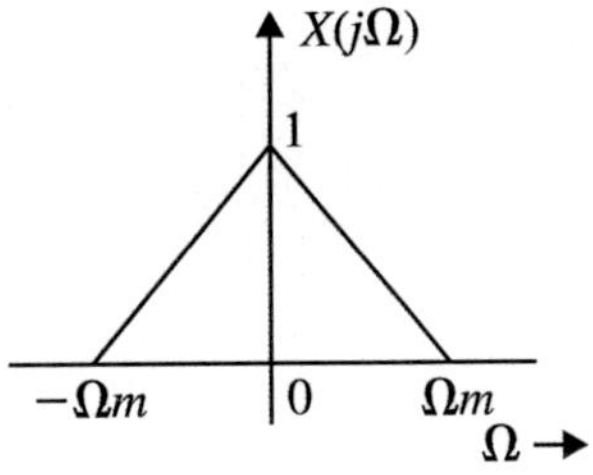

Fig. 1.35 Spectrum of continuous time signal

Equation (1.22) is a Fourier series representation of impulse, $\left[c_k = \frac{1}{T}\right]$

$$x_s(t) = x(t)\left[\frac{1}{T}\sum_{m=-\infty}^{\infty} e^{jm\Omega_0 t}\right]$$

$$x_s(t) = \frac{1}{T}\sum_{m=-\infty}^{\infty} x(t)e^{jm\Omega_0 t} \tag{1.23}$$

$$F[x_s(t)] = \frac{1}{T}\sum_{m=-\infty}^{\infty} F[x(t)e^{jm\Omega_0 t}] \tag{1.24}$$

If $F[x(t)] = X(i\Omega)$ then

$$F[x(t)e^{jm\Omega_0 t}] = X(j(\Omega - m\Omega_0)) \tag{1.25}$$

$$\therefore\ F[x_s(t)] = \frac{1}{T}\sum_{m=-\infty}^{\infty} X(j\Omega - m\Omega_0) \tag{1.26}$$

$$X_s(i\Omega) = F[x_s(t)] = \frac{1}{T}\sum_{m=-\infty}^{\infty} X\left[j\left(\Omega - \frac{2\pi}{T}m\right)\right] \tag{1.27}$$

Thus, the Fourier transform of the sampled signal is given by an infinite sum of shifted replicas of the Fourier transform of the original signal.

Consider the signal $x(t)$ is band limited to f_m, i.e. the highest frequency component of $x(t)$ is f_m. Then $X(j\Omega) = 0$ for $|\Omega| > \Omega_m$ is shown in Fig. 1.35.

In Eq. (1.27), the term $X[j(\Omega - \frac{2\pi m}{T})]$ is the shifting of $X(j\Omega)$ from $\Omega = 0$ to $\Omega = \frac{2\pi m}{T}$ where $m = 0 \pm 1, \pm 2, \ldots$

Case (i) If $\frac{\pi}{T} = \Omega m$, the spectrum of the sampled signal is shown in Fig. 1.36. The replicas will not overlap.

As a result, the frequency spectrum of $TX_s(j\Omega)$ in the frequency range $\left(-\frac{\pi}{T}, \frac{\pi}{T}\right)$ is identical to $X(j\Omega)$. The frequency spectrum $X(j\Omega)$ can be recovered from $X_s(i\Omega)$, by using a lowpass filter which has a sharp cut off at $\Omega = \frac{\pi}{T}$.

Case (ii) If $\frac{\pi}{T} > \Omega m$, the spectrum of the sampled signal is shown in Fig. 1.37. The replicas will not overlap.

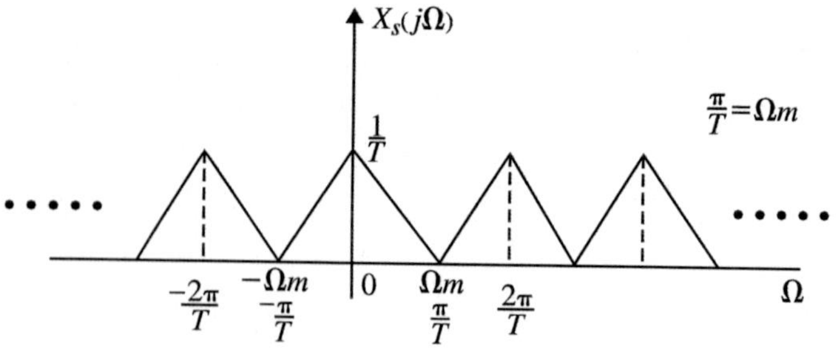

Fig. 1.36 Spectrum of sampled signal for $\frac{\pi}{T} = \Omega m$

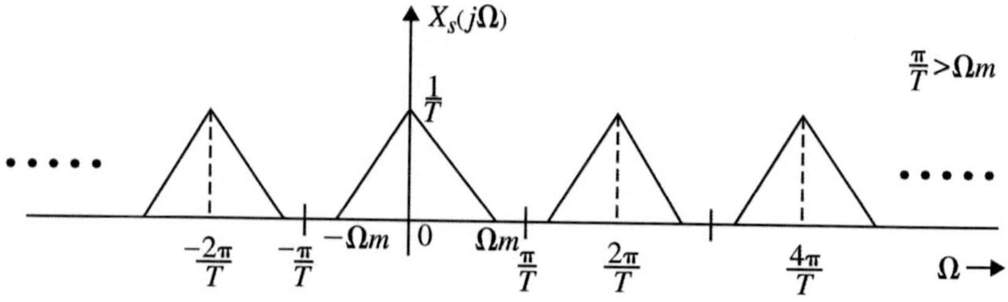

Fig. 1.37 Spectrum of sampled signal for $\frac{\pi}{T} > \Omega m$

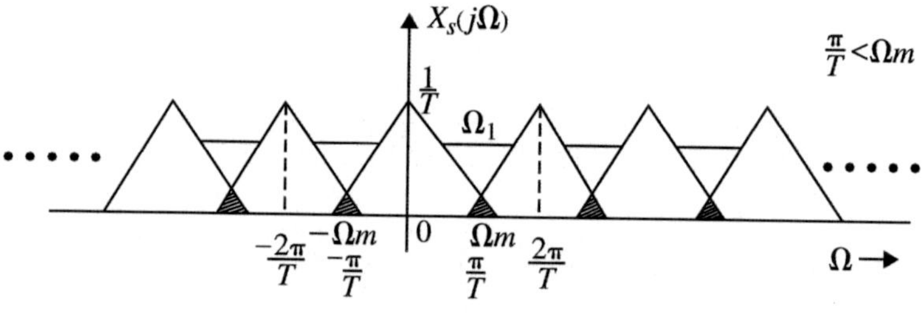

Fig. 1.38 Spectrum of sampled signal for $\frac{\pi}{T} < \Omega m$

The frequency spectrum $X(j\Omega)$ can be recovered from $X_s(j\Omega)$.

Case (iii) If $\frac{\pi}{T} < \Omega m$, the overlapping of successive frequency replicas is shown in Fig. 1.38.

As a result, the frequency spectrum $X(j\Omega)$ will not be recovered from the frequency spectrum of $TX(j\Omega)$.

For example, consider an arbitrary point at Ω_1, which falls in the region of the overlap. The frequency spectrum at $\Omega = \Omega_1$ is the sum of two components, one component is $X_s(j\Omega_1)$ and the other component is $X_s[j(\frac{2\pi}{T} - \Omega_1)]$. This high frequency component $(\frac{2\pi}{T} - \Omega_1)$ is folded in about the folding frequency $\frac{\pi}{T}$ and appears as a low frequency at Ω_1. The super imposition of high frequency component on the low frequency is known as "frequency aliasing". Because of aliasing, the spectrum

$X(j\Omega)$ is no longer recoverable from the spectrum of $X_s(j\Omega)$. The aliasing error can be prevented if the highest frequency component Ωm in the signal $x(t)$ is less than or equal to $\frac{\pi}{T}$. That is

$$\frac{\pi}{T} \geq \Omega m \tag{1.28}$$

If $f_s = \frac{1}{T}$. Then

$$\pi f_s \geq \Omega m \tag{1.29}$$

$$\pi f_s \geq 2\pi f_m \tag{1.30}$$

$$\boxed{f_s \geq 2 f_m} \tag{1.31}$$

To avoid aliasing, the sampling frequency must be greater than twice the highest frequency present in the signal.

1.9.2 Sampling Theorem

A band limited signal $x(t)$ with $X(j\Omega) = 0$ for $|\Omega| > \Omega_m$ is uniquely determined from its sample $x(nT)$, if the sampling frequency $f_s \geq 2 f_m$, i.e. sampling frequency must be at least twice the highest frequency present in the signal.

1.9.3 Nyquist Rate

The minimum rate at which a signal can be sampled and still be reconstructed from its samples is called the Nyquist rate. It is always equal to $2 f_m$, where f_m is the maximum frequency component present in the signal.

$$\boxed{\text{Nyquist rate} = 2 f_m.} \tag{1.32}$$

A signal sampled at greater than the Nyquist rate is said to be over sampled and a signal sampled at less than the Nyquist rate is said to be under sampled.

1.9.4 Anti-aliasing Filter

Sampling theorem states that a signal can be perfectly reconstructed if it is band limited. In practice, communication signals have frequency spectra consisting of low frequency components as well as high frequency noise components. When a signal

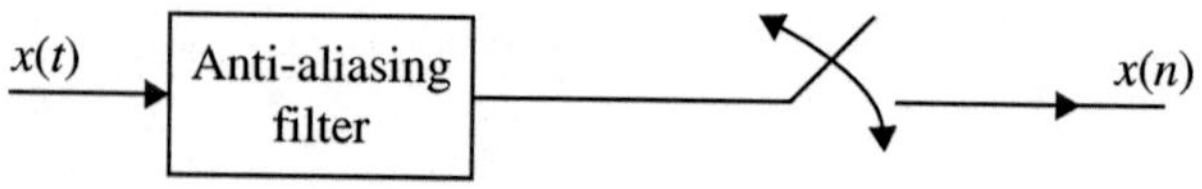

Fig. 1.39 Anti-aliasing filter before the sampler

is sampled, with sampling frequency f_s, all signals with frequency range higher than f_s appear as a signal frequency between 0 and $\frac{\Omega_s}{2}$ creating aliasing.

Therefore, to avoid aliasing errors caused by the undesired high frequency signals, an analog lowpass filter called an anti-aliasing filter is used prior to the sampler.

A filter that is used to reject high frequency signals before it is sampled to reduce the aliasing is called an anti-aliasing filter and is shown in Fig. 1.39.

1.9.5 Signal Reconstruction

The band limited signal $x(t)$ can be reconstructed from its samples if the sampling rate is above the Nyquist rate.

Let the sampled signal be $x_s(t)$.

$$x_s(t) = \sum_{n=-\infty}^{\infty} x(nT)\delta(t - nT) \tag{1.33}$$

$$F[x_s(t)] = \sum_{n=-\infty}^{\infty} F[x(nT)\delta(t - nT)] \tag{1.34}$$

$$X(j\omega) = \frac{1}{T}\sum_{m=-\infty}^{\infty} X\left(j\left(\Omega - \frac{2\pi m}{T}\right)\right) \tag{1.35}$$

To get the original spectrum, the signal $x_s(t)$ is to be passed through a reconstruction (lowpass) filter with an impulse response $h(t)$.

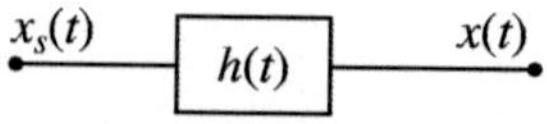

The original signal can be obtained by convolving the sampled signal $x_s(t)$ and the impulse response $h(t)$ of the filter

$$x(t) = x_s(t) * h(t) \tag{1.36}$$

$$= \left[\sum_{n=-\infty}^{\infty} x(nT)\delta(t - nT)\right] * h(t) \tag{1.37}$$

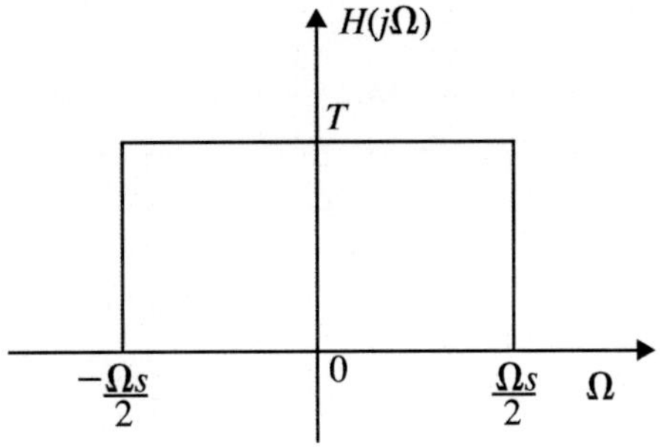

Fig. 1.40 Spectrum of lowpass filter

$$x(t) = \int_{-\infty}^{\infty} \sum_{n=-\infty}^{\infty} x(nT)\delta(\lambda - nT)h(t - \lambda)d\lambda \tag{1.38}$$

$$= \sum_{n=-\infty}^{\infty} x(nT) \int_{-\infty}^{\infty} \delta(\lambda - nT)h(t - \lambda)d\lambda \tag{1.39}$$

$$\therefore\ \delta(\lambda - nT) = \begin{cases} 1, & \lambda = nT \\ 0, & \lambda \neq nT \end{cases}$$

$$x(t) = \sum_{n=-\infty}^{\infty} x(nT)h(t - nT) \tag{1.40}$$

The frequency response of the lowpass filter can be expressed by

$$H(i\Omega) = \begin{cases} T, & \text{for } |\Omega| \leq \frac{\Omega_s}{2} \\ 0, & \text{for } |\Omega| > \frac{\Omega_s}{2} \end{cases} \tag{1.41}$$

The spectrum of ideal lowpass filter is shown in Fig. 1.40.

The impulse response of the lowpass filter is

$$\begin{aligned} h(t) &= F^{-1}[H(j\Omega)] \\ &= \frac{1}{2\pi} \int_{\frac{-\Omega_s}{2}}^{\frac{\Omega_s}{2}} H(j\Omega)e^{j\Omega t}d\Omega \\ &= \frac{T}{2\pi} \int_{\frac{-\Omega_s}{2}}^{\frac{\Omega_s}{2}} e^{j\Omega t}d\Omega = \frac{T}{\pi t} \sin\frac{\Omega_s t}{2} \\ &= \frac{T}{\pi T} \sin\frac{2\pi f_s t}{2} = \frac{T}{\pi t} \sin\frac{\pi t}{T} \end{aligned} \tag{1.42}$$

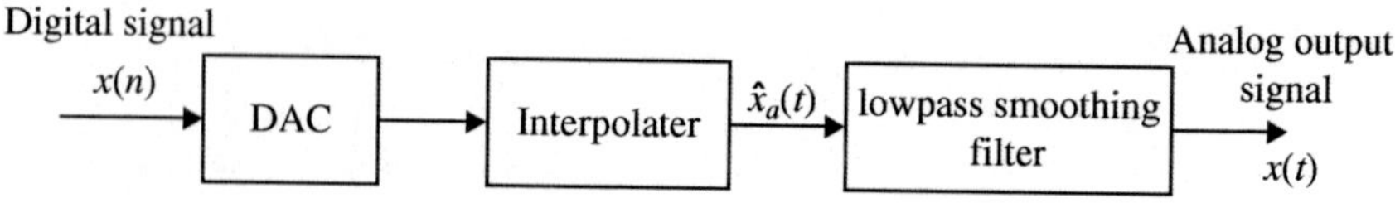

Fig. 1.41 Digital to analog conversion

$$h(t) = \frac{T}{\pi}\frac{\sin\frac{\pi t}{T}}{t} \tag{1.43}$$

$$h(t-nT) = \frac{T}{\pi}\frac{\sin\frac{\pi}{T}(t-nT)}{t-nT} \tag{1.44}$$

Substitute Eq. (1.44) in Eq. (1.39)

$$x(t) = \sum_{n=-\infty}^{\infty} x(nT) \times \frac{\sin\frac{\pi}{T}(t-nT)}{\frac{\pi}{T}(t-nT)} \tag{1.45}$$

$$x(t) = \sum_{n=-\infty}^{\infty} x(nT)\sin c\frac{\pi}{T}(t-nT) \tag{1.46}$$

The response $x(t)$ given by the formula is suitable for practical implementation. The $\sin c$ function spreads over the entire time axis. To reconstruct the signal at the time $t = t_0$, we must know all the sample values including those for $nT > t_0$.

That is, $x(t_0)$ requires the knowledge of future values of $x(nT)$ so it represents a non-casual system. Hence, this type of reconstruction is not suitable for real time applications. Therefore, we need a different approach to convert digital signal to analog signal which is shown in Fig. 1.41.

The digital to analog converter produces an output voltage that is proportional to the value of the binary word. The output of DAC is applied to an interpolator which converts the output samples of DAC into an analog signal. There are several interpolators. One of the interpolators is zero order hold.

1.9.6 Sampling with Zero Order Hold

One of the most widely used interpolators is zero order hold. The block diagram and input–output of zero order hold is shown in Fig. 1.42. In this technique, a given sample is held for an interval until the next sample is received (Fig. 1.43).

$$\hat{x}_a(t) = x(n) \quad \text{for } nT \le n \le (n+1)T \quad n = 0, 1, 2, \ldots \tag{1.47}$$

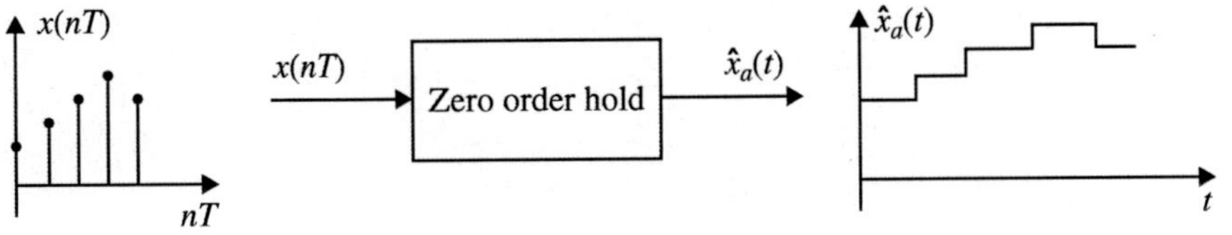

Fig. 1.42 Input–output of zero order hold

Fig. 1.43 Output of a zero order hold

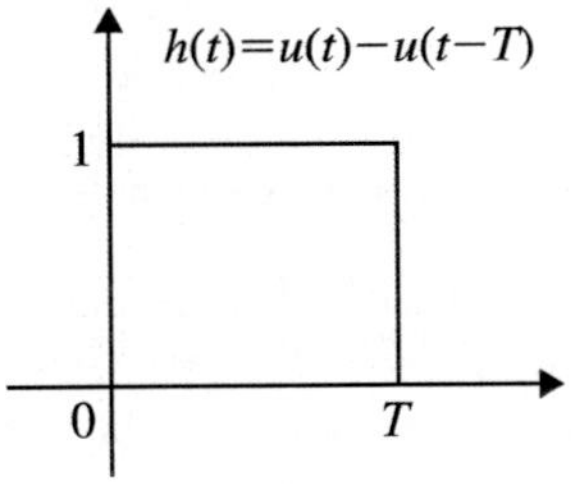

$$\begin{aligned}\hat{x}_a(t) &= x(0) \quad \text{for } 0 \le t < T\\ &= x(T) \quad \text{for } t \le t < 2T\\ &= x(2T) \quad \text{for } 2T \le t < 3T\end{aligned}$$

The impulse response of zero order hold is given by

$$h(t) = \begin{cases} 1, & 0 \le t \le T \\ 0, & \text{otherwise} \end{cases} \tag{1.48}$$

The impulse response function of a zero order hold circuit is shown in Fig. 1.43.

1.9.6.1 Transfer Function of a Zero Order Hold

The output of zero order hold is expressed in terms of $h(t)$. That is the output $\hat{x}_a(t)$ is the convolution of $x(nT)$ and $h(t)$

$$\hat{x}_a(t) = \sum_{n=-\infty}^{\infty} x(nT)h(t - nT) \tag{1.49}$$

$h(t)$ can be expressed as

$$h(t) = u(t) - u(t - T) \tag{1.50}$$

Taking Laplace transform on both sides, we get

$$H(s) = U(s) - U(s)e^{-st}$$

$$\boxed{H(s) = \frac{1}{s}[1 - e^{-st}]} \tag{1.51}$$

where $U(s) = \frac{1}{s}$.

The output of zero order hold consists of higher order harmonics. To remove these harmonics, the output of zero order hold is applied to a lowpass filter. This filter is often called a smoothing filter because it tends to smooth the corners on the step approximations generated by ZOH.

Example 1.34 A signal having a spectrum ranging from DC to 10 kHz is to be sampled and converted into a discrete form. What is the minimum number of samples per second that must be taken to ensure recovery?

Solution Given $f_m = 10\,\text{kHz}$. From the Nyquist rate, the minimum number of samples per second that must be taken to ensure recovery is

$$f_s = 2f_m$$
$$f_s = 20,000 \text{ samples/s}$$

Example 1.35 Consider the following analog signal:

$$x_a(t) = 3\cos 100\pi t$$

(*a*) Determine the minimum sampling rate required to avoid aliasing.
(*b*) Suppose that the signal is sampled at the rate $F_s = 200$ Hz, what is the discrete time signal obtained after sampling?
(*c*) Repeat (*b*) it $F_s = 75$ Hz.
(*d*) What is the frequency $0 < F < \frac{F_s}{2}$ of a sinusoid that yields samples identical to those obtained in part (*c*).

Solution (*a*) Given

$$\Omega_m = 100\pi$$
$$2\pi f_m = 100\pi$$
$$f_m = 100/2$$

Hence, the minimum sampling rate required to avoid aliasing $f_s = 2f_m = 100\,\text{Hz}$.

(*b*) If the signal is sampled at $f_s = 200$ Hz, the discrete time signal is $3\cos 100\pi nT$

$$x(n) = x(t)|_{t=nT}$$
$$x(n) = 3\cos\frac{100\pi}{200}n$$
$$x(n) = 3\cos\frac{n\pi}{2}$$

(*c*) If the signal is sampled at $F_s = 75$ Hz, the discrete time signal is

$$\begin{aligned} x(n) &= x(t)\Big|_{t=nT} \\ &= 3\cos 100\pi nT \\ &= 3\cos\frac{100\pi}{75}n \\ &= 3\cos\left(\frac{4\pi}{3}\right)n \\ &= 3\cos\left(2\pi - \frac{2\pi}{3}\right)n \\ x(n) &= 3\cos\frac{4\pi}{3}n \end{aligned}$$

(*d*) For the sampling rate of $F_s = 75$ Hz, we have

$$F = \frac{F_0}{F_s}$$
$$F_0 = F \cdot F_s$$

Therefore, in general, the sampling of CT sinusoidal signal $x(t) = A\cos(2\pi F_0 t + \theta)$ with sampling rate $F_s = (1/T)$ results in a discrete time signal $x(n) = A\cos(2\pi f_0 n + \theta)$. Therefore, relative frequency $f_0 = (F_0/F_s)$.
The frequency of the sinusoidal in part (*c*) is

$$F = \frac{1}{3}$$
$$\omega = \left(\frac{2\pi}{3}\right)$$
$$T = \left(\frac{2\pi}{\omega}\right)m$$
$$T = \left(\frac{2\pi}{2\pi/3}\right)m = 3m = 3$$

Hence

$$F_0 = \frac{1}{3} \cdot 75$$
$$F_0 = 25 \text{ Hz}$$

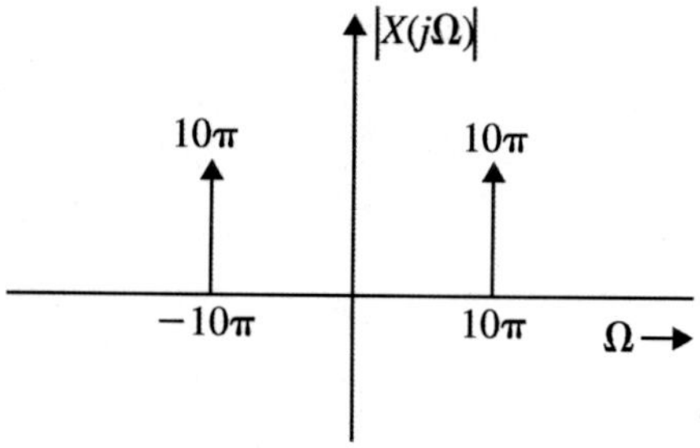

Fig. 1.44 Amplitude spectrum of continuous time signal

Clearly, the sinusoidal signal

$$y_a(t) = 3\cos 2\pi F_0 t$$
$$y_a(t) = 3\cos 50\pi t$$

Example 1.36 The signal $x(t) = 10\cos(10\pi t)$ is sampled at a rate of 8 samples per second. Plot the amplitude spectrum for $|\Omega| \leq 30\pi$. Can the original signal be recovered from samples? Explain.

Solution Given

$$x(t) = 10\cos(10\pi t)$$
$$\Omega m = 10\pi$$
$$F[x(t)] = X(j\Omega) = 10[\pi[\delta(\Omega + \Omega_0) + \delta(\Omega - \Omega_0)]]$$
$$X(j\Omega) = 10\pi[\delta(\Omega + 10\pi) + \delta(\Omega - 10\pi)]$$

The amplitude spectrum of $x(t)$ is shown in Fig. 1.44.

$$\text{The sampling rate } f_s = 8 \text{ Hz} \quad \Rightarrow T = \frac{1}{8}$$
$$\Omega_s = 2\pi f_s = 16\pi$$
$$\Omega m = 10\pi = 2\pi f_m$$
$$f_m = 5 \text{ Hz}$$
$$\text{Nquist rate} = 2f_m = 10 \text{ Hz} \qquad (\text{Here } f_s < 2f_m)$$

Therefore, the original signal cannot be recovered from the samples. The frequency spectrum of sampled signal $x(t)$ is given by

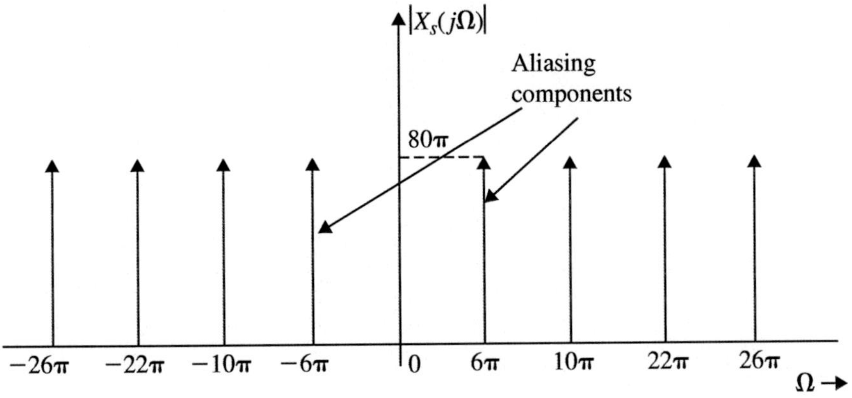

Fig. 1.45 Amplitude spectrum of sampled signal

$$X_s(j\Omega) = \frac{1}{T}\sum_{n-\infty}^{\infty} X\left[j\left(\Omega - \frac{2\pi n}{T}\right)\right]$$

$$= \frac{1}{T}\sum_{n=-\infty}^{\infty} 10\pi\left[\delta\left(\Omega + 10\pi - \frac{2\pi}{T}n\right) + \delta\left(\Omega - 10\pi - \frac{2\pi}{T}n\right)\right]$$

$$X_s(j\Omega) = 80\pi \sum_{n=-\infty}^{\infty} \left[\delta\left(\Omega + 10\pi - 16\pi n\right) + \delta\left(\Omega - 10\pi - 16\pi n\right)\right]$$

The plot of amplitude spectrum for $|\Omega| \leq 30\pi$ is shown in Fig. 1.45.

Example 1.37 A signal $x(t) = \sin c(150\pi t)$ is sampled at a rate of

(*a*) 100 Hz;
(*b*) 200 Hz and
(*c*) 300 Hz.

For each of these three cases, can you recover the signal $x(t)$ from the sampled signal.

Solution Given

$$x(t) = \sin c(150\pi t)$$
$$\Omega_m = 150\pi$$
$$2\pi f_m = 150\pi$$
$$f_m = 75$$
$$f_s = 2f_m \Rightarrow f_s = 150, \quad T = \frac{1}{150}$$

The spectrum of the signal $x(t)$ is a rectangular pulse with a band width (maximum frequency components of 150π rad/s. is shown in Fig. 1.46.

Fig. 1.46 Spectrum of $x(t)$

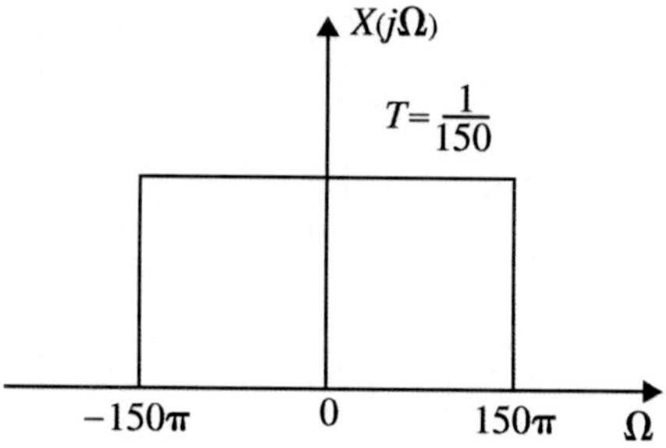

Fig. 1.47 Spectrum of $x(t)$ for Example 1.38

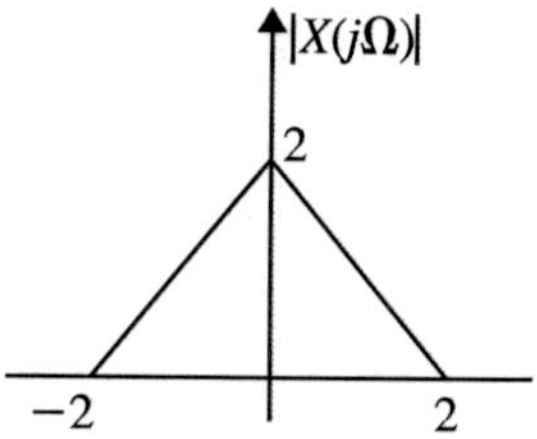

(a) In the first case, the sampling rate is 100 Hz which is less than the Nyquist rate. Therefore, $x(t)$ can be recovered from its samples (Fig. 1.47).

$$f_s = 100, \quad 2f_m = 150 \Rightarrow f_s < 2f_m$$

For (b) and (c) cases, the sampling rate is greater than the Nyquist rate. Therefore, $x(t)$ cannot be recovered from its samples. For $f_s = 200$ and $f_s = 300$

$$2fm = 150 \qquad \text{and} \qquad f_s > 2f_m$$

Example 1.38 Draw $|X_s(i\Omega)|$ for the following cases when $x_s(t) = x(t)\delta_T(t)$ with sampling period T, where

$$\delta_T(t) = \sum_{n=-\infty}^{\infty} \delta(t - nT)$$

The spectrum of its signal for Example 1.38 is shown in Fig. 1.47.

(a) $T = \frac{\pi}{3}$ s
(b) $T = \frac{\pi}{2}$
(c) $T = \frac{2\pi}{3}$

Solution In general, for band limited signal, the amplitude spectrum is shown in Fig. 1.48.

$$\therefore \ \Omega_m = 2\text{rad/s}.$$
$$2\pi f_m = 2$$
$$f_m = \frac{1}{\pi}\ \text{Hz}$$
$$\text{Nyquist rate} = 2f_m$$
$$= \frac{2}{\pi}\ \text{Hz}$$

(*a*) Given

$$T = \frac{\pi}{3}\ \text{s}$$
$$f_s = \frac{1}{T}$$
$$= \frac{3}{\pi}\ \text{Hz}$$
$$\Omega_s = 2\pi f_s$$
$$= 6\ \text{rad/s}.$$
$$f_s > 2f_m \quad \left[\therefore \ \text{since}\ \frac{2}{\pi} > \frac{3}{\pi}\right]$$

Therefore, the spectrum $|X_s(j\Omega)|$ is free from aliasing and is shown in Fig. 1.49.

(*b*) Given

$$T = \frac{\pi}{2}\ \text{s}$$
$$f_s = \frac{2}{\pi}\ \text{Hz}$$
$$\Omega_s = 2\pi f_m = 4\ \text{rad/s}.$$
$$2f_m = \frac{2}{\pi}\ \text{Hz}$$
$$f_s = 2f_m.$$

Therefore, the spectrum $|X_s(i\Omega)|$ in free from aliasing and is shown in Fig. 1.50.

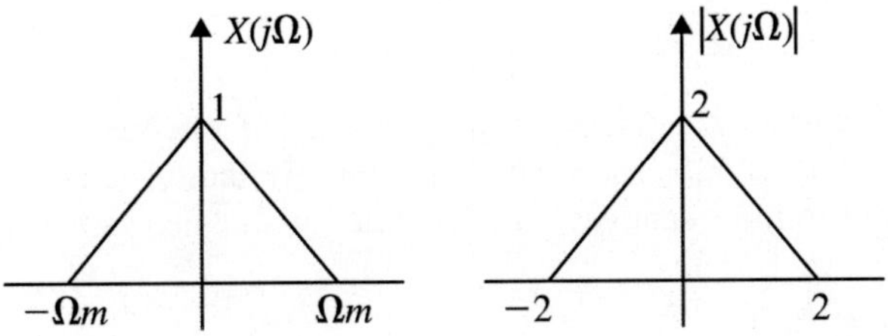

Fig. 1.48 Spectrum of bandlimited signal

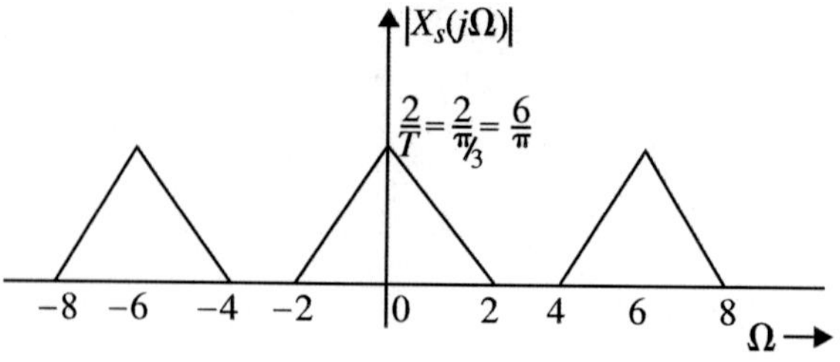

Fig. 1.49 Spectrum of sampled signal for $T = \pi/3$ s

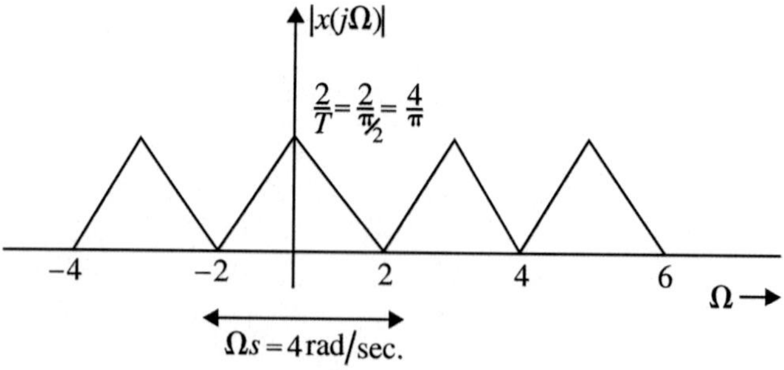

Fig. 1.50 Spectrum of sampled signal for $T = \pi/2$ s

Fig. 1.51 Spectrum of sampled signal for $T = \frac{2\pi}{3}$ s

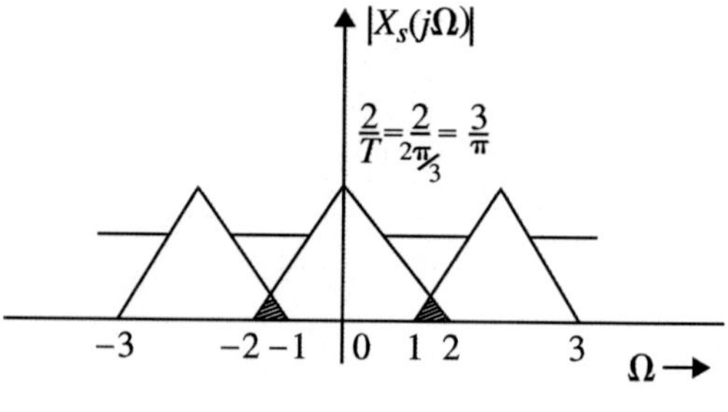

(*c*) Given

$$T = \frac{2\pi}{3}, \quad f_s = \frac{3}{2\pi} \text{ s}, \quad \Omega_s = 3 \text{ rad/s.}$$
$$f_s < 2f_m$$

The sampling frequency is less than the Nyquist rate. The spectra $X(j\Omega)$ repeats for every 3 rad/s and the successive spectrum overlaps. Therefore, $x(t)$ cannot be recovered from its samples. The spectrum of sampled signal for $T = 2\pi/3$ is shown in Fig. 1.51.

Example 1.39 Consider the following sampling and reconstruction block as shown in Fig. 1.52.

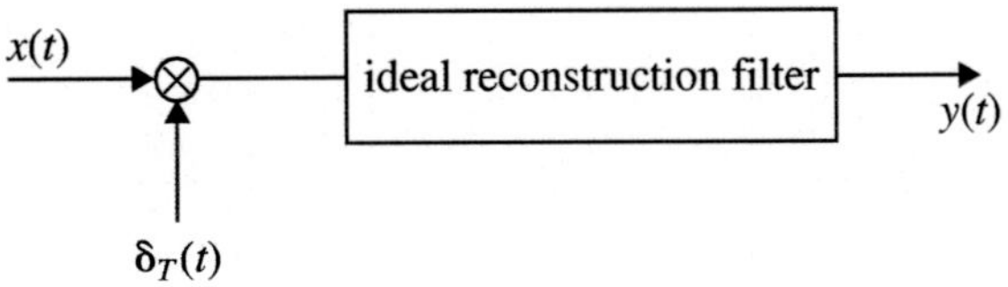

Fig. 1.52 Sampling and reconstruction block

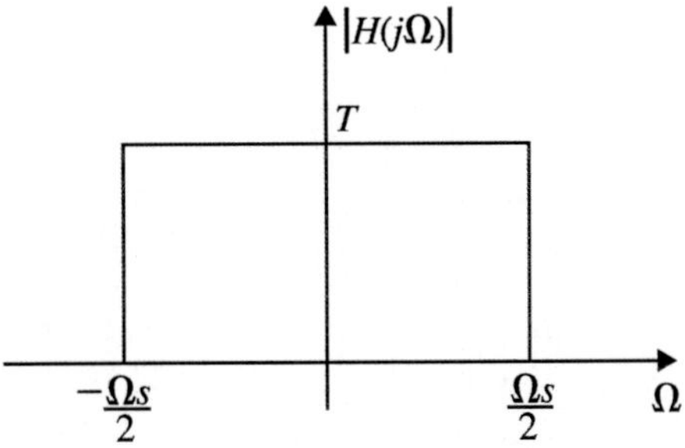

Fig. 1.53 Frequency response of lowpass filter

The output of the ideal reconstruction filter can be found by sending signal $x_s(t)$ through an ideal lowpass filter having characteristics as shown in Fig. 1.53.

For the following signals, draw the spectrum of $|X_s(j\Omega)|$, and find expression for $x(nT)$ and $y(t)$:

(*i*) $x(t) = 2 + \cos(100\pi t)$ for $T = 0.0125$ s.
(*ii*) $x(t) = 2 + \cos(100\pi t)$ for $T = \frac{1}{150}$ s.
(*iii*) $x(t) = 1 + \cos(10\pi t) + \cos 30\pi t$ for $T = 0.04$ s.
(*iv*) If $X(j\Omega) = \frac{1}{2+j\Omega}$ and $T = 2$ s.

Draw $|X_s(j\Omega)|$. Test for aliasing.

Solution (*i*) $x(t) = 2 + \cos(100\pi t)$ for $T = 0.0125$ s.

$$
\begin{aligned}
x(t) &= 2 + \cos(100\pi t) \\
&= 2 + \frac{1}{2}\left[e^{j100\pi t} + e^{-j100\pi t}\right] \\
X(i\Omega) &= 4\pi\delta(\Omega) + \pi\left[\delta(\Omega + 100\pi) + \delta(\Omega - 100\pi)\right] \\
\because\ F[1] &= 2\pi\delta(\Omega) \\
F[e^{j\Omega_0 t}] &= 2\pi\delta(\Omega + \Omega_0)
\end{aligned}
$$

The sketch of the amplitude spectrum of $x(t)$ is shown in Fig. 1.54.
The sampling time

$$
\begin{aligned}
T &= 0.0125 \\
f_s &= \frac{1}{T}
\end{aligned}
$$

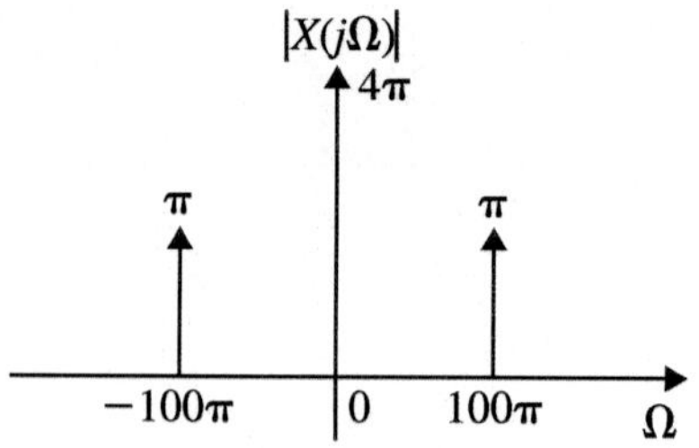

Fig. 1.54 Spectrum of $x(t)$

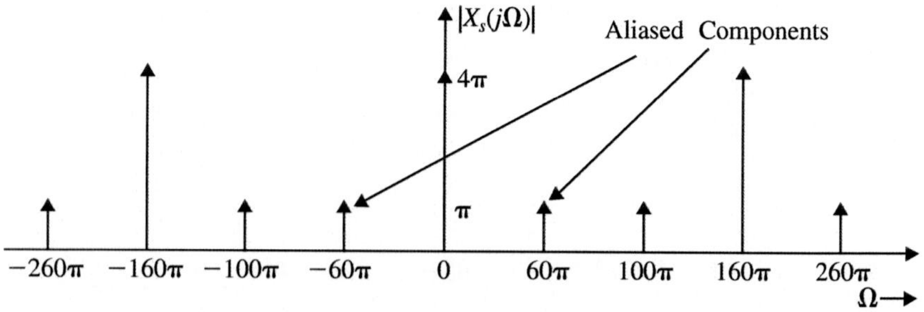

Fig. 1.55 Spectrum of sampled signal

$$
\begin{aligned}
&= \frac{1}{0.0125} \\
&= 80 \text{ Hz} \\
\Omega_s &= 2\pi f_s \\
&= 160 \text{ rad/s} \\
\Omega_m &= 100\pi
\end{aligned}
$$

Given

$$f_m = 50 \text{ Hz}$$

Nyquist rate is $2f_m = 100$ Hz

$$f_s < 2fm$$

Therefore, the successive spectrum $X_s(j\Omega)$ overlap and the signal $x(t)$ cannot be recovered from its samples.
The spectrum $X_s(j\Omega)$ of the sampled signal $x(t)$ is a periodic repetition of $X(j\Omega)$ for every 160π rad/s. (80 Hz) is shown in Fig. 1.55.
The frequency characteristics of the reconstruction filter (LPF) is

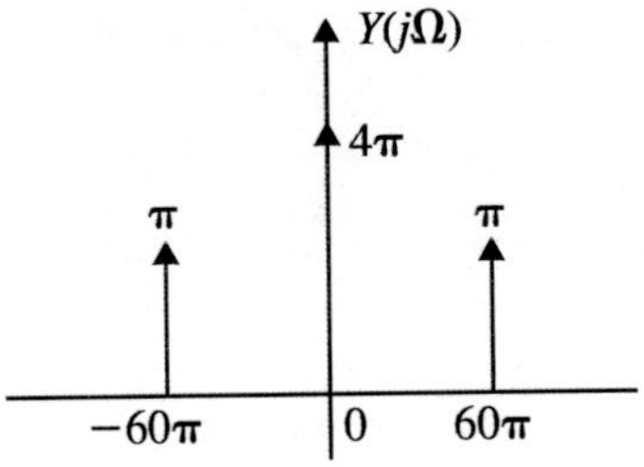

Fig. 1.56 Spectrum of output $y(t)$

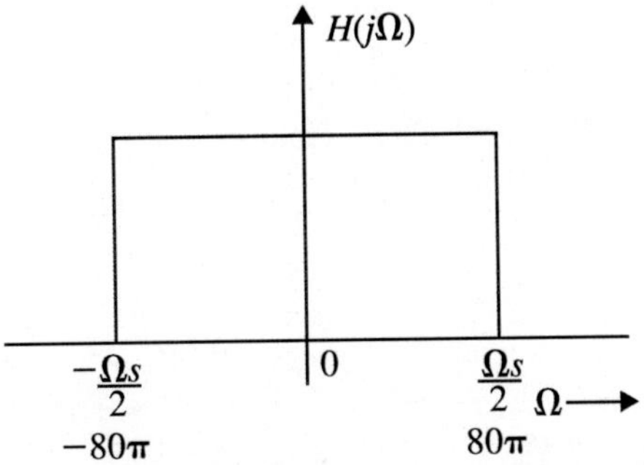

When the spectrum $|X_s(j\Omega)|$ is passed through the reconstruction filter (LPF), the output of the filter consists of the frequency components given by

$$Y(j\Omega) = 4\pi 8(\Omega) + \pi[\delta(\Omega + 60\pi) + \delta(\Omega - 60\pi)]$$

The spectrum of output $y(t)$ is shown in Fig. 1.56.

$$y(t) = 2 + \cos 60\pi t$$
$$x(nT) = x(t)|_{t=nT}$$

$$x(nT) = 2 + \cos 100\pi t|_{t=0.0125n}$$
$$= 2 + \cos(1.25)n\pi$$
$$x(nT) = 2 + \cos(0.75n\pi)$$

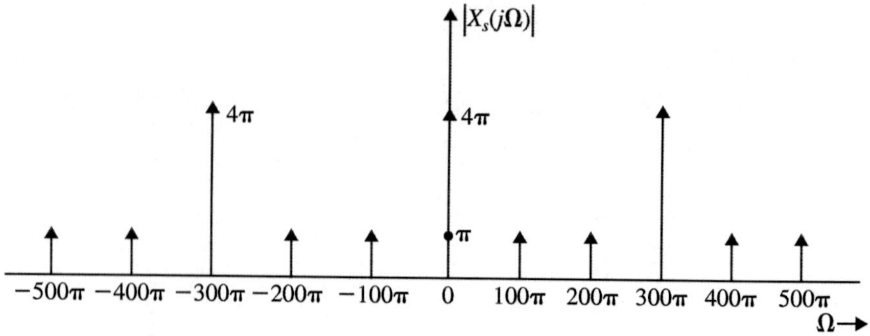

Fig. 1.57 Spectrum of sampled signal

(*ii*) Given $x(t) = 2 + \cos(100\pi t)$

$$T = \frac{1}{150}\ \text{s}$$
$$f_s = \frac{1}{T} = 150\ \text{Hz}$$
$$\Omega_s = 2\pi f_s = 300\pi\ \text{rad/s}.$$

The Amplitude spectrum ($|X(i\Omega)|$) of the signal is same as Fig. 1.57. The spectrum of sampled signal of $x(t)$, is a periodic extension of $X(j\Omega)$ for every 300π (rad/s) and is shown in Fig. 1.55.
The frequency characteristics of a reconstruction filter are shown below.

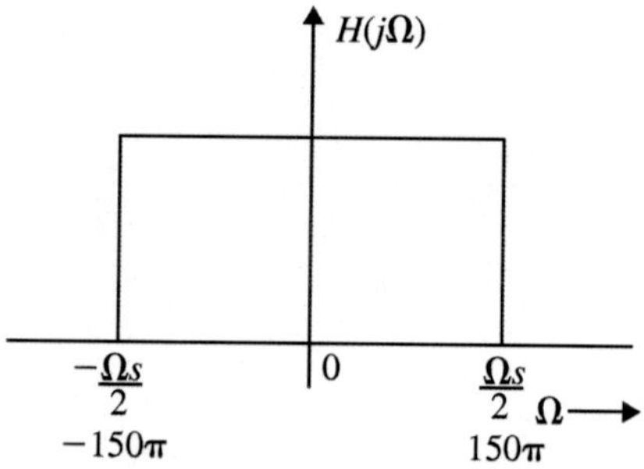

The resultant output spectrum is shown in Fig. 1.55.

$$
\begin{aligned}
Y(j\Omega) &= 4\pi\delta(\Omega) + \pi[\delta(\Omega - 100\pi) + \delta(\Omega + 100\pi)] \\
y(t) &= 2 + \cos 100\pi t \\
x(nT) &= x(t)\Big|_{t=nT} \\
&= 2 + \cos 100\pi t\Big|_{t=\frac{1}{150}n} \\
x(nT) &= 2 + \cos(0.666n\pi)
\end{aligned}
$$

(*iii*) Given

$$
\begin{aligned}
x(t) &= 1 + (\cos 10\pi t) + (\cos 30\pi t); \qquad T = 0.04 \text{ s} \\
x(nT) &= x(t)\Big|_{t=nT} \quad = 0.04n \\
x(nT) &= 1 + (\cos 0.4\pi t) + (\cos 0.2\pi t) \\
x(nT) &= 1 + (\cos 0.4n\pi) + (\cos 1.2n\pi) \\
f_s &= \frac{1}{T} = 25 \text{ Hz}, \quad \Omega_s = 2\pi f_s = 50\pi \text{ rad/s.} \\
\Omega_m &= 2\pi f_m = 30\pi \\
f_m &= 15 \text{ Hz} \\
2f_m &= 30 \text{ Hz} \qquad (\therefore\ f_s < 2f_m)
\end{aligned}
$$

Therefore, the successive spectrum of $X(j\Omega)$ in $X_s(j\Omega)$ overlaps and $x(t)$ cannot be recovered from $x(nT)$ (Fig. 1.58)

$$
\begin{aligned}
X(j\Omega) = 2\pi\delta(\Omega) &+ \pi[\delta(\Omega + 10\pi) + \delta(\Omega - 10\pi)] \\
&+ \pi[8(\Omega + 30\pi) + (\Omega - 30\pi)]
\end{aligned}
$$

The sketch of the spectrum of the sampled signal is shown in Fig. 1.59.
The frequency spectrum of the lowpass filter is given as

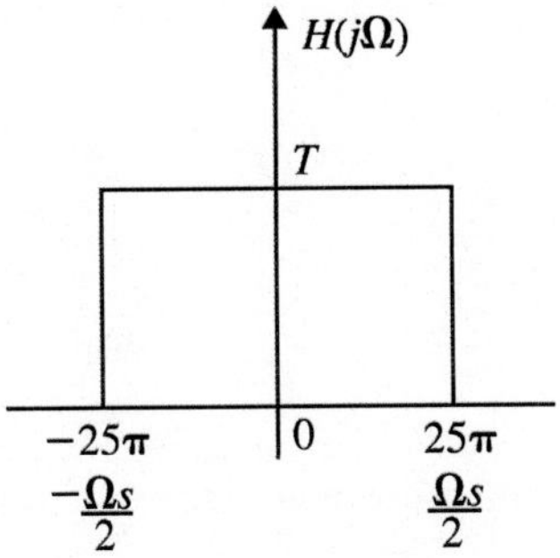

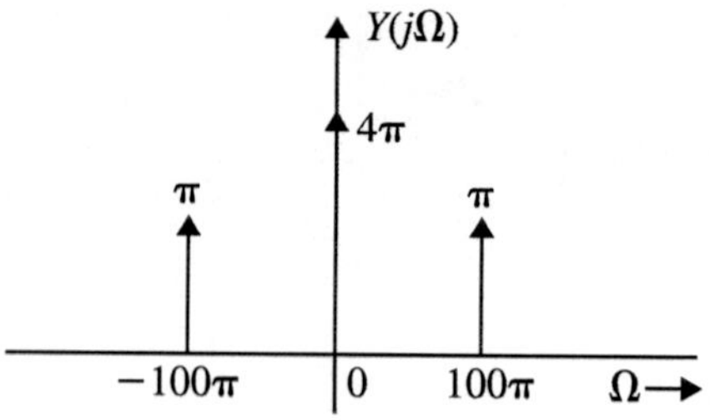

Fig. 1.58 Spectrum of output $y(t)$

The spectrum of $y(\Omega)$ is shown in Fig. 1.58

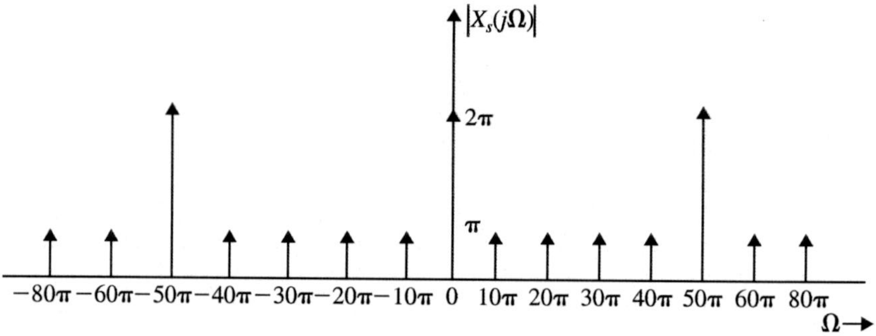

Fig. 1.59 Spectrum of sampled signal

Fig. 1.60 Spectrum of output $y(t)$

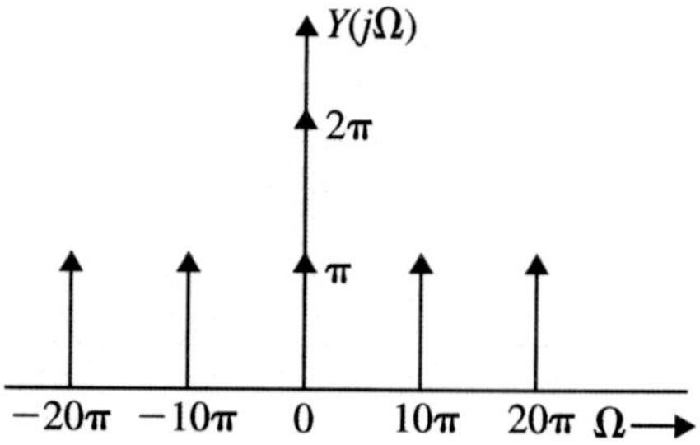

The output spectrum is shown in Fig. 1.60. From that, we can obtain

$$Y(j\Omega) = 2\pi\delta(\Omega) + \pi[\delta(\Omega + 10\pi) + \delta(\Omega - 10\pi)]$$
$$+\pi[\delta(\Omega + 20\pi) + \delta(\Omega - 20\pi)]$$
$$y(t) = 1 + \cos(10\pi t) + \cos(20\pi t)$$

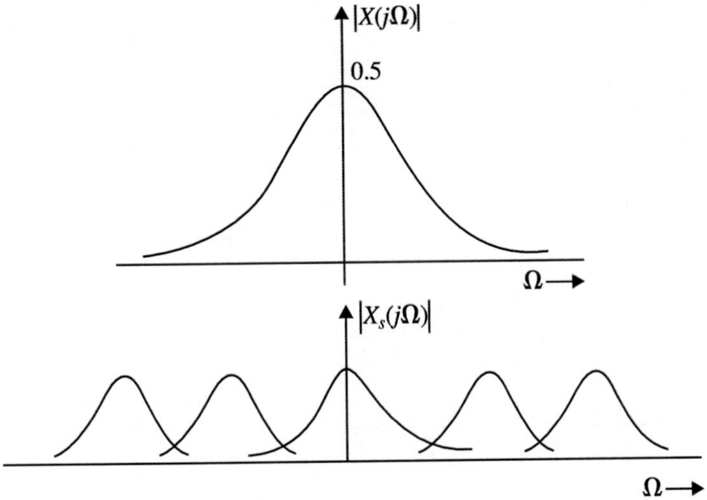

Fig. 1.61 Spectrum of sampled signal

(*iv*) Given $T = 2$ s, $f_s = 0.5$ Hz

$$\Omega_s = \pi \text{ rad/s}$$

$$X(j\Omega) = \frac{1}{2 + j\Omega}$$

$$|X(j\Omega)| = \frac{1}{\sqrt{4 + \Omega^2}}$$

The spectrum of $|X(j\Omega)|$ is not band limited. Therefore, aliasing of successive frequency components occurs even Ω is large. The spectra $|X_s(j\Omega)|$ is shown in Fig. 1.61.

Example 1.40 Determine the Nyquist sampling rate and Nyquist sampling intervals for the following signals:

$$(i) \quad \sin c^2(200\pi t)$$

$$(ii) \quad 0.5 \sin c^2(200\pi t)$$

$$(iii) \quad \sin c(200\pi t) + 3 \sin^2 c(120\pi t)$$

$$(iv) \quad \sin c(100\pi t) \sin c(200\pi t)$$

Solution The Fourier transform of the triangular function is the square of sin *c* function.

Similarly, Fourier transform of a rectangular pulse is a sin *c* function. This is shown in Figs. 1.62 and 1.63, respectively.

(*i*) Given,

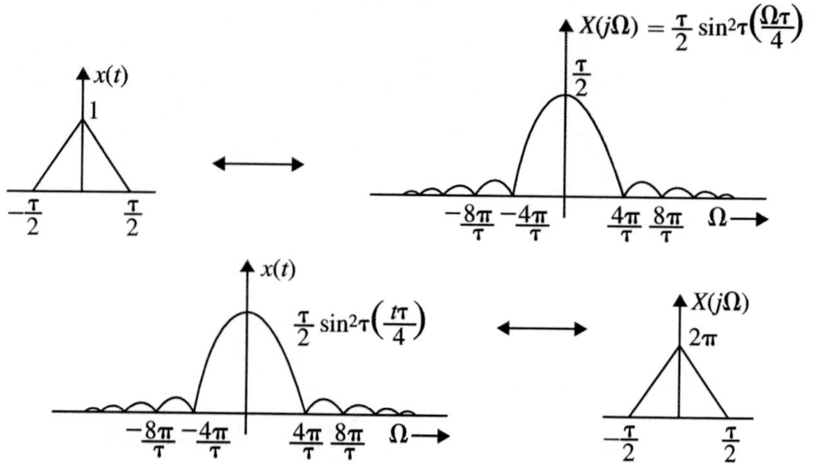

Fig. 1.62 **a** Fourier transform of triangular function. **b** Dual property of figure **a**

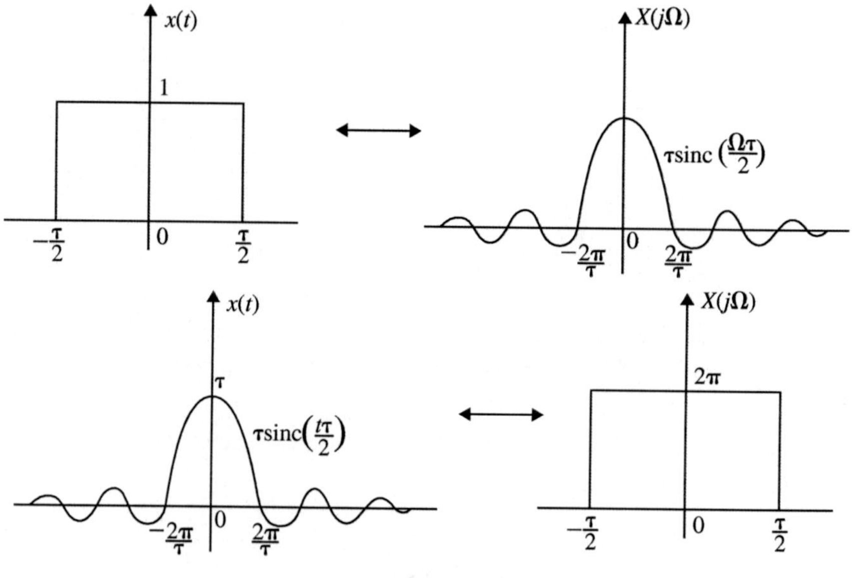

Fig. 1.63 **a** Fourier transform of rectangular pulse. **b** Dual property of figure **a**

$$x(t) = \sin c^2(200\pi t)$$

Comparing the signal $x(t)$ with the above signal, we get

$$x(t) = \frac{\tau}{2}\sin c^2\left(\frac{t\tau}{4}\right)$$
$$\frac{t\tau}{4} = 200\pi t$$
$$\tau = 800\pi$$
$$\frac{\tau}{2} = 400\pi$$

That is the maximum frequency component present in the signal is 400π rad/s (or) the signal is band limited to 400π rad/s.

$$\Omega_m = 400\pi \quad \text{and} \quad f_m = 200 \text{ Hz}$$
$$\text{Nyquist rate } f_s = 2f_m = 400 \text{ Hz.}$$

Sampling interval $T = \frac{1}{f_s} = \frac{1}{400} = 2.5$ ms.

(*ii*) Same values as above.

(*iii*) Given

$$x(t) = \sin c(200\pi t) + 3\sin c^2(120\pi t)$$

Let

$$x_1(t) = \sin c(200\pi t)$$
$$x_2(t) = 3\sin^2 c(120\pi t)$$

Compare the signal $x_1(t)$ with the signal $x(t) = \tau \sin c\left(\frac{\tau t}{2}\right)$

$$\frac{\tau t}{2} = 200\pi t$$
$$\frac{\tau}{2} = 200\pi \text{ rad/s}$$
$$\therefore\ \Omega m_1 = 200 \text{ rad/s}$$
$$f m_1 = 100 \text{ Hz}$$

Compare the signal $x_2(t)$ with the signal $x(t) = \frac{\tau}{2}\sin c^2\left(\frac{\tau t}{4}\right)$,

$$\frac{\tau t}{4} = 120\pi t$$
$$\frac{\tau}{4} = 120\pi$$
$$\frac{\tau}{2} = 240\pi$$
$$\therefore \quad \Omega m_2 = 240\pi \text{ rad/s.}$$

When both signals are added, the maximum frequency components is

$$\Omega_m = \Omega m_2 = 240\pi \text{ rad/s}$$
$$fm = 120 \text{ Hz}$$
$$\text{Nyquist rate } f_s = 2fm = 240 \text{ Hz}$$
$$\text{Nyquist interval } = \frac{1}{f_s} = 4.167 \text{ m s.}$$

(*iv*) $x(t) = \sin c(100\pi t) \sin c(200\pi t)$ Let

$$x_1(t) = \sin c(100\pi t); x_2(t) = \sin c(200\pi t)$$

The signal $x_1(t)$ is band limited to frequency 100π rad/s and the signal $x_2(t)$ is bandlimited to frequency 200π rad/s and are shown in Fig. 1.64.

$$F[x_1(t) \cdot x_2(t)] = X_i(j\Omega) * X_2(j\Omega)$$

The convolution of $X_1(j\Omega)$ and $X_2(j\Omega)$ results is a triangular shape frequency spectra with $\Omega_m = 300\pi$ rad/s and is shown in Fig. 1.65.

$$fm = \frac{\Omega m}{2\pi}$$
$$fm = 150 \text{ Hz}$$

$$\text{Nyquist rate } = 2fm = 300 \text{ Hz}$$
$$\text{Nyquist interval } = \frac{1}{300} = 3.33 \text{ m s.}$$

Example 1.41 A signal $x(t)$, whose spectrum is shown in Fig. 1.66, is sampled at a rate of 300 samples/s. What is the spectrum of the sampled discrete time signal?

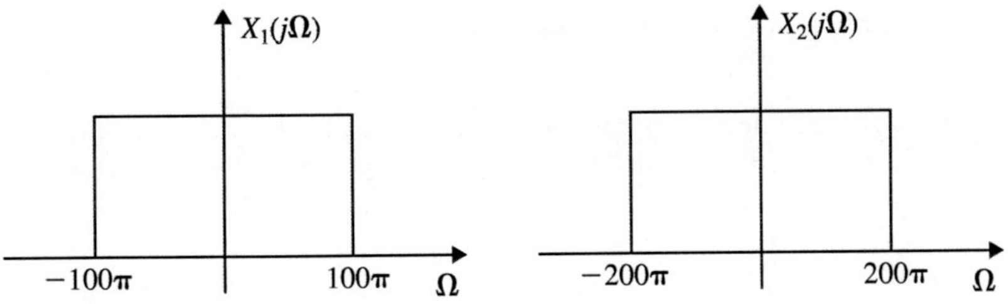

Fig. 1.64 Spectrum of $x_1(t)$ and $x_2(t)$

Fig. 1.65 Spectrum of $x(t)$

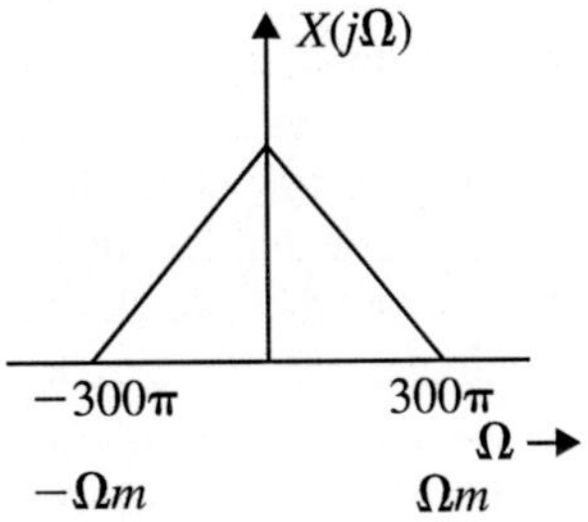

Fig. 1.66 Spectrum of $x(t)$

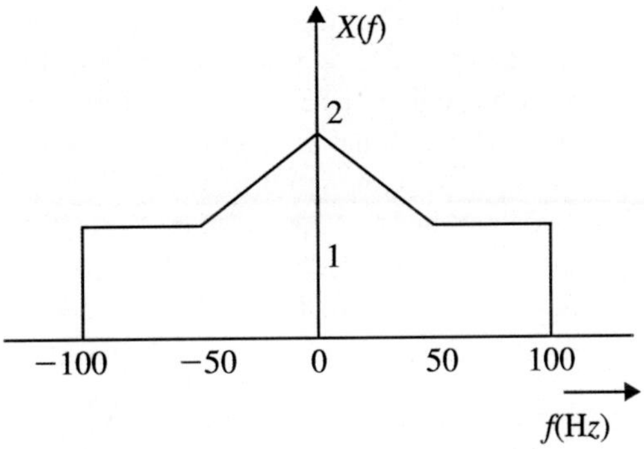

Solution

$$fm = 100 \text{ Hz}$$
$$\text{Nyquist rate} = 2fm = 200 \text{ Hz}$$
$$\text{Sampling frequency } (f_s) = 300 \text{ Hz}$$
$$f_s > 2fm.$$

Therefore, there is no aliasing. The sketch of the spectrum of sampled signal is shown in Fig. 1.67.

1.10 Analog to Digital Conversion

A signal that is specified for a continuum of values of time is a continuous time signal and a signal that is specified only at a discrete value of t is a discrete time signal. Continuous time signals are even though often referred to as analog signals, it is not strictly true. Similarly, discrete time signals are not the same as the digital signals. A signal whose amplitude can take on any value in a continuous range is an analog signal. This means an analog signal amplitude can take on an infinite number of values. On the other hand, a digital signal is one whose amplitude can take on only a finite number of values. The signals which are associated with a microprocessor or a digital signal processor or a digital computer are digital signals and they take on only two values (binary). The terms continuous time and discrete time indicate the nature of a signal along the time axis, whereas the terms analog and digital indicates the nature of the amplitude of the signal which is represented in the vertical axis. Thus, the signals are classified as analog continuous time, digital continuous time, analog discrete time and digital discrete time signal. The necessity of converting an analog signal into a digital signal is explained as given below.

The block diagram of a typical DSP system in its simplest form is shown in Fig. 1.68. The analog signal $x(t)$ is applied to ADC that converts into $x(n)$ which is in digital form. The digital signal $x(n)$ is processed by DSP and the output $y(n)$ is obtained. The ADC normally has a built in sample and hold circuit. The DAC processes the signal back into analog form as $y(t)$, which is applied to a continuous time system. Signal processing using a digital processor implies that the input signal $x(n)$ must be in digital form before it can be processed. The analog to digital conversion requires that the signal is first sampled and converted into a discrete time continuous time amplitude signal. The amplitude of each sample is then quantized into one of 2^b levels where b is the number of bits used to represent a sample in

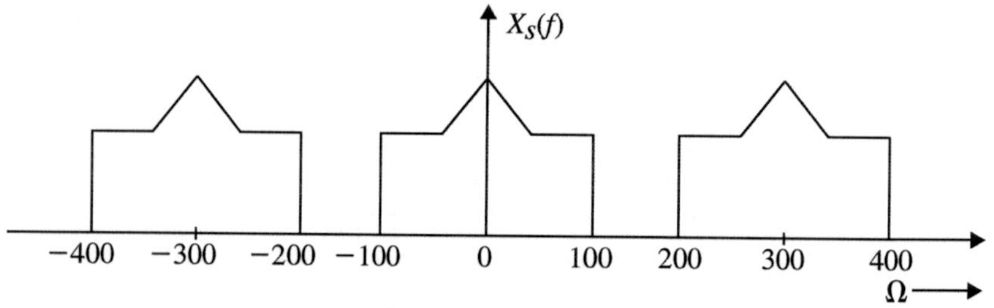

Fig. 1.67 Spectrum of sampled signal

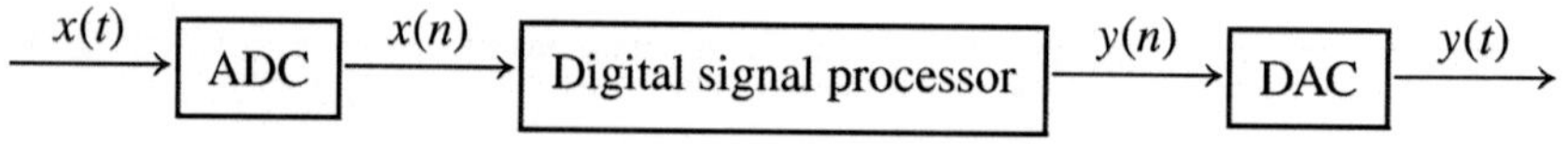

Fig. 1.68 Block diagram of digital signal processing system

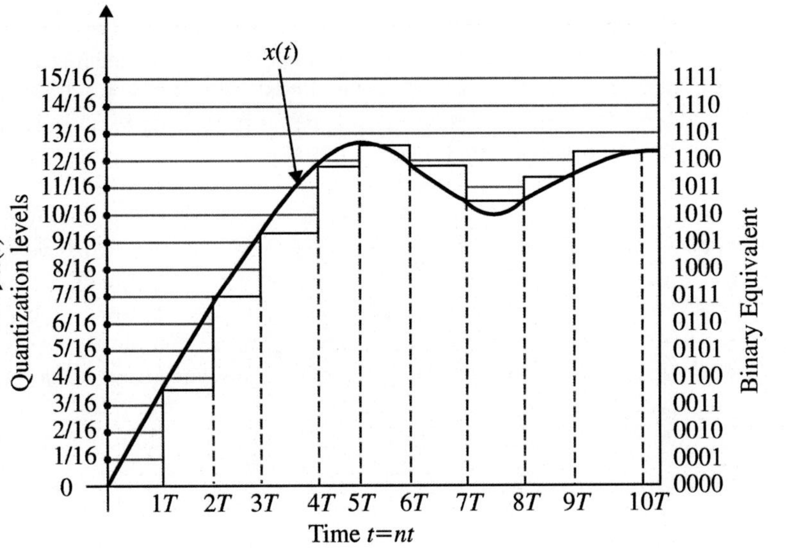

Fig. 1.69 Quantizing analog signal $x(t)$

ADC. The discrete amplitude levels are represented or encoded into distinct binary words each of length 'b' bits.

1.11 Quantization

Quantization is the process of converting analog signal to digital signal. The signal $x(t)$ is converted to a sampled signal at $t = nT$, where n is an integer and T is the sampling period. The sampling rate is at least twice the bandwidth of the signal which is called the Nyquist sampling rate or else analog signal recovery after it was processed by a digital computer or DSP is not possible without distortion. Figure 1.69 shows the continuous signal $x(t)$ which is to be digitized and quantized. The signal $x(t)$ is sampled at discrete time intervals $t = nT$ and held over the sampling interval by a device called zero order hold circuit (ZOH) that yields staircase approximation to the analog signal. The ZOH circuit keeps its output level equal to the magnitude of an input pulse and then resets itself when a new pulse arrives. On the arrival of the new pulse, the signal jumps to the new height of the continuous signal and holds the output at that level. After sampling and holding, the analog to digital converter converts the sampled signal to a digital number as explained below.

The analog voltage of $x(t)$ is divided into equal discrete levels and each level is assigned a digital number. These levels are called quantization levels. A 16 level discritization is shown in Fig. 1.69. A four bit digital number can represent each of the 16 levels. If v is the maximum value of the analog signal, then each level will represent

$\frac{v}{16}$ v. In general, the difference between any two quantization levels is $\frac{v}{2^b}$ v, where b is the binary bit used for analog to digital conversion. However, the quantization level has to be converted to binary bits because the digital computer or the digital signal processor will accept only binary bits as the input for processing the signal. Thus, it necessitates that each sample amplitude is approximated to the nearest quantization level. Each sample can be represented by pulses. Dealing with a large number of distinct pulses is difficult, and therefore, the use of smallest possible number of distinct pulses is preferable, the number two being the best choice. Thus, we need to assign a distinct electrical pulse to each of the two binary states. In practice, a positive pulse is used to represent the binary 1 and a negative pulse is used to represent a binary 0. Therefore, the sample is represented by a group of four binary impulses for 16 levels of quantization, and thus the binary 0110 represents quantization six.

1.11.1 Quantization Error

The analog signal $x(t)$ is sampled at regular interval to get $x(n)$ at $t = nT$, where $n = 0, 1, 2, \ldots$. The amplitude of the sample should be assigned to the nearest quantization level to express it in terms of numeric equivalent $x_q(n)$. This is shown in Fig. 1.70. The difference between $x(n)$, amplitude of the sample signal and $x_q(n)$ the quantized value of $x(n)$ is expressed as

$$e(n) = x_q(n) - x(n)$$

$e(n)$ is called quantization error.

Let a sinusoidal signal amplitude vary from $+1$ to -1 v. Let the ADC use $(b + 1)$ bits which include sign bit. The number of levels available for quantization of $x(n)$ is 2^{b+1}. Therefore, the interval between any quantization level is given by

$$q = \frac{2}{2^{b+1}} = 2^{-b}$$

For $b = 4$, $q = 0.0625$. The sampled signal $x(n)$ is quantized as $x_q(n)$ by the following methods:

(a) Truncation.
(b) Rounding.

By truncation, the sampled signal is approximated to the nearest higher quantization level. By rounding the b bit, sampled signal is rounded to the original number unrounded.

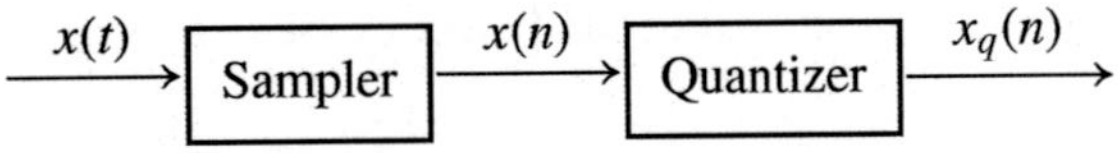

Fig. 1.70 Block diagram of sampler and quantizer

1.12 Energy Spectral Density of CT Signals

Consider two energy signals $x(t)$ and $y(t)$. The convolution of these two signals is written as

$$q(t) = x(t) * y(t)$$

Taking Fourier transform on both sides, we get

$$Q(\omega) = X(\omega)Y(\omega)$$

From equation for cross-correlation, we get

$$R_{xy}(\tau) = [x(t) * y(-t)]$$

Taking Fourier transform on both sides, we get

$$\begin{aligned} F[R_{xy}(\tau)] &= F[x(t) * y(-t)] \\ &= X(\omega)Y(-\omega) \end{aligned}$$

For auto-correlation, the above equation becomes

$$\begin{aligned} F[R_{xx}(\tau)] &= X(\omega)X(-\omega) \\ &= |X(\omega)|^2 \end{aligned} \tag{1.52}$$

Equation (1.52) represents Parseval's theorem for Fourier transform which gives the energy of the signal. $|X(\omega)|^2$ gives the energy spectral density and is denoted by $S_{xx}(\omega)$. It gives the energy density of $x(t)$ as the frequency ω is varied. Thus, from Eq. (1.52), the energy spectral density of a CT signal is nothing but the Fourier transform of its auto-correlation function. Thus,

$$\boxed{R_{xx}(\tau) \stackrel{\text{FT}}{\longleftrightarrow} S_{xx}(\omega)} \tag{1.53}$$

Equation (1.53) implies that by taking Fourier transform of the auto-correlation function, one can get energy spectral density of $x(t)$. Similarly by taking inverse Fourier transform of energy spectral density, the auto-correlation of $x(t)$ is obtained.

1.13 Power Spectral Density of CT Signals

Periodical signals do not have finite energy for the interval $-\infty < t < \infty$. However, they can be considered as energy signal over a period. Consider the auto-correlation of a signal $x(t)$ whose time period is T.

$$R_{xx}(\tau) = \underset{T\to\infty}{\text{Lt}} \frac{1}{T} \int_{-\infty}^{\infty} x(t)x(t-\tau)dt \tag{1.54}$$

Taking Fourier transform on both sides, we get

$$\begin{aligned}
F[R_{xx}(\tau)] &= F\left[\underset{T\to\infty}{\text{Lt}} \frac{1}{T} \int_{-\infty}^{\infty} x(t)x(t-\tau)dt\right] \\
&= \underset{T\to\infty}{\text{Lt}} F\left[\frac{1}{T} \int_{-\infty}^{\infty} x(t)x(t-\tau)dt\right] \\
&= \underset{T\to\infty}{\text{Lt}} \frac{1}{T} F[R_{xx}(\tau)] \\
&= \underset{T\to\infty}{\text{Lt}} \frac{1}{T} |X(\omega)|^2 \\
&= Pxx(\omega)
\end{aligned}$$

$$\boxed{R_{xx}(\tau) \overset{\text{FT}}{\longleftrightarrow} P_{xx}(\omega)} \tag{1.55}$$

The power spectrum density which is a function of ω is obtained by taking the Fourier transform of the auto-correlation function of $x(t)$ which is periodical. Conversely, by taking inverse Fourier transform of power spectral density, the auto-correlation function of $x(t)$ is obtained.

Example 1.42 Find the Energy Spectral Density (ESD) and hence the auto-correlation function for the following signal:

$$x(t) = e^{-2t}u(t)$$

Solution

$$\begin{aligned}
x(t) &= e^{-2t}u(t) \\
X(\omega) = F[x(t)] &= \frac{1}{j\omega + 2} \\
|X(\omega)| &= \frac{1}{\sqrt{\omega^2 + 4}}
\end{aligned}$$

Energy spectral density is given by the following equation:

$$S_{xx} = |X(\omega)|^2$$

$$\boxed{S_{xx} = \frac{1}{\omega^2 + 4}}$$

$$\begin{aligned}
R_{xx}(\tau) &= X(\omega)X(-\omega) \\
&= F\left[\frac{1}{(j\omega+2)}\frac{1}{(-j\omega+2)}\right] \\
&= F^{-1}\frac{1}{4}\left[\frac{1}{j\omega+2} + \frac{1}{2-j\omega}\right] \\
&= F^{-1}\frac{1}{4}\left[\frac{1}{j\omega+2} - \frac{1}{j\omega-2}\right]
\end{aligned}$$

$$\boxed{R_{xx}(\tau) = \frac{1}{4}[e^{-2\tau}u(\tau) + e^{2\tau}u(-\tau)]}$$

Example 1.43 Find the energy spectral density if the auto-correlation function is given by

$$R_{xx}(\tau) = e^{-|\tau|} \qquad \text{for } -\infty < \tau < \infty$$

Solution The energy density spectrum and auto-correlation are related by the following equation:

$$\begin{aligned}
S_{xx} &= \int_{-\infty}^{\infty} R_{xx}(\tau)e^{-j\omega\tau}d\tau \\
&= \int_{-\infty}^{\infty} e^{-4|\tau|}e^{-j\omega\tau}d\tau \\
&= \int_{-\infty}^{0} e^{+4\tau}e^{-j\omega\tau}d\tau + \int_{0}^{\infty} e^{-4\tau}e^{-j\omega\tau}d\tau \\
&= \int_{-\infty}^{\infty} e^{(4-j\omega)\tau}d\tau + \int_{0}^{\infty} e^{-(4+j\omega)\tau}d\tau \\
&= \frac{1}{(4-j\omega)}\left[e^{(4-j\omega)\tau}\right]_{-\infty}^{0} - \frac{1}{(4+j\omega)}\left[e^{-(4+j\omega)\tau}\right]_{0}^{\infty} \\
&= \frac{1}{(4-j\omega)} + \frac{1}{(4+j\omega)}
\end{aligned}$$

$$\boxed{S_{xx} = \frac{8}{(\omega^2 + 16)}}$$

Example 1.44 Find the energy density spectrum of the following discrete time sequence:

$$x[n] = \left(\frac{1}{2}\right)^n u[n]$$

Solution The energy density spectrum of $x[n]$ is the Fourier transform of its autocorrelation function and is related by the following equation:

$$S_{xx} = |X(j\omega)|^2$$

where $X(j\omega)$ is the F.T. of $x[n]$

$$X(j\omega) = \sum_{n=-\infty}^{\infty} \left(\frac{1}{2}\right)^n u[n]e^{-j\omega n}$$

$$X(j\omega) = \sum_{n=0}^{\infty} \left(\frac{1}{2}\right)^n e^{j\omega n}$$

$$= \sum_{n=0}^{\infty} \left(\frac{1}{2}e^{j\omega}\right)^n$$

Using the summation formula, we get

$$X(j\omega) = \frac{1}{\left(1 - \frac{1}{2}e^{j\omega}\right)}$$

$$= \frac{1}{(1 - 0.5\cos\omega - 0.5j\sin\omega)}$$

$$|X(j\omega)|^2 = \frac{1}{(1 - 0.5\cos\omega)^2 + 0.25\sin^2\omega}$$

$$= \frac{1}{(1.25 - \cos\omega)}$$

Energy density spectrum

$$\boxed{S_{xx} = \frac{1}{(1.25 - \cos\omega)}}$$

Example 1.45 Consider the following discrete time sequence:

$$x[n] = \cos\frac{n\pi}{3}$$

Determine the power density spectrum.

Solution Given

$$x[n] = \cos\frac{n\pi}{3}$$

where $\omega_0 = \frac{\pi}{3}$. The fundamental period of $x[n]$ is

$$\begin{aligned} N_0 &= \frac{2\pi}{\omega_0}m \\ &= \frac{2\pi}{\pi/3} \\ &= 6 \qquad \text{for } m = 1 \end{aligned}$$

The discrete time signal $x[n]$ can be written in the following form using Euler's formula:

$$\cos\frac{n\pi}{3} = \frac{1}{2}e^{j(n\pi/3)} + \frac{1}{2}e^{-j(n\pi/3)} \qquad (a)$$

The Fourier series representation of $x[n]$ is given by the following formula:

$$\begin{aligned} x[n] &= \sum_{k=(N_0)} C_k e^{(j2\pi kn)/N_0} \\ &= \sum_{k=-2}^{3} C_k e^{(j2\pi kn)/N_0} \\ &= \sum_{k=-2}^{3} C_k e^{(j\pi kn)/3} \end{aligned}$$

$$x[n] = C_{-2}e^{-(2j\pi n)/3} + C_{-1}e^{-(j\pi n)/3} + C_0 + C_1 e^{(j\pi n)/3} + C_2 e^{(2j\pi n)/3} + C_3 e^{j\pi n} \qquad (b)$$

Comparing equations (a) and (b), we get

$$C_1 = \frac{1}{2} \quad \text{and} \quad C_{-1} = \frac{1}{2}$$

with other coefficients being zero. Power spectral density is given by

$$P_{ss} = \sum_{k=0}^{N_0-1} |C_k|^2 = \left(\frac{1}{2}\right)^2 + \left(\frac{1}{2}\right)^2$$

$$\boxed{P_{xx} = \frac{1}{2}}$$

Example 1.46 Consider the following signal:

$$x[n] = \cos\frac{\pi}{5}n + \sin\frac{\pi}{6}n$$

Determine the power density spectrum.

Solution Given

$$x[n] = \cos\frac{\pi}{5}n + \sin\frac{\pi}{6}n$$

where

$$\begin{aligned}
\omega_{01} &= \frac{\pi}{5} \\
\omega_{02} &= \frac{\pi}{6} \\
N_{01} &= \frac{2\pi}{\omega_{01}} = 10 \\
N_{02} &= \frac{2\pi}{\omega_{02}} = 12
\end{aligned}$$

The periodicity of the combined signal is obtained by taking LCM of 10 and 12, which is 60.

$$\begin{aligned}
N_0 &= 60 \\
\omega_0 &= \frac{2\pi}{N_0} = \frac{\pi}{30}
\end{aligned}$$

Using Euler's formula, the given signal is expressed as

$$x[n] = \frac{1}{2}\left[e^{j(\pi/5)n} + e^{-j(\pi/5)n}\right] + \frac{1}{2j}\left[e^{j(\pi/6)n} - e^{-j(\pi/6)n}\right]$$

The above sequence can be expressed in terms of fundamental frequency as

$$x[n] = \frac{1}{2}\left[e^{j6\omega_0 n} + e^{-j6\omega_0 n}\right] + \frac{1}{2j}\left[e^{j5\omega_0 n} - e^{-j5\Omega_0 n}\right]$$

From the above equations, the Fourier series coefficients are obtained as

$$D_6 = \frac{1}{2}; \quad D_{-6} = \frac{1}{2}; \quad D_5 = \frac{1}{2j}; \quad D_{-5} = \frac{1}{2j}$$

The power spectrum density is given by

$$\begin{aligned}
P_{xx} &= |D_6|^2 + |D_{-6}|^2 + |D_5|^2 + |D_{-5}|^2 \\
&= \frac{1}{4} + \frac{1}{4} + \frac{1}{4} + \frac{1}{4}
\end{aligned}$$

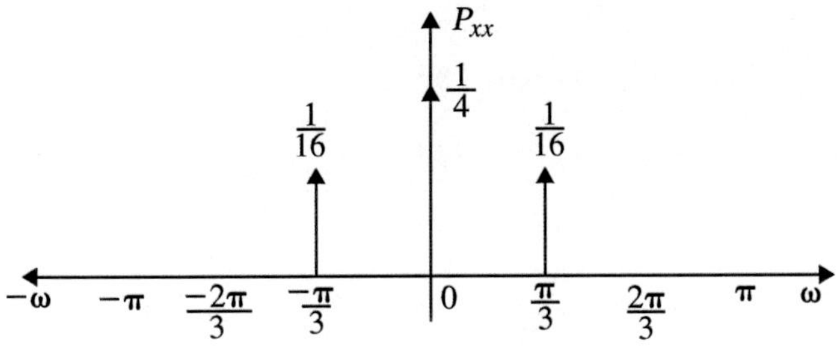

Fig. 1.71 Frequency spectrum of Example 1.47

$$\boxed{P_{xx} = 1}$$

Example 1.47 Find the power spectrum density for the following DT signal:

$$x[n] = \sin^2\left(\frac{\pi}{6}n\right)$$

Sketch the frequency spectrum (Fig. 1.71).

Solution

$$\begin{aligned}
\sin^2\left(\frac{\pi}{6}n\right) &= \frac{1}{2} - \frac{1}{2}\cos\frac{\pi n}{3} \\
\omega_0 &= \frac{\pi}{3} \\
N_0 &= \frac{2\pi m}{6} \\
N_0 &= 6 \qquad \text{for } m = 1
\end{aligned}$$

Using Euler's formula, we can write

$$\begin{aligned}
\left[\sin\frac{\pi}{6}n\right]^2 &= \left[\frac{1}{2j}\left(e^{j(\pi/6)n} - e^{-j(\pi/6)n}\right)\right]^2 \\
&= \frac{1}{2} - \frac{1}{4}\left[e^{j(\pi/3)n} + e^{-j(\pi/3)n}\right] \\
&= \frac{1}{2} - \frac{1}{4}\left[e^{j\omega_0 n} + e^{-j\omega_0 n}\right] \\
D_0 &= \frac{1}{2} \\
D_1 &= -\frac{1}{4} \\
D_{-1} &= D_{-1+6} = D_5 = -\frac{1}{4}
\end{aligned}$$

The frequency spectrum of Example 1.47 is shown in Fig. 1.71
The power density function is given by

$$\begin{aligned} P_{xx} &= |D_0|^2 + |D_1|^2 + |D_5|^2 \\ &= \frac{1}{4} + \frac{1}{16} + \frac{1}{16} \\ &= \frac{3}{8} \end{aligned}$$

1.13.1 Properties of Power Spectral Density

1. **The total area under the power spectrum density curve equals the average value of the signal $x(t)$.**
   Proof: From Eq. (1.53) we have,

$$\begin{aligned} R_{xx}(\tau) &\overset{\text{FT}}{\longleftrightarrow} P_{xx}(\omega) \\ F^{-1}[P_{xx}(\omega)] &= R_{xx}(\tau) \\ &= \int_{-\infty}^{\infty} P_{xx}(\omega) e^{j\omega\tau} d\omega \end{aligned}$$

   For $\tau = 0$,

$$R_{xx}(0) = \int_{-\infty}^{\infty} P_{xx}(\omega) d\omega$$

   From Eq. (1.54), the average power is obtained for $\tau = 0$, as

$$\begin{aligned} R_{xx}(0) &= \underset{T\to\infty}{\text{LT}} \frac{1}{T} \int_{-\infty}^{\infty} |x(t)|^2 dt \\ &= \text{average power} \end{aligned}$$

$$\begin{aligned} \int_{-\infty}^{\infty} P_{xx}(\omega) d\omega &= R_{xx}(0) \\ &= \text{Area under the power spectrum density curve} \end{aligned}$$

2. **The power spectral density of a real valued power signal $x(t)$ is an even function of frequency which satisfies**

$$P_{xx}(-\omega) = P_{xx}(\omega)$$

   Proof: Let $x_T(t)$ be the segment of $x(t)$ which is periodic with period T. It is defined as follows:

$$x_T(t) = \begin{cases} x(t), & |t| \geq T/2 \\ 0, & \text{otherwise} \end{cases}$$

The power density of a periodical signal is given by

$$P_{xx}(\omega) = \operatorname*{Lt}_{T\to\infty}\left[\frac{|x_T(\omega)|^2}{T}\right]$$

$$P_{xx}(-\omega) = \operatorname*{Lt}_{T\to\infty}\left[\frac{|x_T(-\omega)|^2}{T}\right]$$

$$\begin{aligned} P_{xx}(-\omega) &= \operatorname*{Lt}_{T\to\infty}\left[\frac{|x_T(\omega)|^2}{T}\right] \\ &= P_{xx}(\omega) \end{aligned}$$

Hence, the power density function is an even function.

3. **Power density** $P_{xx}(\omega)$ of a signal $x(t)$ is the Fourier transform of the auto-correlation function of $x(t)$.
   Proof: The proof is given in Eq. (1.55).
4. **Power spectral density** $P_{xx}(\omega)$ **of a signal** $x(t)$ **is always non-negative**.
   Proof: From Eq. (1.55), we get

$$P_{xx}(\omega) = \operatorname*{LT}_{T\to\infty}\frac{1}{T}|X(\omega)|^2$$

Since $|X(\omega)|^2$ is always positive, $P_{xx}(\omega)$ is non-negative.

1.14 Recursive Systems

A discrete time system with the Nth order linear constant coefficient can be described by the following difference equation:

$$\sum_{k=0}^{N} a_k y(n-k) = \sum_{k=0}^{M} b_k x(n-k) \tag{1.56}$$

the above equation is written in the following form as:

$$y(n) = \frac{1}{a_0}\left[\sum_{k=0}^{M} b_k x(n-k) - \sum_{k=1}^{N} a_k y(n-k)\right] \tag{1.57}$$

Equation (1.57) gives the value of the output $y(n)$ at any time n in terms of the previous values of the input and output. A discrete time system described by difference equation of the type given in Eq. (1.57), where the output $y(n)$ is fed back and compared with the input, is called a **Recursive System**. In such systems, $y(n-1),\ y(n-2), \ldots, y(n-N)$ need to be known to calculate $y(n)$. Therefore, if the input for all n together with a set of auxiliary conditions $y(-N)$, $y(-N+1), \ldots, y(-1)$, Eq. (1.57) can be solved for $y(n)$. Equation (1.57) is called a recursive equation because it requires a recursive procedure for determining the output in terms of the input and the previous outputs.

Now consider the case when $N = 0$. Equation (1.57) reduces to the following form:

$$y(n) = \sum_{k=0}^{M} \left(\frac{b_k}{a_0}\right) x(n-k) \tag{1.58}$$

Equation (1.58) gives the output $y(n)$ in terms of the present and previous values of the input. Discrete time systems described by difference equation (1.58) is called **Non-Recursive System**. Here we do not use the previously computed values of the output recursively, the system described by Eq. (1.58) is called a non-recursive system. Further, for a non-recursive system, we do not require auxiliary conditions to determine the output $y(n)$ at any time n.

Equation (1.58) also describes the LTI system whose impulse response is given as

$$h(n) = \begin{cases} \frac{b_n}{a_0}, & 0 \le n \le M \\ 0, & \text{otherwise} \end{cases} \tag{1.59}$$

From Eq. (1.59), it is evident that the impulse response is non-zero only over a finite interval. Hence system described by Eq. (1.58) is called **Finite Impulse Response (FIR) System**. On the other hand, the impulse response of equation (1.57) is infinite length and hence it is also called **Infinite Impulse Response (IIR) System**.

Example 1.48 Determine the output of the recursive system described by the following difference equation:

$$y(n+2) + 3y(n+1) + 2y(n) = u(n)$$

where $y(0) = 0$; $y(1) = 1$ and $u(n)$ is unit step response.

Solution

$$y(n+2) + 3y(n+1) + 2y(n) = u(n)$$

For $n = 0$,

$$\begin{aligned} y(2) + 3y(1) + 2y(0) &= u(0) \\ y(2) + 3 &= 1 \\ y(2) &= -2 \end{aligned}$$

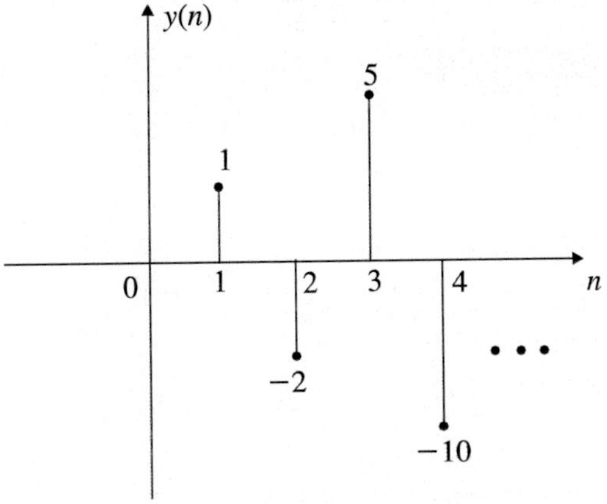

Fig. 1.72 Output response of Example 1.48

For $n = 1$,

$$y(3) + 3y(2) + 2y(1) = u(1)$$
$$y(3) = 5$$

For $n = 2$,

$$y(4) + 3y(3) + 2y(2) = u(2)$$
$$y(4) = -10$$

the output response $y(n)$ is sketched in Fig. 1.72.

Summary

- Signals are broadly classified as continuous time (CT) and discrete time (DT) signals. They are further classified as deterministic and stochastic, periodic and non-periodic, odd and even and energy and power signals.
- Basic DT signals include impulse, step, ramp, parabolic, rectangular pulse, triangular pulse, signum function, sinc function, sinusoid, real and complex exponentials.
- Basic operations on DT signals include addition, multiplication, amplitude scaling, time scaling, time shifting, reflection or folding and amplitude inverted signals.
- In time shifting of DT signal, for $x(n + n_0)$ and $x(-n - n_0)$, the time shift is made to the left of $x(n)$ and $x(-n)$ respectively by n_0. For $x(n - n_0)$ and $x(-n + n_0)$, the time shift is made to the right of the $x(n)$ and $x(-n)$ respectively by n_0.

- To plot DT signals, the operation performed is in the following sequence. The signal is folded (if necessary), time shifted, time scaled, amplitude scaled and inverted.
- Signals are classified as even signals and odd signals. Even signals are **symmetric** about the vertical axis, whereas odd signals are **anti-symmetric** about the time origin. Odd signals pass through the origin. The product of two even signals or two odd signals is an even signal. The product of an even and an odd signal is an odd signal.
- A DT signal which repeats itself every N sequence is called a periodic signal. If the signal is not periodic, it is called an aperiodic or non-periodic signal. The necessary condition for the composite of two or more signals to be periodic is that the individual signal should be periodic.
- A signal is an energy signal if the total energy of the signal satisfies the condition $0 < E < \infty$. A signal is called a power signal if the average power of the signal satisfies the condition $0 < P < \infty$. If the energy of a signal is finite, the average power is zero. If the power of the signal is finite, the signal has infinite energy. All periodic signals are power signals. However, all power signals need not be periodic. Signals which are deterministic and non-periodic are usually energy signals. Some signals are neither energy signals nor power signals.
- The system is broadly classified as a continuous and discrete time system.
- The DT systems are further classified based on the property of causality, linearity, time invariancy, invertibility, memory and stability.
- A discrete time system is said to be causal if the impulse response $h(n) = 0$ for $n < 0$.
- If the impulse response of a discrete time system is absolutely summable, then the system is said to be BIBO stable.
- Step response $s(n)$ is obtained from impulse response $h(n)$ using the mathematical expression

$$s(n) = \sum_{k=0}^{n} h(k)$$

Short Questions and Answers

1. **How are signals classified?** signals are generally classified as CT and DT signals. They are further classified as deterministic and non-deterministic, odd and even, periodic and non-periodic and power and energy signals.
2. **What are odd and even signals?**
   A continuous CT signal is said to be an even signal if it satisfies the condition $x(-t) = x(t)$ for all t. It is said to be an odd signal if $x(-t) = -x(t)$ for all t. For a DT signal if $x[-n] = x[n]$ condition is satisfied, it is an even sequence (signal). If $x[-n] = -x[n]$, the sequence is called an odd sequence.
3. **How even and odd components of a signal are mathematically expressed?**

$$x_e[n] = \frac{1}{2}\{x[n] + x[-n]\}$$
$$x_0[n] = \frac{1}{2}\{x[n] - x[-n]\}$$

4. **What are periodic and non-periodic signals?**
   A discrete time signal is said to be a period signal if it satisfies the condition $x[n] = x[n+N]$ for all n. A signal which is not periodic is said to be non-periodic.
5. **What is the fundamental period of a periodic signal? What is the fundamental frequency?**
   A DT signal is said to be periodic if it satisfies the condition $x(n) = x(n+N)$. If this condition is satisfied for $N = N_0$, it is also satisfied for $N = 2N_0, 3N_0, \ldots$. The smallest value of N that satisfies the above condition is called the fundamental period. The fundamental frequency $f_0 = \frac{1}{N_0}$ Hz. It is also expressed as $\omega_0 = \frac{2\pi}{N_0}$ rad.
6. **What are power and energy signals?**
   For a DT signal $x[n]$, the total energy is defined as

$$E = \sum_{n=-\infty}^{\infty} x^2[n]$$

   The average power is defined as

$$P = \underset{T\to\infty}{Lt} \frac{1}{2N+1} \sum_{n=-N}^{N} x^2[n]$$

7. **Determine whether the signal $x[n] = \cos[0.1\pi n]$ is periodic?**
   The signal $x[n]$ is periodic with fundamental period $N_0 = 20$.
8. **Find whether the signal $x[n] = 5\cos[6\pi n]$ is periodic?**
   The signal is periodic with fundamental period $N_0 = 1$.
9. **Find the average power of the signal.**

$$x[n] = u[n] - u[n-N]$$

   The average power $P = 1$.
10. **Find the total energy of**

$$x[n] = \{1, \underset{\uparrow}{1}, 1\}$$

   The total energy $E = 3$.

11. **If the discrete time signal $x[n] = \{0, 0, 0, 3, 2, 1, -1, -7, 6\}$ then find $y[n] = x[2n - 3]$?**

$$y[n] = \{0, 0, 0, 3, 1, -7\}$$

12. **What is the energy of the signal $x[n] = u[n] - u[n - 6]$?**

$$E = 6$$

13. **What are the properties of systems?**
Systems are generally classified as continuous and discrete time systems. Further classifications of these systems are done based on their properties, which include

   (*a*) linear and non-linear,
   (*b*) time invariant and time variant,
   (*c*) static and dynamic,
   (*d*) causal and non-causal,
   (*e*) stable and unstable and
   (*f*) Invertible and non-invertible.

14. **Define system. What is linear system?**
A system is defined as the interconnection of objects with a definite relationship between objects and attributes.
A system is said to be linear if the weighted sum of several inputs produces a weighted sum of outputs. In other words, the system should satisfy the homogeneity and additivity of the superposition theorem if it is to be linear. Otherwise, it is a non-linear system.
15. **What is time invariant and time variant system?**
A system is said to be time invariant if the output due to the delayed input is the same as the delayed output due to the input. If the continuous time system is described by the differential equation, its coefficients should be time independent for the system to be time invariant. In the case of discrete time system, the coefficients of the difference equation describing the system should be time independent (constant) for the system to be time invariant. If the above conditions are not satisfied, the system (CT as well as DT) is said to be time variant.
16. **What are static and dynamic systems?**
If the output of the system depends only on the present input, the system is said to be static or instantaneous. If the output of the system depends on the past and future input, the system is not static and it is called a dynamic system. A static system does not require memory, and so it is called a memoryless system. A dynamic system requires memory, and hence it is called a system with memory. Systems which are described by differential and difference equations are dynamic systems.
17. **What are causal and non-causal systems?**
If the system output depends on present and on past inputs, it is called a causal

system. If the system output depends on future input it is called a non-causal system.

18. **What are stable and unstable systems?**
If the input is bounded and output is also bounded, the system is called BIBO stable system. If the input is bounded and the output is unbounded, the system is unstable. A system whose impulse response curve has a finite area is also called a stable system.
19. **What are invertible and non-invertible systems?**
A system is said to be invertible if the distinct inputs give distinct outputs.
20. **State the condition for a discrete time LTI system to be causal and stable.**

A discrete time LTI system is said to be causal and stable if the poles of the transfer function all lie in the left half s-plane and the Region of Convergence (ROC) is to the right of the right-most pole.
21. **A certain LTID time system has the following impulse response.**

$$\boldsymbol{h[n] = \left(\frac{1}{3}\right)^n u[n-2]}$$

Is the system both causal and stable?
The response depends on the past input $u[n-2]$, and hence it is causal.

$$\begin{aligned} h[n] &= \sum_{n=2}^{\infty}\left(\frac{1}{3}\right)^n \\ &= \frac{1}{6} < \infty \end{aligned}$$

The system is stable. The system is both causal and stable.
22. **What is convolution sum?**
If $h(n)$ is the impulse response of a linear time invariant discrete time system and if $x(n)$ is input, then the output $y(n)$ is obtained from the following equation:

$$y(n) = \sum_{k=-\infty}^{\infty} x(k)h(n-k)$$

The above equation is called convolution sum. Symbolically, it is represented as

$$y(n) = x(n) * h(n)$$

23. **Evaluate $x(n) * \delta(n)$ and $x(n) * \delta(n-n_0)$?**

$$\begin{aligned} x(n) * \delta(n) &= x(n) \\ x(n) * \delta(n-n_0) &= x(n-n_0) \end{aligned}$$

24. **If $x(n) = \delta(n) - 2\delta(n-1) + 3\delta(n-2) - \delta(n+1)$. Express $x(n-1)$ in terms of sequences?**

$$x(n-1) = \delta(n-1) - 2\delta(n-2) + 3\delta(n-3) - \delta(n)$$

25. **$h(n) = \left(\frac{1}{2}\right)^n u(n)$. Find A so that $h(n) - Ah(n-1) = \delta(n)$?**

$$h(n) - h(n-1) = \left(\frac{1}{2}\right)^n u(n) - A\left(\frac{1}{2}\right)^{n-1} u(n-1)$$

If $A = 1/2$

$$h(n) - h(n-1) = \left(\frac{1}{2}\right)^n u(n) - \left(\frac{1}{2}\right)^n u(n-1) = \delta(n)$$

26. **An LTID system has the following impulse response**

$$h(n) = \left(\frac{1}{2}\right)^n u(n+1)$$

Is the system causal and BIBO stable?
The system is non-zero for $n = -1$. Hence, it is non-causal

$$\begin{aligned}\sum_{-\infty}^{\infty} h(n) &= \sum_{-1}^{\infty}\left(\frac{1}{2}\right)^n \\ &= \left(\frac{1}{2}\right)^{-1} + \sum_{0}^{\infty}\left(\frac{1}{2}\right)^n \\ &= 2 + \frac{1}{1-1/2} = 4 < \infty\end{aligned}$$

The system is BIBO stable.

Long Answer Type Questions

1. **For the following DT signal, find even and odd components**:

$$x[n] = \{1,\ -3,\ 2,\ 5,\ 4\}$$

$$x_e[n] = \{2,\ 2.5,\ 1,\ -1.5,\ \underset{\uparrow}{1},\ -1.5,\ 1,\ 2.5,\ 2\}$$

$$x_0[n] = \{-2,\ -2.5,\ -1,\ 1.5,\ \underset{\uparrow}{0},\ -1.5,\ 1,\ 2.5,\ 2\}$$

2. **Find whether the following signals are periodic. If periodic, find the fundamental period**: (a) $x[n] = \cos(\frac{n}{8} - \pi)$; ($b$) $x[n] = \cos(\frac{\pi}{8} + \frac{\pi}{2}) + \cos(\frac{\pi}{6} - \frac{\pi}{2})$; ($c$) $x[n] = \cos(\frac{5\pi N}{12} + \frac{\pi}{2}) + \sin\frac{10\pi n}{8}$; ($d$) $x[n] = e^{j3n} + e^{j4\pi n}$.
(a) Not periodic. (b) Periodic with fundamental period $N_0 = 48$ samples/s. (c) Periodic with fundamental period $N_0 = 24$ samples/s. (d) Non-periodic.
3. **Given** $x[n]$ and $y[n]$

$$\begin{aligned} x[-1] &= 2 \\ x[n] &= 1 \qquad 1 \le n \le 5 \\ x[6] &= \frac{1}{2} \\ &= 0 \qquad \text{for other } n \end{aligned}$$

Plot (a) $x[\frac{n}{2}]$ **and** (b) $E_v x[n]$.

(a)

$$x[n] = \{2,\ 1,\ 1,\ 1,\ 1,\ 1,\ 1,\ 0.5\}$$
$$\uparrow$$
$$x\left[\frac{n}{2}\right] = \{2,\ 0,\ 1,\ 0,\ 1,\ 0,\ 1, 0,\ 1,\ 0,\ 1,\ 0,\ 1,\ 0.5\}$$
$$\uparrow$$

(b)

$$x_e[n] = \{0.25,\ 0.5,\ 0.5,\ 0.5,\ 0.5,\ 0.5,\ 1,\ 0.5,\ 0.5,\ 0.5,\ 0.5,\ 0.5,\ 0.25\}$$
$$\uparrow$$

4. **Find whether the following signal is periodic. If periodic, find the fundamental period**

$$x[n] = \cos\left(2\pi n + \frac{\pi}{2}\right)\sin\left(5\pi n - \frac{\pi}{4}\right)\sin\left(8\pi n + \frac{\pi}{2}\right)$$

The signal is periodic. Their fundamental period $N_0 = 2$ samples/s.
5. **For the systems given below, determine whether each of them is** (a) **static,** (b) **causal,** (c) **time invariant,** (d) **linear and** (e) **stable.**

$$\boldsymbol{y[n] = x[5n]}$$

(a) The system response depends on present, past and future inputs. Hence, it is dynamic.
(b) Non-causal.
(c) The output due to the delayed input is not the same as the delayed output. Hence, it is time variant.

(*d*) The weighted sum of the output is the same as the output due to the weighted sum of the input. The system is linear.
(*e*) If the input $x[5n]$ is bounded, the output $y[n]$ is also bounded. The system is stable.

6.

$$y[k+2]+3y[k+1]+4y[k]=x[k]$$

(*a*) The system is dynamic.
(*b*) The system is causal.
(*c*) The system is time invariant.
(*d*) The system is linear.
(*e*) The system is stable.

7.

$$y[n]=5x[3^n]$$

(*a*) The system is dynamic.
(*b*) The system is non-causal.
(*c*) The system is time invariant.
(*d*) The system is linear.
(*e*) The system is stable.

8.

$$y[n]=\sin(2\pi x[n])+x[n+1]$$

(*a*) The system is dynamic.
(*b*) The system is non-causal.
(*c*) The system is time invariant.
(*d*) The system is non-linear.
(*e*) The system is stable.

9. **$x[n]=u[n-4]-u[n-10]$ and $h[n]=u[n-5]-u[n-16]$. Find $y[n]=x[n]*h[n]$?**

$$\begin{aligned} y[n] &= (n-8) && 9\le n\le 13 \\ &= 6 && 10\le n\le 19 \\ &= (25-n) && 20\le n\le 24 \\ &= 0 && n>24 \end{aligned}$$

10. **$x[n]=4^n u[-n-2]$ and $h[n]=u[n-2]$. Find $y[n]=x[n]*h[n]$?**

$$y[n] = \frac{1}{3}\left[\frac{1}{4}\right]^{n-1} \qquad n < 0$$
$$= \frac{1}{12} \qquad n > 0$$

11. **Determine whether the following LTID time systems whose impulse response given below are stable.** (*a*) $h[n] = n \sin 2\pi n\, u[n]$, (*b*) $h[n] = 5^n u[-n]$ **and** (*c*) $h[n] = 2^{-n} u[n-5]$?

(a) $y[n] = \infty$ B.I.B.O. unstable.

(b) $y[n] = \frac{1}{4} < \infty$ B.I.B.O. stable.

(c) $y[n] = \frac{1}{6} < \infty$ B.I.B.O. stable.

12. $x[n] = e^{-2n}u[n]$ **and** $h[n] = 3^{-n}u[n]$. **Find** $y[n] = x[n] * h[n]$?

$$y[n] = \frac{[3e^2 - (3e^2)^{-n}]u[n]}{[3e^2 - 1]}$$

13. $x[n] = (0.5)^n\, u[n]$ **and** $h[n] = (0.9)^n u[n]$. **Find** $y[n] = x[n] * h[n]$?

$$y[n] = 2.5[0.9^{n+1} - 0.5^{n+1}]u[n]$$

14. **Determine the power and energy of the following signals**:

(*a*) $x[n] = \left(\frac{1}{3}\right)^n$.

(*b*) $x[n] = e^{j\left(\frac{n\pi}{2} + \frac{\pi}{4}\right)}$.

(*c*) $x[n] = \sin\left(\frac{n\pi}{4}\right)$.

(*d*) $x[n] = e^{2n}u(n)$.

Ans:

$$(a) \quad E = \frac{9}{8}J; \; P = 0.$$
$$(b) \quad E = \infty; \; P = 1W.$$
$$(c) \quad E = \infty; \; P = \frac{1}{2}W.$$
$$(d) \quad E = \infty; \; P = \infty \text{ (Neither power nor energy)}.$$

15. **Determine whether each of the following signals is periodic. Find the fundamental period**

$$(a) \quad x[n] = e^{j6\pi n}.$$
$$(b) \quad x[n] = e^{j\frac{3}{5}(n+\frac{1}{2})}.$$
$$(c) \quad x[n] = \cos\left(\frac{2\pi}{3}\right)n.$$
$$(d) \quad x[n] = \cos\frac{n\pi}{3} + \cos\frac{3n\pi}{4}.$$

Ans:

(a) Periodic. $N = 1$sample

(b) Not Periodic.

(c) Periodic. $N = 3$samples

(d) Periodic. $N = 24$samples

16. **Find the odd and even components of the following sequence.**

$$x(n) = \{2, 3, \underset{\uparrow}{4}, 5, 6\}$$

Ans:

$$x_e(n) = \{4, 4, \underset{\uparrow}{4}, 4, 4\}$$
$$x_0(n) = \{-2, -1, 0, 1, 2\}$$

17. **Test whether the following systems are time variant or time invariant:**

(a) $y(n) = x(n) - x(n-1)$.
(b) $y(n) = nx(n)$.
(c) $y(n) = x(-n)$.

(*d*) $y(n) = x(n) \cos \omega_0 n$.
(*e*) $y(n) = x(n) - bx(n-1)$.
(*f*) $y(n) = x(n) + c$.
(*g*) $y(n) = e^{x(n)}$.
(*h*) $y(n) = nx^2(n)$.
(*i*) $y(n) = \sum_{k=0}^{m} a_k(n-k) - \sum_{k=1}^{m} b_k y(n-k)$.

Ans:

(*a*) Time invariant.
(*b*) Time varying.
(*c*) Time varying.
(*d*) Time varying.
(*e*) Time varying.
(*f*) Time invariant.
(*g*) Time invariant.
(*h*) Time varying.
(*i*) Time invariant.

18. **Test the linearity of the following systems**:

(*a*) $y(n) = nx(n)$.
(*b*) $y(n) = x(n^2)$.
(*c*) $y(n) = x^2(n)$.
(*d*) $y(n) = Ax(n) + B$.
(*e*) $y(n) = e^{x(n)}$.

Ans:

(*a*) Linear.
(*b*) Linear.
(*c*) Non-linear.
(*d*) Non-linear.
(*e*) Non-linear.

19. **Test the causality of the following systems**:

(*a*) $y(n) = x(n) - x(n-1)$.
(*b*) $y(n) = \sum_{k=\infty}^{n} x(k)$.
(*c*) $y(n) = nx(n)$.

Ans:

(*a*) Causal.
(*b*) Causal.
(*c*) Causal.

20. **Test whether the following systems are stable or not**:

(*a*) $y(n) = x(-n-2)$.
(*b*) $y(n) = x(n) + by(n-1)$.

(c) $y(n) = \begin{cases} a^n & n \leq 0, \\ b^n & n < 0. \end{cases}$

Ans:

(a) Stable.
(b) Stable if $|b| < 1$.
(c) Stable if $|a| < 1$ and $|b| > 1$.

21. **Determine the response of the relaxed system characterized by the impulse response $h(n) = (\frac{1}{2})^n u(n)$ to the input signal $x(n) = 2^n u(n)$.**
Ans:

$$y(n) = \left(\frac{1}{2}\right)^n \left(\frac{4^{n+1} - 1}{3}\right) u(n)$$

Chapter 2
The z-transform Analysis of Discrete Time Systems

Learning Objectives

After completing this chapter, you should be able to
- ✠ define the z-transform and the inverse z-transform;
- ✠ find the z-transform and ROC of typical DT signals;
- ✠ find the properties of ROC;
- ✠ find the properties of z-transform;
- ✠ find the inverse z-transform;
- ✠ solve difference equation using the z-transform;
- ✠ establish the relationship between the z-transform, Fourier transform and the Laplace transform.
- ✠ find the causality and stability of DT system;
- ✠ find the discrete time Fourier transform;
- ✠ find the linear convolution of two DT sequences.

2.1 Introduction

The z-transform is the discrete counterpart of the Laplace transform. The Laplace transform converts integro-differential equations into algebraic equations. In the same way, the z-transform converts difference equations of discrete time system to algebraic equations which simplifies the discrete time system analysis. There are many connections between Laplace and z-transforms except for some minor differences. DTFT represents discrete time signal in terms of complex sinusoids. When this sort of representation is generalized and represented in terms of complex exponential, it is termed as z-transform. This sort of representation has a broader

S. Palani, *Discrete Time Systems and Signal Processing*,
https://doi.org/10.1007/978-3-031-32421-5_2

characterization of system with signals. Further, the DTFT is applicable only for stable system whereas z-transform can be applied even to unstable systems which means that z-transform can be used to larger class of systems and signals. It is to be noted that many of the properties in DTFT, Laplace transform and z-transform are common except that the Laplace transform deals with continuous time signals and systems.

2.2 The z-transform

Let z^n be an everlasting exponential. Let $h(n)$ be the impulse response of the discrete time system. The response of a linear, time invariant discrete time system to the everlasting exponential z^n is given as $H(z)z^n$. That is, it is the same exponential within a multiplicative constant. Thus, the system response to the excitation $x[n]$ is the sum of the system's responses to all these exponentials. The tool that is used to represent an arbitrary discrete signal $x[n]$ as a sum of everlasting exponential of the form z^n is called the z-transform.

Let $x[n] = z^n$ be the input signal applied to an LTI discrete time system whose impulse response is $h[n]$. The system output $y[n]$ is given by

$$\begin{aligned} y[n] &= x[n] * h[n] \\ &= \sum_{k=-\infty}^{\infty} h[k]x[n-k] \end{aligned}$$

Substitute $x[n] = z^n$

$$y[n] = \sum_{k=-\infty}^{\infty} h[k]z^{n-k} = z^n \left[\sum_{k=-\infty}^{\infty} h[k]z^{-k} \right]$$

Define the transfer function

$$H[z] = \sum_{k=-\infty}^{\infty} h[k]z^{-k} \tag{2.1}$$

Equation (2.1) may be written as

$$H[z^n] = H[z]z^n$$

To represent any arbitrary signals as a weighted superposition of the Eigen function z^n, let us substitute $z = re^{j\Omega}$ into Eq. (2.1)

$$\begin{aligned} H[re^{j\Omega}] &= \sum_{n=-\infty}^{\infty} h[n][re^{j\Omega}]^{-n} \\ &= \sum_{n=-\infty}^{\infty} \left(h[n]r^{-n}\right) e^{-j\Omega n} \end{aligned} \tag{2.2}$$

Equation (2.2) corresponds to the DTFT of the signal $h[n]r^{-n}$. The inverse of $H[re^{j\Omega}]$, by mathematical manipulation of Eq. (2.2), can be obtained as

$$h[n] = \frac{1}{2\pi j} \oint H(z)z^{n-1}dz \tag{2.3}$$

More generally, Eqs. (2.2) and (2.3) can be written as

$$X[z] = \sum_{n=-\infty}^{\infty} x[n]z^{-n} \tag{2.4}$$

$$x[n] = \frac{1}{2\pi j} \oint X(z)z^{n-1}dz \tag{2.5}$$

The above equations are called z-transform pair. Equation (2.4) is the z-transform of $x[n]$ and Eq. (2.5) is called inverse z-transform. In Eq. (2.4) the range of n is $-\infty < n < \infty$ and hence it is called bilateral z-transform. If $x[n] = 0$ for $n < 0$, Eq. (2.4) can be written as

$$X[z] = \sum_{n=0}^{\infty} x[n]z^{-n} \tag{2.6}$$

Equation (2.6) is called unilateral or right-sided z-transform. Bilateral z-transform has limited practical applications. Unless otherwise it is specifically mentioned, z-transform means unilateral. z-transform and inverse z-transform are symbolically represented as given below.

$$\begin{aligned} Z[x[n]] &= X[z] \\ x[n] &\overset{Z}{\longleftrightarrow} X[z] \\ z^{-1}[X[z]] &= x[n] \\ X[z] &\overset{Z^{-1}}{\longleftrightarrow} x[n]. \end{aligned} \tag{2.7}$$

2.3 Existence of the z-transform

Consider the unilateral z-transform given by Eq. (2.6)

$$X[z] = \sum_{n=0}^{\infty} x[n]z^{-n} = \sum_{n=0}^{\infty} \frac{x[n]}{z^n}$$

For the existence of $X[z]$,

$$|X[z]| \leq \sum_{n=0}^{\infty} \frac{|x[n]|}{|z|^n} < \infty \tag{2.8}$$

If the signal $x[n]$ is expressed in terms of an exponential signal r^n, then if $x[n] \leq r^n$ for some r, then

$$|x[n]| \leq r^n \tag{2.9}$$

Substitute Eq. (2.9) in Eq. (2.8)

$$\begin{aligned} |X[z]| &\leq \sum_{n=0}^{\infty} \left(\frac{r}{z}\right)^n \\ &= \frac{1}{\left[1 - \frac{r}{|z|}\right]} \qquad \text{iff } |z| > r \end{aligned} \tag{2.10}$$

From Eq. (2.10), it is evident that the z-transform of $x[n]$ which is $X(z)$ exists for $|z| > r$ and the signal is z-transformable. If the signal $x[n]$ grows faster than the exponential signal r^n for any r, Eq. (2.10) is not convergence and $x[n]$ is not z-transformable.

2.4 Connection Between Laplace Transform, z-transform and Fourier Transform

Consider the Laplace transform of $x(t)$ which is represented below

$$X(s) = \int_{-\infty}^{\infty} x(t)e^{-st}dt \tag{2.11}$$

When $s = j\omega$, Eq. (2.11) becomes

$$X(j\omega) = \int_{-\infty}^{\infty} x(t)e^{-j\omega t} dt \tag{2.12}$$

Equation (2.12) represents the Fourier transform. **The Laplace transform reduces to the Fourier transform on the imaginary axis where s = $j\omega$.** The relationship between these two transforms can also be interpreted as follows. The complex variable s can be written as $(\sigma + j\omega)$. Equation (2.11) is written as

$$\begin{aligned} X(\sigma + j\omega) &= \int_{-\infty}^{\infty} x(t)e^{-(\sigma+j\omega)t} dt \\ &= \int_{-\infty}^{\infty} \left[x(t)e^{-\sigma t}\right] e^{-j\omega t} dt \end{aligned} \tag{2.13}$$

Equation (2.13) can be recognized as the Fourier transform of $[x(t)e^{-\sigma t}]$. **Thus, the Laplace transform of $x(t)$ is the Fourier transform of $x(t)$ after multiplication by the real exponential $e^{-\sigma t}$ which may be growing or decaying with respect to time.**

The complex variable z can be expressed in polar form as

$$z = re^{j\omega} \tag{2.14}$$

where r is the magnitude of z and ω is the angle of z.

Substitute $z = re^{j\omega}$ in Eq. (2.6)

$$\begin{aligned} X(re^{j\omega}) &= \sum_{n=-\infty}^{\infty} x[n](re^{j\omega})^{-n} \\ &= \sum_{n=-\infty}^{\infty} \{x[n]r^{-n}\}e^{-j\omega n} \\ &= F[x[n]r^{-n}] \end{aligned} \tag{2.15}$$

Thus, $X(re^{j\omega})$ is the Fourier transform of the sequence $x[n]$ which is multiplied by a real exponential r^{-n} which may be growing or decaying with increasing n depending on whether r is greater or less than unity. If $r = 1$, then $|z| = 1$ and equation becomes

$$X(e^{j\omega}) = \sum_{n=-\infty}^{\infty} x[n]e^{-j\omega n} = F[x[n]]$$

The z-transform reduces to Fourier transform in the complex z-plane on the contour of a circle with unit radius. The circle which is called unit circle plays the role in the z-transform similar to the role of the imaginary axis in the s-plane for Laplace transform. The unit circle in the z-plane is shown in Fig. 2.1.

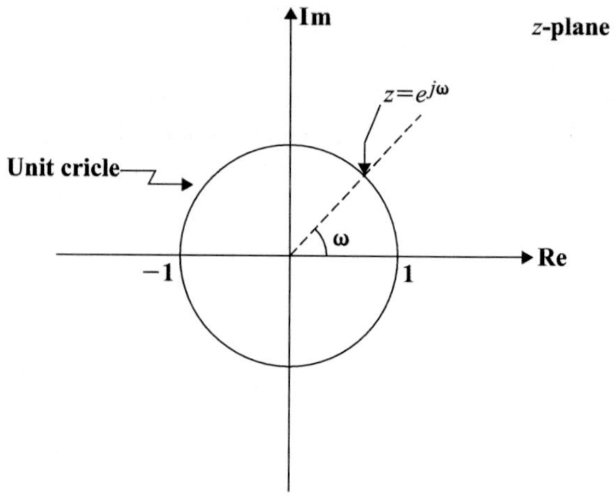

Fig. 2.1 z-transform reduces to FT on the unit circle

2.5 The Region of Convergence (ROC)

In Eq. (2.4) which defines the z-transform $X(z)$ the sum may not cover for all values of z. The values of z in the complex z-plane for which the sum in the z-transform equation converges is called the region of convergence which is written in abbreviated form as ROC. The concept of ROC is illustrated in the following examples.

Example 2.1 Consider the following discrete time signals:

$$(a) \qquad x[n] = a^n u[n] \qquad a < 1$$

$$(b) \qquad x[n] = -a^n u(-n-1) \qquad a < 1$$

$$(c) \qquad x[n] = a^n u[n] - b^n u(-n-1) \qquad b > a \text{ and } a > b$$

Find the z-transform and the ROC in the z-plane.

Solution (*a*) $\boldsymbol{x[n] = a^n u[n]}$ The signal $x[n]$ is shown in Fig. 2.2a which is a right-sided signal.

$$\begin{aligned} X(z) &= \sum_{n=0}^{\infty} a^n u[n] z^{-n} \\ &= \sum_{n=0}^{\infty} a^n z^{-n} \qquad [\because\ u[n] = 1 \text{ all } n \geq 0] \\ &= \sum_{n=0}^{\infty} \left(\frac{a}{z}\right)^n \end{aligned}$$

Using the power series, we get

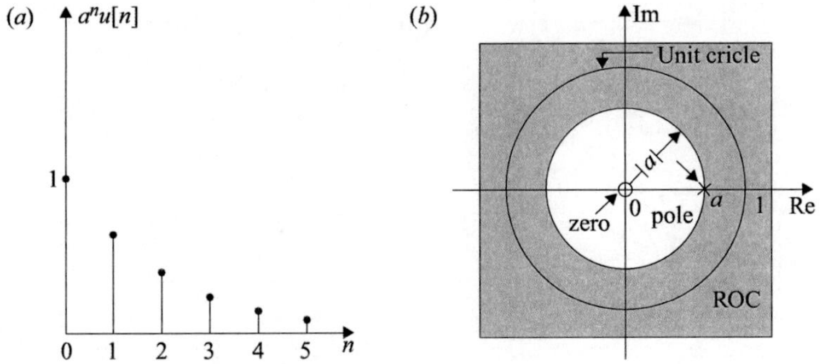

Fig. 2.2 **a** $x[n] = a^n u[n]$ and **b** ROC: $0 < a < 1$

$$X(z) = \frac{1}{\left[1 - \frac{a}{z}\right]}$$

where $\frac{a}{z} < 1$ or $|z| > |a|$.

$$\boxed{X(z) = \frac{z}{(z-a)}} \quad (2.16)$$

$$\boxed{X(z) = \frac{1}{1 - az^{-1}}} \quad (2.17)$$

Fourier transform is represented in the form as shown in Eq. (2.16) to identify poles and zero and system transfer function. Equation (2.17) form is used when inverse z-transform is taken and also for structure realization. z^{-1} is used as time delay operation. z-transform for the causal real exponential converges if $|z| > |a|$. Thus, the ROC of $X(z)$ is to the exterior of the circle of radius a, which is shown in Fig. 2.2b in shaded area. The ROC includes the unit circle for $|a| < 1$.

(*b*) $\boldsymbol{x[n] = -a^n[u[-n-1]]}$ The signal $x[n]$ is shown in Fig. 2.3a which is a left-sided signal

$$Z[-a^n u[-n-1]] = \sum_{n=-\infty}^{-1} -a^n z^{-n} \qquad \because [u(-n-1)] = 1 \text{ for all } -n$$

$$= \sum_{n=-\infty}^{-1} -\left[\frac{a}{z}\right]^n = \sum_{n=1}^{\infty} -\left[\frac{z}{a}\right]^n$$

$$= -\left[\frac{z}{a} + \frac{z^2}{a^2} + \frac{z^3}{a^3} + \cdots\right]$$

$$= 1 - \left[1 + \frac{z}{a} + \left(\frac{z}{a}\right)^2 + \left(\frac{z}{a}\right)^3 + \cdots\right]$$

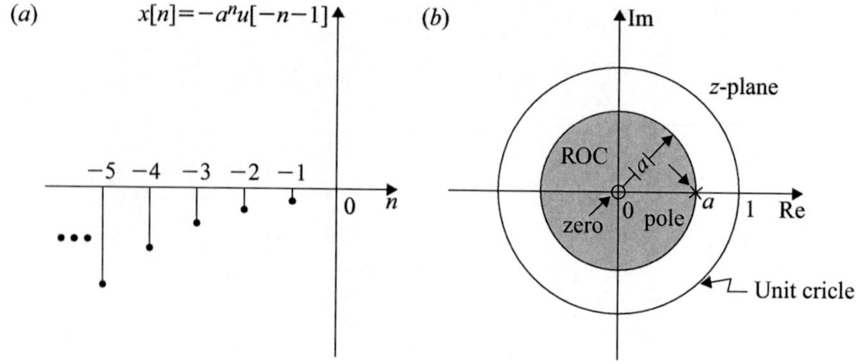

Fig. 2.3 **a** $x[n] = -a^n u[-n-1]$ and **b**) ROC: $0 < a < 1$

$$= 1 - \frac{1}{1 - \frac{z}{a}} \quad \text{if } \left|\frac{z}{a}\right| < 1$$

$$\boxed{Z[-a^n u[-n-1]] = \frac{z}{(z-a)}} \quad \text{ROC } |z| < a \tag{2.18}$$

The z-transforms of $x[n] = a^n u[n]$ which is causal and that of $x[n] = -a^n u[-n-1]$ which is anti-causal are identical. In the former case, the ROC is to the exterior of the circle passing through the outermost pole, and in the latter case (anti-causal), the ROC is to the interior of the circle passing through the innermost pole. The ROC is shown in Fig. 2.3b.

(c) $\boldsymbol{x[n] = a^n u[n] - b^n u[-n-1]}$

From the results derived in Example 2.1(a) and (b), we can find the z-transform of $x[n]$ as

$$X(z) = \frac{z}{(z-a)} + \frac{z}{(z-b)}$$

The right-sided signal $a^n u[n]$ converges if $|z| > a$ and the left-sided signal $-b^n u[-n-1]$ converges if $|z| < b$. The ROC for $|a| > |b|$ and $|a| < |b|$ are shown in Fig. 2.4a, b, respectively. From Fig. 2.4a, it is observed that the two ROCs do not overlap and hence z-transform does not exist for this signal. Now considering Fig. 2.4b, it is observed that the two ROCs overlap and the overlapping area is shaded in the form of a ring. The z-transform exists in the case with ROC as $|a| < |z| < |b|$.

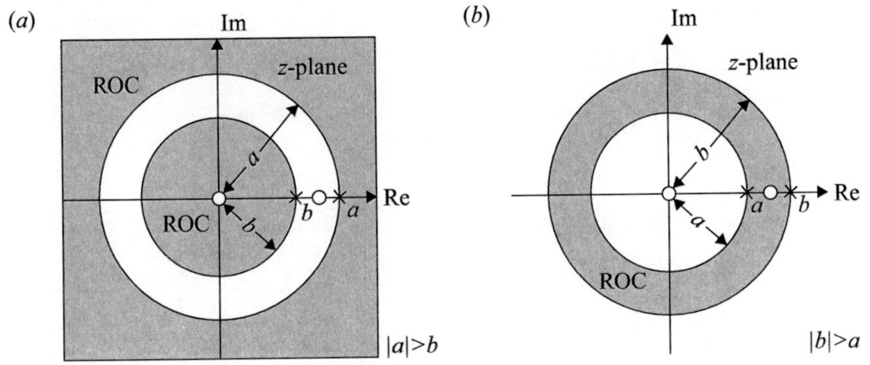

Fig. 2.4 ROC of a two-sided sequence

2.6 Properties of the ROC

Assuming that $X(z)$ is the rational function of z, the properties of the ROC are summed up and given below:

1. The ROC is a concentric ring in the z-plane.
2. The ROC does not contain any pole.
3. If $x[n]$ is a finite sequence in a finite interval $N_1 \leq n \leq N_2$, then the ROC is the entire z-plane except $z = 0$ and $z = \infty$.
4. If $x[n]$ is a right-sided sequence (causal), then the ROC is the exterior of the circle $|z| = r_{\text{max}}$ where r_{max} is the radius of the outermost pole of $X(z)$.
5. If $x[n]$ is a left-sided sequence (non-causal), then the ROC is the interior of the circle $|z| = r_{\text{min}}$ where r_{min} is the radius of the innermost pole of $X(z)$.
6. If $x[n]$ is a two-sided sequence, then the ROC is given by $r_1 < |z| < r_2$ where r_1 and r_2 are the magnitudes of the two poles of $X(z)$. Here ROC is an annular ring between the circle $|z| = r_1$ and $|z| = r_2$ which does not include any poles.

The following examples illustrate the method of finding z-transform $X(z)$ for the discrete time sequence $x[n]$.

Example 2.2 Find the z-transform and the ROC for the sequences $x[n]$ given below.

$$1. \quad x[n] = \{\underset{\uparrow}{2}, -1, 0, 3, 4\}$$

$$2. \quad x[n] = \{1, -2, 3, -1, \underset{\uparrow}{2}\}$$

$$3. \quad x[n] = \{5, 3, \underset{\uparrow}{-2}, 0, 4, -3\}$$

$$4. \quad x[n] = \delta[n]$$

5. $x[n] = u[n]$
6. $x[n] = u[-n]$
7. $x[n] = a^{-n}u[-n]$
8. $x[n] = a^{-n}u[-n-1]$
9. $x[n] = (-a)^n u[-n]$
10. $x[n] = a^{|n|}$ for $|a| < 1$ and $|a| > 1$
11. $x[n] = e^{j\omega_0 n}u[n]$
12. $x[n] = \cos \omega_0 n u[n]$
13. $x[n] = \sin \omega_0 n u[n]$
14. $x[n] = u[n] - u[n-6]$
15. $x[n] = \left[\cos\left(\frac{\pi n}{3} + \frac{\pi}{4}\right)\right]u[n]$

Solution 1. $\boldsymbol{x[n] = \{2, -1, 0, 3, 4\}}$

$$X[z] = \sum_{n=0}^{4} x[n]z^{-n}$$

$$\boxed{X[z] = 2 - z^{-1} + 0 + 3z^{-3} + 4z^{-4}}$$

$X[z]$ will not converge, if $|z| = 0$. Hence, ROC is $|z| > 0$.

2. $\boldsymbol{x[n] = \{1, -2, 3, -1, \underset{\uparrow}{2}\}}$

$$X[z] = \sum_{n=-4}^{0} x[n]z^{-n}$$

$$\boxed{X[z] = z^4 - 2z^3 + 3z^2 - z + 2}$$

$X[z]$ will not converge, if $|z| = \infty$. Hence, ROC is $|z| < \infty$.

3. $\boldsymbol{x[n] = \{5, 3, \underset{\uparrow}{-2}, 0, 4, -3\}}$

$$X[z] = \sum_{n=-2}^{3} x[n]z^{-n}$$

$$\boxed{X[z] = 5z^2 + 3z - 2 + 0 + 4z^{-2} - 3z^{-3}}$$

For $|z| = 0$ and $|z| = \infty$, $X[z]$ is infinity. Hence, ROC is $0 < |z| < \infty$.

4. $\boldsymbol{x[n] = \delta[n]}$

$$X[z] = \sum_{n=-\infty}^{\infty} \delta[n] z^{-n}$$

$$\begin{aligned} \delta[n] &= 1 \quad n = 0 \\ &= 0 \quad n \neq 0 \end{aligned}$$

$$\boxed{X[z] = 1} \quad \text{ROC is entire } z\text{-plane}$$

5. $\boldsymbol{x[n] = u[n]}$

$$\begin{aligned} X[z] &= \sum_{n=0}^{\infty} z^{-n} \\ &= 1 + \frac{1}{z} + \frac{1}{z^2} + \cdots \\ &= \frac{1}{1 - \frac{1}{z}} \quad \text{[By using summation formula]} \end{aligned}$$

$$\boxed{\begin{aligned} X[z] &= \frac{z}{(z-1)} \\ X[z] &= \frac{1}{(1 - z^{-1})} \end{aligned}} \quad \text{ROC: } |z| > 1 \tag{2.19}$$

6. $\boldsymbol{x[n] = u[-n]}$

$$\begin{aligned} X[z] &= \sum_{n=-\infty}^{0} z^{-n} \\ &= \sum_{n=0}^{\infty} z^{n} \\ &= 1 + z + z^2 + \cdots \end{aligned}$$

$$\boxed{X[z] = \frac{1}{1 - z}} \quad \text{ROC: } |z| < 1 \tag{2.20}$$

7. $\boldsymbol{x[n] = a^{-n} u[-n]}$

$$X[z] = \sum_{n=-\infty}^{0} a^{-n} z^{-n}$$

$$= \sum_{n=-\infty}^{0} (az)^{-n}$$

$$= \sum_{n=0}^{\infty} (az)^{n}$$

$$= 1 + (az) + (az)^2 + \cdots$$

$$\boxed{X[z] = \frac{1}{(1 - az)}} \qquad \text{ROC: } |z| < \frac{1}{a} \tag{2.21}$$

8. $\boldsymbol{x[n] = a^{-n}u[-n-1]}$

$$X[z] = \sum_{n=-\infty}^{-1} a^{-n} z^{-n}$$

$$= \sum_{n=-\infty}^{-1} (az)^{-n}$$

$$= \sum_{n=1}^{\infty} (az)^{n}$$

$$= az + (az)^2 + (az)^3 + \cdots$$

$$X[z] = az[1 + az + (az)^2 + \cdots]$$

$$= \frac{az}{1 - az}$$

$$\boxed{X[z] = \frac{-z}{\left(z - \frac{1}{a}\right)}} \qquad \text{ROC: } |z| < \frac{1}{a} \tag{2.22}$$

9. $\boldsymbol{x[n] = (-a)^n u[-n]}$

$$X[z] = \sum_{n=-\infty}^{0} (-a)^n z^{-n}$$

$$= \sum_{n=-\infty}^{0} \left(\frac{z}{-a}\right)^{-n}$$

$$= \sum_{n=0}^{\infty} \left(\frac{z}{-a}\right)^{n}$$

$$= 1 + \left(\frac{z}{-a}\right) + \left(\frac{z}{-a}\right)^2 + \left(\frac{z}{-a}\right)^3 + \cdots$$

$$\boxed{X[z] = \frac{a}{(z+a)}} \qquad \text{ROC: } |z| < |a| \tag{2.23}$$

10. $\boldsymbol{x[n] = a^{|n|};\ a < 1}$

$$x[n] = a^n u[n] + a^{-n} u[-n-1]$$

$$Z[a^n u[n]] = \frac{z}{(z-a)} \qquad \text{ROC: } |z| > a$$

$$Z[a^{-n} u[-n-1]] = \frac{-z}{\left(z - \frac{1}{a}\right)} \qquad \text{ROC: } |z| < \frac{1}{a}$$

$$X[z] = \frac{z}{(z-a)} - \frac{z}{\left(z - \frac{1}{a}\right)}$$

$$\boxed{X[z] = \frac{(a^2-1)}{a} \frac{z}{(z-a)\left(z - \frac{1}{a}\right)}} \tag{2.24}$$

ROC: $a < |z| < \frac{1}{a}$. The ROC is sketched and shown in Fig. 2.5a for $a < 1$.

$$x[n] = a^{|n|} \qquad a > 1$$

The ROC is sketched and shown in Fig. 2.5b. In Fig. 2.5b, the two ROCs do not overlap and there is no common ROC. Hence, $x[n]$ does not have $X[z]$.

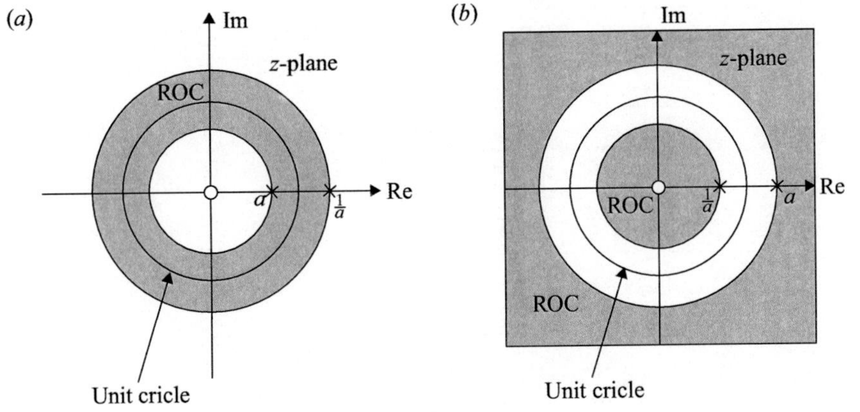

Fig. 2.5 ROC of $x[n] = a^{|n|}$. **a** $a < 1$ and **b** $a > 1$

11. $\boldsymbol{x[n] = e^{j\omega_0 n}u[n]}$

$$\begin{aligned} X[z] &= \sum_{n=0}^{\infty} e^{j\omega_0 n} z^{-n} \\ &= \sum_{n=0}^{\infty} \left(\frac{e^{j\omega_0}}{z}\right)^n \\ &= \left(\frac{1}{1 - \frac{e^{j\omega_0}}{z}}\right) \end{aligned}$$

$$\boxed{X[z] = \frac{z}{(z - e^{j\omega_0})} \qquad \text{ROC: } |z| > |e^{j\omega_0}| \text{ or } |z| > 1} \tag{2.25}$$

12. $\boldsymbol{x[n] = \cos \omega_0 n u[n]}$

$$\begin{aligned} x[n] &= \frac{1}{2}[e^{j\omega_0 n} + e^{-j\omega_0 n}] \\ Z[e^{j\omega_0 n}] &= \frac{z}{(z - e^{j\omega_0})} \\ Z[e^{-j\omega_0 n}] &= \frac{z}{(z - e^{-j\omega_0})} \\ X[z] &= \frac{1}{2}\left[\frac{z}{(z - e^{j\omega_0})} + \frac{z}{(z - e^{-j\omega_0})}\right] \\ &= \frac{z}{2}\frac{\left[z - e^{-j\omega_0} + z - e^{j\omega_0}\right]}{\left[z^2 - z(e^{-j\omega_0} + e^{j\omega_0}) + 1\right]} \end{aligned}$$

$$X[z] = \frac{z}{2}\frac{[2z - 2\cos\omega_0]}{\left[z^2 - 2z\cos\omega_0 + 1\right]}$$

$$\boxed{X[z] = \frac{(1 - z^{-1}\cos\omega_0)}{(1 - z^{-1}2\cos\omega_0 + z^{-2})} \qquad \text{ROC: } |z| > 1} \tag{2.26}$$

13. $\boldsymbol{x[n] = \sin \omega_0 n u[n]}$

$$\begin{aligned} x[n] &= \frac{1}{2j}[e^{j\omega_0 n} - e^{-j\omega_0 n}] \\ Z[e^{j\omega_0 n}u[n]] &= \frac{z}{(z - e^{j\omega_0})} \\ Z[e^{-j\omega_0 n}u[n]] &= \frac{z}{(z - e^{-j\omega_0})} \end{aligned}$$

$$\begin{aligned}
X[z] &= \frac{z}{2j}\left[\frac{1}{(z-e^{j\omega_0})} - \frac{1}{(z-e^{-j\omega_0})}\right] \\
&= \frac{z}{2j}\frac{\left[z-e^{-j\omega_0}-z+e^{j\omega_0}\right]}{\left[z^2-2z\cos\omega_0+1\right]} \\
&= \frac{z\sin\omega_0}{(z^2-2z\cos\omega_0+1)}
\end{aligned}$$

$$\boxed{X[z] = \frac{z^{-1}\sin\omega_0}{(1-2z^{-1}\cos\omega_0+z^{-2})} \qquad \text{ROC: } |z|>1} \tag{2.27}$$

14. $\boldsymbol{x[n] = u[n] - u[n-6]}$

$$\begin{aligned}
x[n] &= \{1,\ 1,\ 1,\ 1,\ 1,\ 1\} \\
X[z] &= 1 + z^{-1} + z^{-2} + z^{-3} + z^{-4} + z^{-5} \\
&= \left[1 + \frac{1}{z} + \frac{1}{z^2} + \frac{1}{z^3} + \frac{1}{z^4} + \frac{1}{z^5}\right]
\end{aligned}$$

$$\boxed{X[z] = \frac{[z^5+z^4+z^3+z^2+z+1]}{[z^5]} \qquad \text{ROC: all } z \text{ except } z \neq 0} \tag{2.28}$$

The above result can be represented in a compact form as (Fig. 2.6)

$$\begin{aligned}
X[z] &= \sum_{n=0}^{5} z^{-n} \\
&= \sum_{n=0}^{5}\left(\frac{1}{z}\right)^n
\end{aligned}$$

The following summation formula is used to simplify this

$$\sum_{k=m}^{n} a^k = \frac{a^{n+1}-a^m}{(a-1)}$$

where $a = \frac{1}{z}$; $k = 0$ and $n = 5$

$$X[z] = \frac{\left(\frac{1}{z}\right)^6 - \left(\frac{1}{z}\right)^0}{\left(\frac{1}{z}-1\right)}$$

$$\boxed{X[z] = \frac{z}{(z-1)}(1-z^{-6})}$$

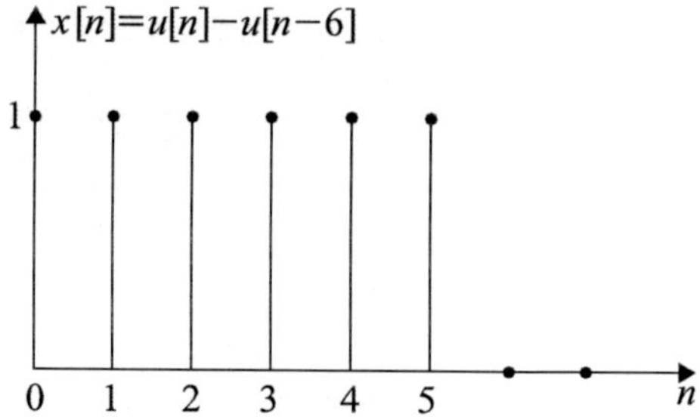

Fig. 2.6 Representation of $x[n] = u[n] - u[n-6]$ for Example 2.2.14

15. $\mathbf{x[n] = \left[\cos\left(\frac{\pi n}{3} + \frac{\pi}{4}\right)\right] u[n]}$

$$\begin{aligned} x[n] &= \frac{1}{2}\left[e^{j\left(\frac{\pi n}{3}+\frac{\pi}{4}\right)} + e^{-j\left(\frac{\pi n}{3}+\frac{\pi}{4}\right)}\right] \\ &= \frac{1}{2}\left[e^{j\frac{\pi}{4}} e^{j\frac{\pi n}{3}} + e^{-j\frac{\pi}{4}} e^{-j\frac{\pi}{4}}\right] \\ X[z] &= \frac{1}{2}\left[e^{j\frac{\pi}{4}} \frac{z}{(z - e^{j\frac{\pi}{3}})} + e^{-j\frac{\pi}{4}} \frac{z}{(z - e^{-j\frac{\pi}{3}})}\right] \end{aligned}$$

$$\begin{aligned} X[z] &= \frac{z}{2} \frac{\left[ze^{j\frac{\pi}{4}} - e^{-j\frac{\pi}{12}} + ze^{-j\frac{\pi}{4}} - e^{-j\frac{\pi}{12}}\right]}{z^2 - z(e^{j\frac{\pi}{3}} + e^{-j\frac{\pi}{3}}) + 1} \\ &= \frac{z}{2} \frac{\left[2z\cos\frac{\pi}{4} - 2\cos\frac{\pi}{12}\right]}{\left(z^2 - 2z\cos\frac{\pi}{3} + 1\right)} \end{aligned}$$

$$\boxed{X[z] = \frac{z[0.707z - 0.966]}{(z^2 - z + 1)} \qquad \text{ROC: } |z| > 1}$$

2.7 Properties of z-transform

The transformations of $x(t)$ and $x[n]$ to $X(s)$ and $X(j\omega)$ using Laplace transform and Fourier transformof become easier if the properties of these transforms are directly applied. Similarly, if the properties of z-transform are applied directly to $x[n]$, then $X[z]$ can be easily derived. Hence, some of the important properties of z-transform which are applied to signals and systems are derived and the applications illustrated. The following properties are derived:

1. Linearity;
2. Time shifting;
3. Time reversal;
4. Multiplication by n;
5. Multiplication by an exponential;

6. Time expansion;
7. Convolution theorem;
8. Initial value theorem;
9. Final value theorem.

2.7.1 Linearity

If

$$x_1[n] \overset{Z}{\longleftrightarrow} X_1[z] \quad \text{and} \quad x_2[n] \overset{Z}{\longleftrightarrow} X_2[z]$$

then

$$\{a_1x_1[n] + a_2x_2[n]\} \overset{Z}{\longleftrightarrow} [a_1X_1[z] + a_2X_2[z]] \tag{2.29}$$

Proof Let

$$\begin{aligned} x[n] &= a_1x_1[n] + a_2x_2[n] \\ X[z] &= \sum_{n=-\infty}^{\infty} [a_1x_1[n] + a_2x_2[n]]z^{-n} \\ &= \sum_{n=-\infty}^{\infty} a_1x_1[n]z^{-n} + \sum_{n=-\infty}^{\infty} a_2x_2[n]z^{-n} \end{aligned}$$

$$\boxed{X[z] = a_1x_1[z] + a_2x_2[z]}$$

2.7.2 Time Shifting

If

$$x[n] \overset{Z}{\longleftrightarrow} X[z]$$

then

$$x[n-k] \overset{Z}{\longleftrightarrow} z^{-k}X[z]$$

Proof Let

$$Z[x[n-k]] = \sum_{n=-\infty}^{\infty} x[n-k]z^{-n}$$

Substitute $(n-k)=m$

$$Z[x[n-k]] = \sum_{m=-\infty}^{\infty} x[m]z^{-(k+m)}$$
$$= \sum z^{-k}x[m]z^{-m}$$

$$\boxed{Z[x[n-k]] = z^{-k}X[z]} \tag{2.30}$$

2.7.3 Time Reversal

If

$$x[n] \stackrel{Z}{\longleftrightarrow} X[z] \qquad \text{ROC: } r_1 < |z| < r_2$$

then

$$x[-n] \stackrel{Z}{\longleftrightarrow} X[z^{-1}] \qquad \text{ROC: } \frac{1}{r_1} < |z| < \frac{1}{r_2}$$

Proof Let

$$Z[x[-n]] = \sum_{n=-\infty}^{\infty} x[-n]z^{-n}$$

Substitute $-n = m$

$$Z[x[-n]] = \sum_{n=\infty}^{-\infty} x[m]z^{m}$$
$$= \sum_{m=-\infty}^{\infty} x[m](z^{-1})^{m}$$

$$\boxed{Z[x[-n]] = X[z^{-1}]} \tag{2.31}$$

Thus, according to time reversal property, folding the signal in the time domain is equivalent to replacing z by z^{-1}. Further the ROC of $X[z]$ which is $r_1 < |z| < r_2$ becomes $r_1 < |z^{-1}| < r_2$ which is $\frac{1}{r_2} < |z| < \frac{1}{r_1}$.

2.7.4 Multiplication by n

If

$$Z[x[n]] = X[z]$$

then

$$Z[nx[n]] = -z\frac{d}{dz}X[z]$$

Proof Let

$$X[z] = \sum_{n=-\infty}^{\infty} x[n]z^{-n}$$

$$Z[nx[n]] = \sum_{n=-\infty}^{\infty} nx[n]z^{-n}$$

$$= z\sum_{n=-\infty}^{\infty} nx[n]z^{-n-1}$$

$$= z\sum_{n=-\infty}^{\infty} x[n][nz^{-n-1}]$$

$$Z[nx[n]] = z\sum_{n=-\infty}^{\infty} -x[n]\frac{d}{dz}[z^{-n}]$$

$$= -z\frac{d}{dz}\sum_{n=-\infty}^{\infty} x[n]z^{-n}$$

$$\boxed{Z[nx[n]] = -z\frac{d}{dz}X[z]} \tag{2.32}$$

2.7.5 Multiplication by an Exponential

If

$$Z[x[n]] = X[z]$$

then

$$Z[a^n x[n]] = X[a^{-1}z]$$

Proof Let

$$Z[a^n x[n]] = \sum_{n=-\infty}^{\infty} a^n x[n] z^{-n}$$

$$= \sum_{n=-\infty}^{\infty} x[n][a^{-1}z]^{-n}$$

$$\boxed{Z[a^n x[n]] = X[a^{-1}z]} \tag{2.33}$$

ROC: $r_1 < |a^{-1}z| < r_2$ or $ar_1 < |z| < ar_2$. In $X[z]$, z is replaced by $\frac{z}{a}$.

2.7.6 Time Expansion

If

$$Z[x[n]] = X[z]$$

then

$$Z[x_k[n]] = X[z^k]$$

Proof

$$Z[x_k[n]] = \sum_{n=-\infty}^{\infty} x\left[\frac{n}{k}\right] z^{-n}$$

where n is multiple of k. Substitute $\frac{n}{k} = l$

$$Z[x_k[n]] = \sum_{l=-\infty}^{\infty} x[l] z^{-kl}$$

$$= \sum_{l=-\infty}^{\infty} x[l][z^k]^{-l} = X[z^k]$$

$$\boxed{Z[x_k[n]] = X[z^k]} \tag{2.34}$$

2.7.7 Convolution Theorem

If

$$y[n] = x[n] * h[n]$$

then

$$Y[z] = X[z]H[z]$$

Proof

$$y[n] = \sum_{k=-\infty}^{\infty} x[k]h[n-k]$$

$$Y[Z] = \sum_{n=-\infty}^{\infty} \left[\sum_{k=-\infty}^{\infty} x[k]h[n-k] \right] z^{-n}$$

$$= \sum_{k=-\infty}^{\infty} x[k]z^{-k} \sum_{n=-\infty}^{\infty} h[n-k]z^{-(n-k)}$$

Substitute $(n-k) = l$

$$Y[z] = \sum_{k=-\infty}^{\infty} x[k]z^{-k} \sum_{l=-\infty}^{\infty} h[l]z^{-l}$$

$$\boxed{Y[z] = X[z]Y[z]} \tag{2.35}$$

2.7.8 Initial Value Theorem

If

$$X[z] = Z[x[n]]$$

where $x[n]$ is causal, then

$$x[0] = \underset{z\to\infty}{Lt}\, X[z]$$

Proof For a causal signal $x[n]$

$$X[z] = \sum_{n=0}^{\infty} x[n]z^{-n}$$
$$= x[0] + x[1]z^{-1} + x[2]z^{-2} + \cdots$$

Taking $z \to \infty$ on both sides, we get

$$\underset{z\to\infty}{Lt}\, X[z] = \underset{z\to\infty}{Lt}\, [x[0] + x[1]z^{-1} + x[2]z^{-2} + \cdots]$$
$$= x[0]$$

$$\boxed{x[0] = \underset{z\to\infty}{Lt}\, X[z]} \tag{2.36}$$

2.7.9 *Final Value Theorem*

If $Z[x[n]] = X[z]$ where $x[n]$ is a causal signal and the ROC of $X[z]$ has no poles on or outside the unit circle, then

$$x[\infty] = \underset{z\to 1}{Lt}\, (z-1)X[z]$$

Proof

$$Z[x[n+1]] - Z[x[n]] = \underset{k\to\infty}{Lt} \sum_{n=0}^{k} [x[n+1] - x[n]]z^{-n}$$

$$x[\infty] = \underset{k\to\infty}{Lt} \sum_{n=0}^{k} [x[n+1] - x[n]]z^{-n}$$

$$zX[z] - x[0] - X[z] = \underset{k\to\infty}{Lt} \sum_{n=0}^{k} [x[n+1] - x[n]]z^{-n}$$

$$(z-1)X[z] - x[0] = \underset{k\to\infty}{Lt} \sum_{n=0}^{k} [x[n+1] - x[n]]z^{-n}$$

Taking $\underset{z\to\infty}{Lt}$ on both sides, we get

$$\begin{aligned}
&\underset{z\to\infty}{Lt}\,(z-1)X[z]-x[0]\\
&=\underset{k\to\infty}{Lt}\,[x[1]-x[0]]+[x[2]-x[-1]]+[x[3]-x[2]]+\cdots+[x[k+1]-x[k]]\\
&=x[\infty]-x[0]
\end{aligned}$$

$$\boxed{x[\infty]=\underset{z\to 1}{Lt}\,(z-1)X[z]} \tag{2.37}$$

Example 2.3 Find the z-transform of the following sequences and also ROC using the properties of z-transform:

1. $x[n]=\delta[n-n_0]$
2. $x[n]=u[n-n_0]$
3. $x[n]=a^{n+1}u[n+1]$
4. $x[n]=a^{n-1}u[n-1]$
5. $x[n]=\left(\frac{1}{2}\right)^n u[-n]$
6. $x[n]=u[n-6]-u[n-10]$
7. $x[n]=nu[n]$
8. $x[n]=n[u[n]-u[n-8]]$
9. $x[n]=a^n\cos\omega_0 nu[n]$
10. $x[n]=a^n\sin\omega_0 nu[n]$
11. Show that $u[n]*u[n-1]=nu[n]$
12. $x[n]=n\left(-\frac{1}{4}\right)^n u[n]*\left(\frac{1}{6}\right)^{-n}u[-n]$
13. $x[n]=\left[\left(\frac{1}{2}\right)^n-\left(\frac{1}{4}\right)^n\right]u[n]$
    Find $X[z]$ and plot the poles and zeros.
14. $x[n]=1 \quad n\geq 0$
    $=z^n \quad n<0$
15. (a) $x[n]=\left[\left(-\frac{1}{3}\right)^n+3\left(\frac{1}{6}\right)^n\right]u[n]$
    (b) $x[n]=\left[\left(-\frac{1}{3}\right)^n u[-n]+3\left(\frac{1}{6}\right)^n\right]u[n]$
    (c) $x[n]=\left[\left(-\frac{1}{3}\right)^n+3\left(\frac{1}{6}\right)^n\right]u[-n]$
16. (a) $x[n]=\left[\left(\frac{1}{4}\right)^n+\left(\frac{1}{5}\right)^n\right]u[n]$
    (b) $x[n]=\left[\left(\frac{1}{5}\right)^n u[n]+\left(\frac{1}{4}\right)^n u[-n-1]\right]$
    (c) $x[n]=\left(\frac{1}{4}\right)^n u[n]+\left(\frac{1}{5}\right)^n u[-n-1]$

17. $x[n] = \delta[n] + \frac{1}{2}\delta(n+1) + \delta(n-3)$
18. $x[n] = 4^n \cos\left[\frac{2\pi n}{6} + \frac{\pi}{4}\right] u[-n-1]$. Sketch the pole–zero plot and indicate the ROC.
19. $x[n] = nu[n-1]$

20. $$\begin{aligned} x[n] &= (4)^n & n &< 0 \\ &= \left(\tfrac{1}{4}\right)^n & n &= 0, 2, 4, \ldots \\ &= \left(\tfrac{1}{5}\right)^n & n &= 1, 3, 5, \ldots \end{aligned}$$

Solution 1. $\boldsymbol{x[n] = \delta[n - n_0]}$

$$\delta[n] \overset{Z}{\longleftrightarrow} 1 \qquad \text{ROC: } |z| > 0$$

By applying time shifting property, we get

$$Z[\delta[n-n_0]] = z^{-n_0} \tag{2.38}$$

ROC: all z excluding $|z| = 0$.

2. $\boldsymbol{x[n] = u[n - n_0]}$

$$u[n] \overset{Z}{\longleftrightarrow} \frac{z}{(z-1)}$$

By applying time shifting (right shifted) property, we get

$$Z[u[n-n_0]] = \frac{z^{-n_0}z}{(z-1)} = \frac{z^{-(n_0-1)}}{(z-1)}$$

$$\boxed{X[z] = \frac{z^{-(n_0-1)}}{(z-1)} \qquad \text{ROC: } 1 < |z| < \infty} \tag{2.39}$$

3. $\boldsymbol{x[n] = a^{n+1}u[n+1]}$

$$a^n u[n] \overset{Z}{\longleftrightarrow} \frac{z}{(z-a)}$$

By applying time shifting (left shifted) property, we get

$$Z[a^{n+1}u[n+1]] = z\frac{z}{(z-a)}$$

$$\boxed{X[z] = \frac{z^2}{(z-a)} \qquad \text{ROC: } |a| < |z| < \infty} \tag{2.40}$$

4. $\boldsymbol{x[n] = a^{n-1}u[n-1]}$

$$a^n u[n] \overset{Z}{\longleftrightarrow} \frac{z}{(z-a)}$$

Applying time shifting (right shifted) property, we get

$$Z[a^{n-1}u[n-1]] = \frac{z^{-1}z}{(z-a)}$$

$$\boxed{X[z] = \frac{1}{(z-a)} \qquad \text{ROC: } a < |z| < \infty} \tag{2.41}$$

5. $\boldsymbol{x[n] = \left(\frac{1}{2}\right)^n u[-n]}$

$$u[-n] \overset{Z}{\longleftrightarrow} \frac{1}{(1-z)}$$

$$x[n] = \left(\frac{1}{2}\right)^n u[-n]$$

By using multiplication property (replacing z by $(\frac{1}{2})^{-1}z$), we get

$$\boxed{X[z] = \frac{1}{(1-2z)}} \qquad \text{ROC: } |z| < \frac{1}{2}$$

6. $\boldsymbol{x[n] = u[n-6] - u[n-10]}$
The signal is represented in Fig. 2.7.

$$\begin{aligned} X[z] &= z^{-6} + z^{-7} + z^{-8} + z^{-9} \\ &= \frac{1}{z^6} + \frac{1}{z^7} + \frac{1}{z^8} + \frac{1}{z^9} \end{aligned}$$

$$\boxed{X[z] = \frac{z^8 + z^2 + z + 1}{z^9}}$$

ROC: all z except $z \neq 0$. The above result can be simplified using summation formula as

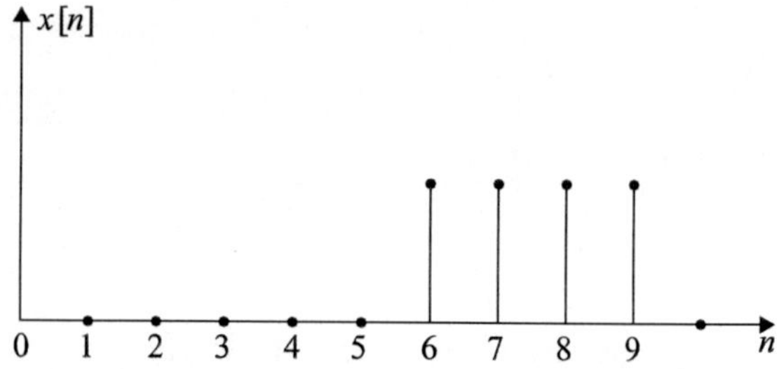

Fig. 2.7 $x[n] = u[n-6] - u[n-10]$

$$X[z] = \sum_{n=6}^{9} \left(\frac{1}{z}\right)^n$$
$$= \frac{\left(\frac{1}{z}\right)^{10} - \left(\frac{1}{z}\right)^6}{\left(\frac{1}{z} - 1\right)}$$

$$\boxed{X[z] = \frac{z}{(z-1)}[z^{-6} - z^{-10}]}$$

The above result can be obtained by time shifting property of unit step sequence.

$$Z[u[n-6]] = \frac{z}{(z-1)}z^{-6}$$
$$Z[u[n-10]] = \frac{z}{(z-1)}z^{-10}$$
$$X[z] = \frac{z}{(z-1)}[z^{-6} - z^{-10}]$$

$$\boxed{X[z] = \frac{(z^{-5} - z^{-9})}{(z-1)}}$$

7. $\boldsymbol{x[n] = nu[n]}$

$$Z[u[n]] = \frac{z}{(z-1)}$$

Applying the differentiation property in z

$$Z[nu[n]] = -z\frac{dX[z]}{dz}$$
$$Z[nu[n]] = -z\frac{d}{dz}\left[\frac{z}{(z-1)}\right]$$

$$\boxed{X[z] = \frac{z}{(z-1)^2}} \tag{2.42}$$

8. $\boldsymbol{x[n] = n[u[n] - u[n-8]]}$
   By using shift theorem, we get

$$Z[u[n] - u[n-8]] = \frac{z}{(z-1)}[1 - z^{-8}]$$
$$= \frac{(z - z^{-7})}{(z-1)}$$
$$Z[n[u[n] - u[n-8]]] = -z\frac{d}{dz}\frac{[z - z^{-7}]}{z-1}$$
$$X[z] = -z\frac{[(z-1)(1+7z^{-8}) - (z - z^{-7})]}{(z-1)^2}$$
$$X[z] = \frac{(-8z^{-6} + 7z^{-7} + z)}{(z-1)^2}$$

$$\boxed{X[z] = \frac{[z^8 - 8z + 7]}{z^7(z-1)^2}}$$

9. $\boldsymbol{x[n] = a^n \cos \omega_0 n u[n]}$
   From Example 2.2.12, we get

$$Z[\cos \omega_0 n u[n]] = \frac{[1 - z^{-1}\cos\omega_0]}{[1 - 2\cos\omega_0 z^{-1} + z^{-2}]}$$

To apply multiplication property, replace z by $|\frac{z}{a}|$ or $z^{-1} = |\frac{z}{a}|^{-1} = az^{-1}$

$$\therefore\ Z[a^n \cos\omega_0 n u[n]] = \frac{[1 - az^{-1}\cos\omega_0]}{[1 - 2a\cos\omega_0 z^{-1} + a^2 z^{-2}]}$$

$$\boxed{X[z] = \frac{[1 - az^{-1}\cos\omega_0]}{[1 - 2a\cos\omega_0 z^{-1} + a^2 z^{-2}]}} \tag{2.43}$$

10. $\boldsymbol{x[n] = a^n \sin \omega_0 n u[n]}$
    From Example 2.2.13, we get

$$Z[\sin\omega_0 n u[n]] = \frac{z^{-1}\sin\omega_0}{[1 - 2\cos\omega_0 z^{-1} + z^{-2}]}$$

To apply multiplication property, as in the previous example, replace z^{-1} by a z^{-1} and $z^{-2} = a^2 z^{-2}$

$$Z[a^n \sin \omega_0 n u[n]] = \frac{a z^{-1} \sin \omega_0}{[1 - 2a \cos \omega_0 z^{-1} + a^2 z^{-2}]}$$

$$\boxed{X[z] = \frac{[a z^{-1} \sin \omega_0]}{[1 - 2a \cos \omega_0 z^{-1} + a^2 z^{-2}]}} \tag{2.44}$$

11. **Show that $u[n] * u[n-1] = nu[n]$**

$$\begin{aligned} Z[u[n]] &= \frac{z}{(z-1)} \\ Z[u[n-1]] &= \frac{1}{(z-1)} \\ Z[u[n] * u[n-1]] &= Z[u[n]]Z[u[n-1]] \\ &= \frac{z}{(z-1)} \frac{1}{(z-1)} \\ &= \frac{z}{(z-1)^2} \end{aligned}$$

Multiplying by Z^{-1} both sides, we get

$$u[n] * u[n-1] = Z^{-1}\left[\frac{z}{(z-1)^2}\right]$$

$$\boxed{u[n] * u[n-1] = n[u[n]]}$$

12. $\boldsymbol{x[n] = n\left(-\frac{1}{4}\right)^n u[n] * \left(-\frac{1}{6}\right)^{-n} u[-n]}$

$$\begin{aligned} x_1[n] = \left(-\frac{1}{4}\right)^n u[n] &\overset{Z}{\longleftrightarrow} \frac{z}{\left(z+\frac{1}{4}\right)} \\ n\left[\left(-\frac{1}{4}\right)^n u[n]\right] &\overset{Z}{\longleftrightarrow} -z\frac{d}{dz}\frac{z}{\left(z+\frac{1}{4}\right)} \\ &= -z\frac{\left[z+\frac{1}{4}-z\right]}{\left(z+\frac{1}{4}\right)^2} \\ &= \frac{\left[-\frac{z}{4}\right]}{\left(z+\frac{1}{4}\right)^2} \qquad \text{ROC: } |z| > \frac{1}{4} \\ x_2[n] = \left(\frac{1}{6}\right)^n u[n] &\overset{Z}{\longleftrightarrow} \frac{z}{\left(z-\frac{1}{6}\right)} \qquad \text{ROC: } |z| > \frac{1}{6} \end{aligned}$$

If time reversal property is used, z is to be replaced by z^{-1}

$$\left(\frac{1}{6}\right)^{-n} u[-n] \overset{Z}{\longleftrightarrow} \frac{z^{-1}}{\left(z^{-1}-\frac{1}{6}\right)}$$

$$\begin{aligned} X_1[z] &= -\frac{6}{z-6} \qquad \text{ROC: } |z| < 6 \\ X[z] &= X_1[z]X_2[z] \\ &= \frac{\frac{z}{4}6}{\left(z+\frac{1}{4}\right)^2 (z-6)} \end{aligned}$$

$$\boxed{X[z] = \frac{1.5z}{\left(z+\frac{1}{4}\right)(z-6)}} \qquad \text{ROC: } \frac{1}{4} < |z| < 6$$

13. $\boldsymbol{x[n] = \left[\left(\frac{1}{2}\right)^n - \left(\frac{1}{4}\right)^n\right] u[n]}$
Find $X[z]$ and plot the poles and zeros.

$$x_1[n] = \left(\frac{1}{2}\right)^n u[n] \overset{Z}{\longleftrightarrow} \frac{z}{\left(z-\frac{1}{2}\right)}$$

$$x_2[n] = \left(\frac{1}{4}\right)^n u[n] \overset{Z}{\longleftrightarrow} \frac{z}{\left(z-\frac{1}{4}\right)}$$

$$\begin{aligned} x[n] &= x_1[n] - x_2[n] \\ X[z] &= X_1[z] - X_2[z] = \frac{z}{(z-0.5)} - \frac{z}{(z-0.25)} \end{aligned}$$

$$\boxed{X[z] = \frac{z0.25}{(z-0.5)(z-0.25)}}$$

The pole–zero plot is shown in Fig. 2.8.

14.

$$\begin{aligned} \boldsymbol{x[n]} &= \boldsymbol{1} \qquad \boldsymbol{n \ge 0} \\ &= \boldsymbol{3^n} \qquad \boldsymbol{n < 0} \end{aligned}$$

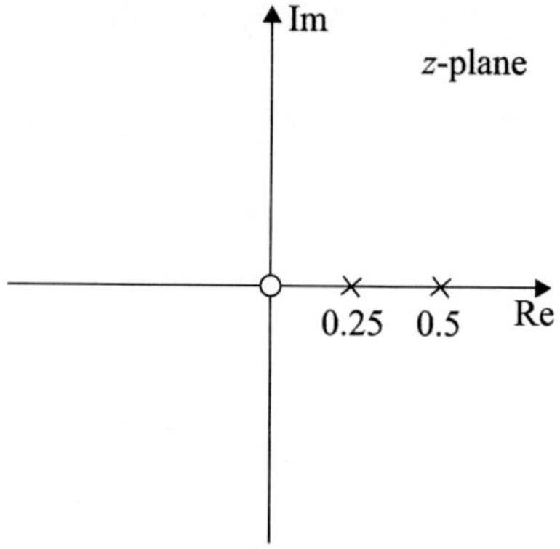

Fig. 2.8 Pole–zero plot

$$\begin{aligned} x[n] &= u[n] + 3^n u[-n-1] \\ &= x_1[n] + x_2[n] \\ X_1[z] &= \frac{z}{(z-1)} \qquad \text{ROC: } |z| > 1 \\ X_2[z] &= (3)^n u[-n-1] \end{aligned}$$

Using time reversal and multiplication properties, we get

$$\begin{aligned} X_2[z] &= -\frac{z}{(z-3)} \qquad \text{ROC: } |z| < 3 \\ X[z] &= X_1(z) + X_2(z) \\ &= \frac{z}{(z-1)} - \frac{z}{(z-3)} \end{aligned}$$

$$\boxed{X[z] = \frac{-2z}{(z-1)(z-3)}} \qquad \text{ROC: } 1 < z < 3$$

15.

$$(a) \quad \boldsymbol{x[n] = \left(-\frac{1}{3}\right)^n u[n] + 3\left(\frac{1}{6}\right)^n u[n]}$$

$$(b) \quad \boldsymbol{x[n] = \left[\left(-\frac{1}{3}\right)^n u[-n] + 3\left(\frac{1}{6}\right)^n\right] u[n]}$$

$$(c) \quad \boldsymbol{x[n] = \left[\left(-\frac{1}{3}\right)^n + 3\left(\frac{1}{6}\right)^n\right] u[-n]}$$

(*a*)

$$x[n] = \left(-\frac{1}{3}\right)^n + 3\left(\frac{1}{6}\right)^n u[n]$$
$$= x_1[n] + x_2[n]$$
$$X_1[z] = \frac{z}{\left(z + \frac{1}{3}\right)} \qquad \text{ROC: } |z| > -\frac{1}{3}$$
$$X_2[z] = 3\frac{z}{\left(z - \frac{1}{6}\right)} \qquad \text{ROC: } |z| > \frac{1}{6}$$

$$X[z] = X_1[z] + X_2[z]$$
$$= z\left[\frac{1}{\left(z + \frac{1}{3}\right)} + \frac{3}{\left(z - \frac{1}{6}\right)}\right] \qquad \text{ROC: } |z| > \frac{1}{6}$$

(*b*)

$$x[n] = \left[\left(-\frac{1}{3}\right)^n u[-n] + 3\left(\frac{1}{6}\right)^n\right] u[n]$$
$$= x_1[n] + x_2[n]$$
$$x_1[n] = \left(-\frac{1}{3}\right)^n u[-n]$$

Applying the properties of time reversal and multiplication, we get

$$X_1[z] = \frac{1}{(1 + 3z)} \qquad \textit{See } \text{Example 2.3.12; ROC: } |z| < \frac{1}{3}$$
$$x_2[n] = 3\left(\frac{1}{6}\right)^n u[n]$$
$$X_2[z] = \frac{3z}{\left(z - \frac{1}{6}\right)} \qquad \text{ROC: } |z| > \frac{1}{6}$$
$$X[z] = X_1[z] + X_2[z]$$

$$\boxed{X[z] = \left[\frac{1}{(1 + 3z)} + \frac{3z}{\left(z - \frac{1}{6}\right)}\right]} \qquad \text{ROC: } \frac{1}{6} < |z| < \frac{1}{3}$$

(*c*)

$$x[n] = \left[\left(-\frac{1}{3}\right)^n + 3\left(\frac{1}{6}\right)^n\right] u[-n]$$
$$= x_1[n] + x_2[n]$$
$$X_1[z] = \frac{1}{(1 + 3z)} \qquad \text{ROC: } |z| < \frac{1}{3}$$

The derivation is given in Example 2.3.15(*b*)

$$x_2[n] \quad = \quad 3\left(\frac{1}{6}\right)^n u[-n]$$

$$u[-n] \overset{Z}{\longleftrightarrow} \frac{1}{(1-z)}$$

Applying multiplication property, we get

$$Z\left(\frac{1}{6}\right)^n u[-n] \overset{Z}{\longleftrightarrow} \frac{1}{(1-6z)} \qquad \text{ROC: } |z| > \frac{1}{6}$$

$$\boxed{X[z] = \left[\frac{1}{(1+3z)} + \frac{3}{(1-6z)}\right]} \quad \text{ROC: } \frac{1}{6} < |z| < \frac{1}{3}$$

16. (*a*)

$$\boldsymbol{x[n] = \left[\left(\frac{1}{4}\right)^n + \left(\frac{1}{5}\right)^n\right] u[n]}$$

Applying results of Eq. (2.16), we get

$$\boxed{X[z] = \left[\frac{z}{\left(z-\frac{1}{4}\right)} + \frac{z}{\left(z-\frac{1}{5}\right)}\right]} \qquad \text{ROC: } |z| > \frac{1}{4}$$

(*b*)

$$\boldsymbol{x[n] = \left[\left(\frac{1}{5}\right)^n u[n] + \left(\frac{1}{4}\right)^n u[-n-1]\right]}$$

$$\left(\frac{1}{5}\right)^n u[n] \overset{Z}{\longleftrightarrow} \frac{z}{\left(z-\frac{1}{5}\right)} \qquad \text{ROC: } |z| > \frac{1}{5}$$

$$\left(\frac{1}{4}\right)^n u[-n-1] \overset{Z}{\longleftrightarrow} \frac{-z}{\left(z-\frac{1}{4}\right)} \qquad \text{ROC: } |z| < \frac{1}{4}$$

$$\left(\frac{1}{5}\right)^n u[n] + \left(\frac{1}{4}\right)^n u[-n-1] \overset{Z}{\longleftrightarrow} \frac{z}{\left(z-\frac{1}{5}\right)} - \frac{z}{\left(z-\frac{1}{4}\right)}$$

$$\boxed{X[z] = \frac{-\frac{z}{20}}{\left(z-\frac{1}{5}\right)\left(z-\frac{1}{4}\right)}} \quad \text{ROC: } \frac{1}{5} < |z| < \frac{1}{4}$$

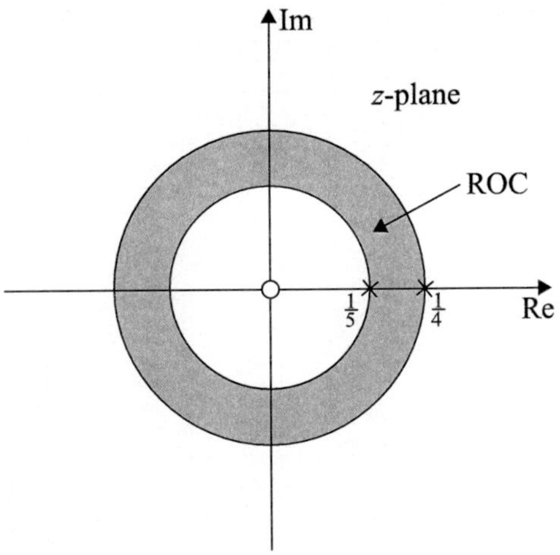

Fig. 2.9 $X[z]$ and its ROC of Example 2.3.16(b)

The poles and zero and the ROC are marked in Fig. 2.9.

(c)

$$\boldsymbol{x[n] = \left[\left(\frac{1}{4}\right)^n u[n] + \left(\frac{1}{5}\right)^n\right] u[-n-1]}$$

$$\left(\frac{1}{4}\right)^n u[n] \overset{Z}{\longleftrightarrow} \frac{z}{\left(z-\frac{1}{4}\right)} \qquad \text{ROC: } |z| > \frac{1}{4}$$

$$\left(\frac{1}{5}\right)^n u[-n-1] \overset{Z}{\longleftrightarrow} \frac{-z}{\left(z-\frac{1}{5}\right)} \qquad \text{ROC: } |z| < \frac{1}{5}$$

The ROCs of the above two equations are shown in Fig. 2.10 and it is seen that they do not overlap and thus the given $x[n]$ does not have $X[z]$.

17. $\boldsymbol{x[n] = \delta[n] + \frac{1}{2}\delta(n+1) + \delta(n-3)}$

$$\delta[n] \overset{Z}{\longleftrightarrow} 1$$

$$\frac{1}{2}\delta[n+1] \overset{Z}{\longleftrightarrow} \frac{1}{2}z$$

$$\delta[n-3] \overset{Z}{\longleftrightarrow} z^{-3}$$

$$\boxed{X[z] = 1 + \frac{1}{2}z + z^{-3}}$$

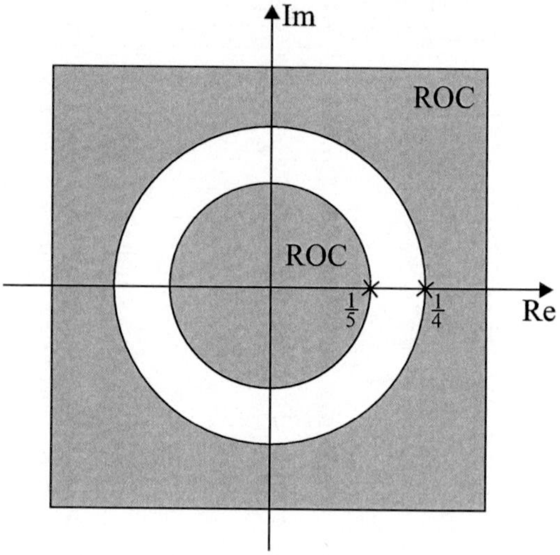

Fig. 2.10 ROC of Example 2.16(c)

18. $\boldsymbol{x[n] = 4^n \cos\left[\frac{2\pi n}{6} + \frac{\pi}{4}\right] u[-n-1]}$

$$\cos\left(\frac{2\pi n}{6}\right) = \cos\frac{\pi n}{3}$$

$$\cos\left(\frac{2\pi n}{6} + \frac{\pi}{4}\right) = \frac{e^{j\left(\frac{\pi}{4}+\frac{\pi n}{3}\right)} + e^{-j\left(\frac{\pi}{4}+\frac{\pi n}{3}\right)}}{2}$$

$$= \frac{1}{2}e^{j\left(\frac{\pi}{4}\right)}e^{j\left(\frac{\pi n}{3}\right)} + \frac{1}{2}e^{-j\left(\frac{\pi}{4}\right)}e^{-j\left(\frac{\pi n}{3}\right)}$$

$$4^n \cos\left(\frac{2\pi n}{6} + \frac{\pi}{n}\right) = \frac{1}{2}e^{j\left(\frac{\pi}{4}\right)}\left(4e^{j\frac{\pi}{3}}\right)^n + \frac{1}{2}e^{-j\frac{\pi}{4}}\left(4e^{-j\frac{\pi}{3}}\right)^n$$

From Eq. (2.18)

$$\left(4e^{j\frac{\pi}{3}}\right)^n u[-n-1] \overset{Z}{\longleftrightarrow} \frac{-z}{\left(z - 4e^{j\frac{\pi}{3}}\right)} \qquad \text{ROC: } |z| < 4$$

$$\left(4e^{-j\frac{\pi}{3}}\right)^n u[-n-1] \overset{Z}{\longleftrightarrow} \frac{-z}{\left(z - 4e^{-j\frac{\pi}{3}}\right)} \qquad \text{ROC: } |z| < 4$$

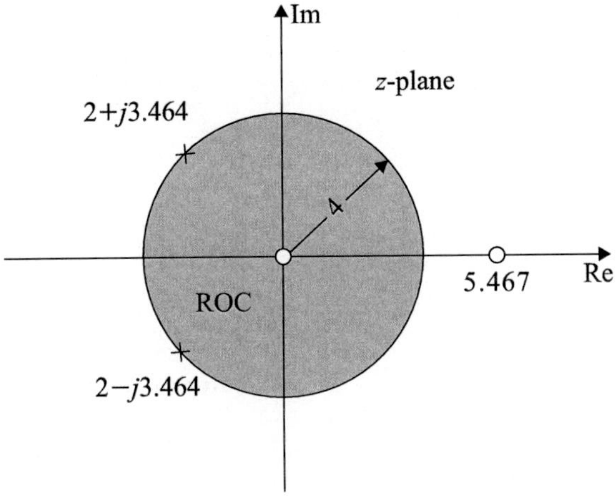

Fig. 2.11 Poles and zeros and ROC of Example 2.3.18

$$X[z] = -\frac{1}{2}z\left[\frac{e^{j\frac{\pi}{4}}}{(z-4e^{j\frac{\pi}{3}})} + \frac{e^{-j\frac{\pi}{4}}}{(z-4e^{-j\frac{\pi}{3}})}\right]$$
$$= -\frac{1}{2}z\left[\frac{ze^{j\frac{\pi}{4}} - 4e^{-j\frac{\pi}{12}} - 4e^{j\frac{\pi}{12}} + ze^{-j\frac{\pi}{4}}}{z^2 - z4(e^{j\frac{\pi}{3}} + e^{-j\frac{\pi}{3}}) + 16}\right]$$
$$= \frac{-\frac{1}{2}z[\sqrt{2}z - 7.73]}{(z^2-4z+16)}$$

$$\boxed{X[z] = \frac{-0.707z[z-5.467]}{(z-2+j3.464)(z-2-j3.464)}} \quad \text{ROC: } |z| < 4$$

The pole–zero diagram is shown in Fig. 2.11. The ROC is the interior of the circle.

19. $\boldsymbol{x[n] = nu[n-1]}$

Method 1

$$u[n-1] \overset{Z}{\longleftrightarrow} \frac{1}{(z-1)}$$

Using differential property, we get

$$nu[n-1] \overset{Z}{\longleftrightarrow} -z\frac{d}{dz}\frac{1}{(z-1)}$$

$$\boxed{z[nu[n-1]] = \frac{z}{(z-1)^2}}$$

Method 2

$$nu[n-1] = (n-1)u[n-1] + u[n-1]$$

$$(n-1)u[n-1] \overset{Z}{\longleftrightarrow} \frac{zz^{-1}}{(z-1)^2} = \frac{1}{(z-1)^2}$$

$$u[n-1] \overset{Z}{\longleftrightarrow} \frac{1}{(z-1)}$$

$$nu[n-1] \overset{Z}{\longleftrightarrow} \frac{1}{(z-1)^2} + \frac{1}{(z-1)}$$

$$\boxed{Z[nu[n-1]] = \frac{z}{(z-1)^2}}$$

20.

$$\boldsymbol{x[n] = \begin{cases} (4)^n & n < 0 \\ \left(\frac{1}{4}\right)^n & n = 0, 2, 4, \ldots \\ \left(\frac{1}{5}\right)^n & n = 1, 3, 5, \ldots \end{cases}}$$

$$\begin{aligned}
X[z] &= \sum_{n=-\infty}^{-1} (4)^n z^{-n} + \sum_{n=0}^{\infty} \left(\frac{1}{4}\right)^n z^{-n} + \sum_{n=0}^{\infty} \left(\frac{1}{5}\right)^n z^{-n} \\
&= X_1[z] + X_2[z] + X_3[z] \\
X_1[z] &= \sum_{n=-\infty}^{-1} (4)^n z^{-n} \\
&= \sum_{n=-\infty}^{-1} \left(\frac{z}{4}\right)^{-n} \\
&= \sum_{n=1}^{\infty} \left(\frac{z}{4}\right)^{n} \\
&= \frac{z}{4} + \left(\frac{z}{4}\right)^2 + \cdots \\
&= \frac{z}{4}\left[1 + \frac{z}{4} + \left(\frac{z}{4}\right)^2 + \cdots\right] \\
&= \frac{z}{4}\frac{1}{\left(1-\frac{z}{4}\right)} = \frac{-z}{(z-4)} \qquad \text{ROC: } |z| < 4 \\
X_2[z] &= \sum_{n=0}^{\infty} \left(\frac{1}{4}\right)^n z^{-n} = \sum_{n=0}^{\infty} (4z)^{-n} \\
&= \sum_{p=0}^{\infty} (4z)^{-2p} \quad \text{where } n = 2p \text{ and } p = 0, 1, 2, \ldots
\end{aligned}$$

$$
\begin{aligned}
X_2[z] &= \sum_{p=0}^{\infty} \left(16z^2\right)^{-p} \\
&= \frac{1}{\left(1-\frac{1}{16z^2}\right)} \\
&= \frac{z^2}{\left(z^2-\frac{1}{16}\right)} \qquad \text{ROC: } |z| > \frac{1}{4} \\
X_3[z] &= \sum_{n=0}^{\infty} \left(\frac{1}{5}\right)^n z^{-n} \\
&= \sum_{n=0}^{\infty} (5z)^{-n} \\
&= \sum_{q=0}^{\infty} (5z)^{-(2q+1)} \quad \text{where } n = 2q+1 \\
&= \frac{1}{5z}\sum_{q=0}^{\infty} \left(25z^2\right)^{-q} = \frac{1}{5z}\frac{1}{\left(1-\frac{1}{25z^2}\right)} \\
&= \frac{z/5}{\left(z^2-\frac{1}{25}\right)} \qquad \text{ROC: } |z| > \frac{1}{5}
\end{aligned}
$$

$$
\boxed{X[z] = \left[-\frac{z}{(z-4)} + \frac{z^2}{\left(z^2-\frac{1}{16}\right)} + \frac{z/5}{\left(z^2-\frac{1}{25}\right)}\right]}
$$

ROC: $\frac{1}{4} < |z| < 4$.

Example 2.4 Find the initial and final values of the following functions:

$$
\begin{aligned}
&(a) \qquad X[z] = \frac{z}{(4z^2-5z-1)} \qquad \text{ROC: } |z| > 1 \\
&(b) \qquad x[z] = \frac{10z(z-0.4)}{(z-0.5)(z-0.3)} \qquad \text{ROC: } |z| > 0.5
\end{aligned}
$$

Solution (*a*) $\mathbf{X[z] = \frac{z}{(4z^2-5z-1)}}$

Initial Value

$$
\begin{aligned}
x[0] &= \underset{z\to\infty}{Lt}\; X[z] \\
&= \underset{z\to\infty}{Lt}\; \frac{z}{z^2\left(4-\frac{5}{z}-\frac{1}{z^2}\right)} \\
&= \underset{z\to\infty}{Lt}\; \frac{1}{z\left(4-\frac{5}{z}-\frac{1}{z^2}\right)}
\end{aligned}
$$

$$\boxed{x[0] = 0}$$

Final Value

$$x[\infty] = \underset{z\to 1}{Lt} \frac{(z-1)}{z}$$

Provided all the poles are inside the unit circle and possibly one pole on the unit circle.

$$(4z^2 - 5z + 1) = 4(z-1)\left(z - \frac{1}{4}\right)$$

$$X[z] = \frac{z}{4(z-1)\left(z-\frac{1}{4}\right)}$$

The poles $(z-1)$ are on the unit circle and $z = \frac{1}{4}$ within unit circle. $X[z]$ is valid to apply final value theorem.

$$x[\infty] = \underset{z\to 1}{Lt} \frac{(z-1)}{z} \frac{z}{4(z-1)\left(z-\frac{1}{4}\right)}$$

$$\boxed{x[\infty] = \frac{1}{3}}$$

(*b*) $\mathbf{X[z] = \frac{10z(z-0.4)}{(z-0.5)(z-0.3)}}$

$$x[0] = \underset{z\to\infty}{Lt} \frac{10z^2\left(1-\frac{0.4}{z}\right)}{z^2\left(1-\frac{0.5}{z}\right)\left(1-\frac{0.3}{z}\right)}$$

$$\boxed{x[0] = 10}$$

To find the final value $x[\infty]$, the poles of $X[z]$ are all inside the unit circle and hence it is valid to apply final value theorem.

$$x[\infty] = \underset{z\to 1}{Lt} \frac{10z(z-1)(z-0.4)}{z(z-0.5)(z-0.3)}$$

$$\boxed{x[\infty] = 0}$$

Example 2.5

$$X[z] = \frac{\left[1 - \frac{1}{4}z^{-2}\right]}{\left[1+\frac{1}{4}z^{-2}\right]\left[1+\frac{5}{4}z^{-1}+\frac{3}{8}z^{-2}\right]}$$

How many different regions of convergence could correspond to $X[z]$?

Solution

$$X[z] = \frac{z^2\left[z^2 - \frac{1}{4}\right]}{\left(z^2 + \frac{1}{4}\right)\left(z^2 + \frac{5}{4}z + \frac{3}{8}\right)} = \frac{z^2\left(z + \frac{1}{2}\right)\left(z - \frac{1}{2}\right)}{\left(z - \frac{j}{2}\right)\left(z + \frac{j}{2}\right)\left(z + \frac{3}{4}\right)\left(z + \frac{1}{2}\right)}$$

$$X[z] = \frac{z^2\left[z - \frac{1}{2}\right]}{\left(z - \frac{j}{2}\right)\left(z + \frac{j}{2}\right)\left(z + \frac{3}{4}\right)}$$

The poles and zeros are located in Fig. 2.12.

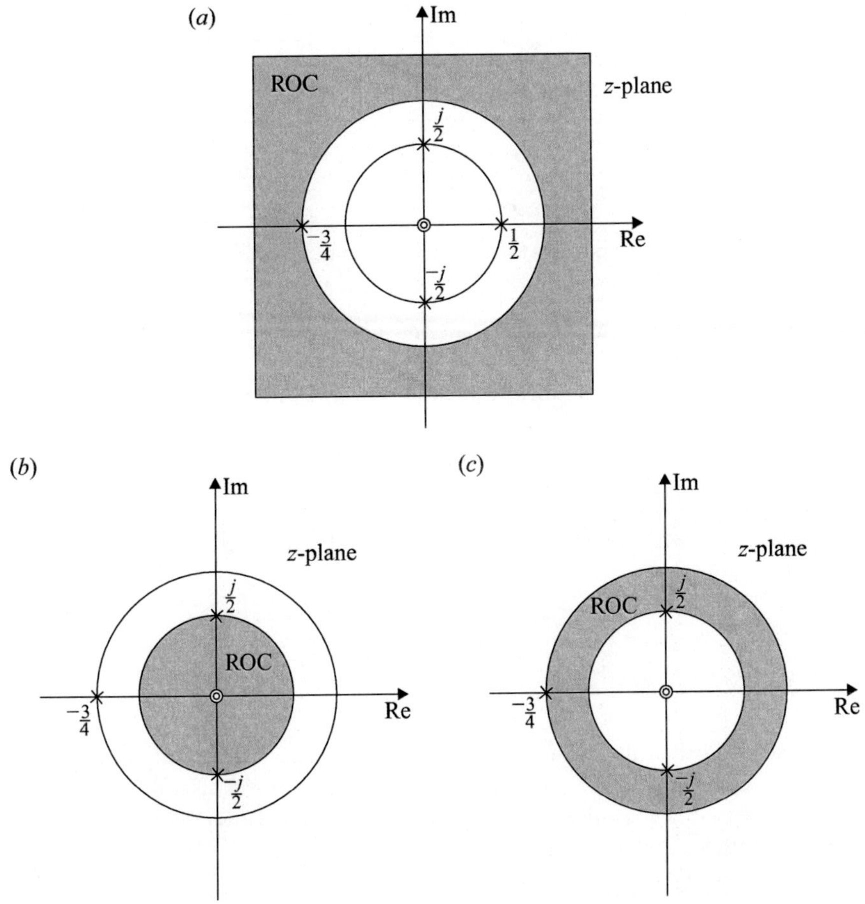

Fig. 2.12 Pole–zero diagram and ROC of $X[z]$

Table 2.1 Unilateral z-transform pairs

No	$x[n]$	$X[z]$
1	$\delta[n]$	1
2	$u[n]$	$\frac{z}{(z-1)}$
3	$nu[n]$	$\frac{z}{(z-1)^2}$
4	$n^2u[n]$	$\frac{z(z+1)}{(z-1)^3}$
5	$a^n u[n]$	$\frac{z}{(z-a)}$
6	$a^{n-1}u[n-1]$	$\frac{1}{(z-a)}$
7	$na^n u[n]$	$\frac{az}{(z-a)^2}$
8	$\cos\omega_0 nu[n]$	$\frac{1-\cos\omega_0 z^{-1}}{1-2\cos\omega_0 z^{-1}+z^{-2}}$
9	$\sin\omega_0 nu[n]$	$\frac{z^{-1}\sin\omega_0}{1-2\cos\omega_0 z^{-1}+z^{-2}}$
10	$a^n\cos\omega_0 nu[n]$	$\frac{1-az^{-1}\cos\omega_0}{1-2a\cos\omega_0 z^{-1}+a^2z^{-2}}$
11	$a^n\sin\omega_0 nu[n]$	$\frac{az^{-1}\sin\omega_0}{1-2a\cos\omega_0 z^{-1}+a^2z^{-2}}$

From Fig. 2.12, circles passing through $|z| = \frac{3}{4}$ and $|z| = \frac{1}{2}$ are drawn. $X[z]$ exists from the following ROCs.

1. $|z| > \frac{3}{4}$. ROC is the exterior of the outermost pole $z = -\frac{3}{4}$. The system is causal and $X[z]$ exits (Fig. 2.12a).
2. $|z| < \frac{1}{2}$. ROC is the interior of the innermost pole $\pm\frac{j}{2}$. The system is anti-causal and $X[z]$ exits (Fig. 2.12b).
3. $\frac{1}{2} < |z| < \frac{3}{4}$. The ROC is a ring between the two circles of radius $r_1 = \frac{3}{4}$ and $r_1 = \frac{1}{2}$. Here $X[z]$ exits. The system is both causal and anti-causal (Fig. 2.12c).

The unilateral z-transform pairs are given in Table 2.1. The properties of z-transform are given in Table 2.2.

2.8 Inverse z-transform

If $X[z]$ is given then the sequence $x[n]$ is determined. This is called inverse z-transform. As in the Laplace transform, in inverse z-transform also, the integration in the complex z-plane using Equation (2.5) is avoided since it is tedious. Instead the following methods are used. They are as follows:

1. Partial fraction method;
2. Power series expansion;
3. Residue method.

Table 2.2 z-transform properties (operations)

Operation	$x[n]$	$X[z]$
Linearity	$a_1x_1[n]+a_2x_2[n]$	$a_1X_1[z]+a_2X_2[z]$
Multiplication by a^n	$a^nx[n]u[n]$	$X\left[\frac{z}{a}\right]$
Multiplication by n	$nx[n]u[n]$	$-z\frac{d}{dz}X[z]$
Time shifting	$x[n-n_0]$	$z^{-n_0}X[z]$
Multiplication by $e^{j\omega_0 n}$	$e^{j\omega_0 n}x[n]$	$X[e^{-j\omega_0}z]$
Time reversal	$x[-n]$	$X\left[\frac{1}{z}\right]$
Accumulation	$\sum_{k=-\infty}^{n}x[n]$	$\frac{z}{(z-1)}X[z]$
Convolution	$x_1[n]*x_2[n]$	$X_1[z]X_2[z]$
Initial value	$x[0]$	$\underset{z\to\infty}{Lt}\ X[z]$
Final value	$x[\infty]$	$\underset{z\to 1}{Lt}\ \frac{(z-1)}{z}X[z]$ poles of $(z-1)X[z]$ are inside the unit circle
Right shifting	$x[n-m]u[n-m]$	$\frac{1}{z^m}X[z]$
	$x[n-m]u[n]$	$\frac{1}{z^m}X[z]+\frac{1}{z^m}\sum_{n=1}^{m}x(-m)z^n$
	$x[n-1]u[n]$	$\frac{1}{z}X[z]+x[-1]$
	$x[n-2]u[n]$	$\frac{1}{z^2}X[z]+\frac{1}{z}x[-1]+x[-2]$
Left shifting	$x[n+m]u[n]$	$z^mX[z]-z^m\sum_{n=0}^{m-1}x[n]z^{-n}$
	$x[n+1]u[n]$	$zX[z]-zx(0)$
	$x[n+2]u[n]$	$z^2X[z]-z^2x[0]-zx[1]$

Of these, the partial fraction method is very easy to apply as was done in determining inverse Laplace transform.

2.8.1 Partial Fraction Method

If $X[z]$ is a rational function of z, then it can be expressed as follows:

$$X[z]=\frac{N[z]}{D[z]}=\frac{K(z-z_1)(z-z_2)\ldots(z-z_m)}{(z-p_1)(z-p_2)\ldots(z-p_n)} \tag{2.45}$$

where $n\geq m$ and all the poles are simple.

$$\begin{aligned}\frac{X[z]}{z}&=\frac{K(z-z_1)(z-z_2)\ldots(z-z_m)}{z(z-p_1)(z-p_2)\ldots(z-p_n)}\\&=\frac{A_0}{z}+\frac{A_1}{z-p_1}+\frac{A_2}{z-p_2}+\ldots+\frac{A_n}{z-p_n}\end{aligned}$$

Table 2.3 z-transform pairs of repeated poles

$X[z]$	$x[n]$ ROC: $\lvert z\rvert > \lvert a\rvert$
1. $\frac{z}{z-a}$	$a^n u[n]$
2. $\frac{z}{(z-a)^2}$	$na^{n-1}u[n]$
3. $\frac{z}{(z-a)^3}$	$\frac{n(n-1)a^{n-2}}{\angle 2}u[n]$
4. $\frac{z}{(z-a)^k}$	$\frac{n(n-1)(n-2)\ldots(n-(k-2))a^{n-k+1}}{\angle(k-1)}u[n]$

where

$$\begin{aligned}
A_0 &= X[z]|_{z=0} \\
A_1 &= (z-p_1)\frac{X[z]}{z}\bigg|_{z=p_1} \\
X[z] &= A_0 + A_1\frac{z}{z-p_1} + \cdots + \frac{A_n z}{z-p_n}
\end{aligned} \tag{2.46}$$

Using z-transform pair table, $x[n]$ can be determined. The following examples illustrate the above method. For repeated poles, the z-transform pairs given in Table 2.3 may be referred to.

Example 2.6 Find the inverse z-transform of

$$X[z] = \frac{1-\frac{1}{3}z^{-1}}{(1-z^{-1})(1+2z^{-1})} \qquad \text{ROC: } |z| > 2$$

Solution

$$\begin{aligned}
X[z] &= \frac{1-\frac{1}{3}z^{-1}}{(1-z^{-1})(1+2z^{-1})} \\
&= \frac{z\left(z-\frac{1}{3}\right)}{(z-1)(z+2)} \\
\frac{X[z]}{z} &= \frac{A_1}{(z-1)} + \frac{A_2}{(z+2)} \\
\left(z-\frac{1}{3}\right) &= A_1(z+2) + A_2(z-1)
\end{aligned}$$

Substitute $z = 1$

$$A_1 = \frac{2}{9}$$

Substitute $z = -2$

$$A_2 = \frac{7}{9}$$

$$X[z] = \frac{1}{9}\left[\frac{2z}{z-1} + \frac{7z}{z+2}\right]$$

$$\boxed{x[n] = \frac{1}{9}\left[2(1)^n + 7(-2)^n\right]u[n]}$$

Example 2.7 Find the inverse z-transform of

$$X[z] = \frac{1}{1024}\left[\frac{1024 - z^{-10}}{1 - \frac{1}{2}z^{-1}}\right] \qquad \text{ROC: } |z| > 0$$

Solution

$$\begin{aligned} X[z] &= \frac{1}{1024}\left[\frac{1024 - z^{-10}}{1 - \frac{1}{2}z^{-1}}\right] \\ &= \frac{z}{\left(z - \frac{1}{2}\right)} - \frac{z}{\left(z - \frac{1}{2}\right)}\frac{z^{-10}}{1024} \end{aligned}$$

Taking inverse z-transform, we get

$$\begin{aligned} x[n] &= \left(\frac{1}{2}\right)^n u[n] - \frac{1}{1024}\left(\frac{1}{2}\right)^{n-10} u[n-10] \\ &= \left(\frac{1}{2}\right)^n u[n] - \frac{1}{1024}\left(\frac{1}{2}\right)^n \left(\frac{1}{2}\right)^{-10} u[n-10] \\ &= \left(\frac{1}{2}\right)^n u[n] - \frac{1}{1024}\left(\frac{1}{2}\right)^n 1024u[n-10] \\ &= \left(\frac{1}{2}\right)^n u[n] - \left(\frac{1}{2}\right)^n u[n-10] \\ x[n] &= \left(\frac{1}{2}\right)^n - 0 \qquad 0 \le n \le 9 \\ &= \left(\frac{1}{2}\right)^n - \left(\frac{1}{2}\right)^n \\ &= 0 \qquad n \ge 10 \end{aligned}$$

$$\boxed{\begin{aligned} x[n] &= \left(\frac{1}{2}\right)^n & 0 \le n \le 9 \\ &= 0 & \text{otherwise} \end{aligned}}$$

Example 2.8 Find the inverse z-transform of

$$X[z] = \frac{z^2}{(1 - az)(z - a)}$$

Solution

$$X[z] = \frac{z^2}{(1 - az)(z - a)}$$

$$\frac{X[z]}{z} = \frac{-z}{a\left[z - \frac{1}{a}\right][z - a]}$$

$$\frac{X[z]}{z} = \frac{A_1}{\left(z - \frac{1}{a}\right)} + \frac{A_2}{(z - a)}$$

$$-\frac{z}{a} = A_1(z - a) + A_2\left(z - \frac{1}{a}\right)$$

Substitute $z = \frac{1}{a}$

$$-\frac{1}{a^2} = A_1\left(\frac{1}{a} - a\right)$$

$$A_1 = \frac{-1}{a(1 - a^2)}$$

Substitute $z = a$

$$-1 = A_2\left(a - \frac{1}{a}\right)$$

$$A_2 = \frac{a}{(1 - a^2)}$$

$$X[z] = \frac{1}{(1 - a^2)}\left[\frac{-1}{a}\frac{z}{\left(z - \frac{1}{a}\right)} + \frac{az}{(z - a)}\right]$$

For $a > 1$, the ROC is shown in Fig. 2.13a. For $a < 1$, the ROC is shown in Fig. 2.13b.

For $a > 1$, the ROC is exterior of the outermost pole. Hence, the function is casual.

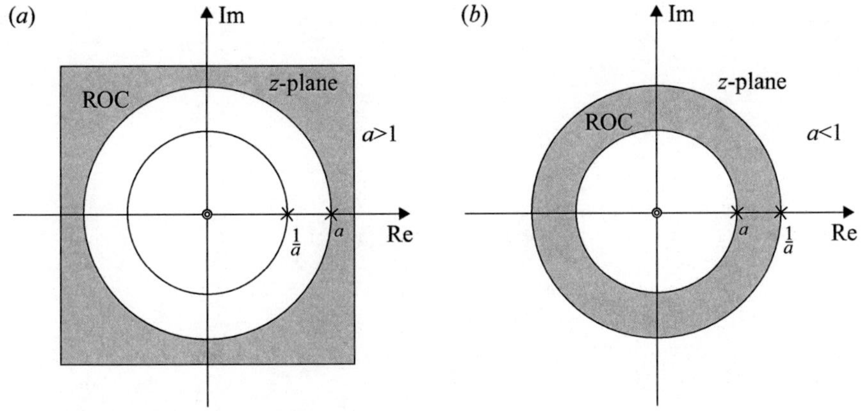

Fig. 2.13 ROC of Example 2.8

$$x[n] = \frac{1}{(1-a^2)}\left[\frac{-1}{a}\frac{1}{(a)^n} + a(a)^n\right]u[n]$$

$$\boxed{x[n] = \frac{1}{(1-a^2)}\left[-\left(\frac{1}{a}\right)^{n+1} + (a)^{n+1}\right]u[n]}$$

For $a < 1$, the ROC is $a < |z| < \frac{1}{a}$ and it is a concentric strip. The pole at $|z| = \frac{1}{a}$ is anti-causal and $z = a$ is causal.

$$\boxed{x[n] = \frac{1}{(1-a^2)}\left[\left(\frac{1}{a}\right)^{n+1} u[-n-1] + (a)^{n+1}u[n]\right]}$$

Example 2.9

$$X[z] = \frac{(7z-23)}{(z-3)(z-4)}$$

Find $x[n]$. ROC: $|z| > 4$.

Solution Method 1:

Dividing both sides by z, we get

$$\begin{aligned}\frac{X[z]}{z} &= \frac{(7z-23)}{z(z-3)(z-4)} \\ &= \frac{A_1}{z} + \frac{A_2}{(z-3)} + \frac{A_3}{(z-4)} \\ (7z-23) &= A_1(z-3)(z-4) + A_2 z(z-4) + A_3 z(z-3)\end{aligned}$$

Substitute $z = 0$

$$-23 = 12A_1; \qquad A_1 = -\frac{23}{12}$$

Substitute $z = 3$

$$-2 = A_2(3)(-1); \qquad A_2 = \frac{2}{3}$$

Substitute $z = 4$

$$5 = 4A_3; \qquad A_3 = \frac{5}{4}$$

$$X[z] = -\frac{23}{12} + \frac{2}{3}\frac{z}{(z-3)} + \frac{5}{4}\frac{z}{(z-4)}$$

$$\boxed{X[n] = \left[-\frac{23}{12}\delta[n] + \frac{2}{3}(3)^n + \frac{5}{4}(4)^n\right]u[n]}$$

Method 2:

$$X[z] = \frac{(7z-23)}{(z-3)(z-4)} = \frac{A_1}{(z-3)} + \frac{A_2}{(z-4)}$$
$$7z - 23 = A_1(z-4) + A_2(z-3)$$

Substitute $z = 3$

$$-2 = -A_1; \qquad A_1 = 2$$

Substitute $z = 4$

$$5 = A_2$$
$$X[z] = \frac{2}{(z-3)} + \frac{5}{(z-4)}$$
$$\frac{2}{(z-3)} \overset{Z^{-1}}{\longleftrightarrow} 2(3)^{n-1}u[n-1]$$
$$\frac{5}{(z-4)} \overset{Z^{-1}}{\longleftrightarrow} 5(4)^{n-1}u[n-1]$$

$$\boxed{x[n] = [2(3)^{n-1} + 5(4)^{n-1}]u[n-1]}$$

The results of the above two methods are the same even though they are expressed in different forms.

Example 2.10

$$X[z] = \frac{10z}{(z+2)(z+4)^2} \qquad \text{ROC: } |z| > 4$$

Find $x[n]$ using partial fraction method.

Solution This is the case with poles repeated twice

$$X[z] = \frac{10z}{(z+2)(z+4)^2}$$

$$\frac{X[z]}{z} = \frac{10}{(z+2)(z+4)^2}$$

$$= \frac{A_1}{(z+2)} + \frac{A_2}{(z+4)} + \frac{A_3}{(z+4)^2}$$

$$10 = A_1(z+4)^2 + A_2(z+2)(z+4) + A_3(z+2)$$

Substitute $z = -2$

$$10 = 4A_1; \qquad A_1 = \frac{5}{2}$$

Substitute $z = -4$

$$10 = -2A_3; \qquad A_3 = -5$$

Compare the coefficients of free terms

$$10 = 16A_1 + 8A_2 + 2A_3$$

$$= 16\frac{5}{2} + 8A_2 - 10$$

$$A_2 = -\frac{5}{2}$$

$$X[z] = \frac{5}{2}\frac{z}{(z+2)} - \frac{5}{2}\frac{z}{(z+4)} - \frac{5}{(z+4)^2}$$

$$\boxed{x[n] = \left[\frac{5}{2}(-2)^n - \frac{5}{2}(-4)^n - 5n(-4)^n\right]u[n]}$$

Example 2.11

$$X[z] = \frac{z(z^2 + z - 30)}{(z-2)(z-4)^3} \qquad \text{ROC: } |z| > 4$$

Find $x[n]$ using partial fraction method.

Solution This is the case with poles repeated thrice

$$X[z] = \frac{z(z^2 + z - 30)}{(z-2)(z-4)^3}$$

$$\frac{X[z]}{z} = \frac{(z^2 + z - 30)}{(z-2)(z-4)^3} = \frac{(z-5)(z+6)}{(z-2)(z-4)^3}$$

$$= \frac{A_1}{(z-2)} + \frac{A_2}{(z-4)^3} + \frac{A_3}{(z-4)^2} + \frac{A_4}{(z-4)}$$

$$(z^2 + z - 30) = A_1(z-4)^3 + A_2(z-2) + A_3(z-2)(z-4) + A_4(z-2)(z-4)^2$$

Substitute $z = 2$

$$(-3)(8) = -8A_1; \qquad A_1 = 3$$

Substitute $z = 4$

$$(-1)(10) = 2A_2; \qquad A_2 = -5$$

$$(z^2 + z - 30) = 3(z^3 - 12z^2 + 48z - 64) - 5(z-2) + A_3(z^2 - 6z + 8) + A_4(z^3 - 10z^2 + 32z - 32)$$

Compare the coefficients of z^2

$$1 = -36 + A_3 - 10A_4$$

$$A_3 - 10A_4 = 37$$

Compare the coefficients of z

$$1 = 144 - 5 - 6A_3 + 32A_4$$

$$6A_3 - 32A_4 = 138$$

Solving the above equation, we get

$$A_3 = 7; \qquad A_4 = -3$$

$$X[z] = \frac{3z}{(z-2)} - \frac{5z}{(z-4)^3} + \frac{7z}{(z-4)^2} - \frac{3z}{(z-4)}$$

$$\frac{z}{(z-4)^3} \overset{Z^{-1}}{\longleftrightarrow} \frac{n(n-1)}{\angle 2}(4)^{n-2}u[n] = \frac{n(n-1)}{32}(4)^n u[n]$$

$$\frac{z}{(z-4)^2} \overset{Z^{-1}}{\longleftrightarrow} n(4)^{n-1}u[n] = \frac{1}{4}n(4)^n u[n]$$

$$\boxed{x[n] = \left[3(2)^n + \left\{-\frac{5}{32}n(n-1) + \frac{7}{4}n - 3\right\}(4)^n\right]u[n]}$$

The values of A_1, A_2 and A_3 determined are checked for their correctness as follows:

$$\frac{X[z]}{z} = \frac{(z-5)(z+6)}{(z-2)(z-4)^3}$$

Substitute $z = 0$

$$\left.\frac{X[z]}{z}\right|_{z=0} = \frac{(-5)(6)}{(-2)(-4)^3} = -\frac{15}{64}$$

$$\frac{X[z]}{z} = \frac{3}{z-2} - \frac{5}{(z-4)^3} + \frac{7}{(z-4)^2} - \frac{3}{(z-4)}$$

Substitute $z = 0$

$$\left.\frac{X[z]}{z}\right|_{z=0} = -\frac{3}{2} + \frac{5}{64} + \frac{7}{16} + \frac{3}{4} = -\frac{15}{64}$$

Hence, the values of A_1, A_2, A_3 and A_4 are found to be correct.

Example 2.12

$$X[z] = \frac{z(z+10)}{(z-1)(z^2-8z+20)}$$

Find $x[n]$ using partial fraction method.

Solution This is the case with complex poles

$$X[z] = \frac{z(z+10)}{(z-1)(z^2-8z+20)}$$

$$\frac{X[z]}{z} = \frac{(z+10)}{(z-1)(z-4+j2)(z-4-j2)}$$

$$= \frac{A_1}{(z-1)} + \frac{A_2}{(z-4+j2)} + \frac{A_3}{(z-4-j2)}$$

$$(z+10) = A_1(z^2-8z+20) + A_2(z-1)(z-4-j2) + A_3(z-1)(z-4+j2)$$

Substitute $z = 1$

$$11 = A_1(13); \qquad A_1 = \frac{11}{13}$$

Substitute $z = 4 + j2$

$$\begin{aligned}
(14 + j2) &= A_3(4 + j2 - 4 + j2)(4 + j2 - 1) \\
A_3 &= \frac{(14 + j2)}{j4(3 + j2)} = \frac{14.142\angle 8.13^\circ}{4\sqrt{13}\angle 123.69^\circ} = 0.98\angle - 115.56^\circ = 0.98e^{-j115.56^\circ} \\
A_2 &= \text{conjugate of } A_3 = 0.98\angle 115.56^\circ = 0.98e^{j115.56^\circ} \\
X[z] &= \frac{11}{13}\frac{z}{(z-1)} + \frac{0.98e^{j115.56^\circ}}{(z - 4 + j2)} + \frac{0.98e^{-j115.56^\circ}}{(z - 4 - j2)}
\end{aligned}$$

Using z-transform pair, we get the following inverse z-transform:

$$x[n] = \frac{11}{13}u[n] + [0.98e^{j115.56^\circ}(4 - j2)^n + 0.98e^{-j115.56^\circ}(4 + j2)^n]u[n]$$

$115.56^\circ = 2$ radians

$$\begin{aligned}
(4 + j2)^n &= (4.47)^n e^{j0.4636n} \\
(4 - j2)^n &= (4.47)^n e^{-j0.4636n}
\end{aligned}$$

$$\begin{aligned}
x[n] &= \frac{11}{13}u[n] + [0.98e^{j2}e^{-j0.4636n}(4.47)^n + 0.98(4.47)^n e^{-j2}e^{j0.4636n}]u[n] \\
&= \frac{11}{13}u[n] + 0.98 * (4.47)^n[e^{j(2-.4636n)} + e^{-j(2-.4636n)}]u[n]
\end{aligned}$$

$$\boxed{x[n] = \left[\frac{11}{13} + 1.96(4.47)^n \cos(2 - 0.4636n)\right]u[n]}$$

Example 2.13

$$X[z] = \frac{(5z^3 - 29z^2 + 8z + 60)}{(z^2 - 7z + 10)}$$

Find $x[n]$ by partial fraction method.

Solution This is the case with irrational system function. The solution of $x[n]$ will have forward and backward shifts. Dividing the numerator polynomial by the denominator polynomial, we get

$$5z + 6$$

$$
\begin{array}{r}
z^2 - 7z + 10\overline{\big)5z^3 - 29z^2 + 8z + 60} \\
\underline{5z^3 - 35z^2 + 50z} \\
6z^2 - 42z + 60 \\
\underline{6z^2 - 42z + 60}
\end{array}
$$

$$
\begin{aligned}
(z^2 - 7z + 10) &= (z-2)(z-5) \\
X[z] &= (5z+6) + \frac{1}{(z-2)(z-5)} \\
&= X_1[z] + X_2[z]
\end{aligned}
$$

where

$$
\begin{aligned}
X_1[z] &= (5z+6) \\
X_2[z] &= \frac{1}{(z-2)(z-5)} \\
\frac{X_2[z]}{z} &= \frac{1}{z(z-2)(z-5)} \\
&= \frac{A_1}{z} + \frac{A_2}{z-2} + \frac{A_3}{z-5} \\
1 &= A_1(z-2)(z-5) + A_2 z(z-5) + A_3 z(z-2)
\end{aligned}
$$

Substitute $z = 0$

$$A_1 = \frac{1}{10}$$

Substitute $z = 2$

$$1 = A_2(2)(-3); \qquad A_2 = -\frac{1}{6}$$

Substitute $z = 5$

$$1 = A_3(5)(3); \qquad A_3 = \frac{1}{15}$$

$$X[z] = 5z + 6 + \frac{1}{10} - \frac{1}{6}\frac{z}{(z-2)} + \frac{1}{15}\frac{z}{(z-5)}$$

$$\boxed{x[n] = \left[\delta(n+1) + 6.1\delta[n] - \frac{1}{6}(2)^n + \frac{1}{15}(5)^n\right] u[n]}$$

Example 2.14 Find the inverse z-transform of

$$X[z] = \frac{(5 + z^{-2} + 4z^{-3})}{(z^2 + 7z + 10)}$$

Solution

$$\begin{aligned} X[z] &= \frac{(5 + z^{-2} + 4z^{-3})}{(z^2 + 7z + 10)} \\ &= \frac{(5 + z^{-2} + 4z^{-3})}{z} \frac{z}{(z+2)(z+5)} \\ &= [5z^{-1} + z^{-3} + 4z^{-4}] \frac{z}{(z+2)(z+5)} \end{aligned}$$

$$\frac{z}{(z+2)(z+5)} = \frac{1}{3}\left[\frac{z}{z+2} - \frac{z}{z+5}\right]$$

$$\frac{z}{(z+2)(z+5)} \overset{Z^{-1}}{\longleftrightarrow} \frac{1}{3}\left[(-2)^n - (-5)^n\right] u[n]$$

Now

$$x[n] = [5z^{-1} + z^{-3} + 4z^{-4}]\frac{1}{3}[(-2)^n - (-5)^n]u[n]$$

Using the time shifting property, we get

$$\boxed{\begin{aligned} x[n] = {} & \frac{5}{3}\left[(-2)^{n-1} - (-5)^{n-1}\right] u[n-1] + \frac{1}{3}\left[(-2)^{n-3} - (-5)^{n-3}\right] u[n-3] \\ & + \frac{4}{3}\left[(-2)^{n-4} - (-5)^{n-4}\right] u[n-4] \end{aligned}}$$

Example 2.15 Find the inverse z-transform for the following system functions:

$$(a) \quad X[z] = \frac{4}{(z-5)} \qquad \text{ROC: } |z| < 5$$

$$(b) \quad X[z] = z(1 - z^{-1})(1 + 2z^{-1}) \qquad \text{ROC: } 0 < |z| < \infty$$

Solution (*a*) $\boldsymbol{X[z] = \frac{4}{(z-5)}}$

$$X[z] = \frac{4}{(z-5)} = 4z^{-1}\frac{z}{z-5}$$

$$x[n] = 4z^{-1}[(5)^n]u[n]]$$

$$\boxed{x[n] = 4(5)^{n-1}u[n-1]}$$

(*b*) $\mathbf{X[z] = z(1 - z^{-1})(1 + 2z^{-1})}$

$$\begin{aligned} X[z] &= z(1 - z^{-1})(1 + 2z^{-1}) \\ X[z] &= z[1 + 2z^{-1} - z^{-1} - 2z^{-2}) \\ &= [z + 1 - 2z^{-1}] \end{aligned}$$

$$\boxed{x[n] = \{1, \underset{\uparrow}{1}, -2\}}$$

2.8.2 Inverse z-transform Using Power Series Expansion

The z-transform Equation (2.4)

$$X[z] = \sum_{n=-\infty}^{\infty} x[n]z^{-n}$$

can be expressed in power series form and the coefficients of $z^{|n|}$ give the values of the sequence. Equation (2.4) can be expressed as

$$\begin{aligned} X[z] = \cdots &+ x[-3]z^3 + x[-2]z^2 \\ &+ x[-1]z + x[0] + x[1]z^{-1} + x[2]z^{-2} + x[3]z^{-3} + \cdots \end{aligned} \tag{2.47}$$

Equation (2.47) does not give closed form. However, if $X[z]$ is not in a simpler form other than the polynomial in z^{-1}, using power series method, $x[n]$ is easily obtained. If $X[z]$ is rational, the power series is obtained by long division. The following examples illustrate the above method.

Example 2.16 Using power series expansion, find the inverse z-transform of the following $X[z]$:

$$(a) \quad X[z] = \frac{4z}{(z^2 - 3z + 2)} \qquad \text{ROC: } |z| > 2$$

$$(b) \quad X[z] = \frac{4z}{(z^2 - 3z + 2)} \qquad \text{ROC: } |z| < 1$$

$$(c) \quad X[z] = \frac{1}{(1 - az^{-1})} \qquad \text{ROC: } |z| > |a| \text{ and ROC: } |z| < |a|$$

Solution (*a*) $\mathbf{X[z] = \frac{4z}{(z^2-3z+2)}}$; ***ROC*: $|z| > 2$**

$$X[z] = \frac{4z}{(z^2 - 3z + 2)} = \frac{4z}{(z-1)(z-2)}$$

For ROC: $|z| > 2$, $x[n]$ is a right-sided sequence where $n \geq 0$. Hence, the long division is done in such a way that $X[z]$ is expressed in power of z^{-1}.

$$\begin{array}{r|l}
 & 4z^{-1} + 12z^{-2} + 28z^{-3} + \cdots \\
\hline
z^2 - 3z + 2 & 4z \\
 & \underline{4z - 12 + 8z^{-1}} \\
 & \quad 12 - 8z^{-1} \\
 & \quad \underline{12 - 36z^{-1} + 24z^{-2}} \\
 & \qquad 28z^{-1} - 24z^{-2} \\
 & \qquad \underline{28z^{-1} - 84z^{-2} + 56z^{-3}}
\end{array}$$

$$X[z] = 4z^{-1} + 12z^{-2} + 28z^{-3} + \cdots$$

$$\boxed{x[n] = \{\underset{\uparrow}{0},\ 4,\ 12,\ 28, \ldots\}}$$

(*b*) $\mathbf{X[z] = \frac{4z}{(z^2-3z+2)}}$; ***ROC*: $|z| < 1$**

For ROC: $|z| < 1$, $x[n]$ sequence is negative where $n \leq 0$. The long division is done in such a way that $X[z]$ is expressed in power of z.

$$\begin{array}{r|l}
 & 2z + 3z^2 + \frac{7}{2}z^3 \\
\hline
2 - 3z + z^2 & 4z \\
 & \underline{4z - 6z^2 + 2z^3} \\
 & \quad 6z^2 - 2z^3 \\
 & \quad \underline{6z^2 - 9z^3 + 3z^4} \\
 & \qquad 7z^3 - 3z^4 \\
 & \qquad \underline{7z^3 - \frac{21}{2}z^4 + \frac{7}{2}z^5}
\end{array}$$

$$X[z] = 2z + 3z^2 + \frac{7}{2}z^3 + \cdots$$

$$x[n] = \left\{\cdots \frac{7}{2},\ 3,\ 2,\ \&0\right\}$$
$$\uparrow$$

(c) $\boldsymbol{X[z] = \frac{1}{(1-az^{-1})}}$; $\boldsymbol{ROC: |z| > |a|}$

$$X[z] = \frac{z}{(z-a)}$$

The ROC: $|z| > a$, and it is exterior of the circle of radius $|a|$. Hence, $x[n]$ is a right-sided sequence where $n \geq 0$. The long division is done in such a way that $X[z]$ is expressed in terms of power of z^{-1} as shown below:

$$\begin{array}{r l}
 & 1 + az^{-1} + a^2z^{-2} + a^3a^{-3} + \cdots \\
z - a) & z \\
 & \underline{z - a} \\
 & a \\
 & \underline{a - a^2z^{-1}} \\
 & a^2z^{-1} \\
 & \underline{a^2z^{-1} - a^3z^{-2}} \\
 & a^3z^{-2} \\
 & \underline{a^3z^{-2} - a^4z^{-3}}
\end{array}$$

$$X[z] = 1 + az^{-1} + a^2z^{-2} + a^3z^{-3} + \cdots$$

$$x[n] = \{1,\ a,\ a^2,\ a^3, \cdots\}$$
$$\uparrow$$
$$x[n] = a^n u[n]$$

For ROC: $|z| < |a|$, $x[n]$ sequence is left-sided

$$\begin{array}{r l}
 & -a^{-1}z - a^{-2}z^2 - a^{-3}z^3 \cdots \\
-a + z) & z \\
 & \underline{z - a^{-1}z^2} \\
 & a^{-1}z^2 \\
 & \underline{a^{-1}z^2 - a^{-2}z^3} \\
 & a^{-2}z^3 \\
 & \underline{a^{-2}z^3 - a^{-3}z^4}
\end{array}$$

$$X[z] = -a^{-1}z - a^{-2}z^2 - a^{-3}z^3 + \cdots$$

$$x[n] = \left\{ \cdots, \frac{1}{a^3}, -\frac{1}{a^2}, -\frac{1}{a}, \&0 \right\}$$
$$\uparrow$$
$$x[n] = -a^n u[-n-1]$$

Example 2.17 Determine the inverse z-transform of

$$X[z] = \log(1 - 2z), \qquad |z| < \frac{1}{2}$$

by using the power series

$$\log(1 - x) = -\sum_{n=1}^{\infty} \frac{x^n}{n}, \qquad |x| < 1$$

and by first differentiating $X[z]$ and then using this to recover $x[n]$.

Solution (*a*) **Using Power Series**

$$\begin{aligned} X[z] &= \log(1 - 2z) \\ &= -\sum_{n=1}^{\infty} \frac{1}{n}(2z)^n \end{aligned}$$

Replace $n = -n$

$$\begin{aligned} X[z] &= \sum_{n=-1}^{-\infty} \frac{1}{n}(2z)^{-n} \\ &= \sum_{n=-1}^{-\infty} \left(\frac{1}{2}\right)^n \frac{1}{n} z^{-n} \end{aligned}$$

By z-transform definition, it is a left-sided signal

$$\begin{aligned} X[n] &= \frac{1}{n}\left(\frac{1}{2}\right)^n u(-n-1) \qquad n \leq -1 \\ &= 0 \qquad n \geq 0. \end{aligned}$$

(b) Using Differentiation Property

$$X[z] = \log(1 - 2z)$$
$$\frac{d}{dz}X[z] = \frac{-2}{(1 - 2z)}$$

Multiplying both sides by $-z$, we get

$$-z\frac{d}{dz}X[z] = \frac{2z}{(1 - 2z)}$$
$$= \frac{-z}{z - \frac{1}{2}}$$

$$-z\frac{d}{dz}X[z] \overset{Z^{-1}}{\longleftrightarrow} nx[n]$$
$$\frac{-z}{z - \frac{1}{2}} \overset{Z^{-1}}{\longleftrightarrow} \left(\frac{1}{2}\right)^n u(-n-1) \qquad \text{ROC: } |z| < \frac{1}{2}$$
$$nx[n] = \left(\frac{1}{2}\right)^n u(-n-1)$$

$$\boxed{x[n] = \left(\frac{1}{2}\right)^n \frac{1}{n} u(-n-1)}$$

Example 2.18 Find the inverse z-transform of

$$(a) \qquad X[z] = \log(1 + az^{-1}) \qquad |z| > |a|$$
$$(b) \qquad X[z] = \log(1 - az^{-1}) \qquad |z| > |a|$$

Solution (a) The power series expansion for $\log(1 + x)$ is

$$\log(1 + x) = \sum_{n=1}^{\infty} \frac{(-1)^{n+1}}{n} x^n \qquad \text{for } x < 1$$
$$\log(1 + az^{-1}) = \sum_{n=1}^{\infty} \frac{(-1)^{n+1}(az^{-1})^n}{n} \qquad |az^{-1}| < 1 \text{ or } |z| > |a|$$
$$= \sum_{n=1}^{\infty} \frac{(-1)^{n+1}a^n z^{-n}}{n}$$

Since the summation is from $n = 1$, using time shifting property, we get

$$\boxed{x[n] = \frac{(-1)^{n+1}a^n}{n} u[n-1]}$$

(*b*) The power series expansion for $\log(1-x)$ is

$$\log(1-x) = -\sum_{n=1}^{\infty} \frac{1}{n} x^n \qquad |x| < 1$$

$$\log(1-az^{-1}) = -\sum_{n=1}^{\infty} \frac{1}{n} (az^{-1})^n$$

$$\log(1+az^{-1}) = -\sum_{n=1}^{\infty} \frac{a^n}{n} z^{-n}$$

$$\boxed{x[n] = -\frac{a^n}{n} u[n-1]}$$

2.8.3 *Inverse z-transform Using Contour Integration or the Method of Residue*

The inverse z-transform can be obtained from Eq. (2.2) which is given by

$$x[n] = \frac{1}{2\pi j} \oint_c X[z] z^{n-1} dz \tag{2.48}$$

The above integral can be evaluated by summing up all the residues of the poles which are inside the circle c of Eq. (2.48) which can be expressed as

$$\begin{aligned} x[n] &= \sum \text{ (Residues of } X[z]z^{-n} \text{ at the poles inside } (c) \\ &= \sum_i (z - z_i) X[z] z^{-n-1} \Big|_{z=z_i} \end{aligned} \tag{2.49}$$

For multiples poles of order k, and $z = \alpha$, the residue is written as

$$\text{Residue} = \frac{1}{\angle(k-1)} \underset{z\to\alpha}{Lt} \left\{ \frac{d^{k-1}}{dz^{k-1}} (z-\alpha)^k X[z] z^{n-1} \right\} \tag{2.50}$$

Example 2.19 Find the inverse z-transform of the following $X[z]$ using Residue method:

$$(a) \quad X[z] = \frac{(1+z^{-1})}{(1+8z^{-1}+15z^{-2})} \qquad |z| > 5$$

$$(b) \quad X[z] = \frac{z^{-1}}{(1-10z^{-1}+24z^{-2})} \qquad 4 < |z| < 6$$

$$(c) \quad X[z] = \frac{z}{\left(z-\frac{1}{2}\right)^2}$$

Solution (*a*) $X[z] = \frac{(1+z^{-1})}{(1+8z^{-1}+15z^{-2})}$; $|z| > 5$

$$X[z] = \frac{z(z+1)}{(z^2+8z+15)}$$

$$X[z] = \frac{z(z+1)}{(z+3)(z+5)}$$

$$x[n] = \sum \text{Residue of } \frac{z(z+1)}{(z+3)(z+5)} z^{n-1}$$

$$= \text{Residue of } (z+3)\frac{z(z+1)}{(z+3)(z+5)} z^{n-1}\Big|_{z=-3}$$

$$+ \text{Residue of } (z+5)\frac{z(z+1)z^{n-1}}{(z+3)(z+5)}\Big|_{z=-5}$$

$$\boxed{x[n] = -(-3)^n + 2(-5)^n}$$

(*b*) $X[z] = \frac{z^{-1}}{(1-10z^{-1}+24z^{-2})}$; $4 < |z| < 6$

$$X[z] = \frac{z}{(z^2-10z+24)} = \frac{z}{(z-4)(z-6)}$$

For $n \geq 0$

$$x[n] = \text{Residue of } X[z]z^{n-1}\Big|_{z=4}$$

$$= (z-4)\frac{z(z^{n-1})}{(z-4)(z-6)}\Big|_{z=4} = -\frac{1}{2}(4)^n u[n]$$

For $n < 0$

$$x[n] = -\left[(z-6)\frac{zz^{n-1}}{(z-4)(z-6)}\right]_{z=6}$$

$$= -\frac{1}{2}(6)^n u(-n-1)$$

$$\boxed{x[n] = -\frac{1}{2}[(4)^n u[n] + (6)^n u(-n-1)]}$$

(c) $X[z] = \frac{z}{(z-\frac{1}{2})^2}$

$$x[n] = \frac{d}{dz}\left[\left(z-\frac{1}{2}\right)^2 \frac{zz^{n-1}}{\left(z-\frac{1}{2}\right)}\right]_{z=\frac{1}{2}}$$

$$= \frac{d}{dz}z^n\Big|_{z=1/2} = nz^{n-1}\Big|_{z=1/2}$$

$$\boxed{x[n] = 2n\left(\frac{1}{2}\right)^n u[n]}$$

2.9 The System Function of DT Systems

Let

1. $x[n]$ = Input of the system;
2. $y[n]$ = Output of the system;
3. $h[n]$ = Impulse response of the system.

The output $y[n]$ can be expressed as the convolution of $x[n]$ with $h[n]$ as

$$y[n] = x[n] * h[n] \tag{2.51}$$

By applying convolution property of z-transform, we obtain

$$Y[z] = X[z]H[z] \tag{2.52}$$

where $Y[z]$, $X[z]$ and $H[z]$ are the z-transforms of $y[n]$, $x[n]$ and $h[n]$, respectively. Equation (2.52) can be expressed as

$$H[z] = \frac{Y[z]}{X[z]} \tag{2.53}$$

In Eq. (2.53), $H[z]$ is referred to as the system function or the transfer function. System function is defined as the ratio of the z-transforms of the output $y[n]$ and the input $x[n]$. The system function completely depends on the system characteristic. Equations (2.51) and (2.52) are illustrated in Fig. 2.14a, b, respectively.

Fig. 2.14 System impulse response and system function

(a) $x[n] \longrightarrow \boxed{h[n]} \longrightarrow y[n] = x[n] * h[n]$

(b) $X[z] \longrightarrow \boxed{H[z]} \longrightarrow Y[z] = X[z]H[z]$

2.10 Causality of DT Systems

An linear time invariant discrete time system is said to be causal if the impulse response $h[n] = 0$ for $n < 0$ and it is therefore right-sided. The ROC of such a system $H[z]$ is the exterior of a circle. If $H[z]$ is rational then the system is said to be causal if the ROC lies exterior of the circle passing through the outermost pole and includes infinity area. A DT system which is linear time invariant with its system function $H[z]$ rational is said to be causal iff the ROC is the exterior of a circle which passes through the outermost pole of $H[z]$. Further, the degree of the numerator polynomial of $H[z]$ should be less than or equal to the degree of the denominator polynomial.

2.11 Stability of DT System

As we discussed in Chap. 2, an LTI discrete time system is said to be BIBO stable if the impulse response $h[n]$ is summable. This is expressed as

$$\sum_{n=-\infty}^{\infty} |h[n]| < \infty \tag{2.54}$$

The corresponding requirement on $H[z]$ is that the ROC of $H[z]$ contains unit circle. By definition of z-transform

$$H[z] = \sum_{n=-\infty}^{\infty} h[n]z^{-n}$$

Let $z = e^{j\Omega}$

$$\begin{aligned} |z| &= |e^{j\Omega}| \\ &= 1 \\ |H[e^{j\Omega}]| &= \left| \sum_{n=-\infty}^{\infty} h[n]e^{-j\Omega n} \right| \\ &\leq \sum_{n=-\infty}^{\infty} \left|h[n]e^{-j\Omega n}\right| \\ &= \sum_{n=-\infty}^{\infty} |h[n]| < \infty \end{aligned} \tag{2.55}$$

From Eq. (2.55), we see that the stability condition given by Eq. (2.54) is satisfied if $z = e^{j\Omega}$. Thus, it implies that $H[z]$ must contain unit circle $|z| = 1$.

An LTI system is stable iff the ROC of its system function $H[z]$ contains the unit circle $|z| = 1$.

2.12 Causality and Stability of DT System

For a causal system whose $H[z]$ is rational, the ROC is outside the outermost pole. For the BIBO stability, the ROC should include the unit circle $|z| = 1$. For the system to be causal and stable, the above requirements are satisfied if all the poles are within the unit circle in the z-plane.

An LTID system with the system function $H[z]$ is said to be both causal and stable iff all the poles of $H[z]$ lie inside the unit circle.

The above characteristics of LTI discrete time systems are illustrated in Fig. 2.15 for a causal system.

Example 2.20 The input to the causal LTI system is

$$x[n] = u[-n-1] + \left(\frac{1}{2}\right)^n u[n]$$

The z-transform of the output of the system is

$$Y[z] = \frac{-\frac{1}{2}z^{-1}}{\left(1 - \frac{1}{2}z^{-1}\right)\left(1 + z^{-1}\right)}$$

Determine $H[z]$, the z-transform of the impulse response, and also, determine the output $y[n]$.

Solution

$$\begin{aligned} X[z] &= -\frac{z}{(z-1)} + \frac{z}{(z-0.5)} \\ &= \frac{-0.5z}{(z-1)(z-0.5)} \end{aligned}$$

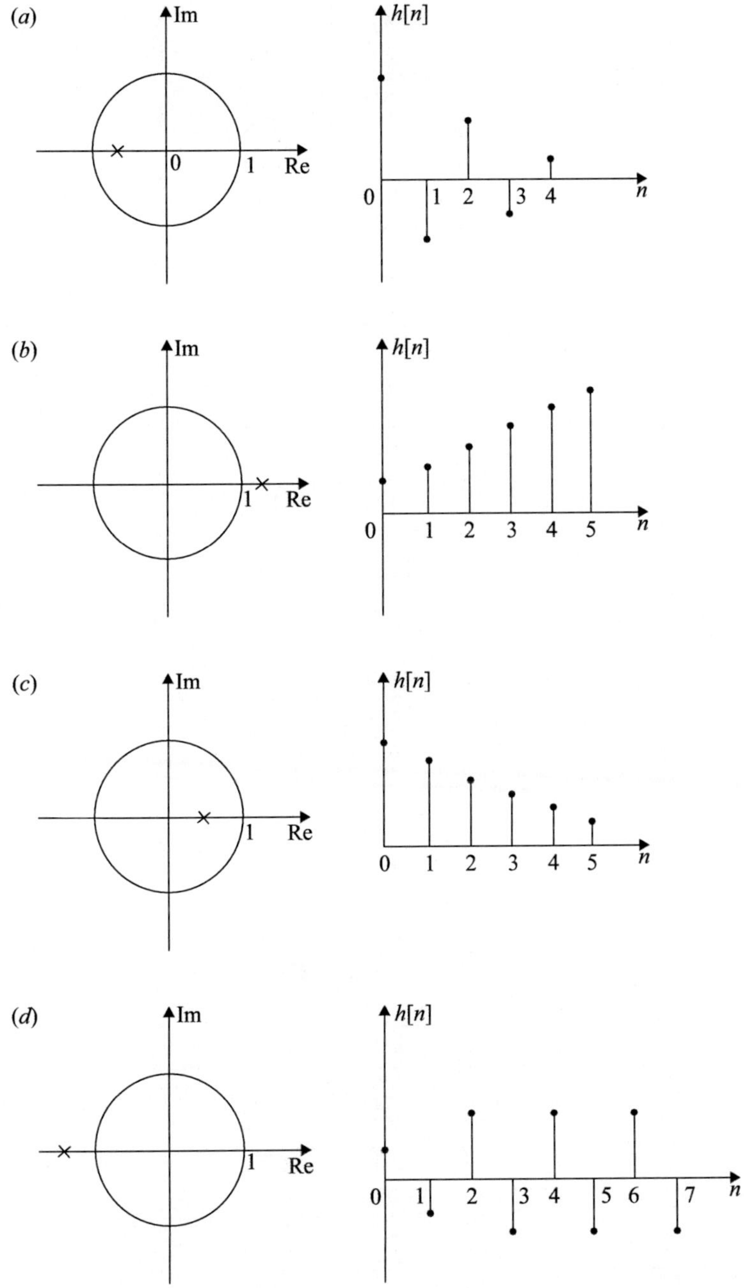

Fig. 2.15 **a to d** Pole location and impulse response of a causal system. **e to f** Pole location and impulse response of a causal system

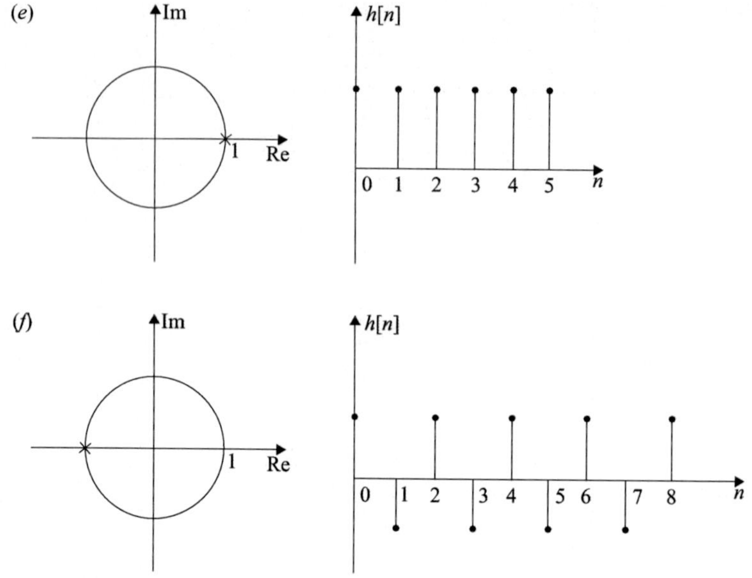

Fig. 2.15 (continued)

$$
\begin{aligned}
Y[z] &= \frac{-\frac{1}{2}z^{-1}}{\left(1-\frac{1}{2}z^{-1}\right)(1+z^{-1})} \\
&= \frac{-\frac{1}{2}z}{(z-0.5)(z+1)} \\
H[z] &= \frac{Y[z]}{X[z]} \\
&= \frac{(-0.5)z(z-1)(z-0.5)}{(z-0.5)(z+1)(-0.5)z} \\
&= \frac{(z-1)}{(z+1)} \\
\frac{H[z]}{z} &= \frac{(z-1)}{z(z+1)} \\
&= \frac{A_1}{z} + \frac{A_2}{z+1} \\
(z-1) &= A_1(z+1) + A_2 z
\end{aligned}
$$

Substitute $z = 0$

$$-1 = A_1$$

Substitute $z = -1$

$$-2 = -A_2; \qquad A_2 = 2$$

$$H[z] = -1 + \frac{2z}{(z+1)}$$

$$\boxed{h[n] = -\delta[n] + (-1)^n 2u[n]}$$

$$\begin{aligned}
Y[z] &= \frac{-\frac{1}{2}z}{(z-0.5)(z+1)} \\
\frac{Y[z]}{z} &= \frac{-\frac{1}{2}}{(z-0.5)(z+1)} \\
&= \frac{A_1}{z-0.5} + \frac{A_2}{z+1} \\
-\frac{1}{2} &= A_1(z+1) + A_2(z-0.5)
\end{aligned}$$

Substitute $z = 0.5$

$$-\frac{1}{2} = \frac{3}{2}A_1; \qquad A_1 = -\frac{1}{3}$$

Substitute $z = -1$

$$-\frac{1}{2} = -\frac{3}{2}A_2; \qquad A_2 = \frac{1}{3}$$

$$Y[z] = \frac{1}{3}\left[-\frac{1}{(z-0.5)} + \frac{1}{(z+1)}\right]$$

$$\boxed{y[n] = \frac{1}{3}\left[-\left(\frac{1}{2}\right)^n + (-1)^n\right]u[n]}$$

Example 2.21 A certain LTI system is described by the following system function:

$$H[z] = \frac{\left(z+\frac{1}{2}\right)}{(z-1)\left(z-\frac{1}{2}\right)}$$

Find the system response to the input $x[n] = 4^{-(n+2)}u[n]$.

Solution

$$
\begin{aligned}
x[n] &= 4^{-(n+2)}u[n] \\
&= \frac{1}{16}(4)^{-n}u[n] \\
X[z] &= \frac{1}{16}\frac{z}{\left(z-\frac{1}{4}\right)} \\
Y[z] &= H[z]X[z] \\
&= \frac{1\left(z+\frac{1}{2}\right)z}{16(z-1)\left(z-\frac{1}{2}\right)\left(z-\frac{1}{4}\right)} \\
\frac{Y[z]}{z} &= \frac{\left(z+\frac{1}{2}\right)}{16(z-1)\left(z-\frac{1}{2}\right)\left(z-\frac{1}{4}\right)} \\
&= \frac{A_1}{z-1}+\frac{A_2}{\left(z-\frac{1}{2}\right)}+\frac{A_3}{\left(z-\frac{1}{4}\right)}
\end{aligned}
$$

$$\frac{1}{16}\left(z+\frac{1}{2}\right) = A_1\left(z-\frac{1}{2}\right)\left(z-\frac{1}{4}\right) + A_2(z-1)\left(z-\frac{1}{4}\right) + A_3(z-1)\left(z-\frac{1}{2}\right)$$

Substitute $z = 1$

$$\left(\frac{1}{16}\right)\left(\frac{3}{2}\right) = \left(\frac{1}{2}\right)\left(\frac{3}{4}\right)A_1; \qquad A_1 = \frac{1}{4}$$

Substitute $z = \frac{1}{2}$

$$\frac{1}{16} = -\left(\frac{1}{2}\right)\left(\frac{1}{4}\right)A_2; \qquad A_2 = -\frac{1}{2}$$

Substitute $z = \frac{1}{4}$

$$\frac{1}{16}\frac{3}{4} = -\left(\frac{3}{4}\right)\left(-\frac{1}{4}\right)A_3; \qquad A_3 = \frac{1}{4}$$

$$Y[z] = \left[\frac{1}{4}\frac{z}{(z-1)} - \frac{1}{2}\frac{z}{\left(z-\frac{1}{2}\right)} + \frac{1}{4}\frac{z}{\left(z-\frac{1}{4}\right)}\right]$$

$$\boxed{y[n] = \left[\frac{1}{4}(1)^n - \frac{1}{2}\left(\frac{1}{2}\right)^n + \frac{1}{4}\left(\frac{1}{4}\right)^n\right]u[n]}$$

Example 2.22 Given

$$x[n] = \{2, -3, 1\}$$
$$h[n] = \{1, 2, -1\}$$

Find $y[n]$ using z-transform.

Solution

$$\begin{aligned}
X[z] &= (2 - 3z^{-1} + z^{-2}) \\
H[z] &= 1 + 2z^{-1} - z^{-2} \\
Y[z] &= X[z]H[z] \\
&= [2 - 3z^{-1} + z^{-2}][1 + 2z^{-1} - z^{-2}] \\
&= 2 + z^{-1} - 7z^{-2} + 5z^{-3} - z^{-4}
\end{aligned}$$

$$\boxed{y[n] = \{2, 1, -7, 5, -1\}}$$

Example 2.23 Given

$$x[n] = u[n]$$
$$y[n] = (2)^n u[n]$$

Find the system function and the impulse response.

Solution

$$\begin{aligned}
x[n] &= u[n] \\
X[z] &= \frac{z}{(z-1)} \qquad |z| > 1 \\
y[n] &= (2)^n u[n] \\
Y[z] &= \frac{z}{(z-2)} \qquad |z| > 2 \\
H[z] &= \frac{Y[z]}{X[z]} = \frac{(z-1)}{(z-2)} \qquad |z| > 2 \\
\frac{H[z]}{z} &= \frac{(z-1)}{z(z-2)} \\
&= \frac{A_1}{z} + \frac{A_2}{(z-2)} \\
z - 1 &= A_1(z-2) + A_2 z
\end{aligned}$$

Substitute $z = 0$

$$-1 = A_1(-2); \qquad A_1 = \frac{1}{2}$$

Substitute $z = 2$

$$1 = 2A_2; \qquad A_2 = \frac{1}{2}$$

$$H[z] = \frac{1}{2}\left[1 + \frac{z}{(z-2)}\right]$$

$$\boxed{y[n] = \frac{1}{2}\left[\delta(n) + (2)^n u[n]\right]}$$

Example 2.24 Given

$$y[n] = \left(\frac{1}{4}\right)^n u[n]$$

$$x[n] = \left(\frac{1}{2}\right)^n u[-n-1]$$

Find the system function and hence the system impulse response.

Solution

$$y[n] = \left(\frac{1}{4}\right)^n u[n]$$

$$Y[z] = \frac{z}{\left(z - \frac{1}{4}\right)} \qquad |z| > \frac{1}{4}$$

$$x[n] = \left(\frac{1}{2}\right)^n u[-n-1]$$

$$X[z] = -\frac{z}{\left(z - \frac{1}{2}\right)} \qquad |z| < \frac{1}{2}$$

$$H[z] = \frac{Y[z]}{X[z]}$$

$$\boxed{H[z] = \frac{-\left(z - \frac{1}{2}\right)}{\left(z - \frac{1}{4}\right)}}$$

$$\frac{H[z]}{z} = \frac{-\left(z-\frac{1}{2}\right)}{z\left(z-\frac{1}{4}\right)}$$
$$= \frac{A_1}{z} + \frac{A_2}{z\left(z-\frac{1}{4}\right)}$$
$$\frac{1}{2} - z = A_1\left(z-\frac{1}{4}\right) + A_2 z$$

Substitute $z = 0$

$$\frac{1}{2} = -\frac{1}{4}A_1; \qquad A_1 = -2$$

Substitute $z = \frac{1}{4}$

$$\frac{1}{2} = \frac{1}{4}A_2; \qquad A_2 = 2$$

$$H[z] = 2\left[-1 + \frac{z}{\left(z-\frac{1}{4}\right)}\right]$$

$$\boxed{h[n] = 2\left[-\delta[n] + \left(\frac{1}{4}\right)^n\right]u[n]}$$

Example 2.25 Consider the following system functions:

$$(a) \qquad H[z] = \frac{(1+4z^{-1}+z^{-2})}{(2z^{-1}+5z^{-2}+z^{-3})}$$
$$(b) \qquad H[z] = \frac{(z-1)(z+2)}{\left(z-\frac{1}{2}\right)\left(z-\frac{3}{4}\right)} \qquad \text{ROC: } |z| > \frac{3}{4}$$
$$(c) \qquad H[z] = \frac{(z-1)(z+2)}{\left(z-\frac{1}{2}\right)\left(z-\frac{3}{4}\right)} \qquad \text{ROC: } |z| < \frac{1}{2}$$

Determine whether these systems are causal or not.

Solution (*a*) $\boldsymbol{H[z] = \frac{(1+4z^{-1}+z^{-2})}{(2z^{-1}+5z^{-2}+z^{-3})}}$

$$H[z] = \frac{(z^3+4z^2+z)}{(2z^2+5z+1)}$$

$H[z]$ is irrational since the degree of the numerator polynomial is greater than the denominator polynomial.

The System is Non-causal.

(*b*) $\boldsymbol{H[z] = \frac{(z-1)(z+2)}{(z-\frac{1}{2})(z-\frac{3}{4})}}$; $\boldsymbol{ROC: |z| > \frac{3}{4}}$
The ROC is the exterior of the circle passing through the outermost pole of $H[z]$. Hence, $h[n]$, the impulse response is right-sided.

The System is Causal.

(*c*) $\boldsymbol{H[z] = \frac{(z-1)(z+2)}{(z-\frac{1}{2})(z-\frac{3}{4})}}$ $\boldsymbol{ROC: |z| < \frac{1}{2}}$
The ROC is the interior of the circle passing through the innermost pole of $H[z]$. Hence, $h[n]$, the impulse response is left-sided.

The System is Non-causal.

Example 2.26 Consider the following system function:

$$H[z] = \frac{\left(2 - \frac{13}{4}z^{-1}\right)}{\left(1 - \frac{1}{4}z^{-1}\right)\left(1 - 3z^{-1}\right)}$$

Determine the causality and stability of the system for the following cases.

(*a*) ROC: $|z| > 3$;
(*b*) ROC: $|z| < \frac{1}{4}$;
(*c*) ROC: $\frac{1}{4} < |z| < 3$.

Solution

$$\begin{aligned} H[z] &= \frac{\left(2 - \frac{13}{4}z^{-1}\right)}{\left(1 - \frac{1}{4}z^{-1}\right)\left(1 - 3z^{-1}\right)} \\ &= \frac{z\left(2z - \frac{13}{4}\right)}{\left(z - \frac{1}{4}\right)(z - 3)} \end{aligned}$$

(*a*) **ROC** : $\boldsymbol{|z| > 3}$
The ROC is the exterior of the circle passing through the outermost pole of $H[z]$ which is rational (the denominator and numerator polynomials have same order). The impulse response $h[n]$ is a right-sided sequence. Hence, $H[z]$ is causal. The ROC does not contain unit circle. Hence, $h[n]$ is not summable. The system is unstable. Refer to Fig. 2.16a.

The System is Causal and Unstable.

(*b*) **ROC :** $|z| < \frac{1}{4}$
The ROC is the interior of the circle passing through the innermost pole of $H[z]$. The impulse response is a left-sided sequence. $H[z]$ is therefore non-causal. The ROC does not include the unit circle. The $h[n]$ is growing exponential negative sequence. The system is unstable. Refer to Fig. 2.16b.

The System is Non-causal and Unstable.

(*c*) **ROC :** $\frac{1}{4} < |z| < 3$
The ROC is to the left of the outermost pole and to the right innermost pole. Hence, $h[n]$ will have right- and left-sided sequences, which is non-causal. The ROC includes unit circle, which means that the right- and left-sided sequences of $h[n]$ will exponentially decay and the system is stable. Refer to Fig. 2.16c.

The System is Non-causal and Stable.

The system cannot be both Causal and Stable.

Example 2.27 Consider the following system function:

$$H[z] = \frac{z}{\left(z - \frac{1}{4}\right)\left(z + \frac{1}{4}\right)\left(z - \frac{1}{2}\right)}$$

For different possible ROCs, determine the causality, stability and the impulse response of the system.

Solution

$$H[z] = \frac{z}{\left(z - \frac{1}{4}\right)\left(z + \frac{1}{4}\right)\left(z - \frac{1}{2}\right)}$$

The possible ROCs for $H[z]$ to exist are (*a*) ROC: $|z| > \frac{1}{2}$, (*b*) ROC: $|z| < \frac{1}{4}$ and (*c*) ROC: $\frac{1}{4} < |z| < \frac{1}{2}$.

$$\begin{aligned} \frac{H[z]}{z} &= \frac{1}{\left(z - \frac{1}{4}\right)\left(z + \frac{1}{4}\right)\left(z - \frac{1}{2}\right)} \\ &= \frac{A_1}{\left(z - \frac{1}{4}\right)} + \frac{A_2}{\left(z + \frac{1}{4}\right)} + \frac{A_3}{\left(z - \frac{1}{2}\right)} \end{aligned}$$

$$1 = A_1\left(z + \frac{1}{4}\right)\left(z - \frac{1}{2}\right) + A_2\left(z - \frac{1}{4}\right)\left(z - \frac{1}{2}\right) + A_3\left(z - \frac{1}{4}\right)\left(z + \frac{1}{4}\right)$$

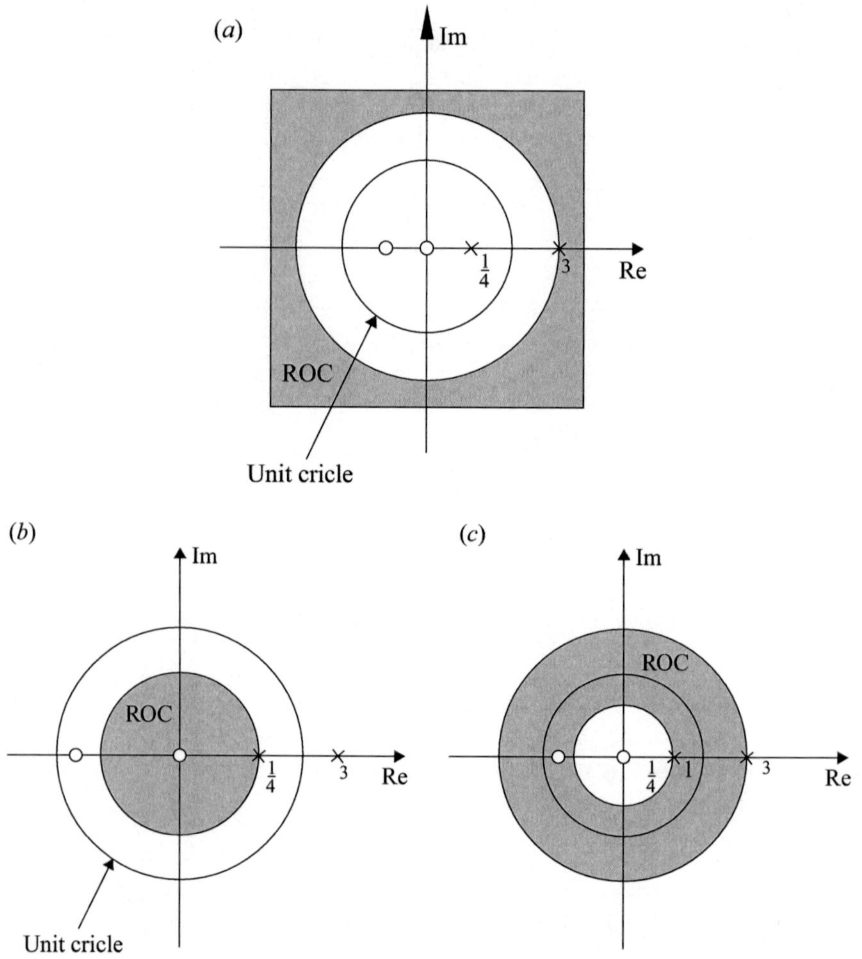

Fig. 2.16 **a** Causal and stable system. **b** Non-causal and unstable system and **c** Non-causal and stable system

Substitute $z = \frac{1}{4}$

$$1 = A_1 \left(\frac{1}{4} + \frac{1}{4}\right)\left(\frac{1}{4} - \frac{1}{2}\right); \qquad A_1 = -8$$

Substitute $z = -\frac{1}{4}$

$$1 = A_2 \left(-\frac{1}{4} - \frac{1}{4}\right)\left(-\frac{1}{4} - \frac{1}{2}\right); \qquad A_2 = \frac{8}{3}$$

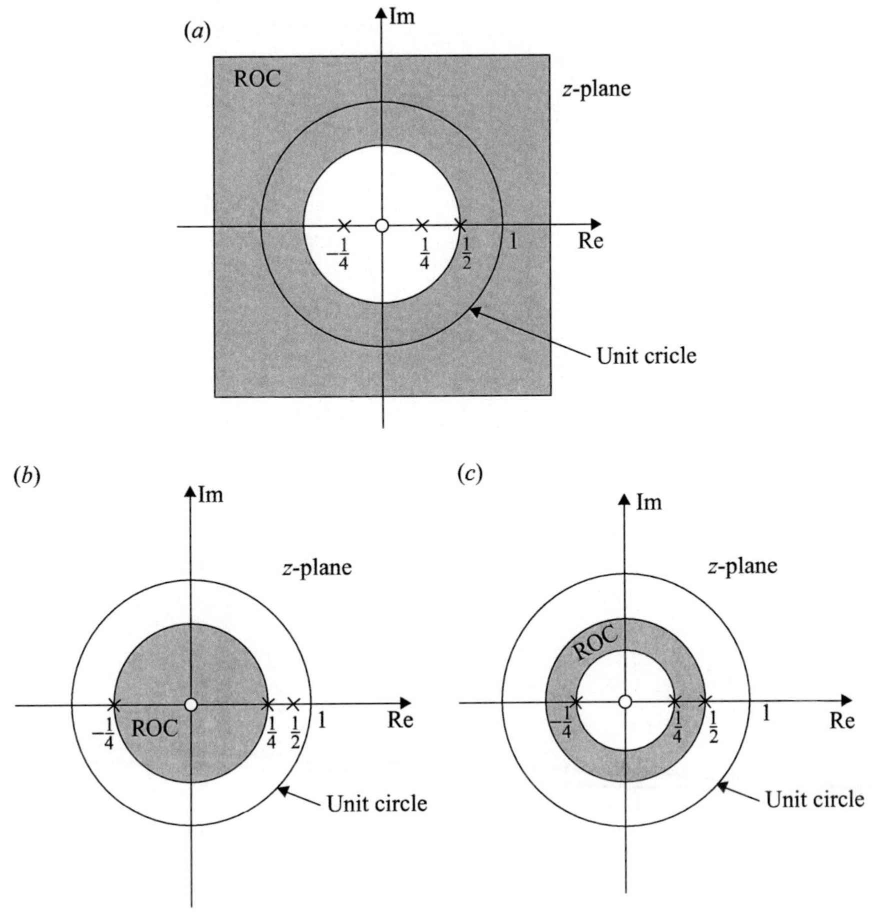

Fig. 2.17 **a** Pole–zero diagram and ROC: $|z| > \frac{1}{2}$ of Example 2.27. **b** Pole–zero diagram and ROC: $|z| < \frac{1}{4}$ and **c** Pole–zero diagram and ROC: $\frac{1}{4} < |z| < \frac{1}{2}$

Substitute $z = \frac{1}{2}$

$$1 = A_3\left(\frac{1}{2} - \frac{1}{4}\right)\left(\frac{1}{2} + \frac{1}{4}\right); \qquad A_3 = \frac{16}{3}$$

$$H[z] = -\frac{8z}{\left(z - \frac{1}{4}\right)} + \frac{8}{3}\frac{z}{\left(z + \frac{1}{4}\right)} + \frac{16}{3}\frac{z}{\left(z - \frac{1}{2}\right)}$$

(*a*) **ROC** : $|z| > \frac{1}{2}$
The pole–zero diagram and the ROC are shown in Fig. 2.17a. From Fig. 2.17a, the ROC is the exterior of the outermost pole $z = \frac{1}{2}$. Further, ROC includes unit

circle. Thus, $h[n]$ is a right-sided sequence and hence $H[z]$ is causal. Since ROC includes unit circle and all the poles are within unit circle, the system is stable. Now,

$$H[z] = -\frac{8z}{\left(z - \frac{1}{4}\right)} + \frac{8}{3}\frac{z}{\left(z + \frac{1}{4}\right)} + \frac{16}{3}\frac{z}{\left(z - \frac{1}{2}\right)}$$

$$\boxed{h[n] = \left[-8\left(\frac{1}{4}\right)^n + \frac{8}{3}\left(-\frac{1}{4}\right)^n + \frac{16}{3}\left(\frac{1}{2}\right)^n\right]u[n]}$$

The System is Causal and Stable.

(*b*) **ROC** : $|z| < \frac{1}{4}$

For ROC: $|z| < \frac{1}{4}$, the pole–zero diagram is shown in Fig. 2.17b. The ROC is interior of the circle passing through the innermost pole. Hence, the system is non-causal. The condition that the ROC does not include unit circle implies that the system is unstable. The sequence $h[n]$ is left-sided. This is obtained as follows:

$$H[z] = -\frac{8z}{\left(z - \frac{1}{4}\right)} + \frac{8}{3}\frac{z}{\left(z + \frac{1}{4}\right)} + \frac{16}{3}\frac{z}{\left(z - \frac{1}{2}\right)}$$

$$h[n] = \left[8\left(\frac{1}{4}\right)^n - \frac{8}{3}\left(-\frac{1}{4}\right)^n - \frac{16}{3}\left(\frac{1}{2}\right)^n\right]u[-n-1]$$

The left-sided sequence $u[-n-1]$ will exponentially increase for $n < 0$ and makes the system unstable.

The System is Non-causal and Unstable.

(*c*) **ROC** : $\frac{1}{4} < |z| < \frac{1}{2}$

The pole–zero diagram and ROC of $H[z]$ are shown in Fig. 2.17c. The ROC is concentric ring for $\frac{1}{4} < |z| < \frac{1}{2}$. The $h[n]$ sequences die to the poles at $z = \frac{1}{4}$ and $z = -\frac{1}{4}$ are right-sided and the sequence due to the pole $z = \frac{1}{2}$ is left-sided. Hence, the system is non-causal. The ROC does not include the unit circle and hence the system is unstable. The impulse response is obtained as follows.

$$H[z] = -\frac{8z}{\left(z - \frac{1}{4}\right)} + \frac{8}{3}\frac{z}{\left(z + \frac{1}{4}\right)} + \frac{16}{3}\frac{z}{\left(z - \frac{1}{2}\right)}$$

$$\boxed{h[n] = \left[-8\left(\frac{1}{4}\right)^n + \frac{8}{3}\left(-\frac{1}{4}\right)^n\right]u[n] - \frac{16}{3}\left(\frac{1}{2}\right)^n u[-n-1]}$$

The term $-\frac{16}{3}(1/2)^n u[-n-1]$ for $n < 0$ yields exponentially increasing sequence.

$$\text{The System is Non} - \text{causal and Unstable.}$$

2.13 z-transform Solution of Linear Difference Equations

As in the case of Laplace transform with differential equation, to get the solution in time domain z-transform is used to solve difference equation to get the output sequence as a function of n. By using the time shift property of z-transform, the difference equation is converted into algebraic equation, taking into account the initial conditions, by taking z-inverse transform, the time domain solution is obtained.

2.13.1 *Right Shift (Delay)*

If

$$x[n]u[n] \stackrel{Z}{\longleftrightarrow} X[z]$$

then

$$\begin{aligned} x[n-1]u[n-1] &\stackrel{Z}{\longleftrightarrow} \frac{1}{z}X[z] \\ x[n-1]u[n] &\stackrel{Z}{\longleftrightarrow} \frac{1}{z}X[z] + x[-1] \\ x[n-2]u[n] &\stackrel{Z}{\longleftrightarrow} \frac{1}{z^2}X[z] + \frac{1}{z}x[-1] + x[-2] \end{aligned}$$

In general,

$$\boxed{x[n-m]u[n] \stackrel{Z}{\longleftrightarrow} z^{-m}X[z] + z^{-m}\sum_{n=1}^{m} x[-n]z^n} \tag{2.56}$$

2.13.2 *Left Shift (Advance)*

If

$$\begin{aligned} x[n]u[n] &\stackrel{Z}{\longleftrightarrow} X[z] \\ x[n+1]u[n] &\stackrel{Z}{\longleftrightarrow} zX[z] - zx[0] \\ x[n+2]u[n] &\stackrel{Z}{\longleftrightarrow} z^2X[z] - z^2x[0] - zx[1] \end{aligned}$$

In general,

$$\boxed{x[n+m]u[n] \overset{Z}{\longleftrightarrow} z^m x[z] - z^m \sum_{n=0}^{m-1} x[n]z^{-n}} \quad (2.57)$$

Equations (2.56) and (2.57) are used to convert difference equations with initial conditions to algebraic equations in z. By application of Eq. (2.56), the delay shift is more common. The following examples illustrate the above procedure.

Example 2.28 Consider the following linear constant coefficient difference equation:

$$y[n] - \frac{3}{4}y[n-1] + \frac{1}{8}y[n-2] = 2x[n-1]$$

Determine $y[n]$ when $x[n] = \delta[n]$ and $y[n] = 0,\ n < 0$.

Solution If $y[n] = 0,\ n = 0$ implies the initial conditions are zero. Taking z-transform on both sides of the given equation, we get

$$\left[1 - \frac{3}{4}z^{-1} + \frac{1}{8}z^{-2}\right] Y[z] = 2z^{-1}X[z]$$

For $\delta[n],\ X[z] = 1$

$$\begin{aligned}
Y[z] &= \frac{2z^{-1}}{1 - \frac{3}{4}z^{-1} + \frac{1}{8}z^{-2}} \\
&= \frac{2z}{z^2 - \frac{3}{4}z + \frac{1}{8}} \\
\frac{Y[z]}{z} &= \frac{2}{\left(z - \frac{1}{2}\right)\left(z - \frac{1}{4}\right)} \\
&= \frac{A_1}{\left(z - \frac{1}{2}\right)} + \frac{A_2}{\left(z - \frac{1}{4}\right)} \\
2 &= A_1\left(z - \frac{1}{4}\right) + A_2\left(z - \frac{1}{2}\right)
\end{aligned}$$

Substitute $z = \frac{1}{2}$

$$2 = A_1\frac{1}{4}; \qquad A_1 = 8$$

Substitute $z = \frac{1}{4}$

$$2 = A_2\left(-\frac{1}{4}\right); \qquad A_2 = -8$$

$$Y[z] = 8\left[\frac{z}{\left(z-\frac{1}{2}\right)} - \frac{z}{\left(z-\frac{1}{4}\right)}\right]$$

$$\boxed{y[n] = 8\left[\left(\frac{1}{2}\right)^n - \left(\frac{1}{4}\right)^n\right]u[n] \qquad \text{ROC: } |z| > \frac{1}{2}}$$

Example 2.29

$$y[n+2] + 1.1y[n+1] + 0.3y[n] = x[n+1] + x[n]$$

where $x[n] = (-4)^{-n}u[n]$. Find $y[n]$ if the initial conditions are zero.

Solution Taking z-transform using left shift property, we get

$$[z^2 + 1.1z + 0.3]Y[z] = [z+1]X[z]$$
$$x[n] = (-4)^{-n}u[n]$$

$$\begin{aligned}
X[z] &= \frac{z}{\left(z+\frac{1}{4}\right)} \\
Y[z] &= \frac{z(z+1)}{\left(z+\frac{1}{4}\right)(z^2+1.1z+0.3)} \\
&= \frac{z(z+1)}{\left(z+\frac{1}{4}\right)(z+0.5)(z+0.6)} \\
\frac{Y[z]}{z} &= \frac{(z+1)}{\left(z+\frac{1}{4}\right)(z+0.5)(z+0.6)} \\
&= \frac{A_1}{\left(z+\frac{1}{4}\right)} + \frac{A_2}{(z+0.5)} + \frac{A_3}{(z+0.6)}
\end{aligned}$$

$$(z+1) = A_1(z+0.5)(z+0.6) + A_2\left(z+\frac{1}{4}\right)(z+0.6) + A_3\left(z+\frac{1}{4}\right)(z+0.5)$$

Substitute $z = -\frac{1}{4}$

$$\left(-\frac{1}{4}+1\right) = A_1\left(-\frac{1}{4}+0.5\right)\left(-\frac{1}{4}+0.6\right); \qquad A_1 = 8.57$$

Substitute $z = -0.5$

$$(-0.5+1) = A_2\left(-0.5+\frac{1}{4}\right)(-0.5+0.6); \qquad A_2 = -20$$

Substitute $z = -0.6$

$$(-0.6+1) = A_3\left(-0.6+\frac{1}{4}\right)(-0.6+0.5); \qquad A_3 = 11.43$$

$$Y[z] = \frac{8.57z}{\left(z+\frac{1}{4}\right)} - \frac{20z}{(z+0.5)} + \frac{11.43}{(z+0.6)}$$

$$\boxed{y[n] = \left[8.57\left(-\frac{1}{4}\right)^n - 20\,(-0.5)^n + 11.43(-0.6)^n\right]u[n]}$$

Example 2.30 A causal LTI system is described by the difference equation

$$y[n] = y[n-1] + y[n-2] + x[n-1]$$

Find (a) System function for this system and (b) Unit impulse response of the system.

Solution Taking z-transform on both sides of the equation and making use of right shift property, we get

$$[1 - z^{-1} - z^{-2}]Y[z] = z^{-1}X[z]$$

(a)

$$H[z] = \frac{Y[z]}{X[z]}$$

$$\boxed{H[z] = \frac{z^{-1}}{(1 - z^{-1} - z^{-2})}}$$

(*b*)

$$
\begin{aligned}
H[z] &= \frac{z}{(z^{-2} - z - 1)} \\
\frac{H[z]}{z} &= \frac{1}{(z - 1.618)(z + 0.618)} \\
&= \frac{A_1}{(z - 1.618)} + \frac{A_2}{(z + 1.618)} \\
1 &= A_1(z + 0.618) + A_2(z - 1.618)
\end{aligned}
$$

Substitute $z = 1.618$

$$1 = A_1\,(1.618 + 0.618)\,; \qquad A_1 = 0.447$$

Substitute $z = -0.618$

$$1 = A_2\,(-0.618 - 1.618)\,; \qquad A_2 = -0.447$$

$$H[z] = 0.447\left[\frac{z}{z - 1.618} - \frac{z}{z + 0.618}\right]$$

$$\boxed{h[n] = 0.447\left[(1.618)^n - (-0.618)^n\right]u[n]}$$

Example 2.31 Find the impulse response of the discrete time system described by the difference equation

$$y[n-2] - 3y[n-1] + 2y[n] = x[n-1]$$

Solution

$$
\begin{aligned}
[z^{-2} - 3z^{-1} + 2]Y[z] &= z^{-1}X[z] \\
H[z] &= \frac{Y[z]}{X[z]} \\
&= \frac{z^{-1}}{(z^{-2} - 3z^{-1} + 2)} \\
&= \frac{z}{(2z^2 - 3z + 1)} \\
\frac{H[z]}{z} &= \frac{0.5}{(z-1)(z-0.5)} \\
&= \frac{1}{(z-1)} - \frac{1}{(z-0.5)} \\
H[z] &= \frac{z}{z-1} - \frac{z}{z-0.5}
\end{aligned}
$$

$$\boxed{h[n] = \left[(1)^n - \left(\frac{1}{2}\right)^n\right] u[n]}$$

Example 2.32 Determine the impulse response and frequency response of the system described by the difference equation

$$y[n] - \left(\frac{1}{6}\right) y[n-1] - \frac{1}{6} y[n-2] = x[n-1]$$

Solution To obtain Impulse Response

$$\left[1 - \frac{1}{6}z^{-1} - \frac{1}{6}z^{-2}\right] Y[z] = z^{-1} X[z]$$

$$H[z] = \frac{Y[z]}{X[z]}$$

$$= \frac{z}{\left(z^2 - \frac{1}{6}z - \frac{1}{6}\right)}$$

$$= \frac{z}{\left(z - \frac{1}{2}\right)\left(z + \frac{1}{3}\right)}$$

$$\frac{H[z]}{z} = \frac{1}{\left(z - \frac{1}{2}\right)\left(z + \frac{1}{3}\right)}$$

$$= \frac{6}{5}\left[\frac{1}{\left(z - \frac{1}{2}\right)} - \frac{1}{\left(z + \frac{1}{3}\right)}\right]$$

$$H[z] = \frac{6}{5}\left[\frac{z}{\left(z - \frac{1}{2}\right)} - \frac{1}{\left(z + \frac{1}{3}\right)}\right]$$

$$\boxed{h[n] = \frac{6}{5}\left[\left(\frac{1}{2}\right)^n - \left(\frac{1}{3}\right)^n\right] u[n]}$$

To obtain Frequency Response

Substitute $z = e^{j\omega}$ in $H[z]$

$$H[e^{j\omega}] = \frac{e^{j\omega}}{\left(e^{j\omega} - \frac{1}{2}\right)\left(e^{j\omega} + \frac{1}{3}\right)}$$

This can be expressed in terms of amplitude and phase as follows:

$$H[e^{j\omega}] = \frac{e^{j\omega}}{\left(\cos\omega + j\sin\omega - \frac{1}{2}\right)\left(\cos\omega + j\sin\omega + \frac{1}{3}\right)}$$

since $|e^{j\omega}| = 1$

$$|H(e^{j\omega})| = \frac{1}{\left[\left\{\left(\cos\omega - \frac{1}{2}\right)^2 + \sin^2\omega\right\}\left\{\left(\cos\omega + \frac{1}{3}\right)^2 + \sin^2\omega\right\}\right]^{\frac{1}{2}}}$$

Since $\angle e^{j\omega} = \omega$

$$\angle H(e^{j\omega}) = \omega - \tan^{-1}\frac{\sin\omega}{\left(\cos\omega - \frac{1}{2}\right)} - \tan^{-1}\frac{\sin\omega}{\left(\cos\omega + \frac{1}{3}\right)}$$

Example 2.33 A causal system is represented by the difference equation

$$y[n] + \frac{1}{4}y[n-1] = x[n] + \frac{1}{2}x[n-1]$$

use z-transform to determine the

1. System function;
2. Unit sample response of the system;
3. Frequency response of the system.

Solution 1.

$$\left[1 + \frac{1}{4}z^{-1}\right]Y[z] = \left[1 + \frac{1}{2}z^{-1}\right]X[z]$$

$$H[z] = \frac{Y[z]}{X[z]}$$

$$H[z] = \frac{\left[1 + \frac{1}{2}z^{-1}\right]}{\left[1 + \frac{1}{4}z^{-1}\right]}$$

2.

$$H[z] = \frac{\left(z + \frac{1}{2}\right)}{\left(z + \frac{1}{4}\right)}$$

$$\frac{H[z]}{z} = \frac{\left(z + \frac{1}{2}\right)}{z\left(z + \frac{1}{4}\right)} = \frac{A_1}{z} + \frac{A_2}{\left(z + \frac{1}{4}\right)}$$

$$\left(z + \frac{1}{2}\right) = A_1\left(z + \frac{1}{4}\right) + A_2 z$$

Substituting $z = 0$ we get,

$$A_1 = 2$$

Substituting $z = -\frac{1}{4}$ we get,

$$\left[-\frac{1}{4} + \frac{1}{2}\right] = A_2\left(-\frac{1}{4}\right); \qquad A_2 = -1$$

$$H[z] = 2 - \frac{z}{\left(z + \frac{1}{4}\right)}$$

$$\boxed{h[n] = 2\delta[n] - \left(\frac{1}{4}\right)^n u[n]}$$

3.

$$H[z] = \frac{\left(z + \frac{1}{2}\right)}{\left(z + \frac{1}{4}\right)}$$

$$H[e^{j\omega}] = \frac{\left(e^{j\omega} + \frac{1}{2}\right)}{\left(e^{j\omega} + \frac{1}{4}\right)} = \frac{\left(\cos\omega + \frac{1}{2}\right) + j\sin\omega}{\left(\cos\omega + \frac{1}{4}\right) + j\sin\omega}$$

$$\boxed{|H(e^{j\omega})| = \frac{\left[\left(\cos\omega + \frac{1}{2}\right)^2 + \sin^2\omega\right]^{1/2}}{\left[\left(\cos\omega + \frac{1}{4}\right)^2 + \sin\omega\right]^{1/2}}}$$

$$\boxed{\angle H(j\omega) = \tan^{-1}\frac{\sin\omega}{\left(\cos\omega + \frac{1}{2}\right)} - \tan^{-1}\frac{\sin\omega}{\left(\cos\omega + \frac{1}{4}\right)}}$$

Example 2.34 Find the output of the system whose input–output is related by the difference equation:

$$y[n] - \frac{5}{6}y[n-1] + \frac{1}{6}y[n-2] = x[n] - \frac{1}{2}x[n-1]$$

for the step input. Assume initial conditions to be zero.

Solution

$$\left[1 - \frac{5}{6}z^{-1} + \frac{1}{6}z^{-2}\right] Y[z] = \left[1 - \frac{1}{2}z^{-1}\right] X[z]$$

$$Y[z] = \frac{\left[1 - \frac{1}{2}z^{-1}\right]}{\left[1 - \frac{5}{6}z^{-1} + \frac{1}{6}z^{-2}\right]} X[z]$$

For unit step input, $X[z] = \frac{z}{z-1}$

$$Y[z] = \frac{z^2\left[z - \frac{1}{2}\right]}{(z-1)\left(z^2 - \frac{5}{6}z + \frac{1}{6}\right)}$$

$$\frac{Y[z]}{z} = \frac{z\left[z - \frac{1}{2}\right]}{(z-1)\left(z - \frac{1}{2}\right)\left(z - \frac{1}{3}\right)}$$

$$= \frac{z}{(z-1)\left(z - \frac{1}{3}\right)}$$

$$= \frac{A_1}{(z-1)} + \frac{A_2}{\left(z - \frac{1}{3}\right)}$$

$$z = A_1\left(z - \frac{1}{3}\right) + A_2(z-1)$$

Substituting $z = 1$ we get,

$$1 = A_1\left(1 - \frac{1}{3}\right); \quad A_1 = \frac{3}{2}$$

Substituting $z = \frac{1}{3}$ we get,

$$\frac{1}{3} = A_2\left(\frac{1}{3} - 1\right); \quad A_2 = -\frac{1}{2}$$

$$Y[z] = \frac{3}{2}\frac{z}{(z-1)} - \frac{1}{2}\frac{z}{\left(z - \frac{1}{3}\right)}$$

$$\boxed{y[n] = \left[\frac{3}{2}(1)^n - \frac{1}{2}\left(\frac{1}{3}\right)^n\right] u[n]}$$

Example 2.35 Find the output response of the discrete time system described by the following difference equation:

$$y[n] - \frac{3}{4}y[n-1] + \frac{1}{8}y[n-2] = x[n]$$

The initial conditions are $y[-1] = 0$ and $y = [-2] = 1$. The input $x[n] = \left(\frac{1}{5}\right)^n u[n]$.

Solution Taking z-transform on both sides of the above equation, we get

$$\begin{aligned} Y[z] - \frac{3}{4}[z^{-1}Y[z] + y[-1]] + \frac{1}{8}[z^{-2}Y[z] \\ +z^{-1}y[-1] + y[-2]] = X[z] \\ \left[1 - \frac{3}{4}z^{-1} + \frac{1}{8}z^{-2}\right]Y[z] = -\frac{1}{8} + \frac{z}{\left(z - \frac{1}{5}\right)} \\ \frac{\left[z^2 - \frac{3}{4}z + \frac{1}{8}\right]}{z^2}Y[z] = -\frac{1}{8} + \frac{z}{\left(z - \frac{1}{5}\right)} \end{aligned}$$

$$\begin{aligned} \frac{Y[z]}{z} &= -\frac{z}{8\left(z - \frac{1}{4}\right)\left(z - \frac{1}{2}\right)} + \frac{z^2}{\left(z - \frac{1}{5}\right)\left(z - \frac{1}{4}\right)\left(z - \frac{1}{2}\right)} \\ &= Y_1[z] + Y_2[z] \\ Y_1[z] &= -\frac{z}{8\left(z - \frac{1}{4}\right)\left(z - \frac{1}{2}\right)} \\ &= -\frac{A_1}{\left(z - \frac{1}{4}\right)} + \frac{A_2}{\left(z - \frac{1}{2}\right)} \\ -\frac{z}{8} &= A_1\left(z - \frac{1}{2}\right) + A_2\left(z - \frac{1}{4}\right) \end{aligned}$$

Substituting $z = \frac{1}{4}$ we get,

$$-\frac{1}{4}\frac{1}{8} = -A_1\frac{1}{4}; \qquad A_1 = \frac{1}{8}$$

Substituting $z = \frac{1}{2}$ we get,

$$-\frac{1}{2}\frac{1}{8} = A_2\frac{1}{4}; \qquad A_2 = -\frac{1}{4}$$

$$Y_1[z] = \frac{1}{8\left(z-\frac{1}{4}\right)} - \frac{1}{4\left(z-\frac{1}{2}\right)}$$

$$Y_2[z] = \frac{z^2}{\left(z-\frac{1}{5}\right)\left(z-\frac{1}{4}\right)\left(z-\frac{1}{2}\right)}$$

$$z^2 = A_1\left(z-\frac{1}{4}\right)\left(z-\frac{1}{2}\right) + A_2\left(z-\frac{1}{5}\right)\left(z-\frac{1}{2}\right) + A_3\left(z-\frac{1}{5}\right)\left(z-\frac{1}{4}\right)$$

Substituting $z = \frac{1}{5}$ we get,

$$\frac{1}{25} = A_1\left(\frac{1}{5}-\frac{1}{4}\right)\left(\frac{1}{5}-\frac{1}{2}\right); \qquad A_1 = \frac{8}{3}$$

Substituting $z = \frac{1}{4}$ we get,

$$\frac{1}{16} = A_2\left(\frac{1}{4}-\frac{1}{5}\right)\left(\frac{1}{4}-\frac{1}{2}\right); \qquad A_2 = -5$$

Substituting $z = \frac{1}{2}$ we get,

$$\frac{1}{4} = A_3\left(\frac{1}{2}-\frac{1}{5}\right)\left(\frac{1}{2}-\frac{1}{4}\right); \qquad A_3 = \frac{10}{3}$$

$$Y_2[z] = \frac{8}{3}\frac{1}{\left(z-\frac{1}{5}\right)} - \frac{5}{\left(z-\frac{1}{4}\right)} + \frac{10}{3}\frac{1}{\left(z-\frac{1}{2}\right)}$$

$$Y[z] = \frac{z}{8\left(z-\frac{1}{4}\right)} - \frac{z}{4\left(z-\frac{1}{2}\right)} + \frac{8}{3}\frac{z}{\left(z-\frac{1}{5}\right)} - \frac{5z}{\left(z-\frac{1}{4}\right)} + \frac{10}{3}\frac{z}{\left(z-\frac{1}{2}\right)}$$

$$= -\frac{39}{8}\frac{z}{\left(z-\frac{1}{4}\right)} + \frac{37}{12}\frac{z}{\left(z-\frac{1}{2}\right)} + \frac{8}{3}\frac{z}{\left(z-\frac{1}{5}\right)}$$

$$\boxed{y[n] = \left[-\frac{39}{8}\left(\frac{1}{4}\right)^n + \frac{37}{12}\left(\frac{1}{2}\right)^n + \frac{8}{3}\left(\frac{1}{5}\right)^n\right]u[n]}$$

Example 2.36 Consider the following difference equation:

$$y[n] + 2y[n-1] + 2y[n-2] = x[n]$$

The initial conditions are $y[-1] = 0$ and $y = [-2] = 2$. Find the step response of the system.

Solution Taking z-transform on both sides of the above equation, we get

$$\begin{aligned} X[z] &= Y[z] + 2[z^{-1}Y[z] + y[-1]] + 2[z^{-2}Y[z] + z^{-1}y[-1] + y[-2]] \\ -4 + X[z] &= [1 + 2z^{-1} + 2z^{-2}]Y[z] \\ -4 + X[z] &= \frac{(z^2 + 2z + 2)}{z^2}Y[z] \end{aligned}$$

For step input $X[z] = \frac{z}{(z-1)}$

$$\begin{aligned} z^2 + 2z + 2 &= (z + 1 + j)(z + 1 - j) \\ \frac{(z + 1 + j)(z + 1 - j)}{z^2}Y[z] &= -4 + \frac{z}{z - 1} \\ &= \frac{(4 - 3z)}{z - 1} \\ \frac{Y[z]}{z} &= \frac{z(4 - 3z)}{(z - 1)(z + 1 + j)(z + 1 - j)} \\ &= \frac{A_1}{(z - 1)} + \frac{A_2}{(z + 1 + j)} + \frac{A_3}{(z + 1 - j)} \\ z[4 - 3z] &= A_1(z^2 + 2z + 2) + A_2(z - 1)(z + 1 - j) \\ &\quad + A_3(z - 1)(z + 1 + j) \end{aligned}$$

Substitute $z = 1$

$$1 = A_1 5; \qquad A_1 = \frac{1}{5}$$

Substitute $z = -1 + j$

$$\begin{aligned} (-1 + j)(4 - 3 + j3) &= A_3(-1 + j - 1)(-1 + 1 + j + j) \\ (-1 + j)(1 + j3) &= A_3(-2 + j)j2 \\ \sqrt{2}\angle 135^\circ \sqrt{10}\angle 71.56^\circ &= A_3\sqrt{5}\angle 153.43^\circ \sqrt{2}\angle 90^\circ \\ A_3 &= \frac{\sqrt{2}\angle 135^\circ \sqrt{10}\angle 71.56^\circ}{\sqrt{5}\angle 153.43^\circ \sqrt{2}\angle 90^\circ} \\ &= 1\angle -36.87^\circ = 1e^{-j0.643} \end{aligned}$$

$$A_2 = \text{conjugate of } A_3 = 1e^{j0.643}$$

The exponentials of A_1 and A_2 are expressed in radians using $57.3^\circ = 1$ radian.

$$Y[z] = \frac{1}{5}\frac{z}{(z-1)} + \frac{e^{j0.643}z}{z+1+j} + \frac{e^{-j0.643}z}{z+1-j}$$

$$Y[z] = \frac{1}{5}\frac{z}{(z-1)} + \frac{e^{j0.643}z}{(z+\sqrt{2}e^{j\frac{\pi}{4}})} + \frac{e^{-j0.643}z}{(z+\sqrt{2}e^{-j\frac{\pi}{2}})}$$

Taking inverse z-transform, we get

$$\begin{aligned} y[n] &= \frac{1}{5} + e^{j0.643}(-\sqrt{2}e^{j\frac{\pi}{4}})^n + e^{-j0.643}(-\sqrt{2}e^{-j\frac{\pi}{4}})^n \\ &= \frac{1}{5} + (-\sqrt{2})^n\left[e^{j(0.643+\frac{\pi}{4}n)} + e^{-j(0.643+\frac{\pi}{4}n)}\right] \end{aligned}$$

$$\boxed{y[n] = \left[\frac{1}{5} + 2(-\sqrt{2})^n\cos\left(\frac{\pi}{4}n + 0.643\right)\right]u[n]}$$

Example 2.37 Solve the following difference equation:

$$y[n] + 6y[n-1] + 8y[n-2] = 5x[n-1] + x[n-2]$$

The initial conditions are $y[-1] = 1$ and $y[-2] = 2$. The input $x[n] = u[n]$.

Solution Taking z-transform on both sides, we get

$$1 + 6(z^{-1}Y[z] + y[-1]) + 8(z^{2}Y[z] + z^{-1}y[-1] + y[-2]) = [5z^{-1} + z^{-2}]X[z]$$

For a causal signal $u[n]$, $x[-2]$, $x[-1]$ are zero.

$$[1 + 6z^{-1} + 8z^{-2}]Y[z] + (6 + 8z^{-1} + 16) = [5z^{-1} + z^{-2}]\frac{z}{(z-1)}$$

$$\begin{aligned} \frac{(z+2)(z+4)}{z^2}Y[z] &= -(22 + 8z^{-1}) + (5z^{-1} + z^{-2})\frac{z}{(z-1)} \\ &= \frac{(-22z^2 + 19z + 9)}{z(z-1)} \\ \frac{Y[z]}{z} &= \frac{(-22z^2 + 19z + 9)}{(z-1)(z+2)(z+4)} \\ &= \frac{A_1}{(z-1)} + \frac{A_2}{(z+2)} + \frac{A_3}{(z+4)} \\ -22z^2 + 19z + 9 &= A_1(z+2)(z+4) + A_2(z-1)(z+4) \\ &\quad + A_3(z-1)(z+2) \end{aligned}$$

Substitute $z = 1$

$$-22 + 19 + 9 = A_1(3)(5); \qquad A_1 = 0.4$$

Substitute $z = -2$

$$-88 - 38 + 9 = A_2(-3)(2); \qquad A_2 = 19.5$$

Substitute $z = -4$

$$-352 - 76 + 9 = A_3(-5)(-2); \qquad A_3 = -41.9$$

$$Y[z] = \frac{0.4z}{(z-1)} + 19.5\frac{z}{(z+2)} - 41.9\frac{z}{(z+4)}$$

$$\boxed{y[n] = \left[0.4 + 19.5(-2)^n - 41.9(-4)^n\right] u[n]}$$

Example 2.38 Find the response of the LTID system described by the following difference equation:

$$y[n+2] + y[n+1] + 0.24y[n] = x[n+1] + 2x[n]$$

where $x[n] = (\frac{1}{2})^n u[n]$ and all the initial conditions are zero.

Solution When the initial conditions are zero,

$$y[n+2] \overset{Z}{\longleftrightarrow} z^2 Y[z]$$
$$y[n+1] \overset{Z}{\longleftrightarrow} zY[z]$$
$$x[n+1] \overset{Z}{\longleftrightarrow} zX[z]$$
$$\left(\frac{1}{2}\right)^2 u[n] \overset{Z}{\longleftrightarrow} \frac{z}{(z-0.5)}$$

The given difference equation can be written in the following form after taking z-transform on both sides.

$$[z^2 + z + 0.24]Y[z] = [z+2]\frac{z}{(z-0.5)}$$
$$(z^2 + z + 0.24) = (z+0.6)(z+0.4)$$
$$\frac{Y[z]}{z} = \frac{(z+2)}{(z-0.5)(z+0.6)(z+0.4)}$$
$$= \frac{A_1}{(z-0.5)} + \frac{A_2}{(z+0.6)} + \frac{A_3}{(z+0.4)}$$
$$(z+2) = A_1(z+0.6)(z+0.4) + A_2(z-0.5)(z+0.4) + A_3(z-0.5)(z+0.6)$$

Substitute $z = 0.5$

$$2.5 = A_1(1.1)(0.9); \qquad A_1 = 2.525$$

Substitute $z = -0.6$

$$1.4 = A_2(-1.1)(-0.2); \qquad A_2 = 6.36$$

Substitute $z = -0.4$

$$1.6 = A_3(-0.9)(-0.2); \qquad A_3 = -8.89$$

$$Y[z] = 2.525\frac{z}{(z-0.5)} + 6.36\frac{z}{(z+0.6)} - 8.89\frac{z}{(z+0.4)}$$

$$\boxed{y[n] = \left[2.525(0.5)^n + 6.36(-0.6)^n - 8.89(-0.4)^n\right]u[n]}$$

Example 2.39 Consider the following difference equation:

$$y[n+2] - 5y[n+1] + 6y[n] = x[n+1] + 4x[n]$$

The auxiliary conditions are as follows $y[0] = 1$ and $y[1] = 2$ and the input $x[n] = u[n]$. Solve for $y[n]$.

Solution

$$\begin{aligned} y[n+2] &\overset{Z}{\longleftrightarrow} z^2Y[z] - z^2y(0) - zy(1) \\ &= z^2Y[z] - z^2 - 2z \\ y[n+1] &\overset{Z}{\longleftrightarrow} zY[z] - zy[0] \\ &= zY[z] - z \\ x[n+1] &\overset{Z}{\longleftrightarrow} zX[z] - zx[0] \\ &= zX[z] - z \end{aligned}$$

Taking z-transform on both sides of the above equation and substituting $X[z] = \frac{z}{(z-1)}$, we get

$$\begin{aligned} [z^2 - 5z + 6]Y[z] &= z^2 + 2z - 5z + (z+4)\frac{z}{(z-1)} - z \\ (z-2)(z-3)Y[z] &= \frac{z(z-4)(z-1) + z(z+4)}{(z-1)} \\ \frac{Y[z]}{z} &= \frac{(z^2 - 4z + 8)}{(z-1)(z-2)(z-3)} \\ &= \frac{A_1}{(z-1)} + \frac{A_2}{(z-2)} + \frac{A_3}{(z-3)} \end{aligned}$$

$$(z^{-2} - 4z + 8) = A_1(z-2)(z-3) + A_2(z-1)(z-3) + A_3(z-1)(z-2)$$

Substitute $z = 1$

$$1 - 4 + 8 = A_1(-1)(-2); \qquad A_1 = 2.5$$

Substitute $z = 2$

$$4 - 8 + 8 = A_2(-1); \qquad A_2 = -4$$

Substitute $z = 3$

$$9 - 12 + 8 = A_3(2)(1); \qquad A_3 = 2.5$$

$$Y[z] = 2.5\frac{z}{(z-1)} - 4\frac{z}{(z-2)} + 2.5\frac{z}{(z-3)}$$

$$\boxed{y[n] = \left[2.5 - 4(2)^n + 2.5(3)^n\right]u[n]}$$

Example 2.40 Solve the following difference equation:

$$y[n+2] - 9y[n+1] + 20y[n] = 4x[n+1] + 2x[n]$$

The input $x[n] = (\frac{1}{2})^n u[n]$. The initial conditions are $y[-1] = 2$ and $y[-2] = 1$.

Solution The given difference equation is in advanced operator form which requires the knowledge of $y[1]$ and $y[2]$. Therefore, the given equation is converted into delay operator form as described below and the given initial conditions are applied. Replacing n with $(n-2)$, the given difference equation is converted as

$$y[n] - 9y[n-1] + 20y[n-2] = 4x[n-1] + 2x[n-2]$$

Since the input is causal, $x[-1] = x[-2] = 0$. Taking z-transform on both sides of the above equation, we get

$$\begin{aligned}
&Y[z] - 9[z^{-1}Y[z] + y[-1]] + 20[z^{-2}Y[z] + z^{-1}y[-1] + y[-2]] \\
&= 4[z^{-1}X[z] + x[-1]] + 2[z^{-2}X[z] + z^{-1}x[-1] + z^{-2}x[-2]]] \\
&= [4z^{-1} + 2z^{-2}]X[z] \\
&= [1 - 9z^{-1} + 20z^{-2}]Y[z] - 18 + 40z^{-1} + 20 = (4z^{-1} + 2z^{-2})X[z] \\
&\frac{[z^2 - 9z + 20]}{z^2}Y[z] = -(2 + 40z^{-1}) + (4z^{-1} + 2z^{-2})X[z]
\end{aligned}$$

Substitute $(z^2 - 9z + 20) = (z-4)(z-5)$ and $X[z] = \frac{z}{(z-0.5)}$

$$\frac{Y[z]}{z} = \frac{(-2z^2 - 35z + 22)}{(z-0.5)(z-4)(z-5)}$$
$$= \frac{A_1}{(z-0.5)} + \frac{A_2}{(z-4)} + \frac{A_3}{(z-5)}$$
$$(-2z^{-2} - 35z + 22) = A_1(z-4)(z-5) + A_2(z-0.5)(z-5) + A_3(z-0.5)(z-4)$$

Substitute $z = 0.5$

$$-0.5 - 17.5 + 22 = A_1(-3.5)(-4.5); \qquad A_1 = 0.254$$

Substitute $z = 4$

$$-32 - 140 + 22 = A_2(3.5)(-1); \qquad A_2 = 42.86$$

Substitute $z = 5$

$$-50 - 175 + 22 = A_3(4.5); \qquad A_3 = -45.1$$

$$Y[z] = \frac{0.254z}{(z-0.5)} + \frac{42.86z}{(z-4)} - \frac{45.1z}{(z-5)}$$

$$\boxed{y[n] = \left[0.254(0.5)^n + 42.86(4)^n - 45.1(5)^n\right] u[n]}$$

2.14 Difference Equation from System Function

Let the system function $H[z]$ be expressed as

$$\frac{Y[z]}{X[z]} = H[z] = \frac{b_0 z^N + b_1 z^{N-1} + \cdots + b_{N-1}z + b_N}{z^N + a_1 z^{N-1} + \cdots + a_{N-1}z + a_N}$$

Cross multiplying and operating z on $Y[z]$ and $X[z]$, we get

$$y[n+N] + a_1 y[n+N-1] + \cdots + a_{N-1}y[n+1] + a_N y[n] = b_0 x[n+N] + b_1 x[n+N-1] + \cdots + b_{N-1}x[n+1] + b_N x[n] \tag{2.58}$$

Similar procedure has to be followed if the system frequency response $H(e^{j\omega})$ is given. Here $e^{j\omega}$ has to be treated as z. The following examples demonstrate the above methods.

Example 2.41 For the system functions given below, determine the difference equation

$$(a) \quad H[z] = \frac{(1 - z^{-1})}{\left(1 - \frac{1}{2}z^{-1} + \frac{1}{4}z^{-2}\right)}$$

$$(b) \quad H[z] = \frac{(z - 1)}{(z + 1)(z - 2)}$$

$$(c) \quad H[z] = \frac{1}{\left(1 - \frac{1}{4}z^{-1}\right)}$$

(*d*) Consider the system consisting of the cascade of two LTI systems with frequency responses

$$H_1(e^{j\omega}) = \frac{2 - e^{j\omega}}{\left(1 + \frac{1}{2}e^{-j\omega}\right)}$$

$$H_2(e^{j\omega}) = \frac{1}{\left(1 - \frac{1}{2}e^{-j\omega} + \frac{1}{4}e^{-j2\omega}\right)}$$

Find the difference equation describing the overall system.

(*e*) Write a difference equation that characterizes a system whose frequency response is

$$H(e^{j\omega}) = \frac{\left(1 - e^{-j\omega} + e^{-3j\omega}\right)}{\left(1 + \frac{1}{2}e^{-j\omega} + \frac{3}{4}e^{-2j\omega}\right)}$$

Solution (*a*) $\boldsymbol{H[z] = \frac{(1-z^{-1})}{(1-\frac{1}{2}z^{-1}+\frac{1}{4}z^{-2})}}$

$$\frac{Y[z]}{X[z]} = \frac{\left(1 - z^{-1}\right)}{\left(1 - \frac{1}{2}z^{-1} + \frac{1}{4}z^{-2}\right)}$$

$$Y[z] - \frac{1}{2}z^{-1}Y[z] + \frac{1}{4}z^{-2}Y[z] = X[z] - z^{-1}X[z]$$

$$\boxed{y[n] - \frac{1}{2}y[n-1] + \frac{1}{4}y[n-2] = x[n] - x[n-1]}$$

(*b*) $\boldsymbol{H[z] = \frac{(z-1)}{(z+1)(z-2)}}$

$$\frac{Y[z]}{X[z]} = \frac{(z-1)}{(z+1)(z-2)}$$
$$= \frac{(z-1)}{(z^2-z-2)}$$
$$z^2Y[z] - zY[z] - 2Y[z] = zX[z] - X[z]$$

$$\boxed{y[n+2] - y[n+1] - 2y[n] = x[n+1] - x[n]}$$

(c) $\boldsymbol{H[z] = \frac{Y[z]}{X[z]} = \frac{1}{(1-\frac{1}{4}z^{-1})}}$

$$\left[1 - \frac{1}{4}z^{-1}\right]Y[z] = X[z]$$

$$\boxed{y[n] - \frac{1}{4}y[n-1] = x[n]}$$

(d) $\boldsymbol{H_1(e^{j\omega}) = \frac{2-e^{j\omega}}{(1+\frac{1}{2}e^{-j\omega})}}$ **and** $\boldsymbol{H_2(e^{j\omega}) = \frac{1}{(1-\frac{1}{2}e^{-j\omega}+\frac{1}{4}e^{-j2\omega})}}$

$$H_1H_2(e^{j\omega}) = \frac{Y(j\omega)}{X(j\omega)}$$
$$= \frac{(2-e^{-j\omega})}{\left(1+\frac{1}{2}e^{-j\omega}\right)\left(1-\frac{1}{2}e^{-j\omega}+\frac{1}{4}e^{-j2\omega}\right)}$$
$$= \frac{(2-e^{-j\omega})}{\left(1-\frac{1}{2}e^{-j\omega}+\frac{1}{4}e^{-j2\omega}+\frac{1}{2}e^{-j\omega}-\frac{1}{4}e^{-j2\omega}+\frac{1}{8}e^{-j3\omega}\right)}$$
$$Y[e^{j\omega}]\left[1+\frac{1}{8}e^{-j3\omega}\right] = [2-e^{-j\omega}]X[e^{j\omega}]$$

$$\boxed{y[n] + \frac{1}{8}y[n-3] = 2x[n] - x[n-1]}$$

(e) $\boldsymbol{\frac{Y[e^{j\omega}]}{X[e^{j\omega}]} = H(e^{j\omega}) = \frac{(1-e^{-j\omega}+e^{-3j\omega})}{(1+\frac{1}{2}e^{-j\omega}+\frac{3}{4}e^{-2j\omega})}}$

$$\left[1+\frac{1}{2}e^{-j\omega}+\frac{3}{4}e^{-2j\omega}\right]Y[e^{j\omega}] = \left[1-e^{-j\omega}+e^{-3j\omega}\right]X[e^{j\omega}]$$

$$\boxed{y[n] + \frac{1}{2}y[n-1] + \frac{3}{4}y[n-2] = x[n] - x[n-1] + x[n-3]}$$

Example 2.42 Obtain the difference equation for the block diagram shown in Fig. 2.18.

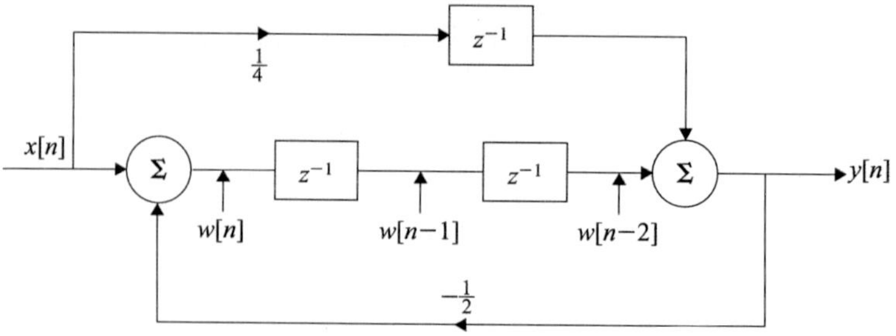

Fig. 2.18 Block diagram of Example 2.42

Solution From Fig. 2.18, the following equations are written:

$$w[n] = x[n] - \frac{1}{2}y[n]$$

Replace n by $(n-2)$

$$\begin{aligned} w[n-2] &= x[n-2] - \frac{1}{2}y[n-2] \\ y[n] &= \frac{1}{4}x[n-1] + w[n-2] \\ &= \frac{1}{4}x[n-1] + x[n-2] - \frac{1}{2}y[n-2] \end{aligned}$$

$$\boxed{y[n] + \frac{1}{2}y[n-2] = \frac{1}{4}x[n-1] + x[n-2]}$$

2.15 Introduction to Fourier Transform

The Fourier series representations of DT signals discussed above are applicable only if the signal is periodic. If the signal is non-periodic, then applying a limiting process the aperiodic continuous time signal can be expressed as a continuous sum of everlasting exponential or sinusoids and this method is termed as Fourier transform of discrete time signal. By applying limiting process to aperiodic signal $x[n]$, it can be expressed as a sum of everlasting exponentials. The spectrum $X[\Omega]$ so obtained is called Discrete Time Fourier Transform (DTFT). If the spectrum obtained by DTFT is sampled for one period of the Fourier transform at a finite number of frequency points, such a transformation is called Discrete Fourier Transform (DFT) which is a very powerful computational tool for the evaluation of FT. Some special algorithms are developed for the easy implementation of DFT which result in saving considerable computation time. Such algorithms are called Fast Fourier Transform (FFT).

The detailed study of DTFT is discussed in this chapter with sufficient illustrated examples.

2.16 Representation of Discrete Time Aperiodic Signals

Consider the aperiodic signal $x[n]$ shown in Fig. 2.19a. The periodic signal $x_{N_0}[n]$ is constructed by repeating the signal $x[n]$ every N_0 units as shown in Fig. 2.19a. The period N_0 is chosen large enough to avoid overlapping. If we put $N_0 \longrightarrow \infty$, the signal repeats after an infinite interval and therefore

$$\underset{N_0 \to \infty}{Lt} x_{N_0}[n] = x[n]$$

For a discrete signal, the FS can be written as

$$x[n] = \sum_{k=[N_0]} D_k e^{jk\Omega_0 n}$$

$$x_{N_0}[n] = \sum_{k=[N_0]} D_k e^{jk\Omega_0 n} \tag{2.59}$$

where $\Omega_0 = \frac{2\pi}{N_0}$ and

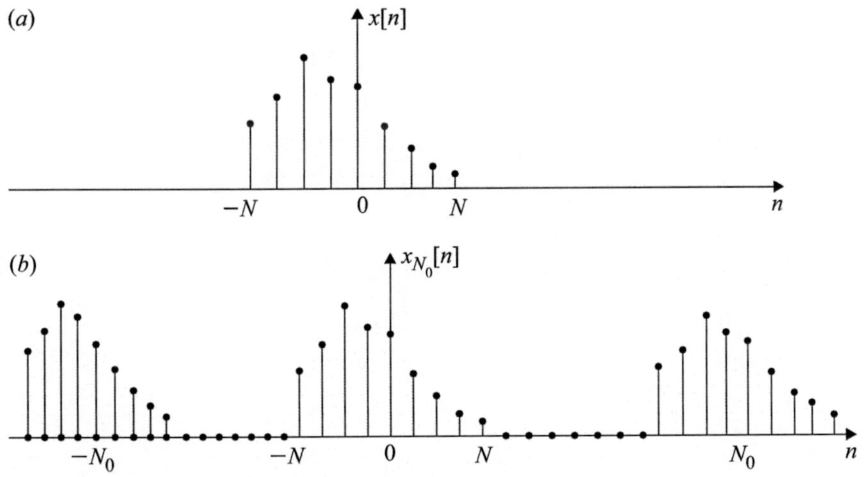

Fig. 2.19 Extension of aperiodic signal to periodic signal

$$D_k = \frac{1}{N_0} \sum_{n=-\infty}^{\infty} x[n]e^{-jk\Omega_0 n} \tag{2.60}$$

With Ω as continuous function, let us define

$$X(\Omega) = \sum_{n=-\infty}^{\infty} x[n]e^{-j\Omega n} \tag{2.61}$$

Substituting Eq. (2.61) in Eq. (2.60), we get

$$D_k = \frac{1}{N_0} X(k\Omega_0) \tag{2.62}$$

Equation (2.62) shows that the Fourier coefficients D_k are $\frac{1}{N_0}$ times the samples of $X(\Omega)$. As $N_0 \to \infty$, the fundamental frequency $\Omega_0 \to 0$ and $D_k \to 0$ and the spectrum becomes continuous. Now consider Eq. (2.61)

$$X(\Omega) = \sum_{n=-\infty}^{\infty} x[n]e^{-jk\Omega_0 n} \tag{2.63}$$

Equation (2.59) can be expressed using Eq. (2.60) as

$$\begin{aligned} x_{N_0}[n] &= \frac{1}{N_0} \sum_{k=[N_0]} X(k\Omega_0)e^{jk\Omega_0 n} \\ &= \sum_{k=[N_0]} X(k\Omega_0)e^{jk\Omega_0 n} \left(\frac{\Omega_0}{2\pi}\right) \end{aligned} \tag{2.64}$$

As $N_0 \to \infty,\ \Omega_0 \to 0$ and $x_{N_0}[n] \to x[n]$

$$x[n] = \underset{\Omega_0 \to 0}{Lt} \sum_{k=N_0} \left[X(k\Omega_0)\frac{\Omega_0}{2\pi}\right] e^{jk\Omega_0 n} \tag{2.65}$$

Since Ω_0 is small, it can be replaced by Ω. Thus,

$$\Delta\Omega = \frac{2\pi}{N_0} \tag{2.66}$$

$$x[n] = \underset{\Delta\Omega_0 \to 0}{Lt} \frac{1}{2\pi} \sum_{k=N_0} X(k\Delta\Omega)e^{jk\Delta\Omega n}\Delta\Omega \tag{2.67}$$

$k = N_0$ implies $N_0\Delta\Omega = 2\pi$. Hence, Eq. (2.67) becomes the integral

$$\boxed{x[n] = \frac{1}{2\pi}\int_{2\pi} X(\Omega)e^{jn\Omega}d\Omega} \tag{2.68}$$

The spectrum $X(\Omega)$ is given by

$$\boxed{X(\Omega) = \sum_{n=-\infty}^{\infty} x[n]e^{-j\Omega n}} \tag{2.69}$$

Equation (2.68) is called the Fourier integral and $X(\Omega)$ is called the Discrete Time Fourier Transform (DTFT). They are called DFTF pair. Symbolically, they are represented as

$$\begin{aligned} x[n] &= \text{IDTFT}\{X(\Omega)\} \\ X(\Omega) &= \text{DTFT}\{x[n]\} \end{aligned} \tag{2.70}$$

Or

$$x[n] \overset{\text{DTFT}}{\longleftrightarrow} X(\Omega)$$

The Fourier transform $X(\Omega)$ is nothing but the description of $x[n]$ in the frequency domain. From Eq. (2.70), it is proved that the spectrum of a discrete time signal is periodic with fundamental period N_0.

As in the case of continuous time signal, the sufficient condition for the convergence of $X(\Omega)$ is that $x[n]$ is either absolutely summable. That is

$$\sum_{n=-\infty}^{\infty} |x[n]| < \infty$$

or the sequence has finite energy, that is

$$\sum_{n=-\infty}^{\infty} |x[n]|^2 < \infty. \tag{2.71}$$

2.17 Connection Between the Fourier Transform and the z-transform

From Eq. (2.70)

$$X(\Omega) = \sum_{n=-\infty}^{\infty} x[n]e^{-j\Omega n} \tag{2.72}$$

The z-transform of $x[n]$ is given by

$$X[z] = \sum_{n=-\infty}^{\infty} x[n]z^{-n} \tag{2.73}$$

From Eqs. (2.72) and (2.73), we see that if the ROC of $X[z]$ contains unit circle, then $X(\Omega)$ equals $X[z]$ evaluated on the unit circle. That is

$$X(\Omega) = X[z]\Big|_{z=e^{j\Omega}} \tag{2.74}$$

Note: Using Eq. (2.74), one can obtain Fourier transform by substituting $z = e^{j\Omega}$ provided $x[n]$ is summable. If $x[n]$ is not summable, as in the case of $u[n]$, one cannot obtain $X(\Omega)$ form $X[z]$.

When Fourier transform and z-transform are connected, $X(\Omega)$ is denoted by $X(e^{j\Omega})$ or $X(e^{j\omega})$.

Example 2.43 Find the FT of the following DT signals:

1. $x[n] = \delta[n]$
2. $x[n] = a^n u[n]$
3. $x[n] = -a^n u[-n-1]$
4. $x[n] = u[n]$
5. $x[n] = (a)^{|n|} \quad |a| < 1$

Solution 1. $\boldsymbol{x[n] = \delta[n]}$

$$F\{\delta[n]\} = \sum_{-\infty}^{\infty} \delta[n]e^{-j\Omega n}$$

$$\delta[n] = \begin{cases} 1 & n = 0 \\ 0 & n \neq 0 \end{cases}$$

$$\boxed{F\{\delta[n]\} = 1}$$

2. $\boldsymbol{x[n] = a^n u[n]}$

$$X[\Omega] = \sum_{n=-\infty}^{\infty} a^n u[n]e^{-j\Omega n}$$

$$u[n] = \begin{cases} 1 & n \geq 0 \\ 0 & n < 0 \end{cases} \quad \text{since } x[n] \text{ is causal}$$

$$X[\Omega] = \sum_{n=0}^{\infty} a^n u e^{-j\Omega n} = \sum_{n=0}^{\infty} [ae^{-j\Omega}]^n$$

By using the summation formula, we get

$$\boxed{X[\Omega] = \frac{1}{(1 - ae^{-j\Omega})}} \quad |a| < 1$$

3. $\boldsymbol{x[n] = -a^n u[-n-1]}$

$$X[\Omega] = \sum_{n=-\infty}^{-1} -a^n u[-n-1] e^{-j\Omega n} = \sum_{n=-\infty}^{-1} -a^n e^{-j\Omega n}$$
$$= \sum_{n=1}^{\infty} -a^{-n} e^{j\Omega n} = -\sum_{n=1}^{\infty} (a^{-1} e^{j\Omega})^n$$
$$= -\{a^{-1}e^{j\Omega} + (a^{-1}e^{j\Omega})^2 + (a^{-1}e^{j\Omega})^3 + \cdots\}$$
$$= -a^{-1}e^{j\Omega}\left[1 + (a^{-1}e^{j\Omega}) + (a^{-1}e^{j\Omega})^2 + \cdots\right]$$
$$= -\frac{a^{-1}e^{j\Omega}}{(1 - a^{-1}e^{j\Omega})}$$

$$\boxed{X(\Omega) = \frac{1}{(1 - ae^{-j\Omega})}} \quad |a| > 1$$

4. $\boldsymbol{x[n] = u[n]}$
Let $x[n] = u[n] \longleftrightarrow X(\Omega)$

$$X(\Omega) = \sum_{n=0}^{\infty} u[n] e^{-j\Omega n}$$
$$X(\Omega) = \sum_{n=0}^{\infty} (e^{-j\Omega})^n$$
$$\neq \frac{1}{(1 - e^{-j\Omega})}$$

which is obtained using summation formula because $|e^{j\Omega_0}| = 1$ and $X(\Omega)$ is not summable. The following procedure is followed to evaluate the FT of a causal step sequence.

$$\delta[n] = u[n] - u[n-1]$$
$$1 = [1 - e^{-j\Omega}]X(\Omega)$$

For $\Omega = 0$, $(1 - e^{-j\Omega}) = 0$. Therefore, $X(\Omega)$ must be written in the following form:

$$X(\Omega) = C\delta(\Omega) + \frac{1}{(1 - e^{-j\Omega})}$$

where C is any constant. The step sequence $u[n]$ can be expressed in terms of odd and even components as

$$x_e[n] = \frac{1}{2} + \frac{1}{2}\delta[n]$$
$$x_O[n] = x[n] - x_e[n]$$
$$= x[n] - \frac{1}{2} - \frac{1}{2}\delta[n]$$
$$F[x_O[n]] = C\delta(\Omega) + \frac{1}{(1 - e^{-j\Omega})} - \pi\delta(\Omega) - \frac{1}{2}$$

From the property of FT, the FT of an odd sequence must be imaginary. To satisfy the condition, $C = \pi$

$$F[x_O[n]] = \frac{1}{(1 - e^{-j\Omega})} - \frac{1}{2}$$
$$F[x_e[n]] = \pi\delta[\Omega] + \frac{1}{2}$$
$$F[x[n]] = F[x_e[n]] + F[x_O[n]]$$
$$= \pi\delta(\Omega) + \frac{1}{2} + \frac{1}{(1 - e^{-j\Omega})} - \frac{1}{2}$$

$$\boxed{u[n] \longleftrightarrow \pi\delta(\Omega) + \frac{1}{(1 - e^{-j\Omega})}}$$

5. $\boldsymbol{x[n] = (a)^{|n|}}$ $\boldsymbol{|a| < 1}$

$$X(\Omega) = \sum_{n=-\infty}^{\infty} (a)^{|n|} e^{-j\Omega n}$$
$$= \sum_{n=-\infty}^{1} a^{-n} e^{-j\Omega n} + \sum_{n=0}^{\infty} a^{n} e^{-j\Omega n}$$
$$= X_1(\Omega) + X_2(\Omega)$$
$$X_2(\Omega) = \frac{1}{(1 - ae^{-j\Omega})} \qquad a < 1$$

$$X_1(\Omega) = \sum_{n=1}^{\infty}(ae^{j\Omega})^n$$
$$= ae^{j\Omega} + (ae^{j\Omega})^2 + (ae^{j\Omega})^3 + \cdots$$
$$= ae^{j\Omega}[1 + ae^{j\Omega} + (ae^{j\Omega})^2 + \cdots]$$
$$= \frac{ae^{j\Omega}}{(1 - ae^{j\Omega})} \qquad a < 1$$
$$X(\Omega) = \frac{ae^{j\Omega}}{(1 - ae^{j\Omega})} + \frac{1}{(1 - ae^{-j\Omega})}$$

$$\boxed{X(\Omega) = \frac{(1 - a^2)}{(1 - 2a\cos\Omega + a^2)}}$$

2.18 Properties of Discrete Time Fourier Transform

The properties of DTFT are very similar to those of CTFT and these properties are very useful to determine the Fourier transform and inverse Fourier transforms very quickly. They are discussed below without proofs.

2.18.1 Linearity

If

$$x_1[n] \overset{\text{DTFT}}{\longleftrightarrow} X_1(\Omega) \quad \text{and} \quad x_2[n] \overset{\text{DTFT}}{\longleftrightarrow} X_2(\Omega)$$

then

$$Ax_1[n] + Bx_2[n] \overset{\text{DTFT}}{\longleftrightarrow} AX_1(\Omega) + BX_2(\Omega)$$

$$\boxed{\{Ax_1[n] + Bx_2[n]\} \overset{\text{DTFT}}{\longleftrightarrow} AX_1(\Omega) + BX_2(\Omega)}$$

2.18.2 Time Shifting Property

If

$$x[n] \overset{\text{DTFT}}{\longleftrightarrow} X(\Omega)$$

then

$$x[n-n_0] \overset{\text{DTFT}}{\longleftrightarrow} X(\Omega)e^{-jn_0\Omega}$$

$$\boxed{x[n-n_0] \overset{\text{DTFT}}{\longleftrightarrow} e^{-j\Omega n_0}X(\Omega)}$$

2.18.3 Frequency Shifting

If

$$x[n] \overset{\text{DTFT}}{\longleftrightarrow} X(\Omega)$$

then

$$e^{j\Omega_0 n}x[n] \overset{\text{DTFT}}{\longleftrightarrow} X(\Omega-\Omega_0)$$

$$\boxed{e^{j\Omega_0 n}x[n] \overset{\text{DTFT}}{\longleftrightarrow} X(\Omega-\Omega_0)}$$

2.18.4 Time Reversal

If

$$x[n] \overset{\text{DTFT}}{\longleftrightarrow} X(\Omega)$$

then

$$x[-n] \overset{\text{DTFT}}{\longleftrightarrow} X(-\Omega)$$

$$\boxed{x[-n] \overset{\text{DTFT}}{\longleftrightarrow} X(-\Omega)}$$

Folding in the time domain corresponds to folding in the frequency domain.

2.18.5 Time Scaling

If

$$x[n] \overset{\text{DTFT}}{\longleftrightarrow} X(\Omega)$$

then

$$x[an] \overset{\text{DTFT}}{\longleftrightarrow} X\left(\frac{\Omega}{a}\right)$$

$$\boxed{x[an] \overset{\text{DTFT}}{\longleftrightarrow} X\left(\frac{\Omega}{a}\right)}$$

2.18.6 *Multiplication by n*

If

$$x[n] \overset{\text{DTFT}}{\longleftrightarrow} X(\Omega)$$

then

$$nx[n] \overset{\text{DTFT}}{\longleftrightarrow} j\frac{dX(\Omega)}{d\Omega}$$

$$\boxed{\{nx[n]\} \overset{\text{DTFT}}{\longleftrightarrow} j\frac{dX(\Omega)}{d\Omega}}$$

2.18.7 *Conjugation*

If

$$x[n] \overset{\text{DTFT}}{\longleftrightarrow} X(\Omega)$$

then

$$x^*[n] \overset{\text{DTFT}}{\longleftrightarrow} X^*(-\Omega)$$

$$\boxed{x^*[n] \overset{\text{DTFT}}{\longleftrightarrow} X^*(-\Omega)}$$

2.18.8 *Time Convolution*

If

$$x_1[n] \overset{\text{DTFT}}{\longleftrightarrow} X_1(\Omega) \quad \text{and} \quad x_2[n] \overset{\text{DTFT}}{\longleftrightarrow} X_2(\Omega)$$

Table 2.4 Properties of DTFT

Operation	$x[n]$	$X(\Omega)$
1. Linearity	$Ax_1[n] + Bx_2[n]$	$AX_1(\Omega) + BX_2(\Omega)$
2. Time shifting	$x[n - n_0]$	$X(\Omega)e^{-jn_0}$
3. Frequency shifting	$e^{j\Omega_0 n}x[n]$	$X(\Omega - \Omega_0)$
4. Time reversal	$x[-n]$	$X(-\Omega)$
5. Time scaling	$x[an]$	$X\left(\frac{\Omega}{a}\right)$
6. Multiplication by n	$nx[n]$	$j\frac{dX(\Omega)}{d\Omega}$
7. Conjugation	$x^*[n]$	$X^*(\Omega)$
8. Time convolution	$x_1[n] * x_2[n]$	$X_1(\Omega)X_2(\Omega)$
9. Parseval's theorem	$\sum_{n=-\infty}^{\infty} \lvert x[n]\rvert^2$	$\frac{1}{2\pi}\int_{2\pi} \lvert X(\Omega)\rvert^2 d\Omega$
10. Modulation	$x[n]\cos(\Omega_c n + \theta)$	$\frac{1}{2}\{X(\Omega - \Omega_c)e^{j\theta}$ $+X(\Omega + \Omega_c)e^{-j\theta}\}$

then

$$x_1[n] * x_2[n] \overset{\text{DTFT}}{\longleftrightarrow} X_1(\Omega)X_2(\Omega)$$

$$\boxed{x_1[n] * x_2[n] \overset{\text{DTFT}}{\longleftrightarrow} X_1(\Omega)X_2(\Omega)}$$

2.18.9 *Parseval's Theorem*

$$\sum_{n=-\infty}^{\infty} |x[n]|^2 \overset{\text{DTFT}}{\longleftrightarrow} \frac{1}{2\pi}\int_{2\pi} |X(\Omega)|^2 d\Omega$$

The above relation states that the average power in a DT periodical signal is equal to the squared magnitude of $X(\Omega)$.

Proof

$$X^*(\Omega) = \sum_{n=-\infty}^{\infty} x^*[n]e^{j\Omega n}$$

Now

$$\begin{aligned}
\sum_{n=-\infty}^{\infty} |x[n]|^2 &= \sum_{n=-\infty}^{\infty} x^*[n]x[n] \\
&= \sum_{n=-\infty}^{\infty} x^*[n]\left[\frac{1}{2\pi}\int_{2\pi} X(\Omega)e^{j\Omega n}d\Omega\right] \\
&= \frac{1}{2\pi}\int_{2\pi} X(\Omega)\left[\sum_{n=-\infty}^{\infty} x^*[n]e^{j\Omega n}\right]d\Omega \\
&= \frac{1}{2\pi}\int_{2\pi} X(\Omega)X^*(\Omega)d\Omega = \frac{1}{2\pi}\int_{2\pi} |X(\Omega)|^2 d\Omega
\end{aligned}$$

2.18.10 Modulation Property

$$x[n]\cos(\Omega_c n + \theta) \overset{\text{DTFT}}{\longleftrightarrow} \frac{1}{2}\{X(\Omega - \Omega_c)e^{j\theta} + X(\Omega + \Omega_c)e^{-j\theta}\}$$

$$\boxed{x[n]\cos(\Omega_c n + \theta) \overset{\text{DTFT}}{\longleftrightarrow} \frac{1}{2}\{X(\Omega - \Omega_c)e^{j\theta} + X(\Omega + \Omega_c)e^{-j\theta}\}}$$

The properties of DTFT are given in Table 2.2.

Example 2.44 Find the discrete time Fourier transform of the following sequences:

1. $x[n] = e^{j\Omega_0 n}$
2. $x[n] = 1 \quad \text{all } n$
3. $x[n] = \cos\Omega_0 n \quad |\Omega_0| \le \pi$
4. $x[n] = u[n] - u[n-N]$
5. $x[n] = \begin{cases} 1 & |n| \le N \\ 0 & |n| > N \end{cases}$
6. $x[n] = a^{-n}u[-n] \quad |a| > 1$
7. $x[n] = 10\left(\frac{1}{6}\right)^{-n} u[-n]$
8. $x[n] = 10\left(\frac{1}{6}\right)^{n} u[n]$
9. $x[n] = n\left(\frac{1}{2}\right)^{n} u[n]$

Table 2.5 DTFT pair

No.	$x[n]$	$X(\Omega)$
1.	$\delta[n]$	1
2.	$u[n]$	$\pi\delta(\Omega) + \frac{1}{(1-e^{-j\Omega})}$
3.	$a^n u[n]$	$\frac{1}{(1-ae^{-j\Omega})}$ $\quad |a| < 1$
4.	$-a^n u[-n-1]$	$\frac{1}{(1-ae^{-j\Omega})}$ $\quad |a| > 1$
5.	$a^{|n|}$	$\frac{(1-a^2)}{(1-2a\cos\Omega + a^2)}$
6.	1	$2\pi\delta(\Omega)$
7.	$\cos\Omega_0 n$	$\pi[\delta(\Omega-\Omega_0) + \delta(\Omega+\Omega_0)]$ $\quad |\Omega|, |\Omega_0| \leq \pi$
8.	$\sin\Omega_0 n$	$j\pi[\delta(\Omega+\Omega_0) - \delta(\Omega-\Omega_0)]$ $\quad |\Omega|, |\Omega_0| \leq \pi$
9.	$u[n] - u[n-N]$	$e^{-j\Omega\left(\frac{N-1}{2}\right)} \frac{\sin(\Omega(N/2))}{\sin(\Omega/2)}$
10.	rect pulse	$\frac{\sin\left(\left(N+\frac{1}{2}\right)\Omega\right)}{\sin(\Omega/2)}$

10. $x[n] = \{2, -1, 2, -2\}$
11. $x[n] = \begin{cases} 1 & 0 \leq n \leq 5 \\ 0 & \text{otherwise} \end{cases}$
12. $y[n] = \left(\frac{1}{2}\right)^n u[n] * \left(\frac{1}{3}\right)^n h[n]$
13. $x[n] = (n+1)a^n u[n]$
14. $x[n] = u[n-1] - u[n-4]$
15. $x[n] = \left(\frac{1}{3}\right)^{n-1} u[n-1]$
16. $x[n] = \left(\frac{1}{4}\right)^{|n-1|} u[n-1]$
17. $x[n] = \delta[n-2] + \delta[n+2]$
18. $x[n] = \delta[n+2] - \delta[n-2]$
19. $x[n] = \sin\left(\frac{\pi}{4}n + \frac{\pi}{3}\right)$
20. $x[n] = \left(\frac{1}{4}\right)^{-n} u[-n-1]$
21. $x[n] = 10 + \cos\left(\frac{\pi}{4}n - \frac{\pi}{5}\right)$
22. $x[n] = x[2-n] + x[-2-n]$
23. $y[n] = (n-1)^2 x[n]$
24. $x[n] = \left(\frac{1}{2}\right)^n u[n+1]$

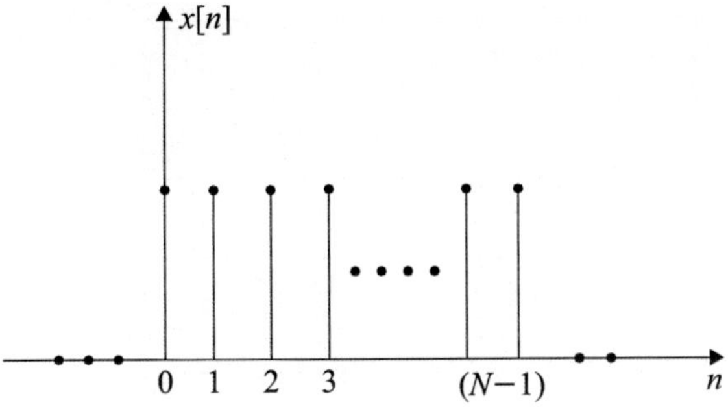

Fig. 2.20 Representation of the discrete sequence $x[n] = e^{j\Omega_0 n}$

Solution 1. $x[n] = e^{j\Omega_0 n}$

Consider the following DTFT:

$$X(\Omega) = 2\pi\delta(\Omega - \Omega_0) \qquad |\Omega|,\ |\Omega_0| \le \pi$$

The inverse DTFT is obtained from (Fig. 2.20)

$$\begin{aligned} x[n] &= \frac{1}{2\pi}\int_{-\pi}^{\pi} X(\Omega)e^{j\Omega n}d\Omega \\ &= \frac{1}{2\pi}\int_{-\pi}^{\pi} 2\pi\delta(\Omega - \Omega_0)e^{j\Omega n}d\Omega \end{aligned}$$

Using the property $\int_{-\infty}^{\infty} \Phi(\Omega)\delta(\Omega - \Omega_0)d\Omega = \Phi(\Omega_0)$ we get

$$\begin{aligned} x[n] &= \frac{1}{2\pi}[2\pi e^{j\Omega_0 n}] \\ &= e^{j\Omega_0 n} \end{aligned}$$

$$\boxed{e^{j\Omega_0 n} \overset{\text{DTFT}}{\longleftrightarrow} 2\pi\delta(\Omega - \Omega_0) \qquad |\Omega|,\ |\Omega_0| \le \pi}$$

2. $x[n] = 1$ ***all n***

$$e^{j\Omega_0 n} \overset{\text{DTFT}}{\longleftrightarrow} 2\pi\delta(\Omega - \Omega_0)$$

Substitute $\Omega_0 = 0$ and $x[n] = 1$

$$\boxed{1 \overset{\text{DTFT}}{\longleftrightarrow} 2\pi\delta(\Omega)}$$

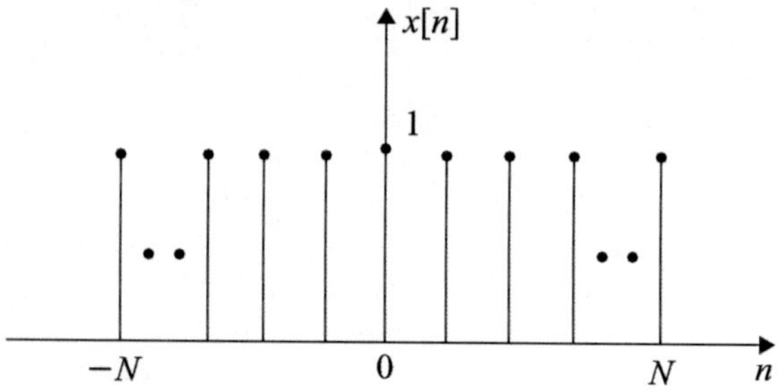

Fig. 2.21 Representation of a rectangular discrete sequence for Example 2.44.5

3. $\boldsymbol{x[n] = \cos\Omega_0 n \qquad |\Omega_0| \le \pi}$

$$\cos\Omega_0 n = \frac{1}{2}[e^{j\Omega_0 n} + e^{-j\Omega_0 n}]$$
$$e^{j\Omega_0 n} \overset{\text{DTFT}}{\longleftrightarrow} 2\pi\delta(\Omega - \Omega_0)$$
$$e^{-j\Omega_0 n} \overset{\text{DTFT}}{\longleftrightarrow} 2\pi\delta(\Omega + \Omega_0)$$

$$\boxed{\cos\Omega_0 n \overset{\text{DTFT}}{\longleftrightarrow} \pi[\delta(\Omega - \Omega_0) + \delta(\Omega + \Omega_0)] \qquad |\Omega|,\ |\Omega_0| \le \pi}$$

4. $\boldsymbol{x[n] = u[n] - u[n - N]}$

$$X(\Omega) = \sum_{n=0}^{N-1} x[n]e^{-j\Omega n}$$

Using the summation formula (Fig. 2.21)

$$\sum_{n=0}^{N-1} a^n = \frac{(1 - a^N)}{(1 - a)}$$

we get

$$X(\Omega) = \frac{1 - e^{-j\Omega N}}{1 - e^{-j\Omega}}$$
$$= \frac{e^{-j\frac{\Omega N}{2}}\left[e^{j\frac{\Omega N}{2}} - e^{-j\frac{\Omega N}{2}}\right]}{e^{-j\frac{\Omega}{2}}\left[e^{j\frac{\Omega}{2}} - e^{-j\frac{\Omega}{2}}\right]}$$

$$\boxed{X(\Omega) = e^{-j\Omega\frac{(N-1)}{2}} \frac{\sin\left(\frac{\Omega N}{2}\right)}{\sin\left(\frac{\Omega}{2}\right)}}$$

5. $x[n] = \begin{cases} 1 & |n| \leq N \\ 0 & |n| > N \end{cases}$

$$
\begin{aligned}
X(\Omega) &= \sum_{-N}^{N} e^{-j\Omega n} \\
&= \sum_{-N}^{-1} e^{-j\Omega n} + \sum_{0}^{N} e^{-j\Omega n} \\
&= e^{j\Omega}\left[\frac{(1-e^{j\Omega N})}{(1-e^{j\Omega})}\right] + \left[\frac{(1-e^{-j\Omega(N+1)})}{(1-e^{-j\Omega})}\right] \\
&= \left[\frac{e^{j\Omega}-e^{j\Omega(N+1)}}{(1-e^{-j\Omega})}\right] + \left[\frac{1-e^{j\Omega(N+1)}}{(1-e^{-j\Omega})}\right] \\
&= \frac{[e^{j\Omega}-1-e^{j\Omega(N+1)}+e^{j\Omega N}+1-e^{-j\Omega(N+1)}-e^{j\Omega}+e^{-j\Omega N}]}{1-(e^{j\Omega}+e^{-j\Omega})+1}
\end{aligned}
$$

$$
\begin{aligned}
X(\Omega) &= \frac{2\cos\Omega N - 2\cos\Omega(N+1)}{2(1-\cos\Omega)} \\
&= \frac{\sin\left(N+\frac{1}{2}\right)\Omega\sin\frac{\Omega}{2}}{\sin^2\frac{\Omega}{2}}
\end{aligned}
$$

$$
\boxed{X(\Omega) = \frac{\sin\left(N+\frac{1}{2}\right)\Omega}{\sin\left(\frac{\Omega}{2}\right)}}
$$

6. $x[n] = a^{-n}u[-n] \quad |a| < 1$

$$
\begin{aligned}
X(\Omega) &= \sum_{n=-\infty}^{0} a^{-n}e^{-j\Omega n} \\
&= \sum_{n=0}^{\infty}(ae^{j\Omega})^n
\end{aligned}
$$

Using summation formula, we get

$$
\boxed{X(\Omega) = \frac{1}{(1-ae^{j\Omega})}}
$$

7. $\boldsymbol{x[n] = 10\left(\frac{1}{6}\right)^{-n} u[-n]}$

$$X(\Omega) = \sum_{n=-\infty}^{0} 10\left(\frac{1}{6}\right)^{-n} e^{-j\Omega n}$$
$$= 10\sum_{n=0}^{\infty}\left(\frac{1}{6}e^{j\Omega n}\right)^{n}$$

$$\boxed{X(\Omega) = \frac{10}{\left(1 - \frac{1}{6}e^{j\Omega}\right)}}$$

8. $\boldsymbol{x[n] = 10\left(\frac{1}{6}\right)^{-n} u[n]}$

$$X(\Omega) = \sum_{n=0}^{\infty} 10\left(\frac{1}{6}e^{j\Omega}\right)^{-n}$$
$$= \sum_{n=0}^{\infty} 10\left(6e^{-j\Omega}\right)^{n}$$

RHS of the equation is not summable and $x[n]$ does not have DTFT.

9. $\boldsymbol{x[n] = n\left(\frac{1}{2}\right)^{n} u[n]}$ **(Using multiplication property by $\boldsymbol{n}$)**

$$\left(\frac{1}{2}\right)^{n} \overset{\text{DTFT}}{\longleftrightarrow} \frac{1}{1 - \left(\frac{1}{2}\right)e^{-j\Omega}}$$
$$= \frac{e^{j\Omega}}{(e^{j\Omega} - 0.5)}$$
$$n\left(\frac{1}{2}\right)^{n} u[n] \overset{\text{DTFT}}{\longleftrightarrow} j\frac{d}{d\Omega}\left[\frac{e^{j\Omega}}{(e^{j\Omega} - 0.5)}\right]$$
$$= \frac{j(e^{j\Omega} - 0.5)e^{j\Omega}(j) - e^{j\Omega}e^{j\Omega}(j)}{(e^{j\Omega} - 0.5)^2}$$
$$= \frac{0.5e^{j\Omega}}{(e^{j\Omega} - 0.5)^2}$$

$$\boxed{n\left(\frac{1}{2}\right)^{n} u[n] \overset{\text{DTFT}}{\longleftrightarrow} \frac{0.5e^{j\Omega}}{(e^{j\Omega} - 0.5)^2}}$$

10. $\boldsymbol{x[n] = \{2, -1, 2, -2\}}$

$$\boxed{x(\Omega) = \{2 - e^{-j\Omega} + 2e^{-j2\Omega} - 2e^{-j3\Omega}\}}$$

11. $\boldsymbol{x[n] = \begin{cases} 1 & 0 \leq n \leq 5 \\ 0 & otherwise \end{cases}}$

$$X(\Omega) = \frac{\sin \Omega \left(N + \frac{1}{2}\right)}{\sin \frac{\Omega}{2}}$$

where $N = 5$

$$\boxed{X(\Omega) = \frac{\sin 5.5\Omega}{\sin 0.5\Omega}}$$

12. $\boldsymbol{y[n] = \left(\frac{1}{2}\right)^n u[n] * \left(\frac{1}{3}\right)^n u[n]}$

$$\begin{aligned} X(\Omega) &= \frac{1}{\left(1 - \frac{1}{2}e^{-j\Omega}\right)} \\ H(\Omega) &= \frac{1}{\left(1 - \frac{1}{3}e^{-j\Omega}\right)} \\ Y(\Omega) &= X(\Omega)H(\Omega) \end{aligned}$$

$$\boxed{Y(\Omega) = \frac{1}{\left(1 - \frac{1}{2}e^{-j\Omega}\right)\left(1 - \frac{1}{3}e^{-j\Omega}\right)}}$$

To find $y[n]$, put $Y(\Omega)$ in partial fraction and take IDTFT.

13. $\boldsymbol{x[n] = (n + 1)a^n u[n]}$

$$\begin{aligned} x[n] &= na^n u[n] + a^n u[n] \\ na^n u[n] &\overset{\text{DTFT}}{\longleftrightarrow} \frac{ae^{j\Omega}}{(e^{j\Omega} - a)^2} \\ a^n u[n] &\overset{\text{DTFT}}{\longleftrightarrow} \frac{e^{j\Omega}}{(e^{j\Omega} - a)} \\ x[n] &\overset{\text{DTFT}}{\longleftrightarrow} \frac{ae^{j\Omega}}{(e^{j\Omega} - a)^2} + \frac{e^{j\Omega}}{(e^{j\Omega} - a)} \\ &= \frac{e^{j\Omega}(a + e^{j\Omega} - a)}{(e^{j\Omega} - a)^2} \\ &= \frac{e^{j2\Omega}}{(e^{j\Omega} - a)^2} \\ &= \frac{1}{(1 - ae^{-j\Omega})^2} \end{aligned}$$

$$\boxed{(n+1)a^n u[n] \overset{\text{DTFT}}{\longleftrightarrow} \frac{1}{(1-ae^{-j\Omega})^2}}$$

14. $\boldsymbol{x[n] = u[n-1] - u[n-4]}$

$$\boxed{X(\Omega) = e^{-j\Omega} + e^{-j2\Omega} + e^{-j3\Omega}}$$

15. $\boldsymbol{x[n] = \left(\frac{1}{3}\right)^{n-1} u[n-1]}$

$$\left(\frac{1}{3}\right)^n u[n] \overset{\text{DTFT}}{\longleftrightarrow} \frac{1}{1-\frac{1}{3}e^{-j\Omega}}$$

Using right shift time shifting property, we get

$$\left(\frac{1}{3}\right)^{n-1} u[n-1] \overset{\text{DTFT}}{\longleftrightarrow} \frac{e^{-j\Omega}}{\left(1-\frac{1}{3}e^{-j\Omega}\right)} = \frac{1}{\left(e^{j\Omega}-\frac{1}{3}\right)}$$

$$\boxed{\left(\frac{1}{3}\right)^{n-1} u[n-1] \overset{\text{DTFT}}{\longleftrightarrow} \frac{1}{\left(e^{j\Omega}-\frac{1}{3}\right)}}$$

16. $\boldsymbol{x[n] = \left(\frac{1}{4}\right)^{|n-1|} u[n-1]}$
From Example 2.43.5,

$$x[n] = (a)^{|n|} \overset{\text{DTFT}}{\longleftrightarrow} \frac{(1-a^2)}{(1-2a\cos\Omega + a^2)}$$

Substitute $|a| = \frac{1}{4}$

$$\left(\frac{1}{4}\right)^{|n|} \overset{\text{DTFT}}{\longleftrightarrow} \frac{(15/16)}{(17/16) - 0.5\cos\Omega} = \frac{15}{17 - 8\cos\Omega}$$

Using right shift time shifting property, we get

$$\boxed{\left(\frac{1}{4}\right)^{|n-1|} \overset{\text{DTFT}}{\longleftrightarrow} \frac{15e^{-j\Omega}}{17 - 8\cos\Omega}}$$

17. $\boldsymbol{x[n] = \delta[n-2] + \delta[n+2]}$

$$\delta[n-2] + \delta[n+2] \overset{\text{DTFT}}{\longleftrightarrow} e^{-j2\Omega} + e^{j2\Omega}$$

$$\boxed{\delta[n-2]+\delta[n+2] \overset{\text{DTFT}}{\longleftrightarrow} 2\cos 2\Omega}$$

18. $\boldsymbol{x[n] = \delta[n+2] + \delta[n-2]}$

$$\begin{aligned}\delta[n+2]-\delta[n-2] \overset{\text{DTFT}}{\longleftrightarrow}\ & e^{j2\Omega} - e^{-j2\Omega} \\ =\ & j2\sin 2\Omega\end{aligned}$$

$$\boxed{\delta[n+2]-\delta[n-2] \overset{\text{DTFT}}{\longleftrightarrow} j2\sin 2\Omega}$$

19. $\boldsymbol{x[n] = \sin\left(\frac{\pi}{4}n + \frac{\pi}{3}\right)}$

$$\begin{aligned}\sin\left(\frac{\pi}{4}n + \frac{\pi}{3}\right) &= \frac{e^{j\left(\frac{\pi}{4}n+\frac{\pi}{3}\right)} - e^{-j\left(\frac{\pi}{4}n+\frac{\pi}{3}\right)}}{2j} \\ &= \frac{1}{2j}[e^{j\frac{\pi}{3}}e^{j\frac{\pi}{4}n} - e^{-j\frac{\pi}{3}}e^{-j\frac{\pi}{4}n}]\end{aligned}$$

From Example 2.44.1, it is derived that

$$\begin{aligned}e^{j\Omega_0 n} &\overset{\text{DTFT}}{\longleftrightarrow} 2\pi\delta(\Omega-\Omega_0) \\ \therefore \quad e^{j\frac{\pi}{4}n} &\overset{\text{DTFT}}{\longleftrightarrow} 2\pi\delta\left(\Omega - \frac{\pi}{4}\right) \\ e^{-j\frac{\pi}{4}n} &\overset{\text{DTFT}}{\longleftrightarrow} 2\pi\delta\left(\Omega + \frac{\pi}{4}\right) \\ \sin\left(\frac{\pi}{4}n + \frac{\pi}{3}\right) &\overset{\text{DTFT}}{\longleftrightarrow} \frac{2\pi}{2j}\left[e^{j\frac{\pi}{3}}\delta\left(\Omega-\frac{\pi}{4}\right) - e^{-j\frac{\pi}{3}}\delta\left(\Omega+\frac{\pi}{4}\right)\right]\end{aligned}$$

$$\boxed{\sin\left(\frac{\pi}{4}n + \frac{\pi}{3}\right) \overset{\text{DTFT}}{\longleftrightarrow} \frac{\pi}{j}\left[e^{j\frac{\pi}{3}}\delta\left(\Omega-\frac{\pi}{4}\right) - e^{-j\frac{\pi}{3}}\delta\left(\Omega+\frac{\pi}{4}\right)\right]}$$

20. $\boldsymbol{x[n] = \left(\frac{1}{4}\right)^{-n} u[-n-1]}$

$$x[n] = \frac{1}{4}\left(\frac{1}{4}\right)^{-n-1} u[-n-1]$$

Using the reversal and left time shift, we get

$$\begin{aligned}\left(\frac{1}{4}\right)^{-n-1} u[-n-1] \overset{\text{DTFT}}{\longleftrightarrow}\ & \frac{e^{j\Omega}}{1-\frac{1}{4}e^{j\Omega}} \\ =\ & \frac{1}{\left(e^{-j\Omega} - \frac{1}{4}\right)}\end{aligned}$$

$$\boxed{\left(\frac{1}{4}\right)^{-n} u[-n-1] \overset{\text{DTFT}}{\longleftrightarrow} \frac{1}{4} \frac{1}{\left(e^{-j\Omega} - \frac{1}{4}\right)}}$$

21. $\boldsymbol{x[n] = 10 + \cos\left(\frac{\pi}{4}n - \frac{\pi}{5}\right)}$

$$\begin{aligned}
\cos\left(\frac{\pi}{4}n - \frac{\pi}{5}\right) &= \frac{1}{2}[e^{-j\frac{\pi}{5}} e^{j\frac{\pi}{4}n} + e^{j\frac{\pi}{5}} e^{-j\frac{\pi}{4}n}] \\
e^{j\frac{\pi}{4}n} &\overset{\text{DTFT}}{\longleftrightarrow} 2\pi\delta\left(\Omega - \frac{\pi}{4}\right) \\
e^{-j\frac{\pi}{4}n} &\overset{\text{DTFT}}{\longleftrightarrow} 2\pi\delta\left(\Omega + \frac{\pi}{4}\right) \\
10 &\overset{\text{DTFT}}{\longleftrightarrow} 20\pi\delta(\Omega)
\end{aligned}$$

$$\boxed{10 + \cos\left(\frac{\pi}{4} - \frac{\pi}{5}\right) \overset{\text{DTFT}}{\longleftrightarrow} 20\pi\delta(\Omega) + \pi\left[e^{-j\frac{\pi}{5}}\delta\left(\Omega - \frac{\pi}{4}\right) + e^{j\frac{\pi}{5}n}\delta\left(\Omega + \frac{\pi}{4}\right)\right]}$$

22. $\boldsymbol{x[n] = x[2-n] + x[-2-n]}$

$$\begin{aligned}
x[2-n] &\overset{\text{DTFT}}{\longleftrightarrow} X(\Omega)e^{-j2\Omega} \quad \text{(Right shift)} \\
x[-2-n] &\overset{\text{DTFT}}{\longleftrightarrow} X(\Omega)e^{j2\Omega} \quad \text{(Left shift)} \\
\{x[2-n] + x[-2-n]\} &\overset{\text{DTFT}}{\longleftrightarrow} X(\Omega)[e^{-j2\Omega} + e^{j2\Omega}] \\
&= 2X(\Omega)\cos 2\Omega
\end{aligned}$$

$$\boxed{\{x[2-n] + x[-2-n]\} \overset{\text{DTFT}}{\longleftrightarrow} 2X(\Omega)\cos 2\Omega}$$

23. $\boldsymbol{y[n] = (n-1)^2 x[n]}$

$$\begin{aligned}
y[n] &= (n^2 - 2n + 1)x[n] \\
&= n^2x[n] - 2nx[n] + x[n] \\
n^2x[n] &\overset{\text{DTFT}}{\longleftrightarrow} (j)^2\frac{d^2X(\Omega)}{d\Omega^2} \\
nx[n] &\overset{\text{DTFT}}{\longleftrightarrow} j\frac{dX(\Omega)}{d\Omega}
\end{aligned}$$

$$\boxed{Y(\Omega) = -\frac{d^2X(\Omega)}{d\Omega^2} - 2j\frac{dX(\Omega)}{d\Omega} + X(\Omega)}$$

24. $\boldsymbol{x[n] = \left(\frac{1}{2}\right)^n u[n+1]}$

$$x[n] = 2\left(\frac{1}{2}\right)^{n+1} u[n+1]$$

Making left shift of $(\frac{1}{2})^n$ we get

$$\left(\frac{1}{2}\right)^{n+1} u[n+1] \overset{\text{DTFT}}{\longleftrightarrow} \frac{e^{j\Omega}}{\left(1 - \frac{1}{2}e^{-j\Omega}\right)}$$

$$\boxed{X(\Omega) = \frac{2e^{j\Omega}}{\left(1 - \frac{1}{2}e^{-j\Omega}\right)}}$$

The above result can be obtained from first principle as follows:

$$X(\Omega) = \sum_{n=-1}^{\infty} \left(\frac{1}{2}\right)^n e^{-j\Omega n}$$

$$\begin{aligned} X(\Omega) &= \left(\frac{1}{2}\right)^{-1} e^{j\Omega} + \sum_{n=0}^{\infty} \left(\frac{1}{2}\right)^n e^{-j\Omega n} \\ &= 2e^{j\Omega} + \sum_{n=0}^{\infty} \left(\frac{1}{2e^{j\Omega}}\right)^n \\ &= 2e^{j\Omega} + \frac{1}{\left(1 - \frac{1}{2}e^{-j\Omega}\right)} \\ &= \frac{2e^{j\Omega} - 1 + 1}{\left(1 - \frac{1}{2}e^{-j\Omega}\right)} \end{aligned}$$

$$\boxed{X(\Omega) = \frac{2e^{j\Omega}}{\left(1 - \frac{1}{2}e^{-j\Omega}\right)}}$$

2.19 Inverse Discrete Time Fourier Transform (IDTFT)

If $x[n]$ is given, the discrete time Fourier transform is obtained using Eq. (2.69) which is given below:

$$X(\Omega) = \sum_{n=-\infty}^{\infty} x[n]e^{-j\Omega n} \tag{2.75}$$

If $X(\Omega)$ is given, then the sequence $x[n]$ is obtained from Eq. (2.68) which is given as

$$x[n] = \frac{1}{2\pi}\int_{2\pi} X(\Omega)e^{j\Omega n}d\Omega \tag{2.76}$$

$x[n]$ is also obtained by putting $X(\Omega)$ by partial fraction and making use of DTFT pair in Table 2.3. The process of getting $x[n]$ from $X(\Omega)$ is called IDTFT. This is illustrated in the following examples.

Example 2.45 Find $x[n]$ for the $X(\Omega)$ given below.

1. $X(\Omega) = 8\pi\delta(\Omega) + 10\pi\delta\left(\Omega - \frac{\pi}{4}\right) + 10\pi\delta\left(\Omega + \frac{\pi}{4}\right)$
2. $X(\Omega) = \begin{cases} j & 0 < \Omega \le \pi \\ -j & -\pi < \Omega \le 0 \end{cases}$
3. $X(\Omega) = \begin{cases} 2 & 0 \le |\Omega| < \frac{\pi}{3} \\ 0 & \frac{\pi}{3} \le |\Omega| \le \pi \end{cases}$
   and $\angle X(\Omega) = -\frac{4}{3}\Omega$
4. $X(\Omega) = 6 + 4e^{-j\Omega} + 7e^{-j2\Omega}$
5. $x[n] = \left(\frac{1}{2}\right)^n u[n]$
   $h[n] = \left(\frac{1}{3}\right)^n u[n]$
   $y[n] = x[n] * h[n]$
6. $X(\Omega) = j\Omega$

Solution 1. $\mathbf{X(\Omega) = 8\pi\delta(\Omega) + 10\pi\delta\left(\Omega - \frac{\pi}{4}\right) + 10\pi\delta\left(\Omega + \frac{\pi}{4}\right)}$

$$2\pi\delta(\Omega) \overset{\text{IDTFT}}{\longleftrightarrow} 1$$

$$\pi\delta\left(\Omega - \frac{\pi}{4}\right) + \pi\delta\left(\Omega + \frac{\pi}{4}\right) \overset{\text{DTFT}}{\longleftrightarrow} \cos\frac{\pi}{4}n$$

$$\boxed{x[n] = 4 + 10\cos\frac{\pi}{4}n}$$

2. $\mathbf{X(\Omega) = \begin{cases} j & 0 < \Omega \le \pi \\ -j & -\pi < \Omega \le 0 \end{cases}}$

$$x[n] = \frac{1}{2\pi}\left\{\int_{-\pi}^{0} -je^{j\Omega n}d\Omega + \int_{0}^{\pi} je^{j\Omega n}d\Omega\right\}$$
$$= \frac{j}{2\pi}\left\{\left[-\frac{1}{jn}e^{j\Omega n}\right]_{-\pi}^{0} + \left[\frac{1}{jn}e^{j\Omega n}\right]_{0}^{\pi}\right\}$$
$$= \frac{j}{2\pi}\left\{-\frac{1}{jn} + \frac{1}{jn}e^{-j\pi n} + \frac{1}{jn}e^{j\pi n} - \frac{1}{jn}\right\}$$
$$= \frac{1}{2\pi n}\left\{-2 + e^{j\pi n} + e^{-j\pi n}\right\}$$
$$= \frac{1}{\pi n}\left\{-1 + \cos\pi n\right\}$$

$$\boxed{x[n] = -\frac{2}{n\pi}\sin^2\frac{n\pi}{2}}$$

3. $X(\Omega) = \begin{cases} 2 & 0 \le |\Omega| < \frac{\pi}{3} \\ 0 & \frac{\pi}{3} \le |\Omega| \le \pi \end{cases}$ *and* $\angle X(\Omega) = -\frac{4}{3}\Omega$

$$\angle X(\Omega) = -\frac{4}{3}\Omega$$
$$= e^{-j\frac{4}{3}\Omega}$$
$$x[n] = \frac{2}{2\pi}\left\{\int_{-\frac{\pi}{3}}^{0} e^{-j\frac{4}{3}\Omega}e^{j\Omega n}d\Omega + \int_{0}^{\frac{\pi}{3}} e^{-j\frac{4}{3}\Omega}e^{j\Omega n}d\Omega\right\}$$
$$= \frac{1}{\pi}\left\{\int_{-\frac{\pi}{3}}^{0} e^{j(n-\frac{4}{3})\Omega}d\Omega + \int_{0}^{\frac{4}{3}} e^{j(n-\frac{4}{3})\Omega}d\Omega\right\}$$
$$= \frac{1}{j\pi}\frac{1}{\left(n-\frac{4}{3}\right)}\left[e^{j(n-\frac{4}{3})\Omega}\right]_{-\frac{\pi}{3}}^{0} + \left[e^{j(n-\frac{4}{3})\Omega}\right]_{0}^{\frac{\pi}{3}}$$
$$= \frac{1}{j\pi\left(n-\frac{4}{3}\right)}\left[1 - e^{-j(n-\frac{4}{3})\frac{\pi}{3}} - 1 + e^{j(n-\frac{4}{3})\frac{\pi}{3}}\right]$$

$$\boxed{x[n] = \frac{2}{\pi\left(n-\frac{4}{3}\right)}\sin\left(n-\frac{4}{3}\right)\frac{\pi}{3}}$$

4. $X(\Omega) = 6 + 4e^{-j\Omega} + 7e^{-j2\Omega}$

$$x[n] = \{6,\ 4,\ 7\}$$

5. $x[n] = \left(\frac{1}{2}\right)^n u[n]$
$h[n] = \left(\frac{1}{3}\right)^n u[n]$
$y[n] = x[n] * h[n]$

$$X(\Omega) = \frac{1}{\left(1 - \frac{1}{2}e^{-j\Omega}\right)}$$
$$H(\Omega) = \frac{1}{\left(1 - \frac{1}{3}e^{-j\Omega}\right)}$$
$$Y(\Omega) = \frac{1}{\left(1 - \frac{1}{2}e^{-j\Omega}\right)\left(1 - \frac{1}{3}e^{-j\Omega}\right)}$$
$$= \frac{A_1}{\left(1 - \frac{1}{2}e^{-j\Omega}\right)} + \frac{A_2}{\left(1 - \frac{1}{3}e^{-j\Omega}\right)}$$
$$1 = A_1\left(1 - \frac{1}{3}e^{-j\Omega}\right) + A_2\left(1 - \frac{1}{2}e^{-j\Omega}\right)$$

Substitute $e^{-j\Omega} = 2$; we get $A_1 = 3$ and substitute $e^{-j\Omega} = 3$; we get $A_2 = -2$

$$Y(\Omega) = \frac{3}{\left(1 - \frac{1}{2}e^{-j\Omega}\right)} - \frac{2}{\left(1 - \frac{1}{3}e^{-j\Omega}\right)}$$

Taking inverse Fourier transform, we get

$$\boxed{Y[n] = 3\left(\frac{1}{2}\right)^n u[n] - 2\left(\frac{1}{3}\right)^n u[n]}$$

6. $\boldsymbol{X(\Omega) = j\Omega}$

$$x[n] = \frac{1}{2\pi}\int_{-\pi}^{\pi} j\Omega e^{j\Omega n} d\Omega$$
$$= \frac{1}{2\pi}\left[\frac{j\Omega}{jn}e^{j\Omega n}\right]_{-\pi}^{\pi}$$
$$= \frac{1}{2\pi n}\left[\pi e^{j\pi n} + \pi e^{-j\pi n}\right]$$
$$= \frac{1}{n}\cos \pi n$$

$$\boxed{x[n] = \frac{(-1)^n}{n} \qquad n \neq 0}$$

2.20 LTI System Characterized by Difference Equation

LTIDT systems are described by linear constant coefficients differential equations of the form

$$\sum_{k=0}^{N} a_k y[n-k] = \sum_{k=0}^{M} b_k x[n-k] \tag{2.77}$$

where $M \leq N$. Taking Fourier transform of both sides of Eq. (2.77) and using the time shifting property, we get

$$\sum_{k=0}^{N} a_k e^{-jk\Omega} Y(\Omega) = \sum_{k=0}^{M} b_k e^{-jk\Omega} X(\Omega) \tag{2.78}$$

$$H(\Omega) = \frac{Y(\Omega)}{X(\Omega)} = \frac{\sum_{k=0}^{M} b_k e^{-jk\Omega}}{\sum_{k=0}^{N} a_k e^{-jk\Omega}} \tag{2.79}$$

In Eq. (2.79) $H(\Omega)$ is called the system transfer function and for any input $X(\Omega)$, the output $y[n]$ can be obtained by taking IDTFT. Further $H(\Omega)$ gives the frequency response of DT system which is periodic and is expressed as

$$H(\Omega) = H(\Omega + 2\pi) \tag{2.80}$$

Unlike continuous time system, for a DT system the frequency response is observed for the frequency range $0 \leq \Omega < 2\pi$ or $-\pi \leq \Omega < \pi$.

Example 2.46 Consider the system consisting of the cascade of two LTI systems with frequency responses

$$H_1(e^{j\omega}) = \frac{2 - e^{j\omega}}{\left(1 + \frac{1}{2}e^{-j\omega}\right)}$$

and

$$H_2(e^{j\omega}) = \frac{1}{\left(1 - \frac{1}{2}e^{-j\omega} + \frac{1}{4}e^{-j2\omega}\right)}$$

Find the difference equation describing the overall system.

Solution Note: As stated earlier, symbols used here are different from the symbols used in this textbook. Their equivalences are

$$\Omega = \omega$$
$$H(\Omega) = H(e^{j\omega})$$
$$H(e^{j\omega}) = H_1(e^{j\omega})H_2(e^{j\omega})$$
$$= \frac{(2 - e^{-j\omega})}{\left(1 + \frac{1}{2}e^{-j\omega}\right)\left(1 - \frac{1}{2}e^{-j\omega} + \frac{1}{4}e^{-j2\omega}\right)}$$
$$\frac{Y(e^{j\omega})}{X(e^{j\omega})} = \frac{2 - e^{-j\omega}}{\left(1 + \frac{1}{8}e^{-j3\omega}\right)}$$
$$\left(1 + \frac{1}{8}e^{-j3\omega}\right)Y(e^{j\omega}) = (2 - e^{-j\omega})X(e^{j\omega})$$

$$\boxed{y[n] + \frac{1}{8}y[n-3] = 2x[n] - x[n-1]}$$

Example 2.47 Find the impulse response of the discrete time system described by the difference equation

$$y[n-2] - 3y[n-1] + 2y[n] = x(n-1)$$

Solution Taking Fourier transform for the both sides of given difference equation, we get

$$[e^{-j2\Omega} - 3e^{-j\Omega} + 2]Y(\Omega) = e^{-j\Omega}X(\Omega)$$
$$Y(\Omega) = \frac{e^{-j\Omega}X(\Omega)}{[e^{-j2\Omega} - 3e^{-j\Omega} + 2]}$$

For an impulse $X(\Omega) = 1$

$$[e^{-j2\Omega} - 3e^{-j\Omega} + 2]Y(\Omega) = (e^{-j\Omega} - 1)(e^{-j\Omega} - 2)$$
$$Y(\Omega) = \frac{e^{-j\Omega}}{(e^{-j\Omega} - 1)(e^{-j\Omega} - 2)}$$
$$= \frac{A_1}{(e^{-j\Omega} - 1)} + \frac{A_2}{(e^{-j\Omega} - 2)}$$
$$e^{-j\Omega} = A_1(e^{-j\Omega} - 2) + A_2(e^{-j\Omega} - 1)$$

Put $e^{-j\Omega} = 1$

$$1 = A_1(1 - 2); \quad A_1 = -1$$

Put $e^{-j\Omega} = 2$

$$2 = A_2(2 - 1); \quad A_2 = 2$$

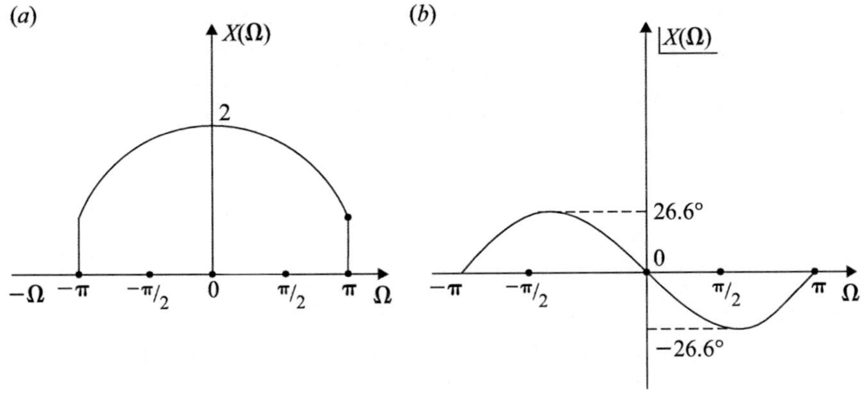

Fig. 2.22 Frequency spectra for Example 2.48. **a** Amplitude spectrum and **b** Phase spectrum

$$\begin{aligned} Y(\Omega) &= \frac{-1}{(e^{-j\Omega}-1)} + \frac{2}{(e^{-j\Omega}-2)} \\ &= \frac{1}{(1-e^{-j\Omega})} + \frac{1}{\left(1-\frac{1}{2}e^{-j\Omega}\right)} \end{aligned}$$

Taking inverse discrete Fourier transform, we get

$$\boxed{y[n] = \left[1 - \left(\frac{1}{2}\right)^n\right] u[n]}$$

Example 2.48 Find the DTFT of

$$x[n] = \left(\frac{1}{2}\right)^n u[n]$$

and plot the spectrum.

Solution

$$\begin{aligned} X(\Omega) &= \frac{1}{\left(1-\frac{1}{2}e^{-j\Omega}\right)} \\ &= \frac{1}{\left(1-\frac{1}{2}\cos\Omega + j\sin\Omega\right)} \\ &= \frac{1\angle -\tan^{-1}\sin\Omega/(1-\frac{1}{2}\cos\Omega)}{\sqrt{\left(1-\frac{1}{2}\cos\Omega\right)^2 + \frac{1}{4}\sin^2\Omega}} \end{aligned}$$

$$= \frac{1\angle - \tan^{-1} \sin\Omega/(1 - \frac{1}{2}\cos\Omega)}{\sqrt{\left(\frac{5}{4} - \cos\Omega\right)}}$$

Ω	$-\pi$	$-\frac{\pi}{2}$	0	$\frac{\pi}{2}$	π
$\lvert X(\Omega)\rvert$	0.667	0.894	2	0.894	0.667
$\angle X(\Omega)$	0	26.6°	0	−26.6°	0

The frequency spectrum is shown in Fig. 2.22.

Example 2.49 Use Fourier transform to find the output of the system whose impulse response $h[n] = \left(\frac{1}{3}\right)^n u[n]$ and the input to the system is $x[n] = \left(\frac{1}{2}\right)^n u[n]$.

Solution

$$H(\Omega) = \frac{1}{\left(1 - \frac{1}{3}e^{-j\Omega}\right)}$$

$$X(\Omega) = \frac{1}{\left(1 - \frac{1}{2}e^{-j\Omega}\right)}$$

$$H(\Omega) = \frac{Y(\Omega)}{X(\Omega)}$$

$$Y(\Omega) = H(\Omega)X(\Omega)$$

$$= \frac{1}{\left(1 - \frac{1}{3}e^{-j\Omega}\right)\left(1 - \frac{1}{2}e^{-j\Omega}\right)}$$

$$= \frac{A_1}{\left(1 - \frac{1}{3}e^{-j\Omega}\right)} + \frac{A_2}{\left(1 - \frac{1}{2}e^{-j\Omega}\right)}$$

$$1 = A_1\left(1 - \frac{1}{2}e^{-j\Omega}\right) + A_2\left(1 - \frac{1}{3}e^{-j\Omega}\right)$$

Put $e^{-j\Omega} = 3$

$$1 = A_1\left(-\frac{1}{2}\right); \qquad A_1 = -2$$

Put $e^{-j\Omega} = 2$

$$1 = A_2\left(\frac{1}{3}\right); \qquad A_2 = 3$$

$$Y(\Omega) = \frac{-2}{\left(1 - \frac{1}{3}e^{-j\Omega}\right)} + \frac{3}{\left(1 - \frac{1}{2}e^{-j\Omega}\right)}$$

$$\boxed{y[n] = \left[-2\left(\frac{1}{3}\right)^n + 3\left(\frac{1}{2}\right)^n\right]u[n]}$$

Example 2.50 Given

$$x[n] = \{1,\ 2,\ 0,\ 2\} \longrightarrow \text{Input}$$
$$\uparrow$$
$$h[n] = \{5,\ A,\ 3\} \longrightarrow \text{Impulse response}$$
$$\uparrow$$
$$\begin{aligned} y[n] &= x[n] * h[n] \\ &= \{5,\ 12,\ 7,\ 16,\ 4,\ 6\} \longrightarrow \text{Output} \end{aligned}$$
$$\uparrow$$

Find the value of A.

Solution

$$x[n] = \{1,\ 2,\ 0,\ 2\}$$
$$\uparrow$$
$$X(\Omega) = [e^{j2\Omega} + 2e^{j\Omega} + 0 + 2e^{-j\Omega}]$$
$$h[n] = \{5,\ A,\ 3\}$$
$$\uparrow$$
$$\begin{aligned} H(\Omega) &= [5e^{j\Omega} + A + 3e^{-j\Omega}] \\ y[n] &= x[n] * h[n] \\ Y(\Omega) &= X(\Omega)H(\Omega) \\ &= [e^{j2\Omega} + 2e^{j\Omega} + 0 + 2e^{-j\Omega}][5e^{j\Omega} + A + 3e^{-j\Omega}] \\ &= 5e^{j3\Omega} + (10 + A)e^{j2\Omega} + (3 + 2A)e^{j\Omega} \\ &\quad +10 + 2Ae^{-j\Omega} + 6e^{-j2\Omega}] \end{aligned} \tag{2.81}$$

Given

$$y[n] = \{5,\ 12,\ 7,\ 16,\ 4,\ 6\}$$
$$\uparrow$$
$$y(\Omega) = 5e^{j3\Omega} + 12e^{j2\Omega} + 7e^{j\Omega} + 16 + 4e^{-j\Omega} + 6e^{-j2\Omega} \tag{2.82}$$

Equate Eqs. (2.81) and (2.82). Equating the coefficients of $e^{j2\Omega}$, we get

$$10 + A = 12; \qquad A = 2$$

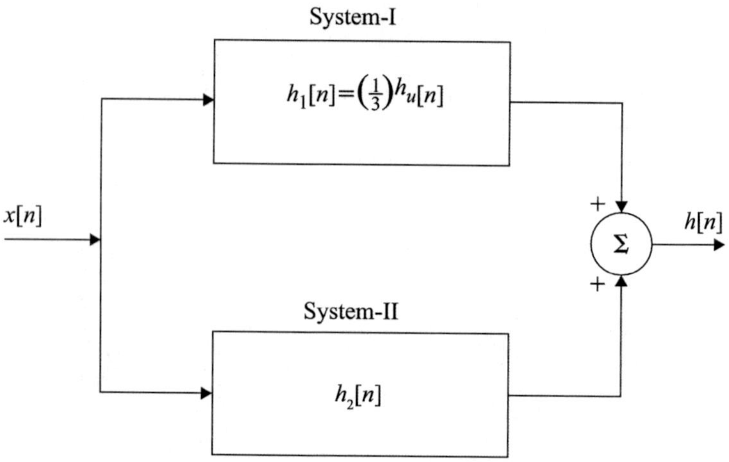

Fig. 2.23 Block diagram representation for Example 2.51

Equating the coefficients of $e^{j\Omega}$, we get

$$3 + 2A = 7; \qquad A = 2$$

Equating the coefficients of $e^{-j\Omega}$, we get

$$2A = 4; \qquad A = 2$$

The value of A is

$$\boxed{A = 2}$$

Example 2.51 Two systems connected in parallel are shown in Fig. 2.23. The impulse response of System-I is

$$h_1[n] = \left(\frac{1}{3}\right)^n u[n]$$

The impulse response function of the combined system

$$H(\Omega) = \frac{(-18 + 5e^{-j\Omega})}{(6 - 5e^{-j\Omega} + e^{-j2\Omega})}$$

Find the transfer function of System-II.

Solution

$$\begin{aligned} H(\Omega) &= \frac{(-18+5e^{-j\Omega})}{(6-5e^{-j\Omega}+e^{-j2\Omega})} \\ &= \frac{\left(-3+\frac{5}{6}e^{-j\Omega}\right)}{\left(1-\frac{5}{6}e^{-j\Omega}+\frac{1}{6}e^{-j2\Omega}\right)} \\ &= \frac{\left(-3+\frac{5}{6}e^{-j\Omega}\right)}{\left(1-\frac{1}{2}e^{-j\Omega}\right)\left(1-\frac{1}{3}e^{-j\Omega}\right)} \end{aligned}$$

$$\begin{aligned} h_1[n] &= \left(\frac{1}{3}\right)^n u[n] \\ H_1(\Omega) &= \frac{1}{\left(1-\frac{1}{3}e^{-j\Omega}\right)} \end{aligned}$$

From Fig. 2.23, $H_2(\Omega)$ can be written as

$$\begin{aligned} H_2(\Omega) &= H(\Omega) - H_1(\Omega) \\ &= \frac{\left(-3+\frac{5}{6}e^{-j\Omega}\right)}{\left(1-\frac{1}{2}e^{-j\Omega}\right)\left(1-\frac{1}{3}e^{-j\Omega}\right)} - \frac{1}{\left(1-\frac{1}{3}e^{-j\Omega}\right)} \\ &= \frac{\left(-4+\frac{4}{3}e^{-j\Omega}\right)}{\left(1-\frac{1}{2}e^{-j\Omega}\right)\left(1-\frac{1}{3}e^{-j\Omega}\right)} \\ &= \frac{-4\left(1-\frac{1}{3}e^{-j\Omega}\right)}{\left(1-\frac{1}{2}e^{-j\Omega}\right)\left(1-\frac{1}{3}e^{-j\Omega}\right)} = \frac{-4}{\left(1-\frac{1}{2}e^{-j\Omega}\right)} \end{aligned}$$

$$\boxed{h_2[n] = -4\left(\frac{1}{2}\right)^n u[n]}$$

2.21 The Convolution Sum

In discrete time system **the convolution sum or superposition sum** is used to determine the output sequence. For any arbitrary input $x(n)$, knowing the impulse response sequence $h_k(n)$, the output response $y(n)$ is obtained as explained below.

Let $h_{-1}(n)$, $h_0(n)$, $h_1(n)$ be the impulse responses to the inputs $x(n)\delta(n+1)$, $x(n)\delta(n)$ and $x(n)\delta(n-1)$. By applying superposition theorem, it is possible to get $y(n)$ which is nothing but the linear combination of the responses due to the individual shifted impulses. Consider the signals $x(n)$, $h_{-1}(n)$, $h_0(n)$ and $h_1(n)$ represented in Fig. 2.24a–d, respectively. The signal $x(-1)\delta(n+1)$ is obtained from $x(n)$ at

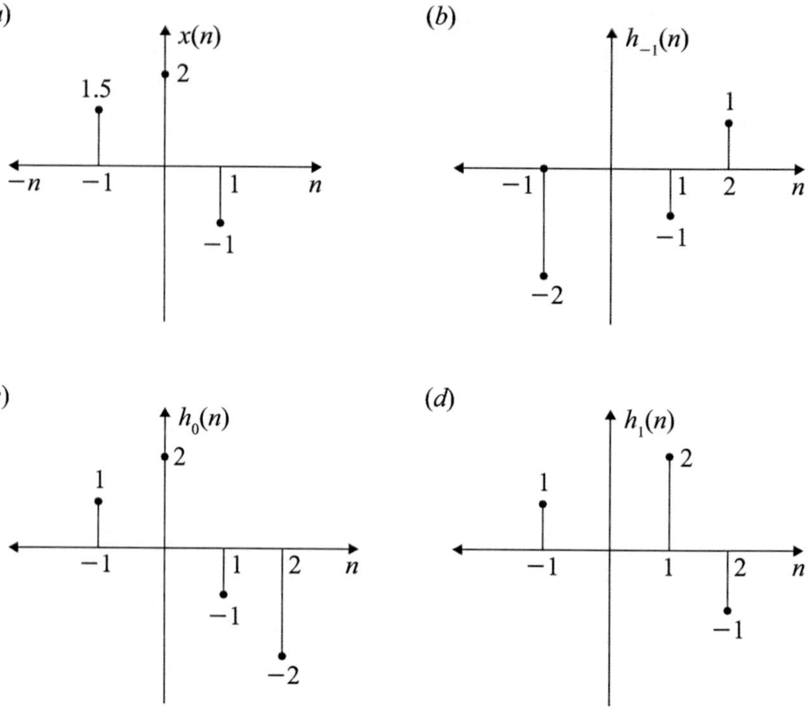

Fig. 2.24 Input and impulse response sequences

$n = -1$. Similarly $x(2)\delta(n)$ is obtained from $x(n)$ at $n = 0$. Similarly $x(1)\delta(n-1)$, $x(2)\delta(n-2)$ can be obtained. The product of $x(-1)\delta(n+1)h_{-1}(n)$ gives $y(-1)$. Similarly $x(0)\delta(n)h_0(n)$ gives $y(0)$ and $x(1)\delta(n-1)h_1(n)$ gives $y(1)$ evaluated at $n = 1$. These signals are represented in Fig. 2.25.

From Fig. 2.25a–k, the values of $y(n)$ for $-\infty < n < \infty$ are calculated from $x(n)h(n)$ by summing up all these values for a given n. Thus,

$$n = -1, \qquad y(-1) = -3 + 2 - 1 = -2$$
$$n = 0, \qquad y(0) = 4 = 4$$

$$n = 1, \qquad y(1) = -1.5 - 2 - 2 = -5.5$$
$$n = 2, \qquad y(2) = 1.5 - 4 + 1 = -1.5$$

The sketch of $y(n)$ is shown in Fig. 2.25k. Thus, the response $y(n)$ at any instant n is nothing but the superposition of the input at every point of n. Since $\delta(n-k)$ is the time shifted version of $\delta(n)$, the response $h_k(n)$ is the time shifted version of $h_0(n)$

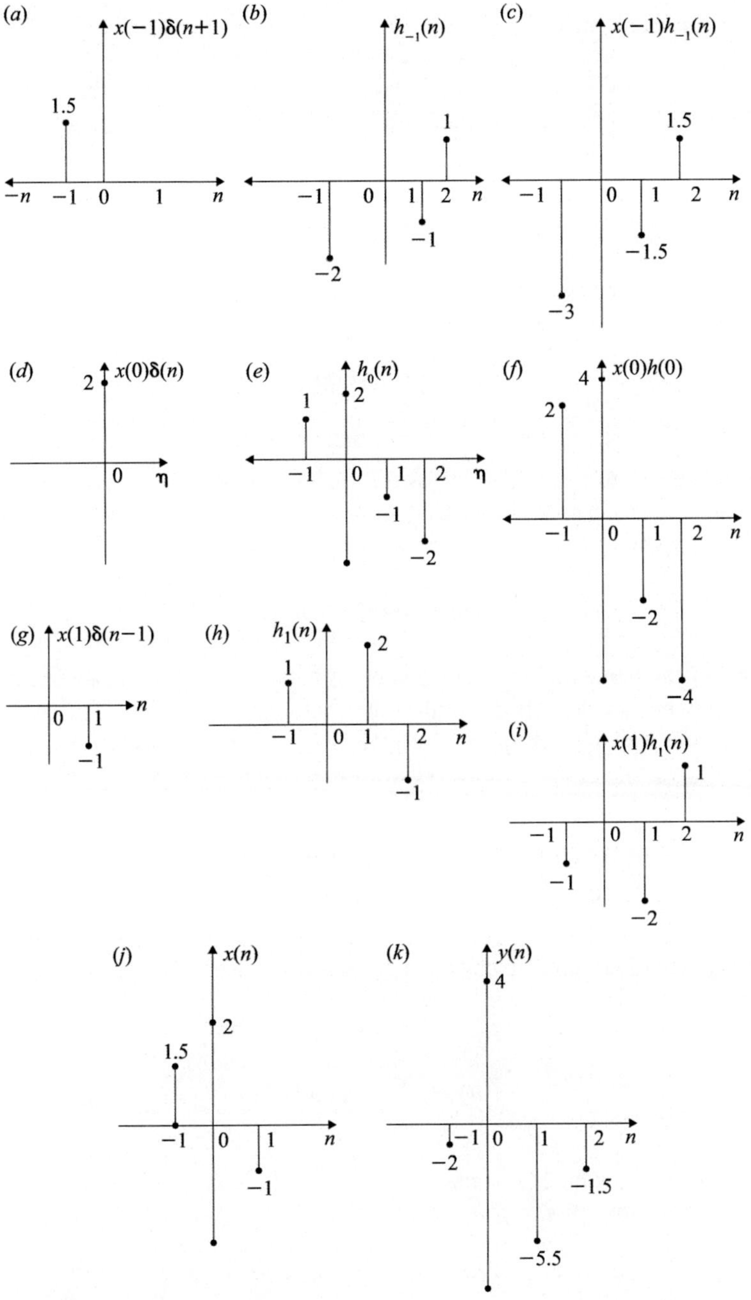

Fig. 2.25 Convolution sum of discrete signals

which can be represented as

$$h_k(n) = h_0(n - k)$$

Now the equation for the convolution sum can be written as

$$y(n) = \sum_{k=-\infty}^{\infty} x(k)h_0(n - k)$$

For convenience if we drop the subscript on $h_0(n - k)$, the above equation is written as

$$\boxed{y(n) = \sum_{k=-\infty}^{\infty} x(k)h(n - k)} \tag{2.83}$$

The above equation is referred to as the convolution sum or superposition sum. $h(n)$ is the impulse response of the LTID time system for the input $\delta(n)$. The convolution sum is symbolically represented as

$$\boxed{y(n) = x(n) * h(n)} \tag{2.84}$$

There are different methods available to get the solution of $y(n)$ by the convolution method, and they are described and illustrated below with examples. The properties of convolution which will be useful for solving convolution sum are discussed first as follows.

Important Points to Remember in Connection with Convolution Sum

1. The convolution sum is mathematically expressed as

$$y(n) = x(n) * h(n)$$

$$y(n) = \sum_{k=-\infty}^{\infty} x(k)h(n - k)$$

   '$*$' denotes convolution operation.
2. The commutative property of convolution is

$$x(n) * h(n) = h(n) * x(n)$$

3. The distributive property of convolution is

$$y(n) = x(n) * h(n)$$

where

$$h(n) = h_1(n) + h_2(n)$$

4. The associative property of convolution is

$$[x(n) * h_1(n)] * h_2(n) = x(n) * [h_1(n) * h_2(n)]$$

5. The shifting property of convolution is

$$x(n-p) * h(n-q) = y(n-p-q)$$

6. The width property of convolution is

$$[x(n) * h(n)] = T_1 + T_2$$

where T_1=width of $x(n)$ sequence and T_2=width of $h(n)$ sequence.
7. The convolution with an impulse is

$$x(n) * \delta(n) = x(n)$$

8. The convolution with delayed impulse is

$$x(n) * \delta(n - n_0) = x(n - n_0)$$

9. Convolution with unit step is

$$x(n) * u(n) = \sum_{k=-\infty}^{\infty} x(k)$$

10. Convolution with a delayed step is

$$x(n) * u(n - n_0) = \sum_{k=-\infty}^{\infty} x(k - n - n_0)$$

11. System causality in terms of impulse response is

$$h(n) = 0 \qquad \text{for } n < 0$$

12. The necessary and sufficient BIBO stability condition is

$$\sum_{n=-\infty}^{\infty} |h(n)| < \infty$$

In other words $|h(n)|$ should be absolutely summable.

13. Step response $s(n)$ is obtained from impulse response using the following mathematical expression:

$$s(n) = \sum_{k=0}^{n} h(k)$$

2.22 Response Using Convolution Sum

If the impulse response $h(n)$ is known, the output response $y(n)$ can be obtained for any input sequence $x(n)$ by using the following methods: (1) Analytical method and (2) Graphical method. In analytical method, $y(n)$ is obtained by

— Using mathematical expression for the convolution sum
— Multiplication method

2.22.1 Analytical Method Using Convolution Sum

Example 2.52 The impulse response $h(n)$ of a certain LTID system is given by $h(n) = a^n u(n)$ where $0 < a < 1$. The system is excited by $x(n) = u(n)$, a step sequence. Find $y(n)$ using convolution sum.

Solution The impulse response and input sequences are represented in Fig. 2.26a, b, respectively. They are causal sequences. Hence, the convolution sum given in Equation (2.83) is written as follows:

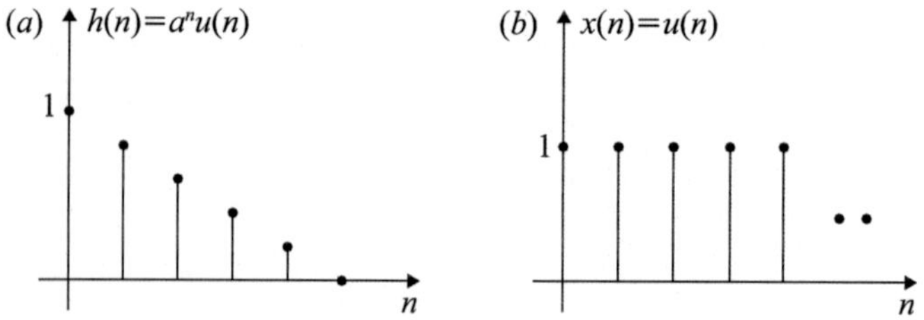

Fig. 2.26 Representation of $h(n)$ and $x(n)$ sequences

$$
\begin{aligned}
x(n) &= u(n) \\
x(k) &= u(k) = 1 \\
h(n) &= a^n u(n) \\
h(n-k) &= a^{n-k} u(n-k) \\
u(k) &= 1 \\
u(n-k) &= 1 \\
y(n) &= \sum_{k=-0}^{n} x(k)h(n-k) \\
&= \sum_{k=-0}^{n} (a^{n-k}) \\
&= a^n \sum_{k=-0}^{n} \left(\frac{1}{a}\right)^k
\end{aligned}
$$

The above expression is simplified using the finite summation formula

$$
\boxed{\sum_{k=-0}^{n} A^k = \frac{(1-A^{n+1})}{(1-A)}}
$$

Therefore

$$
y(n) = \frac{a^n\left(1 - \frac{1}{a^{n+1}}\right)}{\left(1 - \frac{1}{a}\right)}
$$

$$
\boxed{y(n) = \frac{(1-a^{n+1})}{(1-a)} u(n)}
$$

Example 2.53

$$
\begin{aligned}
x(n) &= u(n) \\
h(n) &= u(n)
\end{aligned}
$$

Find

$$
y(n) = x(n) * h(n)
$$

Solution The excitation sequence $x(n)$ and the impulse response sequence $h(n)$ are shown in Fig. 2.27a, b, respectively. Since these two signals are causal, the convolution sum can be written as follows:

$$
y(n) = \sum_{k=0}^{n} (1)^k (1)^{n-k} = (1)^n \sum_{k=0}^{n} (1) = \sum_{k=0}^{n} 1
$$

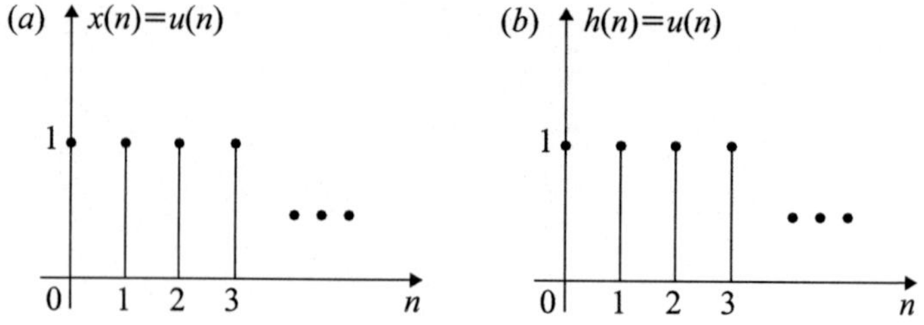

Fig. 2.27 Representation of $x(n) = u(n)$ and $h(n) = u(n)$

According to finite summation formula

$$\boxed{\sum_{k=0}^{n} 1 = (1+n)}$$

Therefore

$$\boxed{y(n) = (1+n)u(n)}$$

Example 2.54

$$x(n) = u(n)$$
$$h(n) = (0.6)^n u(n)$$

Find

$$y(n) = x(n) * h(n)$$

Solution

$$\begin{aligned}
x(n) &= 1^n \\
x(k) &= 1^k \\
h(n) &= (0.6)^n u(n) \\
h(n-k) &= (0.6)^{n-k} u(n-k) \\
y(n) &= \sum_{k=-0}^{n} 1^k (0.6)^{n-k} \\
&= (0.6)^n \sum_{k=0}^{n} 1^k (0.6)^{-k} \\
&= (0.6)^n \sum_{k=0}^{n} \left(\frac{1}{0.6}\right)^k
\end{aligned}$$

$$= (0.6)^n \frac{\left[1 - \frac{1}{(0.6)^{n+1}}\right]}{\left[1 - \frac{1}{0.6}\right]}$$

[Using Finite Summation Formula]

$$= \frac{[(0.6)^{n+1} - 1]}{(0.6) - 1}$$

$$\boxed{y(n) = 2.5[1 - (0.6)^{n+1}]u(n)}$$

Example 2.55

$$x(n) = a^n u(n)$$
$$h(n) = b^n u(n)$$

Find

$$y(n) = x(n) * h(n)$$

Solution $x(n)$ and $h(n)$ are causal and hence the range of convolution sum is $0 < k < n$.

$$x(n) = a^n u(n)$$
$$x(k) = a^k u(k)$$
$$h(n) = b^n u(n)$$
$$h(n-k) = b^{n-k} u(n-k)$$

$$y(n) = \sum_{k=0}^{n} x(k)h(n-k)$$

$$= b^n \sum_{k=0}^{n} \left(\frac{a}{b}\right)^k$$

$$= b^n \frac{\left[1 - (\frac{a}{b})^{n+1}\right]}{(1 - \frac{a}{b})}$$

[Using Finite Summation Formula]

$$\boxed{y(n) = \frac{(b^{n+1} - a^{n+1})}{(b - a)} u(n)}$$

Example 2.56

$$x(n) = \left(\frac{1}{5}\right)^n u(n)$$
$$h(n) = 3^n u(n)$$

Find

$$y(n) = x(n) * h(n)$$

Solution

$$x(n) = \left(\frac{1}{5}\right)^n u(n)$$
$$x(k) = \left(\frac{1}{5}\right)^k u(k)$$
$$h(n) = 3^n u(n)$$
$$h(n-k) = 3^{n-k} u(n-k)$$

For causal $x(n) * h(n)$

$$y(n) = \sum_{k=0}^{n} x(k)h(n-k)$$
$$= \sum_{k=0}^{n} \left(\frac{1}{5}\right)^k 3^{n-k}$$
$$= 3^n \sum_{k=0}^{n} \left(\frac{1}{15}\right)^k$$

Using finite summation formula, we get

$$y(n) = 3^n \frac{\left[1 - \frac{1}{15^{n+1}}\right]}{\left(1 - \frac{1}{15}\right)}$$
$$= 3^n \frac{[15^{n+1} - 1]}{(14)} \frac{15}{15^{n+1}}$$

$$\boxed{y(n) = \frac{1}{14(5)^n}[15^{n+1} - 1]u(n)}$$

Example 2.57

$$x(n) = (0.6)^n u(n)$$
$$h(n) = (0.2)^n u(n)$$

Find

$$y(n) = x(n) * h(n)$$

Solution

$$\begin{aligned} x(n) &= (0.6)^n u(n) \\ x(n-k) &= (0.6)^{n-k} u(n-k) \\ h(n) &= (0.2)^n u(n) \\ h(k) &= (0.2)^k u(k) \end{aligned}$$

Both $x(n)$ and $h(n)$ are causal signals. Hence, the summation of convolution is $0 < k < n$. In this problem, we use commutative property of convolution.

$$\begin{aligned} x(n) * h(n) &= h(k) * x(n) \\ y(n) &= \sum_{k=0}^{n} h(k)x(n-k) \\ &= \sum_{k=0}^{n} (0.2)^k (0.6)^{n-k} \\ &= (0.6)^n \sum_{k=0}^{n} \left(\frac{1}{3}\right)^k \end{aligned}$$

Using finite summation formula, we get

$$y(n) = (0.6)^n \frac{\left[1 - \frac{1}{3^{n+1}}\right]}{\left(1 - \frac{1}{3}\right)}$$

$$y(n) = (0.6)^n \frac{\left[3^{n+1} - 1\right]}{2} \frac{3}{3^{n+1}}$$

$$\boxed{y(n) = 0.5(0.2)^n [3^{n+1} - 1] u(n)}$$

Example 2.58 Two discrete time systems with impulse responses $h_1(n)$ and $h_2(n)$ are connected in cascade as shown in Fig. 2.28. Determine the unit sample response of the interconnected system.

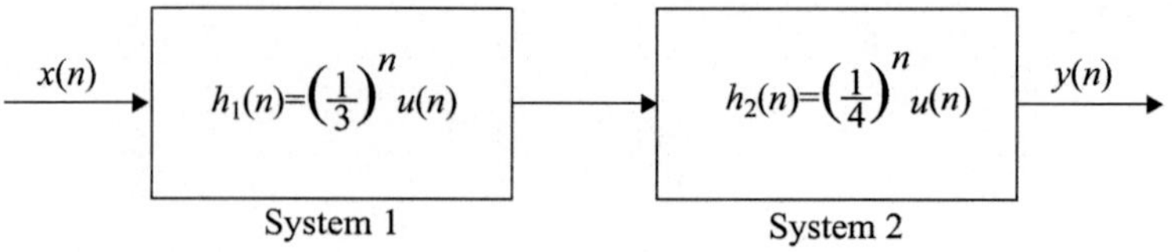

Fig. 2.28 Impulse response of two systems connected in cascade

Solution

$$y(n) = x(n) * h(n)$$

where

$$h(n) = h_1(n) * h_2(n)$$

For a sample

$$\begin{aligned} x(n) &= \delta(n) \\ x(n) * h(n) &= \delta(n) * h(n) \\ &= h(n) \\ &= h_1(n) * h_2(n) \end{aligned}$$

Therefore

$$\begin{aligned} y(n) &= h_1(n) * h_2(n) \\ h_1(k) &= \left(\frac{1}{3}\right)^k u(k) \end{aligned}$$

$$\begin{aligned} h_1(n-k) &= \left(\frac{1}{4}\right)^{n-k} u(n-k) \\ y(n) &= \sum_{k=0}^{n} \left(\frac{1}{3}\right)^k \left(\frac{1}{4}\right)^{n-k} \\ &= \left(\frac{1}{4}\right)^n \sum_{k=0}^{n} \left(\frac{4}{3}\right)^k \\ &= \left(\frac{1}{4}\right)^n \frac{\left[1-\left(\frac{4}{3}\right)^{n+1}\right]}{\left(1-\frac{4}{3}\right)} \end{aligned}$$

$$\boxed{y(n) = \left(\frac{1}{4}\right)^n 3\left[\left(\frac{4}{3}\right)^{n+1} - 1\right] u(n)}$$

Example 2.59 Determine the convolution of the signals

$$x(n) = \cos \pi n \, u(n)$$

$$h(n) = \left(\frac{1}{2}\right)^n u(n)$$

Solution

$$\begin{aligned}
x(n) &= \cos \pi n \, u(n) \\
&= (-1)^n u(n) \\
x(k) &= (-1)^k u(k) \\
h(n) &= \left(\frac{1}{2}\right)^n u(n) \\
h(n-k) &= \left(\frac{1}{2}\right)^{n-k} u(n-k)
\end{aligned}$$

Both $x(n)$ and $h(n)$ are casual. Hence, the convolution sum takes the range of $0 < k < n$.

$$y(n) = \sum_{k=0}^{n} x(k)h(n-k)$$

$$\begin{aligned}
y(n) &= \sum_{k=0}^{n} (-1)^k \left(\frac{1}{2}\right)^{n-k} \\
&= \left(\frac{1}{2}\right)^n \sum_{k=0}^{n} [(-1)(2)]^k \\
&= \left(\frac{1}{2}\right)^n \sum_{k=0}^{n} (-2)^k \\
&= \left(\frac{1}{2}\right)^n \frac{[1-(-2)^{n+1}]}{[1-(-2)]} \\
&= \frac{1}{3}\left(\frac{1}{2}\right)^n [1-(-2)^{n+1}]
\end{aligned}$$

$$\boxed{y(n) = \frac{1}{3}\left(\frac{1}{2}\right)^n [1+2(-2)^n]u(n)}$$

Example 2.60

$$x(n) = (0.2)^n u(n)$$
$$h(n) = (0.2)^{-n} u(-n)$$

Find

$$y(n) = x(n) * h(n)$$

Solution The signal $x(n)$ is a casual signal and $h(n)$ is an anti-causal signal. $x(k)$ is shown in Fig. 2.29a and $h(k)$ is shown in Fig. 2.29b, respectively. $h(-k)$ is shown in Fig. 2.29c and $h(n-k)$ shifted to the extreme left for $n < 0$ and to the right for $n > 0$. They are shown in Fig. 2.29d, e, respectively.

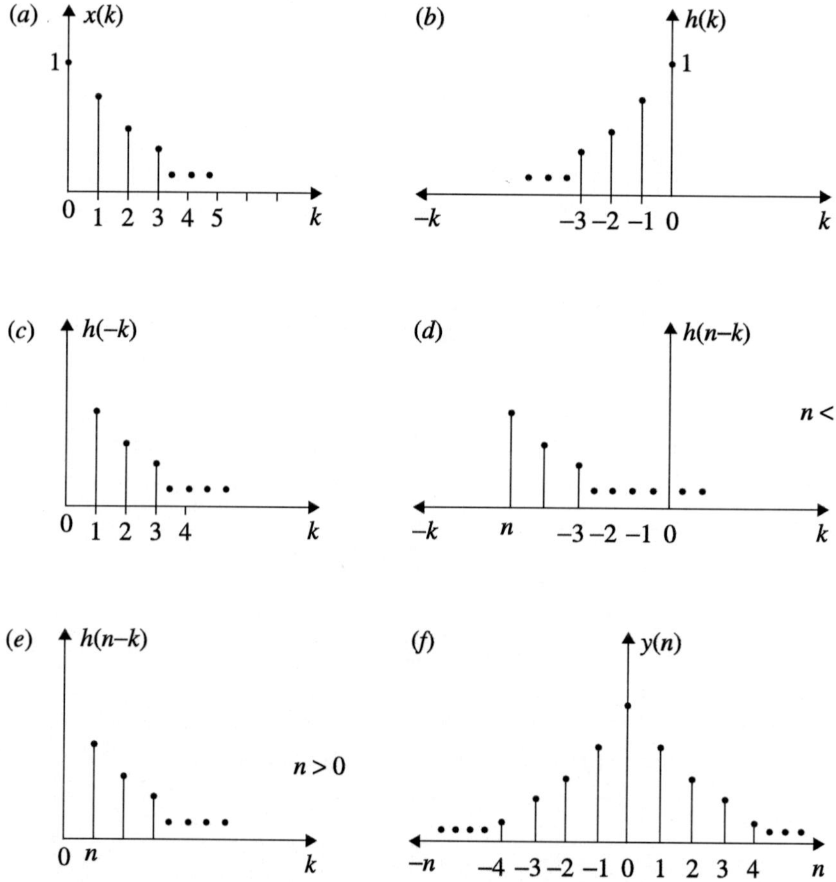

Fig. 2.29 Convolution of casual and anti-causal signals

$$\begin{aligned}
y(n) &= x(n) * h(n) \\
&= \sum_{k=-\infty}^{\infty} x(k)h(n-k) \\
&= \sum_{k=-\infty}^{\infty} (0.2)^k u(k)(0.2)^{-(n-k)} u(-(n-k)) \\
&= (0.2)^{-n} \sum_{k=-\infty}^{\infty} (0.2)^{2k} u(k)u(k-n)
\end{aligned}$$

When $h(n-k)$ is moved towards the right, it overlaps with $x(k)$ for $n < 0$ and for $n > 0$. For $n < 0$

$$u(k)u(k-n) = \begin{cases} 1 & k > 0 \\ 0 & \text{otherwise} \end{cases}$$

Now the limit of summation of $y(n)$ is $0 < k < \infty$. Therefore

$$\begin{aligned}
y(n) &= (0.2)^{-n} \sum_{k=0}^{\infty} [(0.2)^2]^k \\
&= (0.2)^{-n} \left[\frac{1}{1-(0.2)^2} \right] \\
y(n) &= \frac{(0.2)^{-n}}{0.96} \qquad n < 0
\end{aligned}$$

For $n > 0$

$$u(k)u(k-n) = \begin{cases} 1 & \text{for } k \geq n \\ 0 & \text{otherwise} \end{cases}$$

The limit of the summation of $y(n)$ is $n \leq k \leq \infty$

$$\begin{aligned}
y(n) &= \sum_{k=n}^{\infty} (0.2)^k u(k)(0.2)^{-(n-k)} u(-n+k) \\
&= (0.2)^{-n} \sum_{k=n}^{\infty} [(0.2)^2]^k
\end{aligned}$$

Using the finite summation formula

$$\sum_{k=n}^{\infty} a^k = \frac{a^n}{1-a}$$

we get

$$y(n) = (0.2)^{-n}\left[\frac{(0.2)^{2n}}{(1-0.2^2)}\right]$$
$$= \frac{(0.2)^n}{0.96} \qquad n \geq 0.$$

In general

$$\boxed{y(n) = \frac{(0.2)^{|n|}}{0.96} \qquad \text{for all } n}$$

The response $y(n)$ is plotted and is shown in Fig. 2.29f.

Example 2.61

$$x(n) = \begin{cases} 0 & n < -6 \\ \left(\frac{1}{4}\right)^n & n \geq -6 \end{cases}$$

$$h(n) = \begin{cases} 0 & n < 4 \\ \left(\frac{1}{5}\right)^n & n \geq 4 \end{cases}$$

Find

$$y(n) = x(n) * h(n)$$

Solution Figure 2.30(*a*), (*b*) and (*c*) shows $x(h)$, $h(k)$ and $h(n-k)$, respectively.

$$y(n) = x(n) * h(n)$$
$$= \sum_{k=-6}^{n-4} x(k)h(n-k)$$

From Fig. 2.30a, the following equation is written.

$$x(k) = \begin{cases} 0 & k < -6 \\ \left(\frac{1}{4}\right)^k & k \geq -6 \end{cases}$$

From Fig. 2.30b, the following equation is written.

$$h(n-k) = \begin{cases} 0 & (n-k) < 4 \text{ or } k > n-4 \\ \left(\frac{1}{5}\right)^k & (n-k) > 4 \text{ or } k < n-4 \end{cases}$$

When $h(n-k)$ is moved towards the right, $h(n-k)$ overlaps with $x(k)$ at $n \geq -2$

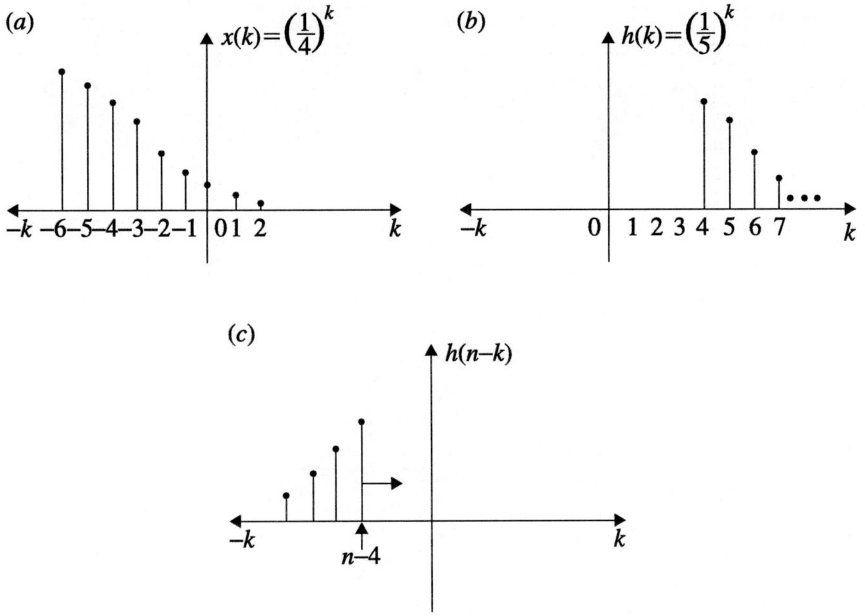

Fig. 2.30 Representation of signals for Example 2.61

$$y(n) = \begin{cases} 0 & n < -2 \\ \sum\limits_{k=-6}^{n-4} \left(\frac{1}{4}\right)^k \left(\frac{1}{5}\right)^{n-k} & n \geq -2 \end{cases}$$

Put $p = k + 6$ for $n \geq -2$

$$y(n) = \sum_{p=0}^{n+2} \left(\frac{1}{4}\right)^{p-6} \left(\frac{1}{5}\right)^{n-p+6}$$

$$= \left(\frac{4}{5}\right)^6 \left(\frac{1}{5}\right)^n \sum_{p=0}^{n+2} \left(\frac{5}{4}\right)^p = \left(\frac{4}{5}\right)^6 \left(\frac{1}{5}\right)^n \left[\frac{1 - \left(\frac{5}{4}\right)^{n+3}}{\left(1 - \frac{5}{4}\right)}\right]$$

$$\boxed{y(n) = 4^7 \left(\frac{1}{5}\right)^{n+6} \left[\left(\frac{5}{4}\right)^{n+3} - 1\right]} \quad n \geq -2$$

Example 2.62 A linear time invariant discrete time system has the following impulse response.

$$h(n) = \left(\frac{1}{2}\right)^n u(n)$$

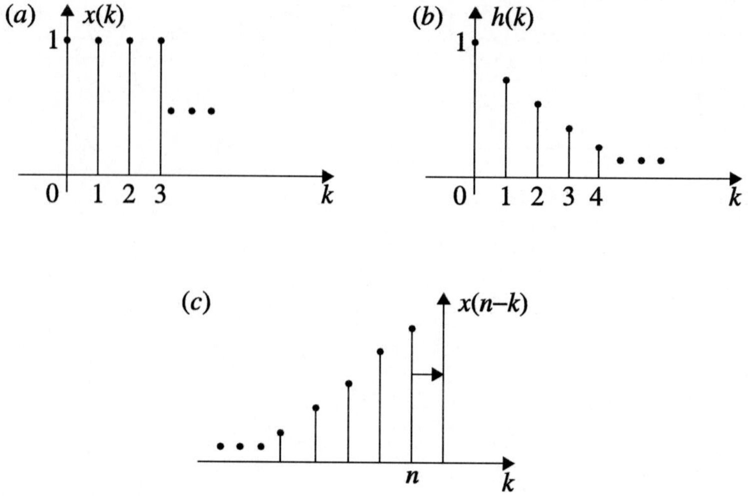

Fig. 2.31 Representation of signals for Example 2.62

The system is excited by the signal $x(n)=u(n)$ determine the output of the system at $n = -6$, $n = 6$ and $n = 12$.

Solution $x(k)$, $h(k)$ and $h(n-k)$ are shown in Fig. 2.31a–c, respectively. At $n = -6$, $h(n-k) < 0$ and it does not overlap with $x(k)$ and hence $y(n) = 0$. For $n > 0$, the following equation is written:

$$
\begin{aligned}
y(n) &= x(n) * h(n) \\
&= \sum_{k=0}^{n} x(n)h(k) \\
&= \sum_{k=0}^{n} \left(\frac{1}{2}\right)^{n-k} \\
&= \left(\frac{1}{2}\right)^{n} \sum_{k=0}^{n} (2)^{k} \\
&= \left(\frac{1}{2}\right)^{n} \frac{1-2^{n+1}}{(1-2)} \\
&= \left(\frac{1}{2}\right)^{n} (2^{n+1}-1)u(n)
\end{aligned}
$$

Substituting $n = 6$

$$
y(6) = \left(\frac{1}{2}\right)^{6} [2^{7}-1]
$$

$$\boxed{y(6) = \frac{127}{64}}$$

Substituting $n = 12$

$$y(12) = \left(\frac{1}{2}\right)^{12} [2^{13} - 1]$$

$$\boxed{y(12) = \frac{8191}{4096}}$$

$$\boxed{y(-6) = 0}$$

Example 2.63

$$x(n) = u(n)$$
$$h(n) = u(n-4)$$

Find

$$y(n) = x(n) * h(n)$$

Solution The signals $x(k)$, $h(k)$ and $h(n-k)$ are shown in Fig. 2.32a–c, respectively. Figure 2.32c is moved towards right so that it overlaps with $x(k)$. Overlap-

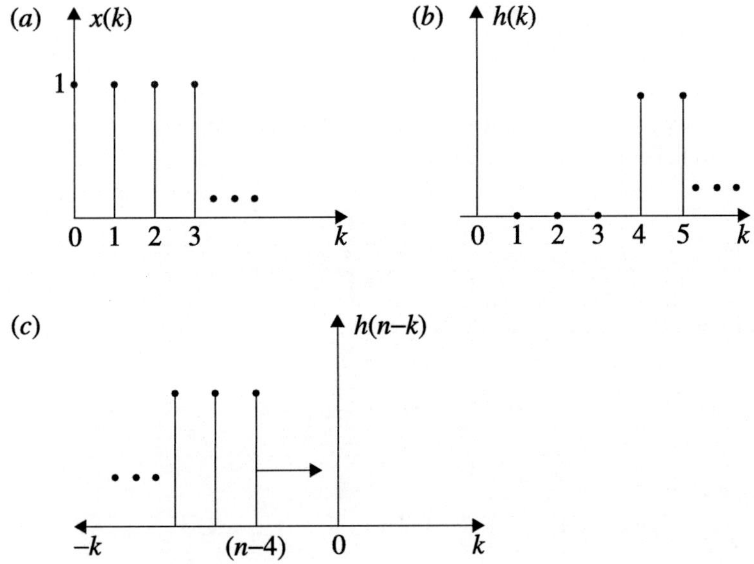

Fig. 2.32 Signals of Example 2.63

ping occurs for $n - 4 \leq k \leq \infty$. Therefore, the limit of convolution sum is $k = 0$ to $k = (n - 4)$

$$y(n) = \sum_{k=0}^{n-4} x(k)h(n-k) \qquad n \geq 4$$

$$= \sum_{k=0}^{n-4} 1 = (n-3)$$

$$\boxed{y(n) = (n-3)} \qquad n \geq 4$$

Example 2.64 A linear time invariant system has the following impulse response:

$$h(n) = [u(n) - u(n-6)]$$

The system is excited by

$$x(n) = [u(n-1) - u(n-5)]$$

Determine the output of the system.

Solution Method 1

1. $x(k)$, $h(k)$ and $h(n-k)$ are represented in Fig. 2.33a–c, respectively. For $n = 0$, there is no overlapping and hence $y(n) = 0$.
2. **Overlapping interval 1**:
   $h(n-k)$ when moved towards right overlaps with $x(k)$ when the leading edge (right edge) of $h(n-k)$ crosses past the left edge $x(k)$. This happens when $n > 1$. This overlapping does not change until $n < 4$. Hence, the overlapping interval is $1 \leq n \leq 4$. The summation interval is from $k = 1$ to $k = n$.

   $$y_1(n) = \sum_{k=1}^{n} 1$$

   The finite summation formula used is

   $$\sum_{k=N_1}^{N_2} 1 = N_2 - N_1 + 1$$

   Therefore,

   $$y_1(n) = n - 1 + 1 = n \qquad 1 \leq n \leq 4$$

   This is shown in Fig. 2.33d.

3. **Overlapping interval 2**:
For $n > 4$, the right edge of $h(n-k)$ crosses past the right edge of $x(k)$. This change of overlapping occurs during the time interval $4 < n < 6$. The output during this interval is $y_2(n)$

$$y_2(n) = \sum_{k=1}^{4} 1 \quad 4 \le n \le 6$$
$$= 4 - 1 + 1$$
$$= 4$$

This is shown in Fig. 2.33e.

4. **Overlapping interval 3**:
When the left edge of $h(n-k)$ crosses past the left edge of $x(k)$, there is overlapping and it continues until $n = 4$. The overlapping interval is therefore $6 \le n < 9$. The output is $y_3(n)$. The limit of summation is from $k = n - 5$ to $k = 4$

$$y_3(n) = \sum_{k=n-5}^{4} 1$$
$$= -n + 5 + 4 + 1$$
$$= 10 - n \quad 6 < n < 9$$

This is shown in Fig. 2.33f.

5. For $n > 9$, there is no overlapping and hence $y(n) = 0$. This is shown in Fig. 2.33g. Hence, the output response $y_1(n)$, $y_2(n)$ and $y_3(n)$ are summed up with their respective time interval to get the total response $y(n)$.
6.

$$y(n) = n \qquad 1 < n < 4$$
$$y(n) = 4 \qquad 4 < n < 6$$
$$y(n) = 10 - n \qquad 6 < n < 9$$

The values of $y(n)$ for $1 \le n \le 10$ is shown in the following table. $y(n) = \{0, 1, 2, 3, 4, 4, 4, 3, 2, 1\}$

n	0	1	2	3	4	5	6	7	8	9	10
$y(n)$	0	1	2	3	4	4	4	3	2	1	0

7. The plot of $y(n)$ with respect to n is shown in Fig. 2.33h.

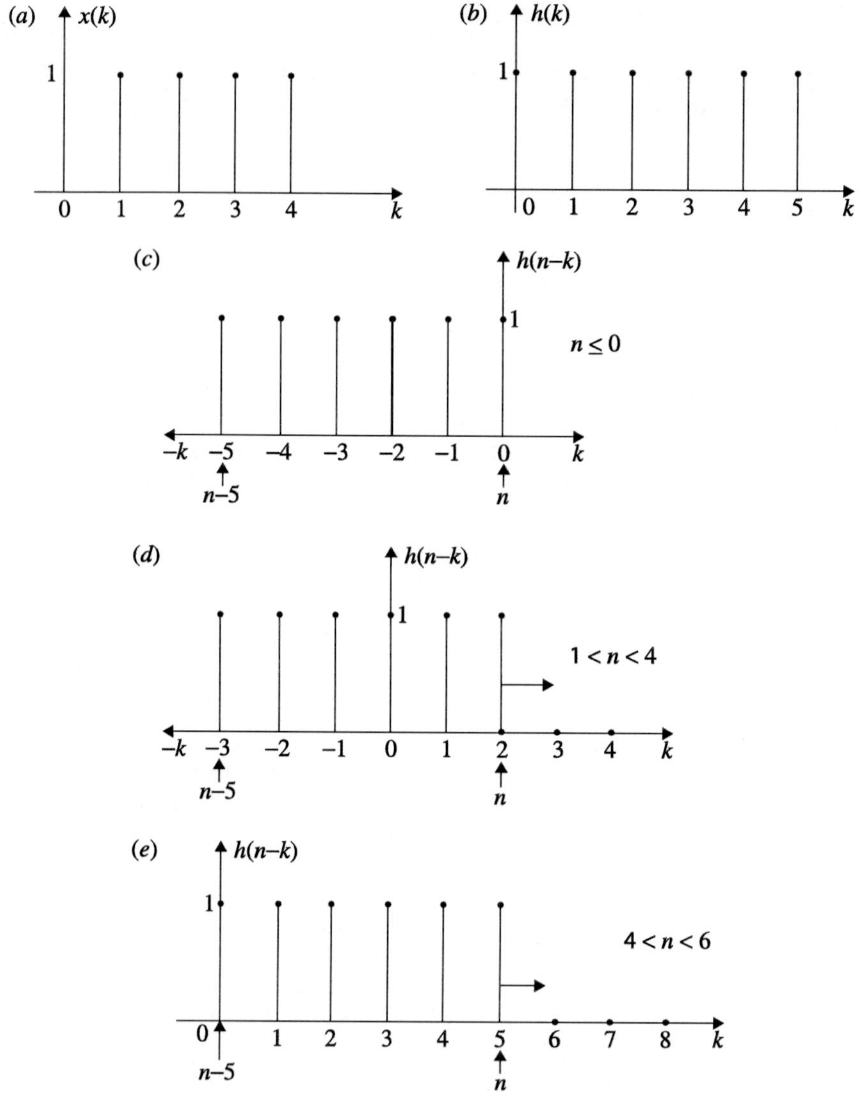

Fig. 2.33 Convolution sum for Example 2.64. Response plot of Example 2.64

8. To check the width property of convolution. The width property can be easily checked. The number of elements of $h(n)$ is $T_1 = 4$. The number of elements of $h(n)$ is $T_2 = 6$. The width property of $y(n)$ is $T = T_1 + T_2 - 1 = 6 + 4 - 1 = 9$. From Fig. 2.33h, it is seen that $y(n)$ has 9 elements.

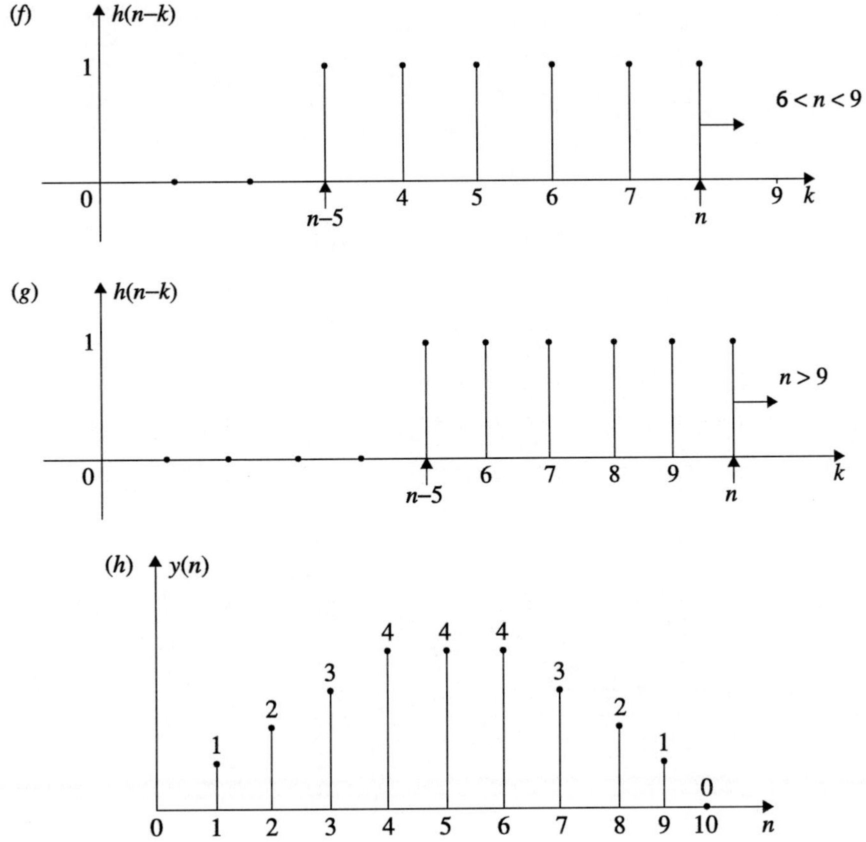

Fig. 2.33 (continued)

2.22.2 *Convolution Sum of Two Sequences by Multiplication Method*

Let

$$
\begin{aligned}
x(n) &= \{x_1,\ x_2,\ \underset{\uparrow}{x_3},\ x_4\} \\
h(n) &= \{h_1,\ h_2,\ \underset{\uparrow}{h_3}\}
\end{aligned}
$$

The convolution of these two sequences $x(n) * h(n)$ is obtained as given below:

1. Write down $x(n)$ and $h(n)$ one below the other:

$$\begin{array}{cccc} x_1 & x_2 & x_3 & x_4 \\ & h_1 & h_2 & h_3 \end{array}$$

2. Carry out the multiplication of the first row by the second row as given below:

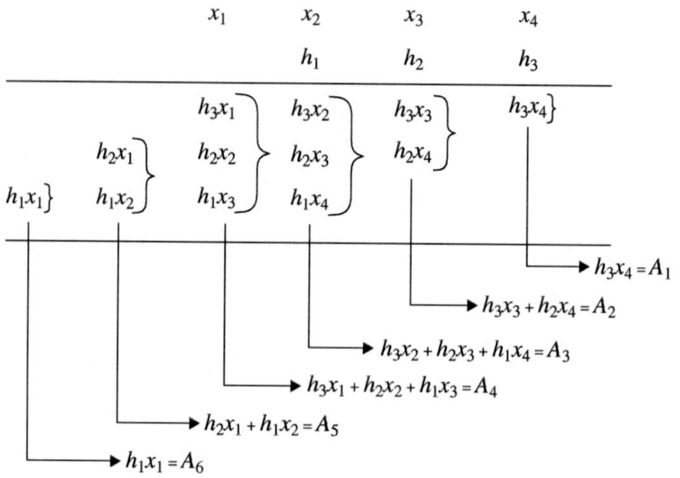

3. Arrange the sequences so obtained as given below:

$$y(n) = \{A_6,\ A_5,\ A_4,\ A_3,\ \underset{\uparrow}{A_2},\ A_1\}$$

4. Let N_1 = width of the $x(n)$ sequence to the left and N_2 = width of the $h(n)$ sequence to the left.
   The width of the output sequence $y(n)$ to the left is $N = N_1 + N_2$. In the example illustrated above $N_1 = 2$ and $N_2 = 2$. In the $y(n)$ sequence the arrow corresponding to $n = 0$ is marked such that to the left of the arrow $N = N_1 + N_2 = 4$ is marked which is at A_2. From $y(n)$ sequence $y(0) = A_2$; $y(1) = A_1$; $y(-1) = A_3$; $y(-2) = A_4$; $y(-3) = A_5$ and $y(-4) = A_6$.
   The width property can also be easily checked. For $x(n)$ the width $T_1 = 4$. For $h(n)$, the width $T_2 = 3$. The width of $y(n) = T = T_1 + T_2 - 1 = 4 + 3 - 1 = 6$.

The following examples illustrate the above method.

Example 2.65 What is the response of an LTID system with impulse response $h(n) = \delta(n) + 2\delta(n-1)$ for the input $x(n) = \{1, 2, 3\}$.

Solution

$$
\begin{aligned}
h(n) &= \delta(n) + 2\delta(n-1) \\
&= \{\underset{\uparrow}{1},\ 2\}
\end{aligned}
$$

$$
x(n) = \{\underset{\uparrow}{1},\ 2,\ 3\}
$$

where $N_1 = 0$, $N_2 = 0$, $N = N_1 + N_2 = 0$ and $T_1 = 2$, $T_2 = 3$, $T = 3 + 2 - 1 = 4$

		1	2
	1	2	3
		3	6
	2	4	
1	2		
1	4	7	6

$$
\boxed{y(n) = \{\underset{\uparrow}{1},\ 4,\ 7,\ 6\}}
$$

The same result as in Example 2.65 is obtained.

Example 2.66 Find the convolution of $x(n) = \{1, 2, 3, 4, 5\}$ with $h(n) = \{1, 2, 3, 3, 2, 1\}$.

Solution When no arrow is marked in $x(n)$ or $h(n)$, the signals are to be taken as causal and therefore $N_1 = 0$, $N_2 = 0$ and $N = N_1 + N_2 = 0$; $T_1 = 6$, $T_2 = 5$, $T = 6 + 5 - 1 = 10$.

The following multiplication is done.

				1	2	3	3	2	1
					1	2	3	4	5
				5	10	15	15	10	5
			4	8	12	12	8	4	
		3	6	9	9	6	3		
	2	4	6	6	4	2			
1	2	3	3	2	1				
1	4	10	19	30	36	35	26	14	5

$$
\boxed{\begin{array}{l} y(n) = \{1,\ 4,\ 10,\ 19,\ 30,\ 36,\ 35,\ 26,\ 14,\ 5\} \\ \qquad\quad \uparrow \end{array}}
$$

Example 2.67 Find the linear convolution of

$$
\begin{array}{l} x(n) = \{\ 1,\ 2,\ 3,\ 4,\ 5,\ 6\} \\ \qquad\qquad\ \ \uparrow \end{array}
$$

and

$$
\begin{array}{l} h(n) = \{\ 2,\ -4,\ 6,\ -8\} \\ \qquad\qquad\ \ \uparrow \end{array}
$$

			1	2	3	4	5	6
					2	−4	6	−8
			−8	−16	−24	−32	−40	−48
		6	12	18	24	30	36	
	−4	−8	−12	−16	−20	−24		
2	4	6	8	10	12			
2	0	4	0	−4	−8	−26	−4	−48

Solution here $N_1 = 0$; $N_2 = 0$ and $N = 0$; $T_1 = 6$, $T_2 = 4$ and $T = 6 + 4 - 1 = 9$.

$$
\boxed{\begin{array}{l} y(n) = \{2,\ 0,\ 4,\ 0,\ -4,\ -8,\ -26,\ -4,\ -48\} \\ \qquad\quad \uparrow \end{array}}
$$

Example 2.68

$$
\begin{array}{l} x(n) = \{-1/2,\ \ 2,\ 1/3,\ 3/2\} \\ \qquad\qquad\qquad\quad\ \ \uparrow \\ h(n) = \{1,\ -1/2,\ \ 2/3\} \\ \qquad\qquad\qquad\quad\ \ \uparrow \end{array}
$$

Find $y(n)$ by convolution method.

Solution Here $N_1 = 1$; $N_2 = 2$ and $N_1 + N_2 = 3$; $T_1 = 4$, $T_2 = 3$ and $T = 4 + 3 - 1 = 6$.

$$
\begin{array}{cccccc}
 & & -\frac{1}{2} & 2 & \frac{1}{3} & \frac{3}{2} \\
 & & & 1 & -\frac{1}{2} & \frac{2}{3} \\
\hline
 & & -\frac{1}{3} & \frac{4}{3} & \frac{2}{9} & 1 \\
 & \frac{1}{4} & -1 & -\frac{1}{6} & -\frac{3}{4} & \\
-\frac{1}{2} & 2 & \frac{1}{3} & \frac{3}{2} & & \\
\hline
-\frac{1}{2} & \frac{9}{4} & -1 & \frac{8}{3} & -\frac{19}{36} & 1 \\
\hline
\end{array}
$$

$$
y(n) = \left\{ -\frac{1}{2}, \ \frac{9}{4}, \ -1, \ \underset{\uparrow}{\frac{8}{3}}, \ -\frac{19}{36}, \ 1 \right\}
$$

Example 2.69 Find the linear convolution of

$$x(n) = \{1, 2, 3, 4, 5\} \quad \text{and} \quad h(n) = \{1, 2, 3, 3, 2, 1\}$$

Use graphical methods.

Solution 1. $x(k)$ is shown in Fig. 2.34a and $h(k)$ is shown in Fig. 2.34b. $h(n-k)$ is shown in Fig. 2.34c. $h(n-k)$ is moved towards right so that it overlaps with $x(k)$. This happens at $n = 0$. The sample with strength 1 in $x(k)$ and the sample $h(k)$ with strength 1 at $n = 0$ overlap as in Fig. 2.34d. The product of these two samples is $y(0) = 1 \times 1 = 1$.

2. $h(n-k)$ is moved further towards right by one more sample $n = 1$. Now this is shown in Fig. 2.34e. The product of the overlapping samples for $n = 1$ is shown by dotted lines $y(1) = 1 \times 2 + 2 \times 1 = 4$.
3. For $n = 2$, $x(k)$ and $h(n-k)$ are shown in Fig. 2.34f. 3 samples in each of $x(k)$ and $h(n-k)$ overlap. Hence, $y(2) = 1 \times 3 + 2 \times 2 + 3 \times 1 = 10$.
4. For $n = 3$, $x(k)$ and $h(n-k)$ are shown in Fig. 2.34g. Here 4 samples overlap. Hence, $y(3) = 1 \times 3 + 2 \times 3 + 3 \times 2 + 4 \times 1 = 19$.
5. For $n = 4$, $x(k)$ and $h(n-k)$ are shown in Fig. 2.34h. $y(4) = 1 \times 2 + 2 \times 3 + 3 \times 3 + 4 \times 2 + 5 \times 1 = 30$.
6. For $n = 5$, $x(k)$ and $h(n-k)$ are shown in Fig. 2.34i. Here 5 samples overlap. $y(5) = 1 \times 1 + 2 \times 2 + 3 \times 3 + 4 \times 3 + 5 \times 2 = 36$.
7. For $n = 6$, $x(k)$ and $h(n-k)$ are shown in Fig. 2.34j. Here 4 samples overlap. $y(6) = 2 \times 1 + 3 \times 2 + 4 \times 3 + 5 \times 3 = 35$.

8. For $n = 7$, $x(k)$ and $h(n-k)$ are shown in Fig. 2.34k. Here 3 samples overlap. Hence, $y(7) = 3 \times 1 + 4 \times 2 + 5 \times 3 = 26$.
9. For $n = 8$, $x(k)$ and $h(n-k)$ are shown in Fig. 2.34l. Here 2 samples overlap. Hence, $y(8) = 4 \times 1 + 5 \times 2 = 14$.
10. For $n = 9$, $x(k)$ and $h(n-k)$ are shown in Fig. 2.34m. Here one sample overlaps. $y(9) = 5 \times 1 = 5$.
11. For $n = 10$, $x(k)$ and $h(n-k)$ do not overlap with any of the samples of $x(k)$. Hence, $y(10) = 0$.
12. For $x(n)$, $N_1 = 0$ and for $h(n)$, $N_2 = 0$. Hence, $N = N_1 + N_2 = 0$. For $x(n)$, $T_1 = 5$ and for $h(n)$, $T_2 = 6$. Hence width of $y(n) = T_1 + T_2 - 1 = 5 + 6 - 1 = 10$.

$$y(n) = \{\underset{\uparrow}{1},\ 4,\ 10,\ 19,\ 30,\ 36,\ 35,\ 26,\ 14,\ 5\}$$

Example 2.70

$$x(n) = 2u(n+2) - 2u(n) - 3u(n-1) + 3u(n-3)$$
$$h(n) = -u(n+1) + 3u(n) - 5u(n-1) + 5u(n-2)$$

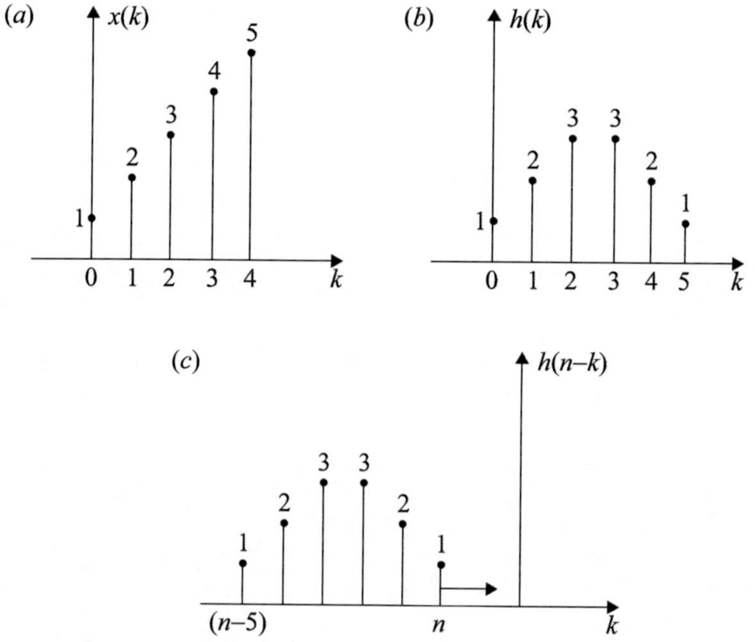

Fig. 2.34 Convolution of Example 2.69(*a*) to (*c*). Convolution of Example 2.69(*d*) to (*g*). Convolution of Example 2.69(*h*) to (*l*). Convolution of Example 2.69(*m*) to (*n*)

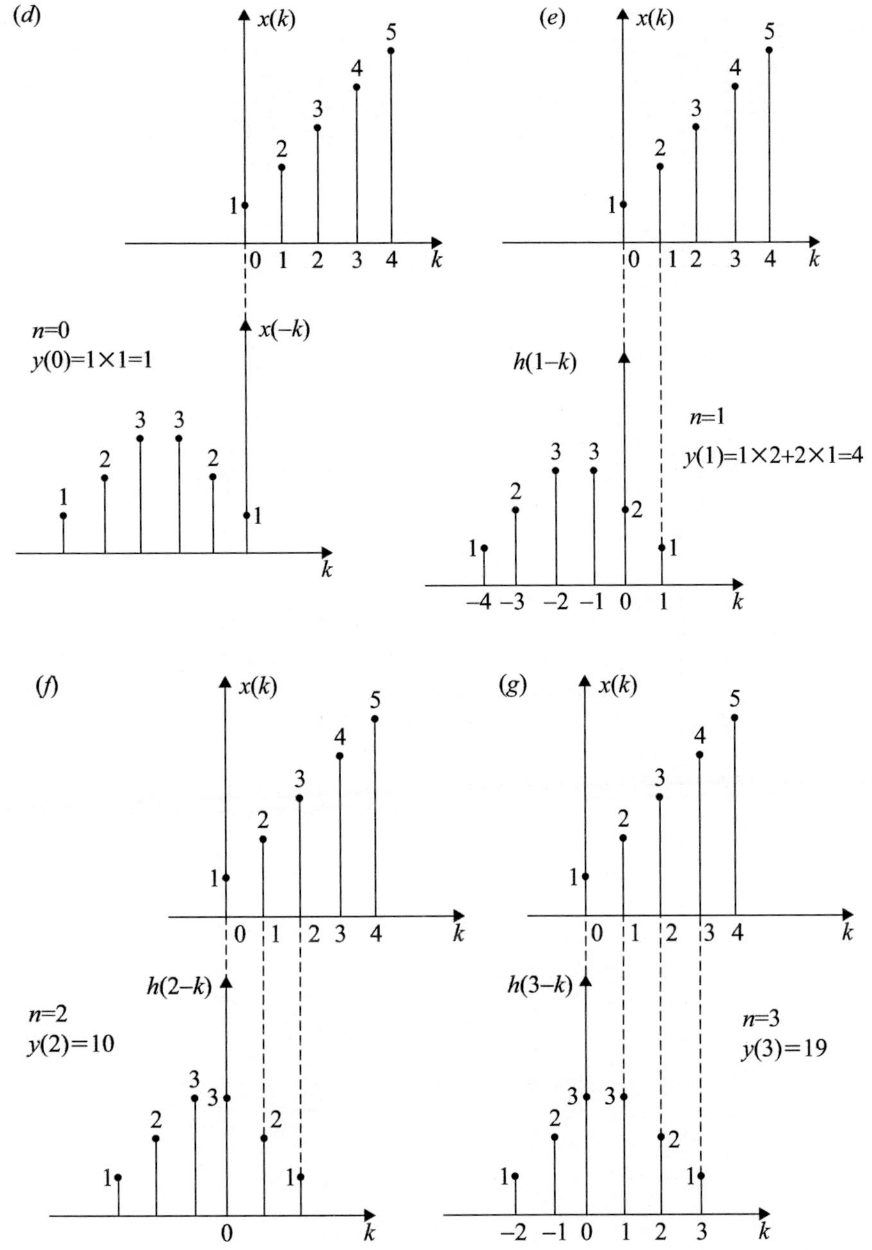

Fig. 2.34 (continued)

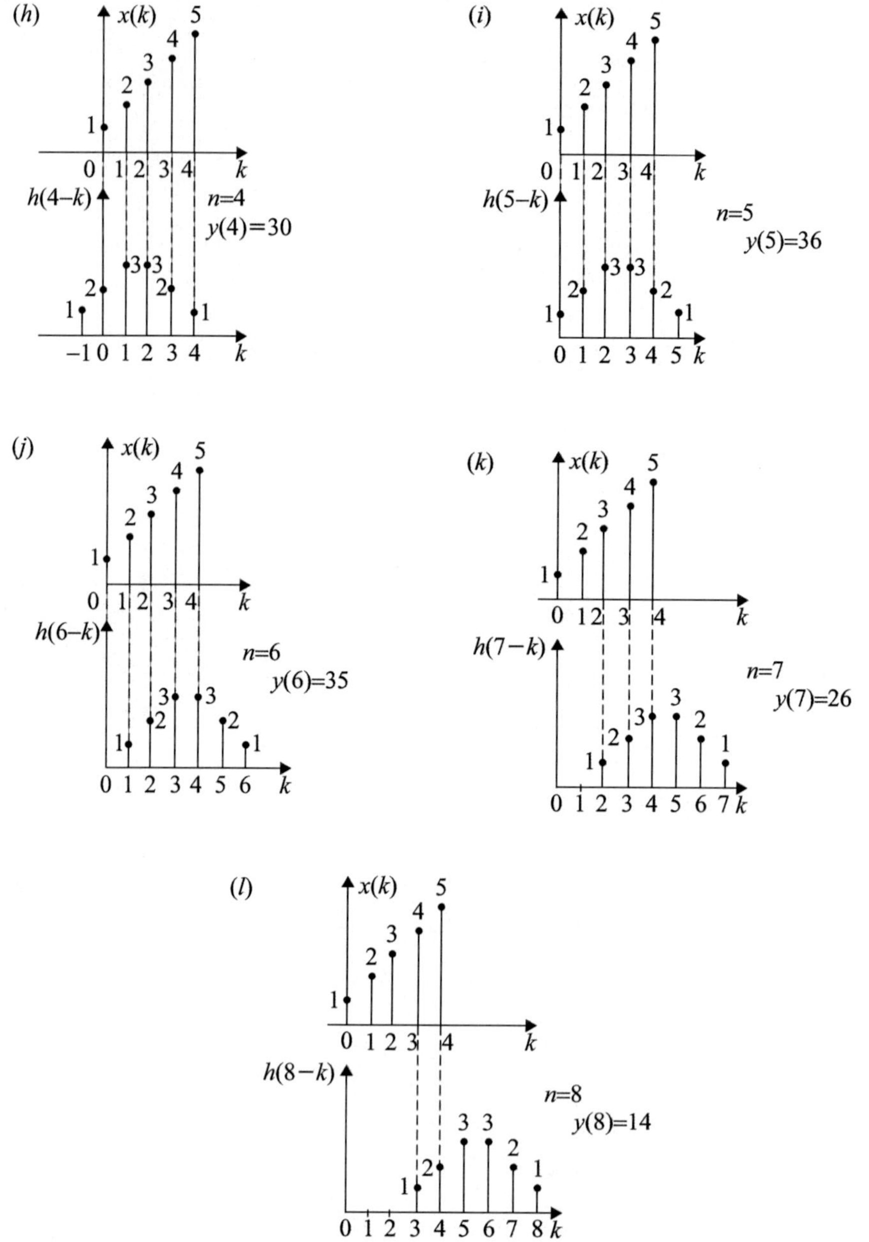

Fig. 2.34 (continued)

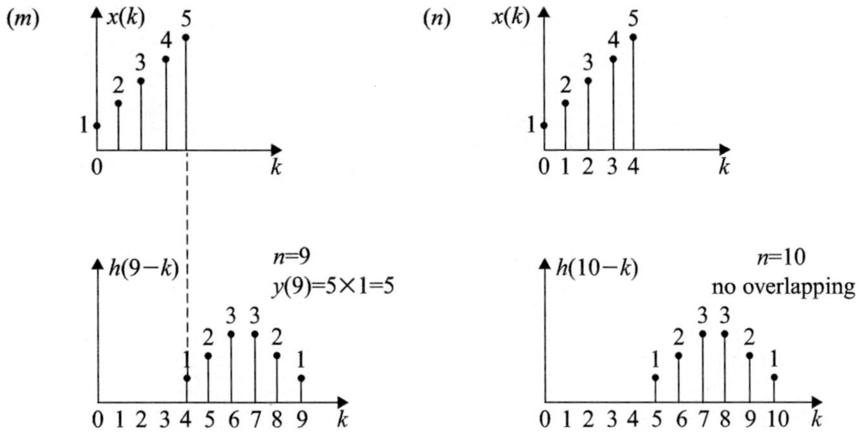

Fig. 2.34 (continued)

Find

$$y(n) = x(n) * h(n)$$

by graphical convolution.

Solution 1.

$$\begin{aligned} x(n) &= 2u(n+2) - 2u(n) - 3u(n-1) + 3u(n-3) \\ &= \{2,\ 2,\ 0,\ -3,\ -3\} \\ &\qquad\qquad \uparrow \end{aligned}$$

This is represented in Fig. 2.35a with n replaced by k.

$$\begin{aligned} h(n) &= -u(n+1) + 3u(n) - 5u(n-1) + 3u(n-2) \\ &= \{-1,\ 2,\ -3\} \\ &\qquad\quad \uparrow \end{aligned}$$

This is represented in Fig. 2.35b with n replaced by k.

2. $h(n-k)$ is created and put to the extreme left and is shown in Fig. 2.35c.
3. $h(n-k)$ is moved towards right sample by sample. The first overlapping between $x(k)$ and $h(n-k)$ occurs when $n = -3$ which is shown in Fig. 2.35d. The dotted line shows the overlapping with $x(k)$. At $n = -3$, $y(-3)$ is obtained as $y(-3) = 2 \times (-1) = -2$.
4. For $n = -2$, $x(k)$ and $h(-2-k)$ are shown in Fig. 2.35e. The overlaps are shown by dotted lines.

$$y(-2) = 2 \times (-1) + 2 \times 2 = 2$$

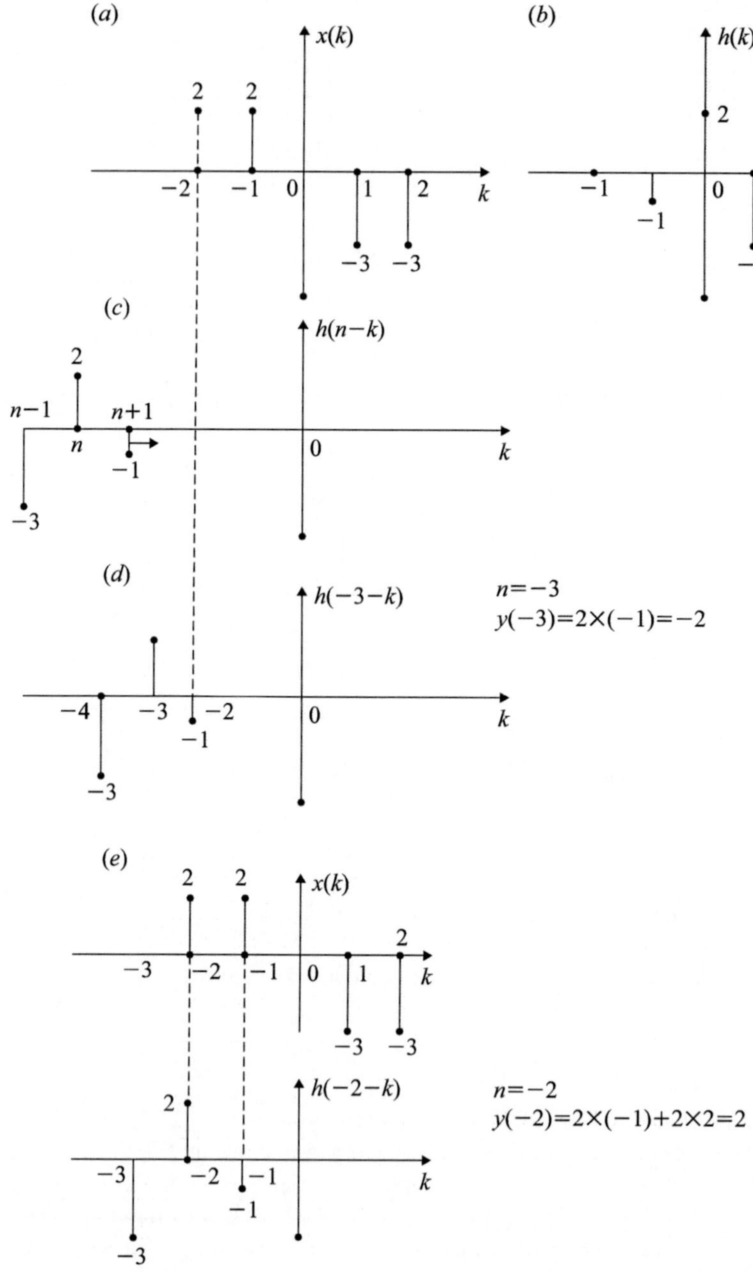

Fig. 2.35 Graphical method of obtaining $y(n)$ of Example 2.70(a) to (e). Graphical method of obtaining $y(n)$ of Example 2.70(f) to (k)

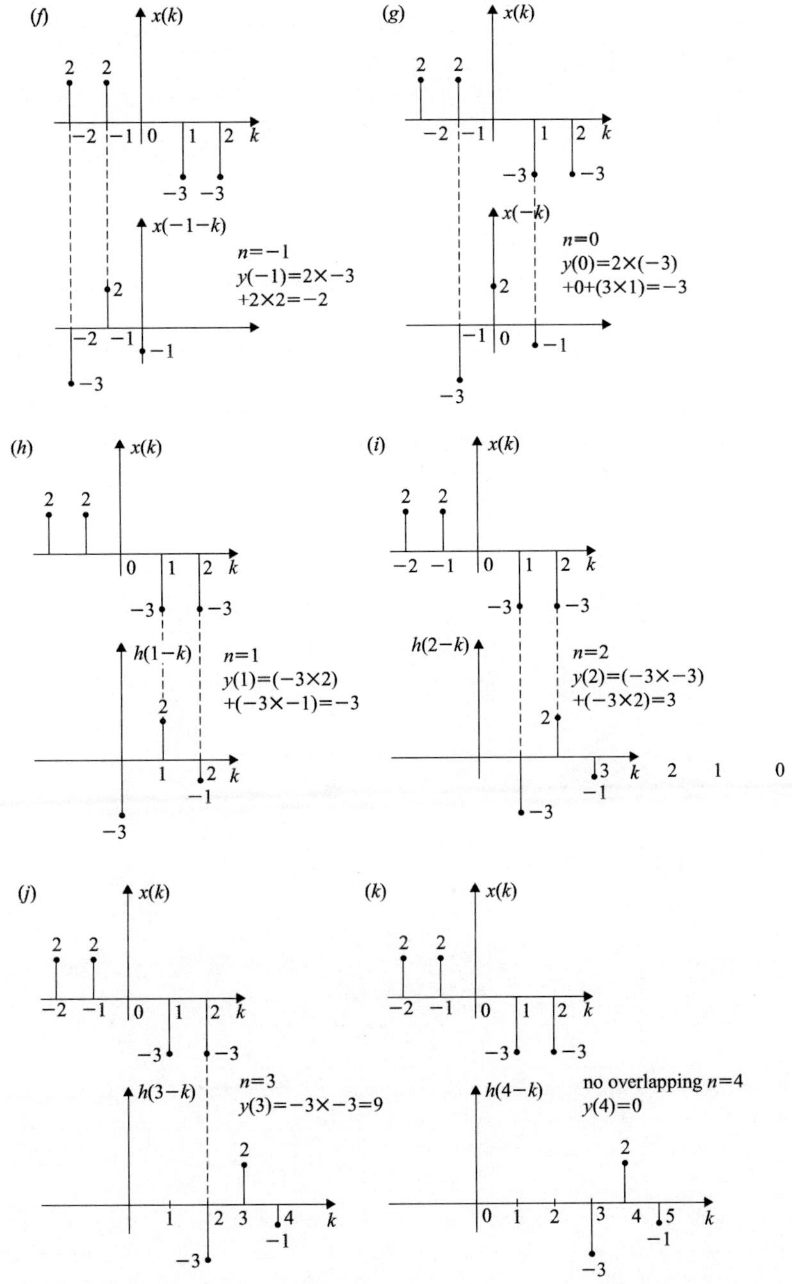

Fig. 2.35 (continued)

5. For $n = -1, x(k)$ and $h(-1-k)$ are shown in Fig. 2.35f. The overlaps are shown in dotted lines. $y(-1) = 2 \times (-3) + 2 \times 2 = -2$.
6. For $n = 0$, $x(k)$ and $h(-k)$ are shown in Fig. 2.35g. The overlaps are shown in dotted lines. $y(0) = 2 \times (-3) + 0 + (3 \times 1) = -3$.
7. For $n = 1$, $x(k)$ and $h(1-k)$ are shown in Fig. 2.35h. The overlaps are shown in dotted lines. $y(1) = (-3 \times 2) + (-3 \times -1) = -3$.
8. For $n = 2$, $x(k)$ and $h(2-k)$ are shown in Fig. 2.35i. The overlaps are shown in dotted lines. $y(2) = (-3 \times -3) + (-3 \times 2) = 3$.
9. For $n = 3, x(k)$ and $h(3-k)$ are shown in Fig. 2.35j. The overlapping is shown in dotted line. $y(3) = (-3 \times -3) = 9$.
10. For $n = 4$, $x(k)$ and $h(4-k)$ are shown in Fig. 2.35k. There is no overlapping and $y(3) = 0$.
11. For the given $x(n)$, $N_1 = 2$ and for $h(n)$, $N_2 = 1$. Hence for $y(n)$, $N = N_1 + N_2 = 2 + 1 = 3$. The width of $x(n)$ is $T_1 = 5$ and that of $h(n)$ is $T_2 = 3$. Hence the width of $y(n)$ is $T = T_1 + T_2 - 1 = 5 + 3 - 1 = 7$.
12.

$$\boxed{\begin{array}{r} y(n) = \{-2,\ 2,\ -2,\ -3,\ -3,\ 3,\ 9\} \\ \leftarrow N{=}3 \rightarrow \quad \uparrow \\ n{=}0 \\ \longleftarrow ----T = 7---- \longrightarrow \end{array}}$$

Summary

- The z-transform for discrete time signals and systems has been developed. This resembles corresponding treatment of the Laplace transform for continuous time system.
- A definite connection exists between Laplace transform, Fourier transform and z-transform. The Laplace transform reduces to Fourier transform on the imaginary axis in the s-plane. Then z-transform reduces to Fourier transform on the unit circle in the complex z-plane.
- For the causal signals and systems (right-sided), the z-transform exists if the ROC is exterior of the circle which passes through the outermost pole of the system function. For the anti-causal signal and system (left-sided) the z-transform exists if the ROC is interior of the circle which passes through the innermost pole of the system. For the right- and left-sided signals, the ROC is a ring which does not include any pole of system function.
- The application of the properties of z-transform very much simplifies the procedure to determine z-transform and inverse z-transform.

- For an LTID system to be causal, the system function should be rational and the ROC is the exterior of the circle which passes through the outermost pole of the system function $H[z]$.
- An LTID system is said to be stable if the ROC of the system function $H[z]$ includes the unit circle.
- An LTID system is said to be causal and stable if all the poles of the system function $H[z]$ lie inside the unit circle in the z-plane.
- Using the properties of z-transform, LTID systems described by constant coefficient difference equation can be converted into algebraic equations and easily analysed. The solution obtained is classified as zero state response, zero input response, natural response and forced response.
- Using Parseval's theorem, the average power of discrete time signal $x[n]$ can be determined by summing up the squared values of discrete Fourier series coefficients over one period using the following formula:

$$P = \sum_{k=0}^{(N_0-1)} |D_k|^2$$

- The Fourier transform is nothing but the description of $x[n]$ in the frequency domain. The transformed spectrum is periodic.
- Discrete Time Fourier Transform (DTFT) and Inverse Discrete Time Fourier Transform (IDTFT) are defined.
- The connection between the DTFT and the z-transform is that if the ROC of z-transform contains unit circle, the DTFT is evaluated on the unit circle. This is equivalent to substituting $z = e^{j\Omega}$ in the z-transform to get DTFT provided $x[n]$ is summable.
- Certain properties of DTFT have been derived. These properties are useful to determine the DTFT and IDTFT very quickly.
- Using DTFT and IDTFT it is possible to solve LTI systems characterized by difference equation.
- The DTFT is used to transform $x[n]$ to a continuous periodic spectrum. This sort of representation is not very much suited for use in digital computer. The DTFT is sampled at finite number of frequency points in one period. This is called Discrete Fourier Transform (DFT) which has many advantages over that of DTFT.

Short Questions and Answers

1. **Define z-transform.**
The z-transform of a discrete time signal $x[n]$ is defined as

$$X[z] = \sum_{n=-\infty}^{\infty} x[n]z^{-n}$$

where z is a complex variable.

2. **Define z-transform pair.**
When the discrete time signal $x[n]$ is z-transformed it is expressed as

$$X[z] = \sum_{n=-\infty}^{\infty} x[n]z^{-n}$$

If we want to recover $x[n]$ form $X[z]$, it is obtained using the following integration:

$$x[n] = \frac{1}{2\pi j} \oint X[z]z^{n-1} dz$$

This equation is called as inverse z-transform. The above two equations for z-transform and inverse z-transform are called z-transform pair.

3. **What do you understand by ROC of z-transform?**
The range of values of z for which the function $X[z]$ converges is called region of convergence which is expressed in abbreviated form as ROC.

4. **Mention the properties of ROC.**

   1. The ROC of $X[z]$ is in the form of a ring in the z-plane which is centred about the origin.
   2. The ROC does not include any poles.
   3. For the right-sided sequence $x[n]$, the ROC is the exterior of the outermost pole.
   4. For the left-sided sequence $x[n]$, the ROC is the interior of the innermost pole.
   5. If the sequence $x[n]$ is two-sided, then the ROC consists of a ring in the z-plane.

5. **What is the scaling property of z-transform?**
If

$$x[n] \stackrel{Z}{\longleftrightarrow} X[z] \qquad \text{ROC: } R$$

then

$$a^n x[n] \longleftrightarrow X\left[\frac{z}{a}\right] \qquad \text{ROC: } aR$$

By using multiplication property, the z-transform is obtained by replacing z by $\frac{z}{a}$ with ROC R replaced by aR.

6. **What is the convolution property of z-transform?**
If

$$x_1[n] \overset{Z}{\longleftrightarrow} X_1[z] \quad \text{ROC: } R_1$$
$$x_2[n] \overset{Z}{\longleftrightarrow} X_2[z] \quad \text{ROC: } R_2$$

then

$$x_1[n] * x_2[n] \overset{Z}{\longleftrightarrow} X_1[z]X_2[z] \quad \text{ROC: } R_1 \cap R_2$$

7. **What is difference property in the z-transform?**
If

$$x[n] \overset{Z}{\longleftrightarrow} X[z] \quad \text{ROC: } R$$

then

$$nx[n] \overset{Z}{\longleftrightarrow} -z\frac{dx[z]}{dz} \quad \text{ROC: } R$$

8. **What are initial and final value theorems?**
If $x[n] = 0$ for $n < 0$, then

$$x[0] = \underset{z\to\infty}{Lt}\ X[z]$$

is called initial value theorem. According to the final value theorem if $X[z]$ is the z-transform $x[n]$ and if all the poles of $X[z]$ are inside the unit circle, then the final value of $x[n] = x[\infty]$ is obtained from

$$x[\infty] = \underset{z\to 1}{Lt}\,(z-1)X[z]$$

9. **What do you understand by the time reversal property of z-transform?**
If

$$x[n] \overset{Z}{\longleftrightarrow} X[z] \quad \text{ROC: } R$$

then

$$x[-n] \overset{Z}{\longleftrightarrow} X\left[\frac{1}{z}\right] \quad \text{ROC: } \frac{1}{R}$$

Thus, the z-transform of the time reversal signal is obtained by replacing z by its reciprocal and also its ROC by its reciprocal.

10. **What do you understand by causality of an LTID system?**
    An linear time invariant discrete time system is said to be causal if the ROC of the system function $H[z]$ is the exterior of the circle containing all the poles of $H[z]$.
11. **What do you understand by stability of an LTID system?**
    An LTID system is said to be stable if the ROC of the system function $H[z]$ includes the unit circle in the z-plane.
12. **When the system is said to be both causal and stable?**
    An LTID system is said to be both causal and stable if all the poles of the system function $H[z]$ are inside the unit circle in the z-plane.
13. **Define system function.**
    System function or transfer function $H[z]$ is defined as the ratio of the z-transform of output sequence $y[n]$ and the input sequence $x[n]$

$$H[z] = \frac{Y[z]}{X[z]}$$

14. **What is the z-transform of $\delta[n-2]$?**

$$\delta[n-2] \overset{Z}{\longleftrightarrow} z^{-2}$$

15. **What is the z-transform of $u[n]$ and $\delta[n]$?**

$$u[n] \overset{Z}{\longleftrightarrow} \frac{z}{(z-1)}$$

$$\delta[n] \overset{Z}{\longleftrightarrow} 1$$

16. **Find the z-transform of $x[n] = u[n] - u[n-5]$.**

$$X[z] = \frac{z}{(z-1)}[1 - z^{-5}]$$

17. **Write the relationship between z-transform and Fourier transform.**
    The z-transform reduces to Fourier transform on the unit circle in the complex z-plane.
18. **Write the relationship between z-transform and Laplace transform.**
    The Laplace transform and z-transform are related as

$$e^s = z$$

$$X[s] = X[z]\Big|_{z=e^s}$$

19. **What is the inverse z-transform of $X[\frac{z}{a}]$?**

$$X\left[\frac{z}{a}\right] \overset{Z^{-1}}{\longleftrightarrow} a^n x[n]$$

20. **Find the system function of the following first-order difference equation $y[n] - 2y[n-1] = x[n] + x[n-1]$.**

$$\begin{aligned} H[z] &= \frac{Y[z]}{X[z]} = \frac{[1+z^{-1}]}{[1-2z^{-1}]} \\ &= \frac{[z+1]}{[z-2]} \end{aligned}$$

21. **Define DTFT pair.**
The discrete time Fourier transform is defined as

$$X(\Omega) = \sum_{n=-\infty}^{\infty} x[n]e^{-j\Omega n}$$

The inverse discrete time Fourier transform is defined as

$$x[n] = \frac{1}{2\pi}\int_{2\pi} X(\Omega)e^{jn\Omega}d\Omega$$

22. **What are the necessary and sufficient condition for the existence of DTFT?**
The necessary and sufficient conditions for the existence of DTFT is that the sequence $x[n]$ should be absolutely summable. That is

$$\sum_{n=-\infty}^{\infty} |x[n]| < \infty$$

Alternatively, if the sequence $x[n]$ has finite energy

$$E = \sum_{n=-\infty}^{\infty} |x[n]|^2 < \infty$$

then DTFT exists.

23. **Define DFT pair.**
The discrete Fourier transform pair is defined as follows. The DFT is given by

$$X(k) = \sum_{n=0}^{N-1} x[n]e^{-j2\pi kn/N}$$

The IDTFT is given by

$$x[n] = \frac{1}{N}\sum_{k=0}^{N-1} X(k)e^{j2\pi kn/N}$$

24. **Is the IDFT of DFT periodic or not? Give your comment.**

$$x[n] = \frac{1}{N}\sum_{k=0}^{N-1} X(\Omega)W_N^{-kn}$$

where $W_N = 1\angle -2\pi/N$. W_N lies on the unit circle in the complex plane and gets repeated for every 2π. Hence, IDFT is periodic with period 2π.

25. **What is the DTFT of a^n?**

$$a^n \overset{\text{DTFT}}{\longleftrightarrow} \frac{(1-a^N)}{(1-ae^{-j\frac{2\pi k}{N}})}$$

26. **What is the connection between the z-transform and the DTFT?**
When the ROC for $X[z]$ includes the unit circle, substituting $z = e^{j\Omega}$ in $X[z]$ gives the DTFT $X(\Omega)$.
27. **Find the system transfer function for the following difference equation.**

$$\mathbf{y[n] - 0.8y[n-1] = x[n]}$$

$$H[\Omega] = \frac{1}{(1-0.8e^{-j\Omega})}$$

28. **Find the frequency responses for the following difference equation.**

$$\mathbf{y[n] - 0.1y[n-1] = x[n]}$$

$$H[\Omega] = \frac{1\angle -\tan^{-1}\frac{0.1\sin\Omega}{(1-0.1\cos\Omega)}}{\sqrt{(1-0.1\cos\Omega)^2 + 0.01\sin^2\Omega}}$$

29. **What is the DTFT of unit step response?**
The DTFT of unit step response is

$$X(\Omega) = \pi\delta(\Omega) + \frac{1}{(1-e^{-j\Omega})}$$

30. **What is the DTFT of $x[-n]$?**
    The DTFT of $x[-n]$ is $X(-\Omega)$
31. **What do you understand by zero padding?**
    In evaluating DFT we assume that the length of the DFT which is N is equal to the length L of the sequence $x[n]$. If $N < L$, time domain aliasing occurs. To avoid this, dummy samples are added with value 0 in $X[k]$. This is called zero padding.
32. **Find DTFT of $\delta[n] = 1$.**

$$\delta[n] \overset{\text{DTFT}}{\longleftrightarrow} 1$$

33. **Find the inverse DTFT of $X[\Omega] = \delta[\Omega]$.**

$$\delta[\Omega] \overset{\text{IDTFT}}{\longleftrightarrow} \frac{1}{2\pi} \qquad -\pi < \Omega < \pi$$

34. **Find $Y(\Omega)$.**

$$x[n] = \left(-\frac{1}{3}\right)^n u[n]$$

$$h[n] = \left(\frac{1}{3}\right)^n u[n]$$

$$y[n] = x[n] * h[n]$$

$$Y(\Omega) = \frac{1}{\left(1 + \frac{1}{3}e^{-j\Omega}\right)\left(1 - \frac{1}{3}e^{-j\Omega}\right)}$$

35. **Find $X(\Omega)$.**

$$x[n] = u[n] - u[n-3]$$

$$X(\Omega) = 1 + e^{-j\Omega} + e^{-j2\Omega}$$

Long Answer Type Questions

1. **Find the z-transform of the following sequence:**

$$x[n] = [3^{n-1} - (-3)^{n-1}]u[n]$$

$$X[z] = \frac{2z^2}{3(z^2 - 9)} \qquad \text{ROC: } |z| > 3$$

2. **Find the z-transform of**

$$x[n] = \sum_{n=0}^{\infty} \frac{1}{3} z^{-n} + \frac{1}{4}(-2)^n z^{-n}$$

$$X[z] = \frac{1}{3}\frac{1}{(1 - z^{-1})} + \frac{1}{3}\frac{1}{(1 + 2z^{-1})} \qquad \text{ROC: } |z| > 2$$

3. **Find the z-transform of**

$$x[n] = \sum_{n=-1}^{\infty} \left(\frac{1}{4}\right)^{n+1} z^{-n}$$

$$X[z] = z + \frac{1}{4}\frac{1}{\left(1 - \frac{1}{4}z^{-1}\right)} \qquad \text{ROC: } |z| > \frac{1}{4}$$

4. **Find the z-transform of**

$$x[n] = \sum_{n=1}^{\infty} \left(\frac{1}{4}\right)^{-n+1} z^{-n}$$

$$X[z] = \frac{1}{4} + \frac{4}{(1 - 4z)} \qquad \text{ROC: } |z| < \frac{1}{4}$$

5. **Find the z-transform of**

$$x[n] = \left(\frac{1}{10}\right)^n u[n-4]$$

$$X[z] = 10^{-4}\frac{z^{-3}}{\left(z - \frac{1}{10}\right)} \qquad \text{ROC: } |z| > \frac{1}{10}$$

6. **Find the z-transform of**

$$\begin{aligned} (a) \quad x[n] &= 1 \quad 0 \le n \le 9 \\ &= 0 \quad \text{otherwise} \\ (b) \quad y[n] &= x[n] - x[n-1] \end{aligned}$$

$$\begin{aligned} (a) \quad & X[z] = \frac{(1 - z^{-10})}{(1 - z^{-1})} \quad \text{ROC: } |z| > 0 \\ (b) \quad & Y[z] = 1 - z^{-10} \quad \text{ROC: } |z| > 0 \end{aligned}$$

7. **Find the unilateral z-transform and the ROC for the following sequences:**

$$\begin{aligned} (a) \quad x[n] &= \left(\frac{1}{6}\right)^n u[n+6] \\ (b) \quad x[n] &= 3\delta[n+4] + \delta[n] + (3)^n u[-n] \\ (c) \quad x[n] &= \left(\frac{1}{4}\right)^{|n|} \end{aligned}$$

$$\begin{aligned} (a) \quad & X[z] = \frac{1}{(1 - 6z^{-1})} \quad \text{ROC: } |z| > 6 \\ (b) \quad & X[z] = 4 \quad \text{ROC: all } z \\ (c) \quad & X[z] = \frac{1}{\left(1 - \frac{1}{4}z^{-1}\right)} \quad \text{ROC: } |z| > \frac{1}{4} \end{aligned}$$

8. **By applying properties of z-transform find the z-transform of the following sequences given $x[n] \overset{Z}{\longleftrightarrow} \frac{z}{(z^2+2)}$**

$$\begin{aligned} (a) \quad y[n] &= x[n-3] \\ (b) \quad y[n] &= nx[n] \\ (c) \quad y[n] &= x[n+1] + x[n-1] \\ (d) \quad x[n] &= 2^n x[n] \\ (e) \quad x[n] &= (n-2)x[n-1] \\ (f) \quad x[n] &= x[-n] \end{aligned}$$

$$(a)\quad Y[z] = \frac{z^{-2}}{(z^2+2)} \quad \text{ROC: } |z| < 2 \text{ (Time shifting property)}$$

$$(b)\quad Y[z] = \frac{z[z^2-2]}{(z^2+2)^2} \quad \text{(Differentiation property)}$$

$$(c)\quad Y[z] = \frac{(z^2+1)}{(z^2+2)} \quad \text{(Time advancing and time delaying)}$$

$$(d)\quad Y[z] = \frac{2z}{(z^2+8)} \quad \text{(Multiplying property)}$$

$$(e)\quad Y[z] = \frac{-4}{(z^2+2)^2} \quad \text{(Time differentiation and time shifting)}$$

$$(f)\quad Y[z] = \frac{z}{(1+2z^2)} \quad \text{(Time reversal)}$$

9. **Find the z-transform of**

$$\mathbf{(a)}\quad \boldsymbol{x[n] = 2^n u[n-2]}$$

$$\mathbf{(b)}\quad \boldsymbol{x[n] = \left(\frac{1}{4}\right)^n u[-n]}$$

$$(a)\quad X[z] = \frac{4z^{-2}}{(1-2z^{-1})}$$

$$(b)\quad X[z] = \frac{1}{(1-4z)}$$

10. **Find the z-transform of**

$$\mathbf{(a)}\quad \boldsymbol{x[n] = (n-4)u[n-4]}$$

$$\mathbf{(b)}\quad \boldsymbol{x[n] = u[n] - u[n-4]}$$

$$\mathbf{(c)}\quad \boldsymbol{x[n] = (n-4)u[n]}$$

$$\mathbf{(d)}\quad \boldsymbol{x[n] = n[u[n] - u[n-4]]}$$

$$(a)\quad X[z] = \frac{z^{-3}}{(z-1)^2} \quad \text{ROC: } |z| > 1$$

$$(b)\quad X[z] = \frac{z}{(z-1)}[1-z^{-4}] \quad \text{ROC: } |z| > 1$$

$$(c)\quad X[z] = \frac{(5z-4z^2)}{(z-1)^2} \quad \text{ROC: } |z| > 1$$

$$(d)\quad X[z] = \frac{(z-3z^{-2}+2z^{-3})}{(z-1)^2} \quad \text{ROC: } |z| > 1$$

11. **Find the z-transform of the following sequence:**

$$\begin{aligned} \boldsymbol{x[n]} &= \left(\frac{1}{4}\right)^n \boldsymbol{u[n]} \\ &= \left(-\frac{1}{2}\right)^n \boldsymbol{u[-n-1]} \end{aligned}$$

$$X[z] = \frac{3}{4}\left[\frac{z}{\left(z-\frac{1}{4}\right)\left(z+\frac{1}{2}\right)}\right] \qquad \text{ROC: } \frac{1}{4} < |z| < \frac{1}{2}$$

12. **Using convolution find $y[n]$ given**

$$\begin{aligned} \boldsymbol{x[n]} &= \left(\frac{1}{2}\right)^n \boldsymbol{u[n]} \\ \boldsymbol{h[n]} &= \left(\frac{1}{3}\right)^n \boldsymbol{u[n]} \\ \boldsymbol{y[n]} &= \boldsymbol{x[n] * h[n]} \end{aligned}$$

$$y[n] = \left[3\left(\frac{1}{2}\right)^n - 2\left(\frac{1}{3}\right)^n\right] u[n] \qquad \text{ROC: } |z| < \frac{1}{2}$$

13. **Using partial function, find the inverse z-transform of**

$$\boldsymbol{H[z] = \frac{(1-z^{-1}+z^{-2})}{(1-z^{-1})(1-2z^{-1})(1-4z^{-1})}} \qquad \boldsymbol{ROC: 2 < |z| < 4}$$

$$h[n] = \left[\frac{1}{3} - \frac{3}{2}(2)^n\right] u[n] + \frac{3}{16}(4)^n u[-n-1]$$

14. **Find the inverse z-transform of**

$$\boldsymbol{H[z] = \frac{4z+1}{z-\frac{1}{4}}}$$

using power series expansion.

$$(a) \qquad ROC : |z| > \frac{1}{4}$$

$$(b) \qquad ROC : |z| < \frac{1}{4}$$

$$(a) \qquad x[n] = \left\{ \underset{\uparrow}{4},\ 2,\ \frac{1}{2},\ \frac{1}{8},\ \frac{1}{32},\ \ldots \right\}$$

$$(b) \qquad x[n] = \{\ldots, 2048,\ -512,\ -128,\ -32,\ \underset{\uparrow}{-4}\}$$

15. **Consider the algebraic expression for the z-transform of $x[n]$**

$$x[n] = \frac{\left(1 - \frac{1}{4}z^{-2}\right)}{\left(1 + \frac{1}{4}z^{-1}\right)\left(1 - \frac{5}{6}z^{-1} + \frac{1}{6}z^{-2}\right)}$$

How many different ROCs could correspond the $X[z]$?

$$(a) \qquad \text{ROC: } |z| > \frac{1}{2}$$

$$(b) \qquad \text{ROC: } 0 < |z| < \frac{1}{4}$$

$$(c) \qquad \text{ROC: } \frac{1}{3} < |z| < \frac{1}{2}$$

16. **Consider the algebraic expression for the z-transform of $x[n]$**

$$x[n] = \frac{\left(1 + z^{-1} + 4z^{-2}\right)}{\left(1 - \frac{1}{4}z^{-1}\right)\left(1 - \frac{7}{24}z^{-1} + \frac{1}{48}z^{-2}\right)}$$

ROC: $|z| > \frac{1}{4}$. Find whether the system is causal and stable.
$X[z]$ is rational and the poles are at $z = \frac{1}{4}$, $z = \frac{1}{6}$ and $z = \frac{1}{8}$. Since the ROC is exterior of the outermost pole, the system is causal. The ROC includes unit circle and the poles are inside the unit circle. The system is stable. Therefore, the system is causal and stable.

17. **A system with impulse response $h[n] = 5(3)^n u[n-1]$ produces on output $y[n] = (-4)^n u[n-1]$. Determine the input $x[n]$.**

$$x[n] = \frac{1}{5}\left[(-4)^n u[n] - 3(-4)^{n-1} u[n-1]\right] \qquad \text{ROC: } |z| > 4$$

18. **Consider the following difference equation.**

$$y[n] - y[n-1] - 2y[n-2] = x[n] + 2x[n-1]$$

The initial conditions are $y[-1] = 1$ and $y[-2] = 2$. The input $x[n] = u[n]$. Find (*a*) Zero input response, (*b*) Zero state response, (*c*) Natural response, (*d*) Forced response and (*e*) Total response.

(*a*) $y_{0i}[n] = [(-1)^n + 4(2)^n]u[n]$

(*b*) $y_{0s}[n] = \left[-\frac{1}{6}(-1)^n + \frac{8}{3}(2)^n - \frac{3}{2}\right]u[n]$

(*c*) $y_n[n] = -\frac{3}{2}u[n]$

(*d*) $y_f[n] = \left[\frac{5}{6}(-1)^n + \frac{20}{3}(2)^n\right]u[n]$

(*e*) $y_{\text{total}}[n] = \left[-\frac{3}{2} + \frac{5}{6}(-1)^n + \frac{20}{3}(2)^n\right]u[n]$

19. **For each of the following difference equations, determine the output response $y[n]$.**

(*a*) $y[n] - 4y[n-1] = x[n]$ **with** $y[-1] = 2$ **and** $x[n] = \left(\frac{1}{3}\right)^n u[n]$

(*b*) $y[n] + \frac{1}{3}y[n-1] = x[n] - \frac{1}{3}x[n-1]$
with $y[-1] = 1$ **and** $x[n] = u[n]$

(*a*) $y[n] = \left[\frac{100}{11}(4)^n - \frac{1}{11}\left(\frac{1}{3}\right)^n\right]u[n]$

(*b*) $y[n] = \left[\frac{3}{2} - \frac{5}{6}\left(-\frac{1}{3}\right)^n\right]u[n]$

20. **Find $x(\infty)$ if**

(*a*) $X(z) = \dfrac{(z+2)}{(z-0.8)^2}$

(*b*) $X(z) = \dfrac{(3z+4)}{(z-1)(z+0.9)}$

$$(a)\ \ x(\infty) = 0; \qquad (b)\ \ x(\infty) = \frac{7}{1.9}$$

21. **By using long direction, determine the inverse z-transform of**

$$(a)\quad X(z) = \frac{(1 + 2z^{-1})}{(1 - 2z^{-1} + z^{-2})}$$

$$(b)\quad X(z) = \frac{(z + 1)}{(z^2 + 3z + 2)}$$

for $x(n)$ being causal and non-causal.

$$(a)\quad x(n) = \{\underset{\uparrow}{1}, 4, 7, 10, 13, 16, 19, \ldots\}\ \text{(causal)}$$

$$x(n) = \{\ldots, 17, 14, 11, 8, 5, 2, \underset{\uparrow}{0}\}\ \text{(non-causal)}$$

$$(b)\quad x(n) = \{\underset{\uparrow}{0}, 1, 4, 10, 22, \ldots\}\ \text{(causal)}$$

$$x(n) = \{\ldots, 1.8125, 1.625, 1.25, \underset{\uparrow}{0.5}\}\ \text{(non-causal)}$$

22. **Find the inverse z-transform of the following:**

$$(a)\quad X(z) = \frac{\frac{1}{4}z^{-1}}{\left(1 - \frac{1}{2}z^{-1}\right)\left(1 - \frac{1}{4}z^{-1}\right)} \qquad ROC : |z| > \frac{1}{2}$$

$$(b)\quad X(z) = \frac{z(z - 1)}{(z + 2)^3(z + 1)}$$

$$(a)\quad x(n) = \left[\left(\frac{1}{2}\right)^n - \left(\frac{1}{4}\right)^n\right]u(n)$$

$$(b)\quad x(n) = \left[2(-2)^n + 2n(-2)^{n-1} + \frac{3n(n-1)}{\angle 2}(-2)^{n-2} - 2(-1)^n\right]u(n)$$

23. **By using residue method, find the inverse z-transform of**

$$(a)\quad X(z) = \frac{(1 + 3z^{-1})}{1 + 3z^{-1}2z^{-2}}; \qquad ROC : |z| > 2$$

$$(b)\quad X(z) = \frac{2z^{-1}}{\left(1 - \left(\frac{1}{4}\right)z^{-1}\right)^2}$$

$$(a) \qquad x(n) = \left[2(-1)^n - (-2)^n\right] u(n)$$

$$(b) \qquad x(n) = 2n\left(\frac{1}{4}\right)^{n-1} u(n)$$

24. **Find the DTFT of**

$$\mathbf{x[n] = (0.5)^n u[n] - 2^{-n} u[-n-1]}$$

DTFT for $2^{-n}u[-n-1]$ does not converge and hence does not exist.

25. **Find the DTFT of**

$$\mathbf{x[n] = (0.5)^n u[n] - (.5)^{-n} u[-n-1]}$$

$$X(\Omega) = \frac{(2 - 2.5e^{-j\Omega})}{(1 - 0.5e^{-j\Omega})(1 - 2e^{-j\Omega})}$$

26. **Find the DTFT of**

$$\mathbf{x[n] = \left(\frac{1}{3}\right)^n u[n+2]}$$

$$X(\Omega) = \frac{9e^{j2\Omega}}{\left(1 - \frac{1}{3}e^{-j\Omega}\right)}$$

27. **Find the DTFT of**

$$\mathbf{x[n] = n\left(\frac{1}{4}\right)^n u[n]}$$

$$X(\Omega) = \frac{0.25e^{j\Omega}}{\left(e^{j\Omega} - 0.25\right)^2}$$

28. **Find the DTFT of $x[n]$**

$$\mathbf{x[n] = \begin{cases} 1 & 0 \le |n| \le 8 \\ 0 & otherwise \end{cases}}$$

$$X(\Omega) = \frac{\sin 8.5\Omega}{\sin 0.5\Omega}$$

29. **Find $Y(\Omega)$.**

$$x[n] = \left(\frac{1}{6}\right)^n u[n]$$
$$h[n] = -(3)^n u[-n-1]$$
$$y[n] = x[n] * h[n]$$

$$Y(\Omega) = \frac{1}{\left(1 - \frac{1}{6}e^{-j\Omega}\right)\left(3e^{-j\Omega} - 1\right)}$$

30. **Find $X(\Omega)$.**

$$x[n] = (n+2)\left(\frac{1}{2}\right)^n u[n]$$

$$X(\Omega) = \frac{2(1 + 1.25e^{-j\Omega})}{(1 - 0.5e^{-j\Omega})^2}$$

31. **Find $X(\Omega)$.**

$$x[n] = \left(\frac{1}{3}\right)^{|n-1|} u[n-1]$$

$$X(\Omega) = \frac{8e^{-j\Omega}}{9 - \cos\Omega}$$

32. **Find $X(\Omega)$.**

$$x[n] = (n-2)^2 x[n]$$

$$X(\Omega) = -\frac{d^2X(\Omega)}{d\Omega^2} - 4j\frac{dX(\Omega)}{d\Omega} + 4X(\Omega)$$

33. **Solve the following difference equation.**

$$y[n-2] - \frac{1}{6}y[n-1] - \frac{1}{6}y[n] = x[n]$$

where $x[n] = \left(\frac{1}{4}\right)^n u[n]$

$$y[n] = \left[\frac{6}{5}\left(\frac{1}{2}\right)^n + \frac{8}{35}\left(-\frac{1}{3}\right)^n - \frac{3}{7}\left(\frac{1}{4}\right)^n\right]u[n]$$

34. **For $X(\Omega)$ given below, find $x[n]$.**

$$X(\Omega) = \begin{cases} 5 & 0 \le |\Omega| < \pi \\ 0 & |\Omega| > \pi \end{cases}$$

$$\begin{aligned} x[n] &= \frac{5}{\pi(n-1)} \sin(n-1)\pi \\ &= 5 \operatorname{sinc}(n-1)\pi \end{aligned}$$

35.

$$\begin{aligned} x[n] &= u[n] - u[n-4] \\ h[n] &= a^n u[n] \qquad 0 < a < 1 \end{aligned}$$

Find $y[n] = x[n] * h[n]$.

$$y[n] = \begin{cases} \frac{(1-a^{n+1})}{(1-a)} & 0 < n < 1 \\ a^{n-3}\frac{(1-a^4)}{(1-a)} & n \ge 3 \end{cases}$$

36.

$$\begin{aligned} x[n] &= u[n] - u[n-6] \\ h[n] &= a^n\,(u[n] - u[n-3]) \end{aligned}$$

where $0 < a < 1$. Find the convolution of $x[n]$ and $h[n]$.

$$y[n] = \begin{cases} \frac{(1-a^{n+1})}{(1-a)} & 0 \le n \le 2 \\ \frac{(1-a^3)}{(1-a)} & 2 \le n \le 5 \\ a^{n-5}\frac{(1-a^{8-n})}{(1-a)} & 5 \le n \le 8 \end{cases}$$

37.

$$\begin{aligned} x[n] &= 2^n u[-n-2] \\ h[n] &= u[n-1] \end{aligned}$$

Find $y[n] = x[n] * h[n]$.

$$y[n] = \begin{cases} 2^n & -\infty < n < -1 \\ 0.5 & n > -1 \end{cases}$$

38.

$$\boldsymbol{x[n] = u[n] - 3u[n-2] + 2u[n-4]}$$
$$\boldsymbol{h[n] = u[n+1] - u[n-8]}$$

Find $y[n] = x[n] * h[n]$.

$$y[n] = \{1, \underset{\uparrow}{2}, 0, -2, -2, -2, -2, -2, -2, -3, -4, -2\}$$

39.

$$\boldsymbol{x[n] = \{1, 2, 3, 4\}}$$
$$\boldsymbol{h[n] = \{2, -4, 6, -8\}}$$

Find $y[n] = x[n] * h[n]$.

$$y[n] = \{2, 0, 4, 0, -14, \underset{\uparrow}{0}, -32\}$$

40. **Find the convolution of $x[n]$ with $h[n]$**

$$\boldsymbol{x[n] = a^n u[n] \quad 0 < a < 1}$$
$$\boldsymbol{h[n] = \begin{cases} 1 & 0 \le n \le 9 \\ 0 & n \ge 10 \end{cases}}$$

$$y[n] = \begin{cases} 0 & n < 0 \\ \frac{(1-a^{n+1})}{(1-a)} & 0 \le n \le 9 \\ a^{n-9}\left[\frac{(1-a^{10})}{(1-a)}\right] & n > 9. \end{cases}$$

Chapter 3
Discrete Fourier Transform and Computation

Learning Objectives

After completing this chapter, you should be able to

- ✠ define Discrete Fourier Transform (DFT) and Inverse Discrete Fourier Transform (IDFT).
- ✠ establish the properties of DFT.
- ✠ study different methods of circular convolution and solve numerical problems.
- ✠ study different Fast Fourier Transform (FFT) algorithms and their applications.
- ✠ study the use of FFT algorithms in linear filtering and correlation.

3.1 Introduction

In the study of signals and systems, the discrete time periodic signals are represented by Discrete time Fourier Series (DFS) using a parallel development of continuous system. The Fourier series representation in these cases was applicable only if the signals are periodic. If the signal is non-periodic, then by applying a limiting process, the aperiodic discrete time signal $x[n]$ can be expressed as a sum of everlasting exponentials or sinusoids. The spectrum of $x(\omega)$ so obtained is called Discrete Time Fourier Transform (DTFT). If the spectrum obtained by DTFT is sampled for one period of the Fourier transform, such a transformation is called Discrete Fourier Transform (DFT) which is a very powerful computational tool for the evaluation

S. Palani, *Discrete Time Systems and Signal Processing*,
https://doi.org/10.1007/978-3-031-32421-5_3

of Fourier transform. DFT finds wide applications in linear filtering, correlation analysis and spectrum analysis. Some special algorithms are developed for the easy implementation of DFT which result in the saving of considerable computation time. Such algorithms are called Fast Fourier Transform (FFT). By the divide-and-conquer approach, the DFT which has a size N, where N is a composite number, is reduced to the smaller DFTs and computation is performed. The computational algorithms are developed when the size of N is a power of 2 and a power of 4.

3.2 Discrete Fourier Transform (DFT)

By Fourier transform, the sequence $x[n]$ is transformed as $X(\omega)$ in the frequency domain. If the time sequence $x[n]$ is continuous and periodic, the transformation is called Discrete Time Fourier Transform (DTFT). For non-periodic signal $x[n]$, a slightly modified transform technique which is known as Discrete Fourier Transform (DFT) is used which transforms $x[n]$ to $X(\omega)$. DFT is a very powerful tool for the analysis and synthesis of discrete signals and systems. The method is ideally suited for use in digital computers or specially designed digital hardware. The DFT is obtained by sampling one period of DTFT only at a finite number of frequency points. It has the following features:

1. The original finite duration signal can be easily recovered from its DFT since there exists a one-to-one correspondence between $x[n]$ and $X(\omega)$.
2. For the calculation of the DFT of finite duration sequences, a very efficient and fast technique called FFT has been developed.
3. As far as realization in digital computers is concerned, DFT is the appropriate representation since it is discrete and of finite length in both the time and frequency domains.
4. DFT is closely related to discrete Fourier series, the Fourier transform, convolution, correlation and filtering.

3.2.1 *The Discrete Fourier Transform Pairs*

Consider the sequence $x[n]$ of length N. The Fourier transform of $x[n]$ is given by

$$X(\Omega) = \sum_{n=-\infty}^{\infty} x[n]e^{-j\Omega n} \tag{3.1}$$

In Eq. (3.1), $X(\Omega)$ is the continuous function of Ω. The range of Ω is from $-\pi$ to π or 0 to 2π. Hence, calculating $X(\Omega)$ on a digital computer or DSP is impossible. It is, therefore, necessary to compute $X(\Omega)$ at discrete values of Ω. When Fourier transform $X(\Omega)$ is calculated at only discrete points k, it is called Discrete Fourier

Transform (DFT). The DFT is denoted by $X(k)$. For finite discrete points N, Eq. (3.1) is written as

$$X(k) = \sum_{n=0}^{N-1} x[n]e^{-j2\pi kn/N} \tag{3.2}$$

where $k = 0, 1, 2, \ldots, (N-1)$. $X(k)$ is computed at $k = 0, 1, 2, \ldots, (N-1)$ discrete points. $X(k)$ is the sequence of N samples. The sequence $x[n]$ is obtained back from the following equation.

$$X[n] = \frac{1}{N}\sum_{k=0}^{N-1} X(k)e^{j2\pi kn/N} \tag{3.3}$$

Let us define $W_N = e^{-j2\pi/N}$, where W_N is called the Twiddle factor. Equations (3.2) and (3.3) are called DFT and IDFT or simply a discrete Fourier transform pair. They can be represented in terms of the twiddle factor as given below

$$X(k) = \sum_{n=0}^{N-1} x[n]W_N^{kn} \tag{3.4}$$

$$x[n] = \frac{1}{N}\sum_{k=0}^{N-1} X(k)W_N^{-kn} \tag{3.5}$$

Let the sequence $x[n]$ be resentenced as a vector x_N of N samples as

$$x_N = \begin{matrix} n=0 \\ n=1 \\ \vdots \\ n=N-1 \end{matrix} \begin{bmatrix} x(0) \\ x(1) \\ \vdots \\ x(N-1) \end{bmatrix}_{N\times 1} \tag{3.6}$$

and $X(k)$ be represented as a vector X_N of N samples as

$$X_N = \begin{matrix} k=0 \\ k=1 \\ \vdots \\ k=N-1 \end{matrix} \begin{bmatrix} X(0) \\ X(1) \\ \vdots \\ X(N-1) \end{bmatrix}_{N\times 1} \tag{3.7}$$

The twiddle factor W_N^{kn} is represented as a matrix with k rows and N columns as

$$W_N = \begin{matrix} k=0 \\ \\ \\ \\ k=N-1 \end{matrix} \begin{bmatrix} W_N^0 & W_N^0 & W_N^0 & \cdots & W_N^0 \\ W_N^0 & W_N^1 & W_N^2 & \cdots & W_N^{N-1} \\ W_N^0 & W_N^2 & W_N^4 & \cdots & W_N^{2(N-1)} \\ \vdots & \vdots & \vdots & \vdots & \vdots \\ W_N^0 & W_N^{N-1} & W_N^{2(N-1)} & \cdots & W_N^{(N-1)(N-1)} \end{bmatrix}_{N \times N} \tag{3.8}$$

Thus, Eqs. (3.4) and (3.5) can be written in matrix form as

$$X_N = [W_N]x_N \tag{3.9}$$

$$x_N = \frac{1}{N}[W_N^*]X_N \tag{3.10}$$

where $W_N^* = W_N^{-kn}$

$$\begin{aligned} W_N &= e^{-j\frac{2\pi}{N}} \\ &= 1\angle{-2\pi/N} \end{aligned} \tag{3.11}$$

From Eq. (3.11), the magnitude of the twiddle factor is 1, and the phase angle is $-\frac{2\pi}{N}$. It lies on the unit circle in the complex plane from 0 to 2π angle, and it gets repeated for every cycle.

3.2.2 Four-Point, Six-Point and Eight-Point Twiddle Factors

As in Eq. (3.11), the magnitude of the twiddle factor is 1, and the angle -2π is equally divided in the interval N. The most commonly used intervals are $N = 4$ and $N = 8$. For $N = 4$, the angle between any $N = 0$ and $N = 1$ is $\frac{\pi}{2}$.

3.2.2.1 Four-Point Twiddle Factor

For $N = 4$

$$W_N = \begin{matrix} & n=0 & 1 & 2 & 3 \\ k=0 & W_4^0 & W_4^0 & W_4^0 & W_4^0 \\ 1 & W_4^0 & W_4^1 & W_4^2 & W_4^3 \\ 2 & W_4^0 & W_4^2 & W_4^4 & W_4^6 \\ 3 & W_4^0 & W_4^3 & W_4^6 & W_4^9 \end{matrix} \tag{3.12}$$

Note: $W_4^4 = W_4^0$; $W_4^6 = W_4^2$ and $W_4^9 = W_4^1$. From Eq. (3.11)

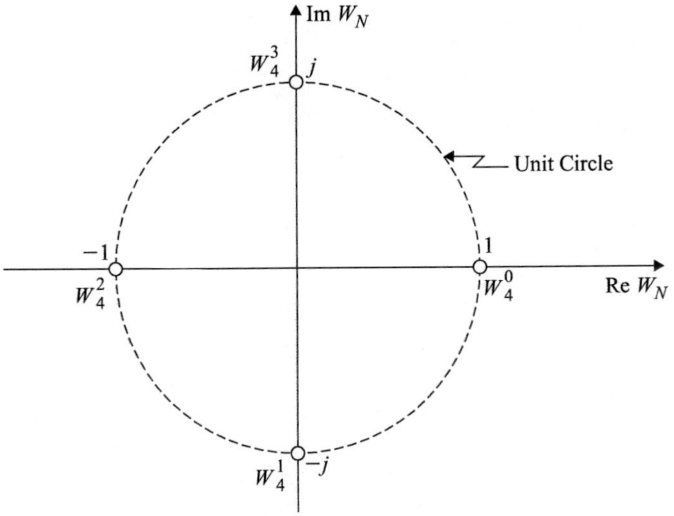

Fig. 3.1 Representation of W_4^{-nk}

$$W_4^1 = 1\angle{-\pi/2}$$

For $N = 4$, the unit circle is divided into four equal segments in the clockwise sequence and labelled as W_4^0, W_4^1, W_4^2 and W_4^3. From Fig. 3.1, the twiddle factors are obtained as

$$W_4^0 = 1; \quad W_4^1 = -j; \quad W_4^2 = -1; \quad W_4^3 = -j$$

Equation (3.12) is written as

$$W_N = \begin{bmatrix} 1 & 1 & 1 & 1 \\ 1 & -j & -1 & j \\ 1 & -1 & 1 & -1 \\ 1 & j & -1 & -j \end{bmatrix} \tag{3.13}$$

Equation (3.13) represents the twiddle factor to express DFT of any sequence $x[n]$. Twiddle factors for 6-point DFT and 8-point DFT are derived below.

3.2.2.2 Six-Point Twiddle Factor

For $N = 6$, the unit circle is divided into six equal segments and in the clockwise sequence labelled as W_6^0, W_6^1, W_6^2, W_6^3, W_6^4 and W_6^5 noting that $W_6^6 = W_6^0$, $W_6^7 = W_6^1$ and so on. This is shown in Fig. 3.2. Each segment is shifted by $-60°$ on the unit

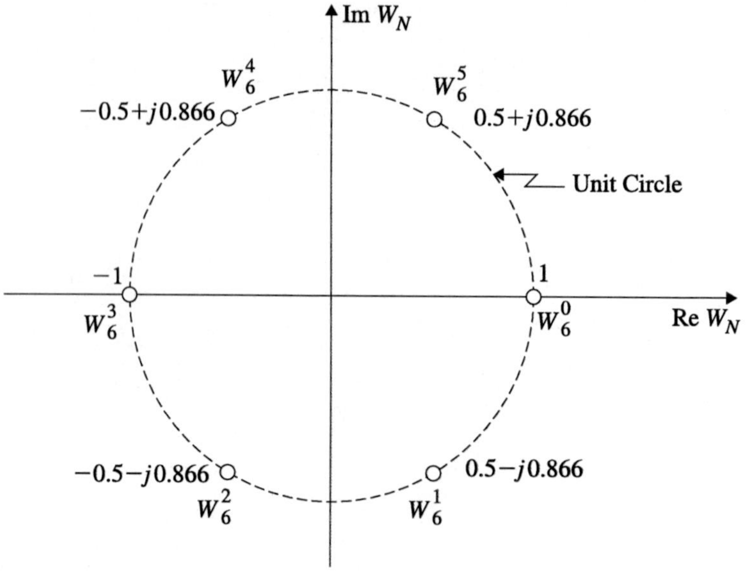

Fig. 3.2 Representation of W_6^{-nk}

circle. For $N = 6$, W_6 is obtained by multiplying the rows and columns of W_6 and is given below

$$W_N = \begin{bmatrix} W_6^0 & W_6^0 & W_6^0 & W_6^0 & W_6^0 & W_6^0 \\ W_6^0 & W_6^1 & W_6^2 & W_6^3 & W_6^4 & W_6^5 \\ W_6^0 & W_6^2 & W_6^4 & W_6^6 & W_6^8 & W_6^{10} \\ W_6^0 & W_6^3 & W_6^6 & W_6^9 & W_6^{12} & W_6^{15} \\ W_6^0 & W_6^4 & W_6^8 & W_6^{12} & W_6^{16} & W_6^{20} \\ W_6^0 & W_6^5 & W_6^{10} & W_6^{15} & W_6^{20} & W_6^{25} \end{bmatrix} \tag{3.14}$$

$$W_6^0 = W_6^6 = W_6^{12} = W_6^{18} = W_6^{24} = 1$$
$$W_6^1 = W_6^7 = W_6^{13} = W_6^{19} = W_6^{25} = 1e^{-j\frac{\pi}{3}} = 0.5 - j0.866$$
$$W_6^2 = W_6^8 = W_6^{14} = W_6^{20} = W_6^{26} = 1e^{-j\frac{2\pi}{3}} = -0.5 - j0.866$$
$$W_6^3 = W_6^9 = W_6^{15} = W_6^{21} = W_6^{27} = -1$$
$$W_6^4 = W_6^{10} = W_6^{16} = W_6^{22} = W_6^{28} = 1e^{j\frac{2\pi}{3}} = -0.5 + j0.866$$
$$W_6^5 = W_6^{11} = W_6^{17} = W_6^{23} = W_6^{29} = 1e^{j\frac{\pi}{3}} = 0.5 + j0.866$$

Substituting the values of the elements of the matrix W_6, we get

$$W_6 = \begin{bmatrix} 1 & 1 & 1 & 1 & 1 & 1 \\ 1 & 0.5 - j0.866 & -0.5 - j0.866 & -1 & -0.5 + j0.866 & 0.5 + j0.866 \\ 1 & -0.5 - j0.866 & -0.5 + j0.866 & 1 & -0.5 - j0.866 & -0.5 + j0.866 \\ 1 & -1 & 1 & -1 & 1 & -1 \\ 1 & -0.5 + j0.866 & -0.5 - j0.866 & 1 & -0.5 + j0.866 & -0.5 - j0.866 \\ 1 & 0.5 + j0.866 & -0.5 + j0.866 & -1 & -0.5 - j0.866 & 0.5 - j0.866 \end{bmatrix} \tag{3.15}$$

3.2.2.3 Eight-Point Twiddle Factor

The 8-point twiddle factor is represented in Fig. 3.3. The twiddle factor W_8 is obtained from Fig. 3.3 and is shown below

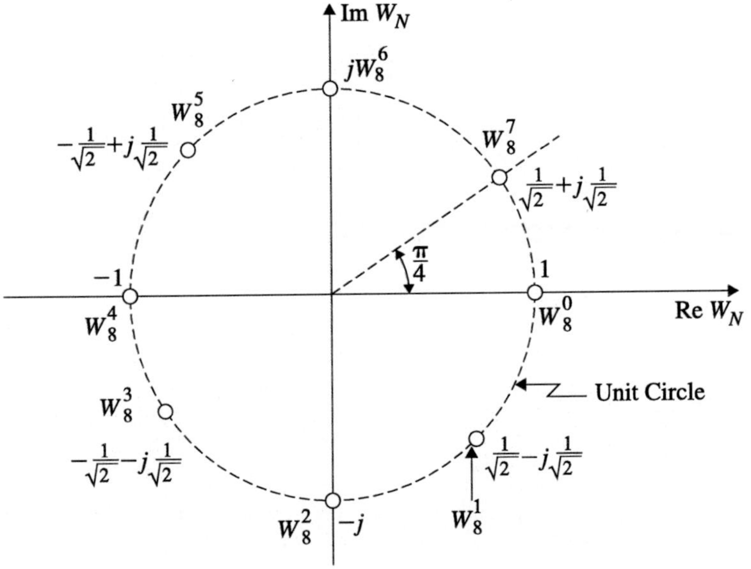

Fig. 3.3 Representation of W_8^{-kn}

$$W_8 = \begin{bmatrix} W_8^0 & W_8^0 & W_8^0 & W_8^0 & W_8^0 & W_8^0 & W_8^0 & W_8^0 \\ W_8^0 & W_8^1 & W_8^2 & W_8^3 & W_8^4 & W_8^5 & W_8^6 & W_8^7 \\ W_8^0 & W_8^2 & W_8^4 & W_8^6 & W_8^8 & W_8^{10} & W_8^{12} & W_8^{14} \\ W_8^0 & W_8^3 & W_8^6 & W_8^9 & W_8^{12} & W_8^{15} & W_8^{18} & W_8^{21} \\ W_8^0 & W_8^4 & W_8^8 & W_8^{12} & W_8^{16} & W_8^{20} & W_8^{24} & W_8^{28} \\ W_8^0 & W_8^5 & W_8^{10} & W_8^{15} & W_8^{20} & W_8^{25} & W_8^{30} & W_8^{35} \\ W_8^0 & W_8^6 & W_8^{12} & W_8^{18} & W_8^{24} & W_8^{30} & W_8^{36} & W_8^{42} \\ W_8^0 & W_8^7 & W_8^{14} & W_8^{21} & W_8^{28} & W_8^{35} & W_8^{42} & W_8^{49} \end{bmatrix} \tag{3.16}$$

$$W_8 = \begin{bmatrix} 1 & 1 & 1 & 1 & 1 & 1 & 1 & 1 \\ 1 & \frac{1}{\sqrt{2}} - j\frac{1}{\sqrt{2}} & -j & -\frac{1}{\sqrt{2}} - j\frac{1}{\sqrt{2}} & -1 & -\frac{1}{\sqrt{2}} + j\frac{1}{\sqrt{2}} & j & \frac{1}{\sqrt{2}} + j\frac{1}{\sqrt{2}} \\ 1 & -j & -1 & j & 1 & -j & -1 & j \\ 1 & -\frac{1}{\sqrt{2}} - j\frac{1}{\sqrt{2}} & j & \frac{1}{\sqrt{2}} - j\frac{1}{\sqrt{2}} & -1 & \frac{1}{\sqrt{2}} + j\frac{1}{\sqrt{2}} & -j & -\frac{1}{\sqrt{2}} + j\frac{1}{\sqrt{2}} \\ 1 & -1 & 1 & -1 & 1 & -1 & 1 & -1 \\ 1 & -\frac{1}{\sqrt{2}} + j\frac{1}{\sqrt{2}} & -j & \frac{1}{\sqrt{2}} + j\frac{1}{\sqrt{2}} & -1 & \frac{1}{\sqrt{2}} - j\frac{1}{\sqrt{2}} & j & -\frac{1}{\sqrt{2}} - j\frac{1}{\sqrt{2}} \\ 1 & j & -1 & -j & 1 & j & -1 & -j \\ 1 & \frac{1}{\sqrt{2}} + j\frac{1}{\sqrt{2}} & j & -\frac{1}{\sqrt{2}} + j\frac{1}{\sqrt{2}} & -1 & -\frac{1}{\sqrt{2}} - j\frac{1}{\sqrt{2}} & -j & \frac{1}{\sqrt{2}} - j\frac{1}{\sqrt{2}} \end{bmatrix} \tag{3.17}$$

Example 3.1 Compute the DFT of the sequence $x[n] = \{1,\ j,\ -1,\ -j\}$ for $N = 4$.

Solution

Method 1

$$W_4 = \begin{bmatrix} 1 & 1 & 1 & 1 \\ 1 & -j & -1 & j \\ 1 & -1 & 1 & -1 \\ 1 & j & -1 & -j \end{bmatrix}; \qquad X_4 = \begin{bmatrix} 1 \\ j \\ -1 \\ -j \end{bmatrix}$$

From Eq. (3.9)

$$X_4 = W_4 x_4$$

$$X_4 = \begin{bmatrix} 1 & 1 & 1 & 1 \\ 1 & -j & -1 & j \\ 1 & -1 & 1 & -1 \\ 1 & j & -1 & -j \end{bmatrix} \begin{bmatrix} 1 \\ j \\ -1 \\ -j \end{bmatrix}$$
$$= \begin{bmatrix} 1+j-1-j \\ 1+1+1+1 \\ 1-j-1+j \\ 1-1+1-1 \end{bmatrix} = \begin{bmatrix} 0 \\ 4 \\ 0 \\ 0 \end{bmatrix}$$

$$\boxed{\begin{array}{l} X(0) = 0 \\ X(1) = 4 \\ X(2) = 0 \\ X(3) = 0 \end{array}}$$

Method 2

$$X(k) = \sum_{n=0}^{3} x[n]e^{-j\frac{2\pi kn}{4}}; \quad k = 0, 1, 2, 3, \ldots$$

For $k = 0$

$$\begin{aligned} X(0) &= \sum_{n=0}^{3} x[n] \\ &= x[0] + x[1] + x[2] + x[3] \\ &= 1 + j - 1 - j = 0 \end{aligned}$$

For $k = 1$

$$\begin{aligned} X(1) &= \sum_{n=0}^{3} x[n]e^{-j\frac{\pi n}{2}} \\ &= x[0] + x[1]e^{-j\frac{\pi}{2}} + x[2]e^{-j\pi} + x[3]e^{-j\frac{3\pi}{2}} \\ &= 1 + j(-j) + (-1)(-1) + (-j)(j) \\ &= 1 + 1 + 1 + 1 = 4 \end{aligned}$$

For $k = 2$

$$
\begin{aligned}
X(2) &= \sum_{n=0}^{3} x[n]e^{-j\pi n} \\
&= x[0] + x[1]e^{-j\pi} + x[2]e^{-j2\pi} + x[3]e^{-j3\pi} \\
&= 1 + j(-1) + (-1)(1) + (-j)(-1) \\
&= 1 - j - 1 + j = 0
\end{aligned}
$$

For $k = 3$

$$
\begin{aligned}
X(3) &= \sum_{n=0}^{3} x[n]e^{-j\frac{3\pi n}{2}} \\
&= x[0] + x[1]e^{-\frac{3\pi}{2}} + x[2]e^{-j3\pi} + x[3]e^{-j\frac{9\pi}{2}} \\
&= 1 + j(j) + (-1)(-1) + (-j)(-j) \\
&= 1 - 1 + 1 - 1 = 0
\end{aligned}
$$

$$
\boxed{\begin{aligned}
X(0) &= 0 \\
X(1) &= 4 \\
X(2) &= 0 \\
X(3) &= 0
\end{aligned}}
$$

Method 1 is simpler and quicker.

Example 3.2 Find the 8-point DFT of $x[n] = \{1,\ -1,\ 1,\ -1,\ 1,\ -1,\ 1,\ -1\}$.

Solution

$$
\begin{aligned}
X_N &= W_N x_N \\
&= W_8 x_8
\end{aligned} \tag{3.18}
$$

W_8 is given in Eq. (3.17):

$$
x_8 = \begin{bmatrix} 1 \\ -1 \\ 1 \\ -1 \\ 1 \\ -1 \\ 1 \\ -1 \end{bmatrix}
$$

$X(0)$ is obtained by multiplying x_8 with the first row of W_8. Thus

$$X(0) = 1 - 1 + 1 - 1 + 1 - 1 + 1 - 1 = 0$$

$$\begin{aligned} X(1) &= \text{2nd row of } W_8 \text{ to multiply } x_8 \\ &= 1 - \frac{1}{\sqrt{2}} + j\frac{1}{\sqrt{2}} - j + \frac{1}{\sqrt{2}} + j\frac{1}{\sqrt{2}} - 1 + \frac{1}{\sqrt{2}} - j\frac{1}{\sqrt{2}} + j = 0 \end{aligned}$$

$$\begin{aligned} X(2) &= \text{3rd row of } W_8 \text{ to multiply } x_8 \\ &= 1 + j - 1 - j + 1 + j - 1 - j = 0 \\ X(3) &= 1 + \frac{1}{\sqrt{2}} + j\frac{1}{\sqrt{2}} + j - \frac{1}{\sqrt{2}} + j\frac{1}{\sqrt{2}} - 1 - \frac{1}{\sqrt{2}} - j\frac{1}{\sqrt{2}} - j + \frac{1}{\sqrt{2}} - j\frac{1}{\sqrt{2}} \\ &= 0 \\ X(4) &= \text{5th row of } W_8 \text{ to multiply } x_8 \\ &= 1 + 1 + 1 + 1 + 1 + 1 + 1 + 1 = 8 \\ X(5) &= \text{6th row of } W_8 \text{ to multiply } x_8 \\ &= 1 + \frac{1}{\sqrt{2}} - j\frac{1}{\sqrt{2}} - j - \frac{1}{\sqrt{2}} - j\frac{1}{\sqrt{2}} - 1 - \frac{1}{\sqrt{2}} + j\frac{1}{\sqrt{2}} + j + \frac{1}{\sqrt{2}} + j\frac{1}{\sqrt{2}} \\ &= 0 \\ X(6) &= 1 - j - 1 + j + 1 - j - 1 + j = 0 \\ X(7) &= 1 - \frac{1}{\sqrt{2}} - j\frac{1}{\sqrt{2}} + j + \frac{1}{\sqrt{2}} - j\frac{1}{\sqrt{2}} - 1 + \frac{1}{\sqrt{2}} + j\frac{1}{\sqrt{2}} - j - \frac{1}{\sqrt{2}} + j\frac{1}{\sqrt{2}} \\ &= 0 \end{aligned}$$

$$X_8 = \begin{bmatrix} 0 \\ 0 \\ 0 \\ 0 \\ 8 \\ 0 \\ 0 \\ 0 \end{bmatrix}$$

3.2.3 Zero Padding

In evaluating the DFT, we assumed that the length of the DFT which is N is equal to the length L of the sequence $x[n]$. If $N < L$, time domain aliasing occurs due to under-sampling and in the process, we could miss out some important details and get misleading information. To avoid this N, the number of samples of $x[n]$ is increased by adding some dummy sample of 0 value. This addition of dummy samples is known as zero padding. Zero padding not only increases the number of samples but also helps in getting a better idea of the frequency spectrum of $X(\Omega)$.

Example 3.3 Compute the 4-point DFT of the sequence

$$x[n] = 1 \qquad 0 \le n < 2$$

Solution For the given sequence, $L = 3$ and $N = 4$. By adding dummy samples of 0 values (zero padding), the given sequence becomes

$$x[n] = \{1,\ 1,\ 1,\ 0\}$$

$$x_N = \begin{bmatrix} 1 \\ 1 \\ 1 \\ 0 \end{bmatrix}$$

W_4 is given in Eq. (3.13):

$$\begin{aligned} X_4 &= W_4 x_4 \\ &= \begin{bmatrix} 1 & 1 & 1 & 1 \\ 1 & -j & -1 & j \\ 1 & -1 & 1 & -1 \\ 1 & j & -1 & -j \end{bmatrix} \begin{bmatrix} 1 \\ 1 \\ 1 \\ 0 \end{bmatrix} \\ X(0) &= [1 + 1 + 1 + 0] = 3 \\ X(1) &= [1 - j - 1 + 0] = -j \\ X(2) &= [1 - 1 + 1 + 0] = 1 \\ X(3) &= [1 + j - 1 + 0] = j \\ X_4 &= \begin{bmatrix} 3 \\ -j \\ 1 \\ j \end{bmatrix} \end{aligned}$$

Example 3.4 Compute the 4-point DFT of the following sequences:

$$\begin{aligned} &(1) \quad x[n] = \{1,\ 1,\ 1,\ 1\} \\ &(2) \quad x[n] = \{1,\ 1,\ 0,\ 0\} \\ &(3) \quad x[n] = \cos \pi n \\ &(4) \quad x[n] = \sin \frac{n\pi}{2} \end{aligned}$$

Solution (1) $\boldsymbol{x[n] = \{1,\ 1,\ 1,\ 1\}}$

$$x_4 = \begin{bmatrix} 1 \\ 1 \\ 1 \\ 1 \end{bmatrix}$$

$$X_4 = W_4 x_4$$

$$= \begin{bmatrix} 1 & 1 & 1 & 1 \\ 1 & -j & -1 & j \\ 1 & -1 & 1 & -1 \\ 1 & j & -1 & -j \end{bmatrix} \begin{bmatrix} 1 \\ 1 \\ 1 \\ 1 \end{bmatrix}$$

$$X(0) = 1 + 1 + 1 + 1$$
$$= 4$$
$$X(1) = 1 - j - 1 + j$$
$$= 0$$
$$X(2) = 1 - 1 + 1 - 1$$
$$= 0$$
$$X(3) = 1 + j - 1 - j$$
$$= 0$$

$$X_4 = \begin{bmatrix} 4 \\ 0 \\ 0 \\ 0 \end{bmatrix}$$

(2) $\boldsymbol{x[n] = \{1,\ 1,\ 0,\ 0\}}$

$$X_4 = W_4 x_4$$

$$X_4 = \begin{bmatrix} 1 & 1 & 1 & 1 \\ 1 & -j & -1 & j \\ 1 & -1 & 1 & -1 \\ 1 & j & -1 & -j \end{bmatrix} \begin{bmatrix} 1 \\ 1 \\ 0 \\ 0 \end{bmatrix}$$

$$X(0) = 1 + 1 + 0 + 0$$
$$= 2$$
$$X(1) = 1 - j + 0 + 0$$
$$= (1 - j)$$

$$
\begin{aligned}
X(2) &= 1 - 1 + 0 + 0 \\
&= 0 \\
X(3) &= 1 + j + 0 + 0 \\
&= (1 + j)
\end{aligned}
$$

$$
X_4 = \begin{bmatrix} 2 \\ 1 - j \\ 0 \\ 1 + j \end{bmatrix}
$$

(3) **$x[n] = \cos \pi n$; where $n = 0, 1, 2, 3, \ldots$**

$$
\begin{aligned}
x[n] &= \{1,\ -1,\ 1,\ -1\} \\
X_4 &= W_4 x_4 \\
&= \begin{bmatrix} 1 & 1 & 1 & 1 \\ 1 & -j & -1 & j \\ 1 & -1 & 1 & -1 \\ 1 & j & -1 & -j \end{bmatrix} \begin{bmatrix} 1 \\ -1 \\ 1 \\ -1 \end{bmatrix} \\
X(0) &= 1 - 1 + 1 - 1 \\
&= 0 \\
X(1) &= 1 + j - 1 - j \\
&= 0 \\
X(2) &= 1 + 1 + 1 + 1 \\
&= 4 \\
X(3) &= 1 - j - 1 + j \\
&= 0
\end{aligned}
$$

$$
X_4 = \begin{bmatrix} 0 \\ 0 \\ 4 \\ 0 \end{bmatrix}
$$

(4) **$x[n] = \sin \frac{n\pi}{2}$; where $n = 0, 1, 2, 3, \ldots$**

$$
\begin{aligned}
x[n] &= \{0,\ 1,\ 0,\ -1\} \\
X_4 &= W_4 x_4
\end{aligned}
$$

$$X_4 = \begin{bmatrix} 1 & 1 & 1 & 1 \\ 1 & -j & -1 & j \\ 1 & -1 & 1 & -1 \\ 1 & j & -1 & -j \end{bmatrix} \begin{bmatrix} 0 \\ 1 \\ 0 \\ -1 \end{bmatrix}$$

$$\begin{aligned} X(0) &= 0 + 1 + 0 - 1 \\ &= 0 \\ X(1) &= 0 - j + 0 - j \\ &= -j2 \\ X(2) &= 0 - 1 + 0 + 1 \\ &= 0 \\ X(3) &= 0 + j + 0 + j \\ &= j2 \end{aligned}$$

$$X_4 = \begin{bmatrix} 0 \\ -j2 \\ 0 \\ j2 \end{bmatrix}$$

Example 3.5 Find the N-point DFT of the following sequences for $0 \leq n \leq N - 1$:

$$x[n] = \delta[n]$$

Solution $x[n] = \delta[n]$

$$X(k) = \sum_{n=0}^{N-1} x[n] e^{-j\frac{2\pi kn}{N}}$$

$$x[n] = \begin{cases} 1 & n = 0 \\ 0 & n \neq 0 \end{cases}$$

$$\boxed{X(k) = 1}$$

Example 3.6 Find the IDFT of the following functions with $N = 4$:

(1) $X(k) = \{1,\ 0,\ 1,\ 0\}$

(2) $X(k) = \{6,\ (-2 + j2),\ -2,\ (-2 - j2)\}$

Solution (1) $\boldsymbol{X(k) = \{1,\ 0,\ 1,\ 0\}}$

From Eq. (3.10)

$$x_N = \frac{1}{N}[W_N^*]X_N$$

$$X_N = \begin{bmatrix} 1 \\ 0 \\ 1 \\ 0 \end{bmatrix}$$

$$W_N^* = \begin{bmatrix} 1 & 1 & 1 & 1 \\ 1 & j & -1 & -j \\ 1 & -1 & 1 & -1 \\ 1 & -j & -1 & j \end{bmatrix}$$

For $N = 4$

$$x_N = \frac{1}{4}\begin{bmatrix} 1 & 1 & 1 & 1 \\ 1 & j & -1 & -j \\ 1 & -1 & 1 & -1 \\ 1 & -j & -1 & j \end{bmatrix}\begin{bmatrix} 1 \\ 0 \\ 1 \\ 0 \end{bmatrix}$$

$$x[0] = \frac{1}{4}[1 + 0 + 1 + 0] = 0.5$$

$$x[1] = \frac{1}{4}[1 + 0 - 1 + 0] = 0$$

$$x[2] = \frac{1}{4}[1 + 0 + 1 + 0] = 0.5$$

$$x[3] = \frac{1}{4}[1 + 0 - 1 + 0] = 0$$

$$\boxed{x[n] = \{0.5,\ 0,\ 0.5,\ 0\}}$$

(2) $\boldsymbol{x[n] = \{6,\ (-2 + j2),\ -2,\ (-2 - j2)\}}$

$$x_N = \frac{1}{4}\begin{bmatrix} 1 & 1 & 1 & 1 \\ 1 & j & -1 & -j \\ 1 & -1 & 1 & -1 \\ 1 & -j & -1 & j \end{bmatrix}\begin{bmatrix} 6 \\ -2 + j2 \\ -2 \\ -2 - j2 \end{bmatrix}$$

$$x[0] = \frac{1}{4}[6 - 2 + j2 - 2 - 2 - j2] = 0$$

$$x[1] = \frac{1}{4}[6 - j2 - 2 + 2 + j2 - 2] = 1$$

$$x[2] = \frac{1}{4}[6 + 2 - j2 - 2 + 2 + j2] = 2$$

$$x[3] = \frac{1}{4}[6 + j2 + 2 + 2 - j2 + 2] = 3$$

$$\boxed{x[n] = \{0,\ 1,\ 2,\ 3\}}$$

Example 3.7 Compute the 4-point DFT of a causal three sample sequence given by

$$x(n) = \begin{cases} \frac{1}{3}, & 0 \le n \le 2 \\ 0, & \text{else} \end{cases}$$

Plot the magnitude and phase spectrum.

Solution N-point DFT of $x(n)$ is

$$X(k) = \sum_{n=0}^{N-1} x(n) e^{-j\frac{2\pi kn}{N}}; \qquad k = 0, \ldots, N-1$$

Here $N = 4$

$$X(k) = \sum_{n=0}^{3} x(n) e^{-j\frac{2\pi kn}{4}} = \sum_{n=0}^{3} x(n) e^{-j\frac{\pi kn}{2}}; \qquad k = 0, 1, 2, 3, \ldots$$

$$X(k) = x(0)e^{0} + x(1)e^{-\frac{j\pi k}{2}} + x(2)e^{-j\pi k} + x(3)e^{-\frac{j3\pi k}{2}}$$

$$= \frac{1}{3} + \frac{1}{3}e^{-\frac{j\pi k}{2}} + \frac{1}{3}e^{-j\pi k} + 0$$

$$= \frac{1}{3}\left[1 + e^{-\frac{j\pi k}{2}} + e^{-j\pi k}\right]$$

$$X(k) = \frac{1}{3}\left[1 + \cos\frac{\pi k}{2} - j\sin\frac{\pi k}{2} + \cos\pi k - j\sin\pi k\right]; \quad k = 0, 1, 2, 3, \ldots$$

For $k = 0$

$$X(0) = \frac{1}{3}\left[1 + \cos 0 - j\sin 0 + \cos 0 - j\sin 0\right]$$

$$= \frac{1}{3}[1 + 1 + 1] = 1\angle 0$$

For $k = 1$

$$X(1) = \frac{1}{3}\left[1 + \cos\frac{\pi}{2} - j\sin\frac{\pi}{2} + \cos\pi - j\sin\pi\right]$$
$$= \frac{1}{3}[1 - j - 1] = -\frac{1}{3}j = \frac{1}{3}\angle -\pi/2$$

For $k = 2$

$$X(2) = \frac{1}{3}\,[1 + \cos\pi - j\sin\pi + \cos 2\pi - j\sin 2\pi]$$
$$= \frac{1}{3}[1 - 1 + 1] = \frac{1}{3}\angle 0$$

For $k = 3$

$$X(3) = \frac{1}{3}\left[1 + \cos\frac{3\pi}{2} - j\sin\frac{3\pi}{2} + \cos 3\pi - j\sin 3\pi\right]$$
$$= \frac{1}{3}[1 + j - 1] = \frac{j}{3} = \frac{1}{3}\angle\pi/2$$

Therefore

$$X(k) = \left\{1\angle 0,\ \frac{1}{3}\angle -\pi/2,\ \frac{1}{3}\angle 0,\ \frac{1}{3}\angle\pi/2\right\}$$

$$\text{Magnitude function } |X(k)| = \left\{1,\ \frac{1}{3},\ \frac{1}{3},\ \frac{1}{3}\right\}$$
$$\text{Phase function } \angle X(k) = \left\{0,\ -\frac{\pi}{2},\ 0,\ \frac{\pi}{2}\right\}$$

The magnitude and phase spectrum are shown in Fig. 3.4a and b, respectively.

Example 3.8 Compute the 4-point DFT of the sequence

$$x(n) = \{0,\ 1,\ 2,\ 3\}$$

Sketch the magnitude and phase spectrum.

Solution Given

$$x(n) = \{0, 1, 2, 3\}$$

Here, the length of the sequence $L = 4$. Hence, we can compute the 4-point DFT since $N \geq L$. The 4-point DFT of the sequence $x(n)$ is given by

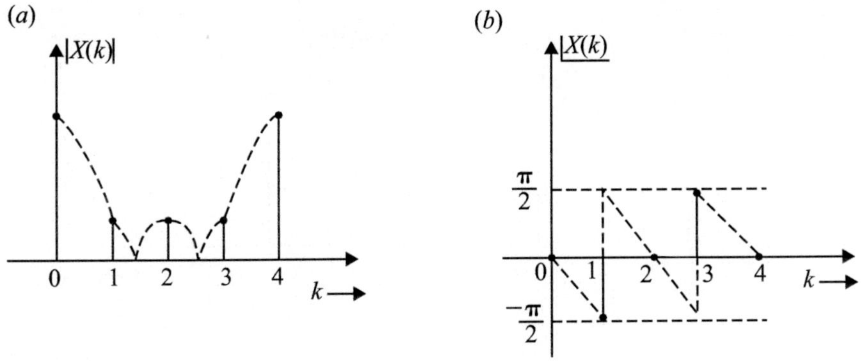

Fig. 3.4 Magnitude and phase spectrum of Example 3.7

$$
\begin{aligned}
X(k) &= \sum_{n=0}^{3} x(n)e^{-j\frac{2\pi kn}{4}} = \sum_{n=0}^{3} x(n)e^{-j\frac{\pi kn}{2}} \\
&= x(0)e^{0} + x(1)e^{-j\frac{\pi k}{2}} + x(2)e^{-j\pi k} + x(3)e^{-j\frac{3\pi k}{2}} \\
&= 0 + e^{-j\frac{\pi k}{2}} + 2e^{-j\pi k} + 3e^{-j\frac{3\pi k}{2}} \\
X(4) &= \left(\cos\frac{\pi k}{2} - j\sin\frac{\pi k}{2}\right) + 2\left(\cos\pi k - j\sin\pi k\right) + 3\left(\cos\frac{3\pi k}{2} - j\sin\frac{3\pi k}{2}\right)
\end{aligned}
$$

For $k = 0$

$$
\begin{aligned}
X(0) &= (\cos 0 - j\sin 0) + 2\,(\cos 0 - j\sin 0) + 3\,(\cos 0 - j\sin 0) \\
&= 1 + 2 + 3 = 6 = 6\angle 0
\end{aligned}
$$

For $k = 1$

$$
\begin{aligned}
X(1) &= \left(\cos\frac{\pi}{2} - j\sin\frac{\pi}{2}\right) + 2(\cos\pi - j\sin\pi) + 3\left(\cos\frac{3\pi}{2} - j\sin\frac{3\pi}{2}\right) \\
&= -j - 2 + 3j = -2 + 2j = 2.8\angle 135^\circ = 2.8\angle 135 \times \frac{\pi}{180} \\
&= 2.8\angle 0.75\pi
\end{aligned}
$$

For $k = 2$

$$
\begin{aligned}
X(2) &= (\cos\pi - j\sin\pi) + 2(\cos 2\pi - j\sin 2\pi) + 3(\cos 3\pi - j\sin 3\pi) \\
&= -1 + 2 - 3 = -2 = 2\angle 180 = 2\angle 180 \times \frac{\pi}{180} \\
&= 2\angle\pi
\end{aligned}
$$

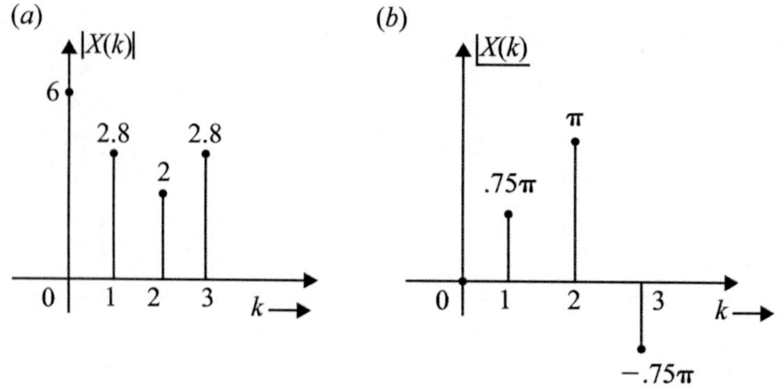

Fig. 3.5 Magnitude and phase spectrum of Example 3.8

For $k = 3$

$$X(3) = \left(\cos\frac{3\pi}{2} - j\sin\frac{3\pi}{2}\right) + 2(\cos 3\pi - j\sin 3\pi) + 3\left(\cos\frac{9\pi}{2} - j\sin\frac{9\pi}{2}\right)$$
$$= j - 2 - 3j = -2 - 2j = 2.8\angle{-135^\circ} = 2.8\angle{-135} \times \frac{\pi}{180}$$
$$= 2.8\angle{-0.75\pi}$$

Therefore,

$$X(k) = \{6\angle 0,\ 2.8\angle 0.75\pi,\ 2\angle\pi,\ 2.8\angle{-0.75\pi}\}$$

$$\text{Magnitude function } |X(k)| = \{6,\ 2.8,\ 2,\ 2.8\}$$
$$\text{Phase function } \angle X(k) = \{0,\ 0.75\pi,\ \pi,\ -0.75\pi\}$$

The magnitude and phase spectrum are shown in Fig. 3.5a and b, respectively.

Example 3.9 Find the DFT of a sequence $x(n) = \{1, 1, 0, 0\}$ and find the IDFT of $Y(k) = \{1, 0, 1, 0\}$.

Solution Let us assume $L = N = 4$. The 4-point DFT of $x(n)$ is

$$X(k) = \sum_{n=0}^{N-1} x(n)e^{-j\frac{2\pi kn}{N}}, \quad k = 0, 1, \ldots, N-1$$

$$X(k) = \sum_{n=0}^{3} x(n)e^{-j\frac{2\pi kn}{N}} = \sum_{n=0}^{3} x(n)e^{-j\frac{\pi kn}{2}}$$

$$X(0) = \sum_{n=0}^{3} x(n) = x(0) + x(1) + x(2) + x(3)$$
$$= 1 + 1 + 0 + 0 = 2$$
$$X(1) = \sum_{n=0}^{3} x(n)e^{-j\frac{\pi n}{2}} = x(0) + x(1)e^{-j\frac{\pi}{2}} + x(2)e^{-j\pi} + x(3)e^{-j\frac{3\pi}{2}}$$
$$= 1 + \cos\frac{\pi}{2} - j\sin\frac{\pi}{2} = 1 - j$$
$$X(2) = \sum_{n=0}^{3} x(n)e^{-j\pi n} = x(0) + x(1)e^{-j\pi} + x(2)e^{-j2\pi} + x(3)e^{-j3\pi}$$
$$= 1 + \cos\pi - j\sin\pi = 1 - 1 = 0$$
$$X(3) = \sum_{n=0}^{3} x(n)e^{-j\frac{3\pi n}{2}} = x(0) + x(1)e^{-j\frac{3\pi}{2}} + x(2)e^{-j3\pi} + x(3)e^{-j\frac{9\pi}{2}}$$
$$= 1 + \cos\frac{3\pi}{2} - j\sin\frac{3\pi}{2} = 1 + j$$

Therefore

$$\boxed{X(k) = \{2,\ 1 - j,\ 0,\ 1 + j\}}$$

Given $Y(k) = \{1, 0, 1, 0\}$.

The 4-point IDFT of $Y(k)$ is

$$y(n) = \frac{1}{N}\sum_{k=0}^{N-1} Y(k)e^{j\frac{2\pi kn}{N}}, \qquad n = 0, 1, \ldots, N-1$$
$$y(n) = \frac{1}{4}\sum_{k=0}^{3} Y(k)e^{j\frac{2\pi kn}{3}}$$
$$y(0) = \frac{1}{4}\sum_{k=0}^{3} Y(k)$$
$$= \frac{1}{4}[Y(0) + Y(1) + Y(2) + Y(3)]$$

$$y(0) = \frac{1}{4}[1 + 0 + 1 + 0] = 0.5$$

$$y(1) = \frac{1}{4}\sum_{k=0}^{3} Y(k)e^{\frac{j\pi k}{2}}$$

$$= \frac{1}{4}\left[Y(0) + Y(1)e^{\frac{j\pi}{2}} + Y(2)e^{j\pi} + Y(3)e^{\frac{j3\pi}{2}}\right]$$

$$= \frac{1}{4}\left[Y(0) + Y(1)j\sin\frac{\pi}{2} + Y(2)\cos\pi + Y(3)j\sin\frac{3\pi}{2}\right]$$

$$= \frac{1}{4}[1 + 0 - 1 + 0] = 0$$

$$y(2) = \frac{1}{4}\sum_{k=0}^{3} Y(k)e^{\frac{j4\pi k}{4}}$$

$$= \frac{1}{4}\left[Y(0) + Y(1)e^{j\pi} + Y(2)e^{j2\pi} + Y(3)e^{j3\pi}\right]$$

$$= \frac{1}{4}[1 + 0 + \cos 2\pi + 0]$$

$$= \frac{1}{4}[1 + 0 + 1 + 0] = 0.5$$

$$y(3) = \frac{1}{4}\sum_{k=0}^{3} Y(k)e^{\frac{j3\pi k}{2}}$$

$$= \frac{1}{4}\left[Y(0) + Y(1)e^{j\frac{3\pi}{2}} + Y(2)e^{j3\pi} + Y(3)e^{j\frac{9\pi}{2}}\right]$$

$$= \frac{1}{4}[1 + 0 + \cos 3\pi + 0]$$

$$= \frac{1}{4}[1 + 0 - 1 + 0] = 0$$

Therefore,

$$\boxed{y(n) = \{0.5, 0, 0.5, 0\}}$$

Example 3.10 Find the DFT of a sequence

$$x(n) = \begin{cases} 1, & \text{for } 0 \le n \le 2 \\ 0, & \text{otherwise} \end{cases}$$

for (*i*) $N = 4$ and (*ii*) $N = 8$. Plot $|X(k)|$ and $\angle X(k)$. Comment on the result.

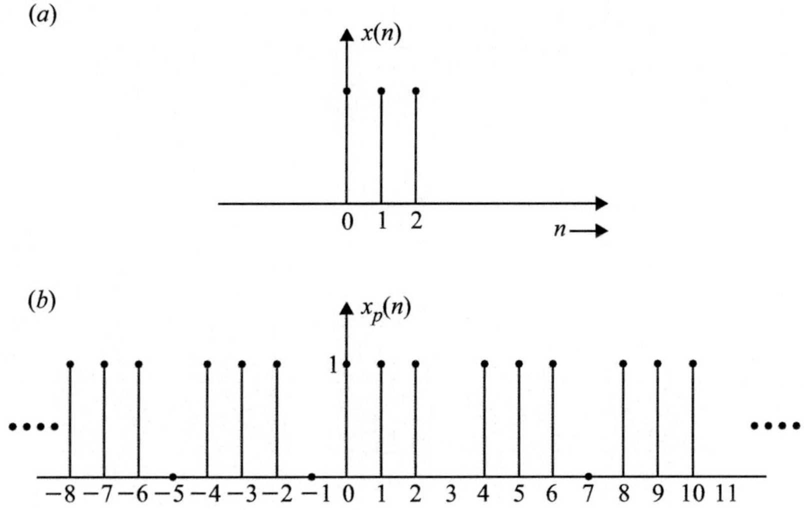

Fig. 3.6 DT signal and its periodic extension

Solution Given length of the sequence $L = 3$. $x[n]$ is represented in Fig. 3.6a.

(*i*) For $N = 4$, the periodic extension of the sequence is shown in Fig. 3.6b. Consider

$$
\begin{aligned}
x[n] &= \{1,\ 1,\ 1,\ 0\} \\
X(k) &= \sum_{n=0}^{N-1} x(n) e^{-j\frac{2\pi kn}{N}} \\
&= \sum_{n=0}^{3} x(n) e^{-j\frac{2\pi kn}{4}} \\
X(0) &= \sum_{n=0}^{3} x(n) = x(0) + x(1) + x(2) + x(3) \\
&= 3 = 3\angle 0 \\
X(1) &= \sum_{n=0}^{3} x(n) e^{-j\frac{\pi n}{2}} \\
&= x(0) + x(1)e^{-j\frac{\pi}{2}} + x(2)e^{-j\pi} + x(3)e^{-j\frac{3\pi}{2}} \\
&= 1 + (-j) + (-1) \\
&= -j \\
&= 1\angle \pi/2
\end{aligned}
$$

$$
\begin{aligned}
X(2) &= \sum_{n=0}^{3} x(n)e^{-j\pi n} \\
&= x(0) + x(1)e^{-j\pi} + x(2)e^{-j2\pi} + x(3)e^{-j3\pi} \\
&= 1 + (-1) + 1 = 1 = 1\angle 0 \\
X(3) &= \sum_{n=0}^{3} x(n)e^{-j\frac{3\pi n}{2}} \\
&= x(0) + x(1)e^{-j\frac{3\pi}{2}} + x(2)e^{-j3\pi} + x(3)e^{-j\frac{9\pi}{2}} \\
&= 1 + j - 1 = j = 1\angle \pi/2
\end{aligned}
$$

Therefore,

$$
\begin{aligned}
X(k) &= \{3\angle 0,\ 1\angle \pi/2,\ 1\angle 0,\ 1\angle \pi/2\} \\
|X(k)| &= \{3, 1, 1, 1\} \\
\angle X(k) &= \left\{0, -\frac{\pi}{2}, 0, \frac{\pi}{2}\right\}
\end{aligned}
$$

Magnitude and phase spectrum are shown in Fig. 3.7a and b, respectively.

(*ii*) **For** $N = 8$

$$x(n) = \{1, 1, 1, 0, 0, 0, 0, 0\}$$

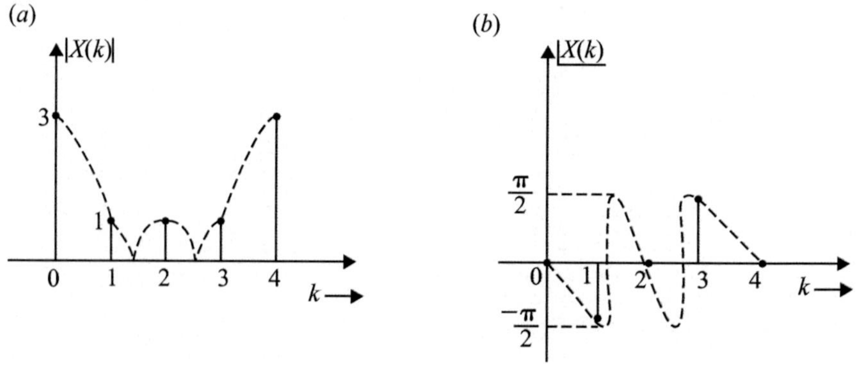

Fig. 3.7 Magnitude and phase spectrum for $N = 4$

$$
\begin{aligned}
X(k) &= \sum_{n=0}^{7} x(n)e^{-j\frac{2\pi kn}{8}}, \quad k = 0, \ldots, 7 \\
X(0) &= \sum_{n=i}^{7} x(n) = 1+1+1+0+0+0+0+0 = 3 \\
X(1) &= \sum_{n=0}^{7} x(n)e^{-j\frac{\pi n}{4}} \\
&= x(0) + x(1)e^{-j\frac{\pi}{4}} + x(2)e^{-j\frac{2\pi}{4}} + 0 + 0 \\
&= 1 + \frac{1}{\sqrt{2}} - j\frac{1}{\sqrt{2}} - j = 1.707 - j1.707 \\
X(1) &= 2.414\angle -\pi/4 \\
X(2) &= \sum_{n=0}^{7} x(n)e^{-j\frac{\pi n}{2}} \\
&= x(0) + x(1)e^{-j\frac{\pi}{2}} + x(2)e^{-j\pi} + 0 + 0 \\
&= 1 + \cos\frac{\pi}{2} - j\sin\frac{\pi}{2} + \cos\pi - j\sin\pi \\
&= 1 - j - 1 = -j \\
&= 1\angle -\pi/2 \\
X(3) &= \sum_{n=0}^{7} x(n)e^{-j\frac{3\pi n}{4}} \\
&= x(0) + x(1)e^{-j\frac{3\pi}{4}} + x(2)e^{-j\frac{3\pi}{2}} + 0 + 0 \\
&= 1 + \cos\frac{3\pi}{4} - j\sin\frac{3\pi}{4} + \cos\frac{3\pi}{2} - j\sin\frac{3\pi}{2} \\
&= 1 - 0.707 - j0.707 + j = 0.293 + j0.293 \\
&= 0.414\angle \pi/3 \\
X(4) &= \sum_{n=0}^{7} x(n)e^{-jn\pi} \\
&= x(0) + x(1)e^{-j\pi} + x(2)e^{-j2\pi} + 0 + 0 \\
&= 1 + \cos\pi - j\sin\pi + \cos 2\pi - j\sin 2\pi \\
&= 1 - 1 + 1 = 1\angle 0 \\
X(5) &= \sum_{n=0}^{7} x(n)e^{-j\frac{5\pi n}{4}} \\
&= x(0) + x(1)e^{-j\frac{5\pi}{4}} + x(2)e^{-j\frac{5\pi}{2}} \\
&= 1 + \cos\frac{5\pi}{4} - j\sin\frac{5\pi}{7} + \cos\frac{5\pi}{2} - j\sin\frac{5\pi}{2} \\
&= 1 - 0.707 + j0.707 - j
\end{aligned}
$$

$$= 0.293 - j0.293 = 0.414\angle -\pi/4$$

$$X(6) = \sum_{n=0}^{7} x(n)e^{-j\frac{3\pi n}{2}}$$
$$= x(0) + x(1)e^{-j\frac{3\pi}{2}} + x(2)e^{-j3\pi} + 0$$
$$= 1 + \cos\frac{3\pi}{2} - j\sin\frac{3\pi}{2} + \cos 3\pi - j\sin 3\pi$$
$$= 1 + j - 1 = j$$
$$= 1\angle \pi/2$$

$$X(7) = \sum_{n=0}^{7} x(n)e^{-j\frac{7\pi nk}{4}}$$
$$= 1 + e^{-j\frac{7\pi}{4}} + e^{-j\frac{7\pi}{2}} + 0$$
$$= 1 + \cos\frac{7\pi}{4} - j\sin\frac{7\pi}{4} + \cos\frac{7\pi}{2} - j\sin\frac{7\pi}{2} + 0$$
$$= 1 + 0.707 + j0.707 + j$$
$$= 1.707 + j1.707$$
$$= 2.414\angle \pi/4$$

Therefore

$$X(k) = \{3, 2.414, 1, 0.414, 1, 0.414, 1, 2.414\}$$
$$\angle X(k) = \left\{0, -\frac{\pi}{4}, -\frac{\pi}{2}, \frac{\pi}{4}, 0, -\frac{\pi}{4}, \frac{\pi}{2}, \frac{\pi}{4}\right\}$$

The magnitude and phase spectrum are shown in Fig. 3.8a and b, respectively.

Comments: It is very difficult to extrapolate the entire frequency spectrum with $N = 4$, i.e. the resolution of the spectrum is very poor. In order to increase the resolution, we must increase N. It is possible to extrapolate the frequency spectrum with $N = 8$. Thus, the zero padding gives us a high-density spectrum and provides better displayed magnitude and phase plots.

Example 3.11 Find IDFT of the sequence $X(k) = \{5, 0, 1 - j, 0, 1, 0, 1 + j, 0\}$.

Solution Given $N = 8$

$$x(n) = \frac{1}{N}\sum_{k=0}^{N-1} X(k)e^{j\frac{2\pi kn}{N}}, \quad n = 0, \ldots, N-1$$

$$x(n) = \frac{1}{8}\sum_{k=0}^{7} X(k)e^{j\frac{\pi kn}{4}}, \quad n = 0, \ldots, 7$$

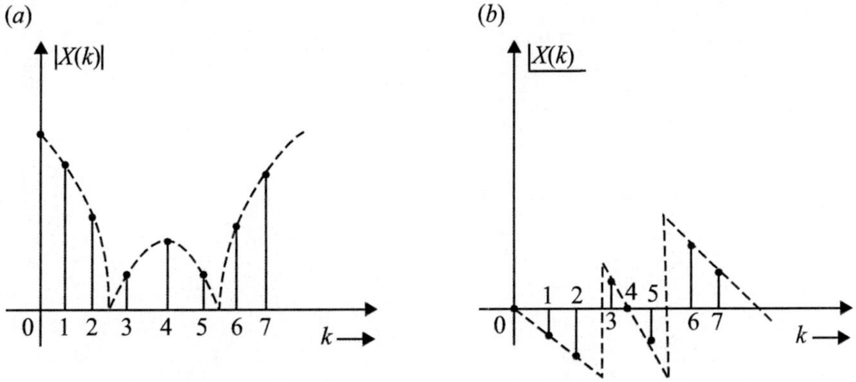

Fig. 3.8 Magnitude and phase spectrum for $N = 8$

$$
\begin{aligned}
x(0) &= \frac{1}{8}\sum_{k=0}^{7} X(k) \\
&= \frac{1}{8}[5+0+1-j+0+1+0+1+j+0] = 1 \\
x(1) &= \frac{1}{8}\sum_{k=0}^{7} X(k)e^{j\frac{\pi k}{4}} \\
&= \frac{1}{8}[5+(1-j)j+1(-1)+(1+j)(-j)] \\
&= \frac{1}{8}[6] = 0.75 \\
x(2) &= \frac{1}{8}\sum_{k=0}^{7} X(k)e^{j\frac{\pi k}{2}} \\
&= \frac{1}{8}[5+(1-j)(-1)+1(1)+(1+j)(-1)] \\
&= \frac{1}{8}[4] = 0.5 \\
x(3) &= \frac{1}{8}\sum_{k=0}^{7} X(k)e^{j\frac{3\pi k}{4}} \\
&= \frac{1}{8}[5+(1-j)(-j)+1(-1)+(1+j)(j)] \\
&= \frac{1}{8}[2] = 0.25
\end{aligned}
$$

$$x(4) = \frac{1}{8}\sum_{k=0}^{7} X(k)e^{j\pi k}$$
$$= \frac{1}{8}[5 + (1 - j)(1) + 1(1) + (1 + j)(1)] = 1$$
$$x(5) = \frac{1}{8}\sum_{k=0}^{7} X(k)e^{j\frac{5\pi k}{4}}$$
$$= \frac{1}{8}[5 + (1 - j)(j) + (1)(1) + (1 + j)(-j)]$$
$$= \frac{1}{8}[6] = 0.75$$
$$x(6) = \frac{1}{8}\sum_{k=0}^{7} X(k)e^{j\frac{3\pi k}{2}}$$
$$= \frac{1}{8}[5 + (1 - j)(-1) + 1(1) + (1 + j)(-1)]$$
$$= \frac{1}{8}[4] = 0.5$$
$$x(7) = \frac{1}{8}\sum_{k=0}^{7} X(k)e^{j\frac{7\pi k}{4}}$$
$$= \frac{1}{8}[5 + (1 - j)(-j) + 1(-1) + (1 + j)(j)]$$
$$= \frac{1}{8}[2] = 0.25$$

Therefore,

$$\boxed{x(n) = \{1, 0.75, 0.5, 0.25, 1, 0.75, 0.5, 0.25\}}$$

Example 3.12 Two finite duration sequences are given by

$$x[n] = \sin\left(\frac{n\pi}{2}\right) \qquad \text{for } n = 0, 1, 2, 3$$
$$h[n] = 2^n \qquad \text{for } n = 0, 1, 2, 3$$

(a) Calculate the 4-point DFT $X(k)$. (b) Calculate the 4-point DFT $H(k)$ and (c) if $Y(k) = X(k)H(k)$, determine the inverse DFT $y(n)$ of $Y(k)$ and sketch it.

Solution (*a*) To calculate the 4-point DFT $X(k)$

Given

$$\begin{aligned}
x[n] &= \sin\left(\frac{n\pi}{2}\right) \qquad \text{for } n = 0, 1, 2, 3 \\
x[n] &= \{0,\ 1,\ 0,\ -1\} \\
X_4 &= W_4 x_4 \\
&= \begin{bmatrix} 1 & 1 & 1 & 1 \\ 1 & -j & -1 & j \\ 1 & -1 & 1 & -1 \\ 1 & j & -1 & -j \end{bmatrix} \begin{bmatrix} 0 \\ 1 \\ 0 \\ -1 \end{bmatrix} \\
X(0) &= 0 + 1 + 0 - 1 = 0 \\
X(1) &= 0 - j + 0 - j = -j2 \\
X(2) &= 0 - 1 + 0 + 1 = 0 \\
X(3) &= 0 + j + 0 + j = j2 \\
X_4 &= \begin{bmatrix} 0 \\ -2j \\ 0 \\ 2j \end{bmatrix}
\end{aligned}$$

$$\boxed{X(k) = \{0,\ -2j,\ 0,\ 2j\}}$$

(*b*) Calculate the 4-point DFT $H(k)$

Given

$$\begin{aligned}
h[n] &= 2^n \qquad \text{for } n = 0, 1, 2, 3 \\
h[n] &= \{1,\ 2,\ 4,\ 8\} \\
H_4 &= W_4 h_4 \\
&= \begin{bmatrix} 1 & 1 & 1 & 1 \\ 1 & -j & -1 & j \\ 1 & -1 & 1 & -1 \\ 1 & j & -1 & -j \end{bmatrix} \begin{bmatrix} 1 \\ 2 \\ 4 \\ 8 \end{bmatrix} \\
H(0) &= 1 + 2 + 4 + 8 = 15 \\
H(1) &= 1 - 2j - 4 + 8j = -3 + 6j \\
H(2) &= 1 - 2 + 4 - 8 = -5 \\
H(3) &= 1 + 2j - 4 - 8j = -3 - 6j
\end{aligned}$$

$$H_4 = \begin{bmatrix} 15 \\ -3+6j \\ -5 \\ -3-6j \end{bmatrix}$$

$$\boxed{H(k) = \{15,\ -3+6j,\ -5,\ -3-6j\}}$$

(c)

$$\begin{aligned} Y(k) &= X(k)H(k) \\ &= \{0,\ -j2,\ 0,\ j2\}\{15,\ -3+j6,\ -5,\ -3-j6\} \\ &= \{0\times 15,\ (-j2)(-3+j6),\ 0\times(-5),\ j2(-3-j6)\} \\ &= \{0,\ 6j+12,\ 0,\ -6j+12\} \\ Y(k) &= \{0,\ 12+6j,\ 0,\ 12-6j\} \end{aligned}$$

From Eq. (3.10)

$$y_n = \frac{1}{N}[W_N^*]Y_N$$

$$Y_n = \begin{bmatrix} 0 \\ 12+6j \\ 0 \\ 12-6j \end{bmatrix}$$

$$W_N^* = \begin{bmatrix} 1 & 1 & 1 & 1 \\ 1 & j & -1 & -j \\ 1 & -1 & 1 & -1 \\ 1 & -j & -1 & j \end{bmatrix}$$

For $N = 4$

$$y_N = \frac{1}{4}\begin{bmatrix} 1 & 1 & 1 & 1 \\ 1 & j & -1 & -j \\ 1 & -1 & 1 & -1 \\ 1 & -j & -1 & j \end{bmatrix}\begin{bmatrix} 0 \\ 12+6j \\ 0 \\ 12-6j \end{bmatrix}$$

$$y[0] = \frac{1}{4}[0+12+6j+0+12-6j] = 6$$

$$y[1] = \frac{1}{4}[0+12j-6+0-12j-6] = -3$$

$$y[2] = \frac{1}{4}[0-12-6j+0-12+6j] = -6$$

$$y[3] = \frac{1}{4}[0-12j+6+0+12j+6] = 3$$

$$\boxed{y[n] = \{6,\ -3,\ -6,\ 3\}}$$

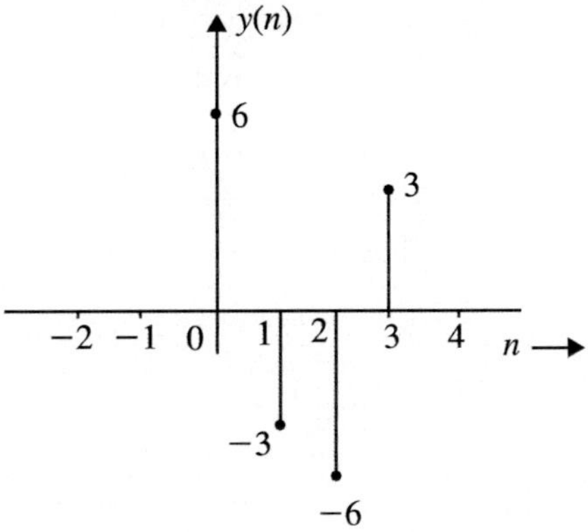

Example 3.13 Compute the DFT of

$$x[n] = e^{-0.5n}, \qquad 0 \le n \le 5$$

Solution For the given sequence $L = 5$. In general $N \ge L$.

Let $N = 8$

$$X(k) = \sum_{n=0}^{N-1} x(n)e^{\frac{-j2\pi kn}{8}}$$

$$X(k) = \sum_{n=0}^{5} e^{-0.5n}e^{\frac{-j2\pi kn}{8}}$$

$$= \sum_{n=0}^{5} e^{-0.5n}e^{\frac{-j\pi kn}{4}}$$

$$= \sum_{n=0}^{5} \left(e^{-0.5}e^{\frac{-j\pi k}{4}}\right)^n$$

$$= \sum_{n=0}^{5} \left(0.607e^{\frac{-j\pi k}{4}}\right)^n$$

$$= \frac{1 - \left(0.607e^{\frac{-j\pi k}{4}}\right)^6}{1 - 0.607e^{\frac{-j\pi k}{4}}}$$

$$\boxed{X(k) = \frac{1 - 0.05e^{\frac{-j3\pi k}{2}}}{1 - 0.607e^{\frac{-j\pi k}{4}}}}$$

The following summation formula is used above:

$$\sum_{n=0}^{N-1} x^n = \frac{1 - x^N}{1 - x}$$

Example 3.14 Calculate the DFT of the sequence

$$x[n] = \{1,\ 1,\ -2,\ -2\}$$

Solution Given

$$x[n] = \{1,\ 1,\ -2,\ -2\}$$

$$x_4 = \begin{bmatrix} 1 \\ 1 \\ -2 \\ -2 \end{bmatrix}$$

W_4 is given in Eq. (3.13):

$$\begin{aligned}
X_4 &= W_4 x_4 \\
&= \begin{bmatrix} 1 & 1 & 1 & 1 \\ 1 & -j & -1 & j \\ 1 & -1 & 1 & -1 \\ 1 & j & -1 & -j \end{bmatrix} \begin{bmatrix} 1 \\ 1 \\ -2 \\ -2 \end{bmatrix} \\
X(1) &= 1 + 1 - 2 + 2 = 2 \\
X(2) &= 1 - j + 2 + 2j = 3 + j \\
X(3) &= 1 - 1 - 2 - 2 = -4 \\
X(4) &= 1 + j + 2 - 2j = 3 - j \\
X_4 &= \begin{bmatrix} 2 \\ 3 + j \\ -4 \\ 3 - j \end{bmatrix}
\end{aligned}$$

4-point DFT of $x(n)$ is

$$\boxed{X(k) = \{2,\ 3+j,\ -4,\ 3-j\}}$$

Example 3.15 A finite duration sequence of length L is given as

$$x(n) = \begin{cases} 1, & 0 \le n \le L-1, \\ 0, & \text{otherwise} \end{cases}$$

Determine the N-point DFT of the sequence for $N = L$.

Solution For $N = L$, a finite duration $x(n)$ is

$$x(n) = \begin{cases} 1, & 0 \le n \le N-1, \\ 0, & \text{otherwise} \end{cases}$$

N-point DFT of $x(n)$ is

$$\begin{aligned} X(k) &= \sum_{n=0}^{N-1} x(n) e^{\frac{-j2\pi kn}{N}} \\ &= \sum_{n=0}^{N-1} 1 \cdot e^{\frac{-j2\pi kn}{N}} \\ &= \sum_{n=0}^{N-1} \left(e^{\frac{-j2\pi k}{N}}\right)^n \\ &= \frac{1 - e^{\frac{-j2\pi kN}{N}}}{1 - e^{\frac{-j2\pi k}{N}}} \qquad \because \quad e^{\frac{-j2\pi kN}{N}} = e^{-j2\pi k} = 1 \\ &= 0 \end{aligned}$$

Example 3.16 Compute DFT of $x[n]$

$$x(n) = \begin{cases} 0, & 0 \le n \le 4, \\ 1, & 5 \le n < 7 \end{cases}$$

Solution Given

$$x(n) = \begin{cases} 0, & 0 \le n \le 4, \\ 1, & 5 \le n < 7 \end{cases}$$

$$x[n] = \{0,\ 0,\ 0,\ 0,\ 0,\ 1,\ 1\}$$

Let us assume $N = L = 8$. Therefore, $x[n] = \{0, 0, 0, 0, 0, 1, 1, 0\}$

$$X(k) = \sum_{n=0}^{N-1} x(n)e^{\frac{-j2\pi nk}{N}}, \quad k = 0, \ldots, N-1$$

$$X(k) = \sum_{n=0}^{7} x(n)e^{\frac{-j2\pi nk}{8}}, \quad k = 0, 1, \ldots, 7$$

$$X(k) = \sum_{n=0}^{7} x(n)e^{\frac{-j\pi nk}{4}}, \quad k = 0, 1, \ldots, 7$$

For $k = 0$

$$X(0) = \sum_{n=0}^{7} x(n)$$

$$X(0) = 0 + 0 + 0 + 0 + 0 + 1 + 1 + 0 = 2$$

For $k = 1$

$$\begin{aligned} X(1) &= \sum_{n=0}^{7} x(n)e^{\frac{-j\pi n}{4}} \\ &= x(0) + x(1)e^{\frac{-j\pi}{4}} + x(2)e^{\frac{-j\pi}{2}} + x(3)e^{\frac{-j3\pi}{4}} + x(4)e^{-j\pi} \\ &\quad + x(5)e^{\frac{-j5\pi}{4}} + x(6)e^{\frac{-j3\pi}{2}} + x(7)e^{\frac{-j7\pi}{4}} \\ &= 0 + 0 + 0 + 0 + 0 + e^{\frac{-j5\pi}{4}} + e^{\frac{-j3\pi}{2}} + 0 \\ X(1) &= -0.707 + 0.707j + j \\ &= -0.707 + 1.707j \end{aligned}$$

For $k = 2$

$$\begin{aligned} X(2) &= \sum_{n=0}^{7} x(n)e^{\frac{-j\pi n}{2}} \\ &= x(5)e^{\frac{-j5\pi}{2}} + x(6)e^{-j3\pi} \\ &= -j - 1 = -1 - j \end{aligned}$$

For $k = 3$

$$
\begin{aligned}
X(3) &= \sum_{n=0}^{7} x(n)e^{\frac{-j3\pi n}{4}} \\
&= x(5)e^{\frac{-j15\pi}{4}} + x(6)e^{\frac{-j9\pi}{2}} \\
&= 0.707 + 0.707j - j \\
&= 0.707 - 0.293j
\end{aligned}
$$

For $k = 4$

$$
\begin{aligned}
X(4) &= \sum_{n=0}^{7} x(n)e^{-j\pi n} \\
&= x(5)e^{-j5\pi} + x(6)e^{-j6\pi} \\
&= -1 + 1 = 0
\end{aligned}
$$

For $k = 5$

$$
\begin{aligned}
X(5) &= \sum_{n=0}^{7} x(n)e^{\frac{-j5\pi n}{4}} \\
&= x(5)e^{\frac{-j25\pi}{4}} + x(6)e^{\frac{-j15\pi}{2}} \\
&= 0.707 - 0.707j + j \\
&= 0.707 + 0.293j
\end{aligned}
$$

For $k = 6$

$$
\begin{aligned}
X(6) &= \sum_{n=0}^{7} x(n)e^{\frac{-j3\pi n}{2}} \\
&= x(5)e^{\frac{-j15\pi}{2}} + x(6)e^{-j9\pi} \\
&= j - 1 = -1 + j
\end{aligned}
$$

For $k = 7$

$$
\begin{aligned}
X(7) &= \sum_{n=0}^{7} x(n)e^{\frac{-j7\pi n}{4}} \\
&= x(5)e^{\frac{-j35\pi}{2}} + x(6)e^{\frac{-j21\pi}{2}} \\
&= -0.707 - 0.707j - j \\
&= -0.707 - 1.707j
\end{aligned}
$$

$$X(k) = \Big\{2,\ -0.707 + 1.707j,\ -1 - j,\ 0.707 - 0.293j,\ 0,\ 0.707 + 0.293j,\ -1 + j,\ -0.707 - 1.707j\Big\}$$

Example 3.17 Compute the DFT of the sequence $x[n] = e^{-n}$ where $0 \le n \le 4$.

Solution Let $N = 8$, therefore

$$x(n) = \begin{cases} e^{n}, & 0 \le n \le 4, \\ 0, & 5 \le n \le 7 \end{cases}$$

$$\begin{aligned} X(k) &= \sum_{n=0}^{7} x(n) e^{\frac{-j2\pi nk}{8}} \\ &= \sum_{n=0}^{4} x(n) e^{\frac{-j2\pi nk}{8}} \\ &= \sum_{n=0}^{4} e^{-n} e^{\frac{-j2\pi nk}{8}} \\ &= \sum_{n=0}^{4} \left(e^{-1} \cdot e^{\frac{-j2\pi k}{8}}\right)^{n} \\ &= \sum_{n=0}^{4} \left(0.3679 e^{\frac{-j\pi k}{4}}\right)^{n} \end{aligned}$$

$$X(k) = \frac{1 - \left(0.3679 e^{\frac{-j\pi k}{4}}\right)^{5}}{1 - 0.3679 e^{\frac{-j\pi k}{4}}}, \qquad k = 0, \ldots, N-1$$

because

$$\sum_{n=0}^{N} x^{n} = \frac{1 - x^{N+1}}{1 - x}$$

Example 3.18 Determine the 8-point DFT of the sequence

$$x[n] = \{0,\ 0,\ 1,\ 1,\ 1,\ 0,\ 0,\ 0\}$$

Solution Given

$$X_N = W_N x_N$$
$$X_8 = W_8 x_8$$

W_8 is given in Eq. (3.17):

$$x_8 = \begin{bmatrix} 0 \\ 0 \\ 1 \\ 1 \\ 1 \\ 0 \\ 0 \\ 0 \end{bmatrix}$$

$X(0)$ is obtained by multiplying x_8 with the first row of W_8. Thus

$$\begin{aligned}
X(0) &= 0 + 0 + 1 + 1 + 1 + 0 + 0 + 0 = 3 \\
X(1) &= \text{2nd row of } W_8 \text{ to multiply } x_8 \\
&= -j - \frac{1}{\sqrt{2}} - \frac{j}{\sqrt{2}} - 1 = -1.707 - j1.707 \\
X(2) &= \text{3rd row of } W_8 \text{ to multiply } x_8 \\
&= -1 + j + 1 = j \\
X(3) &= \text{4th row of } W_8 \text{ to multiply } x_8 \\
&= j + \frac{1}{\sqrt{2}} - j\frac{1}{\sqrt{2}} - 1 = -0.293 + j0.293 \\
X(4) &= \text{5th row of } W_8 \text{ to multiply } x_8 \\
&= 1 - 1 + 1 = 1 \\
X(5) &= \text{6th row of } W_8 \text{ to multiply } x_8 \\
&= -j + \frac{1}{\sqrt{2}} + \frac{j}{\sqrt{2}} - 1 = -0.293 - j0.293 \\
X(6) &= \text{7th row of } W_8 \text{ to multiply } x_8 \\
&= -1 - j + 1 = -j \\
X(7) &= \text{8th row of } W_8 \text{ to multiply } x_8 \\
&= j - \frac{1}{\sqrt{2}} + \frac{j}{\sqrt{2}} - 1 = -1.707 + j1.707
\end{aligned}$$

Therefore,

$$\boxed{X(k) = \left\{3,\ -1.707 - j1.707,\ j,\ -0.293 + j0.293,\ 1,\ -0.293 - j0.293,\ -j,\ -1.707 + j1.707\right\}}$$

Example 3.19 Find the N-point DFT of the following signals:

$$(a) \quad x[n] = \delta(n - n_0)$$
$$(b) \quad x[n] = a^n$$

Solution (a) $x[n] = \delta(n - n_0)$

$$X(k) = \sum_{n=0}^{N-1} x(n) e^{\frac{-j2\pi kn}{N}}$$

$$x[n] = \begin{cases} 1, & n = n_0, \\ 0, & n \neq n_0 \end{cases}$$

$$\boxed{X(k) = e^{\frac{-j2\pi kn_0}{N}}}$$

(b) $x[n] = a^n$

$$X(k) = \sum_{n=0}^{N-1} a^n e^{\frac{-j2\pi kn}{N}}$$
$$= \sum_{n=0}^{N-1} \left(ae^{\frac{-j2\pi k}{N}}\right)^n$$

Using the summation formula

$$\sum_{n=0}^{N} x^n = \frac{1 - x^{N+1}}{1 - x}$$

we get

$$X(k) = \frac{1 - \left(ae^{\frac{-j2\pi k}{N}}\right)^N}{1 - ae^{\frac{-j2\pi k}{N}}}$$

$$\boxed{X(k) = \frac{1 - a^N}{1 - ae^{\frac{-j2\pi k}{N}}}}$$

3.3 Relationship of the DFT to Other Transforms

3.3.1 Relationship to the Fourier Series Coefficients of a Periodic Sequence

Fourier series of the discrete time signal is written as

$$x(n) = \sum_{k=0}^{N-1} C_k e^{j\frac{2\pi nk}{N}} \qquad -\infty < n < \infty \tag{3.19}$$

where the Fourier series coefficients are represented as

$$C_k = \frac{1}{N}\sum_{n=0}^{N-1} x(n)e^{-j\frac{2\pi nk}{N}} \qquad k = 0, \ldots, N-1 \tag{3.20}$$

$$\text{DFT} \Longrightarrow \quad X(k) = \sum_{n=0}^{N-1} x(n)e^{-j\frac{2\pi nk}{N}} \tag{3.21}$$

Comparing Eqs. (3.20) and (3.21), we get

$$C_k = \frac{1}{N}X(k)$$

$$\boxed{X(k) = NC_k} \tag{3.22}$$

3.3.2 Relationship to the Fourier Transform of an Aperiodic Sequence

Fourier transform of $x(n)$ is

$$X(\omega) = \sum_{n=-\infty}^{\infty} x(n)e^{-j\omega n} \tag{3.23}$$

$$X(k) = X(\omega)\Big|_{\omega=\frac{2\pi k}{N}}, \quad k = 0, \ldots, N-1 \tag{3.24}$$

The finite duration sequence

$$X(k) = \sum_{n=0}^{N-1} x(n) e^{-j\frac{2\pi kn}{N}}$$

3.3.3 Relationship to the z-Transform

The z-transform of an N-point sequence $x(n)$ is given by

$$X(z) = \sum_{n=0}^{N-1} x(n) z^{-n} \tag{3.25}$$

Let us evaluate $X(z)$ at N equally spaced points on a unit circle that is at $z = e^{j\frac{2\pi k}{N}}$

$$\begin{aligned} X(z)\Big|_{z=e^{j\frac{2\pi k}{N}}} &= \sum_{n=0}^{N-1} x(n) e^{-j\frac{2\pi kn}{N}} \\ &= X(k) \end{aligned}$$

$$X(k) = X(z)\Big|_{z=e^{j\frac{2\pi k}{N}}} \tag{3.26}$$

From Eq. (3.26), we can conclude that the N-point DFT of a finite duration sequence can be obtained from the z-transform of the sequence at N equally spaced points around the unit circle.

Example 3.20 Consider the finite-length sequence $x(n)$ as represented below

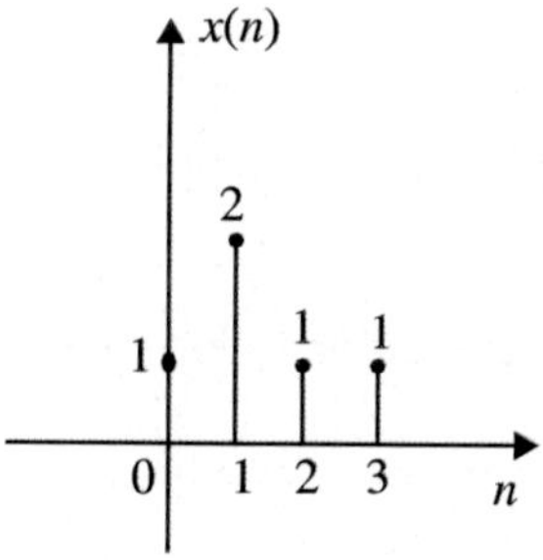

Representation of $x(n)$ sequence.

Let $X(z)$ be the z-transform of $x(n)$. If we sample $X(z)$ at

$$z = e^{j\left(\frac{2\pi}{4}\right)k}, \qquad k = 0, 1, 2, 3$$

using the relation

$$X_1(k) = X(z)\Big|_{z=e^{j\left(\frac{2\pi}{4}\right)k}}$$

find the sequence $x_1(n)$.

Solution

$$X_1(k) = X(z)\Big|_{z=e^{j\left(\frac{2\pi}{4}\right)k}}$$

$$\begin{aligned} X(z) &= \sum_{n=0}^{3} x(n)z^{-n} = x(0) + x(1)z^{-1} + x(2)z^{-2} + x(3)z^{-3} \\ &= 1 + 2z^{-1} + z^{-2} + z^{-3} \end{aligned}$$

$$X_1(k) = \left(1 + 2z^{-1} + z^{-2} + z^{-3}\right)\Big|_{z=e^{j\left(\frac{2\pi}{4}\right)k}}$$

$$X_1(k) = 1 + 2e^{\frac{-j2\pi k}{4}} + e^{\frac{-j2\pi k\cdot 2}{4}} + e^{\frac{-j2\pi k\cdot 3}{4}}$$

$$X_1(k) = x_1(0) + x_1(1)e^{\frac{-j2\pi k}{4}} + x_1(2)e^{\frac{-j2\pi k\cdot 2}{4}} + x_1(3)e^{\frac{-j2\pi k\cdot 3}{4}}$$

Therefore,

$$\boxed{x_1(n) = \{1,\ 2,\ 1,\ 1\}}$$

3.4 Properties of DFT

3.4.1 Periodicity

If a sequence $x(n)$ is periodic with periodicity of N samples, then N-point DFT of the sequence is also periodic with periodicity of N samples.

Hence, if $x(n)$ and $X(k)$ are an N point DFT pair, then

$$\begin{aligned} x(n+N) &= x(n) \qquad \text{for all } n \\ X(k+N) &= X(k) \qquad \text{for all } k \end{aligned} \tag{3.27}$$

Proof

$$X(k) = \sum_{n=0}^{N-1} x(n)e^{-j\frac{2\pi kn}{N}}$$

$$X(k+N) = \sum_{n=0}^{N-1} x(n)e^{-j\frac{2\pi (k+N)n}{N}}$$

$$= \sum_{n=0}^{N-1} x(n)e^{-j\frac{2\pi kn}{N}}e^{-j2\pi n}$$

$$e^{-j2\pi n} = 1 \quad \text{for all } n, \text{(Here } n \text{ is an integer)}$$

$$X(k+N) = \sum_{n=0}^{N-1} x(n)e^{-j\frac{2\pi kn}{N}}$$

$$\boxed{X(k+N) = X(k)}$$

3.4.2 Linearity

If

$$x_1(n) \overset{\text{DFT}}{\underset{N}{\longleftrightarrow}} X_1(k) \quad \text{and} \quad x_2(n) \overset{\text{DFT}}{\underset{N}{\longleftrightarrow}} X_2(k)$$

then for any real-valued or complex-valued constants a_1 and a_2

$$\boxed{a_1x_1(n) + a_2x_2(n) \overset{\text{DFT}}{\underset{N}{\longleftrightarrow}} a_1X_1(k) + a_2X_2(k)} \tag{3.28}$$

Proof

$$\text{DFT}\{a_1x_1(n) + a_2x_2(n)\} = \sum_{n=0}^{N-1}(a_1x_1(n) + a_2x_2(n))e^{-j\frac{2\pi kn}{N}}$$

$$= a_1\sum_{n=0}^{N-1} x_1(n)e^{-j\frac{2\pi kn}{N}} + a_2\sum_{n=0}^{N-1} x_2(n)e^{-j\frac{2\pi kn}{N}}$$

$$= a_1X_1(k) + a_2X_2(k)$$

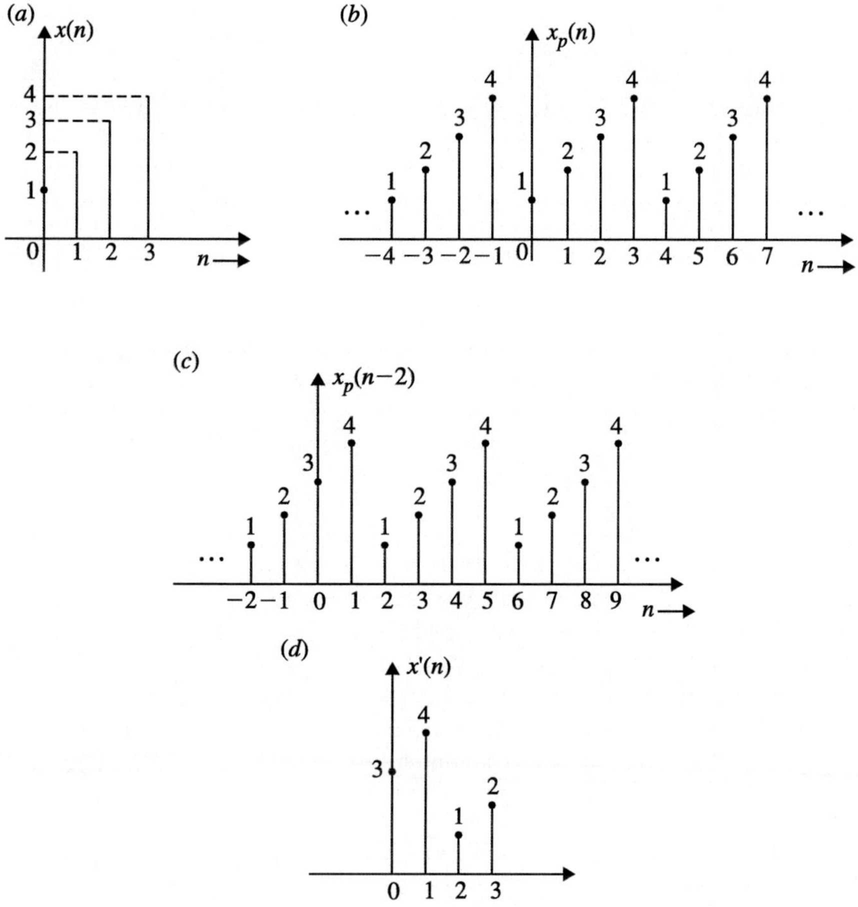

Fig. 3.9 **a** Discrete time signal $x(n)$, **b** periodic extension of $x(n)$, **c** shifted periodic extension of $x(n)$ and **d** Discrete time signal $x'(n)$

3.4.3 Circular Shift and Circular Symmetry of a Sequence

Consider a finite duration sequence $x(n)$ and its periodic extension $x_p(n)$. The periodic extension of $x(n)$ can be expressed as $x_p(n) = x(n + N)$ where N is the periodicity. For example, let

$$x(n) = \{\underset{\uparrow}{1}, 2, 3, 4\} \quad \text{and} \quad N = 4$$

The sequence and its periodic extension are shown in Fig. 3.9a and b, respectively. Let us shift the periodic sequence $x_p(n)$ by two units of time to the right.

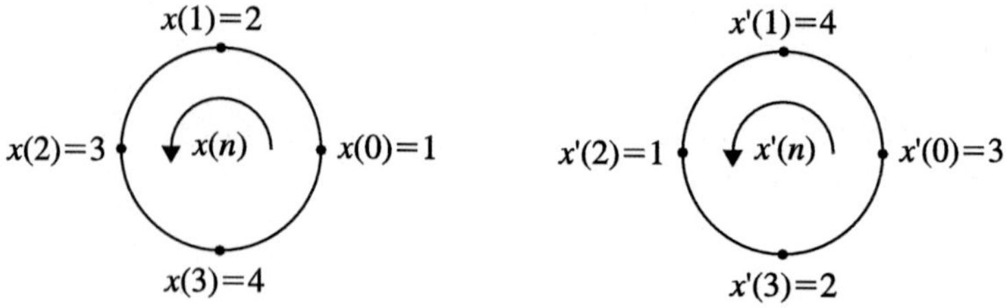

Fig. 3.10 Circular representation

Let us denote one period of this shifted sequence by $x'(n)$. The sequence $x'(n)$ can be represented by $x(n-2, (\text{mod}4))$ where mod 4 indicates that the sequence repeats after 4 samples:

$$\begin{aligned} x'(n) &= x(n-2, \text{ mod } 4) \\ x'(0) &= x(-2, \text{ mod } 4) = x(2) = 3 \\ x'(1) &= x(-1, \text{ mod } 4) = x(3) = 4 \\ x'(2) &= x(0, \text{ mod } 4) = x(0) = 1 \\ x'(3) &= x(1, \text{ mod } 4) = x(1) = 2 \end{aligned}$$

Circular representation of $x(n)$ and $x'(n)$ are shown in Fig. 3.10.

From this, $x'(n)$ is simply $x(n)$ shifted circularly by two units in time where the counterclockwise direction has been arbitrarily selected as the positive direction. From this, we conclude that a circular shift of an N-point sequence is equivalent to the linear shift of its periodic extension.

Let $x(n)$ be an N-point sequence represented on a circle and $x'(n)$ be its shifted sequence by k units of time

$$x'(n) = x(n-k, \text{ mod } 4) = x(n-k)_N \tag{3.29}$$

Even Symmetry

An N-point sequence is called even, if it is symmetric about zero on the circle

$$x(N-n) = x(n) \quad \text{for} \quad 0 \leq n \leq N-1 \tag{3.30}$$

Odd Symmetry

An N-point sequence is called odd, if it is asymmetric about zero on the circle

$$x(N-n) = -x(n) \quad \text{for} \quad 0 \leq n \leq N-1 \tag{3.31}$$

Time Reversal

Time reversal of an N-point sequence is obtained as

$$x(-n,\ \mathrm{mod}\ N) = x(N-n) \tag{3.32}$$

$$x(-n)_N = x(N-n), \quad 0 \le n \le N-1 \tag{3.33}$$

Time reversal is equivalent to plotting $x(n)$ in a clockwise direction on a circle.

Even and odd sequences for a periodic sequence $x_p(n)$ is given as

$$\text{Even: } x_p(n) = x_p(-n) = x_p(N-n) \tag{3.34}$$

$$\text{Odd: } x_p(n) = -x_p(-n) = -x_p(N-n) \tag{3.35}$$

If the periodic sequence is complex-valued, then

$$\text{Conjugate Even: } x_p(n) = x_p^*(N-n) \tag{3.36}$$

$$\text{Conjugate Odd: } x_p(n) = -x_p^*(N-n) \tag{3.37}$$

$$x_p(n) = x_{pe}(n) + x_{po}(n) \tag{3.38}$$

$$x_{pe}(n) = \frac{1}{2}[x_p(n) + x_p^*(N-n)] \tag{3.39}$$

$$x_{po}(n) = \frac{1}{2}[x_p(n) + x_p^*(N-n)] \tag{3.40}$$

3.4.4 Symmetry Properties of the DFT

Let us assume that N-point sequence $x(n)$ and its DFT are complex-valued, then

$$\begin{aligned}
x(n) &= x_R(n) + jx_I(n) \qquad 0 \le n \le N-1 \\
X(k) &= X_R(k) + jX_I(k) \qquad 0 \le k \le N-1 \\
&= \sum_{n=0}^{N-1}(x_R(n) + jx_I(n))e^{-j\frac{2\pi kn}{N}} \\
&= \sum_{n=0}^{N-1}(x_R(n) + jx_I(n))\left(\cos\frac{2\pi kn}{N} - j\sin\frac{2\pi kn}{N}\right) \\
X(k) &= \sum_{n=0}^{N-1}\left[x_R(n)\cos\frac{2\pi kn}{N} + x_I(n)\sin\frac{2\pi kn}{N}\right] \\
&\quad -j\sum_{n=0}^{N-1}\left[x_R(n)\sin\frac{2\pi kn}{N} - x_I(n)\cos\frac{2\pi kn}{N}\right] \\
&= X_R(k) + jX_I(k)
\end{aligned} \tag{3.41}$$

where

$$X_R(k) = \sum_{n=0}^{N-1} \left[x_R(n) \cos \frac{2\pi kn}{N} + x_I(n) \sin \frac{2\pi kn}{N} \right]$$

$$X_I(k) = -\sum_{n=0}^{N-1} \left[x_R(n) \sin \frac{2\pi kn}{N} - x_I(n) \cos \frac{2\pi kn}{N} \right]$$

Similarly

$$x_R(n) = \frac{1}{N} \sum_{k=0}^{N-1} \left[X_R(k) \cos \frac{2\pi kn}{N} - X_I(k) \sin \frac{2\pi kn}{N} \right] \tag{3.42}$$

$$x_I(n) = \frac{1}{N} \sum_{k=0}^{N-1} \left[X_R(k) \sin \frac{2\pi kn}{N} + X_I(k) \cos \frac{2\pi kn}{N} \right] \tag{3.43}$$

Case (*i*) Real Value Sequence

If $x(n)$ is real,

$$X(N-k) = X^*(k) = X(-k)$$
$$X(k) = X^*(N-k),\ X(k) \text{ has complex conjugate symmetry}$$

$$|X(N-k)| = |X(k)|$$
$$\angle X(N-k) = -\angle X(k) \tag{3.44}$$

Case (*ii*) Real and Even Value Sequence

If $x(n)$ is real and even, that is

$$x(n) = x(N-n), \quad 0 \le n \le N-1$$
$$X_I(k) = 0, \quad x_I(n) = 0$$

$$\text{DFT} \Longrightarrow X(k) = X_R(k) = \sum_{n=0}^{N-1} x(n) \cos \frac{2\pi kn}{N}; \quad 0 \le n \le N-1 \tag{3.45}$$

$$\text{IDFT} \Longrightarrow x(n) = \frac{1}{N} \sum_{k=0}^{N-1} X(k) \cos \frac{2\pi kn}{N}; \quad 0 \le n \le N-1 \tag{3.46}$$

Case (*iii*) Real and Odd Sequence

If $x(n)$ is real and odd, that is

$$x(n) = -x(N-n)$$
$$X_R(k) = 0, \quad x_I(n) = 0$$

$$\text{DFT} \implies X(k) = jX_I(k) = -j\sum_{n=0}^{N-1} x(n)\sin\frac{2\pi kn}{N}; \quad 0 \le n \le N-1 \quad (3.47)$$

$$\text{IDFT} \implies x(n) = j\frac{1}{N}\sum_{k=0}^{N-1} X(k)\sin\frac{2\pi kn}{N}; \quad 0 \le k \le N-1 \quad (3.48)$$

Case (*iv*) Purely Imaginary Sequence

$$x(n) = jx_I(n)$$

$$X_R(k) = \sum_{n=0}^{N-1} x_I(n)\sin\frac{2\pi kn}{N} \quad (3.49)$$

$$X_I(k) = \sum_{n=0}^{N-1} x_I(n)\cos\frac{2\pi kn}{N} \quad (3.50)$$

$$X(k) = X_R(k) + X_I(k) \quad (3.51)$$

Example 3.21 The first five points of the 8-point DFT of real-valued sequences are

$$\{28, \ -4+j9.565, \ -4+j4, \ -4+j1.656, \ -4\}$$

Determine the remaining three points.

Solution Give $N = 8$,

$$X(0) = 28 \qquad X(1) = -4+j9.565$$
$$X(2) = -4+j4 \qquad X(3) = -4+j1.656$$
$$X(4) = -4$$

We have

$$X(k) = X^*(N-k)$$

The remaining three DFT points are

$$X(5) = X^*(8-5) = X^*(3) = -4-j1.656$$
$$X(6) = X^*(8-6) = X^*(2) = -4-j4$$
$$X(7) = X^*(8-7) = X^*(1) = -4-j9.565$$

3.4.5 Multiplication of Two DFTs and Circular Convolution

Let $x_1(n)$ and $x_2(n)$ be finite duration sequences of length N. Their respective N-point DFTs are

$$X_1(k) = \sum_{n=0}^{N-1} x_1(n)e^{-j\frac{2\pi nk}{N}}; \quad k = 0, 1, 2, \ldots, N-1 \tag{3.52}$$

$$X_2(k) = \sum_{n=0}^{N-1} x_2(l)e^{-j\frac{2\pi lk}{N}}; \quad k = 0, 1, 2, \ldots, N-1 \tag{3.53}$$

If we multiply the two DFTs together, the result is a DFT, say $X_3(k)$ of a sequence $x_3(n)$ of length N

$$X_3(k) = X_1(k)X_2(k); \quad k = 0, 1, 2, \ldots, N-1 \tag{3.54}$$

$$\text{IDFT}\{X_3(k)\} = x_3(m)$$

By the definition of IDFT

$$x_3(m) = \frac{1}{N}\sum_{k=0}^{N-1} X_3(k)e^{j\frac{2\pi km}{N}} \qquad m = 0, \ldots, (N-1) \tag{3.55}$$

$$= \frac{1}{N}\sum_{k=0}^{N-1} X_1(k)X_2(k)e^{j\frac{2\pi km}{N}}$$

$$= \frac{1}{N}\sum_{k=0}^{N-1}\left[\sum_{n=0}^{N-1} x_1(n)e^{-j\frac{2\pi nk}{N}}\right]\left[\sum_{l=0}^{N-1} x_2(l)e^{-j\frac{2\pi kl}{N}}\right]e^{j\frac{2\pi km}{N}}$$

$$x_3(m) = \frac{1}{N}\sum_{n=0}^{N-1} x_1(n)\sum_{l=0}^{N-1} x_2(l)\sum_{k=0}^{N-1} e^{j\frac{2\pi k[m-n-l]}{N}}$$

Let $(m-n-l) = PN$, where P is an integer.

$$e^{\frac{j2\pi k(m-n-l)}{N}} = e^{\frac{j2\pi kPN}{N}}$$
$$= e^{j2\pi kp} = (e^{j2\pi p})^k$$

From finite geometric series sum formula

$$\sum_{n=0}^{N-1} a^n = \begin{cases} N, & \text{for } a = 1 \\ \frac{1-a^N}{1-a}, & \text{for } a \neq 1 \end{cases}$$

Therefore,

$$\sum_{k=0}^{N-1} e^{\frac{j2\pi k(m-n-l)}{N}} = \sum_{k=0}^{N-1} (e^{j2\pi p})^k = \sum_{k=0}^{N-1} 1^k = N$$

$$\therefore \quad x_3(m) = \frac{1}{N} \sum_{n=0}^{N-1} x_1(n) \sum_{l=0}^{N-1} x_2(l) N$$

$$= \sum_{n=0}^{N-1} x_1(n) \sum_{l=0}^{N-1} x_2(l)$$

If $x_2(l)$ is a periodic sequence with periodicity of N samples, then $x_2(l \pm PN) = x_2(l)$. Here $m - n - l = PN$, and therefore $l = m - n - PN$

$$x_2(l) = x_2(m - n - PN) = x_2(m - n)_N$$

$$= x_2((m - n), \quad \text{mod } N)$$

$$\therefore\ x_3(n) = \sum_{n=0}^{N-1} x_1(n) \sum_{n=0}^{N-1} x_2(m - n)_N$$

$$= \sum_{n=0}^{N-1} x_1(n) x_2(m - n)_N$$

$$x_3(n) = x_1(n) \, Ⓝ \, x_2(n)$$

$$\text{IDFT } [X_1(k) X_2(k)] = x_1(n) \, Ⓝ \, x_2(n)$$

$$X_1(k) X_2(k) = \text{DFT } [x_1(n) \, Ⓝ \, x_2(n)] \tag{3.56}$$

Therefore,

$$\boxed{x_1(n) \, Ⓝ \, x_2(n) \overset{\text{DFT}}{\longleftrightarrow} X_1(k) X_2(k)} \tag{3.57}$$

Thus, we conclude that the multiplication of the DFTs of two sequences is equivalent to the DFT of the circular convolution of the two sequences.

3.4.6 Time Reversal of a Sequence

If

$$x(n) \underset{N}{\overset{\text{DFT}}{\longleftrightarrow}} X(k)$$

then

$$x((-n))_N = x(N-n) \underset{N}{\overset{\text{DFT}}{\longleftrightarrow}} X((-k))_N = X(N-k) \tag{3.58}$$

Hence, reversing the N-point sequence in time is equivalent to reversing the DFT values (enter the sequence is clockwise direction in the circle).

Proof

$$\text{DFT}\,\{x(n-N)\} = \sum_{n=0}^{N-1} x(N-m)e^{-j\frac{2\pi kn}{N}}$$

Let $m = N - n$

$$\begin{aligned}
\text{DFT}\,\{x(n-N)\} &= \sum_{n=0}^{N-1} x(m)e^{-j\frac{2\pi k(N-m)}{N}} \\
&= \sum_{m=0}^{N-1} x(m)e^{j\frac{2\pi km}{N}} \cdot e^{-j\frac{2\pi kN}{N}} \\
&= \sum_{m=0}^{N-1} x(m)e^{-j\frac{2\pi(N-k)m}{N}} = X(N-k)
\end{aligned}$$

$$X(N-k) = X((-k))_N, \quad 0 \le n \le N-1$$

3.4.7 Circular Time Shift of a Sequence

If

$$x(n) \underset{N}{\overset{\text{DFT}}{\longleftrightarrow}} X(k)$$

then

$$x(n-l)_N \underset{N}{\overset{\text{DFT}}{\longleftrightarrow}} X(k)e^{-j\frac{2\pi kl}{N}} \tag{3.59}$$

Proof

$$\text{DFT}\{x((n-l))_N\} = \sum_{n=0}^{N-1} x((n-l))_N e^{-j\frac{2\pi kn}{N}}$$

$$\text{DFT}\{x((n-l))_N\} = \sum_{n=0}^{l-1} x((n-l))_N e^{-j\frac{2\pi kn}{N}} + \sum_{n=0}^{N-1} x(n-l)_N e^{-j\frac{2\pi kn}{N}}$$

let

$$\begin{aligned} x((n-l))_N &= x(N+(n-l)) \\ &= x(N+n-l) \\ m &= N+n-l \end{aligned}$$

$$\begin{aligned} \therefore \sum_{n=0}^{l-1} x((n-l))_N e^{-j\frac{2\pi kn}{N}} &= \sum_{m=N-l}^{N-1} x(N+n-l) e^{-j\frac{2\pi kn}{N}} \\ &= \sum_{m=N-l}^{N-1} x(m) e^{-j\frac{2\pi k(m+l-N)}{N}} \\ &= \sum_{m=N-l}^{N-1} x(m) e^{-j\frac{2\pi km}{N}} \cdot e^{-j\frac{2\pi kl}{N}} \\ \sum_{n=l}^{N-1} x(n-l) e^{-j\frac{2\pi kn}{N}} &= \sum_{m=0}^{N-1-l} x(m) e^{-j\frac{2\pi k(m+l)}{N}} \end{aligned}$$

Therefore,

$$\begin{aligned} \text{DFT}\{x(n-l)_N\} &= \sum_{m=N-l}^{N-1} x(m) e^{-j\frac{2\pi k(m+l)}{N}} + \sum_{m=0}^{N-1-l} x(m) e^{-j\frac{2\pi k(m+l)}{N}} \\ &= \sum_{m=0}^{N-1} x(m) e^{-j\frac{2\pi k(m+l)}{N}} \\ &= e^{-j\frac{2\pi kl}{N}} \sum_{m=0}^{N-1} x(m) e^{-j\frac{2\pi km}{N}} \\ &= e^{-j\frac{2\pi kl}{N}} \cdot X(k) \end{aligned}$$

3.4.8 Circular Frequency Shift

If

$$x(n) \overset{\text{DFT}}{\underset{N}{\longleftrightarrow}} X(k)$$

then

$$x(n)e^{j\frac{2\pi ln}{N}} \overset{\text{DFT}}{\underset{N}{\longleftrightarrow}} X((k-l))_N \tag{3.60}$$

Proof

$$\begin{aligned}
\text{DFT}\left\{X((k-l))_N\right\} &= \frac{1}{N}\sum_{k=0}^{N-1} X((k-l))_N e^{j\frac{2\pi kn}{N}} \\
&= \frac{1}{N}\sum_{k=0}^{l-1} X((k-l))_N e^{j\frac{2\pi kn}{N}} + \frac{1}{N}\sum_{k=l}^{N-1} X((k-l))_N e^{j\frac{2\pi kn}{N}}
\end{aligned}$$

Let $X((k-l))_N = X(N+k-l)$

$$\begin{aligned}
N+k-l &= m \\
\text{when } k = 0 \Longrightarrow \quad m &= N-l \\
\text{when } k = l-1 \Longrightarrow \quad m &= N-1
\end{aligned}$$

$$\begin{aligned}
\frac{1}{N}\sum_{k=0}^{l-1} X((k-l))_N e^{j\frac{2\pi kn}{N}} &= \frac{1}{N}\sum_{m=N-l}^{N-1} x(N+k-l)e^{j\frac{2\pi kn(m+l-N)}{N}} \\
&= \frac{1}{N}\sum_{m=N-l}^{N-1} X(m)e^{j\frac{2\pi(m+l)n}{N}} \\
\frac{1}{N}\sum_{k=l}^{N-1} X((k-l))e^{j\frac{2\pi kn}{N}} &= \frac{1}{N}\sum_{m=0}^{N-1-l} X(m)e^{j\frac{2\pi(m+l)n}{N}} \\
\therefore\ \text{DFT}\left\{X((k-l))_N\right\} &= \frac{1}{N}\sum_{m=0}^{N-1} X(m)e^{j\frac{2\pi(m+l)n}{N}} \\
&= e^{j\frac{2\pi ln}{N}}\frac{1}{N}\sum_{m=0}^{N-1} X(m)e^{j\frac{2\pi mn}{N}} \\
&= e^{j\frac{2\pi ln}{N}}x(n) \\
X((k-l))_N &= \text{DFT}\left\{x(n)e^{j\frac{2\pi ln}{N}}\right\}
\end{aligned}$$

3.4.9 Complex Conjugate Properties

If

$$x(n) \overset{\text{DFT}}{\underset{N}{\longleftrightarrow}} X(k)$$

then

$$x^*(n) \overset{\text{DFT}}{\underset{N}{\longleftrightarrow}} X^*((-k))_N = X^*(n-k) \tag{3.61}$$

Proof

$$\begin{aligned}
\text{DFT}\left\{X^*(n)\right\} &= \sum_{n=0}^{N-1} x^*(n)e^{-j\frac{2\pi kn}{N}} \\
&= \left[\sum_{n=0}^{N-1} x(n)e^{j\frac{2\pi kn}{N}}\right]^* \\
\text{DFT}\left\{X^*(n)\right\} &= \left[\sum_{n=0}^{N-1} x(n)e^{-j\frac{2\pi(N-k)n}{N}}\right]^* \\
&= X^*(n-k)
\end{aligned}$$

3.4.10 Circular Correlation

3.4.10.1 Circular Cross Correlation

In general, for complex-valued sequence $x(n)$ and $y(n)$ if

$$x(n) \overset{\text{DFT}}{\underset{N}{\longleftrightarrow}} X(k) \quad \text{and} \quad y(n) \overset{\text{DFT}}{\underset{N}{\longleftrightarrow}} Y(k)$$

then

$$\gamma_{xy}(l) \overset{\text{DFT}}{\underset{N}{\longleftrightarrow}} R_{xy}(k) = X(k)Y^*(k) \tag{3.62}$$

3.4.10.2 Circular Auto-Correlation

If $x(n) = y(n)$

$$\gamma_{xx} \overset{\text{DFT}}{\underset{N}{\longleftrightarrow}} R_{xx}(k) = |X(k)|^2 \tag{3.63}$$

3.4.11 *Multiplication of Two Sequences*

If

$$x_1(n) \overset{\text{DFT}}{\underset{N}{\longleftrightarrow}} X_1(k) \quad \text{and} \quad x_2(n) \overset{\text{DFT}}{\underset{N}{\longleftrightarrow}} X_2(k)$$

then

$$x_1(n)x_2(n) \overset{\text{DFT}}{\underset{N}{\longleftrightarrow}} \frac{1}{N}\sum_{k=0}^{N-1} X_1(k) \, Ⓝ \, X_2(k) \tag{3.64}$$

3.4.12 *Parseval's Theorem*

For complex-valued sequences $x(n)$ and $y(n)$, if

$$x(n) \overset{\text{DFT}}{\underset{N}{\longleftrightarrow}} X(k) \quad \text{and} \quad y(n) \overset{\text{DFT}}{\underset{N}{\longleftrightarrow}} Y(k)$$

then

$$\sum_{n=0}^{N-1} x(n)y^*(n) = \frac{1}{N}\sum_{k=0}^{N-1} X(k)Y^*(k) \tag{3.65}$$

If $x(n) = y(n)$

$$\sum_{n=0}^{N-1} |x(n)|^2 = \frac{1}{N}\sum_{k=0}^{N-1} |X(k)|^2$$

Proof

$$
\begin{aligned}
\text{IDFT}\,\{R_{xy}(k)\} &= \gamma_{xy}(l) \\
&= \frac{1}{N}\sum_{k=0}^{N-1} R_{xy}(k)e^{j\frac{2\pi kl}{N}}
\end{aligned}
$$

Using cross correlation

$$
\begin{aligned}
\sum_{n=0}^{N-1} x(n)y^*(n-l) &= \frac{1}{N}\sum_{k=0}^{N-1} R_{xy}(k)e^{j\frac{2\pi kl}{N}} \\
\sum_{n=0}^{N-1} x(n)y^*(n) &= \frac{1}{N}\sum_{k=0}^{N-1} R_{xy}(k) \quad [\text{when } l = 0]
\end{aligned}
$$

If $x(n) = y(n)$

$$
\sum_{n=0}^{N-1} |x(n)|^2 = \frac{1}{N}\sum_{k=0}^{N-1} R_{xy}(k)
$$

Now

$$
\begin{aligned}
R_{xy}(k) &= X(k)Y^*(k) \\
X(k) &= Y(k) \\
R_{xy}(k) &= X(k)X^*(k) = |X(k)|^2
\end{aligned}
$$

Therefore

$$
\boxed{\sum_{n=0}^{N-1} |x(n)|^2 = \frac{1}{N}\sum_{k=0}^{N-1} |X(k)|^2}
$$

Summary of properties of the discrete Fourier Transform is given in Table 3.1

3.5 Circular Convolution

The circular convolution of two sequences requires that one of the sequences should be periodic. If both sequences are non-periodic, then periodically extend one of the sequences and then perform circular convolution.

The circular convolution can be performed only if both sequences consist of the same number of samples. If the sequences have a different number of samples, then

Table 3.1 Summary of properties of the discrete Fourier transform

No.	Finite-length sequence (length N)	N point DFT (length N)
1.	$x(n)$	$X(k)$
2.	$x_1(n), x_2(n)$	$X_1(k), X_2(k)$
3.	$ax_1(n) + bx_2(n)$	$aX_1(k) + bX_2(k)$
4.	$X[n]$	$Nx[((-k))_N]$
5.	$x[((n-m))_N]$	$e^{\frac{-j2\pi km}{N}} X[k]$
6.	$e^{\frac{-j2\pi ln}{N}} Nx(n)$	$X[((k-l))_N]$
7.	$\sum_{m=0}^{N-1} x_1(m)x_2((n-m))_N$	$X_1[k]X_2[k]$
8.	$x_1(n)x_2(n)$	$\frac{1}{2}\sum_{l=0}^{N-1} X_1(l)X_2((k-l))_N$
9.	$x^*(n)$	$X^*((-k))_N$
10.	$x^*((-n))_N$	$X^*(k)$
11.	$\text{Re}\{x(n)\}$	$X_e(k) = \frac{1}{2}\{X((k))_N + X^*((-k))_N\}$
12.	$j\text{Im}\{x(n)\}$	$X_o(k) = \frac{1}{2}\{X((k))_N - X^*((-k))_N\}$
13.	$x_e = \frac{1}{2}\{x(n) + x^*((-n))_N\}$	$\text{Re}\{X(k)\}$
14.	$x_o = \frac{1}{2}\{x(n) - x^*((-n))_N\}$	$j\text{Im}\{X(k)\}$
15.	Symmetric properties	$X(k) = X^*((-k))_N$
		$\text{Re}\{X(k)\} = \text{Re}\{X((-k))_N\}$
		$\text{Im}\{X(k)\} = -\text{Im}\{X((-k))_N\}$
		$\lvert X(k)\rvert = \lvert X((-k))_N\rvert$
		$\angle X(k) = -\angle X((-k))_N$
16.	$x_e(n) = \frac{1}{2}\{x(n) + x((-n))_N\}$	$\text{Re}\{X(k)\}$
17.	$x_o(n) = \frac{1}{2}\{x(n) - x((-n))_N\}$	$j\text{Im}\{X(k)\}$

convert the smaller size sequences to the size of a larger size sequence by appending zero padding. The circular convolution has length the same as the length of input sequences, whereas in linear convolution the length is $N_1 + N_2 - 1$. That is,

$$\text{Linear convolution} \Rightarrow \text{output length} = N_1 + N_2 - 1 \tag{3.66}$$

$$\text{Circular convolution} \Rightarrow \text{output length} = \max(N_1, N_2) \tag{3.67}$$

3.5.1 *Method of Performing Circular Convolution*

3.5.1.1 Graphical Method (Concentric Circle Method) (Method 1)

Let $x_3(m)$ be the sequence obtained by circular convolution of $x_1(n)$ and $x_2(n)$

$$x_3(m) = \sum_{n=0}^{N-1} x_1(n)x_2(m-n) = x_1(n) \circledN x_2(n) \tag{3.68}$$

Procedure

1. Graph N samples of $x_1(n)$ at equally spaced points around the outer circle in counterclockwise direction.
2. Start at the same point as was done for $x_1(n)$. Graph N samples of $x_2(n)$ at equally spaced points around an inner circle in clockwise direction.
3. Multiply corresponding samples on the two circles and sum the products to produce the output.
4. Rotate the inner circle keeping the outer circle fixed by one sample in the counterclockwise direction and repeat step 3 to obtain the next value of output.
5. Repeat step 4 until the inner circle first sample coincides with the first sample of the exterior circle once again. This completes one complete rotation.

Example 3.22 Perform circular convolution of two sequences

$$x_1(n) = \{\underset{\uparrow}{2}, 1, 2, 1\} \quad \text{and} \quad x_2(n) = \{\underset{\uparrow}{1}, 2, 3, 4\}$$

The circular convolution of Example 3.22 is shown in Fig. 3.11 for different values of m.

Solution Given (Fig. 3.11)

$$x_1(n) = \{\underset{\uparrow}{2}, 1, 2, 1\} \quad \text{and} \quad x_2(n) = \{\underset{\uparrow}{1}, 2, 3, 4\}$$

$$\therefore\ x_3(m) = \{\underset{\uparrow}{14}, 16, 14, 16\}$$

3.5.1.2 Circular Convolution Using DFT and IDFT (Method 2)

Procedure

1. Take N-point DFT $x_1(n)$ and $x_2(n)$:

$$X_1(k) = \text{DFT}(x_1(n)) \quad \text{and} \quad X_2(k) = \text{DFT}[x_2(n)]$$

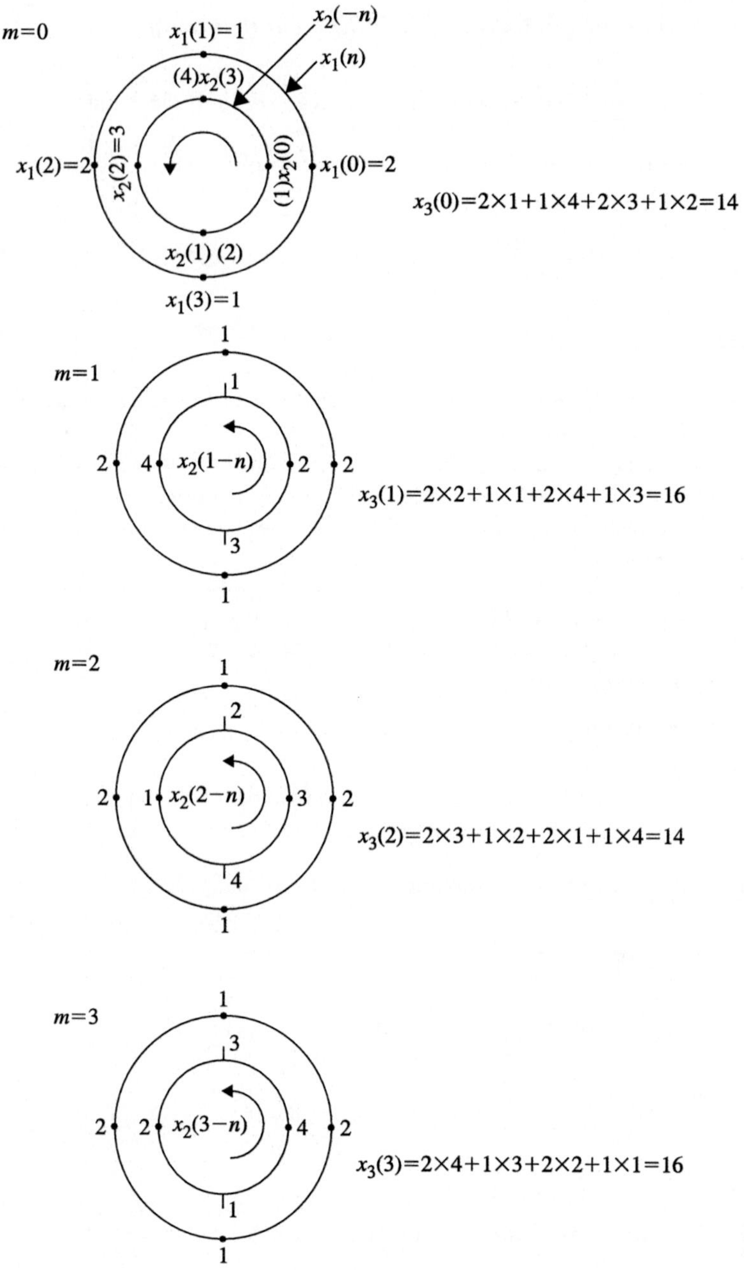

Fig. 3.11 Circular convolution of Example 3.22 for different values of m

2. Determine the product of $X_1(k)$ and $X_2(k)$. Let

$$X_3(k) = X_1(k)X_2(k)$$

3.

$$\begin{aligned} x_3(m) &= \text{IDFT}[X_3(k)] \\ &= \text{IDFT}[X_1(k)X_2(k)] \end{aligned}$$

Example 3.23 Find $x_3(m)$ using DFT and IDFT methods:

$$x_1(n) = \{\ 2, 1, 2, 1\} \quad \text{and} \quad x_2(n) = \{1, 2, 3, 4\}$$

Solution

$$X_1(k) = \sum_{n=0}^{3} x_1(n)e^{-j\frac{2\pi kn}{4}}, \quad k = 0, 1, 2, 3\ldots$$

$$X_1(k) = x_1(0) + x_1(1)e^{-j\frac{\pi kn}{2}} + x_1(2)e^{-j\pi k} + x_1(3)e^{-j\frac{3\pi kn}{2}}$$

For $k = 0$

$$X_1(0) = 2 + 1 + 2 + 1 = 6$$

For $k = 1$

$$X_1(1) = 2 + e^{-j\frac{\pi}{2}} + 2e^{-j\pi} + e^{\frac{-j3\pi}{2}} = 2 - j - 2 + j = 0$$

For $k = 2$

$$X_1(2) = 2 + e^{-j\pi} + 2e^{-j2\pi} + e^{-j3\pi} = 2 - 1 + 2 - 1 = 2$$

For $k = 3$

$$X_1(3) = 2 + e^{-j\frac{3\pi}{2}} + 2e^{-j3\pi} + e^{-j\frac{9\pi}{2}} = 2 + j - 2 - j = 0$$

$$X_1(k) = \{\underset{\uparrow}{6}, 0, 2, 0\}$$

$$X_2(k) = \sum_{n=0}^{3} X_2(n)e^{-j\frac{2\pi nk}{4}}, \quad k = 0, 1, 2, 3\ldots$$

$$X_2(k) = x_2(0) + x_2(1)e^{-j\frac{\pi k}{2}} + x_2(2)e^{-j\pi k} + x_3(3)e^{\frac{-j3\pi k}{2}}$$

For $k = 0$

$$X_2(0) = 1 + 2 + 3 + 4 = 10$$

For $k = 1$

$$X_2(1) = 1 + 2e^{-j\frac{\pi}{2}} + 3e^{-j\pi} + 4e^{-j\frac{3\pi n}{2}} = 1 - 2j - 3 + 4j = -2 + j2$$

For $k = 2$

$$X_2(3) = 1 + 2e^{-j\pi} + 3e^{-j2\pi} + 4e^{-j3\pi} = 1 - 2 + 3 - 4 = -2$$

For $k = 3$

$$X_2(3) = 1 + 2e^{-j\frac{3\pi}{2}} + 3e^{-j3\pi} + 4e^{-j\frac{9\pi}{2}} = 1 + 2j - 3 - 4j = -2 - 2j$$

$$\therefore\ X_2(k) = \{\underset{\uparrow}{10}, -2 + 2j, -2, -2 - 2j\}$$

$$\begin{aligned} X_3(k) &= X_1(k)X_2(k) \\ &= [6,\ 0,\ 2,\ 0][10,\ -2 + j2,\ -2,\ -2 - j2] \\ &= [6 \times 10,\ 0 \times (-2 + j2),\ 2 \times (-2),\ 0(-2 - j2)] \\ &= \{\underset{\uparrow}{60}, 0, -4, 0\} \end{aligned}$$

$$\begin{aligned} x_3(m) &= \text{IDFT}\ [X_3(k)] \\ &= \frac{1}{N}\sum_{k=0}^{N-1} X_3(k)e^{j\frac{2\pi mk}{N}}, \quad m = 0, 1, 2, 3, \ldots \\ &= \frac{1}{4}\sum_{k=0}^{3} X_3(k)e^{j\frac{2\pi mk}{4}} \\ &= \frac{1}{4}[X_3(0) + X_3(2)e^{j\pi m}] \end{aligned}$$

$$x_3(m) = \frac{1}{4}[60 - 4e^{j\pi m}]$$

For $m = 0$

$$X_3(0) = \frac{1}{4}[60 - 4] = \frac{56}{4} = 14$$

For $m = 1$

$$X_3(1) = \frac{1}{4}[60 - 4e^{j\pi}] = \frac{64}{4} = 16$$

For $m = 2$

$$X_3(2) = \frac{1}{4}[60 - 4e^{2j\pi}] = \frac{56}{4} = 14$$

For $m = 3$

$$X_3(3) = \frac{1}{4}[60 - 4e^{3j\pi}] = \frac{64}{4} = 16$$

$$\boxed{x_3(m) = x_1(n) \textcircled{N} x_2(n) = \{\underset{\uparrow}{14}, 16, 14, 16\}}$$

3.5.1.3 Circular Convolution Using Matrices (Method 3)

Circular convolution of $x_1(n)$ and $x_2(n)$ is

$$x(m) = x_1(n) \textcircled{N} x_2(n)$$
$$= \begin{bmatrix} x_2(0) & x_2(N-1) & x_2(N-2) & \cdots & x(1) \\ x_2(1) & x(0) & x_2(N-1) & \cdots & x(2) \\ \vdots & \vdots & \vdots & \vdots & \vdots \\ x_2(N-1) & x_2(N-2) & x_2(N-3) & \cdots & x(0) \end{bmatrix} \begin{bmatrix} x_1(0) \\ x_1(1) \\ \vdots \\ x_1(N-1) \end{bmatrix}$$
$$= \begin{bmatrix} x_3(0) \\ x_3(1) \\ \vdots \\ x_3(m-1) \end{bmatrix}$$

Example 3.24 Perform the circular convolution using the matrices method

$$x_1(n) = \{\ 2, 1, 2, 1\} \quad \text{and} \quad x_2(n) = \{1, 2, 3, 4\}$$

Solution

$$\begin{matrix} x_2(n-k)_N & x_1(n) & x_3(m) \end{matrix}$$

$$\begin{bmatrix} x_2(0) & x_2(3) & x_2(2) & x_2(1) \\ x_2(1) & x_2(0) & x_2(3) & x_2(2) \\ x_2(2) & x_2(1) & x_2(0) & x_2(3) \\ x_2(3) & x_2(2) & x_2(1) & x_2(0) \end{bmatrix} \begin{bmatrix} x_1(0) \\ x_1(1) \\ x_1(2) \\ x_1(3) \end{bmatrix} = \begin{bmatrix} x_3(0) \\ x_3(1) \\ x_3(2) \\ x_3(3) \end{bmatrix}$$

$$\begin{bmatrix} 1 & 4 & 3 & 2 \\ 2 & 1 & 4 & 3 \\ 3 & 2 & 1 & 4 \\ 4 & 3 & 2 & 1 \end{bmatrix} \begin{bmatrix} 2 \\ 1 \\ 2 \\ 1 \end{bmatrix} = \begin{bmatrix} 14 \\ 16 \\ 14 \\ 16 \end{bmatrix}$$

Therefore,

$$\boxed{x_3(m) = x_1(n) \textcircled{N} x_2(n) = \{14, 16, 14, 16\}}$$

3.5.2 *Performing Linear Convolution Using DFT*

To perform linear convolution using DFT, both the sequences should be converted into $N_1 + N_2 - 1$ point sequences by padding with zeros. Then take $N_1 + N_2 - 1$ point DFT of both the sequences and determine the product of their DFTs. The resultant sequence is obtained by taking the inverse DFT of the product of the DFTs.

Example 3.25 Find the response (linear convolution) of the system whose impulse response and input sequence are

$$h(n) = \{\underset{\uparrow}{0.5}, 1\} \quad \text{and} \quad x(n) = \{\underset{\uparrow}{1}, 0.5\}$$

using DFT-IDFT.

Solution Here $N_1 = 2$ and $N_2 = 2$

$$N = N_1 + N_2 - 1 = 2 + 2 - 1 = 3$$

$$\therefore\ x(n) = \{\underset{\uparrow}{1}, 0.5, 0\} \quad \text{and} \quad h(n) = \{\underset{\uparrow}{0.5}, 1, 0\}$$

$$X(k) = \sum_{n=0}^{2} x(n)e^{\frac{-j2\pi kn}{3}}, \qquad k = 0, \ldots, 2$$

$$X(0) = 1 + 0.5 = 1.5$$

$$X(1) = \sum_{n=0}^{2} x(n)e^{\frac{-j2\pi n}{3}} = 1 + 0.5e^{\frac{-j2\pi}{3}} = 0.75 - j0.433$$

$$X(2) = \sum_{n=0}^{3} x(n)e^{\frac{-j2\pi n2}{3}} = \sum_{n=0}^{3} x(n)e^{\frac{-j4\pi n}{3}}$$

$$= 1 + 0.5e^{\frac{-j4\pi}{3}} + 0 = 0.75 + j0.433$$

Therefore,

$$X(k) = \{1.5, 0.75 - j0.433, 0.75 + j0.433\}$$

$$H(k) = \sum_{n=0}^{2} h(n)e^{\frac{-j2\pi kn}{3}}, \qquad k = 0, \ldots, 2$$

$$H(0) = h(0) + h(1) + h(3) = 0.5 + 1 + 0 = 1.5$$

$$H(1) = \sum_{n=0}^{2} h(n)e^{\frac{-j2\pi n}{3}}$$

$$H(1) = 0.5 + e^{\frac{-j2\pi}{3}} + 0 = 0.5 - 0.5 - j0.866$$

$$= -j0.866$$

$$H(2) = \sum_{n=0}^{2} h(n)e^{\frac{-j4\pi n}{3}}$$

$$= 0.5 + e^{\frac{-j4\pi}{3}} + 0 = j0.866$$

$$H(k) = \{1.5,\ -j0.866,\ j.866\}$$

Let

$$\begin{aligned}Y(k) &= X(k)H(k)\\ &= \{2.25, -0.375 - j0.6495, -0.375 + j0.6495\}\end{aligned}$$

The sequence $y(n)$ is obtained by taking IDFT of $Y(k)$

$$y(n) = \frac{1}{N}\sum_{k=0}^{N-1} Y(k)e^{\frac{j2\pi kn}{N}}, \qquad n = 0, \ldots, N-1$$

$$y(n) = \frac{1}{3}\sum_{k=0}^{2} Y(k)e^{\frac{j2\pi kn}{3}}, \qquad n = 0, \ldots, 2$$

For $n = 0$,

$$y(0) = \frac{1}{3}[y(0) + y(1) + y(2)] = \frac{1}{3}[1.5] = 0.5$$

For $n = 1$,

$$\begin{aligned}y(1) &= \frac{1}{3}\sum_{k=0}^{2} Y(k)e^{\frac{j2\pi k}{3}}\\ &= \frac{1}{3}\left[y(0) + y(1)e^{\frac{j2\pi}{3}} + y(2)e^{\frac{j4\pi}{3}}\right]\\ &= \frac{1}{3}[2.25 + 0.75 + j0 + 0.75 + j0] = 1.25\end{aligned}$$

For $n = 2$,

$$\begin{aligned}y(2) &= \frac{1}{3}\left[y(0) + y(1)e^{\frac{j4\pi}{3}} + y(2)e^{\frac{j8\pi}{3}}\right]\\ &= \frac{1}{3}[2.25 - 0.375 + j.6495 - 0.375 - j0.6495]\\ &= 0.5\end{aligned}$$

$$\therefore \quad \boxed{y(n) = \{\underset{\uparrow}{0.5},\ 1.25,\ 0.5\}}$$

Example 3.26 Determine the output response $y(n)$ if

$$h(n) = \{1, 1, 1\} \quad \text{and} \quad x(n) = \{1, 2, 3, 1\}$$

by using (*i*) Linear convolution and (*ii*) Circular convolution.

Solution (*i*) **Linear Convolution**

$$N_1 = 4 \quad \text{and} \quad N_2 = 3$$
$$N = N_1 + N_2 - 1 = 4 + 3 - 1 = 6$$

The linear convolution obtained by the graphical method is shown in Fig. 3.12.

(*ii*) **Circular Convolution With Zero Padding**

To get the result of linear convolution with circular convolution, we have to add appropriate number of zeros to both sequences.

$$N = N_1 + N_2 - 1 = 4 + 3 - 1 = 6$$
$$\therefore\ x(n) = \{1,\ 2,\ 3,\ 1,\ 0,\ 0\}$$
$$h(n) = \{1,\ 1,\ 1,\ 0,\ 0,\ 0\}$$

$$\begin{bmatrix} 1&0&0&0&1&1 \\ 1&1&0&0&0&1 \\ 1&1&1&0&0&0 \\ 0&1&1&1&0&0 \\ 0&0&1&1&1&0 \\ 0&0&0&1&1&1 \end{bmatrix} \begin{bmatrix} 1 \\ 2 \\ 3 \\ 1 \\ 0 \\ 0 \end{bmatrix} = \begin{bmatrix} 1 \\ 3 \\ 6 \\ 6 \\ 4 \\ 1 \end{bmatrix}$$

$$\therefore\ y(n) = \{\underset{\uparrow}{1},\ 3,\ 6,\ 6,\ 4,\ 1\}$$

Example 3.27 For the $x_1(n)$, $x_2(n)$ and N given, compute $x_1(n)$ Ⓝ $x_2(n)$

$$(a) \quad x_1(n) = \delta(n) + \delta(n-1) + \delta(n-2)$$
$$x_2(n) = 2\delta(n) - \delta(n-1) + 2\delta(n-2) \qquad (N = 3)$$
$$(b) \quad x_1(n) = \delta(n) + \delta(n-1) - \delta(n-2) + \delta(n-3)$$
$$x_2(n) = \delta(n) - \delta(n-2) + \delta(n-4) \qquad (N = 5)$$

Solution (*a*) Given

$$x_1(n) = \{\underset{\uparrow}{1}, 1, 1\} \quad \text{and} \quad x_2(n) = \{2, \underset{\uparrow}{-1}, 2\}$$
$$N = 3$$

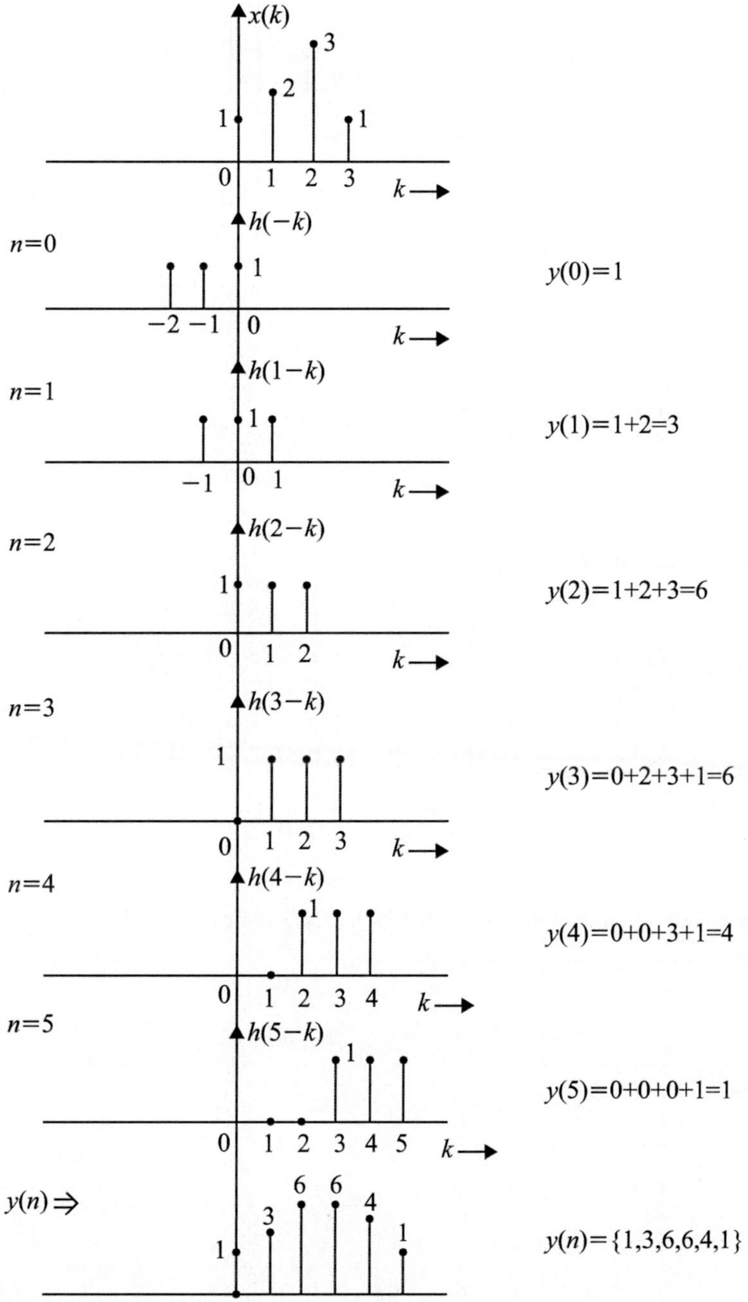

Fig. 3.12 Linear convolution of two sequences for Example 3.26

$$\begin{bmatrix} 2 & 2 & -1 \\ -1 & 2 & 2 \\ 2 & -1 & 2 \end{bmatrix} \begin{bmatrix} 1 \\ 1 \\ 1 \end{bmatrix} = \begin{bmatrix} 3 \\ 3 \\ 3 \end{bmatrix}$$

$$y(n) = x_1(n) \circledN x_2(n) = \{\underset{\uparrow}{3}, 3, 3\}$$

(*b*) Given

$$x_1(n) = \{\underset{\uparrow}{1}, 1, -1, 1\} \quad \text{and} \quad x_2(n) = \{1, \underset{\uparrow}{0}, -1, 0, 1\}$$

$$N = 5$$

Add one zero to the sequence $x_1(n)$

$$x_1(n) = \{\underset{\uparrow}{1}, 1, -1, 1, 0\} \quad \text{and} \quad x_2(n) = \{1, \underset{\uparrow}{0}, -1, 0, 1\}$$

Using matrix method, we get

$$\begin{bmatrix} 1 & 1 & 0 & -1 & 0 \\ 0 & 1 & 1 & 0 & -1 \\ -1 & 0 & 1 & 1 & 0 \\ 0 & -1 & 0 & 1 & 1 \\ 1 & 0 & -1 & 0 & 1 \end{bmatrix} \begin{bmatrix} 1 \\ 1 \\ -1 \\ 1 \\ 0 \end{bmatrix} = \begin{bmatrix} 3 \\ 0 \\ -1 \\ 0 \\ 2 \end{bmatrix}$$

$$\boxed{y(n) = \{\underset{\uparrow}{3}, 0, -1, 0, 2\}}$$

Example 3.28 Find the circular convolution of $x(n) \circledN h(n)$

$$x(n) = 1, \qquad 0 \le n \le 10$$
$$h(n) = \left(\frac{1}{2}\right)^n, \qquad 0 \le n \le 10$$

Solution

$$x(n) = 1, \qquad 0 \le n \le 10$$
$$x(n) = \{1, 1, 1, 1, 1, 11, 1, 1, 1, 1\}$$
$$h(n) = \left(\frac{1}{2}\right)^n, \qquad 0 \le n \le 10$$
$$= \{1, 0.5, 0.25, 0.125, 0.0625, 0.03125 0.0156, 0.008, 0.004, 0.002, 0.001\}$$

$$h((N-k))_N$$

$$\begin{bmatrix} 1 & 0.001 & 0.002 & 0.004 & 0.008 & 0.0156 & 0.03125 & 0.0625 & 0.125 & 0.25 & 0.5 \\ 0.5 & 1 & 0.001 & 0.002 & 0.004 & 0.008 & 0.0156 & 0.03125 & 0.0625 & 0.125 & 0.25 \\ 0.25 & 0.5 & 1 & 0.001 & 0.002 & 0.004 & 0.008 & 0.0156 & 0.03125 & 0.0625 & 0.125 \\ 0.125 & 0.25 & 0.5 & 1 & 0.001 & 0.002 & 0.004 & 0.008 & 0.0156 & 0.03125 & 0.0625 \\ 0.0625 & 0.125 & 0.25 & 0.5 & 1 & 0.001 & 0.002 & 0.004 & 0.008 & 0.0156 & 0.03125 \\ 0.03125 & 0.0625 & 0.125 & 0.25 & 0.5 & 1 & 0.001 & 0.002 & 0.004 & 0.008 & 0.0156 \\ 0.0156 & 0.03125 & 0.0625 & 0.125 & 0.25 & 0.5 & 1 & 0.001 & 0.002 & 0.004 & 0.008 \\ 0.008 & 0.0156 & 0.03125 & 0.0625 & 0.125 & 0.25 & 0.5 & 1 & 0.001 & 0.002 & 0.004 \\ 0.004 & 0.008 & 0.0156 & 0.03125 & 0.0625 & 0.125 & 0.25 & 0.5 & 1 & 0.001 & 0.002 \\ 0.002 & 0.004 & 0.008 & 0.0156 & 0.03125 & 0.0625 & 0.125 & 0.25 & 0.5 & 1 & 0.001 \\ 0.001 & 0.002 & 0.004 & 0.008 & 0.0156 & 0.03125 & 0.0625 & 0.125 & 0.25 & 0.5 & 1 \end{bmatrix}$$

$$\begin{matrix} & x(n) & & x(n) \, \circledN \, h(n) \\ \times & \begin{bmatrix} 1 \\ 1 \\ 1 \\ 1 \\ 1 \\ 1 \\ 1 \\ 1 \\ 1 \\ 1 \\ 1 \end{bmatrix} & = & \begin{bmatrix} 1.999 \approx 2 \\ 1.999 \approx 2 \\ 1.999 \approx 2 \\ 1.999 \approx 2 \\ 1.999 \approx 2 \\ 1.999 \approx 2 \\ 1.999 \approx 2 \\ 1.999 \approx 2 \\ 1.999 \approx 2 \\ 1.999 \approx 2 \\ 1.999 \approx 2 \end{bmatrix} \end{matrix}$$

$$\boxed{x(n) \, \circledN \, h(n) = \{2,\ 2,\ 2,\ 2,\ 2,\ 2,\ 2,\ 2,\ 2,\ 2,\ 2\}}$$

Example 3.29 Find the output sequence $y(n)$ if

$$h(n) = \{1,\ 1,\ 1\} \quad \text{and} \quad x(n) = \{1,\ 2,\ 3\,1\}$$

using circular convolution.

Solution

$$\begin{aligned} \text{Length of output sequence} &= N_1 + N_2 - 1 \\ &= 3 + 4 - 1 = 6 \end{aligned}$$

Therefore

$$h(n) = \{1,\ 1,\ 1,\ 0,\ 0,\ 0\} \quad \text{and} \quad x(n) = \{1,\ 2,\ 3,\ 1,\ 0,\ 0\}$$

Using circular convolution

$$y(n) = \overset{h((N-k))_N}{\begin{bmatrix} 1\,0\,0\,0\,1\,1 \\ 1\,1\,0\,0\,0\,1 \\ 1\,1\,1\,0\,0\,0 \\ 0\,1\,1\,1\,0\,0 \\ 0\,0\,1\,1\,1\,0 \\ 0\,0\,0\,1\,1\,1 \end{bmatrix}} \overset{x(n)}{\begin{bmatrix} 1 \\ 2 \\ 3 \\ 1 \\ 0 \\ 0 \end{bmatrix}}$$

$$= \begin{bmatrix} 1+0+0+0+0+0 \\ 1+2+0+0+0+0 \\ 1+2+3+0+0+0 \\ 0+2+3+1+0+0 \\ 0+0+3+1+0+0 \\ 0+0+0+1+0+0 \end{bmatrix} = \begin{bmatrix} 1 \\ 3 \\ 6 \\ 6 \\ 4 \\ 1 \end{bmatrix}$$

Example 3.30 Find the convolution sum of

$$x(n) = \begin{cases} 1, & n = -2, 0, 1 \\ 2, & n = -1 \\ 0, & \text{otherwise} \end{cases}$$

$$h(n) = \delta(n) - \delta(n-1) + \delta(n-2) - \delta(n-3)$$

Solution $x(k)$ and $h(n-k)$ are represented as shown in Fig. 3.13. $x(n-k)$ is moved such that it overlaps with $x(k)$, and $y(n)$ is obtained as shown in Fig. 3.13:

$$x(n) = \{1, \underset{\uparrow}{2}, 1, 1\} \quad \text{and} \quad h(n) = \{\underset{\uparrow}{1}, -1, 1, -1\}$$

convolution sum

$$y(n) = x(n) * h(n)$$
$$= \sum_{n=0}^{N-1} x(k)h(n-k)$$

Output sequence starts at

$$n = n_1 + n_2 = -2 + 0 = -2$$

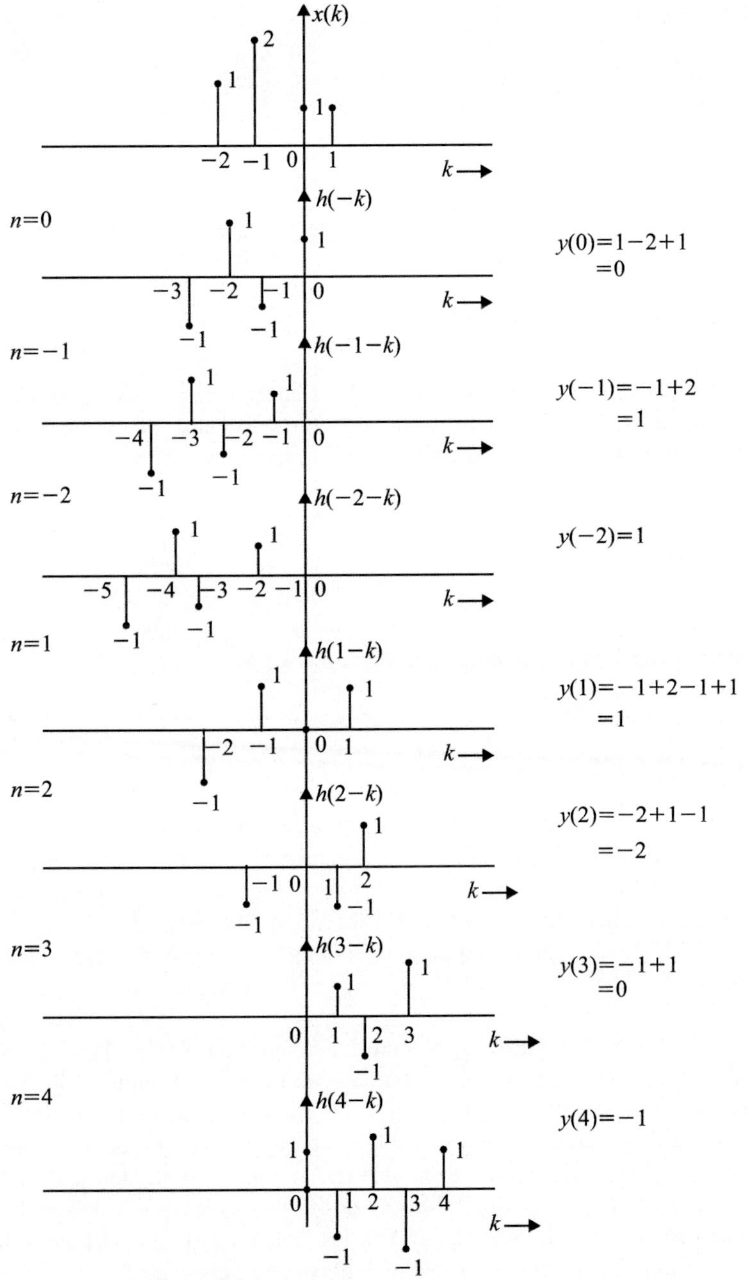

Fig. 3.13 Linear convolution of two sequences for Example 3.30

Length of output (convolution sum) sequence

$$N = N_1 + N_2 - 1 = 4 + 4 - 1 = 7$$

$$\boxed{y(n) = \{1,\ 1,\ \underset{\uparrow}{0},\ 1,\ -2,\ 0,\ -1\}}$$

3.6 Fast Fourier Transform (FFT)

For spectral analysis of discrete signals, the DFT approach is a very straightforward one. For larger values of N, DFT becomes tedious because of the huge number of mathematical operations required to be performed. Consider the following DFT where $N = 8$:

$$X(k) = \sum_{k=0}^{7} x(n)e^{\frac{-jk2\pi n}{8}}, \qquad k = 0, 1, \ldots, 7 \tag{3.69}$$

Substituting $(k2\pi/8) = K$ in the above equation, we get

$$X(k) = x(0)e^{-jK0} + x(1)e^{-jK} + x(2)e^{-jK2} + x(3)e^{-jK3} + x(4)e^{-jK4}$$
$$+x(5)e^{-jK5} + x(6)e^{-jK6} + x(7)e^{-jK7} \qquad k = 0, 1, \ldots, 7 \tag{3.70}$$

Equation (3.70) has eight terms on the right-hand side in which each term contains the multiplication of a real term with complex exponential. Thus, for example $x(1)e^{-jk} = x(1)[\cos K - j \sin K]$ requires two multiplications and one addition for each value of K where $K = \frac{2\pi k}{8}$, $k = 0, 1, 2, \ldots, 7$. Thus, in Eq. (3.70) each term on the right-hand side requires eight complex multiplications and seven additions. The 8-point DFT therefore requires $8 \times 8 = 8^2 = 64$ complex multiplications and $8 \times 7 = 8(8-1) = 56$ additions.

In general, for an N-point DFT, N^2 multiplications and $N(N-1)$ additions are required. For $N = 1024$, about 10^8 multiplications and an equal number of additions are required which results in a computational burden. Further, such a huge number of mathematical operations limit the bandwidth of digital signal processors. Several algorithms have been developed to reduce the computation burden and ease the implementation of DFT. The algorithm developed by Cooley and Tukey in 1965 is the most efficient one and is called Fast Fourier Transform (FFT). The applications of FFT algorithms are discussed below with illustrated examples.

Table 3.2 Comparison of number of computations by DFT and FFT

	Direct Computation		Radix-2 FFT	
Number of points N	Addition $N(N-1)$	Multiplication N^2	Addition $N\log_2^N$	Multiplication $\left(\frac{N}{2}\right)\log_2^N$
4	12	16	8	4
8	56	64	24	12
16	240	256	64	32
32	992	1,024	160	80
64	4,032	4,096	384	192
128	16,256	16,384	896	448

3.6.1 *Radix-2 FFT Algorithm*

For efficient computation of DFT, several algorithms have been developed based on divide-and-conquer methods. However, the method is applicable for N not being a prime number. Consider the case when $N = r_1 r_2 r_3 \ldots r_v$ where the $\{r_j\}$ are prime. If $r_1 = r_2 = r_3 = \cdots = r$, then $N = r^v$. In such a case, the DFTs are of size r. The number r is called the radix of the FFT algorithm. The most widely used FFT algorithms are radix-2 and radix-4 algorithms and are discussed in the following sections.

For performing radix-2 FFT, the value of N should be such that $N = 2^m$. Here, the decimation can be performed m times, where $m = \log_2^N$.

In the direct computation of N-point DFT, the total number of complex additions is $N(N-1)$ and the total number of complex multiplications is N^2. In radix-2 FFT, the total number of complex additions is reduced to $N\log_2^N$ and the total number of complex multiplications is $\left(\frac{N}{2}\right)\log_2^N$. The comparison of the number of computations by DFT and FFT is shown in Table 3.2.

3.6.1.1 Decimation in Time (DIT) Radix-2 FFT

Consider Eqs. (3.71) and (3.72) which are given below

$$X(k) = \sum_{n=0}^{N-1} x(n)e^{(-j2\pi kn)/N} \tag{3.71}$$

$$x(n) = \frac{1}{N}\sum_{k=0}^{N-1} X(k)e^{(j2\pi kn)/N} \tag{3.72}$$

In the direct evaluation of spectral components, the number of complex multiplications and additions required are N^2 and $N(N-1)$ respectively as stated in Sect. 3.6.

Such a huge number of mathematical operations limit the BW of digital signal processors. The classical DFT approach does not use the two important properties of the twiddle factor, namely symmetry and periodicity properties which are given below

$$W_N^{k+N/2} = -W_N^k$$
$$W_N^{k+N} = W_N^k$$

The radix-2 FFT algorithm exploits these two properties thereby removing redundant mathematical operations. However, the results obtained using FFT algorithms is exactly the same as that of DFT. Further, the efficiency of FFT algorithm increases as N is increased. For example, if $N = 512$, DFT requires nearly 110 times more multiplications than the FFT algorithm. The basic principle of the FFT algorithm is therefore to decompose DFT into successively smaller DFTs. The manner in which this decomposition is done leads to different FFT algorithms. The two basic classes of algorithms are

1. Decimation in Time (DIT)
2. Decimation in Frequency (DIF).

In the algorithm developed by DIT, the sequence $x(n)$ is decomposed into successively smaller subsequences. In the DIF algorithm, the sequence of DFT coefficients $x(k)$ is decomposed into smaller subsequences. The DIT radix-2 FFT algorithm is discussed in this section.

In the decimation in Time (DIT) algorithm, the time domain sequence $x(n)$ is decimated and smaller point DFTs are performed. The results of smaller point DFTs are combined to get the result of an N-point DFT.

In DIT radix-2 FFT, the time domain sequence is decimated into 2-point sequences. For each 2-point sequence, the 2-point DFT is computed. The results of 2-point DFTs are used to compute 4-point DFTs. Two 4-point DFTs are combined to get an 8-point DFT. This process is continued until we get N-point DFT. In general we can say that, in decimation in time algorithm, the N-point DFT can be realized from two $\frac{N}{2}$-point DFTs, and the $\frac{N}{2}$-point DFT can be realized from two $\frac{N}{4}$-point DFTs and so on. This is explained as given below.

Let us consider an N-point DFT of $x(n)$ which is written as

$$X(k) = \sum_{n=0}^{N-1} x(n)e^{\frac{-j2\pi kn}{N}}, \qquad k = 0, 1, 2, \ldots, (N-1)$$

Let

$$W_N = e^{\frac{-j2\pi k}{N}}$$

where $W_N =$ twiddle factor (or) phase factor. Therefore

$$e^{\frac{-j2\pi kn}{N}} = (e^{-j2\pi})^{\frac{n\pi}{N}} = W_N^{nk}$$

$$X(k) = \sum_{n=0}^{N-1} x(n) W_N^{nk}, \qquad k = 0, 1, 2, \ldots, (N-1) \tag{3.73}$$

Let $x(n)$ be an N-sample sequence. Decimate $x(n)$ into two sequences of $\frac{N}{2}$ samples. Let the two sequences be $f_1(n)$ and $f_2(n)$. Let $f_1(n)$ be of even-numbered samples of $x(n)$ and $f_2(n)$ that of odd-numbered samples of $x(n)$. Thus

$$\therefore \; f_1(n) = x(2n), \qquad n = 0, 1, \ldots, \left(\frac{N}{2} - 1\right)$$

$$f_2(n) = x(2n+1) \qquad n = 0, 1, 2, \ldots, \left(\frac{N}{2} - 1\right)$$

Equation (3.73) can be written as

$$X(k) = \sum_{n=0}^{N-1} x(n) W_N^{kn}$$

$$X(k) = \sum_{n=\text{even}} x(n) W_N^{kn} + \sum_{n=\text{odd}} x(n) W_N^{kn}, \quad k = 0, 1, 2, \ldots, N-1 \tag{3.74}$$

$$X(k) = \sum_{n=0}^{\frac{N}{2}-1} x(2n) W_N^{k(2n)} + \sum_{n=0}^{\frac{N}{2}-1} x(2n+1) W_N^{k(2n+1)}$$

$$W_N^{k(2n)} = (e^{-j2\pi})^{\frac{k2n}{N}} = (e^{-j2\pi})^{\frac{kn}{\frac{N}{2}}} = W_{\frac{N}{2}}^{kn}$$

$$\begin{aligned} W_N^{k(2n+1)} &= (e^{-j2\pi})^{\frac{k(2n+1)}{N}} = (e^{-j2\pi})^{\frac{k2n}{N}} \cdot (e^{-j2\pi})^{\frac{k}{N}} \\ &= (e^{-j2\pi})^{\frac{kn}{\frac{N}{2}}} \cdot (e^{-j2\pi})^{\frac{k}{N}} \\ &= W_{\frac{N}{2}}^{kn} \cdot W_N^k \end{aligned}$$

Therefore

$$X(k) = \sum_{n=0}^{\frac{N}{2}-1} f_1(n) W_{\frac{N}{2}}^{kn} + \sum_{n=0}^{\frac{N}{2}-1} f_2(n) W_{\frac{N}{2}}^{kn} W_N^k \tag{3.75}$$

$$X(k) = F_1(k) + W_N^k F_2(k) \tag{3.76}$$

where $F_1(k)$ and $F_2(k)$ are $\frac{N}{2}$ point DFT of $f_1(n)$ and $f_2(n)$, respectively

$$F_1(k) = \sum_{n=0}^{\frac{N}{2}-1} f_1(n) W_{\frac{N}{2}}^{kn}, \qquad F_2(k) = \sum_{n=0}^{\frac{N}{2}-1} f_2(n) W_{\frac{N}{2}}^{kn} \tag{3.77}$$

$f_1(n)$ would result in the two $\frac{N}{4}$ point sequences, and $f_2(n)$ would result in another $\frac{N}{4}$ point sequences.

Let the decimated $\frac{N}{4}$ point sequences of $f_1(n)$ be $V_{11}(n)$ and $f_2(n)$ be $V_{12}(n)$. That is,

$$V_{11}(n) = f_1(2n), \quad n = 0, 1, 2, \ldots, \left(\frac{N}{4} - 1\right)$$

$$V_{12}(n) = f_1(2n + 1), \quad n = 0, 1, 2, \ldots, \left(\frac{N}{4} - 1\right)$$

$$V_{21}(n) = f_2(2n), \quad n = 0, 1, 2, \ldots, \left(\frac{N}{4} - 1\right)$$

$$V_{22}(n) = f_2(2n + 1), \quad n = 0, 1, 2, \ldots, \left(\frac{N}{4} - 1\right)$$

Let

$$V_{11}(k) = \frac{N}{4} \text{ point DFT of } V_{11}(n)$$

$$V_{12}(k) = \frac{N}{4} \text{ point DFT of } V_{12}(n)$$

$$V_{21}(k) = \frac{N}{4} \text{ point DFT of } V_{21}(n)$$

$$V_{22}(k) = \frac{N}{4} \text{ point DFT of } V_{22}(n)$$

Therefore

$$F_1(k) = V_{11}(k) + W_{\frac{N}{2}}^{k} V_{12}(k), \quad k = 0, 1, 2, \ldots, \left(\frac{N}{2} - 1\right) \tag{3.78}$$

$$F_1\left(k + \frac{N}{4}\right) = V_{11}(k) - W_{\frac{N}{2}}^{k} V_{12}(k), \quad k = 0, 1, 2, \ldots, \left(\frac{N}{2} - 1\right)$$

$$F_2(k) = V_{21}(k) + W_{\frac{N}{2}}^{k} V_{22}(k), \quad k = 0, 1, 2, \ldots, \left(\frac{N}{2} - 1\right) \tag{3.79}$$

$$F_2\left(k + \frac{N}{4}\right) = V_{21}(k) - W_{\frac{N}{2}}^{k} V_{22}(k), \quad k = 0, 1, 2, \ldots, \left(\frac{N}{2} - 1\right)$$

The decimation of the data sequence can be repeated again and again until the resulting sequences are reduced to a 2-point sequence.

8-point DFT Using Radix-2 DIT FFT

Let us consider the computation of an $N = 8$ point DFT. Here $N = 8 = 2^3$, and therefore the number of stages of computation is equal to 3. The given 8-point sequence is decimated to 2-point sequences. For each 2-point sequence, the 2-point DFT is computed. From the result of a 2-point DFT, a 4-point DFT can be computed. From the result of a 4-point DFT, an 8-point DFT can be computed. The block diagram of the radix-2 DIT FFT for $N = 8$ is shown in Fig. 3.14a.

Step 1: Before Decimation, the Sequences are Arranged in Bit Reversed Order

Let us consider $N = 8$ sequence. In the first level decimation, we have the sequence $x(0), x(2), x(4), x(6), x(1), x(3), x(5)$ and $x(7)$ and in the second level decimation, the sequence is $x(0), x(4), x(2), x(6), x(1), x(5), x(3)$ and $x(7)$. This is represented in Fig. 3.14a.

The shifting of the input data sequences is to be arranged in a well-defined order. The index n of $x(n)$ is expressed in binary form. The data point $x(4)$ is expressed in binary form as $x(100)$ and is placed in position $m = 001$ or $m = 1$ in the decimal array. Now the data $x(n)$ after decimation is stored in bit reversed order as shown in Fig. 3.14b.

The basic computation in Fig. 3.14b involves a butterfly operation which is illustrated in Fig. 3.15. Consider numbers a and b. The number b is multiplied by W_N^r and then added and the product subtracted form a to form a new complex numbers (A, B). This basic computation is called a butterfly because the flow graph resembles a butterfly. In each butterfly, one complex multiplication and two complex additions are performed.

The block diagram of Radix-2 DIT FFT algorithm is shown in Fig. 3.16.

Natural order		Bit reversed order	
$x(0)$	$x(000)$	$x(000)$	$x(0)$
$x(1)$	$x(001)$	$x(100)$	$x(4)$
$x(2)$	$x(010)$	$x(010)$	$x(2)$
$x(3)$	$x(011)$	$x(110)$	$x(6)$
$x(4)$	$x(100)$	$x(001)$	$x(1)$
$x(5)$	$x(101)$	$x(101)$	$x(5)$
$x(6)$	$x(110)$	$x(011)$	$x(3)$
$x(7)$	$x(111)$	$x(111)$	$x(7)$

Let

$$\begin{aligned} x(n) &= \text{8-point sequence} \\ f_1(n),\ f_2(n) &= \text{4-point sequences obtained from } x(n) \\ V_{11}(n),\ V_{12}(n) &= \text{2-point sequences obtained from } f_1(n) \\ V_{21}(n),\ V_{22}(n) &= \text{2-point sequences obtained from } f_2(n) \end{aligned}$$

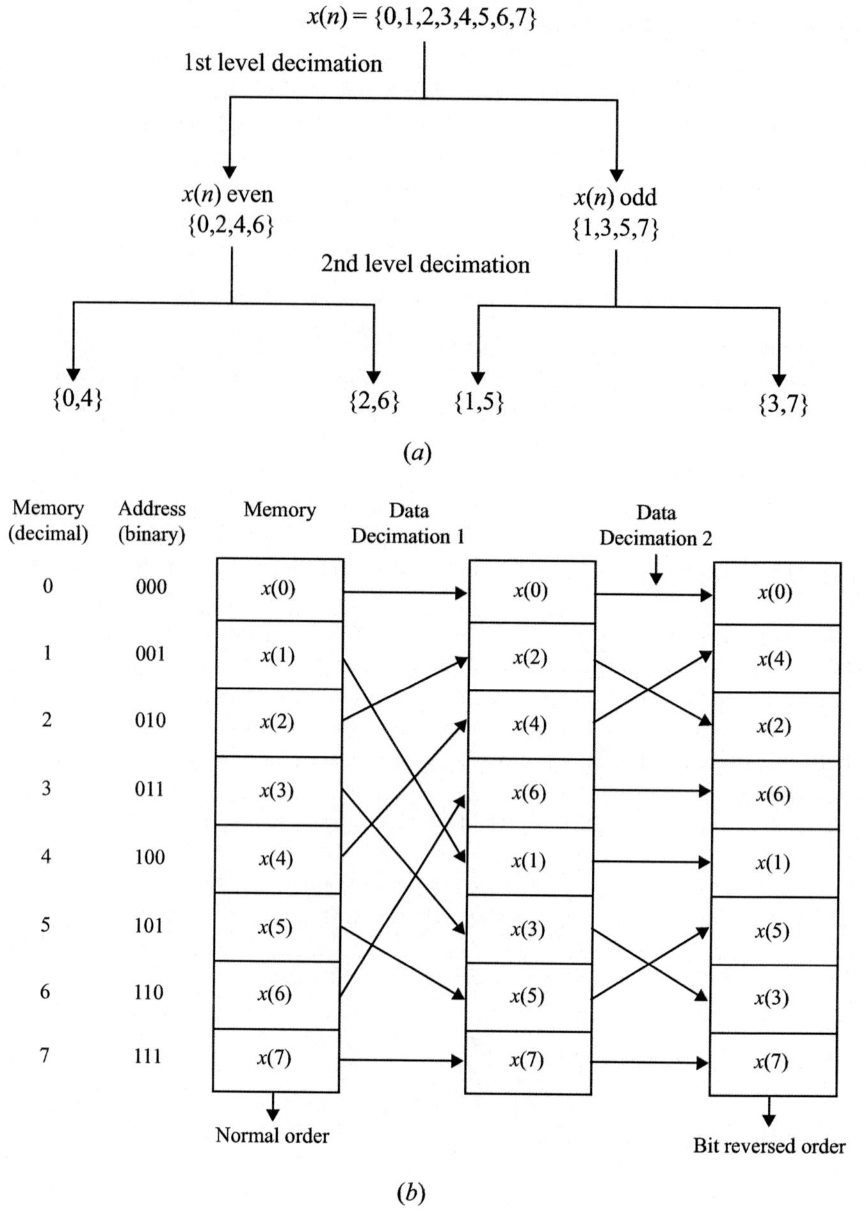

Fig. 3.14 **a** Data sequence decimation by radix-2. **b** Shifting of data and bit reversal

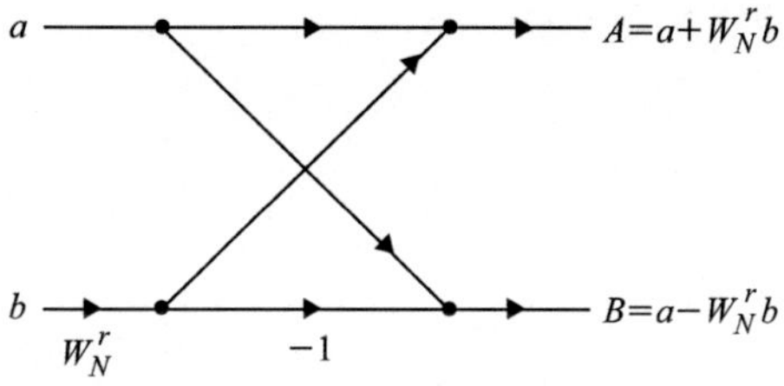

Fig. 3.15 Basic butterfly computation in the DIT FFT algorithm

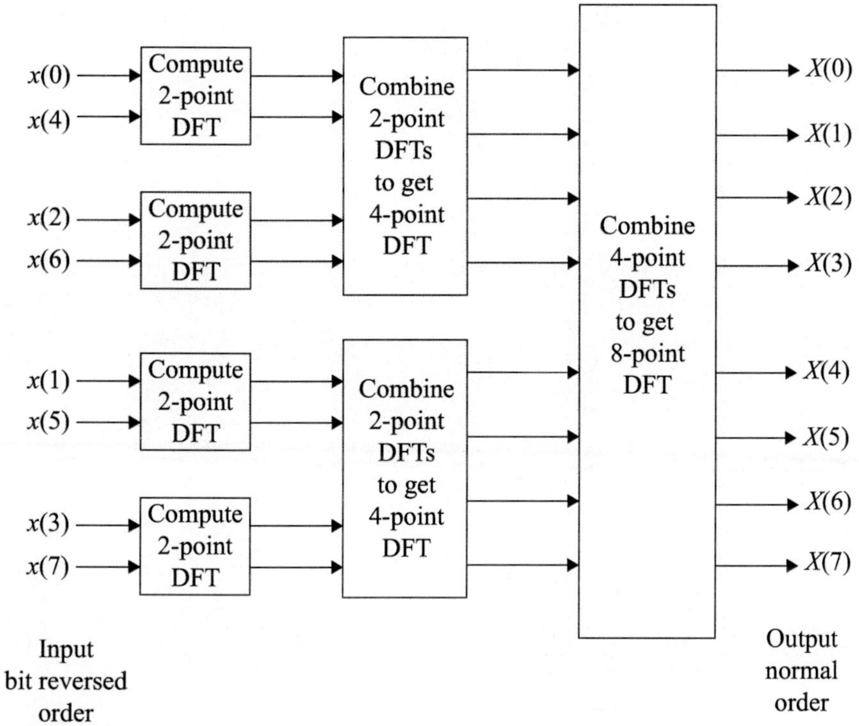

Fig. 3.16 Block diagram of radix-2 DIT FFT

First Stage (2-point DFT Computation)

2-point DFT of $V_{11}(n)$ is

$$V_{11}(k) = \sum_{n=0,1} V_{11}(n) W_{\frac{N}{4}}^{nk} \quad \text{for } k = 0, 1 \tag{3.80}$$

$$V_{11}(0) = \sum_{n=0,1} V_{11}(n) = V_{11}(0) + V_{11}(1) = x(0) + x(4)$$

$$V_{11}(1) = \sum_{n=0,1} V_{11}(n) W_{\frac{N}{4}}^{n} = V_{11}(0) W_{\frac{N}{4}}^{0} + V_{11}(1) W_{\frac{N}{4}}^{1}$$

$$= V_{11}(0) - V_{11}(1) W_{\frac{N}{4}}^{0} = x(0) - x(4)$$

Similarly,

$$V_{12}(k) = \sum_{n=0,1} V_{12}(n) W_{\frac{N}{4}}^{nk}, \qquad k = 0, 1 \tag{3.81}$$

$$V_{12}(0) = \sum_{n=0,1} V_{12}(n) = V_{12}(0) + V_{12}(1) = x(2) + x(6)$$

$$V_{12}(1) = \sum_{n=0,1} V_{12}(n) W_{\frac{N}{4}}^{n} = V_{12}(0) W_{\frac{N}{4}}^{0} + V_{12}(1) W_{\frac{N}{4}}^{1}$$

$$= V_{12}(0) - V_{12}(1)$$

$$= x(2) - x(6)$$

$$V_{21}(k) = \sum_{n=0,1} V_{21}(n) W_{\frac{N}{4}}^{nk}, \qquad k = 0, 1 \tag{3.82}$$

$$V_{21}(0) = \sum_{n=0,1} V_{21}(n) = V_{21}(0) + V_{21}(1) = x(1) + x(5)$$

$$V_{21}(1) = \sum_{n=0,1} V_{21}(n) W_{\frac{N}{4}}^{n} = V_{21}(0) + V_{21}(1) W_{\frac{N}{4}}^{1}$$

$$= V_{21}(0) - V_{21}(1)$$

$$= x(1) - x(5)$$

$$V_{22}(k) = \sum_{n=0,1} V_{22}(n) W_{\frac{N}{4}}^{nk}, \qquad k = 0, 1 \tag{3.83}$$

$$V_{22}(0) = \sum_{n=0,1} V_{22}(n) = V_{22}(0) + V_{22}(1) = x(3) + x(7)$$

$$\begin{aligned} V_{22}(1) &= \sum_{n=0,1} V_{22}(n) W_{\frac{N}{7}}^{n} = V_{22}(0) + V_{22}(1) W_{\frac{N}{4}}^{1} \\ &= V_{22}(0) - V_{22}(1) \end{aligned}$$

The Second Stage Computation (4-point DFT)

Let

$$F_1(k) = \text{DFT}\,[f_1(n)] \quad \text{and} \quad F_2(k) = \text{DFT}\,[f_2(n)]$$

From Eq. (3.81), we get

$$\begin{aligned} F_1(k) &= V_{11}(k) + W_{\frac{N}{2}}^{k} V_{12}(k) \quad \text{for } k = 0, 1, 2, 3 \\ F_1(0) &= V_{11}(0) + W_{\frac{N}{2}}^{0} V_{12}(0) \\ F_1(1) &= V_{11}(1) + W_{\frac{N}{2}}^{1} V_{12}(1) \\ F_1(2) &= V_{11}(2) + W_{\frac{N}{2}}^{2} V_{12}(2) \\ &= V_{11}(0) - W_{\frac{N}{2}}^{0} V_{12}(0) \\ F_1(3) &= V_{11}(3) + W_{\frac{N}{2}}^{3} V_{12}(3) \\ &= V_{11}(1) - W_{\frac{N}{2}}^{1} V_{12}(1) \end{aligned}$$

Here $V_{11}(k)$ and $V_{12}(k)$ are periodic with periodicity 2. Therefore,

$$V_{11}(k+2) = V_{11}(k) \quad \text{and} \quad V_{12}(k+2) = V_{12}(k)$$

Similarly, from Eq. (3.82), we get

$$\begin{aligned} F_2(k) &= V_{21}(k) + W_{\frac{N}{2}}^{k} V_{22}(k) \quad \text{for } k = 0, 1, 2, 3 \\ F_2(0) &= V_{21}(0) + W_{\frac{N}{2}}^{0} V_{22}(0) \\ F_2(1) &= V_{21}(1) + W_{\frac{N}{2}}^{1} V_{22}(1) \\ F_2(2) &= V_{21}(2) + W_{\frac{N}{2}}^{2} V_{22}(2) \\ &= V_{21}(0) - W_{\frac{N}{2}}^{0} V_{22}(0) \\ F_2(3) &= V_{21}(3) + W_{\frac{N}{2}}^{3} V_{22}(3) \\ &= V_{21}(1) - W_{\frac{N}{2}}^{1} V_{22}(1) \end{aligned}$$

The Third Stage of Computation (8-point DFT)

Let

$$
\begin{aligned}
X(k) &= \text{DFT } x(n)\\
\therefore\ X(k) &= F_1(k) + W_N^k F_2(k), \quad \text{for } k = 0, 1, 2, 3, 4, 5, 6, 7\\
X(0) &= F_1(0) + W_N^0 F_2(0)\\
X(1) &= F_1(1) + W_N^1 F_2(1)\\
X(2) &= F_1(2) + W_N^2 F_2(2)\\
X(3) &= F_1(3) + W_N^3 F_2(3)\\
X(4) &= F_1(4) + W_N^4 F_2(4)\\
&= F_1(0) - W_N^0 F_2(0)\\
X(5) &= F_1(5) + W_N^5 F_2(5)\\
&= F_1(1) - W_N^1 F_2(1)\\
X(6) &= F_1(6) + W_N^6 F_2(6)\\
&= F_1(2) - W_N^2 F_2(2)\\
X(7) &= F_1(7) + W_N^7 F_2(7)\\
&= F_1(3) - W_N^3 F_2(3)
\end{aligned}
$$

$F_1(k)$ and $F_2(k)$ are periodic with periodicity 4. Therefore,

$$F_1(k+4) = F_1(k) \quad \text{and} \quad F_2(k+4) = F_2(k)$$

The flow graph or butterfly diagram for an 8-point DIT FFT is shown in Fig. 3.17.

The phase factor values are

$$
\begin{aligned}
W_2^0 &= 1\\
W_4^0 &= e^{-j2\pi\times\frac{0}{4}} = 1\\
W_4^1 &= e^{-j2\pi\times\frac{1}{4}} = e^{-j\frac{\pi}{2}} = -j\\
W_8^0 &= 1\\
W_8^1 &= e^{-j\frac{2\pi}{8}} = e^{\frac{-j\pi}{4}} = \frac{1}{\sqrt{2}} - j\frac{1}{\sqrt{2}}\\
W_8^2 &= e^{-j2\pi\times\frac{2}{8}} = e^{\frac{-j\pi}{2}} = -j\\
W_8^3 &= e^{-j2\pi\times\frac{3}{8}} = e^{-j\frac{3\pi}{4}} = \frac{-1}{\sqrt{2}} - j\frac{1}{\sqrt{2}}
\end{aligned}
$$

Summary of steps of radix-2 DIT FFT algorithm

1. The number of input samples $N = 2^M$, where M is an integer.
2. The input sequence is shuffled through bit reversal.

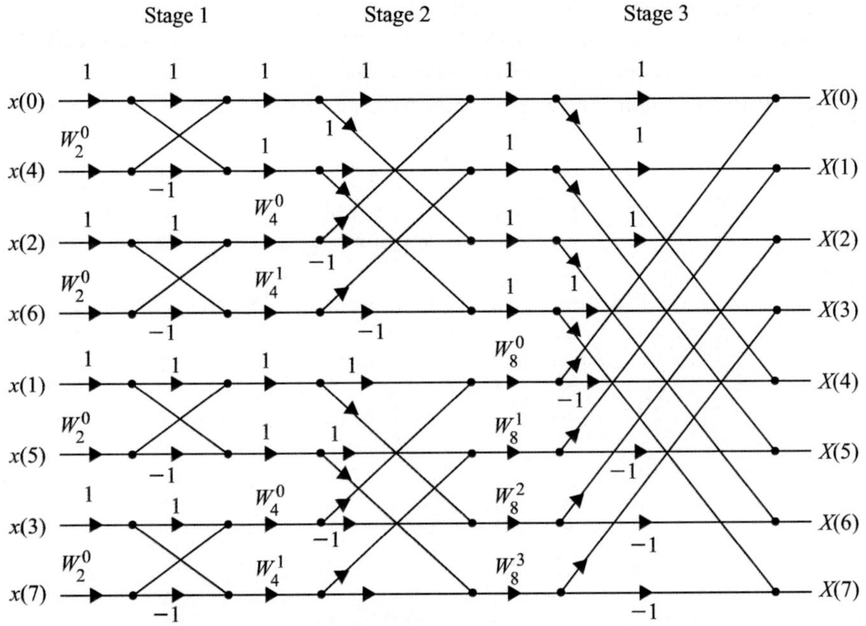

Fig. 3.17 8-point radix-2 DIT FFT Butterfly diagram

3. The number of stages in the flow graph is given by $M = \log_2^N$.
4. Each stage consists of $N/2$ butterflies.
5. Input/output for each butterfly are separated by 2^{m-1} samples, where m represents the stage index, i.e. for first stage $m = 1$ and for second stage $m = 2$ and so on.
6. The number of complex multiplications is given by $\frac{N}{2}\log_2^N$.
7. The number of complex additions is given by $N\log_2^N$.
8. The twiddle factor exponents are a function of the stage index m and are given by

$$k = \frac{Nt}{2^m}, \quad t = 0, 1, 2, \ldots, 2^{m-1} - 1.$$

9. The number of sets or sections of butterflies in each stage is given by the formula 2^{M-m}.
10. The exponent repeat factor (ERF), which is the number of times the exponent sequence associated with m is repeated, is given by 2^{M-m}.

3.6.1.2 Decimation in Frequency (DIF) Radix-2 FFT

In decimation in frequency algorithm the frequency domain sequence $X(k)$ is decimated. In this algorithm, the N-point time domain sequence is converted to two $\frac{N}{2}$ point sequences. Then each $\frac{N}{2}$ point sequence is converted to two $\frac{N}{4}$ point sequences. This process is continued until getting $\frac{N}{2}$ 2-point sequences. Finally, the 2-point DFT of each 2-point sequence is computed. The 2-point DFTs of $\frac{N}{2}$ 2-point sequences will give N-samples, which is the N-point DFT of the time domain sequence.

Consider an N-point sequence $x(n)$. The N-point DFT of $x(n)$ is given by

$$X(k) = \sum_{n=0}^{N-1} x(n) W_N^{kn} = \sum_{n=0}^{\frac{N}{2}-1} x(n) W_N^{kn} + \sum_{n=\frac{N}{2}}^{N-1} x(n) W_N^{kn} \tag{3.84}$$

$$= \sum_{n=0}^{\frac{N}{2}-1} x(n) W_N^{kn} + \sum_{n=0}^{\frac{N}{2}-1} x\left(n+\frac{N}{2}\right) W_N^{k(n+\frac{N}{2})}$$

$$= \sum_{n=0}^{\frac{N}{2}-1} x(n) W_N^{kn} + \sum_{n=0}^{\frac{N}{2}-1} x\left(n+\frac{N}{2}\right) W_N^{kn} \cdot W_N^{\frac{kN}{2}}$$

$$= \sum_{n=0}^{\frac{N}{2}-1} \left[x(n) W_N^{kn} + (-1)^k x\left(n+\frac{N}{2}\right) W_N^{kn}\right] \qquad \left(\because\ W_N^{\frac{kN}{2}} = (-1)^k\right)$$

$$X(k) = \sum_{n=0}^{\frac{N}{2}-1} \left[x(n) + (-1)^k x\left(n+\frac{N}{2}\right)\right] W_N^{kn} \tag{3.85}$$

Let us split $X(k)$ into even and odd number sequences as follows:

$$X(2k)\text{ (even)} = \sum_{n=0}^{\frac{N}{2}-1} \left[x(n) + (-1)^{2k} x\left(n+\frac{N}{2}\right)\right] W_N^{2kn} \tag{3.86}$$

$$= \sum_{n=0}^{\frac{N}{2}-1} \left[x(n) + x\left(n+\frac{N}{2}\right)\right] W_{\frac{N}{2}}^{kn}, \qquad k = 0, 1, \ldots, \left(\frac{N}{2}-1\right)$$

$$X(2k+1)\text{(odd)} = \sum_{n=0}^{\frac{N}{2}-1} \left[x(n) + (-1) x\left(n+\frac{N}{2}\right)\right] W_N^{(2k+1)^n} \tag{3.87}$$

$$= \sum_{n=0}^{\frac{N}{2}-1} \left[x(n) - x\left(n+\frac{N}{2}\right)\right] W_N^{2kn} W_N^n$$

$$= \sum_{n=0}^{\frac{N}{2}-1} \left[x(n) - x\left(n+\frac{N}{2}\right)\right] W_N^n W_{\frac{N}{2}}^{kn}, k = 0, 1, 2, \ldots, \left(\frac{N}{2}-1\right)$$

Let

$$g_1(n) = x(n) + x\left(n + \frac{N}{2}\right), \qquad n = 0, 1, 2 \ldots, \left(\frac{N}{2} - 1\right)$$

$$g_2(n) = \left[x(n) - x\left(n + \frac{N}{2}\right)\right] W_N^n, \qquad n = 0, 1, 2, \ldots, \left(\frac{N}{2} - 1\right)$$

where $g_1(n)$ and $g_2(n)$ are $\frac{N}{2}$ point time domain sequences. Therefore,

$$X(2k) = \sum_{n=0}^{\frac{N}{2}-1} g_1(n) W_{\frac{N}{2}}^{kn}, \qquad k = 0, 1, 2, \ldots, \left(\frac{N}{2} - 1\right) \tag{3.88}$$

$$X(2k+1) = \sum_{n=0}^{\frac{N}{2}-1} g_2(n) W_{\frac{N}{2}}^{kn}, \qquad k = 0, 1, 2, \ldots, \left(\frac{N}{2} - 1\right) \tag{3.89}$$

$X(2k)$ is $\frac{N}{2}$ point DFT of $g_1(n)$, and $X(2k+1)$ is $\frac{N}{2}$ point DFT of $g_2(n)$

$$\therefore\ G_1(k) = \sum_{n=0}^{\frac{N}{2}-1} g_1(n) W_{\frac{N}{2}}^{kn}, \qquad k = 0, 1, 2, \ldots, \left(\frac{N}{2} - 1\right) \tag{3.90}$$

$$G_2(k) = \sum_{n=0}^{\frac{N}{2}-1} g_2(n) W_{\frac{N}{2}}^{kn}, \qquad k = 0, 1, 2, \ldots, \left(\frac{N}{2} - 1\right) \tag{3.91}$$

The $g_1(n)$ is $\frac{N}{2}$ point sequence, and it can be decimated into two $\frac{N}{4}$ point sequences.

$$\begin{aligned}
\therefore\ G_1(k) &= \sum_{n=0}^{\frac{N}{2}-1} g_1(n) W_{\frac{N}{2}}^{kn} \\
&= \sum_{n=0}^{\frac{N}{4}-1} g_1(n) W_{\frac{N}{2}}^{kn} + \sum_{n=0}^{\frac{N}{4}-1} g_1\left(n + \frac{N}{4}\right) W_{\frac{N}{2}}^{k\left(n+\frac{N}{4}\right)} \\
&= \sum_{n=0}^{\frac{N}{4}-1} \left[g_1(n) + g_1\left(n + \frac{N}{4}\right) W_{\frac{N}{2}}^{k\frac{N}{4}}\right] W_{\frac{N}{2}}^{kn}
\end{aligned}$$

Therefore,

$$G_1(2k) = \sum_{n=0}^{\frac{N}{4}-1} d_{11}(n) W_{\frac{N}{4}}^{kn} = D_{11}(k) \tag{3.92}$$

$$G_1(2k+1) = \sum_{n=0}^{\frac{N}{4}-1} d_{12}(n) W_{\frac{N}{4}}^{kn} = D_{12}(k) \tag{3.93}$$

$$G_2(2k) = \sum_{n=0}^{\frac{N}{4}-1} d_{21}(n) W_{\frac{N}{4}}^{kn} = D_{21}(k) \tag{3.94}$$

$$G_2(2k+1) = \sum_{n=0}^{\frac{N}{4}-1} d_{22}(n) W_{\frac{N}{4}}^{kn} = D_{22}(k) \tag{3.95}$$

where $D_{11}(k)$, $D_{12}(k)$, $D_{21}(k)$ and $D_{22}(k)$ are $\frac{N}{4}$ point DFTs of $d_{11}(n), d_{12}(n), d_{21}(n)$ and $d_{22}(n)$, respectively.

The decimation of the frequency domain sequence can be continued until the resulting sequence is reduced to two-point sequences.

The 8-point DFT Using Radix-2 DIF FFT

Let $x(n)$ be an 8-point sequence where $N = 8$.

First Stage of Computation

In the first stage of computation, two 4-point sequences $g_1(n)$ and $g_2(n)$ are obtained:

$$g_1(n) = x(n) + x\left(n + \frac{N}{2}\right)$$

$$g_1(n) = x(n) + x(n+4), \quad n = 0, 1, 2, 3$$

$$g_1(0) = x(0) + x(4)$$
$$g_1(1) = x(1) + x(5)$$
$$g_1(2) = x(2) + x(6)$$
$$g_1(3) = x(3) + x(7)$$

and

$$g_2(n) = \left[x(n) - x\left(n + \frac{N}{2}\right)\right] W_N^n$$
$$g_2(n) = [x(n) - x(n+4)]\, W_8^n, \quad n = 0, 1, 2, 3$$
$$g_2(0) = [x(0) - x(4)]\, W_8^0$$
$$g_2(1) = [x(1) - x(5)]\, W_8^1$$
$$g_2(2) = [x(2) - x(6)]\, W_8^2$$
$$g_2(3) = [x(3) - x(7)]\, W_8^3$$

Second Stage of Computation

In the second stage, 4 2-point sequences are obtained:

$$d_{11}(n) = g_1(n) + g_1\left(n + \frac{N}{4}\right)$$
$$d_{11}(n) = g_1(n) + g_1(n+2), \qquad n = 0, 1$$
$$d_{11}(0) = g_1(0) + g_1(2)$$
$$d_{11}(1) = g_1(1) + g_1(3)$$
$$d_{12}(n) = \left[g_1(n) - g_1\left(n + \frac{N}{4}\right)\right] W_{\frac{N}{2}}^{n}$$
$$= [g_1(n) - g_1(n+2)]\, W_4^n, \qquad n = 0, 1$$
$$d_{12}(0) = [g_1(0) - g_1(2)]\, W_4^0$$
$$d_{12}(1) = [g_1(1) - g_1(3)]\, W_4^1$$
$$d_{21}(n) = g_2(n) + g_2\left(n + \frac{N}{4}\right)$$
$$= g_2(n) + g_2(2), \qquad n = 0, 1$$
$$d_{21}(n) = g_2(0) + g_2(2)$$
$$d_{21}(1) = g_2(1) + g_2(3)$$
$$d_{22}(n) = \left[g_2(n) - g_2\left(n + \frac{N}{4}\right)\right] W_{\frac{N}{2}}^{n}$$
$$= [g_2(n) - g_2(n+2)]\, W_4^n, \qquad n = 0, 1$$

$$d_{22}(0) = [g_2(0) - g_2(2)]\, W_4^0$$
$$d_{22}(1) = [g_2(1) - g_2(3)]\, W_4^1$$

Third Stage Computation

In the third stage, 2-point DFTs are calculated:

$$D_{11}(k) = \sum_{n=0,1} d_{11}(n) W_2^{kn}, \qquad k = 0, 1$$
$$D_{11}(0) = \sum_{n=0,1} d_{11}(n) = d_{11}(0) + d_{11}(1)$$
$$D_{11}(1) = \sum_{n=0,1} d_{11}(n) W_2^{n} = [d_{11}(0) - d_{11}(1)]\, W_2^0$$

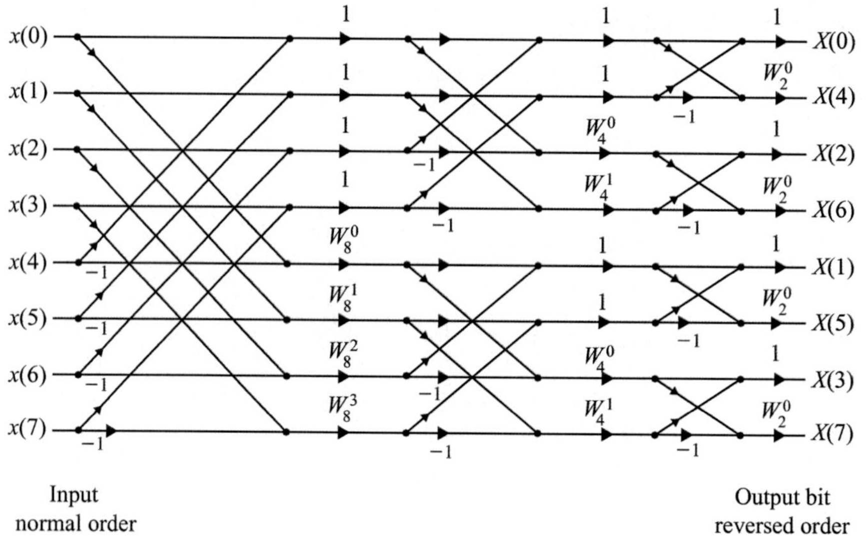

Fig. 3.18 8-point radix-2 DIF-FFT butterfly diagram

Similarly

$$
\begin{aligned}
D_{12}(0) &= d_{12}(0) + d_{12}(1) \\
D_{12}(1) &= [d_{12}(0) - d_{12}(1)]\, W_2^0 \\
D_{21}(0) &= d_{21}(0) + d_{21}(1) \\
D_{21}(1) &= [d_{21}(0) - d_{21}(1)]\, W_2^0 \\
D_{22}(0) &= d_{22}(0) + d_{22}(1) \\
D_{22}(1) &= [d_{22}(0) - d_{22}(1)]\, W_2^0
\end{aligned}
$$

The signal flow graph or butterfly diagram of all three stages is shown in Fig. 3.18.

Summary of Steps of Radix-2 DIF-FFT Algorithm

1. The number of input samples $N = 2^M$, where M is an integer.
2. The input sequence is in normal order.
3. The number of stages in the flow graph is given by $M = \log_2^N$.
4. Each stage consists of $N/2$ butterflies.
5. Input/output for each butterfly are separated by 2^{m-1} samples, where m represents the stage index, i.e. for the first stage $m = 1$ and for the second stage $m = 2$ and so on.
6. The number of complex multiplications is given by $\frac{N}{2}\log_2^N$.
7. The number of complex additions is given by $N\log_2^N$.

8. The twiddle factor exponents are a function of the stage index m and are given by

$$k = \frac{Nt}{2^{M-m+1}}, \quad t = 0, 1, 2, \ldots, 2^{m-1} - 1$$

9. The number of sets or sections of butterflies in each stage is given by the formula 2^{m-1}.
10. The exponent repeat factor (ERF), which is the number of times the exponent sequence associated with m is repeated, is given by 2^{m-1}.

3.6.1.3 Comparison of DIF and DIT

Difference Between DIF and DIT

(*i*) In DIT, the input is bit reversed while the output is in normal order. For DIF, the input is in normal order, while the output is in bit reversed order.
(*ii*) Considering the butterfly diagram, in DIF, the complex multiplication takes place after the add-subtract operation.

Similarities

(*i*) Both algorithms require the same number of operations to compute DFT.
(*ii*) Both algorithms require bit reversal at some place during computation.

3.6.2 Radix-4 FFT Algorithms

The decimation in time (DIT) radix-4 FFT recursively partitions a DFT into four quarter-length DFTs of groups of every fourth time sample. The outputs of these shorter FFTs are reused to compute many outputs, thus greatly reducing the total computation cost. In the decimation in frequency radix-4 FFT groups, every fourth output sample is decimated into shorter length DFTs to save computations.

3.6.2.1 Radix-4 DIT FFT Algorithms

Radix-4 DIT rearranges the DFT equation into four parts: sums over all groups of every fourth discrete time index $n = [0, 4, 8, \ldots, N-4]$, $n = [15, 9, \ldots, N-3]$, $n = [2, 6, 10, \ldots, N-2]$ and $n = [3, 7, 11, \ldots, N-1]$, i.e.

$$
\begin{aligned}
X(k) &= \sum_{n=0}^{N-1} x(n)e^{-j\frac{2\pi nk}{N}} \\
&= \sum_{n=0}^{\frac{N}{4}-1} x(4n)e^{-j\frac{2\pi k(4n)}{N}} + \sum_{n=0}^{\frac{N}{4}-1} x(4n+1)e^{-j\frac{2\pi k(4n+1)}{N}} \\
&\quad + \sum_{n=0}^{\frac{N}{4}-1} x(4n+2)e^{-j\frac{2\pi k(4n+2)}{N}} + \sum_{n=0}^{\frac{N}{4}-1} x(4n+3)e^{-j\frac{2\pi k(4n+3)}{N}} \qquad (3.96)\\
&= \sum_{n=0}^{\frac{N}{4}-1} x(4n)e^{-j\frac{2\pi k4n}{N}} + \sum_{n=0}^{\frac{N}{4}-1} x(4n+1)e^{-j\frac{2\pi k4N}{N}} \cdot e^{-j\frac{2\pi k}{N}} \\
&\quad + \sum_{n=0}^{\frac{N}{4}-1} x(4n+2)e^{-j\frac{2\pi 4n}{N}} \cdot e^{-j\frac{2\pi k\cdot 2}{N}} + \sum_{n=0}^{\frac{N}{4}-1} x(4n+3)e^{-j\frac{2\pi 4n}{N}} \cdot e^{-j\frac{2\pi k\cdot 3}{N}} \qquad (3.97)\\
&= X_1(k) + W_N^k X_2(k) + W_N^{2k} X_3(k) + W_N^{3k} X_4(k)
\end{aligned}
$$

where

$$X_1(k) = \sum_{n=0}^{\frac{N}{4}-1} x(4n) W_N^{4nk} \tag{3.98}$$

$$X_2(k) = \sum_{n=0}^{\frac{N}{4}-1} x(4n+1) W_N^{4nk} \tag{3.99}$$

$$X_3(k) = \sum_{n=0}^{\frac{N}{4}-1} x(4n+2) W_N^{4nk} \tag{3.100}$$

$$X_4(k) = \sum_{n=0}^{\frac{N}{4}-1} x(4n+3) W_N^{4nk} \tag{3.101}$$

16-Point Radix-4 DIT FFT

Here $N = 16 = 4^2$, and the number of stages are two:

$$X(k) = X_1(k) + W_{16}^k X_2(k) + W_{16}^{2k} X_3(k) + W_{16}^{3k} X_4(k) \tag{3.102}$$

Stage I

$$X_1(k) = \sum_{n=0}^{3} x(4n) W_{16}^{4nk}, \quad k = 0, 1, 2, 3$$
$$= x(0) + x(4) W_{16}^{4k} + x(8) W_{16}^{8k} + x(12) W_{16}^{12k} \tag{3.103}$$

$$X_2(k) = \sum_{n=0}^{3} x(4n+1) W_{16}^{4nk}$$
$$= x(1) + x(5) W_{16}^{4k} + x(9) W_{16}^{8k} + x(13) W_{16}^{12k} \tag{3.104}$$

$$X_3(k) = \sum_{n=0}^{3} x(4n+2) W_{16}^{4nk}$$
$$= x(2) + x(6) W_{16}^{4k} + x(10) W_{16}^{8k} + x(14) W_{16}^{12k} \tag{3.105}$$

$$X_4(k) = \sum_{n=0}^{3} x(4n+3) W_{16}^{4nk}$$
$$= x(3) + x(7) W_{16}^{4k} + x(11) W_{16}^{8k} + x(15) W_{16}^{12k} \tag{3.106}$$

Stage II

$$[X_i(k+4) = X_i(k)]$$

$$X(k) = X_1(k) + W_{16}^{k} X_2(k) + W_{16}^{2k} X_3(k) + W_{16}^{3k} X_4(k)$$
$$X(0) = X_1(0) + X_2(0) + X_3(0) + X_4(0)$$
$$X(1) = X_1(1) + W_{16}^{1} X_2(1) + W_{16}^{2} X_3(1) + W_{16}^{3} X_4(1)$$
$$X(2) = X_1(2) + W_{16}^{2} X_2(2) + W_{16}^{4} X_3(2) + W_{16}^{6} X_4(2)$$
$$X(3) = X_1(3) + W_{16}^{3} X_2(3) + W_{16}^{6} X_3(3) + W_{16}^{9} X_4(3)$$
$$X(4) = X_1(0) + W_{16}^{4} X_2(0) + W_{16}^{8} X_3(0) + W_{16}^{12} X_4(0)$$
$$X(5) = X_1(1) + W_{16}^{5} X_2(1) + W_{16}^{10} X_3(1) + W_{16}^{15} X_4(1)$$
$$X(6) = X_1(2) + W_{16}^{6} X_2(2) + W_{16}^{12} X_3(2) + W_{16}^{18} X_4(2)$$
$$X(7) = X_1(3) + W_{16}^{7} X_2(3) + W_{16}^{14} X_3(3) + W_{16}^{21} X_4(3)$$
$$X(8) = X_1(0) + W_{16}^{8} X_2(0) + W_{16}^{16} X_3(0) + W_{16}^{24} X_4(0)$$
$$X(9) = X_1(1) + W_{16}^{9} X_2(1) + W_{16}^{18} X_3(1) + W_{16}^{27} X_4(1)$$
$$X(10) = X_1(2) + W_{16}^{10} X_2(2) + W_{16}^{20} X_3(2) + W_{16}^{30} X_4(2)$$

$$X(11) = X_1(3) + W_{16}^{11} X_2(3) + W_{16}^{22} X_3(3) + W_{16}^{33} X_4(3)$$
$$X(12) = X_1(0) + W_{16}^{12} X_2(0) + W_{16}^{24} X_3(0) + W_{16}^{36} X_4(0)$$

$$X(13) = X_1(1) + W_{16}^{13} X_2(1) + W_{16}^{26} X_3(1) + W_{16}^{39} X_4(1)$$
$$X(14) = X_1(2) + W_{16}^{14} X_2(2) + W_{16}^{28} X_3(2) + W_{16}^{42} X_4(2)$$
$$X(15) = X_1(3) + W_{16}^{15} X_2(3) + W_{16}^{30} X_3(3) + W_{16}^{45} X_4(3)$$

16-point radix-4 DIT FFT butterfly diagram is shown in Fig. 3.19.

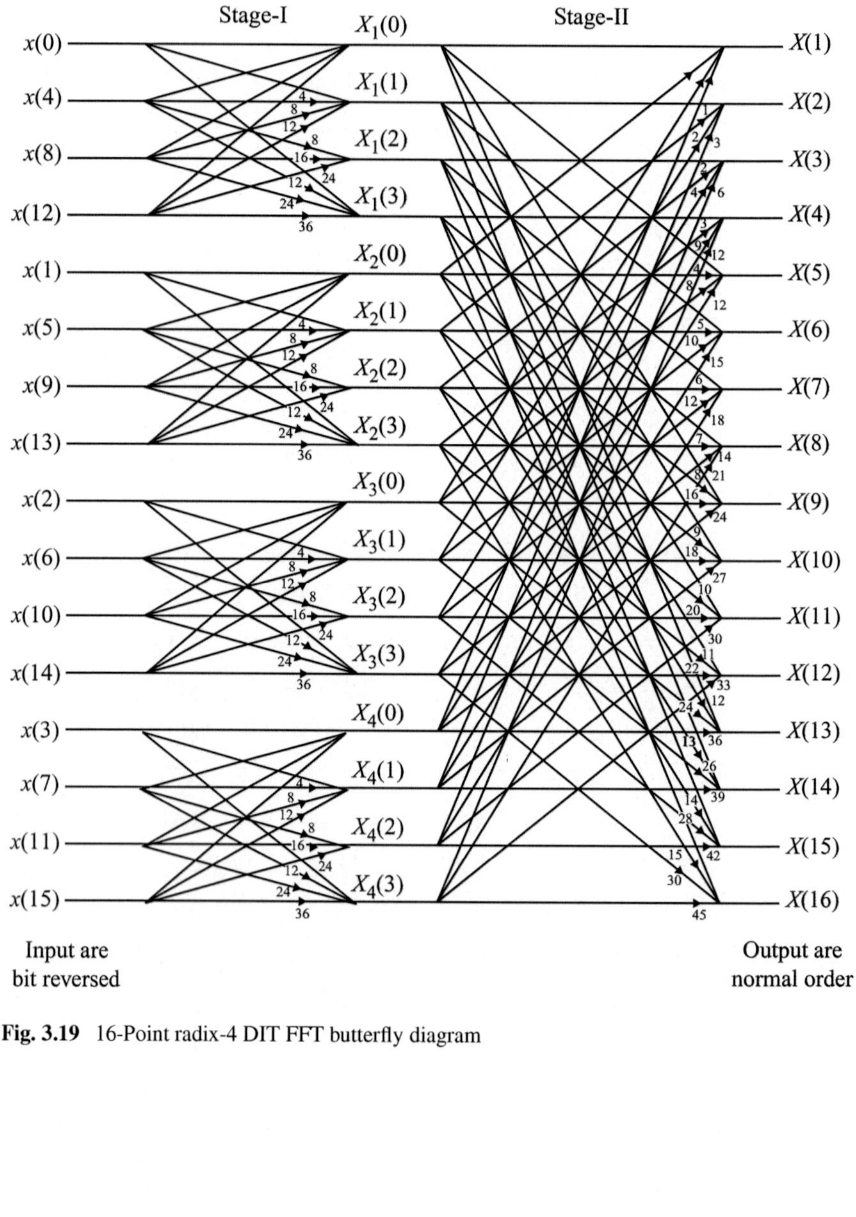

Fig. 3.19 16-Point radix-4 DIT FFT butterfly diagram

In Radix-4 FFT

$$N = 4^m$$
$$M = \log_4^N = \frac{\log_2^N}{2}$$

where $N/4$ implies the number of butterflies per stage.

Number of complex multiplications

$$3\frac{N}{4}\frac{\log_N^2}{2} = \frac{3}{8}N\log_2^N \quad \text{(75\%of radix-2 FFT)}$$

Number of complex additions

$$8\frac{N}{4}\frac{\log_2^N}{2} = N\log_2^N \quad \text{(same as a radix-2 FFT)}$$

3.6.2.2 Radix-4 Decimation in Frequency FFT

N-point DFT formula breaking into four smaller DFTs is given by

$$X(k) = \sum_{n=0}^{N-1} x(n)W_N^{kn}$$

$$= \sum_{n=0}^{\frac{N}{4}-1} x(n)W_N^{kn} + \sum_{n=(\frac{N}{4})}^{\frac{N}{2}-1} x(n)W_N^{kn} + \sum_{n=(\frac{N}{2})}^{3(\frac{N}{4})-1} x(n)W_N^{kn}$$

$$+ \sum_{n=(\frac{3N}{4})}^{N-1} x(n)W_N^{kn} \tag{3.107}$$

$$= \sum_{n=0}^{\frac{N}{4}-1} x(n)W_N^{kn} + \sum_{n=0}^{\frac{N}{2}-1} x\left(n+\frac{N}{4}\right)W_N^{k(n+\frac{N}{4})}$$

$$+ \sum_{n=0}^{\frac{N}{4}-1} x\left(n+\frac{N}{2}\right)W_N^{k(n+\frac{N}{2})} + \sum_{n=0}^{\frac{N}{4}-1} x\left(n+\frac{3N}{4}\right)W_N^{k(n+\frac{3N}{4})} \tag{3.108}$$

$$= \sum_{n=0}^{\frac{N}{4}-1} x(n)W_N^{kn} + W_N^{kn/4}\sum_{n=0}^{\frac{N}{4}-1} x\left(n+\frac{N}{4}\right)W_N^{kn} + W_N^{kn/2}$$

$$+ \sum_{n=0}^{\frac{N}{4}-1} x\left(n+\frac{N}{2}\right)W_N^{nk} + W_N^{3Nk/4}\sum_{n=0}^{\frac{N}{4}-1} x\left(n+\frac{3N}{4}\right)W_N^{kn} \tag{3.109}$$

$$W_N^{kN/4} = e^{-j\frac{2\pi kN}{4N}} = \left(e^{-j\frac{\pi}{2}}\right)^k = (-j)^k$$
$$W_N^{kN/2} = (-1)^k$$
$$W_N^{3kN/4} = j^k$$

Substituting these twiddle factors in Eq. (3.92), we get

$$X(k) = \sum_{n=0}^{\frac{N}{4}-1}\left[x(n)+(-j)^k x\left(n+\frac{N}{4}\right)+(-1)^k x\left(n+\frac{N}{4}\right)+(j)^k x\left(n+\frac{3N}{4}\right)\right]W_N^{kn}$$

The four $N/4$ point DFTs are written as

$$X(4k) = \sum_{n=0}^{\frac{N}{4}-1}\left[x(n)+x\left(n+\frac{N}{4}\right)+x\left(n+\frac{N}{2}\right)+x\left(n+\frac{3N}{4}\right)\right]W_N^{4kn}$$

$$X(4k) = \sum_{n=0}^{\frac{N}{4}-1}\left[x(n)+x\left(n+\frac{N}{4}\right)+x\left(n+\frac{N}{2}\right)+x\left(n+\frac{3N}{4}\right)\right]W_N^{0}\cdot W_{\frac{N}{4}}^{kn} \quad (3.110)$$

$$X(4k+1) = \sum_{n=0}^{\frac{N}{4}-1}\left[x(n)-jx\left(n+\frac{N}{4}\right)-x\left(n+\frac{N}{2}\right)+jx\left(n+\frac{3N}{4}\right)\right]W_N^{n}\cdot W_{\frac{N}{4}}^{kn} \quad (3.111)$$

$$X(4k+2) = \sum_{n=0}^{\frac{N}{4}-1}\left[x(n)-x\left(n+\frac{N}{4}\right)+x\left(n+\frac{N}{2}\right)-x\left(n+\frac{3N}{4}\right)\right]W_N^{2n}\cdot W_{\frac{N}{4}}^{kn} \quad (3.112)$$

$$X(4k+3) = \sum_{n=0}^{\frac{N}{4}-1}\left[x(n)+jx\left(n+\frac{N}{4}\right)-x\left(n+\frac{N}{2}\right)-jx\left(n+\frac{3N}{4}\right)\right]W_N^{3n}\cdot W_{\frac{N}{4}}^{kn} \quad (3.113)$$

16-Point DFT Using Radix-4 DIF FFT

Let $x(n)$ be a 16-point sequence. The samples of $x(n)$ are $x(0), x(1), x(2), \ldots, x(15)$. Here $N = 16$.

Stage I

In the first stage, four 4-point sequences $g_1(n)$, $g_2(n)$, $g_3(n)$ and $g_4(n)$ are obtained from Eqs. (3.93)–(3.96), respectively. Thus,

$$g_1(n) = \left[x(n)+x\left(n+\frac{N}{4}\right)+x\left(n+\frac{N}{2}\right)+x\left(n+\frac{3N}{4}\right)\right]W_N^0$$

$n = 0, 1, 2, 3$ and $N = 16$.

$$g_1(0) = [x(0)+x(4)+x(8)+x(12)]\,W_{16}^0$$
$$g_1(1) = [x(1)+x(5)+x(9)+x(13)]\,W_{16}^0$$

$$g_1(2) = [x(2) + x(6) + x(10) + x(14)]\, W_{16}^0$$
$$g_1(3) = [x(3) + x(7) + x(11) + x(15)]\, W_{16}^0$$

$$g_2(n) = \left[x(n) - jx\left(n + \frac{N}{4}\right) - x\left(n + \frac{N}{2}\right) + jx\left(n + \frac{3N}{4}\right)\right] W_N^n$$

$$g_2(0) = [x(0) - jx(4) - x(8) + jx(12)]\, W_{16}^0$$
$$g_2(1) = [x(1) - jx(5) - x(9) + jx(13)]\, W_{16}^1$$
$$g_2(2) = [x(2) - jx(6) - x(10) + jx(14)]\, W_{16}^2$$
$$g_2(3) = [x(3) - jx(7) - x(11) + jx(15)]\, W_{16}^3$$

$$g_3(n) = \left[x(n) - x\left(n + \frac{N}{4}\right) + x\left(n + \frac{N}{2}\right) - x\left(n + \frac{3N}{4}\right)\right] W_N^{2n}$$

$$g_3(0) = [x(0) - x(4) + x(8) - x(12)]\, W_{16}^0$$
$$g_3(1) = [x(1) - x(5) + x(9) - x(13)]\, W_{16}^2$$
$$g_3(2) = [x(2) - x(6) + x(10) - x(14)]\, W_{16}^4$$
$$g_3(3) = [x(3) - x(7) + x(11) - x(15)]\, W_{16}^6$$

$$g_4(n) = \left[x(n) + jx\left(n + \frac{N}{4}\right) - x\left(n + \frac{N}{2}\right) - jx\left(n + \frac{3N}{4}\right)\right] W_N^{3n}$$

$$g_4(0) = [x(0) - jx(4) - x(8) - jx(12)]\, W_{16}^0$$
$$g_4(1) = [x(1) - jx(5) - x(9) - jx(13)]\, W_{16}^3$$
$$g_4(2) = [x(2) - jx(6) - x(10) - jx(14)]\, W_{16}^6$$
$$g_4(3) = [x(3) - jx(7) - x(11) - jx(15)]\, W_{16}^9$$

Stage II

The four 4-point DFTs are

$$G_1(k) = \sum_{n=0}^{\frac{N}{4}-1} g_1(n) W_{N/4}^{kn} \qquad k = 0, 1, \ldots, 3$$

$$G_1(k) = \sum_{n=0}^{3} g_1(n) W_4^{kn}$$

$$G_1(0) = \sum_{n=0}^{3} g_1(n) = g_1(0) + g_1(1) + g_1(2) + g_1(3)$$

$$G_1(1) = \sum_{n=0}^{3} g_1(n) W_4^n$$
$$= g_1(0)W_4^0 + g_1(1)W_4^1 + g_1(2)W_4^2 + g_1(3)W_4^3$$

$$G_1(2) = \sum_{n=0}^{3} g_1(n) W_4^{2n}$$
$$= g_1(0)W_4^0 + g_1(1)W_4^2 + g_1(2)W_4^0 + g_1(3)W_4^2$$
$$G_1(3) = \sum_{n=0}^{3} g_1(n) W_4^{3n}$$
$$= g_1(0)W_4^0 + g_1(1)W_4^3 + g_1(2)W_4^2 + g_1(3)W_4^1$$

Similarly

$$G_2(k) = \sum_{n=0}^{\frac{N}{4}-1} g_2(n) W_{N/4}^{kn} \qquad k = 0, 1, \ldots, 3$$
$$G_3(k) = \sum_{n=0}^{\frac{N}{4}-1} g_3(n) W_{N/4}^{kn} \qquad k = 0, 1, \ldots, 3$$
$$G_4(k) = \sum_{n=0}^{\frac{N}{4}-1} g_4(n) W_{N/4}^{kn} \qquad k = 0, 1, \ldots, 3$$

16-point radix-4 DIF FFT butterfly diagram is shown in Fig. 3.20:

$$\text{Number of complex multiplications} = \frac{3}{8} N \log_2^N \tag{3.114}$$
$$\text{Number of complex additions} = N \log_2^N \quad \text{(same as DIT algorithm)}$$

3.6.2.3 4-Point DFT Using Radix-4 FFT Algorithm

$$X(k) = \sum_{n=0}^{3} x(n) W_N^{kn} \qquad k = 0, \ldots, 3$$
$$X(0) = \sum_{n=0}^{3} x(n) W_N^0 = x(0) + x(1) + x(2) + x(3) \tag{3.115}$$

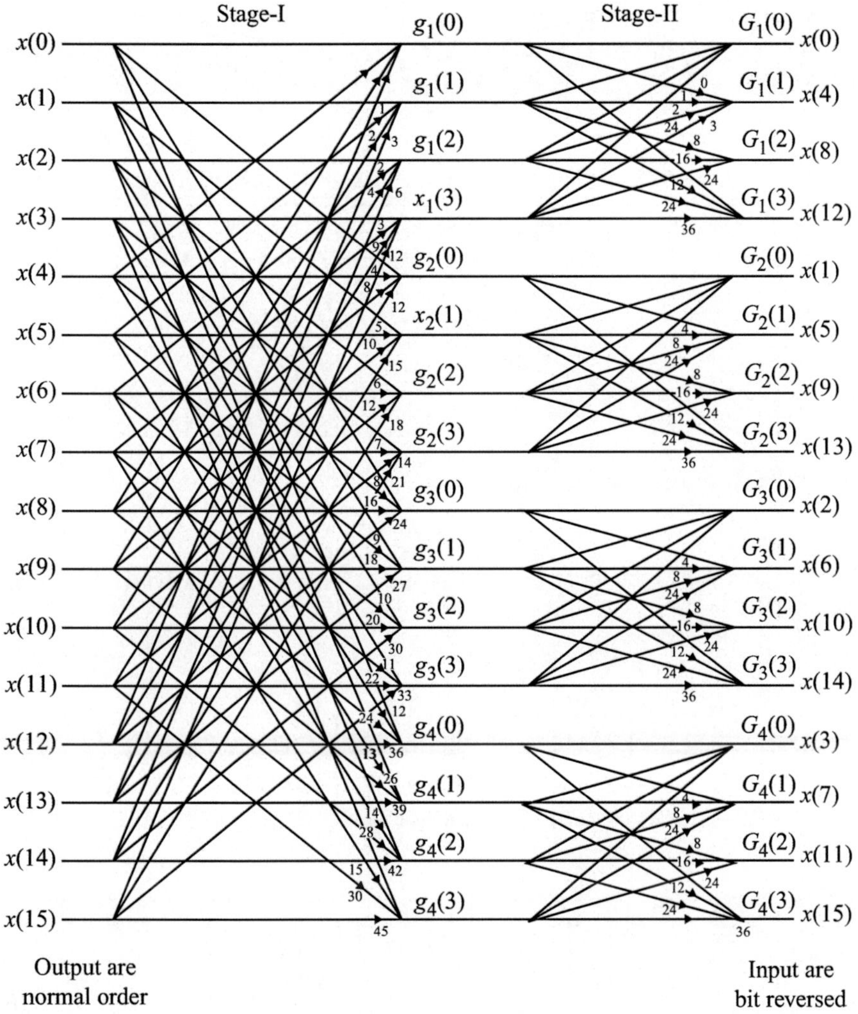

Fig. 3.20 16-Point radix-4 DIF FFT butterfly diagram

$$X(1) = \sum_{n=0}^{3} x(n) W_N^{n} = x(0)W_4^0 - jx(1)W_4^0 - x(2)W_4^0 + jx(3)W_4^0 \tag{3.116}$$

$$X(2) = \sum_{n=0}^{3} x(n) W_N^{2n} = x(0)W_4^0 - x(1)W_4^0 + x(2)W_4^0 - x(3)W_4^0 \tag{3.117}$$

$$X(3) = \sum_{n=0}^{3} x(n) W_N^{3n} = x(0)W_4^0 + jx(1)W_4^0 - x(2)W_4^0 - jx(3)W_4^0 \tag{3.118}$$

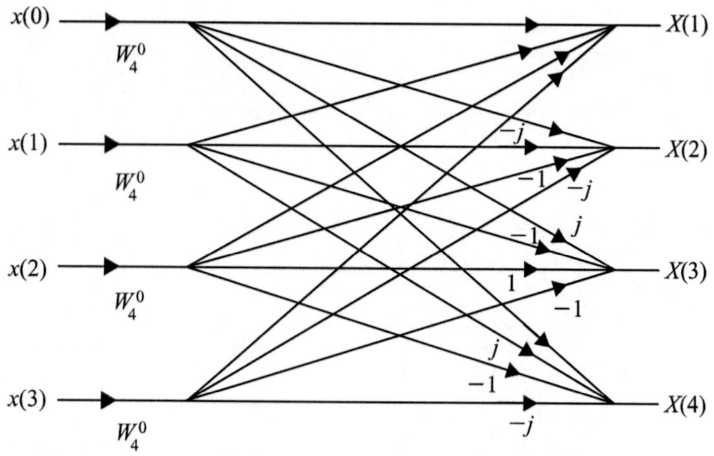

Fig. 3.21 4-Point radix-4 FFT butterfly diagram

4-point radix-4 FFT butterfly diagram is shown in Fig. 3.21. In matrix form, the 4-point DFT is written as

$$\begin{bmatrix} X(0) \\ X(1) \\ X(2) \\ X(3) \end{bmatrix} = \begin{bmatrix} 1 & 1 & 1 & 1 \\ 1 & W_4^1 & W_4^2 & W_4^3 \\ 1 & W_4^2 & W_4^4 & W_4^6 \\ 1 & W_4^3 & W_4^6 & W_4^9 \end{bmatrix} \begin{bmatrix} x(0) \\ x(1) \\ x(2) \\ x(3) \end{bmatrix} \tag{3.119}$$

or

$$\begin{bmatrix} X(0) \\ X(1) \\ X(2) \\ X(3) \end{bmatrix} = \begin{bmatrix} 1 & 1 & 1 & 1 \\ 1 & -j & -1 & j \\ 1 & -1 & 1 & -1 \\ 1 & j & -1 & -j \end{bmatrix} \begin{bmatrix} W_4^0 x(0) \\ W_4^0 x(1) \\ W_4^0 x(2) \\ W_4^0 x(3) \end{bmatrix} \tag{3.120}$$

Therefore, in general, the four $N/4$ point DFTs of $X(k)$ in matrix form are

$$\begin{bmatrix} X(0,q) \\ X(1,q) \\ X(2,q) \\ X(3,q) \end{bmatrix} = \begin{bmatrix} 1 & 1 & 1 & 1 \\ 1 & -j & -1 & j \\ 1 & -1 & 1 & -1 \\ 1 & j & -1 & -j \end{bmatrix} \begin{bmatrix} W_N^0 F(0,q) \\ W_N^q F(1,q) \\ W_N^{2q} F(2,q) \\ W_N^{3q} F(3,q) \end{bmatrix} \tag{3.121}$$

The basic butterfly computation in radix-4 FFT algorithm is shown in Fig. 3.22.

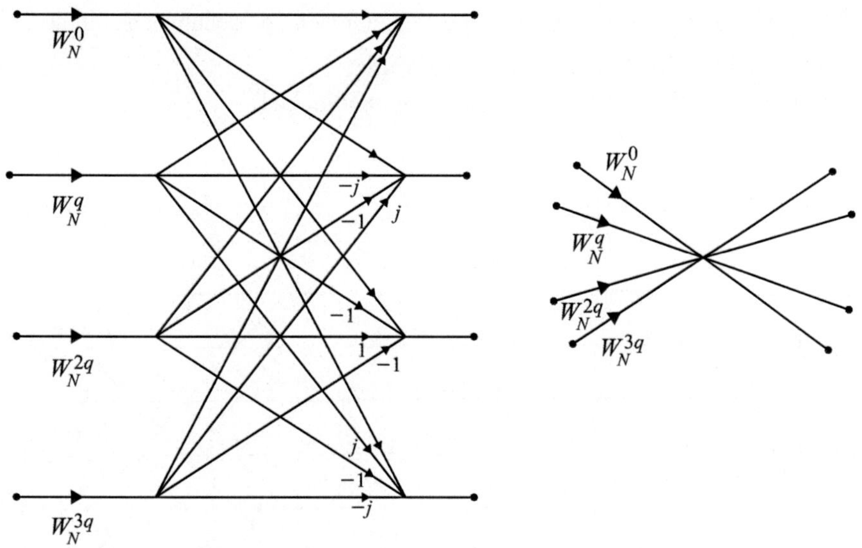

Fig. 3.22 Basic butterfly in a radix-4 FFT algorithm

3.6.3 Computation of IDFT Through FFT

The IDFT of an N-point sequence $X(k)$ is defined as

$$\begin{aligned} x(n) &= \frac{1}{N}\sum_{k=0}^{N-1} X(k)e^{j\frac{2\pi kn}{N}} \\ &= \frac{1}{N}\sum_{k=0}^{N-1} X(k)W_N^{-nk} \qquad \left[\because\ W_N = e^{-j\frac{2\pi}{N}}\right] \end{aligned}$$

Taking the conjugate and multiplying by N, we get

$$\begin{aligned} Nx^*(n) &= \sum_{k=0}^{N-1} X^*(k)W_N^{nk} \\ x^*(n) &= \frac{1}{N}\sum_{k=0}^{N-1} X^*(k)W_N^{nk} \\ x(n) &= \frac{1}{N}\left[\sum_{k=0}^{N-1} X^*(k)W_N^{nk}\right]^* \end{aligned} \tag{3.122}$$

where $\sum_{k=0}^{N-1} X^*(k)W_N^{nk}$ is an N-point DFT of $X^*(k)$.

Hence, in order to compute IDFT of $X(k)$ the following procedure is be followed:

1. Take conjugate of $X(k)$ [i.e. determine $X^*(k)$].
2. Compute the N-point DFT of $X^*(k)$ using radix-2 FFT.
3. Take conjugate of the output sequence of FFT.
4. Divide the sequence obtained in step 3 by N. The resultant sequence is $x(n)$.

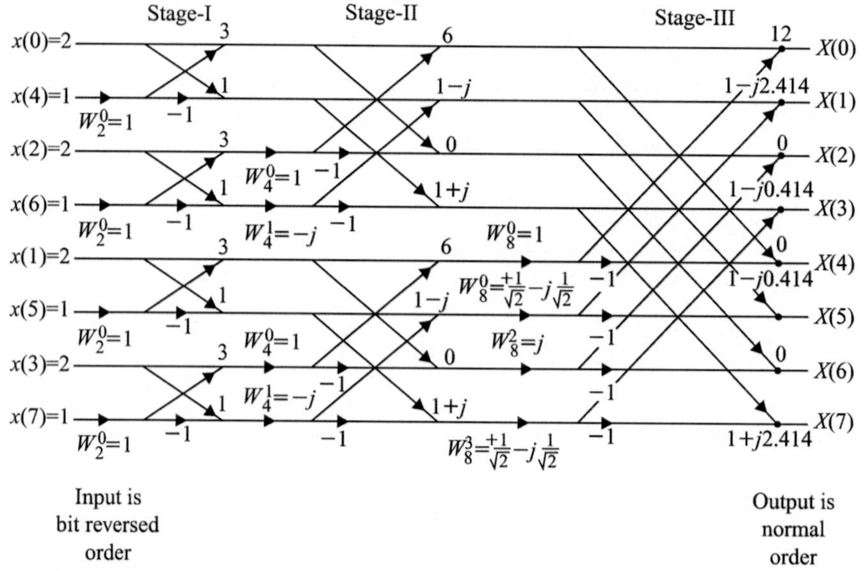

Fig. 3.23 Butterfly diagram of Example 3.31(i)

Butterfly diagram of Example 3.31 (i) shown in Fig. 3.23.

Example 3.31 An 8-point sequence is given by $x(n) = \{2, 2, 2, 2, 1, 1, 1, 1\}$. Compute 8-point DFT of $x(n)$ by (i) radix-2 DIT FFT and (ii) radix-2 DIF-FFT. Also sketch the magnitude and phase spectrum.

Solution Given (Figs. 3.23, 3.24 and 3.25)

$$\begin{array}{rcccccccc} x(n) = \{2, & 2, & 2, & 2, & 1, & 1, & 1, & 1\} \\ x(0) & x(1) & x(2) & x(3) & x(4) & x(5) & x(6) & x(7) \end{array}$$

$$W_2^0 = 1$$
$$W_4^0 = 1$$
$$W_4^1 = e^{\frac{-j2\pi}{4}} = -j$$
$$W_8^0 = 1$$

$$W_8^1 = e^{\frac{-j2\pi}{8}} = \frac{1}{\sqrt{2}} - j\frac{1}{\sqrt{2}}$$

$$W_8^2 = e^{\frac{-j2\pi \times 2}{8}} = -j$$

$$W_8^3 = e^{\frac{-j2\pi \times 3}{8}} = -\frac{1}{\sqrt{2}} - j\frac{1}{\sqrt{2}}$$

(*i*) DIT FFT

$$X(k) = \{12, -j2.414, 0, 1 - j0.414, 0, 1 + j0.414, 0, 1 + j2.414\}$$

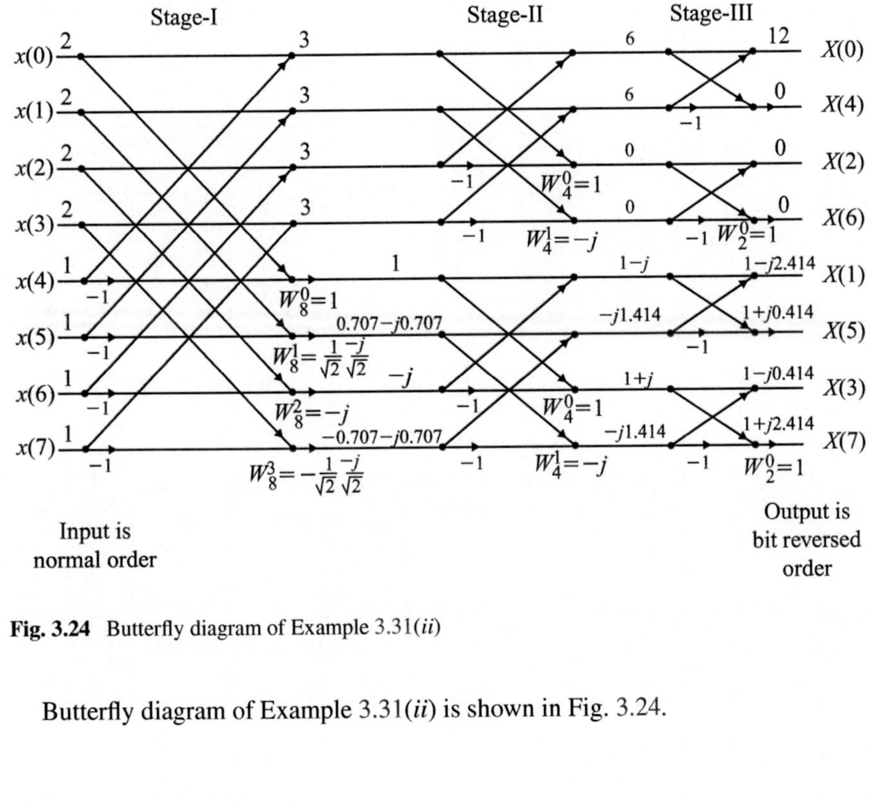

Fig. 3.24 Butterfly diagram of Example 3.31(*ii*)

Butterfly diagram of Example 3.31(*ii*) is shown in Fig. 3.24.

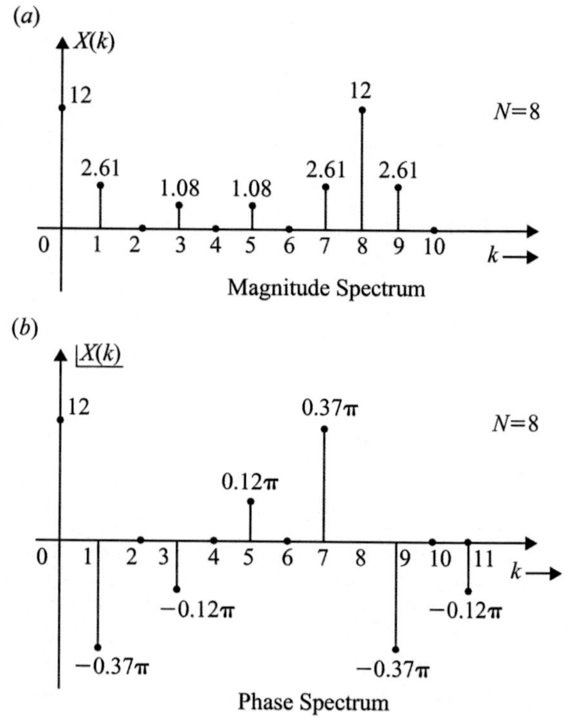

Fig. 3.25 Magnitude and phase spectrum of Example 3.31

Magnitude and phase spectrums of Example 3.31 are shown in Fig. 3.25a and 3.25b respectively.

Input	Stage-I	Stage-II	Stage-III	Output
2	$2+1=3$	$3+3=6$	$6+6=12$	$X(0)$
1	$2-1=1$	$1+(-j)=1-j$	$1-j+(\frac{1}{\sqrt{2}}-j\frac{1}{\sqrt{2}})(1-j)=1-j2.414$	$X(1)$
2	$2+1=3$	$3-3=0$	$0+0=0$	$X(2)$
1	$2-1=1$	$1+(-1)(-j)=1+j$	$1+j+(\frac{1}{\sqrt{2}}-\frac{j}{\sqrt{2}})(1+j)=1-j0.414$	$X(3)$
2	$2+1=3$	$3+3=6$	$6-6=0$	$X(4)$
1	$2-1=1$	$1+(-j)=1-j$	$1-j-(\frac{1}{\sqrt{2}}-\frac{j}{\sqrt{2}})(1-j)=1-j0.414$	$X(5)$
2	$2+1=3$	$3-3=0$	$0-0=0$	$X(6)$
1	$2-1=1$	$1+(-1)(-j)=1+j$	$1+j-(\frac{1}{\sqrt{2}}-\frac{j}{\sqrt{2}})(1+j)=1+j2.414$	$X(7)$

(*ii*) **DIF-FFT**

$$X(k)=\{12, 1-j2.414, 0, 1-j0.414, 0, 1+j0.414, 0, 1+j2.414\}$$

$$|X(k)|=\{12, 2.61, 0, 1.08, 0, 1.08, 0, 2.61\}$$

$$\angle X(k)=\{0, -0.37\pi, 0, -0.12\pi, 0, 0.12\pi, 0, 0.37\pi\}$$

Input	Stage-I	Stage-II	Stage-III	Output
2	$2+1=3$	$3+3=6$	$6+6=12$	$X(0)$
2	$2+1=3$	$3+3=6$	$6-6=0$	$X(4)$
2	$2+1=3$	$(3-3)(j)=0$	$0+0=0$	$X(2)$
2	$2+1=3$	$(3-3)(-j)=0$	$0-0=0$	$X(6)$
1	$(2-1)1=1$	$1-j$	$1-j-j1.414=$ $1-j2.414$	$X(1)$
1	$(2-1)(\frac{1}{\sqrt{2}}-\frac{j}{\sqrt{2}})=$ $\frac{1}{\sqrt{2}}-\frac{j}{\sqrt{2}}$	$\frac{1}{\sqrt{2}}-\frac{j}{\sqrt{2}}-\frac{1}{\sqrt{2}}-\frac{j}{\sqrt{2}}=$ $-j1.414$	$1-j+j1.414=$ $1+j0.414$	$X(5)$
1	$(2-1)(-j)=-j$	$1-(-j)=1+j$	$1+j-j1.414=$ $1-j0.414j$	$X(3)$
1	$(2-1)(\frac{-1}{\sqrt{2}}-\frac{-j}{\sqrt{2}})=$ $\frac{-1}{\sqrt{2}}-\frac{j}{\sqrt{2}}$	$(\frac{1}{\sqrt{2}}-\frac{j}{\sqrt{2}}+\frac{1}{\sqrt{2}}+\frac{j}{\sqrt{2}})-j$ $=-j1.414$	$1+j+j1.414=$ $1+j2.414$	$X(7)$

Example 3.32 In an LTI system, the input $x(n)=\{1,\ 1,\ 1\}$ and the impulse response $h(n)=\{-1,\ -1\}$. Determine the response of the LTI system by radix-2 DIT FFT.

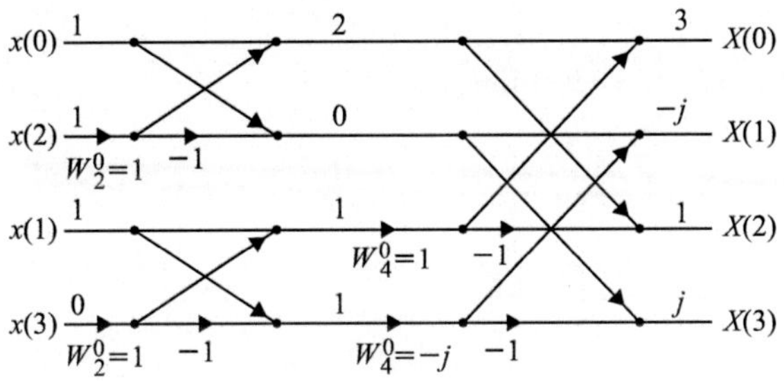

Fig. 3.26 Butterfly diagram of $X(k)$

Butterfly diagram for Example 3.32 is shown in Fig. 3.26.

Solution The response of LTI system is given by linear convolution of input $x(n)$ and impulse response $h(n)$. That is

$$y(n)=x(n)*h(n)$$

where $N_1=3$ and $N_2=2$. The length of output sequence $y(n)$ is $N=N_1+N_2-1=3+2-1=4$. Therefore, the given sequence $x(n)$ and $h(n)$ are converted into 4-point sequences by appending zeros (Fig. 3.26):

$$x(n)=\{1,\ 1,\ 1,\ 0\} \quad \text{and} \quad h(n)=\{-1,\ -1,\ 0,\ 0\}$$
$$y(n)=x(n)*h(n)$$
$$Y(k)=X(k)*H(k)$$
$$y(n)=\text{IDFT}\ \{Y(k)\}$$

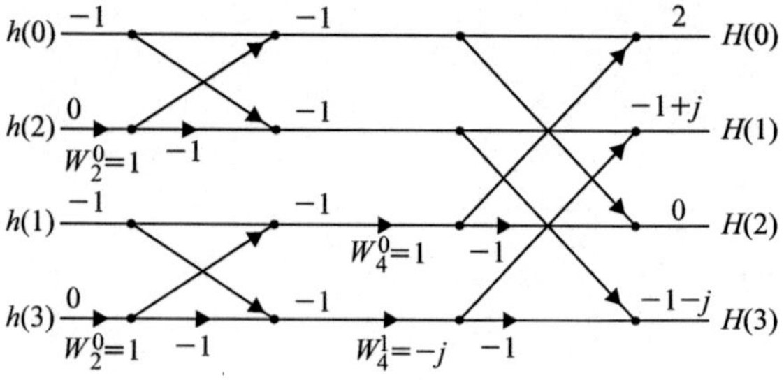

Fig. 3.27 Butterfly diagram of $H(k)$

$$y(n) = \text{IDFT}\,\{X(k)H(k)\}$$

Step 1: To Determine $X(k)$

$$x(n) = \{1,\ 1,\ 1,\ 0\}$$

Therefore, by referring to the butterfly diagram, we get

$$X(k) = \{3,\ -j,\ 1,\ j\}$$

Step 2: To Determine $H(k)$

$$h(n) = \{-1,\ -1,\ 0,\ 0\}$$

Therefore, by referring to Fig. 3.27, we get

$$H(k) = \{2,\ -1 + j,\ 0,\ -1 - j\}$$

Step 3: To Determine the Product $X(k)H(k)$

Let

$$Y(k) = X(k)H(k)$$

where

$$\begin{aligned}
k = 0,\quad & Y(0) = X(0)H(0) = -6\\
k = 1,\quad & Y(1) = X(1)H(1) = 1 + j\\
k = 2,\quad & Y(2) = X(2)H(2) = 0\\
k = 3,\quad & Y(3) = X(3)H(3) = 1 - j
\end{aligned}$$

Therefore

$$Y(k) = \{-6,\ 1 + j,\ 0,\ 1 - j\}$$

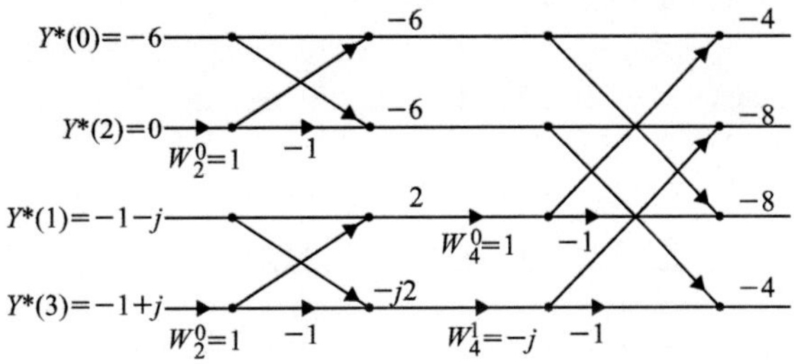

Fig. 3.28 Butterfly diagram of $Y^*(k)$

Step 4: To Determine IDFT of $Y(k)$

The IDFT of $Y(k)$ is given by

$$y(n) = \frac{1}{N}\left[\sum_{k=0}^{N-1} Y^*(k) W_N^{nk}\right]^*$$

$$\therefore\ Y^*(k) = \{-6,\ 1-j,\ 0,\ 1+j\}$$

The 4-point DFT of $Y^*(k)$ is

Therefore, by referring to Fig. 3.28, we get

$$\begin{aligned} y(n) &= \frac{1}{4}\{4 \text{ point DFT } Y^*(k)\}^* \\ &= \frac{1}{4}\{-4,\ -8,\ -8,\ -4\}^* \\ &= \frac{1}{4}\{-4,\ -8,\ -8,\ -4\} \\ y(n) &= \{-1,\ -2,\ -2,\ -1\} \end{aligned}$$

Example 3.33 Find the DFT of sequence

$$x(n) = \{1,\ 2,\ 3,\ 4,\ 4,\ 3,\ 2,\ 1\}$$

using DIT and DIF algorithms.

Solution The twiddle factors are

$$\begin{aligned} W_2^0 &= 1, \quad W_4^0 = 1, \quad W_4^1 = -j \\ W_8^0 &= 0, \quad W_8^1 = e^{-j\frac{2\pi}{8}} = \frac{1}{\sqrt{2}} - \frac{j}{\sqrt{2}} \\ W_8^2 &= e^{-j\frac{2\pi\times 2}{8}} = -j \\ W_8^3 &= e^{-j\frac{2\pi\times 3}{8}} \\ &= -\frac{1}{\sqrt{2}} - \frac{j}{\sqrt{2}} \end{aligned}$$

(*i*) Using DIT Algorithm

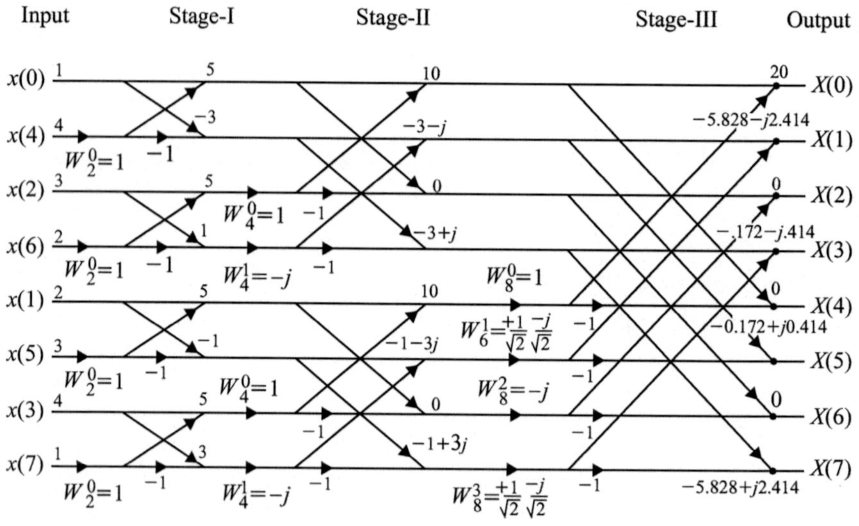

Fig. 3.29 Butterfly diagram of Example 3.33(*i*)

The butterfly diagram is shown in Fig. 3.29.

(*i*) Using DIT Algorithm

Input	Stage-I	Stage-II	Output
1	$1+4=5$	$5+5=10$	$10+10=20$
4	$1-4=-3$	$-3-j$	$-3-j+(-1-3j)(\frac{1}{\sqrt{2}}-\frac{j}{\sqrt{2}})$ $=-5.828-j2.414$
3	$3+2=5$	$5-5=0$	$0+0=0$
2	$3-2=1$	$-3+j$	$-3+j+(-1+3j)(\frac{-1}{\sqrt{2}}-\frac{j}{\sqrt{2}})=$ $-0.172-j0.414$
2	$2+3=5$	$5+5=10$	$10-10=0$
3	$2-3=-1$	$-1+(3)(-j)=-1-3j$	$-3-j-(-1-3j)(\frac{1}{\sqrt{2}}-\frac{j}{\sqrt{2}})=$ $-0.172+j0.414$
4	$4+1=5$	$5-5=0$	$0+0=0$
1	$4-1=3$	$-1+(3)(-j)(-1)=-1+3j$	$-3+j-(-1+3j)(\frac{-1}{\sqrt{2}}-\frac{j}{\sqrt{2}})=$ $-5.828+j2.414$

$X(k)=$
$\{20,\ -5.828-j2.414,\ 0,\ -0.172-j0.414,\ 0,\ -0.172+j0.414,\ 0,\ -5.828+j2.414\}$

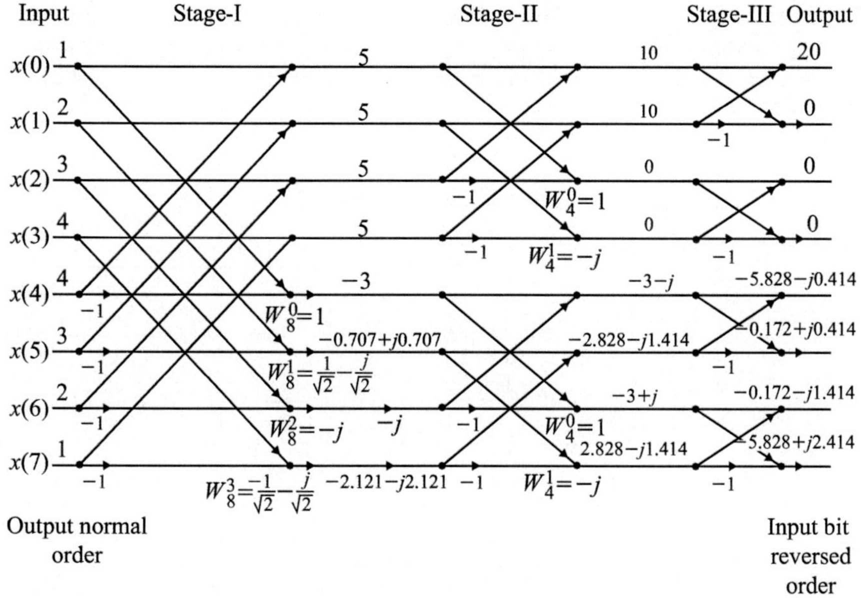

Fig. 3.30 Butterfly diagram of Example 3.33(*ii*)

(*ii*) Using DIF Algorithm

$X(k) =$

$\{20,\ -5.828-j2.414,\ 0,\ -0.172vj0.414,\ 0,\ -0.172+j0.414,\ 0,\ -5.828+j2.414\}$

Butterfly diagram using DIF algorithm of Example 3.33(*ii*) is shown in Fig. 3.30.

Input	Stage-I	Stage-II	Output
1	$1+4=5$	$5+5=10$	$10+10=20$
2	$2+3=5$	$5+5=10$	$10-10=0$
3	$3+2=5$	$(5-5)(1)=0$	$0+0=0$
4	$4+1=5$	$(5-5)(-j)=0$	$0-0=0$
4	$(1-4)1=-3$	$-3-j$	$-3-j-2.828-j1.414$ $=-5.828-j0.414$
3	$(2-3)(\frac{1}{\sqrt{2}}-\frac{j}{\sqrt{2}})$ $=-0.707+j0.707$	$-0.707+j0.707-2.121-j2.121$ $=-2.828-j1.414$	$-3-j+2.828+j1.414$ $=-0.172+j0.414$
2	$(3-2)(-j)=-j$	$-3+j$	$-3+j+2.828-j1.414$ $=-0.172-j0.414$
1	$(4-1)(\frac{-1}{\sqrt{2}}-\frac{j}{\sqrt{2}})$ $=-2.121-j2.121$	$-0.707+j0.707-(-2.121-j2.121)$ $(-j)=2.828-j1.414$	$-3+j-2.828+j1.414$ $=-5.828+j2.414$

Example 3.34 Compute 4-point sequence of

$$x(n) = \{0,\ 1,\ 2,\ 3\}$$

using the DIT algorithm.

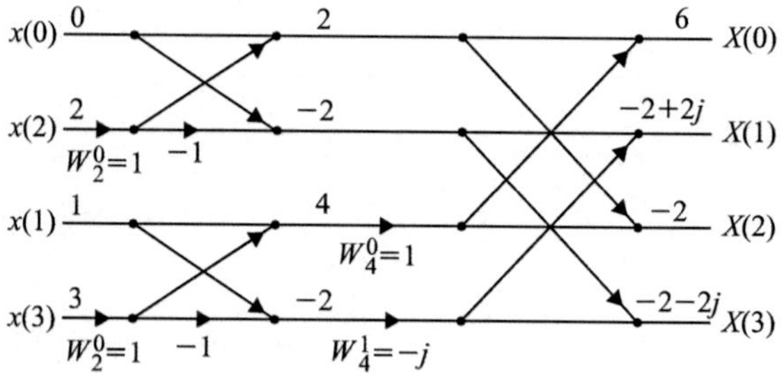

Fig. 3.31 Butterfly diagram of $X(k)$ for Example 3.34

The Butterfly diagram DIT algorithm for Example 3.34 is shown in Fig. 3.31.

Solution The twiddle factors are

$$W_2^0 = 1, \quad W_4^0 = 1, \quad W_4^1 = -j$$

$$\boxed{X(k) = \{6,\ -2+2j,\ -2,\ -2-2j\}}$$

Example 3.35 Find the IDFT of the sequence

$$X(k) = \{4,\ 1 - j2.414,\ 0,\ 1 - j0.414,\ 0,\ 1 + j0.414,\ 0,\ 1 + j2.414\}$$

using the DIT algorithm.

Solution

$$X^*(k) = \{4,\ 1 + j2.414,\ 0,\ 1 + j0.414,\ 0,\ 1 - j0.414,\ 0,\ 1 - j2.414\}$$

The 8-point butterfly diagram of $X^*(k)$ is shown in Fig. 3.32.

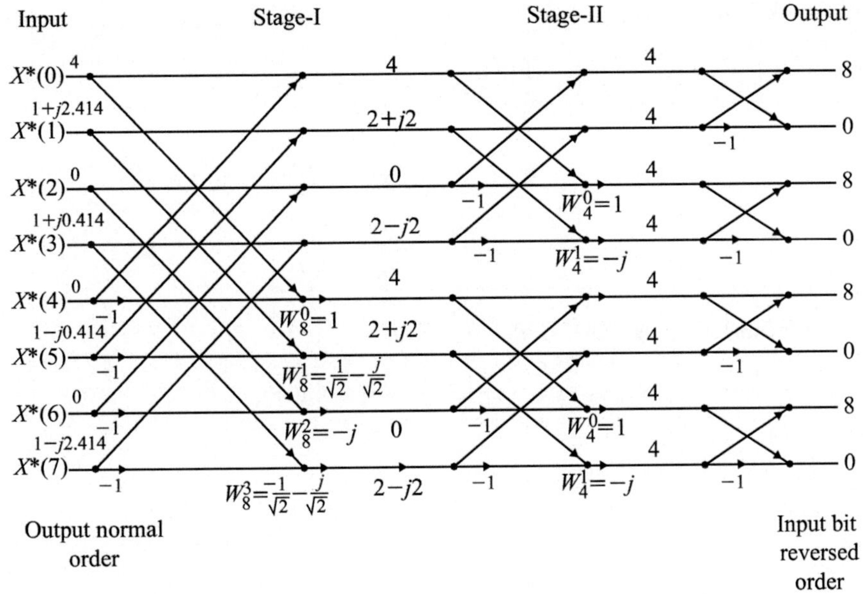

Fig. 3.32 Butterfly diagram of $X^*(k)$

Input	Stage-I	Stage-II	Output
4	$4+0=4$	$4+0=4$	$4+4=8$
$1+j2.414$	$1+j2.414+1-j0.414=2+j2$	$2+j2+2-j2=4$	$4-4=0$
0	$0+0=0$	$4-0=4$	$4+4=8$
$1+j0.414$	$1+j0.414+1-j2.414=2-j2$	$2+j2-(2-j2)(-j)=4$	$4-4=0$
0	$4-0=4$	$4+0=4$	$4+4=8$
$1-j0.414$	$1+j2.414-(1-j0.414)\left(\frac{1}{\sqrt{2}}-\frac{j}{\sqrt{2}}\right)$ $=2+j2$	$2+j2+2-j2=4$	$4-4=0$
0	$0-0=0$	$4-0=4$	$4+4=8$
$1-j2.414$	$1-j0.414-(1-j2.414)\left(\frac{-1}{\sqrt{2}}-\frac{j}{\sqrt{2}}\right)$ $=2-j2$	$2+j2+2-(2-j2)(-j)=4$	$4-4=0$

8-point DFT of $X^*(k)$ is

$$x(n) = \frac{1}{8}\{8 \text{ point DFT of } X^*(k)\}^*$$
$$= \frac{1}{8}\{8,\ 8,\ 8,\ 8,\ 0,\ 0,\ 0,\ 0\}^*$$

$$\boxed{x(n) = \{1,\ 1,\ 1,\ 1,\ 0,\ 0,\ 0,\ 0\}}$$

Example 3.36 Find the IDFT of the sequence

$$X(k) = \{10,\ -2 + j2,\ -2,\ -2 - j2\}$$

using the DIT algorithm.

Solution Given

$$X(k) = \{10,\ -2 + j2,\ -2,\ -2 - j2\}$$
$$X^*(k) = \{10,\ -2 - j2,\ -2,\ -2 + j2\}$$

The butterfly diagram of $X^*(k)$ is shown in Fig. 3.33.

4-point DFT of $X^*(k) = \{4,\ 12,\ 8,\ 16\}$

$$x(n) = \frac{1}{4}\{4 \text{ point DFT of } X^*(k)\}^*$$
$$= \frac{1}{4}\{4,\ 8,\ 12,\ 16\}^*$$

$$\boxed{x(n) = \{1,\ 2,\ 3,\ 4\}}$$

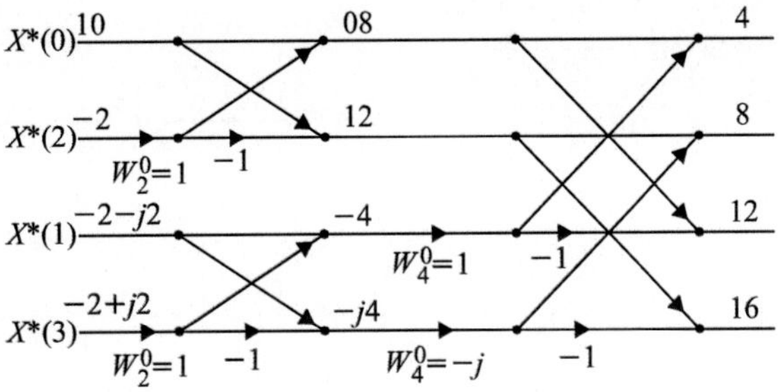

Fig. 3.33 Butterfly diagram of $X^*(k)$

Example 3.37 Compute the DFT of the sequence

$$x(n) = \cos\left(\frac{n\pi}{2}\right)$$

where $N = 4$, using the DIF FFT algorithm.

Solution Given

$$x(n) = \cos\left(\frac{n\pi}{2}\right), \qquad N = 4$$
$$x(0) = 1$$
$$x(1) = 0$$
$$x(2) = -1$$
$$x(3) = 0$$

Therefore,

$$x(n) = \{1, 0, -1, 0\}$$

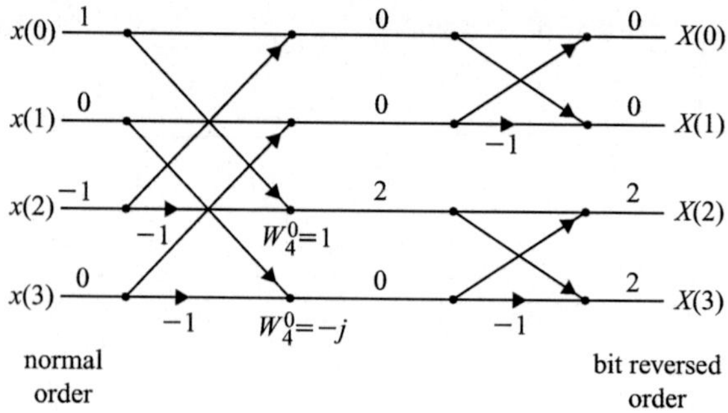

Fig. 3.34 Butterfly diagram of $X(k)$ for Example 3.37

Therefore

$$\boxed{X(k) = \{0,\ 2,\ 0,\ 2\}}$$

Butterfly diagram for Example 3.37 is shown in Fig. 3.34.

Example 3.38 Using the decimation in the frequency FFT flow graph, compute the DFT of

$$x[n] = \cos\left(\frac{n\pi}{4}\right) \qquad \text{for} \qquad 0 \le n \le 7$$

Solution Given

$$x[n] = \cos\left(\frac{n\pi}{4}\right) \qquad \text{for} \qquad 0 \le n \le 7$$
$$x[n] = \{1,\ 0.707,\ 0,\ -0.707,\ -1,\ -0.07,\ 0,\ 0.707\}$$

The butterfly diagram is shown in Fig. 3.35. From Fig. 3.35, $X(k)$ is obtained as

$$\boxed{X(k) = \{0,\ 4,\ 0,\ 0,\ 0,\ 0,\ 0,\ 4\}}$$

Example 3.39 Draw the butterfly diagram using 8-point DIT FFT for the following sequence:

$$x[n] = \{1, 0, 0, 0, 0, 0, 0, 0\}$$

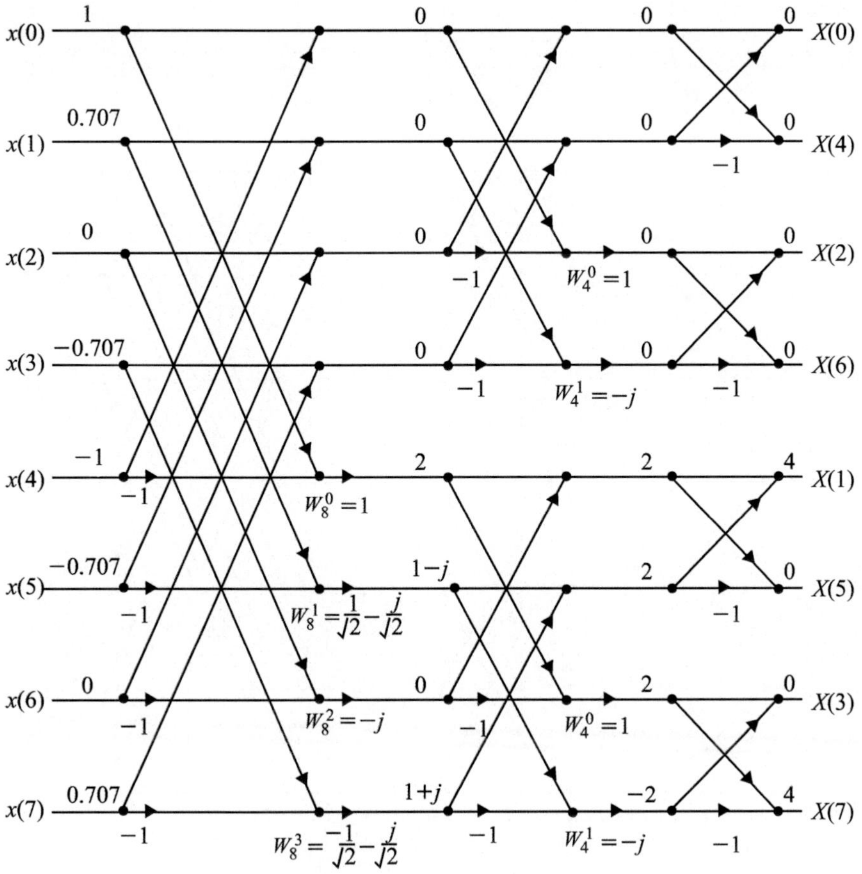

Fig. 3.35 Butterfly diagram for Example 3.38

Solution The butterfly diagram for Example 3.39 is shown in Fig. 3.36.

Input	Stage-I	Stage-II	Output
1	$1+0=1$	$1+0=1$	$1+0=1$
0	$0+1=1$	$1+0=1$	$1+0=1$
0	$0+0=0$	$0+1=1$	$1+0=1$
0	$0+0=0$	$0+1=1$	$1+0=1$
0	$0+0=0$	$0+0=0$	$0+1=1$
0	$0+0=0$	$0+0=0$	$0+1=1$
0	$0+0=0$	$0+0=0$	$0+1=1$
0	$0+0=0$	$0+0=0$	$0+1=1$

$$\boxed{X(k) = \{1,\ 1,\ 1,\ 1,\ 1,\ 1,\ 1,\ 1\}}$$

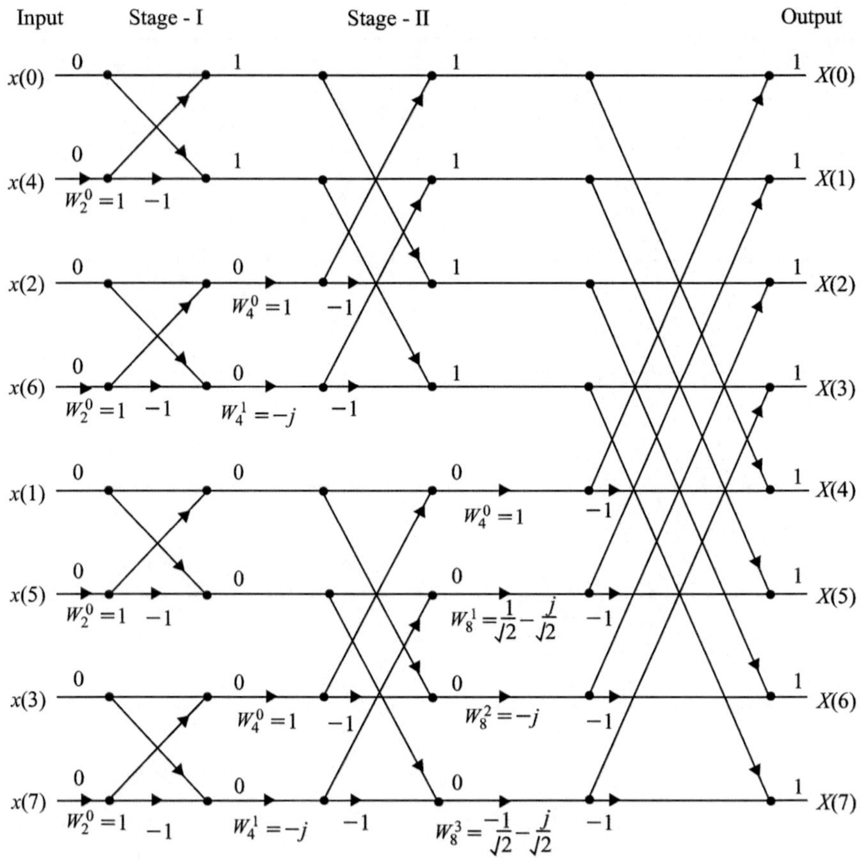

Fig. 3.36 Butterfly diagram for Example 3.39

Example 3.40 Compute the 8-point DFT of the sequence

$$x[n] = \{0.5, 0.5, 0.5, 0.5, 0, 0, 0, 0\}$$

using the in-place radix-2 DIF-FFT algorithm.

Solution The twiddle factors are

$$\begin{aligned} W_8^0 &= 1, \\ W_8^1 &= \frac{1}{\sqrt{2}} - \frac{j}{\sqrt{2}} \\ W_8^2 &= -j \\ W_8^3 &= -\frac{-1}{\sqrt{2}} - \frac{j}{\sqrt{2}} \end{aligned}$$

$$W_4^0 = 1$$
$$W_4^1 = -j$$
$$W_2^0 = 1$$

The butterfly diagram is shown in Fig. 3.37.

Input	Stage-I	Stage-II	Output
0.5	$0.5 + 0 = 0.5$	$0.5 + 0.5 = 1$	$1 + 1 = 2$
0.5	$0.5 + 0 = 0.5$	$0.5 + 0.5 = 1$	$1 - 1 = 0$
0.5	$0.5 + 0 = 0.5$	$(0.5 - 0.5)1 = 0$	$0 + 0 = 0$
0.5	$0.5 + 0 = 0.5$	$(0.5 - 0.5)(-j) = 0$	$0 - 0 = 0$
0	$(0.5 + 0)1 = 0.5$	$0.5 - 0.5j$	$0.5 - 0.5j - 0.707j$ $= 0.5 - 1.207j$
0	$(0.5 + 0)\left(\frac{1}{\sqrt{2}} - \frac{j}{\sqrt{2}}\right)$ $= 0.3535 - j0.3525$	$(0.3525 - j0.3525)$ $+(-0.3525 - j0.3525) = -0.707j$	$0.5 - 0.5j + 0.707j$ $= 0.5 + 0.207j$
0	$(0.5 + 0)(-j) = -0.5j$	$0.5 + 0.5j$	$0.5 + 0.5j - 0.707j$ $= 0.5 - 0.207j$
0	$(0.5 + 0)\left(\frac{-1}{\sqrt{2}} - \frac{j}{\sqrt{2}}\right)$ $= -0.3535 - j0.3525$	$[(0.3525 - j0.3525)$ $-(-0.3525 - j0.3525)](-j)$ $= -0.707j$	$0.5 + 0.5j + 0.707j$ $= 0.5 + 1.207j$

$$\boxed{X(k) = \{2,\ 0.5 - 1.207j,\ 0,\ 0.5 - 0.207j,\ 0,\ 0.5 + 0.207j,\ 0,\ 0.5 + 1.207j\}}$$

Example 3.41 Find the 8-point DFT of the given sequence

$$x[n] = \{0, 1, 2, 3, 4, 5, 6, 7\}$$

using the DIF radix-2 FFT algorithm.

Solution

Input	Stage-I	Stage-II	Output
0	$0 + 4 = 4$	$8 + 4 = 12$	$12 + 16 = 28$
1	$1 + 5 = 6$	$6 + 10 = 16$	$12 - 16 = -4$
2	$2 + 6 = 8$	$4 - 8 = -4$	$-4 + 4j$
3	$3 + 7 = 10$	$(6 - 10) - j = 4j$	$-4 - 4j$
4	$0 - 4 = -4$	$-4 + 4j$	$-4 + 4j + \frac{8j}{\sqrt{2}}$ $= -4 + 9.656j$
5	$(1 - 5)\left(\frac{1}{\sqrt{2}} - \frac{j}{\sqrt{2}}\right)$ $= \frac{-4}{\sqrt{2}} + \frac{4j}{\sqrt{2}}$	$\frac{-4}{\sqrt{2}} + \frac{4j}{\sqrt{2}} + \frac{4}{\sqrt{2}} + \frac{4j}{\sqrt{2}} = \frac{8j}{\sqrt{2}}$	$-4 + 4j - \frac{8j}{\sqrt{2}}$ $= -4 - 1.656j$
6	$(2 - 6)(-j) = 4j$	$-4 - 4j$	$-4 + 4j + \frac{8j}{\sqrt{2}}$ $= -4 + 1.656j$
7	$(3 - 7)\left(-\frac{1}{\sqrt{2}} - \frac{j}{\sqrt{2}}\right)$ $= \frac{4}{\sqrt{2}} + \frac{4j}{\sqrt{2}}$	$\left(-\frac{4}{\sqrt{2}} + \frac{4j}{\sqrt{2}} - \frac{4}{\sqrt{2}} - \frac{4j}{\sqrt{2}}\right)$ $-j = \frac{8j}{\sqrt{2}}$	$-4 - 4j - \frac{8j}{\sqrt{2}}$ $= -4 - 9.656j$

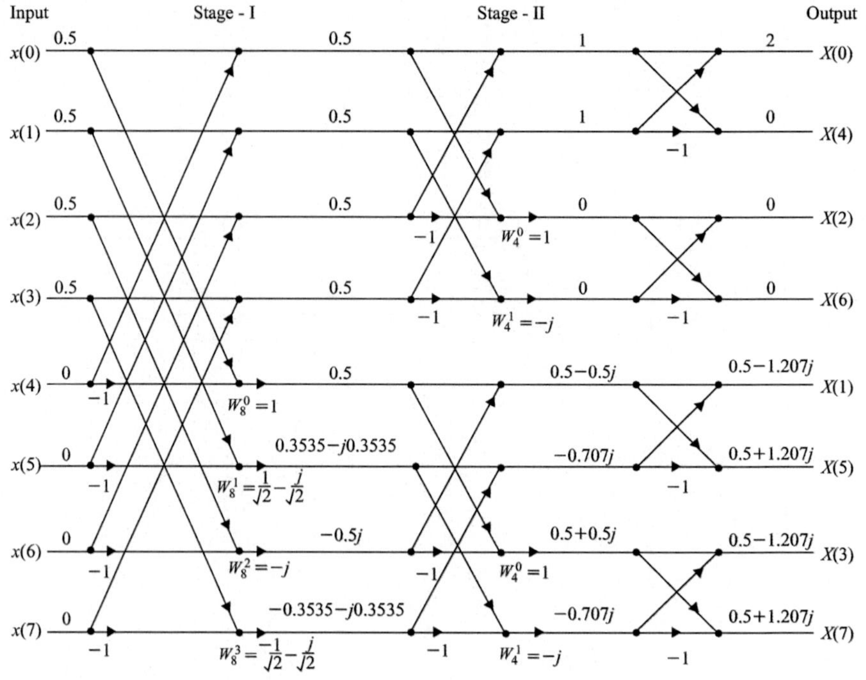

Fig. 3.37 Butterfly diagram for Example 3.40

$$X(k) = \Big\{28, -4 + 9.656j, -4 + 4j, -4 + 1.656j, -4, -4 - 1.656j, -4 - 4j, -4 - 9.656j\Big\}$$

The butterfly diagram is shown in Fig. 3.38.

Example 3.42 Find the response of an LTI system with impulse response $h(n) = \{2, 1, 3\}$ for the input $x(n) = \{1, 2\}$ using the DIT radix-2 FFT algorithm.

Solution The response of the LTI system is given by the linear convolution of input $x(n)$ and impulse $h(n)$ i.e.

$$y(n) = x(n) * h(n)$$

$N_1 = 2$ and $N_2 = 3$. The length of output sequence $y(n)$ is $N = N_1 + N_2 - 1 = 2 + 3 - 1 = 4$. Therefore, the given sequence $x(n)$ and $h(n)$ are converted into a 4-point sequence by appending zeros:

$$x(n) = \{1, 2, 0, 0\} \quad \text{and} \quad h(n) = \{2, 1, 3, 0\}$$

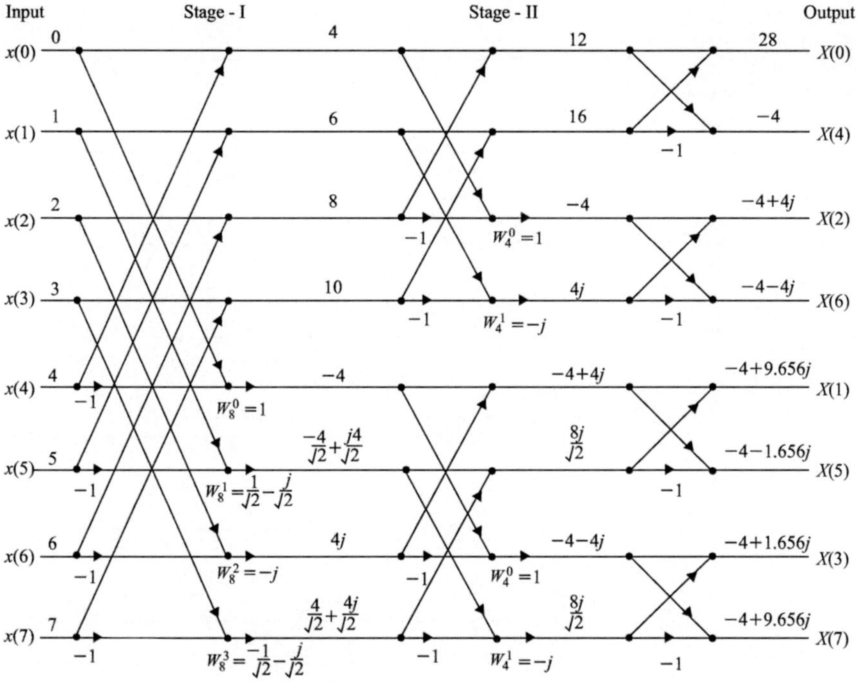

Fig. 3.38 Butterfly diagram for Example 3.41

$$y(n) = \text{IDFT}\{X(k)H(k)\}$$

Butterfly diagram to determine $X(k)$ is shown in Fig. 3.39a. Butterfly diagram to determine $H(k)$ is shown in Fig. 3.39b. Butterfly diagram to determine $Y(k)$ is shown in Fig. 3.39c for Example 3.42.

Step 1: To determine $X(k)$ (Refer to Fig. 3.39a)

$$x(n) = \{1, 2, 0, 0\}$$

$$X(k) = \{3, 1 - 2j, -1, 1 + 2j\}$$

Step 2: To determine $H(k)$ (Refer to Fig. 3.39b)

$$h(n) = \{2, 1, 3, 0\}$$

$$H(k) = \{6, -1 - j, 4, -1 + j\}$$

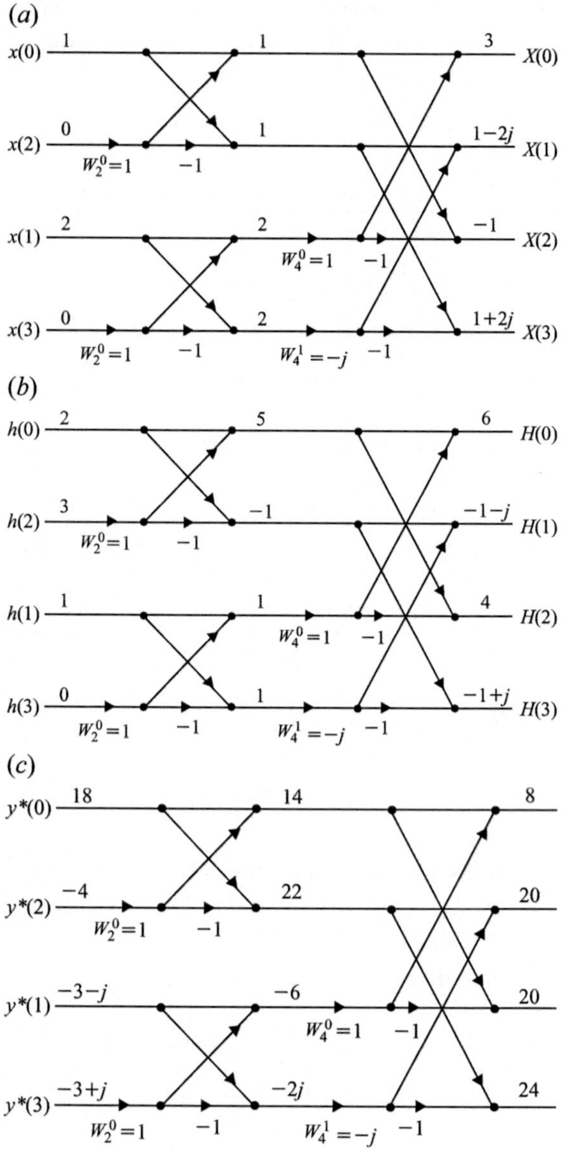

Fig. 3.39 **a** Butterfly diagram for Example 3.42. **b** Butterfly diagram for Example 3.42. **c** Butterfly diagram for Example 3.42

Step 3:

$$\begin{aligned} Y(k) &= X(k)H(k) \\ &= \{3,\ (1-2j),\ -1,\ (1+2j)\}\{6,\ (-1-j),\ 4,\ (-1+j)\} \\ &= \{3\times 6,\ (1-2j)(-1-j),\ -1\times 4,\ (1+2j)(-1+j)\} \\ Y(k) &= \{18, -3+j, -4, -3-j\} \end{aligned}$$

Step 4: To determine IDFT of $Y(k)$ (Refer to Fig. 3.39c)

The IDFT of $Y(k)$ is given by

$$\begin{aligned} y(n) &= \frac{1}{N}\left[\sum_{k=0}^{N-1} Y^*(k)W_N^{nk}\right]^* \\ &= \frac{1}{N}\left[N \text{ point DFT of } Y^*(k)\right]^* \\ Y^*(k) &= \{18, -3-j, -4, -3+j\} \end{aligned}$$

The 4-point DFT of $Y^*(k)$ is therefore,

$$\begin{aligned} y(n) &= \frac{1}{N}\left[4 \text{ point DFT of } Y^*(k)\right]^* \\ &= \frac{1}{4}\{8, 20, 20, 24\}^* \\ &= \frac{1}{4}\{8, 20, 20, 24\} \end{aligned}$$

$$\boxed{y(n) = \{2, 5, 5, 6\}}$$

Example 3.43 Develop a radix-3 DIF-FFT algorithm for evaluating the DFT for $N = 9$.

Solution For $N = 9 = 3\cdot 3$

$$\begin{aligned} X(k) &= \sum_{n=0}^{2} x(3n)W_9^{3nk} + \sum_{n=0}^{2} x(3n+1)W_9^{(3n+1)k} \\ &\quad + \sum_{n=0}^{2} x(3n+2)W_9^{(3n+2)k} \\ X(k) &= X_1(k) + W_9^k X_2(k) + W_9^{2k} X_3(k) \end{aligned}$$

where

$$X_1(k) = \sum_{n=0}^{2} x(3n) W_9^{3nk} = x(0) + x(3) W_9^{3k} + x(6) W_9^{6k} \qquad k = 0, \ldots, 2$$

$$X_2(k) = \sum_{n=0}^{2} x(3n+1) W_9^{3nk} = x(1) + x(4) W_9^{3k} + x(7) W_9^{6k} \qquad k = 0, \ldots, 2$$

$$X_3(k) = \sum_{n=0}^{2} x(3n+2) W_9^{3nk} = x(2) + x(5) W_9^{3k} + x(8) W_9^{6k} \qquad k = 0, \ldots, 2$$

The butterfly diagram is shown in Fig. 3.40.

Stage-I: Computation

$$X_1(k) = \sum_{n=0}^{2} x(3n) W_9^{3nk}$$
$$= x(0) + x(3) W_9^{3k} + x(6) W_9^{6k} \qquad k = 0, \ldots, 2$$

$$X_1(0) = x(0) + x(3) W_9^0 + x(6) W_9^0$$
$$X_1(1) = x(0) + x(3) W_9^3 + x(6) W_9^6$$
$$X_1(2) = x(1) + x(3) W_9^6 + x(6) W_9^{12}$$
$$X_2(k) = x(1) + x(4) W_9^{3k} + x(7) W_9^{6k}, \qquad k = 0, \ldots, 2$$
$$X_2(0) = x(1) + x(3) W_9^0 + x(7) W_9^0$$

$$X_2(1) = x(1) + x(3) W_9^3 + x(7) W_9^6$$
$$X_2(2) = x(1) + x(3) W_9^6 + x(7) W_9^{12}$$
$$X_3(k) = x(2) + x(5) W_9^{3k} + x(8) W_9^{6k}, \qquad k = 0, \ldots, 2$$
$$X_3(0) = x(2) + x(5) W_9^0 + x(8) W_9^0$$
$$X_3(1) = x(2) + x(5) W_9^3 + x(8) W_9^6$$
$$X_3(2) = x(2) + x(5) W_9^6 + x(8) W_9^{12}$$

Stage-II: Computation

$$X(k) = X_1(k) + W_9^k X_2(k) + W_9^{2k} X_3(k) \qquad k = 0, \ldots, 8$$
$$X(0) = X_1(0) + W_9^0 X_2(0) + W_9^0 X_3(0)$$
$$X(1) = X_1(1) + W_9^1 X_2(1) + W_9^2 X_3(1)$$
$$X(2) = X_1(2) + W_9^2 X_2(2) + W_9^4 X_3(2)$$
$$X(3) = X_1(0) + W_9^3 X_2(0) + W_9^6 X_3(0)$$

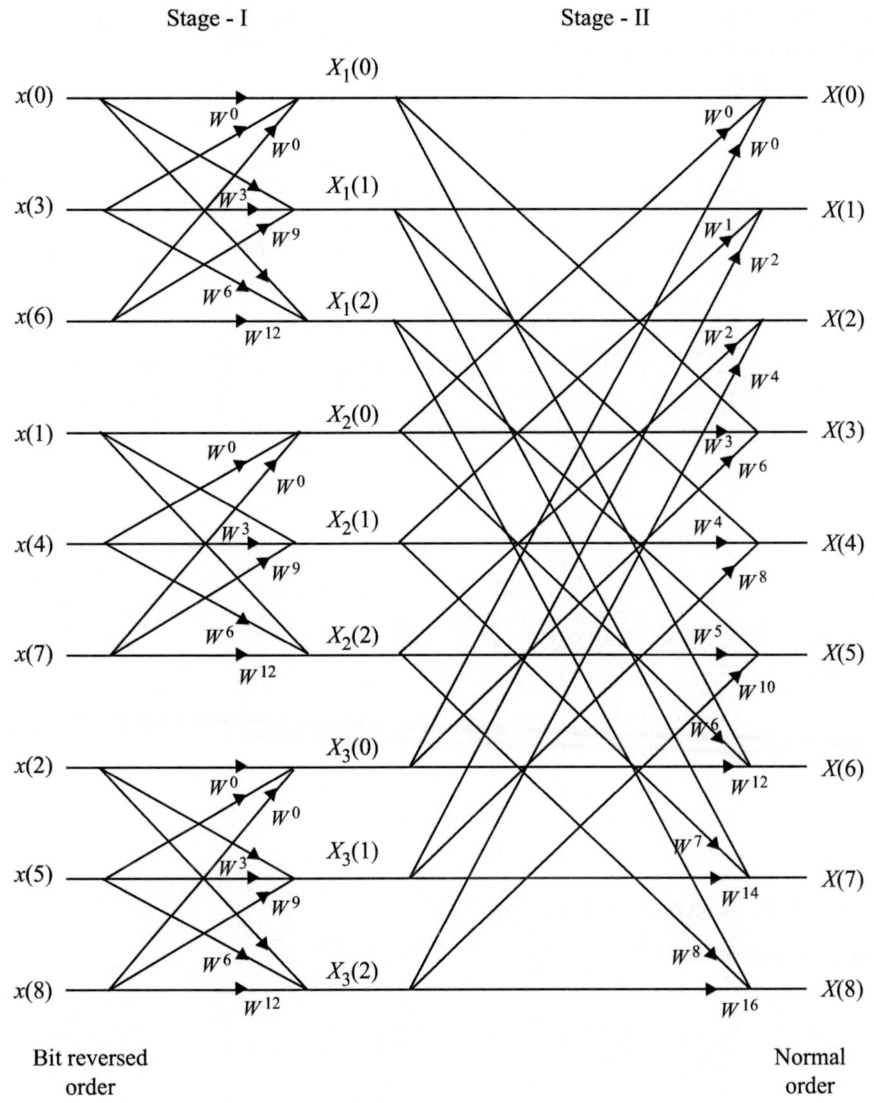

Fig. 3.40 Radix-3 DIT FFT flow diagram for $N = 9$

$$
\begin{aligned}
X(4) &= X_1(1) + W_9^4 X_2(1) + W_9^8 X_3(1)\\
X(5) &= X_1(2) + W_9^5 X_2(2) + W_9^{10} X_3(2)\\
X(6) &= X_1(0) + W_9^6 X_2(0) + W_9^{12} X_3(0)\\
X(7) &= X_1(1) + W_9^7 X_2(1) + W_9^{14} X_3(1)\\
X(8) &= X_1(2) + W_9^8 X_2(2) + W_9^{16} X_3(2)
\end{aligned}
$$

Example 3.44 Develop a DIT FTT algorithm for decomposing the DFT for $N = 6$ and draw the flow diagrams for

(*a*) $N = 2 \cdot 3$
(*b*) $N = 3 \cdot 2$
(*c*) Also, by using the FFT algorithm developed in part (*b*); evaluate the DFT values for $x(n) = \{1, 2, 3, 4, 5, 6\}$.

Solution (*a*) For $N = 6 = 2 \cdot 3$, where number of sequence $m_1 = 2$, number of elements in each subsequences $N_1 = 3$:

$$X(k) = \sum_{n=0}^{2} x(2n)W_6^{2nk} + \sum_{n=0}^{2} x(2n+1)W_6^{(2n+1)k} + \sum_{n=0}^{2} x(2)W_6^{2nk}$$
$$+W_6^k \sum_{n=0}^{2} x(2n+1)W_6^{2nk}$$

Also, $X_i(k+3) = X_i(k)$

$$X(k) = X_1(k) + W_6^k X_2(k), \quad k = 0, \ldots, 5$$

where

$$X_1(k) = \sum_{n=0}^{2} x(2n)W_6^{2nk} \qquad k = 0, \ldots, 2$$
$$X_2(k) = \sum_{n=0}^{2} x(2n+1)W_6^{2nk} \qquad k = 0, \ldots, 2$$

The butterfly diagram is shown in Fig. 3.41.

Stage-I: Computation

$$X_1(k) = \sum_{n=0}^{2} x(2n)W_6^{2nk} \qquad k = 0, \ldots, 2$$
$$X_1(k) = x(0) + x(2)W_6^{2k} + x(4)W_6^{4k} \qquad k = 0, \ldots, 2$$
$$X_1(0) = x(0) + x(2)W_6^0 + x(4)W_6^0$$
$$X_1(1) = x(0) + x(3)W_6^2 + x(4)W_6^4$$
$$X_1(2) = x(1) + x(3)W_9^4 + x(4)W_6^8$$
$$X_1(2) = x(0) + x(3)W_9^4 + x(4)W_6^2$$

$$X_2(k) = \sum_{n=0}^{2} x(2n+1)W_6^{2nk} \qquad k = 0, \ldots, 2$$
$$X_2(k) = x(1) + x(3)W_6^{2k} + x(5)W_6^{4k} \qquad k = 0, \ldots, 2$$

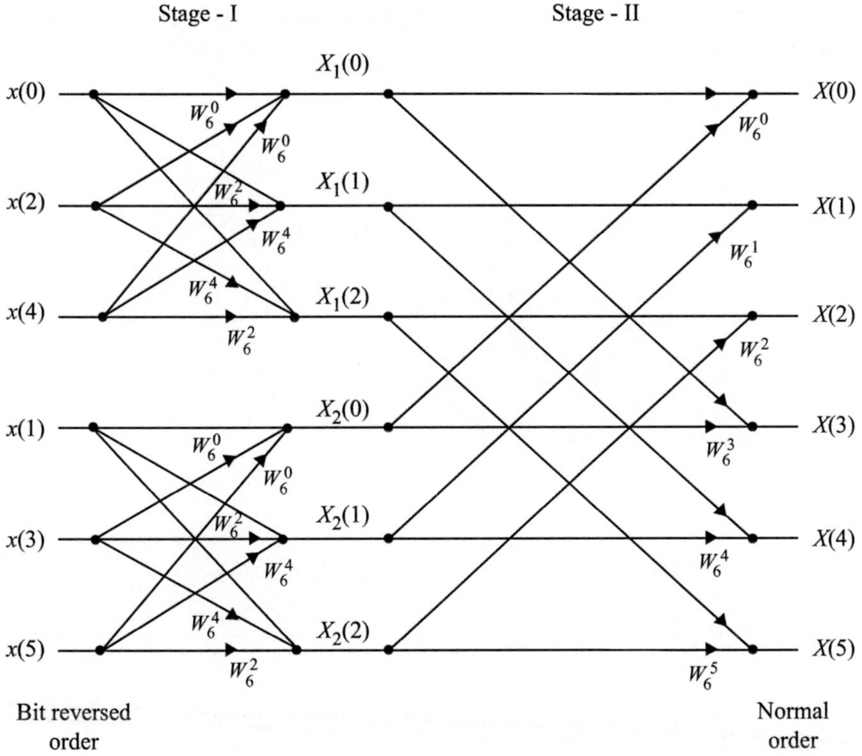

Fig. 3.41 DIT FFT flow diagram for $N = 6 = 2 \cdot 3$

$$X_2(0) = x(1) + x(3)W_6^0 + x(5)W_6^0$$
$$X_2(1) = x(1) + x(3)W_6^2 + x(5)W_6^4$$
$$X_2(2) = x(1) + x(3)W_6^4 + x(5)W_6^8$$
$$X_2(2) = x(1) + x(3)W_6^4 + x(5)W_6^2$$

Stage-II: Computation

$$X(k) = X_1(k) + W_6^k X_2(k), \qquad k = 0, \ldots 5$$
$$X(0) = X_1(0) + W_6^0 X_2(0)$$
$$X(1) = X_1(1) + W_6^1 X_2(1)$$
$$X(2) = X_1(2) + W_6^2 X_2(2)$$
$$X(3) = X_1(3) + W_6^3 X_2(3) = X_1(0) + W_6^3 X_2(0)$$
$$X(4) = X_1(4) + W_6^4 X_2(4) = X_1(1) + W_6^4 X_2(1)$$
$$X(5) = X_1(5) + W_6^5 X_2(5) = X_1(2) + W_6^5 X_2(2)$$

(*b*) For $N = 6 = 2 \cdot 3$, where number of sequence $m_1 = 3$, number of elements in each subsequences $N_1 = 2$:

$$X(k) = \sum_{n=0}^{1} x(3n) W_6^{3nk} + \sum_{n=0}^{1} x(3n+1) W_6^{(3n+1)k} + \sum_{n=0}^{1} x(3n+2) W_6^{(3n+2)k}$$

$$= \sum_{n=0}^{1} x(3n) W_6^{3nk} + W_6^{k} \sum_{n=0}^{1} x(3n+1) W_6^{3nk} + W_6^{2k} \sum_{n=0}^{1} x(3n+2) W_6^{3nk}$$

$$X(k) = X_1(k) + W_6^{k} X_2(k) + W_6^{2k} X_3(k)$$

where

$$X_1(k) = \sum_{n=0}^{1} x(3n) W_6^{3nk} = x(0) + x(3) W_6^{3k}$$

$$X_2(k) = \sum_{n=0}^{1} x(3n+1) W_6^{3nk} = x(1) + x(4) W_6^{3k}$$

$$X_3(k) = \sum_{n=0}^{1} x(3n+2) W_6^{3nk} = x(2) + x(5) W_6^{3k}$$

Also, $X_i(k+2) = X_i(k)$.

Stage-I: Computation

$$\begin{aligned}
X_1(k) &= x(0) + x(3) W_6^{3k} \qquad k = 0, 1 \\
X_1(0) &= x(0) + x(3) W_6^{0} \\
X_1(1) &= x(0) + x(3) W_6^{3} \\
X_2(k) &= x(1) + x(4) W_6^{3k} \qquad k = 0, 1 \\
X_2(0) &= x(1) + x(4) W_6^{0} \\
X_2(1) &= x(1) + x(4) W_6^{3} \\
X_3(k) &= x(2) + x(5) W_6^{3k} \qquad k = 0, 1 \\
X_3(0) &= x(2) + x(5) W_6^{0} \\
X_3(1) &= x(2) + x(5) W_6^{3}
\end{aligned}$$

The butterfly diagram is shown in Fig. 3.42.

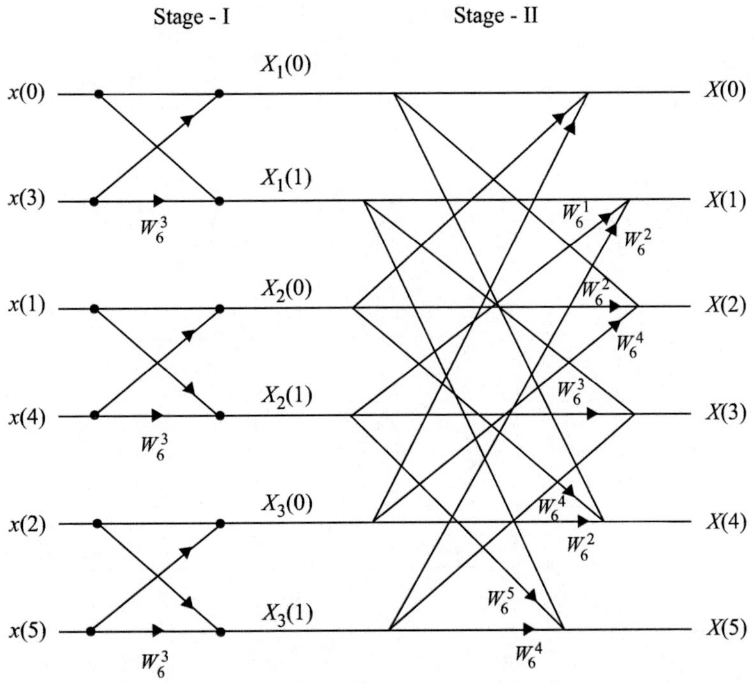

Fig. 3.42 DIT FFT flow diagram for $N = 6 = 3 \cdot 2$

Stage-II: Computation

$$
\begin{aligned}
X(k) &= X_1(k) + W_6^k X_2(k) + W_6^{2k} X_3(k), \qquad k = 0, \ldots, 5 \\
X(0) &= X_1(0) + W_6^0 X_2(0) + W_6^0 X_3(0) \\
X(1) &= X_1(1) + W_6^1 X_2(1) + W_6^2 X_3(1) \\
X(2) &= X_1(2) + W_6^2 X_2(2) + W_6^4 X_3(2) \\
&= X_1(0) + W_6^2 X_2(0) + W_6^4 X_3(0) \\
X(3) &= X_1(3) + W_6^3 X_2(3) + W_6^6 X_3(3) \\
&= X_1(0) + W_6^3 X_2(1) + W_6^0 X_3(0) \\
X(4) &= X_1(4) + W_6^4 X_2(4) + W_6^8 X_3(4) \\
&= X_1(0) + W_6^4 X_2(0) + W_6^2 X_3(0) \\
X(5) &= X_1(5) + W_6^5 X_2(5) + W_6^{10} X_3(5) \\
&= X_1(1) + W_6^5 X_2(1) + W_6^4 X_3(1)
\end{aligned}
$$

(c) To evaluate the DFT values for $x(n)$ (Refer to Fig. 3.43)

$$x(n) = \{1, 2, 3, 4, 5, 6\}$$

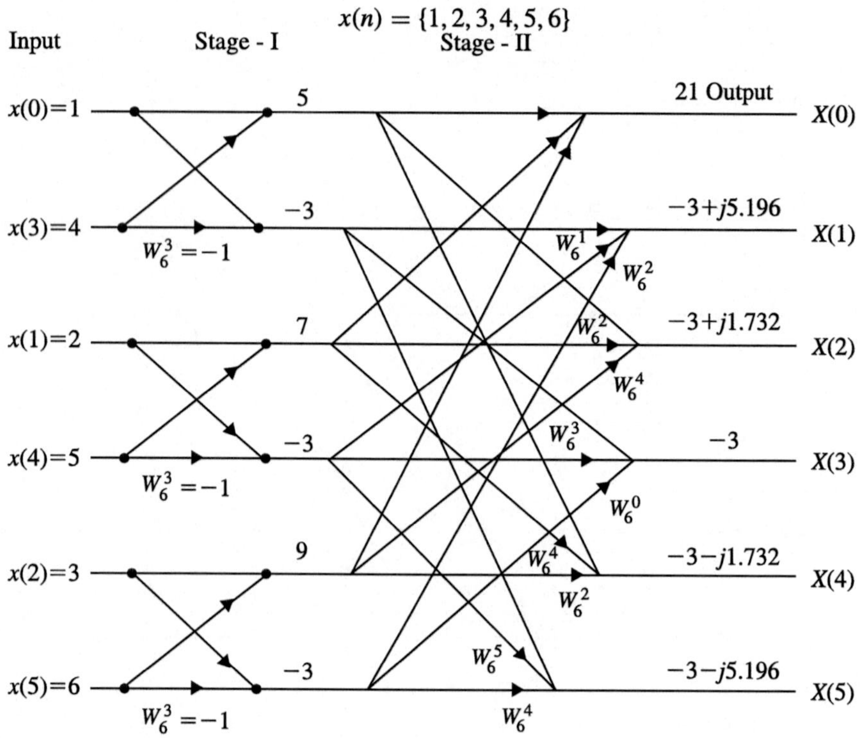

Fig. 3.43 Butterfly diagram for Example 3.44(*c*)

By using the FFT algorithm developed in part (*b*), twiddle factors are

$$
\begin{aligned}
W_6^0 &= 1\\
W_6^1 &= e^{-\frac{j2\pi}{6}} = 0.5 - j0.866\\
W_6^2 &= e^{-\frac{j2\pi\cdot 2}{6}} = -0.5 - j0.866\\
W_6^3 &= e^{-\frac{j2\pi\cdot 3}{6}} = -1\\
W_6^4 &= e^{-\frac{j2\pi\cdot 4}{6}} = -0.5 + j0.866\\
W_6^5 &= e^{-\frac{j2\pi\cdot 5}{6}} = 0.5 + j0.866
\end{aligned}
$$

Input	Stage-I	Stage-II (or) Output
0	$1+4=5$	$5+7+9=21=X(0)$
4	$1-4=-3$	$-3+(0.5-j0.866)(-3)+(-0.5-j0.866)(-3)$ $=-3+j5.196=X(1)$
2	$2+5=7$	$5+(0.5-j0.866)(7)+(-0.5+j0.866)(9)$ $=-3+j1.732=X(2)$
5	$2-5=-3$	$-3+(-1)(-3)+(1)(-3)=-3=X(3)$
3	$3+6=9$	$5+(0.5+j0.866)(7)+(-0.5-j0.866)(9)$ $=-3-j1.732=X(4)$
6	$3-6=-3$	$-3+(0.5+j0.866)(-3)+(-0.5+j0.866)(-3)$ $=-3-j5.196=X(5)$

$$\boxed{X(k)=\{21,-3+j5.196,-3+j1.732,-3,-3-j1.732,-3-j5.196\}}$$

Example 3.45 Develop the DIT FFT algorithm for decomposing the DFT for $N=12$ and draw the flow diagrams.

Solution For $N=12=3\cdot 4$ where $m_1=3$ and $N_1=4$:

$$X(k)=\sum_{n=0}^{3}x(3n)W_{12}^{3nk}+\sum_{n=0}^{3}x(3n+1)W_{12}^{(3n+1)k}+\sum_{n=0}^{3}x(3n+2)W_{12}^{(3n+2)k}$$
$$=\sum_{n=0}^{3}x(3n)W_{12}^{3nk}+W_{12}^{k}\sum_{n=0}^{3}x(3n+1)W_{12}^{3nk}+W_{12}^{2k}\sum_{n=0}^{3}x(3n+2)W_{12}^{3nk}$$
$$X(k)=X_1(k)+W_{12}^{k}X_2(k)+W_{12}^{2k}X_3(k)$$

where

$$X_1(k)=\sum_{n=0}^{3}x(3n)W_{12}^{3nk}\qquad k=0,\ldots,3$$
$$=x(0)+W_{12}^{3k}x(3)+W_{12}^{6k}x(6)+W_{12}^{9k}x(9)$$
$$X_2(k)=\sum_{n=0}^{3}x(3n+1)W_{12}^{3nk}$$
$$=x(1)+x(4)W_{12}^{3k}+x(7)+W_{12}^{6k}+x(0)W_{12}^{7k}$$
$$X_3(k)=\sum_{n=0}^{3}x(3n+2)W_{12}^{3nk}$$
$$=x(2)+W_{12}^{3k}x(5)+W_{12}^{6k}x(8)+W_{11}^{9k}$$

The butterfly diagram is shown in Fig. 3.44.

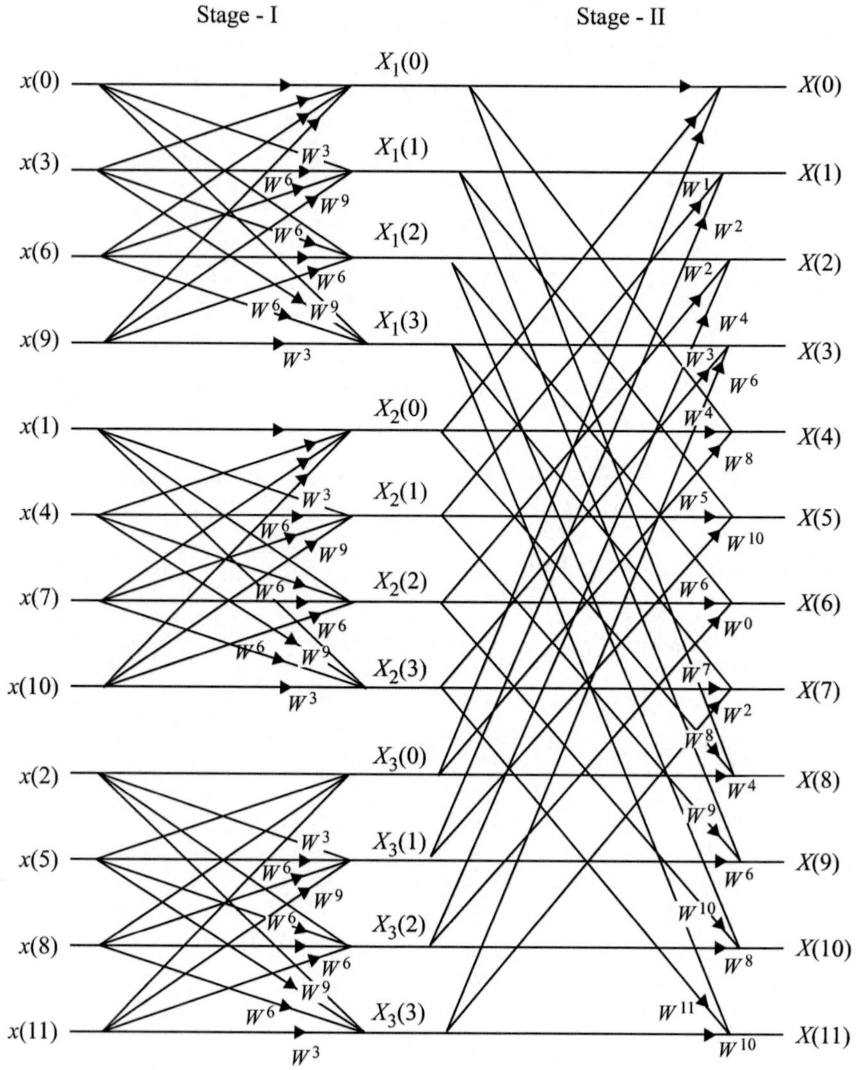

Fig. 3.44 Flow diagram of 12-point composite radix-FFT

Stage-I: Computation

$$
\begin{aligned}
X_1(k) &= x(0) + W_{12}^{3k}x(3) + W_{12}^{6k}x(6) + W_{12}^{9k}x(9) \qquad k = 0, \ldots, 3\\
X_1(0) &= x(0) + x(3) + x(6) + x(9)\\
X_1(1) &= x(0) + W_{12}^{3}x(3) + W_{12}^{6}x(6) + W_{12}^{9}x(9)\\
X_1(2) &= x(0) + W_{12}^{6}x(3) + W_{12}^{0}x(6) + W_{12}^{6}x(9)\\
X_1(3) &= x(0) + W_{12}^{9}x(3) + W_{12}^{6}x(6) + W_{12}^{3}x(9)\\
X_2(k) &= x(1) + x(4)W_{12}^{3k} + x(7)W_{12}^{6k} + x(10)W_{12}^{9k} \qquad k = 0, \ldots, 5\\
X_2(0) &= x(1) + x(4) + x(7) + x(10)\\
X_2(1) &= x(1) + x(4)W_{12}^{3} + x(7)W_{12}^{6} + x(10)W_{12}^{9}\\
X_2(2) &= x(1) + x(4)W_{12}^{6} + x(7)W_{12}^{0} + x(10)W_{12}^{6}\\
X_2(3) &= x(1) + x(4)W_{12}^{9} + x(7) + W_{12}^{6} + x(10)W_{12}^{3}\\
X_3(k) &= x(2) + x(5)W_{12}^{3k} + x(8)W_{12}^{6k} + x(11)W_{12}^{9k} \qquad k = 0, \ldots, 5\\
X_3(0) &= x(2) + x(5) + x(8) + x(11)\\
X_3(1) &= x(2) + x(5)W_{12}^{3} + x(8)W_{12}^{6} + x(11)W_{12}^{9}\\
X_3(2) &= x(2) + x(5)W_{12}^{6} + x(8)W_{12}^{0} + x(11)W_{12}^{6}\\
X_3(3) &= x(2) + x(5)W_{12}^{9} + x(8)W_{12}^{6} + x(11)W_{12}^{3}
\end{aligned}
$$

Stage-II: Computation

$$
\begin{aligned}
X(k) &= X_1(k) + W_{12}^{k}X_2(k) + W_{12}^{2k}X_3(k), \qquad k = 0, \ldots, 11\\
X(0) &= X_1(0) + X_2(0) + X_3(0)\\
X(1) &= X_1(1) + W_{12}^{1}X_2(1) + W_{12}^{2}X_3(1)\\
X(2) &= X_1(2) + W_{12}^{2}X_2(2) + W_{12}^{4}X_3(2)\\
X(3) &= X_1(3) + W_{12}^{3}X_2(3) + W_{12}^{6}X_3(3)\\
X(4) &= X_1(0) + W_{12}^{4}X_2(0) + W_{12}^{8}X_3(0)\\
X(5) &= X_1(1) + W_{12}^{5}X_2(1) + W_{12}^{10}X_3(1)\\
X(6) &= X_1(2) + W_{12}^{6}X_2(2) + W_{12}^{0}X_3(2)\\
X(7) &= X_1(3) + W_{12}^{7}X_2(3) + W_{12}^{2}X_3(3)\\
X(8) &= X_1(0) + W_{12}^{8}X_2(0) + W_{12}^{4}X_3(0)\\
X(9) &= X_1(1) + W_{12}^{9}X_2(1) + W_{12}^{6}X_3(1)\\
X(10) &= X_1(2) + W_{12}^{10}X_2(2) + W_{12}^{8}X_3(2)\\
X(11) &= X_1(3) + W_{12}^{11}X_2(3) + W_{12}^{10}X_3(3)
\end{aligned}
$$

Example 3.46 Develop a radix-4 DIT FFT algorithm for evaluating the DFT for $N = 16$ and hence determine the 16-point DFT of the sequence

$$x(n) = \{0, 1, 0, 1, 0, 1, 0, 1, 0, 1, 0, 1, 0, 1, 0, 1\}$$

Solution For $N = 16 = 4 \cdot 4$ where $m_1 = 4$ and $N_1 = 4$:

$$X(k) = \sum_{n=0}^{3} x(4n) W_{16}^{4nk} + \sum_{n=0}^{3} x(4n+1) W_{16}^{(4n+1)k} + \sum_{n=0}^{3} x(4n+2) W_{16}^{(4n+2)k} + \sum_{n=0}^{3} x(4n+3) W_{16}^{(4n+3)k}$$

$$X_1(k) = \sum_{n=0}^{3} x(4n) W_{16}^{4nk} + W_{16}^{k} \sum_{n=0}^{3} x(4n+1) W_{16}^{4nk} + W_{16}^{2k} \sum_{n=0}^{3} x(4n+2) W_{16}^{4nk} + W_{16}^{3k} \sum_{n=0}^{3} x(4n+3) W_{16}^{4nk}$$

$$= X_1(k) + W_{16}^{k} X_2(k) + W_{16}^{2k} X_3(k) + W_{16}^{3k} X_4(k)$$

The twiddle factors are

$$W_{16}^{0} = 1$$
$$W_{16}^{1} = e^{-\frac{j2\pi}{16}} = (0.923 - j0.382)$$
$$W_{16}^{2} = e^{-\frac{j2\pi \cdot 2}{16}} = 0.707 - j0.707$$
$$W_{16}^{3} = e^{-\frac{j2\pi \cdot 3}{16}} = 0.3826 - j0.923$$
$$W_{16}^{4} = e^{-\frac{j2\pi \cdot 4}{16}} = -j$$
$$W_{16}^{5} = -0.3826 - j0.9238$$
$$W_{16}^{6} = e^{-\frac{j2\pi \cdot 6}{16}} = -0.707 - j0.707$$
$$W_{16}^{7} = e^{-\frac{j2\pi \cdot 7}{16}} = 0.9238 - j0.382$$

The radix-4 DIT FFT algorithm is shown in Fig. 3.45.

Input	Stage-I	Stage-II (or) Output
0	$0+0+0+0=0$	$0+4+0+4=8=X(1)$
0	$0+0+0+0=0$	$0+(0.923-j0.382)(-j2)+(0.707-j0.707)(0)+(0.3826-j0.923)(-j2)$ $=-2.61-j2.61=X(1)$
0	$0+0+0+0=0$	$0+0+0+0=0=X(2)$
0	$0+0+0+0=0$	$0+0+0+0=0=X(3)$
1	$1+1+1+1=4$	$0+(-j)4+(-1)(0)+(j)4=0=X(4)$
1	$1-j-1-j=-2j$	$0+(-j2)(-0.3826-j0.9238)+0+(-j2)(0.9038+j0.382)=-j1.0824-1.0836$ $=X(5)$
1	$1+1(-1)+1(1)+1(-1)=0$	$0+0+0+0=0=X(6)$
1	$1+1(-j)+1(-1)+1(j)=0$	$0+0+0+0=0=X(7)$
0	0	$0+4(-1)+0+4(1)=0=X(8)$
0	0	$0+(-j2)(-0.923+j0.9826)+0+(-j2)(-0.382+j0.923)=2.6112+j2.6112$ $=X(9)$
0	0	$0+0+0+0=0=X(10)$
0	0	$0+0+0+0=0=X(11)$
1	$1+1+1+1=4$	$0+4(-j)+0+j4=0=X(12)$
1	$1+1(-j)+1(-1)+1(-1)=-j2$	$0+(-j2)(0.3826+j0.923)+0+(-j2)(-0.9238-j0.382)=1.0818+j1.0848$ $=X(13)$
1	$1+1(-1)+1(1)+1(-1)=0$	$0+0+0+0=0=X(14)$
1	$1+1(-j)+1(-1)+1(j)=0$	$0+0+0+0=0=X(15)$

$$
\begin{aligned}
W_{16}^{8} &= e^{-\frac{j2\pi\cdot 8}{16}} = -1\\
W_{16}^{9} &= e^{-\frac{j2\pi\cdot 9}{16}} = -0.923+j0.3826\\
W_{16}^{10} &= -0.707+j0.707\\
W_{16}^{11} &= -0.3826+j0.923\\
W_{16}^{12} &= -j\\
W_{16}^{13} &= 0.3826+j0.923\\
W_{16}^{14} &= 0.707+j0.707\\
W_{16}^{15} &= 0.9238+j0.382
\end{aligned}
$$

Example 3.47 Develop a DIF FFT algorithm for decomposing the DFT for $N=6$ and draw the flow diagram for (*a*) $N=3\cdot 2$ and (*b*) $N=2\cdot 3$.

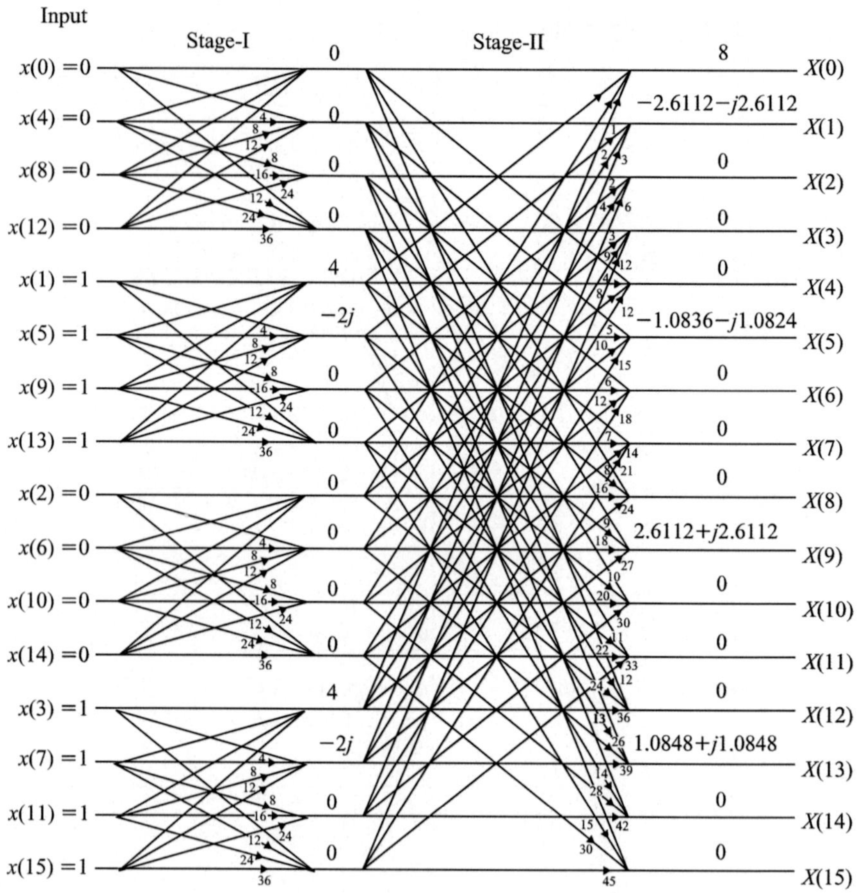

Fig. 3.45 Radix-4 DIT FFT algorithm for Example 3.46

Solution (*a*) To develop DIF FFT algorithm for $N = 3 \cdot 2$

$$X(k) = \sum_{n=0}^{5} x(n) W_6^{nk} = \sum_{n=0}^{2} x(n) W_6^{nk} + \sum_{n=3}^{5} x(n) W_6^{nk}$$

$$= \sum_{n=0}^{2} x(n) W_6^{nk} + \sum_{n=0}^{2} x(n+3) W_6^{(n+3)k}$$

$$X(k) = \sum_{n=0}^{2} [x(n) + x(n+3) W_6^{3k}] W_6^{nk}$$

$$X(2k) = \sum_{n=0}^{2} [x(n) + x(n+3) W_6^{6k}] + W_6^{2nk}$$

$$= \sum_{n=0}^{2} [x(n) + x(n+3)] W_6^{2nk}$$

$$X(2k+1) = \sum_{n=0}^{2} [x(n) + x(n+3) W_6^{3(2k+1)}] W_6^{(2k+1)n}$$

$$= \sum_{n=0}^{2} [x(n) - x(n+3)] W_6^n W_6^{2nk}$$

Let $g(n) = x(n) + x(n+3)$, $h(n) = x(n) - x(n+3)$

$$X(2k) = \sum_{n=0}^{2} g(n) W_6^{2nk}$$

$$X(2k+1) = \sum_{n=0}^{2} h(n) W_6^n W_6^{2nk}$$

The butterfly diagram is shown in Fig. 3.46.

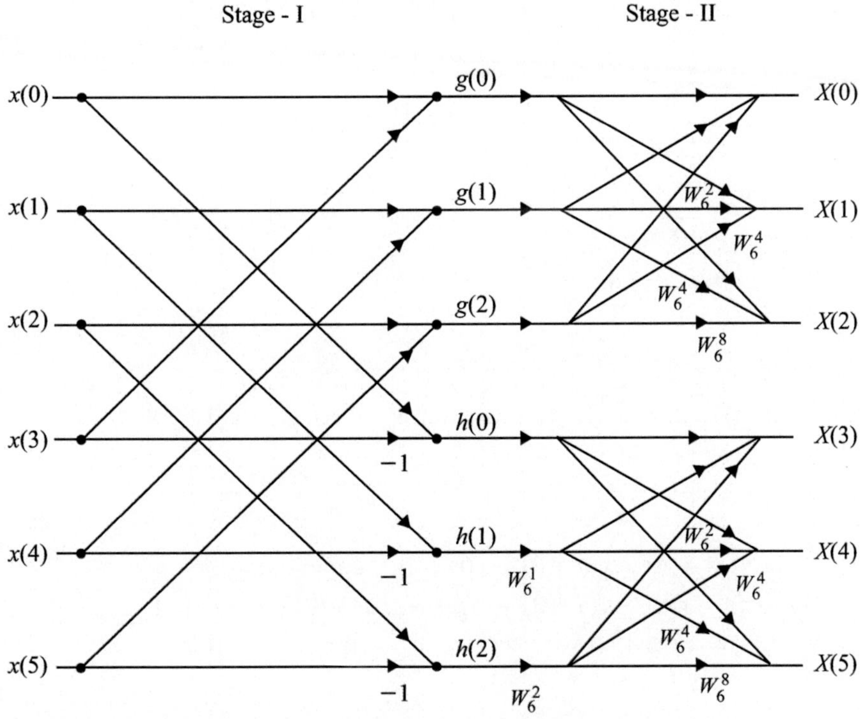

Fig. 3.46 DIF FFT flow diagram for decomposing the DFT for $N = 6 = 3 \cdot 2$

Stage I

$$g(0) = x(0) + x(3) \qquad h(0) = x(0) - x(3)$$
$$g(1) = x(1) + x(4) \qquad h(1) = x(1) - x(4)$$
$$g(2) = x(2) + x(5) \qquad h(2) = x(2) - x(5)$$

Stage II

$$X(0) = g(0) + g(1) + g(2)$$
$$X(2) = g(0) + g(1)W_6^2 + g(2)W_6^4$$
$$X(4) = g(0) + g(1)W_6^4 + g(2)W_6^8$$
$$X(1) = h(0) + h(1)W_6^1 + h(2)W_6^2$$
$$X(3) = h(0) + h(1)W_6^1 W_6^2 + h(2)W_6^2 W_6^4$$
$$X(5) = h(0) + h(1)W_6^1 W_6^4 + h(2)W_6^2 W_6^8$$

The DIF FFT flow diagram for decomposing the DFT is shown in Fig. 3.47 for $N = 6 = 3 \times 2$.
(b) To develop DIF FFT algorithm for $N = 2 \cdot 3$

$$X(k) = \sum_{n=0}^{5} x(n)W_6^{nk} = \sum_{n=0}^{1} x(n)W^{nk} + \sum_{n=2}^{3} x(n)W_6^{nk} + \sum_{n=4}^{5} x(n)W_6^{nk}$$
$$= \sum_{n=0}^{1} [x(n) + x(n+2)W_6^{2k} + x(n+4)W_6^{4k}]W_6^{nk}$$
$$X(3k) = \sum_{n=0}^{1} [x(n) + x(n+2) + x(n+4)]W_6^{3nk}$$

$$X(3k+1) = \sum_{n=0}^{1} [x(n) + x(n+2)W_6^2 + x(n+4)W_6^4]W_6^n W_6^{3nk}$$
$$X(3k+2) = \sum_{n=0}^{1} [x(n) + x(n+2)W_6^4 + x(n+4)W_6^2]W_6^{2n} W_6^{3nk}$$

Let

$$f(n) = x(n) + x(n+2) + x(n+4), \qquad n = 0, 1, \ldots$$
$$g(n) = x(n) + x(n+2) + W_6^2 + x(n+4)W_6^4, \qquad n = 0, 1, \ldots$$
$$h(n) = x(n) + x(n+2)W_6^4 + x(n+4)W_6^2, \qquad n = 0, 1, \ldots$$

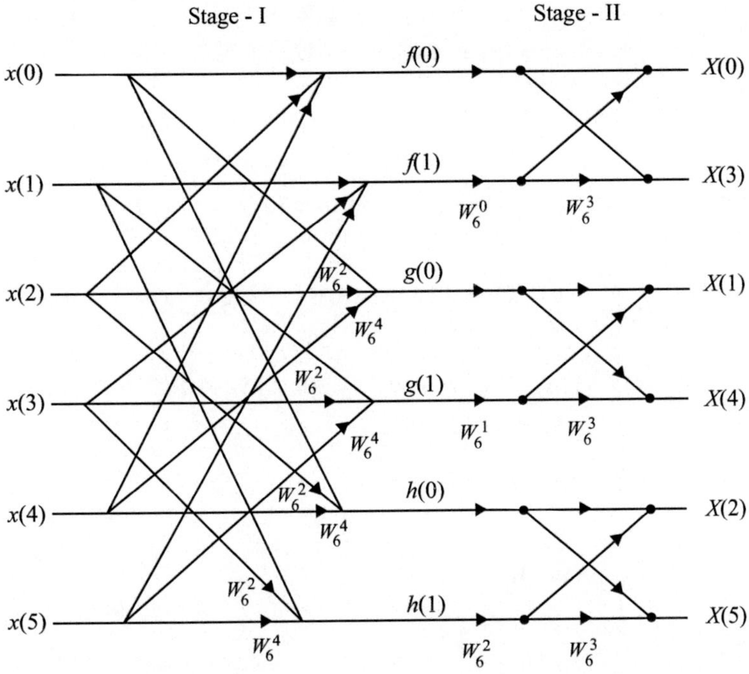

Fig. 3.47 DIF FFT flow diagram for decomposing the DFT for $N = 6 = 2 \cdot 3$

Stage I

$$f(0) = x(0) + x(2) + x(4)$$
$$f(1) = x(1) + x(3) + x(5)$$
$$g(0) = x(0) + x(2)W_6^2 + x(4)W_6^4$$
$$g(1) = x(1) + x(3)W_6^2 + x(5)W_6^4$$

The DIF FFT flow diagram for decomposing the DFT for $N = 6 = 2 \times 3$ is shown in Fig. 3.47.

$$h(0) = x(0) + x(2)W_6^4 + x(4)W_6^2$$
$$h(1) = x(1) + x(3)W_6^4 + x(5)W_6^2$$

Stage II

$$X(0) = f(0) + f(1)W_6^0$$
$$X(1) = g(0) + g(1)W_6^1$$
$$X(2) = h(0) + h(1)W_6^2$$
$$X(3) = f(0) + f(1)W_6^0W_6^3$$

$$X(4) = g(0) + g(1)W_6^1 W_6^3$$
$$X(5) = h(0) + h(1)W_6^2 W_6^3$$

Example 3.48 Develop a radix-3 DIF FFT algorithm for evaluating the DFT for $N = 9$.

Solution To develop a radix-3 DIF FFT algorithm for $N = 9 = 3 \cdot 3$

$$X(k) = \sum_{n=0}^{2} x(n)W_9^{nk} + \sum_{n=0}^{2} x(n+3)W_9^{(n+3)k} + \sum_{n=0}^{2} x(n+6)W_9^{(n+6)k}$$

$$X(k) = \sum_{n=0}^{2} [x(n) + x(n+3)W_9^{3k} + x(n+6)W_9^{6k}]W_9^{nk}$$

$$X(3k) = \sum_{n=0}^{2} [x(n) + x(n+3) + x(n+6)]W_9^{3nk}]$$
$$= \sum_{n=0}^{2} f(n)W_9^{3nk}$$

$$X(3k+1) = \sum_{n=0}^{2} [x(n) + x(n+3)W_9^3 + x(n+6)W_9^6]W_9^n W_9^{3nk}$$
$$= \sum_{n=0}^{2} g(n)W_9^n W_9^{3nk}$$

$$X(3k+2) = \sum_{n=0}^{2} [x(n) + x(n+3)W_9^6 + x(n+6)W_9^3]W_9^{2n} W_9^{3nk}$$
$$= \sum_{n=0}^{2} h(n)W_9^{2n} W_9^{3nk}$$

The butterfly diagram is shown in Fig. 3.48.

Stage I

$$f(0) = x(0) + x(3) + x(6)$$
$$f(1) = x(1) + x(4) + x(7)$$
$$f(2) = x(2) + x(5) + x(8)$$
$$g(0) = x(0) + x(3)W_9^3 + x(6)W_9^6$$
$$g(1) = x(1) + x(4)W_9^3 + x(7)W_9^6$$
$$g(2) = x(2) + x(5)W_9^3 + x(8)W_9^6$$

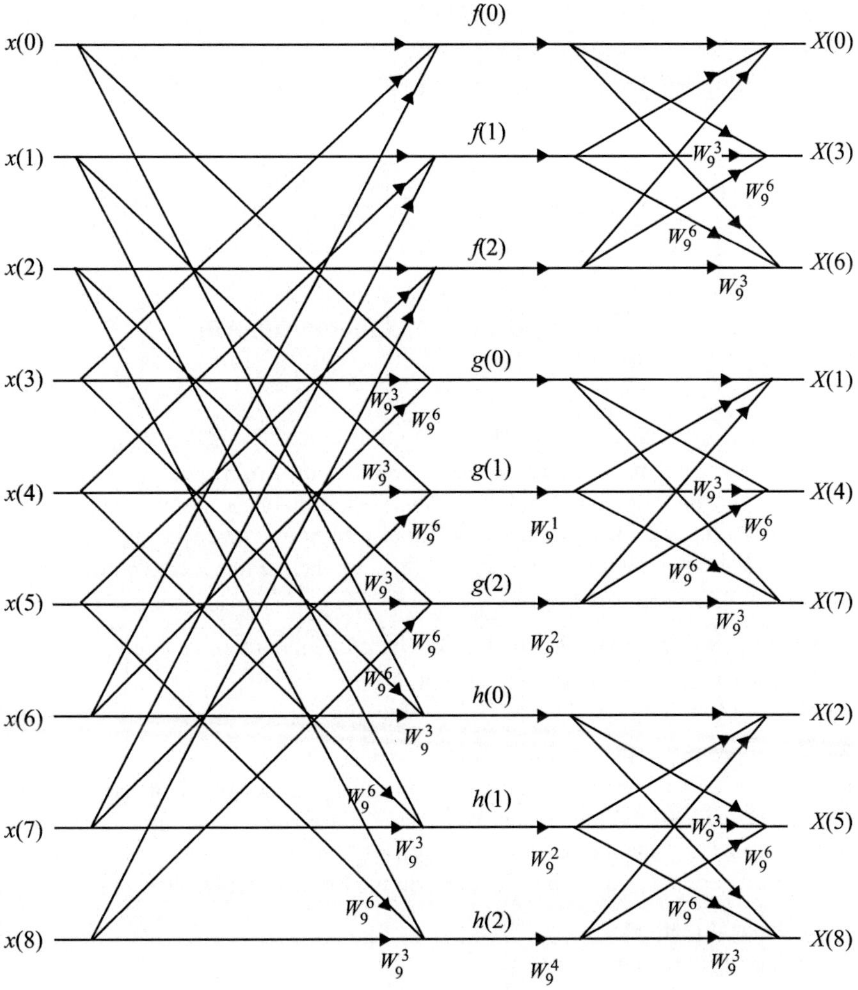

Fig. 3.48 A radix-3 DIF FFT flow diagram for decomposing the DFT for $N = 9 = 3 \cdot 3$

$$h(0) = x(0) + x(3)W_9^6 + x(6)W_9^3$$
$$h(1) = x(1) + x(4)W_9^6 + x(7)W_9^3$$
$$h(2) = x(2) + x(5)W_9^6 + x(8)W_9^3$$

Stage II

$$X(0) = f(0) + f(1) + f(2)$$
$$X(3) = f(0) + f(1)W_9^3 + f(2)W_9^6$$
$$X(6) = f(0) + f(1)W_6^0 + f(2)W_9^3$$

$$X(1) = g(0) + g(1)W_9^1 + g(2)W_9^2$$
$$X(4) = g(0) + g(1)W_9^1 W_9^3 + g(2)W_9^2 W_9^6$$
$$X(7) = g(0) + g(1)W_9^1 W_9^6 + g(2)W_9^2 W_9^3$$
$$X(2) = h(0) + h(1)W_9^2 + h(2)W_9^4$$
$$X(5) = h(0) + h(1)W_9^2 W_9^3 + h(2)W_9^4 W_9^6$$
$$X(8) = h(0) + h(1)W_9^2 W_9^6 + h(2)W_9^4 W_9^3$$

3.6.4 *Use of the FFT Algorithm in Linear Filtering and Correlation*

An important application of the FFT algorithm is in FIR linear filtering of long data sequences. The response of an LTI system for any arbitrary input is given by linear convolution of the input and the impulse response of the system. If one of the sequences (either the input sequence or impulse response sequence) is very much larger than the other, then it is very difficult to compute the linear convolution using DFT for the following reasons:

1. The entire sequence should be available before convolution can be carried out. This makes long delay in getting the output.
2. Large amount of memory is required to store the sequences.

The above problems can be overcome by linear filtering of a longer sequence into the size of smaller sequences. Then the linear convolution of each section of longer sequences and the smaller sequence is performed. The output sequences obtained from the convolutions of the sections are combined to get the overall output sequences. There are two methods to perform the linear filtering. They are

1. Overlap add method.
2. Overlap save method.

3.6.4.1 Overlap Add Method

Let N_1 be the length of a longer sequence and N_2, the length of a smaller sequence. Let the longer sequence be divided into sections of size N_3 samples. (*Note*: Normally, the longer sequence is divided into sections of size the same as that of a smaller sequence.)

The linear convolution of each section with a smaller sequence will produce an output sequence of size $N_3 + N_2 - 1$ samples. In this method, the last $N_2 - 1$ samples of each output sequence overlaps with the first $N_2 - 1$ samples of the next section. While combining the output sequences of the various sectioned convolutions, the corresponding samples of overlapped regions are added and the samples of non-overlapped regions are retained as such.

3.6.4.2 Overlap Save Method

Let N_1 be the length of a longer sequence and N_2, the length of a smaller sequence. Let the longer sequence be divided into sections of size N_3 samples.

In the overlap save method, the results of linear convolution are obtained by circular convolution. Hence, each section of a longer sequence and the smaller sequence are converted to the size of the output sequence of size $N_3 + N_2 - 1$ samples. The smaller sequence is converted to size of $N_3 + N_2 - 1$ samples by appending with zeros. The conversion of each section of a longer sequence to the size $N_3 + N_2 - 1$ samples can be performed in two different methods.

Method I

In this method, the first $N_2 - 1$ samples of a section is appended as last $N_2 - 1$ samples of the previous section. The circular convolution of each section will produce an output sequence of size $N_3 + N_2 - 1$ samples. In this output, the first $N_2 - 1$ samples are discarded and the remaining samples of the output of sectioned convolution are saved as the overall output sequence.

Method II

In this method, the last $N_2 - 1$ samples of a section are appended as last $N_2 - 1$ samples of the next section (i.e. the overlapping samples are placed at the beginning of the section). The circular convolution of each section will produce an output sequence of size $N_3 + N_2 - 1$ samples. In this output, the last $N_2 - 1$ samples are discarded and the remaining samples of the output of sectioned convolution are saved as the overall output sequence.

Example 3.49 Perform the linear convolution of the following sequence by (*i*) overlap add method and (*ii*) overlap save method:

$$x(n) = \{1, -1, 2, -2, 3, -3, 4, -4\} \quad \text{and} \quad h(n) = \{-1, 1\}$$

Solution Overlap Add Method

Here $x(n)$ is a longer sequence when compared to $h(n)$. Hence, $x(n)$ is sectioned into sequences of size equal to $h(n)$.

Let $x(n)$ be sectioned into three sequences, each consisting of two samples of $x(n)$ as

$$x_1(n) = \begin{cases} 1; & n = 0 \\ -1; & n = 1 \\ 2; & n = 2 \end{cases} \qquad x_2(n) = \begin{cases} -2; & n = 3 \\ +3; & n = 4 \\ -3; & n = 5 \end{cases} \qquad x_3(n) = \begin{cases} 4; & n = 6 \\ -4; & n = 7 \\ 0; & n = 8 \end{cases}$$

Let $y_1(n)$, $y_2(n)$, $y_3(n)$ and $y_4(n)$ be the output of linear convolution of $x_1(n)$, $x_2(n)$, $x_3(n)$ and $x_4(n)$ with $h(n)$, respectively.

Butterfly diagram of $X(k)$ for Example 3.49 using overlap add method is shown in Fig. 3.49.

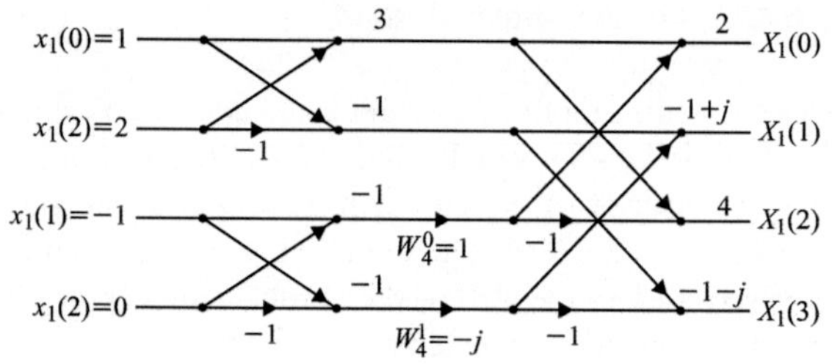

Fig. 3.49 Butterfly diagram of $X(k)$ for Example 3.49, using overlap add method

Convolution of Section I

$$\begin{aligned}y_1(n) &= x_1(n) * h(n)\\ Y_1(k) &= X_1(k)H(k)\\ y_1(n) &= IDFT\,[X_1(k)H(k)]\end{aligned}$$

$$X_1(k) \Rightarrow x_1(n) = \{1, -1, 2\}$$

and

$$h(n) = \{-1, 1\}$$

Length of $y_1(n) = 3 + 2 - 1 = 4;$

$$n_1 = 0 \quad \text{and} \quad n_2 = 0$$

so

$$n = n_1 + n_2 = 0.$$

Therefore,

$$x_1(n) = \{1, -1, 2, 0\}$$

and

$$h(n) = \{-1, 1, 0, 0\}$$

Therefore,

$$X_1(k) = \{2, -1 + j, 4, -1 - j\}$$

$$\begin{aligned}\therefore\ H(k) &= \{0, -1 - j, -2, -1 + j\}\\ X_1(k)H(k) &= \{0, (-1 + j)(-1 - j), -8, (-1 - j)(-1 + j)\}\\ &= \{0, 2, -8, 2\}\end{aligned}$$

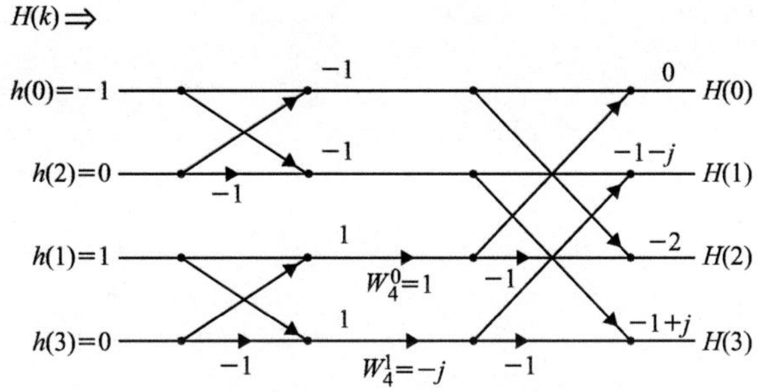

Fig. 3.50 Butterfly diagram of $H(k)$

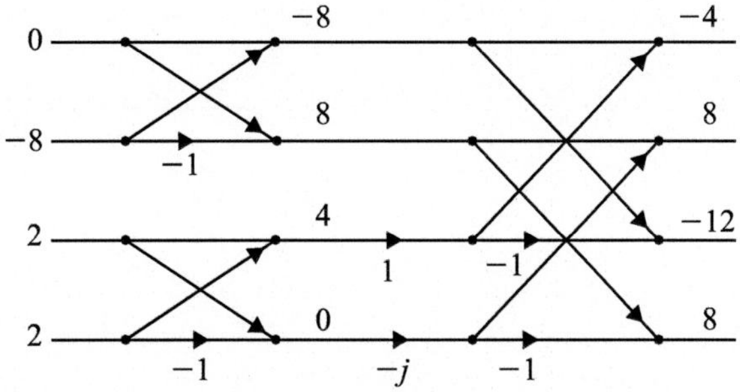

Fig. 3.51 Butterfly diagram of $[X_1(k)H(k)]^*$

IDFT $[X_1(k)H(k)] \Rightarrow$

$$\{X_1(k)H(k)\}^* = \{0, 2, -8, 2\}$$

DFT of $[X_1(k)H(k)]^*$ is (Figs. 3.50 and 3.51)

$$\{-4, 8, -12, 8\}$$

$$\begin{aligned} \therefore\ y_1(n) &= \frac{1}{n}\{-4, 8, -12, 8\} \\ &= \frac{1}{4}\{-4, 8, -12, 8\} = \{\underset{\uparrow}{-1}, 2, -3, 2\} \end{aligned}$$

Convolution of Section II (Fig. 3.52)

$$x_2(n) = \{-2, 3, -3\} \quad \text{and} \quad h(n) = \{-1, 1\}$$

$N = 3 + 2 - 1 = 4$

$$x_2(n) = \{-2, 3, -3, 0\} \quad \text{and} \quad h(n) = \{-1, 1, 0, 0\}$$

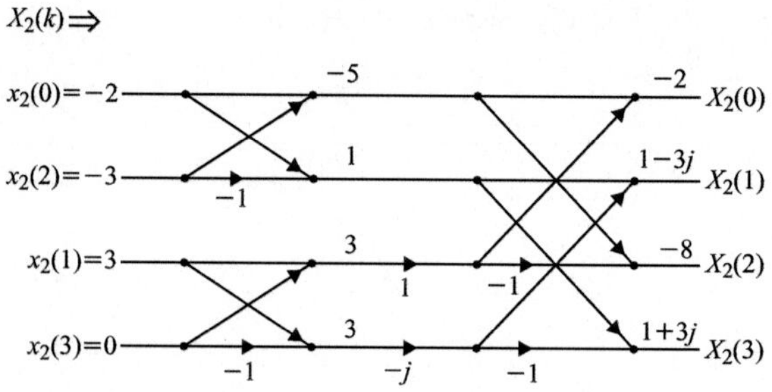

Fig. 3.52 Butterfly diagram of $X_2(k)$

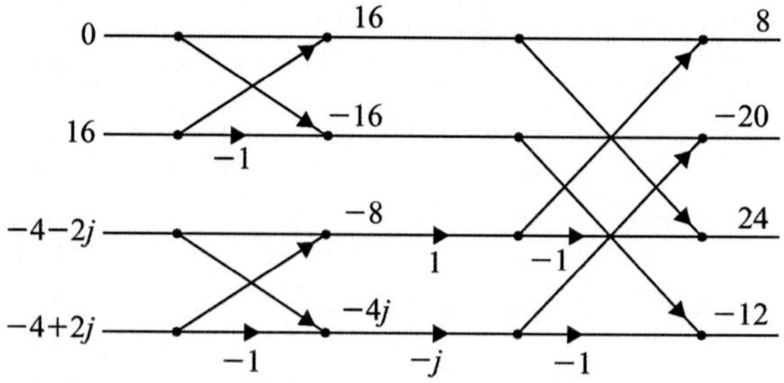

Fig. 3.53 Butterfly diagram of $[X_2(k)H(k)]^*$

$n_1 = 3$, $n_2 = 0$ and $n = n_1 + n_2 = 3$.

$$\begin{aligned} X_2(k) &= \{-2, 1-3j, -8, 1+3j\} \\ X_2(k)H(k) &= \{0, (-1-j)(1-3j), (-2)(-8), (-1+j)(1+3i)\} \\ &= \{0, -4+2j, 16, -4-2j\} \end{aligned}$$

IDFT $[X_2(k)H(k)] \Rightarrow$

$$\{X_2(k)H(k)\}^* = \{0, -4-2j, 16, -4+2j\}$$

DFT $\{X_2(k)H(k)\}^*$ is (Fig. 3.53)
Therefore,

$$\begin{aligned} y_2(n) &= \frac{1}{N}\left\{DFT\,[X_2(k)H(k)]^*\right\}^* = \frac{1}{4}\{8, -20, 24, -12\} \\ &= \{\underset{\substack{\uparrow \\ n=3}}{2}, -5, 6, -3\} \end{aligned}$$

Convolution of Section III

$$x_3(n) = \{4, -4, 0\} \quad \text{and} \quad h(n) = \{-1, 1\}$$

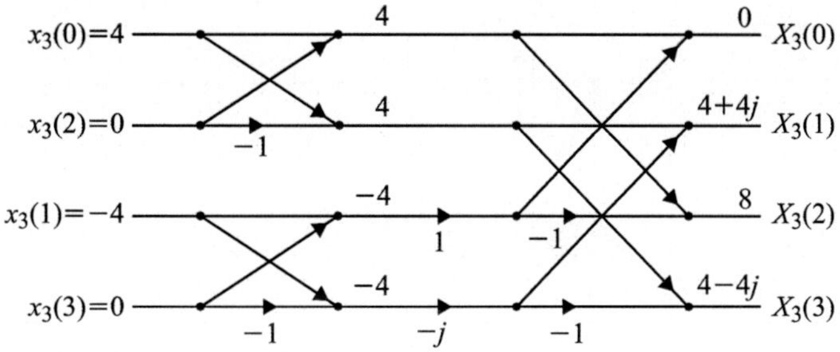

Fig. 3.54 Butterfly diagram of $X_3(k)$

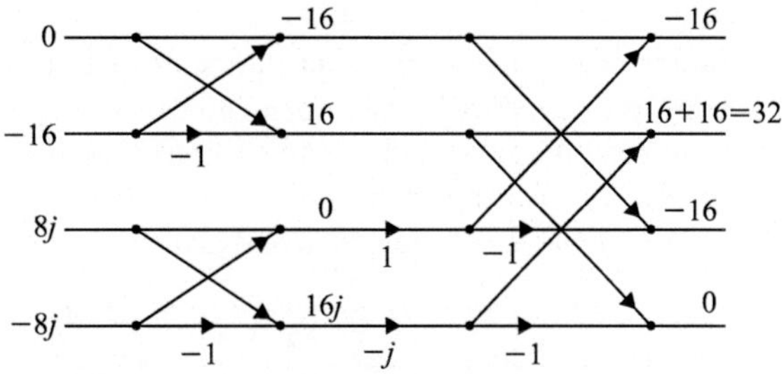

Fig. 3.55 Butterfly diagram of $[x_3(k)H(k)]^*$

$N = 3 + 2 - 1 = 4$, $n_1 = 6$, $n_2 = 0$ so $n = n_1 + n_2 = 6$. Therefore (Figs. 3.54 and 3.55)

$$x_3(n) = \{4, -4, 0, 0\} \quad \text{and} \quad h(n) = \{-1, 1, 0, 0\}$$

$$\begin{aligned} X_3(k) &= \{0, 4 + 4j, 8, 4 - 4j\} \\ X_3(k)H(k) &= \{0, (-1 - j)(4 + 4j), -2(8), (-1 + j)(4 - 4j)\} \\ &= \{0, -8j, -16, 8j\} \end{aligned}$$

IDFT $[X_2(k)H(k)] \Rightarrow$

$$[x_3(k)H(k)]^* = \{0, 8j, -16, -8j\}$$

DFT $\left[\{x_3(k)H(k)\}^*\right]$ is

$$\begin{aligned} y_3(n) &= \frac{1}{N}\left[1DFT\left[x_3(k)H(k)\right]\right] \\ &= \frac{1}{N}\left[DFT\left[[x_3(k)H(k)]^*\right]\right]^* \\ &= \frac{1}{4}\{-16, 32, -16, 0\}^* \end{aligned}$$

Table 3.3 Combined output of each convolution section

n	**0**	**1**	**2**	**3**	**4**	**5**	**6**	**7**	**8**
$y_1(n)$	−1	2	−3	2					
$y_2(n)$				2	−5	6	−3		
$y_3(n)$							−4	8	−4
$y(n)$	−1	2	−3	4	−5	6	−7	8	−4

$$= \{\underset{\substack{\uparrow \\ n=6}}{-4}, 8, -4, 0\}$$

To combine the output of the convolution of each section, $N_2 - 1$ samples are overlapped. The overlapped portion (or samples) are added while combining the output. The combined output response of each convolution is shown in Table 3.3.
Where $N_2 - 1$ are samples that are overlapped. Here, $N_2 = 2$ and $N_2 - 1 = 1$. Samples that are overlapped are added and given below as $y(n)$.

$$\boxed{\therefore\ y(n) = \{-1, 2, -3, 4, -5, 6, -7, 8, -4\}}$$

The above result can be verified as follows:

	1	−1	2	−2	3	−3	4	−4
							−1	1
	1	−1	2	−2	3	−3	4	−4
−1	1	−2	2	−3	3	−4	4	
−1	2	−3	4	−5	6	−7	8	−4

$$\boxed{y(n) = \{-1,\ \underset{\uparrow}{2},\ -3,\ 4,\ -5,\ 6,\ -7,\ 8,\ -4\}}$$

II Overlap Save Method

Method I

In this, the $N_2 - 1$ overlapping samples are placed at the beginning of the section. The number of overlapping samples are $N_2 - 1 = 2 - 1 = 1$ samples:

$$\begin{aligned} x_1(n) &= 1; \quad n = 0 & x_2(n) &= -2; \quad n = 3 & x_3(n) &= 4; \quad n = 6 \\ &-1; \quad n = 1 & &+3; \quad n = 4 & &-4; \quad n = 7 \\ &2; \quad n = 2 & &-3; \quad n = 5 & &0; \quad n = 8 \\ &-2; \quad n = 2 & &4; \quad n = 6 & &0; \quad n = 9 \end{aligned}$$

Convolution of Section-I

$$x_1(n) = \{\underset{\uparrow}{1}, -1, 2, -2\} \quad \text{and} \quad h(n) = \{-1, \underset{\uparrow}{1}, 0, 0\}$$

at $n = 0$ (Figs. 3.56 and 3.57).

$$X_1(k) = \{0, -1 - j, 6, -1 + j\}$$

$$H(k) = \{0, -1 - j, -2, -1 + j\}$$

$$\begin{aligned} X_1(k)H(k) &= \{0, (-1 - j)(-1 - j), (-2)(6), (-1 + j)(-1 + j)\} \\ &= \{0, 2j, -12, -2j\} \\ \{X_1(k)H(k)\}^* &= \{0, -2j, -12, 2j\} \end{aligned}$$

DFT $\left[[X_1(k)H(k)]^*\right]$ is (Fig. 3.58)

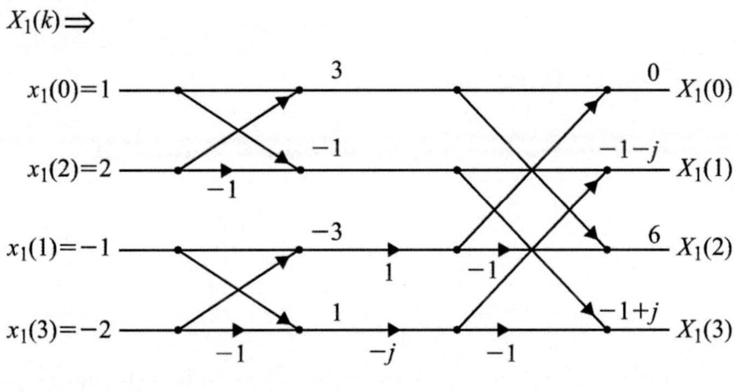

Fig. 3.56 Butterfly diagram of $X_1(k)$

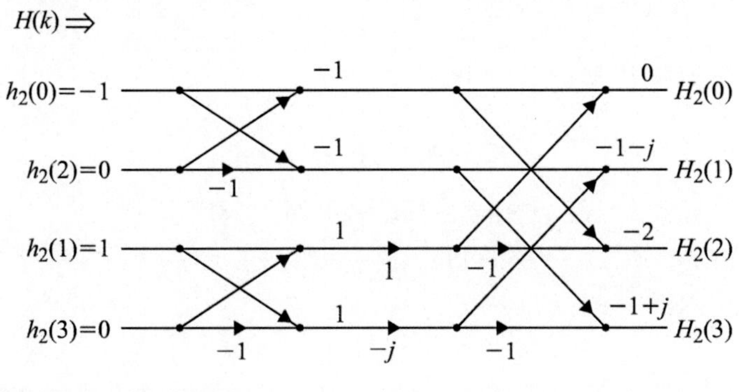

Fig. 3.57 Butterfly diagram of $H(k)$

Fig. 3.58 Butterfly diagram of $[X_1(k)H(k)]^*$

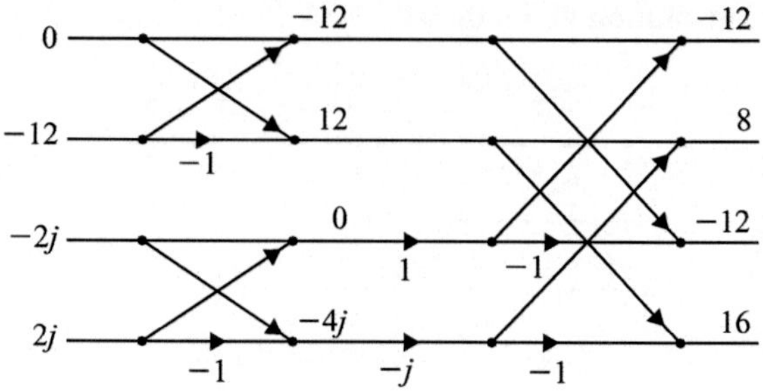

$$\{-12, 8, -12, 16\}$$

$$y_1(n) = \frac{1}{N}\left\{DFT\left[[x(k)H(k)]^*\right]\right\}^* = \frac{1}{4}\{-12, 8, -12, 16\}^*$$

$$y_1(n) = \{\underset{\underset{n=0}{\uparrow}}{-3}, 2, -3, 4\}$$

Convolution of Section-II (Fig. 3.59)

$$x_2(n) = \{\underset{\underset{n=3}{\uparrow}}{-2}, 3, -3, 4\} \quad \text{and} \quad h(n) = \{-1, \underset{\uparrow}{1}, 0, 0\}$$

$$X_2(k) = \{2, 1+j, -12, 1-j\}$$

$$\begin{aligned} X_2(k)H(k) &= \{0, (1+j)(-1-j), (-12)(-2), (-1+j)(1-j)\} \\ &= \{0, -2j, 24, 2j\} \\ [X_2(k)H(k)] &= \{0, 2j, 24, -2j\} \end{aligned}$$

DFT $\left[[X_2(k)H(k)]^*\right]$ is (Fig. 3.60)

$$\{24, -20, 24, -28\}$$

$$\begin{aligned} y_2(n) &= \frac{1}{4}\{24, -20, 24, -28\} \\ &= \{\underset{\underset{n=3}{\uparrow}}{6}, -5, 6, -7\} \end{aligned}$$

Convolution of Section-III (Fig. 3.61)

$X_2(k) \Rightarrow$

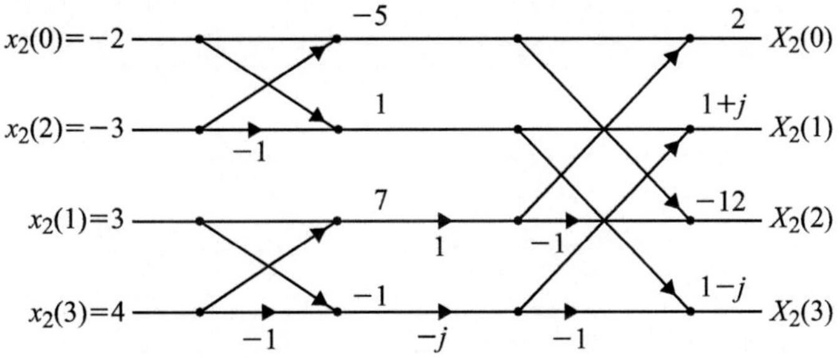

Fig. 3.59 DFT Computation of $X_2(k)$

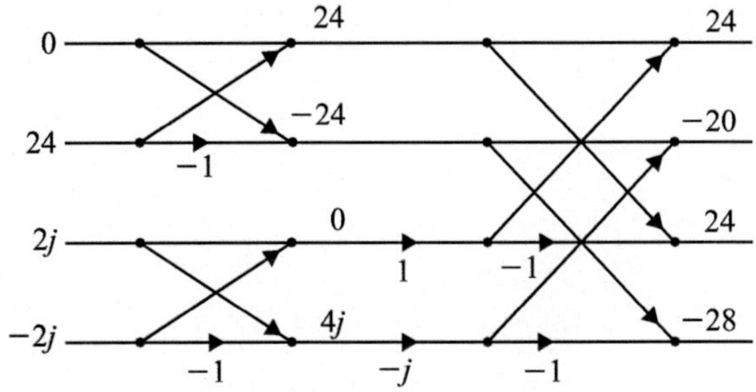

Fig. 3.60 DFT computation of $[X_2(k)H(k)]^*$

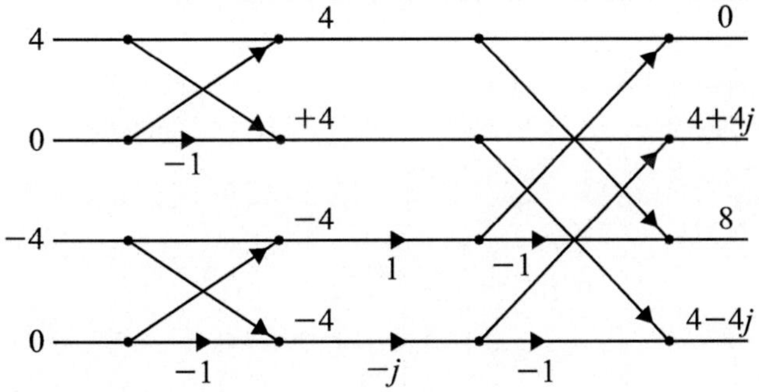

Fig. 3.61 DFT Computation of $x_3(k)$

$$x_3(n) = \{\underset{\substack{\uparrow \\ n=6}}{4}, -4, 0, 0\} \quad \text{and} \quad h(n) = \{\underset{\uparrow}{-1}, 1, 0, 0\}$$

$$X_3(k) = \{0, 4+4j, 8, 4-4j\}$$

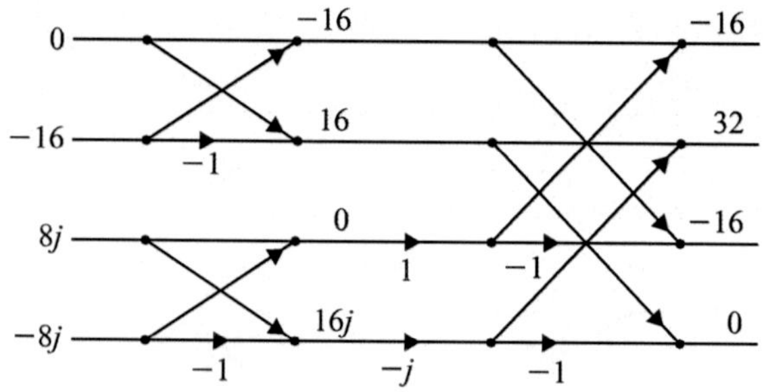

Fig. 3.62 DFT Computation of $[X_3(k)H(k)]^*$

Table 3.4 Combined output of each convolution section in method I

n	0	1	2	3	4	5	6	7	8	9
$y_1(n)$	X	2	−3	4						
$y_2(n)$				X	−5	6	−7			
$y_3(n)$	3						X	8	−4	0
$y(n)$	3	2	−3	4	−5	6	−7	8	−4	0

$$X_3(k)H(k) = \{0, (4+4j)(-1-j), 8(-2), (4-4j)(-1+j)\}$$
$$= \{0, -8j, -16, 8j\}$$
$$\{X_3(k)H(k)\} = \{0, 8j, -16, -8j\}$$

DFT $\left[\{X_3(k)H(k)\}^*\right]$ is (Fig. 3.62)

Butterfly diagram of DFT compulation of $[X_3(k)H(k)]^*$ obtained by overlap save method of Example 3.49 is shown in Fig. 3.62.

$$y_3(n) = \frac{1}{4}\{-16, 32, -16, 0\}^*$$
$$= \{\underset{\substack{\uparrow \\ n=6}}{-4}, 8, -4, 0\}$$

To combine the output of the convolution of each section, the first $N_2 - 1$ samples are overlapped that are discarded. The combined output of each convolution is shown in Table 3.4.

Here $y(n)$ is linear convolution of $x(n)$ and $h(n)$. It can be observed that the results of both the methods are the same, except the first $N_2 - 1$ samples.

Table 3.5 Combined output of each convolution in method II

n	0	1	2	3	4	5	6	7	8	9
$y_1(n)$	−1	2	−3	×						
$y_2(n)$				4	−5	6	×			
$y_3(n)$							−7	8	−4	×
$y(n)$	−1	2	−3	4	−5	6	−7	8	−4	×

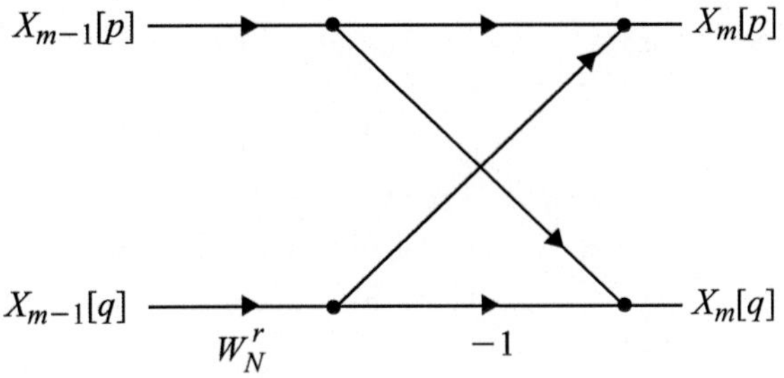

Fig. 3.63 Flow graph

Method II

$$x_1(n) = 1;\ n = 0 \qquad x_2(n) = -2;\ n = 3 \qquad x_3(n) = 4;\ n = 7$$
$$-1;\ n = 1 \qquad +3;\ n = 4 \qquad -4;\ n = 8$$
$$2;\ n = 2 \qquad -3;\ n = 5 \qquad 0;\ n = 9$$
$$0;\ n = 2 \qquad 4;\ n = 6 \qquad 0;\ n = 10$$

$h(n) = \{-1, 1, 0, 0\}$. Find $y_1(n)$, $y_2(n)$ and $y_3(n)$ using the FFT algorithm, the same above method. The result is shown in Table 3.5.

3.7 In-Plane Computation

The Butterfly flow graph of Fig. 3.63 describes an algorithm for the computation of the DFT. In the Butterfly flow graph, the branches connecting the nodes and the transmittance of each of these branches. No matter how the nodes in the flow graph are rearranged, it will always represent the same computation provided that the connection between the nodes and the transmittance of the connection are maintained. The particular form for the flow graph in Fig. 3.63 arose out of deriving the algorithm by separating the original sequences into the even-numbered and odd-numbered points and then continuing to create smaller and smaller subsequences in the same way. An interesting by-product of this derivation is that this flow graph, in addition to describing an efficient procedure for computing the discrete Fourier transform, also suggests a useful way of storing the original data and storing the results of the computation in intermediate arrays.

When implementing the computation depicting in Fig. 3.63, we can imagine the use of two arrays of (complex) storage registers, one for the arrays being computed and one for the data being used in the computation. For example, in computing the first array in Fig. 3.63, one set of storage registers would contain the input data and the second set would contain the computed results for the first stage. We denote the sequence of complex numbers resulting from the nth stage of computation as $X_m(l)$ where $l = 0, 1, \dots N-1$, and $m = 1, 2, \dots, v[v = \log_2^N]$. Furthermore, for convenience we define the set of input samples of $X_0[l]$. We can think of $X_{m-1}[l]$ as the input array and $X_m(l)$ as the output array for the mth stage computation. Thus, for the case $N = 8$ as in Fig. 3.63, we get

$$\begin{aligned} X_0[0] &= x[0] & X_0[4] &= x[1] \\ X_0[1] &= x[4] & X_0[5] &= x[5] \\ X_0[2] &= x[2] & X_0[6] &= x[3] \\ X_0[3] &= x[6] & X_0[7] &= x[7] \end{aligned}$$

Using this notation, the basic butterfly diagram is drawn as shown in Fig. 3.63, with the associated equation as

$$\begin{aligned} X_m[p] &= X_{m-1}[p] + W_N^r X_{m-1}[q] \\ X_m[q] &= X_{m-1}[p] - W_N^r X_{m-1}[q] \end{aligned}$$

In the above equations, p, q and r vary from stage to stage in a manner that is readily inferred from Fig. 3.15. It is clear from Figs. 3.15 and 3.63 that only the complex numbers in locations p and q of the $(m-1)$st array are required to compute the elements p and q of the mth array. Thus, only one complex array of N storage registers is physically necessary to implement the complete computation. That is, $X_m[p]$ and $X_m[q]$ are stored in the same storage registers as $X_{m-1}[p]$ and $X_{m-1}[q]$, respectively. This kind of computation is commonly referred to as an 'in-plane computation'.

If (n_2, n_1, n_0) is the binary representation of the index of the sequence $x[n]$, then the sequence value $x[n_2, n_1, n_0]$ is stored in the array position $X_0[n_0, n_1, n_2]$. That is, in determining the position of $x[n_2, n_1, n_0]$ in the input array, we must reverse the order of the bits of the index n. Let us first consider the process depicted in Fig. 3.64 for sorting a data sequence in normal order by successive examination of the bits representing the data index. If the most significant bit of the data index is zero, $x[n]$ belongs to the top half of the sorted array, otherwise it belongs to the bottom half. Next, the top half and bottom half subsequences can be sorted by examining the second most significant bit, and so on.

Let us now consider the process depicted in Fig. 3.63 for sorting a data sequence in bit reversed order. If the least significant bit is 0, the sequence value corresponds to an even-numbered sample and therefore will appear in the top half and if the least significant bit is 1, the sequence value corresponds to an odd-numbered sample and consequently will appear in the bottom half of the array $X_0[l]$. Next, the even and odd indexed subsequences are sorted into their even and odd indexed parts, and this

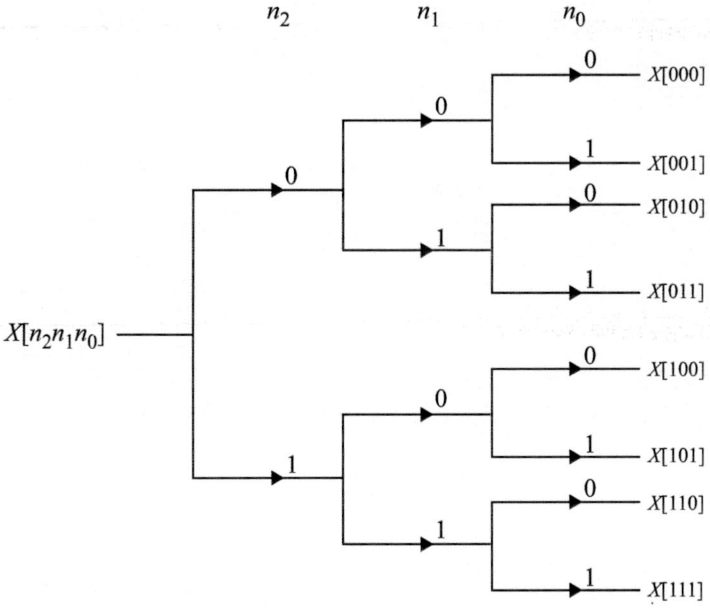

Fig. 3.64 Tree diagram depicting normal-order sorting

can be done by examining the second least significant bit in the index. This process is repeated until N subsequences are depicted by the tree diagram Fig. 3.65.

Summary

- For spectral analysis of discrete signals, the DFT approach is a very straightforward one. However, for larger values of N, DFT becomes tedious because of the huge number of mathematical operations required to perform. Such a huge number of mathematical operations limit the bandwidth of the Digital Signal Processors.
- Several algorithms have been developed to reduce the computational burden and ease the implementation of DFT. The algorithm developed by Cooley and Tukey is the most efficient one and is called Fast Fourier Transform (FFT).
- The most widely used FFT algorithms are radix-2 and radix-4.
- Radix-2 FFT algorithm exploits the two important properties of twiddle factors, namely symmetry and periodicity thereby removing redundant mathematical operations.
- The basic principle of the FFT algorithm is to decompose DFT into successively smaller DFTs. The manner in which this decomposition is done

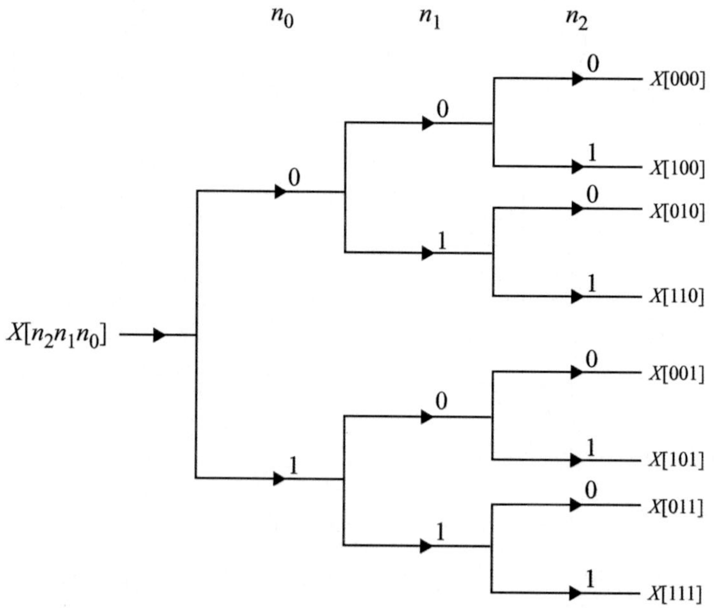

Fig. 3.65 Tree diagram depicting bit reversed-order sorting

leads to different FFT algorithms. Decimation in Time (DIT) and Decimation in Frequency (DIF) are the two basic classes of the algorithm.

- In the algorithm in DIT, the sequence $x(n)$ is decomposed into successively smaller subsequences. Similarly in the DIF algorithm, the sequence of DFT coefficients $X(k)$ is decomposed in smaller subsequences.

Short Questions and Answers

1. **Define DFT of a discrete time sequence (or) analysis equation of DFT.** *(Nov./Dec. 2003)*
Ans: The DFT is used to convert a finite discrete time sequence $x(n)$ to an N-point frequency domain sequence denoted by $X(k)$. The N-point DFT of a finite duration sequence $x(n)$ of length L, where $L \leq N$ is defined as

$$X(k) = \sum_{n=0}^{N-1} x(n)e^{-j\frac{2\pi kn}{N}} \quad \text{for} \quad k = 0, 1, 2, \ldots, (N-1)$$

2. **Define IDFT (or) synthesis equation of DFT.** *(Nov./Dec. 2003)*
Ans: The IDFT is used to convert the N-point frequency domain sequence $X(k)$ to an N-point time domain sequence. The IDFT of the sequence $X(k)$ of

length N is defined as

$$x(n) = \frac{1}{N}\sum_{n=0}^{N-1} X(k)e^{j\frac{2\pi kn}{N}} \quad \text{for} \quad k = 0, 1, 2, \ldots, (N-1)$$

3. **What is the drawback in the Fourier transform and how is it overcome?**
Ans: The drawback in the Fourier transform is that it is a continuous function of ω, and so it cannot be processed by a digital system. This drawback is overcome by using the Discrete Fourier transform. The DFT converts the continuous function of ω to a discrete function of ω.
4. **Give any two applications of DFT (or the importance of DFT)**
Ans:

   (*i*) The DFT is used for spectral analysis of signals using a digital computer.
   (*ii*) The DFT is used to perform filtering operation on signals using digital computer.

5. **What is the relationship between z-transform and DFT?**
Ans: Let an N-point DFT of $x(n)$ be $X(k)$ and the z-transform of $x(n)$ be $X(z)$. The N-point sequence $X(k)$ can be obtained from $X(z)$ by evaluating $X(z)$ at N equally spaced points around the unit circle. That is

$$X(k) = X(z)\Big|_{Z=e^{j\frac{2\pi kn}{N}}} \quad \text{for} \quad k = 0, 1, 2, \ldots, (N-1)$$

6. **Why is linear convolution important in DSP?**
Ans: The response or the output of the LTI discrete time system for any input $x(n)$ is given by linear convolution of the input $x(n)$ and the impulse response $h(n)$ of the system. (This means that if the impulse response of a system is known, then the response of the system for any input can be determined by convolution operation).
7. **Write the properties of linear convolution.**
Ans: The linear convolution has the following properties:

   1. Commutative property: $x(n) * h(n) = h(n) * x(n)$.
   2. Associative property: $[x(n) * h_1(n)] * h_2(n) = x(n) * [h_1(n) * h_2(n)]$.
   3. Distributive property: $x(n) * (h_1(n) + h_2(n)) = x(n) * h_1(n) + x(n) * h_2(n)$.

8. **What is circular convolution?**
Ans: The convolution of two periodic sequences with periodicity N is called circular convolution. If $x_1(n)$ and $x_2(n)$ are two periodic sequences with N samples in a period, then the circular convolution of $x_1(n)$ and $x_2(n)$ is defined as

$$x_1(n) * x_2(n) = \sum_{n=0}^{N-1} x_1(n)x_2(m-n)$$

9. What is zero padding? Why is it needed? *(Nov./Dec. 2003)*
Ans: Appending zeros to a sequence in order to increase the size or length of the sequence is called zero padding. In circular convolution, when the two input sequences are of different size, then they are converted to equal size by zero padding.

10. List the difference between linear convolution and circular convolution.
Ans:

	Linear convolution	Circular convolution
1.	The length of the input sequences can be different	The length of the input sequences should be same
2.	Zero padding is not required	If the length of the input sequences are different, then zero padding is required
3.	The input sequences need not be periodic	At least one of the input sequences should be periodic or should periodically be extended
4.	The output sequence is non-periodic	The output sequence is periodic. The periodicity is same as that of input sequence
5	The length of the output sequence will be greater than the length of input sequences	The length of the input and output sequences are same

11. Why is circular convolution important in DSP?
Ans: The Discrete Fourier Transform is used for the analysis and design of discrete time systems using digital computers. The DFT supports only circular convolution. Hence, when DFT techniques are employed, the results of linear convolution are obtained only via circular convolution.

12. How will you perform linear convolution via circular convolution?
Ans: The linear convolution of two sequences of length N_1 and N_2 produces an output sequence of length $N_1 + N_2 - 1$. To perform linear convolution via circular convolution, convert the input sequences to $N_1 + N_2 - 1$ point sequences by padding with zeros. Now the circular convolution of the $N_1 + N_2 - 1$ point sequences will give an output sequence, which is the same as that of linear convolution of the original two sequences of length N_1 and N_2.

13. What is sectioned convolution?
Ans: In linear convolution of two sequences, if one of the sequences is very much larger than the other, then it is very difficult to compute the linear convolution using DFT. In such cases, the longer sequence is sectioned into sizes of a smaller sequence. Then the linear convolution of each section of a longer sequence and the smaller sequence is performed. The output sequence obtained from the convolution of all the sections are combined to get the overall output sequence. This technique of convolution is called sectioned convolution.

14. Why is sectioned convolution performed?
Ans: In linear convolution of two sequences, if one of the sequences is very

much larger than the other, then it is very difficult to compute the linear convolution using DFT for the following reasons:

(*i*) The entire sequence should be available before convolution can be carried out. This makes long delay in getting the output.
(*ii*) Large amount of memory is required to store the sequences.

15. Compare the overlap add and overlap save methods of sectioned convolutions.

	Overlap add method	Overlap save method
1.	Linear convolution of each section of longer sequence with smaller sequence is performed	Circular convolution of each section of longer sequence with smaller sequence is performed (after converting them to the size of output sequence)
2.	Zero padding is not required	Zero padding is required to convert the input sequence to the size of output sequence
3.	Overlapping of samples of input sections are not required	The $N_2 - 1$ samples of an input section of longer sequence is overlapped with next input section
4	The overlapped samples in the output of sectioned convolutions are added to get the overall output	Depending on the method of overlapping the input samples, either the last $N_2 - 1$ samples or the first $N_2 - 1$ samples of the output sequence of each sectioned convolution are discarded

16. In what way is zero padding implemented in the overlap save method?
Ans: In the overlap save method, the zero padding is employed to convert the small input sequence to the size of the output sequence of each sectioned convolution. The zero padding is also employed to convert either the last section or the first section of the longer input sequence to the size of the output sequence of each sectioned convolution.

17. What is FFT?
Ans: The FFT is a method for computing the DFT with reduced number of calculations. The computational efficiency is achieved by employing the divide-and-conquer approach. This is based on the decomposition of an N-point DFT into successively smaller DFTs.

In an N-point sequence, if N can be expressed as $N = r^m$, then the sequence can be decimated into an r-point sequence, and r-point DFTs are computed. From the results of r-point DFTs, the r^2-point DFTs are computed. From the results of an r^2-point DFT, the r^3-point DFTs are computed and so on, until we get an r^m point DFT. Hence, the number of stages of computation is m. The number r is called the radix of the FFT algorithm.

18. What is radix-2 FFT?
Ans: The radix-2 FFT is an efficient algorithm for computing an N-point DFT of an N-point sequence. In radix-2 FFT, the N-point sequence is decimated into

2-point sequences, and a 2-point DFT for each decimated sequence is computed. From the results of 2-point DFTs, the 4-point DFTs are computed. From the results of 4-point DFTs, the 8-point DFTs are computed and so on until we get an N-point DFT.

19. How many multiplications and additions are involved in radix-2 FFT?

Ans: For performing radix-2 FFT, the value of N should be such that $N = 2^m$. The total number of complex additions are $N \log_2 N$, and the total number of complex multiplications are $(N/2)\log_2 N$.

20. Calculate the percentage saving in calculations in a 512-point radix-2 FFT, when compared to direct DFT.

Ans: Direct computation of DFT:

$$\begin{aligned}\text{Number of complex additions} &= N(N-1) = 512 \times (512-1)\\ &= 2,61,632\\ \text{Number of complex multiplications} &= N^2 = 512 \times 512 = 2,61,144\end{aligned}$$

Radix-2 FFT:

$$\begin{aligned}\text{Number of complex additions} &= N \log_2 N = 4,608\\ \text{Number of complex multiplications} &= (N/2)\log_2 N = 2,304\end{aligned}$$

$$\begin{aligned}&\text{Percentage saving in additions}\\ &= 100 - \frac{\text{Number of additions in radix-2 FFT}}{\text{Number of additions in direct DFT}} \times 100\\ &= 100 - \frac{(4,608)}{(2,61,632)} \times 100 = 98\%\end{aligned}$$

$$\begin{aligned}&\text{Percentage saving in multiplications}\\ &= 100 - \frac{\text{Number of multiplications in radix-2 FFT}}{\text{Number of multiplications in direct DFT}} \times 100\\ &= 100 - \frac{(2,304)}{(2,61,144)} \times 100 = 99.1\%\end{aligned}$$

21. Draw and explain the basic butterfly diagram or flow graph of DIT radix-2 FFT.

Ans:

The basic butterfly diagram of DIT radix-2 FFT is shown in Fig. 3.66. It performs the following operations. Here, a and b are input complex numbers and A and B are output complex numbers.

(*i*) Input complex number b is multiplied by the phase factor W_N^k.

(*ii*) The product bW_N^k is added to the input complex number a to form a new complex number A.

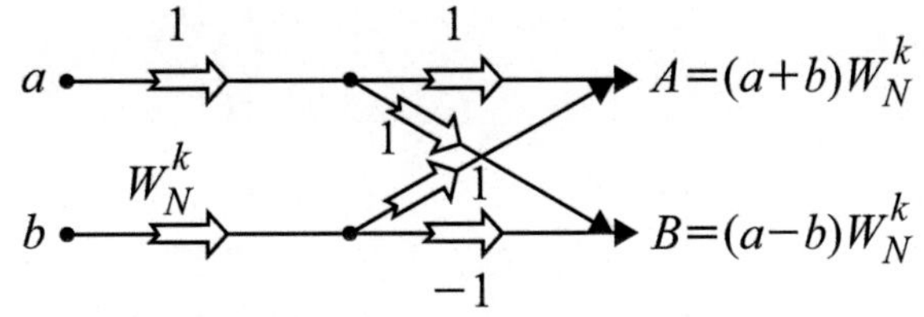

Fig. 3.66 Butterfly diagram for DIT radix-2 FFT

(*iii*) The product bW_N^k is subtracted from the input complex number a to form a new complex number B.

22. What is DIF radix-2 FFT?

Ans: The DIF radix-2 FFT is an efficient algorithm for computing DFT. In this algorithm, the N-point time domain sequence is converted to two $N/2$ point sequences. Then each $N/2$ point sequence is converted to two $N/4$ point sequences. This process is continued until we get $N/2$ 2-point sequences. Now the 2-point DFTs of $N/2$ 2-point sequences will give N samples, which is the N-point DFT of the time domain sequence. Here, the equations for forming $N/2$ point sequences, $N/4$ point sequences, *etc.* are obtained by decimation of frequency domain sequences. Hence this method is called DIF.

23. Compare the DIT and DIF radix-2 FFT.

Ans:

	DIT Radix-2 FFT	DIF Radix-2 FFT
1.	The time domain sequence is decimated	The frequency domain sequence is decimated
2.	When the input is in bit reversed order, the output will be in normal order and *vice versa*	When the input is in normal order, the output will be in bit reversed order and *vice versa*
3.	In each stage of computation, the phase factors are multiplied before add and subtract operations	In each stage of computation, the phase factors are multiplied after add and subtract operations
4.	The value of N should be expressed such that $N - 2m$ and this algorithm consists of m stages of computations	The value of N should be expressed such that $N = 2m$ and this algorithm consists of m stages of computations
5	Total number of arithmetic operations are $N \log_2 N$ complex additions and $(N/2)\log_2 N$ complex multiplications	Total number of arithmetic operations are $N \log_2 N$ complex additions and $(N/2)\log_2 N$ complex multiplications

24. How will you compute IDFT using the radix-2 FFT algorithm?

Ans: Let $x(n)$ be an N-point sequence and $X(k)$ be the N-point DFT of $x(n)$. The sequence $x(n)$ can be computed from the sequence,

$$x(n) = \frac{1}{N}\left(\sum_{k=0}^{N-1} X^*(k) W_N^{nk}\right)^*$$

The following procedure can be used to determine $x(n)$ using radix-2 FFT algorithm:

(*i*) Take conjugate of $X(k)$ (i.e. determine $X^*(k)$).
(*ii*) Compute an N-point DFT of $X^*(k)$ using radix-2 FFT.
(*iii*) Take conjugate of the output sequence from FFT.
(*iv*) Divide the sequence obtained in Step (*iii*) by N. The resultant sequence is $x(n)$.

25. **What is direct or slow convolution and fast convolution.**
Ans: The response of an LTI system is given by convolution of input and impulse response. The computation of the response of an LTI system by convolution sum formula is called slow convolution because it involves very large number of calculations. The number of calculations in DFT computation of the response of LTI system by FFT algorithm is called fast convolution.
26. **Why is FFT needed?**
Ans: The FFT is needed to compute DFT with reduced number of calculations. The DFT is required for spectrum analysis and filtering operations on the signals using digital computers.
27. **How is FFT faster?**
Ans: The FFT algorithm exploits the two important properties of twiddle factors, namely symmetry and periodicity thereby removing redundant mathematical operation. It requires $N/2 \log_2^N$ complex multiplication and $N \log_2^N$ complex addition for computing an N-point DFT. FFT reduces the computation time required to compute a discrete Fourier transform and improves the performance by a factor 100 or more over the direct evaluation of the DFT.
28. **Compute DFT of the sequence** $x(n) = e^{-n},\ 0 \le n \le 4$
Ans:

$$X(k) = \sum_{n=0}^{N-1} x(n) e^{-\frac{j2\pi kn}{N}} = \sum_{n=0}^{4} e^{-n} e^{-\frac{j2\pi kn}{5}} = \sum_{n=0}^{4} \left[e^{-1-\frac{j2\pi k}{5}}\right]^n$$

$$X(k) = \frac{1 - e^{-\left(1+\frac{j2\pi k}{5}\right)^5}}{1 - e^{-\left(1+\frac{j2\pi k}{5}\right)}}$$

29. **Calculate the number of complex multiplications for the direct evaluation of an 8-point DFT.**
Ans: The number of complex multiplications required using the direct computation is

$$N^2 = 8^2 = 64$$

30. Distinguish between DFT and DTFT.
Ans: DFT:

- Obtained by performing sampling operation in both the time and frequency domains.
- Discrete frequency spectrum (discrete function of ω).

DTFT:

- Sampling is performed only in time domain.
- Continuous frequency spectrum (continuous function of ω).

31. Give the computation efficiency of an 1024-point FFT over a 1024-point DFT.
Ans: The number of complex multiplications required using the direct computation is

$$N^2 = (1024)^2 = 1048576$$

The number of complex multiplications required using FFT is

$$\begin{aligned}\frac{N}{2}\log_2^N &= \frac{1024}{2}\log_2^{1024}\\ &= 512\log_2^{2^{10}}\\ &= 5120\end{aligned}$$

Computation efficiency in multiplication

$$\begin{aligned}&= 100 - \frac{5120}{1048576} \times 100\\ &= 99.5\%\end{aligned}$$

The number of complex multiplications required using the direct computation is

$$N(N-1) = 1024(1024-1) = 1047552$$

The number of complex multiplications required using FFT is

$$\begin{aligned}N\log_2^N &= 1024\log_2^{1024}\\ &= 1024\log_2^{2^{10}}\\ &= 10240\end{aligned}$$

Computation efficiency in addition

$$= 100 - \frac{10240}{1048576} \times 100$$
$$= 99\%$$

32. Write the analysis and synthesis equation of DFT.
Ans: Analysis equation of DFT is

$$X(k) = \sum_{n=0}^{N-1} x(n)e^{\frac{-j2\pi kn}{N}}, \quad n = 0, 1, 2, \ldots, N-1$$

Synthesis equation of DFT is

$$x(n) = \frac{1}{N}\sum_{k=0}^{N-1} X(k)e^{\frac{j2\pi kn}{N}j}, \quad n = 0, 1, 2, \ldots, N-1$$

33. Determine the response of the system with $y(n) = x(n-1)$ for the input signal

$$x(n) = \begin{cases} |n|, & -3 \leq n \leq 3 \\ 0, & \text{otherwise} \end{cases}$$

Ans:

$$x(n) = \begin{cases} |n|, & -3 \leq n \leq 3 \\ 0, & \text{otherwise} \end{cases}$$
$$x(n) = \{3, 2, 1, \underset{\uparrow}{0}, 1, 2, 3\}$$

$$y(n) = x(n-1)$$

34. Calculate the multiplication reduction factor α in computing a 1024-point DFT in a radix-2 FFT algorithm.
Ans: The number of complex multiplications required using the direct computation is

$$N^2 = (1024)^2 = 1048576$$

The number of complex multiplications required using FFT is

$$\frac{N}{2}\log_2^N = \frac{1024}{2}\log_2^{1024}$$

$$= 512 \log_2^{2^{10}}$$
$$= 5120$$

Multiplication reduction factor is

$$\alpha = \frac{1048576}{5120} = 204.8$$

35. Find the values of W_N^k when $N = 8$ and $k = 2$ and also for $k = 3$:

$$W_N^k = e^{-\frac{j2\pi k}{N}}$$

Ans: For $N = 8$ and $k = 2$,

$$W_N^k = W_8^2 = e^{-\frac{j2\pi 2}{8}} = -j$$

For $N = 8$ and $k = 3$,

$$W_N^k = W_8^3 = e^{-\frac{j2\pi 3}{8}} = -0.707 - j0.707$$

36. State and prove Parseval's relation to DFT.

Ans: For complex-valued sequence $x(n)$ and $y(n)$, if

$$x(n) \overset{\text{DFT}}{\underset{N}{\longleftrightarrow}} X(k) \quad \text{and} \quad y(n) \overset{\text{DFT}}{\underset{N}{\longleftrightarrow}} Y(k)$$

then

$$\sum_{n=0}^{N-1} x(n) y^*(n) = \frac{1}{N} \sum_{k=0}^{N-1} X(k) Y^*(k)$$

Proof

$$\text{IDFT}\{R_{xy}(k)\} = \Gamma_{xy}(l)$$
$$= \frac{1}{N} \sum_{k=0}^{N-1} R_{xy}(k) e^{\frac{j2\pi kl}{N}}$$

using cross correlation

$$\sum_{k=0}^{N-1} x(n) Y^*(n-l) = \frac{1}{N} \sum_{k=0}^{N-1} R_{xy}(k) e^{\frac{j2\pi kl}{N}}$$

$$\sum_{k=0}^{N-1} x(n)Y^*(n) = \frac{1}{N}\sum_{k=0}^{N-1} R_{xy}(k) \qquad [\text{when } l = 0]$$

$$\sum_{k=0}^{N-1} x(n)Y^*(n) = \frac{1}{N}\sum_{k=0}^{N-1} X(k)Y^*(k)$$

If $x(n) - y(n)$

$$\sum_{k=0}^{N-1} x(n)x^*(n-l) = \frac{1}{N}\sum_{k=0}^{N-1} X(k)X^*(k)$$

$$\sum_{k=0}^{N-1} |x(n)|^2 = \frac{1}{N}\sum_{k=0}^{N-1} |X(k)|^2$$

37. What do you mean by the term 'bit reversal' as applied to FFT?
Ans: In FFT, before decimation the sequences are arranged in bit reversed order. Bit reversal process for $N = 8$ is

Normal order		Bit reversal order	
Input sample	Binary representation	Bit reversal binary	Bit reversed sample input
$x(0)$	000	000	$x(0)$
$x(1)$	001	100	$x(4)$
$x(2)$	010	010	$x(2)$
$x(3)$	011	110	$x(6)$
$x(4)$	100	001	$x(1)$
$x(5)$	101	101	$x(5)$
$x(6)$	110	011	$x(3)$
$x(7)$	111	111	$x(7)$

38. The first five DFT coefficients of a sequence $x(n)$ are $X(0) = 20$, $X(1) = 5 + j2$, $X(2) = 20$, $X(3) = 0.2 + j0.4$ and $X(4) = 0$. Determine the remaining DFT coefficients.
Ans: Let $N = 8$. We have $X(k) = X^*(N - k)$. The first five DFT coefficients are given, and the remaining DFT coefficients are obtained as

$$X(5) = X^*(8-5) = X^*(3) = 0.2 - j0.4$$
$$X(6) = X^*(8-6) = X^*(2) = 20$$
$$X(7) = X^*(8-7) = X^*(1) = 5 - j2$$

39. What are the advantages of the FFT algorithm over the direct computation of DFT?
Ans:

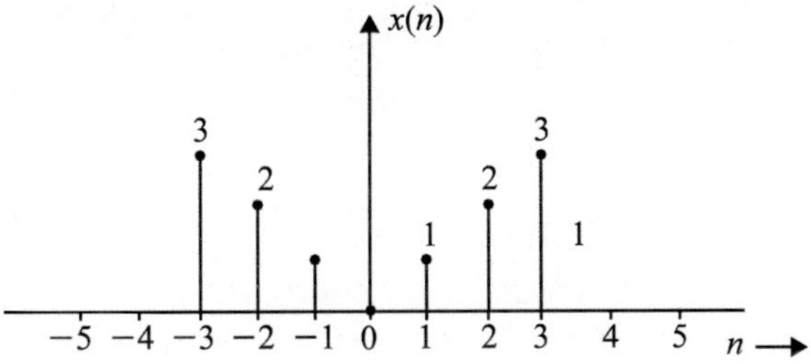

Fig. 3.67 Input sequence for question 33

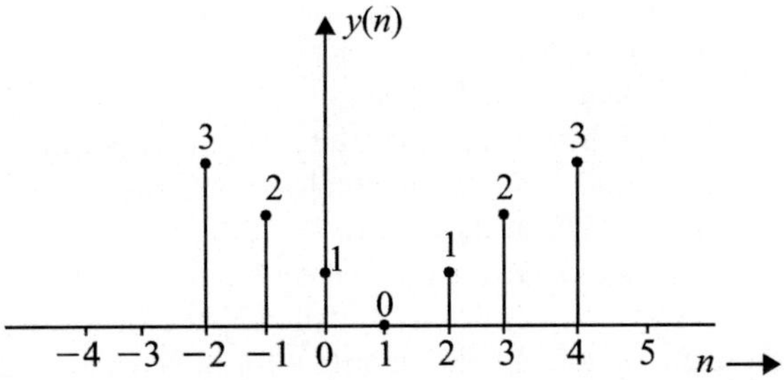

Fig. 3.68 Output response for question 33 with $y(n) = x(n-1)$

1. FFT algorithm reduces the number of complex multiplication and complex addition operations.
2. It reduces the computation time required to compute DFT (Figs. 3.67 and 3.68).

Figure 3.67 for question 33 input sequence

Long Answer Type Questions

1. Find the 8-point DFT of the sequence $x(n) = a^n$ where $a = 3$.
2. Compute the DFTs of the following sequences where $N = 4$ using the DIT algorithm.

   (*a*) $x(n) = 2^n$
   (*b*) $x(n) = 2^{-n}$
   (*c*) $x(n) = \sin\left(\frac{n\pi}{2}\right)$
   (*d*) $x(n) = \cos\left(\frac{n\pi}{2}\right)$.

3. Compute the 16-point DFTs of the following sequences:

$$x(n) = \cos\left(\frac{\pi}{2}\right), \quad 0 \leq n \leq 15$$

   using the DIT algorithm.
4. Compute the FFT for the sequence $x(n) = n^2 + 1$ where $N = 8$ using the DIT algorithm.
5. Compute the DFT coefficients of a finite duration sequence {0,1,2,3,0,0,0,0}.

6. Compute the IDFT of the following sequences using (*a*) the DIT algorithm and (*b*) the DIF algorithm.

$$\begin{array}{ll} (i) & \{5, 0, 1 - j, 0, 1, 0, 1 + j, 0\} \\ (ii) & \{1, 1 + j, 1 - j2, 1, 0, 1 + j2, 1 + j\} \end{array}$$

7. Derive the equation for radix-4 FFT for $N = 4$ and draw the butterfly diagram.
8. Derive the DFT of the sample data sequence $x(n) = \{1, 1, 1, 0\}$ and compute the corresponding amplitude and phase spectrum.
9. Compute the DFT for the sequence $\{1, 2, 0, 0, 0, 2, 1, 1\}$ using the radix-2 DIF-FFT algorithm.
10. Prove that the multiplication of the DFTs of two sequences is equivalent to the DFT of the circular convolution of the two sequences in time domain.
11. Using the DFT-IDFT method, perform circular convolution of the two sequences $x(n) = \{1, 2, 0, 1\}$ and $h(n) = \{2, 2, 1, 1\}$.
12. By means of DFT and IDFT, determine the response of an FIR filter with impulse response $h(n) = \{1, 2, 3\}$ to the input sequence $x(n) = \{1, 2, 2, 1\}$.
13. Develop a DIT FFT algorithm for $N = 12 = 3 \cdot 2 \cdot 2$ and draw the signal flow diagram.
14. The first five DFT coefficient are $\{22, -7.5353 - j3.1213, 1 + j, -0.4645 - j1.1213, 0\}$. Determine the remaining DFT coefficients for $N = 16$.
15. Determine the 8-point DFT of the sequence

$$x(n) = \begin{cases} 1, & -3 \leq n \leq 3 \\ 0, & \text{otherwise} \end{cases}$$

16. Convolve the following sequences using (*i*) overlap add method and (*ii*) overlap save method.

$$x(n) = \{1, -1, 2, 1, 2, -1, 1, 3, 1\} \quad \text{and} \quad h(n) = \{1, 2, 1\}$$

17. An input sequence $x(n) = \{2, 1, 0, 1, 2\}$ is applied to a DSP system having an impulse sequence $h(n) = \{5, 3, 2, 1\}$. Determine the output sequence produced by (*i*) linear convolution and (*ii*) verify the same through circular convolution.
18. Determine the DFT of the sequence

$$x(n) = \begin{cases} 1, & 2 \leq n \leq 6 \\ 0, & \text{otherwise} \end{cases}$$

Assume that $N = 10$.
19. Derive and draw the radix-4 DIT and DFT algorithms for FFT of 8 points.
20. Discuss in detail the use of the FFT algorithm in linear filtering.

Chapter 4
Design of IIR Digital Filters

Learning Objectives

After completing this chapter, you should be able to:

✠ study the characteristics and classify digital filters.

✠ establish the causality and stability conditions necessary when analog filters are converted into digital filters.

✠ design digital filters from analog filters using impulse invariance and bilinear transformation methods.

✠ design digital lowpass filters by Butterworth, Chebyshev approximations and approximation of derivatives.

✠ study analog and digital frequency transformations and design filters of different types.

✠ draw frequency response of digital filters from the system function.

✠ realize the structure of digital filters in different forms.

4.1 Introduction

A filter, is a network or system that changes the frequency response (amplitude *versus* frequency and phase *versus* frequency) characteristics of a signal in a desired manner. The filter extracts information from the signal by removing noise. It is also used to separate two or more signals which can make use of the available

S. Palani, *Discrete Time Systems and Signal Processing*,
https://doi.org/10.1007/978-3-031-32421-5_4

communication channel very efficiently. Filters are broadly classified as analog filters and digital filters. Analog filters operate on analog signals. They are governed by differential equations. They consist of electric components like resistors, capacitors and inductors. On the other hand, digital filters operate on digital samples of the signal. They are governed by linear difference equation. They consist of adders, multipliers and delay implemented in logic.

Digital filters play very important role in Digital Signal Processing. They are extensively used in data transmission, data compression, echo cancellation, image processing, speech processing, telephone, digital audio, biomedical signal processing *etc.* Digital filters have the following advantages over analog filters.

4.1.1 Advantages

1. Digital filters are used in low frequency devices used in biomedical applications where analog filters cannot be used.
2. They have linear phase response characteristic.
3. The performance of digital filters do not vary with environmental changes.
4. Filtered and unfiltered data can be saved for future use.
5. They are cheap and consume low power. Precision is high compared to analog filters.
6. The frequency response can be automatically adjusted using programmable processor.

4.1.2 Disadvantages

The disadvantages of digital filters include the following:

1. The time taken for the design and hardware development is longer compared to analog filters.
2. The maximum bandwidth of signals that can be handled by digital filters is lower compared to analog filters.
3. Digital filters are subject to ADC noise due to quantization and round off noise produced during computation.

4.2 IIR and FIR Filters

Digital filters are broadly classified an Infinite Impulse Response (IIR) and Finite Impulse Response (FIR) filters. The impulse response of the filters can be represented by the sequence $h(k)$, $(k = 0, 1, 2, \ldots)$ as in Fig. 4.1.

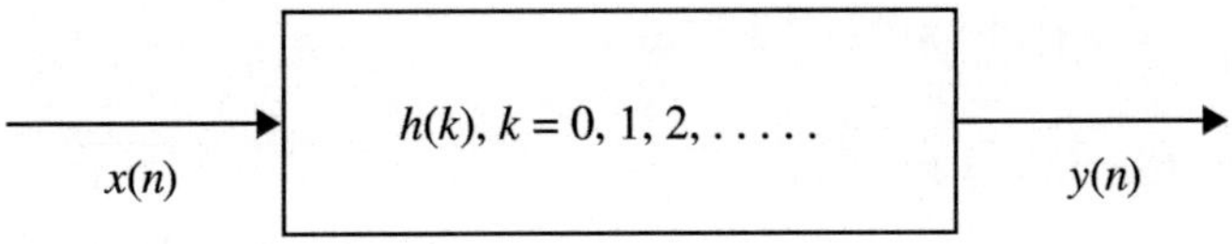

Fig. 4.1 A digital filter

From Fig. 4.1, the following equations are written for the convolution sum.

$$y(n) = \sum_{R=0}^{\infty} h(k)x(n-k) \tag{4.1}$$

$$y(n) = \sum_{R=0}^{N-1} h(k)x(n-k) \tag{4.2}$$

Equation (4.1) gives the output of an IIR filter while Eq. (4.2) gives the output of an FIR filter, where the response is of infinite and finite duration respectively. The output of IIR filter using Eq. (4.1) cannot be computed in practice because the length of the impulse is too long. Equation (4.1) is therefore modified and expressed in recursive form as given below:

$$y(n) = \sum_{R=0}^{N} b_k x(n-k) - \sum_{k=1}^{M} a_k y(n-k) \tag{4.3}$$

In Eq. (4.3) a_k and b_k are the coefficients of IIR filter. Equations (4.2) and (4.3) are used to design FIR and IIR filters respectively. It is to be noted that, the output of the IIR filter (recursive filter) as seen from Eq. (4.3), depends on past outputs as well as present and past input samples. Further, Eq. (4.3) shows the inherent feedback present in the IIR filter. On the other hand, the output of the FIR filter (non-recursive filter) as seen from Eq. (4.2), depends on the past and present values of the input samples. Equations (4.2) and (4.3) can also be represented in terms of the transfer function as given below:

$$H(z) = \sum_{k=0}^{N-1} h(k)z^{-k} \tag{4.4}$$

$$H(z) = \frac{\sum_{k=0}^{N} b_k z^{-k}}{\left[1 + \sum_{k=1}^{M} a_k z^{-k}\right]} \tag{4.5}$$

Equations (4.4) and (4.5) respectively are the transfer functions of FIR and IIR filters.

4.3 Basic Features of IIR Filters

The transfer function of IIR filter is given in Eq. (4.5). For the IIR filter design, it is necessary to find suitable values for the coefficients a_k and b_k so that the desired frequency response is obtained. The design stages for digital IIR filters is as follows:

1. Filter specification: The function of the filter (lowpass, highpass, bandpass, band rejected filter) and the desired performance are given.
2. Calculation of coefficients a_k and b_k.
3. Structure realization.
4. Finite word length effects analysis and solutions.
5. Hardware and software implementation.

4.4 Performance Specifications

For the design of IIR filters, the following specifications are normally followed.

1. Signal characteristics.
2. The frequency response characteristics of the filter.
3. Cost constraints.
4. Mode of implementation.

The frequency response specifications of frequency selective filters such as lowpass, highpass, bandpass and band stop filters are expressed in the form of tolerance scheme as shown in Fig. 4.2.

Analog filter design is a well-established one. Several techniques have been developed to design digital filters which are all based on converting an analog filter into a digital filter. The digital filter is therefore first designed in the analog domain and then the design is converted into the digital domain. For an analog filter, the system function can be represented by the following equation:

$$H_a(s) = \frac{B(s)}{A(s)} = \frac{\sum_{k=0}^{M} \beta_k s^k}{\sum_{k=0}^{N} \alpha_k s^k} \tag{4.6}$$

where $[\alpha_k]$ and $[\beta_k]$ are the coefficients of the filter. The system function $H_a(s)$ can also be written by its impulse response as given below:

$$H_a(s) = \int_{-\infty}^{\infty} h(t)e^{-st}dt \tag{4.7}$$

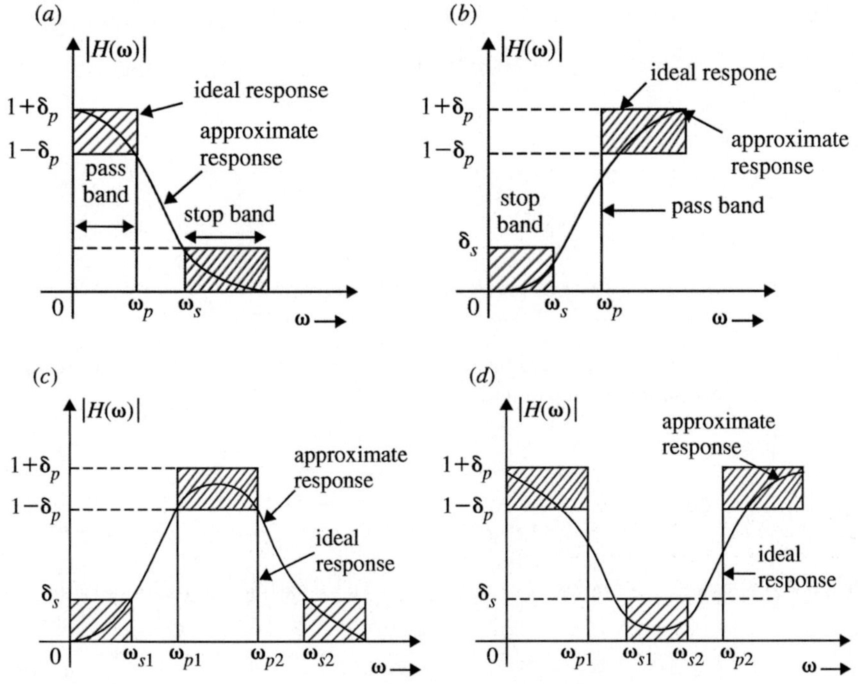

Fig. 4.2 Magnitude response. **a** Lowpass, **b** Highpass, **c** Bandpass and **d** Band stop filter

The analog filter input and output equation can also be represented by the following equations:

$$\sum_{k=0}^{N} \alpha_k \frac{d^k y(t)}{dt^k} = \sum_{k=0}^{M} \beta_k \frac{d^k x(t)}{dt^k} \tag{4.8}$$

where $x(t)$ is input signal and $y(t)$ is the output response. The above equations which represent the characteristics of analog filter in three different forms, are converted to their equivalent forms in digital domain while designing a digital filer. It is to be noted that the conversion becomes valid and effective if the following conditions are satisfied. If $H_a(s)$ is the system function of a linear time invariant stable system, then all its poles should fall in left half of the s-plane (LHP). On that basis, when analog filer is converted to digital filter, then it should satisfy the following stability conditions:

1. The j axis in the s-plane should map into unit circle in the z-plane. This results in a direct relationship between the two frequency variables in s-domain and z-domain.
2. The LHP of the s-plane should map into the interior of the unit circle in the z-plane. This guarantees the stability of the digital filter when converted from analog filter.

3. The T.F.s $H_a(s)$ and $H(z)$ should be rational, so that the filter is causal.

The following methods of digital filter design by converting analog filter are described:

1. Impulse invariance transform method.
2. Bilinear transformation method.

4.5 Impulse Invariance Transform Method

The impulse response of the digital filter is obtained by uniformly sampling the impulse response of the analog filter.

$$h(n) = h_a(t)\Big|_{t=nt} = h_a(nT) \tag{4.9}$$

$$H_a(s) = L[h_a(t)]$$

When $H_a(s)$ has distinct poles, it can be expressed by partial fraction expression as

$$H_a(s) = \sum_{i=1}^{N} \frac{A_i}{s - p_i} = \frac{A_1}{s - p_1} + \frac{A_2}{s - p_2} + \cdots + \frac{A_N}{s - p_N} \tag{4.10}$$

$$h_a(t) = A_1 e^{p_1 t} u_a(t) + A_2 e^{p_2 t} u_a(t) + \cdots + A_N e^{p_N t} u_a(t)$$

$$\therefore\ h(n) = h_a(t)\Big|_{t=nT}$$

$$h(n) = A_1 e^{p_1 nT} u_a(nT) + A_2 e^{p_2 nT} u_a(nT) + \cdots + A_N e^{p_N nT} u_a(nT) \tag{4.11}$$

$$H(z) = Z[h(n)]$$

$$= A_1 \frac{1}{1 - e^{p_1 T} z^{-1}} + A_2 \frac{1}{1 - e^{p_2 T} z^{-1}} + \cdots + A_N \frac{1}{1 - e^{p_N T} z^{-2}}$$

$$= \sum_{i=1}^{N} \frac{A_i}{1 - e^{p_i T} z^{-1}} \tag{4.12}$$

Therefore, by impulse invariant transformation

$$\boxed{\frac{1}{s - p_i} \xrightarrow{\text{is transformed to}} \frac{1}{1 - e^{p_i T} z^{-1}}} \tag{4.13}$$

4.5.1 Relation Between Analog and Digital Filter Poles

Let the analog poles be at $s = p_i$ where $i = 1, 2, \ldots, N$. The digital poles are given by roots of term $(1 - e^{p_i T} z^{-1})$, i.e.,

$$z = e^{p_i T}, \qquad i = 1, 2, \ldots, N \tag{4.14}$$

Any point in the s-plane, can be expressed as

$$\begin{aligned} s &= \sigma_1 + j\Omega_1 \\ z &= e^{p_i T} \\ z &= e^{sT} \end{aligned} \tag{4.15}$$

i.e., analog pole at $s = p_i$ is transformed to digital pole at $z = e^{p_i t}$

$$\begin{aligned} z &= e^{(\sigma_1 + j\Omega_1)T} \\ z &= e^{\sigma_1 T} \cdot e^{j\Omega_1 T} \\ |z|\angle z &= e^{\sigma_1 T} \cdot e^{j\Omega_1 T} \\ |z| &= e^{\sigma_1 T} \quad \text{and} \quad \angle z = \Omega_1 T \end{aligned} \tag{4.16}$$

(*i*) If $\sigma_1 < 0$, then the analog pole s lies on left half of s-plane. In this case, $|z| < 1$, and hence the corresponding digital pole z will lie inside the unit circle in the z-plane.

(*ii*) If $\sigma_1 = 0$, then the analog pole s lies on imaginary axis of s-plane. Therefore, $|z| = 1$, and hence the corresponding digital pole z will lie on the unit circle in the z-plane.

(*iii*) If $\sigma_1 > 0$, then the analog pole s lies on right half of s-plane. In this case, $|z| = 1$, and hence the corresponding digital pole z will lie outside the unit circle in z-plane as shown in Fig. 4.3.

The stability of a filter is related to the location of the poles. For a stable analog filter the poles should lie on the left half of the s-plane. Since the left half of s-plane maps inside the unit circle in z-plane we can say that, for a stable digital filter the poles should lie inside the unit circle in z-plane.

4.5.2 Relation Between Analog and Digital Frequency

Let Ω be the analog frequency in rad/s, and ω is digital frequency in rad/s.

$$\begin{aligned} z &= re^{j\omega} \\ s &= \sigma + j\Omega \end{aligned} \tag{4.17}$$

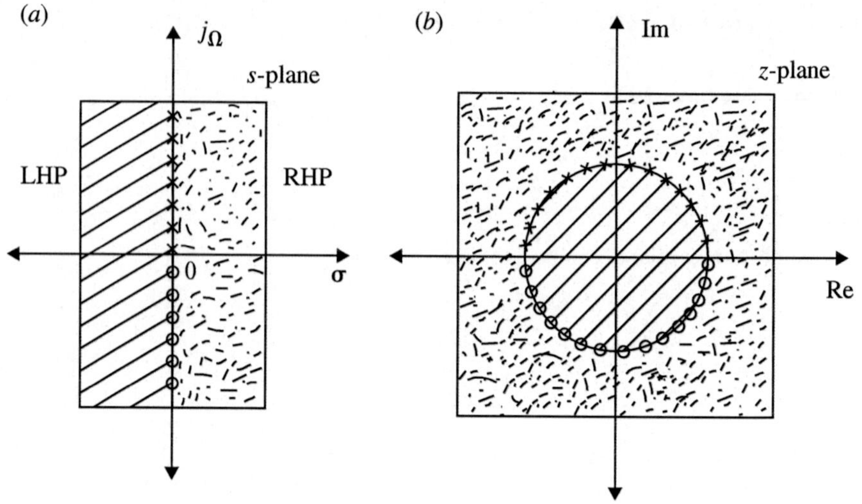

Fig. 4.3 s-plane to z-plane mapping

By impulse invariant transformation, the following equations are written

$$z = e^{sT}$$
$$re^{j\omega} = e^{(\sigma + j\Omega)T}$$
$$re^{j\omega} = e^{\sigma T} \cdot e^{j\Omega T}$$

$$\boxed{\omega = \Omega T \quad \text{or} \quad \Omega = \frac{\omega}{T}} \tag{4.18}$$

The mapping of analog to digital frequency is not one-to-one. Since ω is unique over the range $(-\pi, \pi)$, i.e., $\omega = \Omega T$ implies that the interval $-\pi/T \leq \Omega \leq (\pi/T)$ maps into the corresponding values of $-\pi \leq \omega \leq \pi$. This reflects the effects of aliasing in sampling.

The following impulse invariant transformations from s-plane to z-plane are used for different pole locations:

1. For distinct poles

$$\frac{1}{s - p_i} \longrightarrow \frac{1}{1 - e^{p_i T} z^{-1}} \tag{4.19}$$

2. For multiples poles

$$\frac{1}{(s + p_i)^m} \longrightarrow \frac{(-1)^{m-1}}{(m-1)!} \frac{d^{m-1}}{dp_i^{m-1}} \left(\frac{1}{1 - e^{-p_i T} z^{-1}} \right) \tag{4.20}$$

3. For complex poles

$$\frac{s+a}{(s+a)^2+b^2} \longrightarrow \frac{1-e^{-aT}(\cos bT)z^{-1}}{1-2e^{-aT}(\cos bT)z^{-1}+e^{-2aT}z^{-2}} \tag{4.21}$$

$$\frac{b}{(s+a)^2+b^2} \longrightarrow \frac{1-e^{-aT}(\sin bT)z^{-1}}{1-2e^{-aT}(\cos bT)z^{-1}+e^{-2aT}z^{-2}} \tag{4.22}$$

4.6 Bilinear Transformation

IIR filter design by means of impulse invariance method has a serve limitation in that it can be applied only for lowpass filters and a few limited bandpass filters. This serious limitation is overcome by bilinear transformation which is described below.

The bilinear transformation is a conformal mapping that transforms the imaginary axis of s-plane into the unit circle in the z-plane only once, thus avoiding aliasing of frequency components. In this mapping all points in the left half of s-plane are mapped inside the unit circle in the z-plane and all points in the right half of s-plane are mapped outside the unit circle in the z-plane.

Consider an analog filter with the following system function

$$H_a(s) = \frac{Y(s)}{X(s)} = \frac{b}{(s+a)} \tag{4.23}$$

Equation (4.23) is written in differential equation form as

$$\frac{dy(t)}{dt} + ay(t) = bx(t) \tag{4.24}$$

$y(t)$ can be expressed in terms of its derivative $y'(t)$ with the initial condition $y(t_0)$ as given below:

$$y(t) = \int_{t_0}^{t} y'(\tau)d\tau + y(t_0). \tag{4.25}$$

By using the trapezoidal formula at $t = nT$ and $t_0 = nT - T$, the above integral is approximated as

$$y(nT) = \frac{T}{2}[y'(nT) + y'(nT-T)] + y(nT-T) \tag{4.26}$$

Now Eq. (4.24) is written as follows:

$$y'(nT) = -ay(nT) + bx(nT) \tag{4.27}$$

Substituting Eq. (4.26) in Eq. (4.27), the following difference equation for the discrete system is obtained.

$$\left(1+\frac{aT}{2}\right)y(n)-\left(1-\frac{aT}{2}\right)y(n-1)=\frac{bT}{2}[x(n)+x(n-1)] \quad (4.28)$$

Taking z-transform on both sides of the above equation we get,

$$\left(1+\frac{aT}{2}\right)Y(z)-\left(1-\frac{aT}{2}\right)z^{-1}Y(z)=\frac{bT}{2}[1+z^{-1}]X(z) \quad (4.29)$$

Rearranging the terms, the discrete system function is written as follows.

$$H(z)=\frac{Y(z)}{X(z)}=\frac{\left(\frac{bT}{2}\right)(1+z^{-1})}{\left(1+\frac{aT}{2}\right)-\left(1-\frac{aT}{2}\right)z^{-1}}$$
$$=\frac{b}{\left[\frac{2}{T}\left(\frac{1-z^{-1}}{1+z^{-1}}\right)+a\right]} \quad (4.30)$$

Comparing Eq. (4.23) with Eq. (4.30), the following relationship becomes valid while mapping from s-plane to z-plane.

$$\boxed{s=\frac{2}{T}\left(\frac{1-z^{-1}}{1+z^{-1}}\right)} \quad (4.31)$$

Equation (4.31) is called the bilinear transformation. This holds good for any N^{th} order system. The following examples illustrate the method of transforming $H_a(s)$ to digital system function $H(z)$.

Example 4.1 All-pole analog filters have transfer function

$$H(s)=\frac{1}{(s^2+5s+6)}$$

Find $H(z)$ by impulse invariant technique. Assume $T=1\,\text{s}$.

Solution

$$(s^2 + 5s + 6) = (s + 2)(s + 3)$$

$$H(s) = \frac{1}{(s + 2)(s + 3)}$$

$$= \frac{A}{(s + 2)} + \frac{B}{(s + 3)}$$

$$H(s) = \frac{1}{(s + 2)} - \frac{1}{(s + 3)}$$

By applying impulse invariant transformation we get

$$H(z) = \frac{1}{(1 - e^{-2T} z^{-1})} - \frac{1}{(1 - e^{-3T} z^{-1})}$$

Substituting $T = 1$, we get

$$H(z) = \frac{1}{(1 - 0.135z^{-1})} - \frac{1}{(1 - 0.05z^{-1})}$$

$$\boxed{H(z) = \frac{0.085z^{-1}}{(1 - 0.135z^{-1})(1 - 0.05z^{-1})}}$$

Example 4.2 Design a digital filter using

$$H(s) = \frac{1}{s^2 + 9s + 18}$$

with $T = 0.2\,\mathrm{s}$.

Solution

$$(s^2 + 9s + 18) = (s + 3)(s + 6)$$

$$H(s) = \frac{1}{(s + 3)(s + 6)}$$

$$= \frac{1}{3}\left[\frac{1}{s + 3} - \frac{1}{s + 6}\right]$$

$$H(z) = \frac{1}{3}\left[\frac{1}{(1 - e^{-3T} z^{-1})} - \frac{1}{(1 - e^{-6T} z^{-1})}\right]$$

Substituting $T = 0.2\,\mathrm{s}$ we get

$$H(z) = \frac{1}{3}\left[\frac{1}{(1 - 0.55z^{-1})} - \frac{1}{(1 - 0.3z^{-1})}\right]$$

$$\boxed{H(z) = \frac{0.083z^{-1}}{(1 - 0.55z^{-1})(1 - 0.3z^{-1})}}$$

Example 4.3 Consider the following TF of an analog filter

$$H(s) = \frac{(s+1)}{(s+1)^2 + 1}$$

Convert the analog filter into a digital IIR filter by impulse invariance method for $T = 1$.

Solution Method 1

$$\begin{aligned}
H(s) &= \frac{(s+1)}{(s+1)^2 + 1} \\
&= \frac{(s+1)}{s^2 + 2s + 2} \\
&= \frac{(s+1)}{(s+1+j)(s+1-j)} \\
&= \frac{A_1}{(s+1+j)} + \frac{A_1^*}{(s+1-j)} \\
A_1 &= \left.\frac{A(s+1)}{(s+1+j)}\right|_{s=-1+j} = \frac{1}{2} \\
A_1^* &= \frac{1}{2}
\end{aligned}$$

By impulse invariant transformation the TF of digital IIR filter is obtained as follows:

$$H(z) = \frac{1}{2}\left[\frac{1}{1 - e^{-(1+j)T}z^{-1}}\right] + \frac{1}{2}\left[\frac{1}{1 - e^{(-1+j)T}z^{-1}}\right]$$

Substituting $T = 1$, we get

$$\begin{aligned}
H(z) &= \frac{1}{2}\left[\frac{1}{1 - 0.3678e^{-j}z^{-1}}\right] + \frac{1}{2}\left[\frac{1}{1 - 0.3678e^{j}z^{-1}}\right] \\
&= \frac{1 - 0.3678e^{j}z^{-1} + 1 - 0.3678e^{-j}z^{-1}}{2(1 - 0.3678e^{-j}z^{-1})(1 - 0.3678e^{j}z^{-1})} \\
&= \frac{1 - 0.3678\cos 1z^{-1}}{1 - 0.3678z^{-1}(e^{j} + e^{-j}) + 0.1353z^{-2}}
\end{aligned}$$

$$H(z) = \frac{(1 - 0.2z^{-1})}{(1 - 0.4z^{-1} + 0.1353z^{-2})}$$

Method 2: Using Eq. (4.21), we get

$$\frac{(s+a)}{(s+a)^2 + b^2} = \frac{1 - e^{-aT} \cos bT z^{-1}}{1 - 2e^{-aT} \cos bT z^{-1} + 2e^{-2aT} z^{-2}}.$$

Here, $a = 1; b = 1; T = 1$.

$$H(z) = \frac{(1 - 0.2z^{-1})}{(1 - 0.4z^{-1} + 0.1353z^{-2})}$$

Example 4.4 Using impulse invariant method find $H(z)$ at $T = 1$ s.

$$H(z) = \frac{2}{(s^2 + 8s + 15)}$$

Solution

$$(s^2 + 8s + 15) = (s+3)(s+5)$$
$$H(s) = \frac{2}{(s+3)(s+5)}$$
$$= \frac{1}{s+3} - \frac{1}{s+5}$$

$$H(z) = \frac{1}{(1 - e^{-3T} z^{-1})} - \frac{1}{(1 - e^{-5T} z^{-1})}$$
$$= \frac{1}{(1 - 0.05z^{-1})} - \frac{1}{(1 - 6.73 \times 10^{-3} z^{-1})}$$

$$H(z) = \frac{0.04326z^{-1}}{(1 - 0.05z^{-1})(1 - 6.73 \times 10^{-3} z^{-1})}$$

Hence, in the s-domain transfer function, if s is substituted by the term $\frac{2}{T}\frac{1-z^{-1}}{1+z^{-1}}$, then the resulting transfer function will be z-domain transfer function.

4.6.1 Relation Between Analog and Digital Filter Poles

The mapping of s-domain function to z-domain function by bilinear transformation is a one-to-one mapping i.e., for every point in s-plane, there is exactly a corresponding point in z-plane and *vice versa.*

$$s = \frac{2}{T}\frac{1 - z^{-1}}{1 + z^{-1}} \tag{4.32}$$

$$\frac{Ts}{2} = \frac{1 - \frac{1}{z}}{1 + \frac{1}{z}}$$

$$\frac{Ts}{2} = \frac{z - 1}{z + 1}$$

$$\frac{Ts}{2}(z + 1) = z - 1$$

$$\frac{Ts}{2}z - z = -1 - \frac{Ts}{2}$$

$$-z\left(1 - \frac{Ts}{2}\right) = -\left(1 + \frac{Ts}{2}\right)$$

$$\boxed{z = \frac{1 + \frac{Ts}{2}}{1 - \frac{Ts}{2}}} \tag{4.33}$$

Let $s = \sigma_1 + j\Omega_1$

$$z = \frac{1 + \frac{T}{2}(\sigma_1 + j\Omega_1)}{1 - \frac{T}{2}(\sigma_1 + j\Omega_1)} = \frac{\left(1 + \frac{T}{2}\sigma_1\right) + j\frac{T}{2}\Omega_1}{\left(1 - \frac{T}{2}\sigma_1\right) + j\frac{T}{2}\Omega_1} \tag{4.34}$$

$$|z| = \sqrt{\frac{\left(1 + \frac{T}{2}\sigma_1\right)^2 + \left(\frac{T}{2}\Omega_1\right)^2}{\left(1 - \frac{T}{2}\sigma_1\right)^2 + \left(\frac{T}{2}\Omega_1\right)^2}} \tag{4.35}$$

1. If $\sigma_1 < 0$ (i.e., σ_1 is negative), then the point $s = \sigma_1 + j\Omega_1$, lies on the left half of s-plane. In this case, $|z| < 1$, and hence the corresponding point in z-plane will lie inside the unit circle in z-plane.
2. If $\sigma_1 = 0$ (i.e., real part is zero), then the point $s = \sigma_1 + j\Omega_1$, lies on the imaginary axis in the s-plane. In this case, $|z| = 1$, and hence the corresponding point in z-plane will lie on the unit circle in z-plane.
3. If $\sigma_1 > 0$ (i.e., σ_1 is positive), then the point $s = \sigma_1 + j\Omega_1$, lies on the right half of s-plane. In this case, $|z| > 1$, and hence the corresponding point in z-plane will lie outside the unit circle in z-plane.

The above three transformations are represented in Fig. 4.4.

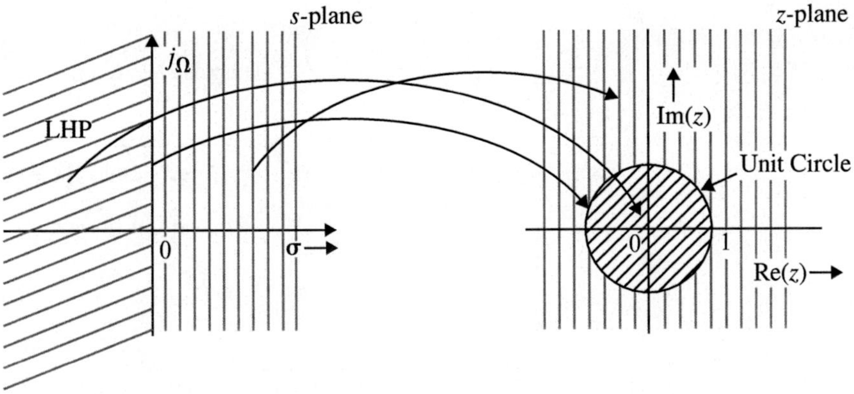

Fig. 4.4 Mapping of s-plane into z-plane in Bilinear transformation

4.6.2 Relation Between Analog and Digital Frequency

Let $s = j\Omega$ be points on imaginary axis and the corresponding points on the z-plane on unit circle are given by $z = e^{j\omega}$ where Ω is the analog frequency and ω is the digital frequency. Using bilinear transformation

$$s = \frac{2}{T}\frac{1 - z^{-1}}{1 + z^{-1}}$$

and substituting $s = j\Omega$ and $z = e^{j\omega}$, we get

$$\begin{aligned} j\Omega &= \frac{2}{T}\frac{1 - e^{-j\omega}}{1 + e^{-j\omega}} = \frac{2}{T}\left[\frac{e^{(j\omega)/2} \cdot e^{-(j\omega)/2} - e^{-j\omega}}{e^{(j\omega)/2} \cdot e^{-(j\omega)/2} + e^{-j\omega}}\right] \\ &= \frac{2}{T}\frac{e^{-(j\omega)/2}(e^{(j\omega)/2} - e^{-(j\omega)/2})}{e^{-(j\omega)/2}(e^{(j\omega)/2} + e^{-(j\omega)/2})} \\ &= \frac{2}{T}\frac{(e^{(j\omega)/2} - e^{-(j\omega)/2})/2}{(e^{(j\omega)/2} + e^{-(j\omega)/2})/2} \end{aligned}$$

$$\begin{aligned} \Omega &= \frac{2}{T}\frac{(e^{(j\omega)/2} - e^{-(j\omega)/2})/2j}{\cos(\omega/2)} \\ &= \frac{2}{T}\frac{\sin(\omega/2)}{\cos(\omega/2)} = \frac{2}{T}\tan\frac{\omega}{2} \end{aligned}$$

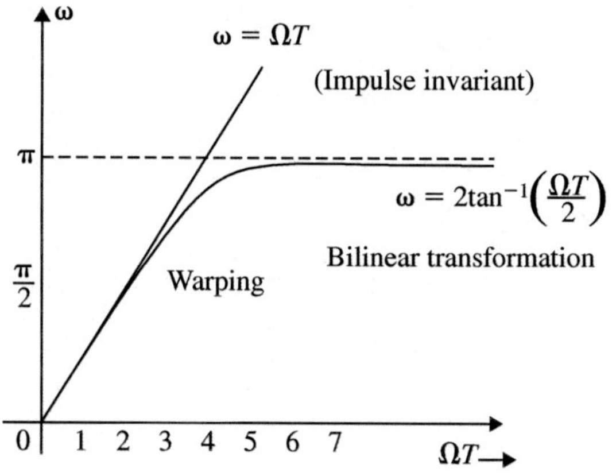

Fig. 4.5 Correspondence between analog and digital frequencies

Analog frequency

$$\boxed{\Omega = \frac{2}{T}\tan\frac{\omega}{2}} \tag{4.36}$$

Digital frequency

$$\boxed{\omega = 2\tan^{-1}\frac{\Omega T}{2}} \tag{4.37}$$

The analog frequency Ω and digital frequency ω have a non-linear relationship and is shown is Fig. 4.5 because the entire negative imaginary axis in the s-plane ($\Omega = -\infty$ to 0) is mapped into the lower half of unit circle in z-plane ($\omega = -\pi$ to 0) and the entire positive imaginary axis in the s-plane ($\Omega = 0$ to ∞) is mapped into upper half of unit circle in z-plane ($\omega = 0$ to π). This non-linear mapping introduces a distortion in the frequency axis which is called frequency warping.

4.6.3 *Effect of Warping on the Magnitude Response*

Consider an analog filter with a number of passbands. The corresponding digital filter will have same number of passbands, but with disproportionate bandwidth as shown in Fig. 4.6. The passband in the analog filter are of constant width with regular intervals. However, the passbands for the digital equivalent is somewhat pushed together.

The effect of warping on amplitude responses can be eliminated by pre-warping the analog filter. In this method, the specified digital frequencies are converted to analog equivalent ($\Omega = \frac{2}{T}\tan(\omega/2)$). These analog frequencies are called pre-warps frequencies. Using the pre-warp frequencies, the analog filter transfer function is designed and then it is transformed to digital filter transfer function.

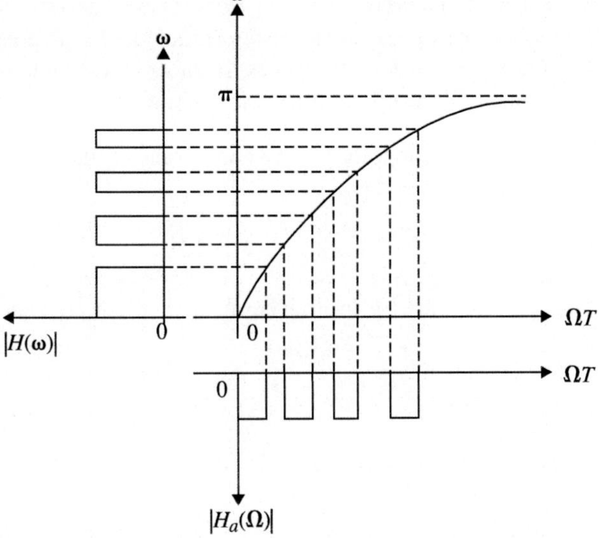

Fig. 4.6 Effect of warping on the magnitude response. (Equally spaced analog passbands are pushed together at high frequency in the digital domain.)

4.6.4 *Effect of Warping on the Phase Response*

Consider an analog filter with linear phase. The phase response of corresponding digital filter will be non-linear as shown in Fig. 4.7.

From the above, it can be state that, the bilinear transformation preserves the magnitude response of an analog filter only if the specification requires piecewise constant

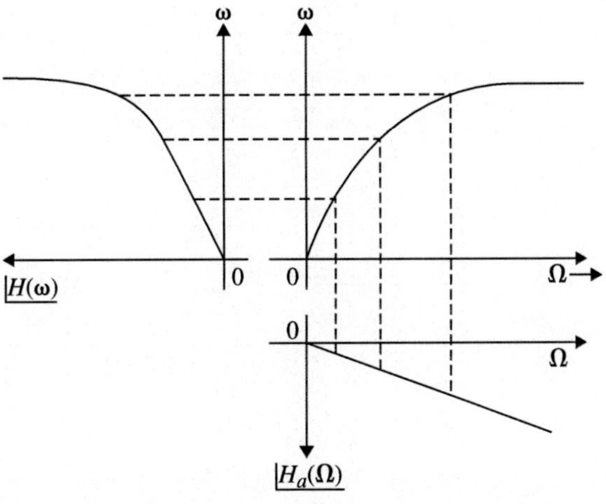

Fig. 4.7 Effect of warping on the phase response

magnitude, but the phase response of the analog filter is not preserved. Therefore, the bilinear transformation can be used only to design digital filters with prescribed magnitude response with piecewise constant values. A linear phase analog filter cannot be transformed to a linear phase digital filter using bilinear transformation.

Example 4.5 Convert the analog filter with system transfer function

$$H_a(s) = \frac{2}{(s+1)(s+2)}$$

into a digital IIR filter by impulse invariance method, when (*i*) $T = 1$ s and (*ii*) $T = 0.1$ s.

Solution Given

$$H_a(s) = \frac{2}{(s+1)(s+2)} = \frac{A}{s+1} + \frac{B}{s+2}$$

$$A = \left.\frac{2}{s+2}\right|_{s=-1} = 2$$

$$B = \left.\frac{2}{s+1}\right|_{s=-2} = -2$$

$$H_a(s) = \frac{2}{(s+1)} - \frac{2}{(s+2)} = \frac{2}{s-(-1)} - \frac{2}{s-(-2)}$$

By impulse invariant transformation, we know that

$$\frac{A_p}{s-p_i} \xrightarrow{\text{transformed to}} \frac{A_i}{1-e^{p_i T} z^{-1}}$$

$$H(z) = \frac{2}{1-e^{p_1 T} z^{-1}} - \frac{2}{1-e^{p_2 T} z^{-1}} \quad (\text{where } p_1 = -1,\ p_2 = -2)$$

$$H(z) = \frac{2}{1-e^{-T} z^{-1}} - \frac{2}{1-e^{-2T} z^{-1}}$$

(*i*) When $T = 1$ s.

$$H(z) = \frac{2}{1-e^{-1} z^{-1}} - \frac{2}{1-e^{-2} z^{-1}}$$

$$= \frac{2}{1-0.3678z^{-1}} - \frac{2}{1-0.1353z^{-1}}$$

$$\boxed{H(z) = \frac{0.465z^{-1}}{(1 - 0.3678z^{-1})(1 - 0.1353z^{-1})}}$$

(*ii*) When $T = 0.1$ s.

$$\begin{aligned} H(z) &= \frac{2}{1 - e^{-0.1}z^{-1}} - \frac{2}{1 - e^{-0.2}z^{-1}} \\ &= \frac{2}{1 - 0.9048z^{-1}} - \frac{2}{1 - 0.8187z^{-1}} \end{aligned}$$

$$\boxed{H(z) = \frac{0.1722z^{-1}}{(1 - 0.9048z^{-1})(1 - 0.8187z^{-1})}}$$

Example 4.6 Convert the analog filter with system transfer function

$$H_a(s) = \frac{(s + 0.1)}{(s + 0.1)^2 + 9}$$

into a digital IIR filter by using the impulse invariant method, when $T = 1$ s.

Solution Given

$$\begin{aligned} H(s) &= \frac{(s + 0.1)}{(s + 0.1)^2 + 9} \\ &= \frac{s + 0.1}{s^2 + 0.2s + 9.01} \\ &= \frac{s + 0.1}{[s - (-0.1 + j3)][s - (-0.1 - j3)]} \\ &= \frac{A_1}{s - (-0.1 + j3)} + \frac{A_1^*}{s - (-0.1 - j3)} \\ A_1 &= \left.\frac{s + 0.1}{s - (-0.1 - j3)}\right|_{s=-0.1+j3} = \frac{1}{2} \\ A_1^* &= \frac{1}{2} \\ H(s) &= \frac{1/2}{s - (-0.1 + j3)} + \frac{1/2}{s - (-0.1 - j3)} \end{aligned}$$

By impulse invariant transformation we get,

$$\begin{aligned}H(z) &= \frac{1/2}{1-e^{(-0.1+j3)T}z^{-1}} + \frac{1/2}{1-e^{(-0.1-j3)T}z^{-1}} \\ &= \frac{0.5-0.5e^{-0.1T}e^{-j3T}z^{-1}+0.5-0.5e^{-0.1T}e^{j3T}z^{-1}}{(1-e^{(-0.1+j3)T}z^{-1})(1-e^{(-0.1-j3)T}z^{-1})} \\ &= \frac{1-0.5e^{-0.1T}z^{-1}(e^{j3T}+e^{-j3T})}{1-e^{-0.1T}-e^{-j3T}z^{-1}e^{-0.1T}e^{j3T}z^{-1}+e^{-0.1T}e^{j3T}e^{-0.1T}e^{-j3T}z^{-2}} \\ &= \frac{1-0.5e^{-0.1T}z^{-1}2\cos 3T}{1-e^{-0.1T}z^{-1}(e^{j3T}+e^{-j3T})+e^{-0.2T}z^{-2}} \\ &= \frac{1-\cos(3T)e^{-0.1T}z^{-1}}{1-2\cos(3T)e^{-0.1T}z^{-1}+e^{-0.2T}z^{-2}}\end{aligned}$$

For $T = 1$ s.

$$H(z) = \frac{1-\cos(3)e^{-0.1}z^{-1}}{1-2\cos(z)e^{-0.1}z^{-1}+e^{-0.2}z^{-2}}$$

where 3 is in radian

$$\boxed{H(z) = \frac{1+0.8959z^{-1}}{1+1.7915z^{-1}+0.8187z^{-2}}}$$

Example 4.7 Apply the bilinear transformation to

$$H_a(s) = \frac{2}{(s+1)(s+2)}$$

with $T = 1$ s and find $H(z)$.

Solution Given

$$H_a(s) = \frac{2}{(s+1)(s+2)}$$

put $s = \frac{2}{T}\frac{1-z^{-1}}{1+z^{-1}}$ in $H_a(s)$ to get $H(z)$

$$\begin{aligned}H(z) &= \frac{2}{\left(\frac{2}{T}\frac{1-z^{-1}}{1+z^{-1}}+1\right)\left(\frac{2}{T}\frac{1-z^{-1}}{1+z^{-1}}+2\right)} \\ &= \frac{2}{\left(\frac{2-2z^{-1}+1+z^{-1}}{1+z^{-1}}\right)\left(\frac{2-2z^{-1}+2+2z^{-1}}{1+z^{-1}}\right)}\end{aligned}$$

$$H(z) = \frac{2(1+z^{-1})^2}{(3-z^{-1})4}$$
$$= \frac{0.5(1+z^{-1})^2}{3-z^{-1}}$$

$$\boxed{H(z) = \frac{-0.5(z^{-1}+1)^2}{(z^{-1}-3)}}$$

Example 4.8 The normalized transfer function of an analog filter is given by

$$Ha(s_n) = \frac{1}{s_n^2 + 1.414s_n + 1}$$

convert the analog filter to a digital filter with a cut-off frequency of 0.4π, using bilinear transformation.

Solution To preserve the magnitude response the pre-warping of analog filter has to be performed. For this analog cut-off frequency is determined. The analog transfer function is un-normalized by replacing $s_n = s/\Omega_c$ using this analog cut-off frequency. Then the analog transfer function is converted to digital filter transfer function using bilinear transformation.

Given $\omega_c = 0.4\pi$ rad/s Let $T = 1$ s

$$\Omega_c = \frac{2}{T}\tan\left(\frac{0.4\pi}{2}\right)$$
$$= 2\tan\left(\frac{0.4\pi}{2}\right) = 1.453 \text{ rad/s.}$$

Normalized analog transfer function is,

$$H_a(s_n) = \frac{1}{s_n^2 + 1.414s_n + 1}$$

The analog transfer function is un-normalized by replacing s_n by s/Ω_c

$$H_a(s) = \frac{1}{\left(\frac{s}{\Omega_c}\right)^2 + 1.414\left(\frac{s}{\Omega_c}\right) + 1}$$
$$= \frac{\Omega_c^2}{s^2 + 1.414\Omega_c s + \Omega_c^2}$$
$$= \frac{2.11}{s^2 + 2.05s + 2.11}$$

$$H(z) = H_a(s)\Big|_{s=\frac{2}{T}\frac{1-z^{-1}}{1+z^{-1}}}$$

$$= \frac{2.11}{\left[\frac{2(1-z^{-1})}{1+z^{-1}}\right]^2 + 2.05\left[\frac{2(1-z^{-1})}{1+z^{-1}}\right] + 2.11}$$

$$= \frac{2.11(1+z^{-1})^2}{4(1-z^{-1})^2 + 4.1(1-z^{-1})(1+z^{-1}) + 2.11(1+z^{-1})^2}$$

$$H(z) = \frac{2.11(1+2z^{-1}+z^{-2})}{10.21 - 3.78z^{-1} + 2.01z^{-2}}$$

$$= \frac{2.11(1+2z^{-1}+z^{-2})}{10.21(1-0.37z^{-1}+0.197z^{-2})}$$

$$\boxed{H(z) = \frac{0.207(1+2z^{-1}+z^{-2})}{1-0.37z^{-1}+0.197z^{-2}}}$$

4.7 Specifications of the Lowpass Filter

Let ω_p be passband digital frequency in rad/s and ω_s be the stopband digital frequency in rad/s. Let

$$A_p = |H(\omega)|_{\omega=\omega_p} \quad \text{and} \quad A_s = |H(\omega)|_{\omega=\omega_s} \tag{4.38}$$

where A_p is the gain at the passband frequency and A_s is the gain at the stopband frequency. Let

$$\alpha_1 = \frac{1}{A_p} \quad \text{and} \quad \alpha_2 = \frac{1}{A_s} \tag{4.39}$$

where α_1 is attenuation at the passband frequency and α_2 is attenuation at the stopband frequency.

The maximum values of normalized gain is unity and so A_p, A_s are less than 1 and α_1, α_2 are greater than 1. The magnitude response of lowpass filter in terms of gain and attenuation are shown in Fig. 4.8.

The magnitude response in dB, is shown in Fig. 4.9. Let k_1 be gain in dB at passband frequency and k_2 the gain in dB at stopband frequency.

$$k_1 = 20\log A_p \quad \text{and} \quad k_2 = 20\log A_s \tag{4.40}$$

$$A_p = 10^{k_1/20} \quad \text{and} \quad A_s = 10^{k_2/20} \tag{4.41}$$

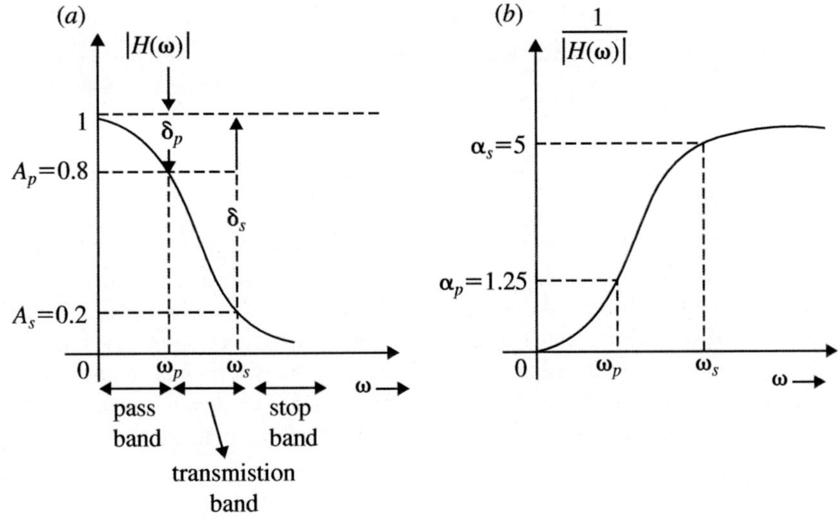

Fig. 4.8 Gain $V_s\omega$ and attenuation $V_s\omega$

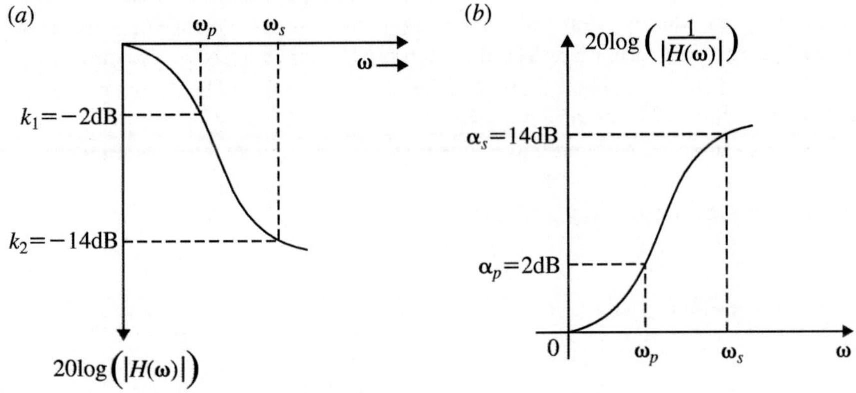

Fig. 4.9 Gain and attenuation $V_s\omega$ in dB

Let α_1 be attenuation in dB at ω_1, α_2 be attenuation in dB at ω_2

$$\alpha_p = 20\log\left(\frac{1}{A_p}\right) \quad \text{and} \quad \alpha_s = 20\log\left(\frac{1}{A_s}\right)$$
$$\alpha_p = -20\log A_p \quad \text{and} \quad \alpha_s = -20\log A_s \tag{4.42}$$

Sometimes the specifications are given in terms of passband ripple δ_p and stopband ripple δ_s. In this case dB gain and attenuation can be estimated as

$$k_1 = 20 \log(1 - \delta_p) \quad \text{and} \quad \alpha_p = -20 \log(1 - \delta_p)$$
$$k_2 = 20 \log \delta_s \quad \text{and} \quad \alpha_s = -20 \log \delta_s. \tag{4.43}$$

4.8 Design of Lowpass Digital Butterworth Filter

On many practical applications, realization of $H(z)$ from analog transfer function $H(z)$ may not be available and it has to be determined from the specifications of the desired digital filters. Normally the cut-off frequency Ω_c, of the filter, the passband attenuation and stopband attenuation are given. The system function $H(s)$ is obtained to satisfy the above specifications. For realization the filter response characteristics have to be approximated. There are several approximation techniques available for the filter design. The most widely used approximation techniques are:

1. Butterworth filter.
2. Chebyshev filter.
3. Approximation of derivatives.

The analog filter transfer function $H(s)$ obtained using any of these methods is converted to digital filter transfer function by using either impulse invariant transformation or bilinear transformation described earlier. The characteristics of Butterworth and Chebyshev filters are ideally suited to design lowpass filter. The design procedure for Butterworth filter is described below.

Let us consider the squared magnitude response of an ideal lowpass filter shown in Fig. 4.10. Ω_c is the cut-off frequency of the filter. This ideal lowpass filter characteristics is however not physically realizable.

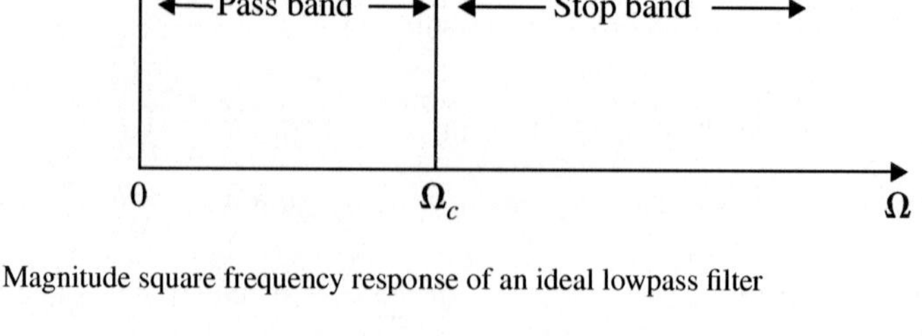

Fig. 4.10 Magnitude square frequency response of an ideal lowpass filter

4.8.1 Analog Butterworth Filter

The analog butterworth filter is designed by approximating the ideal frequency response using an error function. The error function is selected such that magnitude is maximally flat in the passband and monotonically decreasing in the stopband.

The magnitude response of lowpass filter obtained by the approximation is given by

$$|H_a(\Omega)|^2 = \frac{1}{1+\left(\frac{\Omega}{\Omega_c}\right)^{2N}} \tag{4.44}$$

We know that $s = j\Omega$. Substituting Ω by s/j in Eq. (4.44)

$$H_a(s)H_a(-s) = \frac{1}{1+\left(\frac{s/j}{\Omega_c}\right)^{2N}} = \frac{1}{1+\left(-\frac{s^2}{\Omega_c^2}\right)^{N}} \tag{4.45}$$

Let $s/\Omega_c = s_n$ is the normalized function. Therefore, the normalized transfer function is

$$H_a(s_n)H_a(-s_n) = \frac{1}{1+\left(-s_n^2\right)^{N}} \tag{4.46}$$

The normalized transfer function has $2N$ poles. For a stable and causal filter, the poles should lie on the left half of s-plane. when N is even, all poles are complex and exit as conjugate pair. When N is odd, one of the poles is real and all other poles are complex and exit as conjugate pair.

From Eq. (4.46), it is possible to locate the poles of the analog filter in the s-plane. For odd number of N,

$$\begin{aligned} 1+(-s^2)^N &= 0 \\ s^{2N} &= 1 \end{aligned}$$

For $N = 1$,

$$\begin{aligned} s^2 &= 1 \\ s_1 &= 1 \\ s_2 &= -1 \end{aligned}$$

The poles are located as shown in Fig. 4.11.

For a stable filter, only LHP poles along are taken and the denominator of the TF is written as

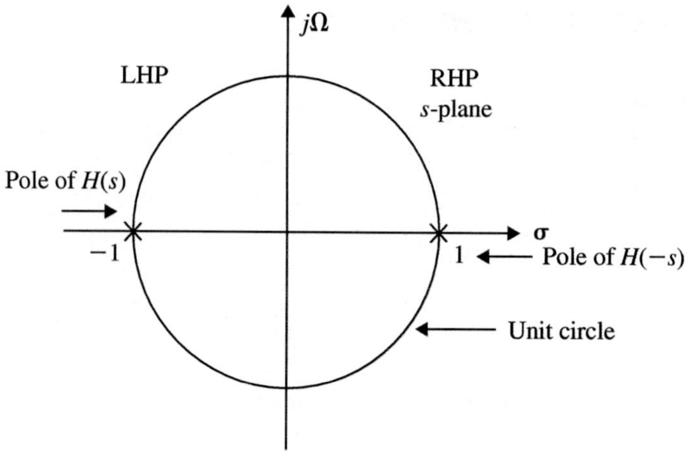

Fig. 4.11 Pole location of first order filter

$$\boxed{H(s) = \frac{1}{(s+1)}}$$

For $N = 3$,

$$s^6 = 1$$

The magnitude of the six poles are 1 and their phase angle is given as

$$\phi_k = \frac{(2k)\pi}{2N}, \quad k = 0, 1, 2, 3, 4, 5$$

Thus,

$$s_1 = 1\angle 0^\circ;\ s_2 = 1\angle 60^\circ;\ s_3 = 1\angle 120^\circ;\ s_4 = 1\angle 180^\circ;\ s_5 = 1\angle 240^\circ;\ s_6 = 1\angle 300^\circ$$

The pole locations are shown in Fig. 4.12.

Considering only LHP poles, the LF of the filter is obtained as

$$H(s) = \frac{1}{(s+1)(s+0.5+j0.866)(s+5-j0.866)}$$

$$\boxed{H(s) = \frac{1}{(s+1)(s^2+s+1)}}$$

Now let us consider N being even. The following numerator polynomial for $H(s)$ is written.

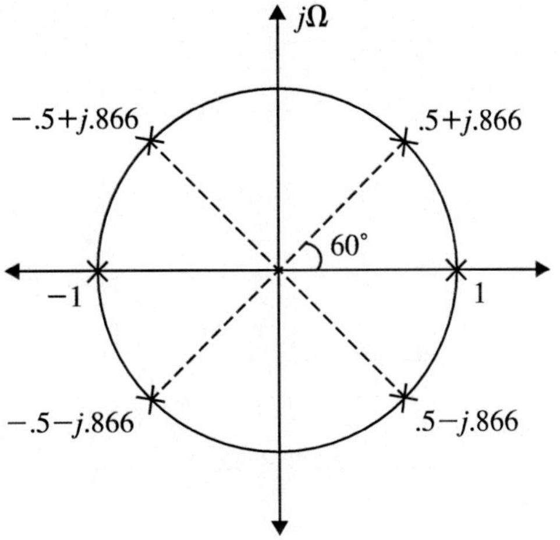

Fig. 4.12 Pole location of third order filter

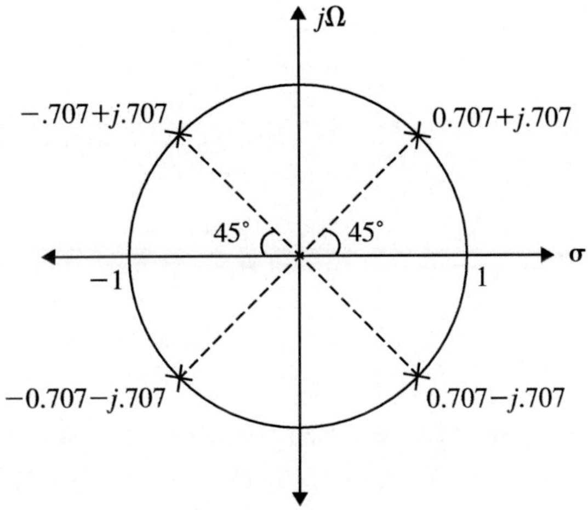

Fig. 4.13 Pole location of a second order filter

$$1 + (-s^2)^N = 0$$
$$s^{2N} = 1$$
$$s_k = 1\angle(2k+1)\pi/2N; \quad k = 0, 1, \ldots, 2N$$

For a second order filter $N = 2$. The poles of $H(s)$ are located as follows:

$$s_1 = 1\angle 45°; \quad s_2 = 1\angle 135°; \quad s_3 = 1\angle 225°; \quad s_4 = 1\angle 315°;$$

The pole locations are shown in Fig. 4.13.

Considering only the LHP poles, the TF of the filter is obtained as given below:

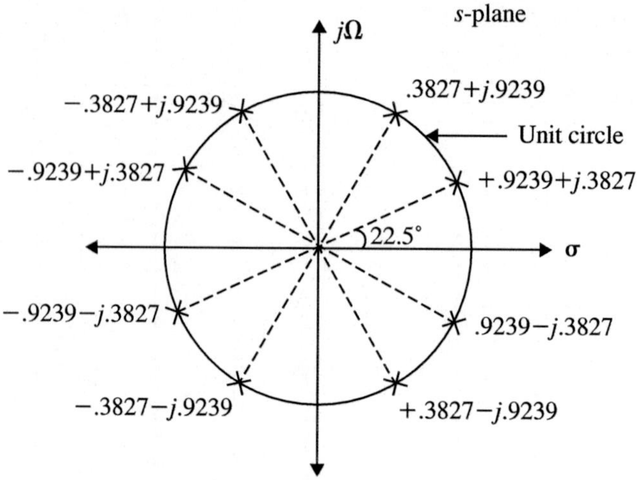

Fig. 4.14 Pole location of fourth order filter

$$H(s) = \frac{1}{(s + 0.707 + j0.707)(s + 0.707 - j.707)}$$

$$\boxed{H(s) = \frac{1}{(s^2 + \sqrt{2}s + 1)}}$$

Now consider $N = 4$,

$$\begin{aligned}
s_k &= 1\angle(2k+1)\pi/2N; \qquad k = 0, 1, 2, 3, 4, 5, 6, 7 \\
s_1 &= 1\angle 22.5°; \quad s_2 = 1\angle 67.5°; \quad s_3 = 1\angle 112.5°; \quad s_4 = 1\angle 157.5°; \\
s_5 &= 1\angle 202.5°; \quad s_6 = 1\angle 247.5°; \quad s_7 = 292.5°; \quad s_8 = 1\angle 337.5° \\
s_1 &= 1\angle 22.5° = \cos 22.5° + j \sin 22.5° = 0.9239 + j.3827 \\
s_2 &= \cos 67.5° + j \sin 67.5° = -0.3827 + j.9239 \\
s_3 &= \cos 112.5° + j \sin 112.5° = -0.3827 + j.9239 \\
s_4 &= \cos 157.5° + j \sin 157.5° = -0.9239 + j.3827 \\
s_5 &= \cos 202.5° + j \sin 202.5° = -0.9239 - j.7071 \\
s_6 &= \cos 247.5° + j \sin 247.5° = -0.3827 - j.9239 \\
s_7 &= \cos 292.5° + j \sin 292.5° = 0.3227 - j.9239 \\
s_8 &= \cos 337.5° + j \sin 337.5° = 0.9239 - j.3827
\end{aligned}$$

The pole locations are given in Fig. 4.14.

Table 4.1 Butterworth polynomial for different N

N	**Denominator of $H(s)$**
1	$(s+1)$
2	$(s^2+\sqrt{2}s+1)$
3	$(s+1)(s^2+s+1)$
4	$(s^2+0.7654s+1)(s+1.8478s+1)$
5	$(s+1)(s^2+0.6180s+1)(s^2+1.6180s+1)$
6	$(s^2+1.9319s+1)(s^2+\sqrt{2}s+1)(s^2+0.517s+s+1)$
7	$(s+1)(s^2+1.8019s+1)(s^2+1.247s+1)(s^2+0.44s+1)$

Considering the LHP poles only, the TF of the filter is obtained as

$$H(s)=\frac{1}{\left[\begin{array}{l}(s+0.3827+j.9239)(s+0.3827-j.9239)\\ \times(s+0.9239+j.3827)(s+0.9239-j.3827)\end{array}\right]}$$

$$\boxed{H(s)=\frac{1}{(s+0.7654s+1)(s+1.8478s+1)}}$$

The Butterworth polynomial for various values of N for $\Omega_c=1$ rad/s is given in Table 4.1.

The Butterworth polynomials given in Table 4.1 are called **Normalized polynomials and the corresponding poles are normalized poles**. Here $\Omega_c=1$ rad/s. If $\Omega_c\neq 1$, the polynomial is called un-normalized polynomial and s is replaced by s/Ω_c in $H(s)$.

4.8.1.1 Determination of Order N and Cut-Off Frequency 'Ω_c' of Butterworth Filter

Determination of order N and cut-off frequency Ω_c is very important in the design of Butterworth filter. Let the following specifications be considered.

$$A_p\leq|H_a(\Omega)|\leq 1 \quad 0\leq\Omega\leq\Omega_p$$
$$|H_a(\Omega)|\leq A_s \quad \Omega_s\leq\Omega$$

Consider the squared magnitude response of Butterworth filter which is given below:

$$|H_a(\Omega)|^2 = \frac{1}{1 + \left(\frac{\Omega}{\Omega_c}\right)^{2N}}$$

Since $1 + \left(\frac{\Omega}{\Omega_c}\right)^{2N} \gg 1$, the above equation is written as

$$|H_a(\Omega)|^2 = \frac{1}{\left(\frac{\Omega}{\Omega_c}\right)^{2N}}$$

According to the given specifications, the following equations are written:

$$\frac{1}{\left(1 + \frac{\Omega_p}{\Omega_c}\right)^{2N}} \geq A_p^2$$

$$\frac{1}{\left(1 + \frac{\Omega_s}{\Omega_c}\right)^{2N}} \leq A_s^2$$

The above equations are rearranged and written as follows:

$$\left(\frac{\Omega_p}{\Omega_c}\right)^{2N} \leq \frac{1}{A_p^2} - 1$$

$$\left(\frac{\Omega_s}{\Omega_c}\right)^{2N} \geq \frac{1}{A_s^2} - 1$$

Dividing the equation one by the other after considering the equalities at Ω_c we get

$$\left(\frac{\Omega_s}{\Omega_p}\right)^{2N} = \frac{\left(\frac{1}{A_s^2} - 1\right)}{\left(\frac{1}{A_p^2} - 1\right)}$$

After taking logarithms on both sides, we solve for N as,

$$\boxed{N = \frac{1}{2} \frac{\log\left[\left(\frac{1}{A_s^2} - 1\right) / \left(\frac{1}{A_p^2} - 1\right)\right]}{\log\left(\frac{\Omega_s}{\Omega_p}\right)}}$$

Note: If the alternations A_p and A_s are given in dB, then A_p and A_s are determined as follows:

$$20 \log A_p = -\text{dB of passband attenuation}$$

$$20 \log A_s = -\text{dB of stopband attenuation}$$

Now to determine Ω_c, any of the equation for passband or stopband attenuation is considered.

$$\left(\frac{\Omega_p}{\Omega_c}\right)^{2N} = \frac{1}{A_p^2} - 1$$

$$\left(\frac{\Omega_p}{\Omega_c}\right) = \left(\frac{1}{A_p^2} - 1\right)^{\frac{1}{2N}}$$

$$\boxed{\Omega_c = \frac{\Omega_p}{\left(\frac{1}{A_p^2} - 1\right)^{1/2N}} \quad \text{or} \quad \Omega_c = \frac{\Omega_s}{\left(\frac{1}{A_s^2} - 1\right)^{1/2N}}}$$

The above equations may not give identical values for Ω_c. Sometimes, the average of these two frequencies are taken to determine Ω_c which is given below:

$$\Omega_c = \frac{1}{2}\left[\frac{\Omega_p}{\left(\frac{1}{A_p^2-1}\right)^{1/2N}} + \frac{\Omega_s}{\left(\frac{1}{A_s^2-1}\right)^{1/2N}}\right]$$

Therefore, the transfer function of the Butterworth filters will be product of second order factors. The analog filter transfer function of normalized and un-normalized Butterworth lowpass filter is also expressed as given below:

Normalized Butterworth Lowpass Filter Transfer Function

Let N be the order of the filter. When N is even

$$H_a(s) = \prod_{k=1}^{N/2} \frac{1}{s_n^2 + b_k s_n + 1} \tag{4.47}$$

When N is odd

$$H_a(s) = \frac{1}{s_n + 1} \prod_{k=1}^{(N-1)/2} \frac{1}{s_n^2 + b_k s_n + 1} \tag{4.48}$$

where

$$b_k = 2\sin\left(\frac{(2k-1)\pi}{2N}\right) \tag{4.49}$$

Un-normalized Butterworth Lowpass Filter Transfer Function

The un-normalized transfer function is obtained by replacing s_n by s/Ω_c, where $\Omega_c = 3$ dB cut-off frequency. When N is even

$$H_a(s) = \prod_{k=1}^{N/2} \frac{\Omega_c^2}{s^2 + b_k \Omega_c s + \Omega_c^2} \tag{4.50}$$

When N is odd

$$H_a(s) = \frac{\Omega_c}{s + \Omega_c} \prod_{k=1}^{(N-1)/2} \frac{\Omega_c^2}{s^2 + b_k \Omega_c s + \Omega_c^2} \tag{4.51}$$

where

$$b_k = 2\sin\left(\frac{(2k-1)\pi}{2N}\right) \tag{4.52}$$

4.8.2 *Frequency Response of Butterworth Filter*

The frequency response of Butterworth filter depends on the order N. The approximated magnitude response approaches the ideal response as the value of N increases.

A_p is gain at the passband frequency ω_p and A_s is the stopband frequency ω_s. The order of the filter is given by

$$N \geq \frac{1}{2} \frac{\log\left[\left(\frac{1}{A_s^2} - 1\right) \Big/ \left(\frac{1}{A_p^2} - 1\right)\right]}{\log\left(\frac{\Omega_s}{\Omega_p}\right)} \tag{4.53}$$

The frequency response of Butterworth filter is shown in Fig. 4.15.

4.8.3 *Properties of Butterworth Filters*

1. The butterworth filters are all-pole designs, (i.e. the zeros of the filter exist at infinity).
2. At the cut-off frequency Ω_c, the magnitude is $1/\sqrt{2}$ or -3 dB.
3. The filter order N completely specifies the filter.
4. The magnitude is flat at the origin.
5. The magnitude is monotonically decreasing function of Ω.
6. The magnitude response approaches the ideal response as the value of N increases.

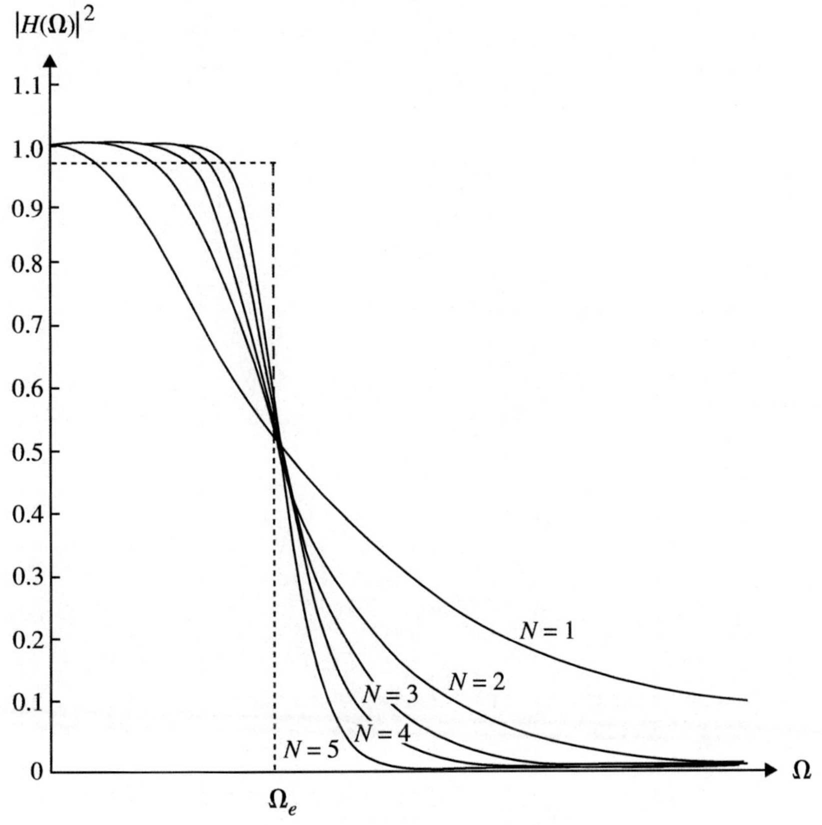

Fig. 4.15 Frequency response of Butterworth filter

4.8.4 Design Procedure for Lowpass Digital Butterworth Filters

Let A_p be gain at passband frequency ω_p and A_s be gain at stopband frequency ω_s. Let Ω_p is analog frequency corresponding to ω_p and Ω_s is analog frequency corresponding to ω_s.

Step 1. Choose either bilinear or impulse invariant transformation.
Step 2. Calculate the ratio Ω_s / Ω_p. For bilinear transformation,

$$\frac{\Omega_s}{\Omega_p} = \frac{\tan(\omega_s/2)}{\tan(\omega_p/2)}$$

For impulse invariant,

$$\frac{\Omega_s}{\Omega_p} = \frac{\omega_s}{\omega_p}$$

Step 3. Determine the order N of the filter using the following equation

$$N \geq \frac{1}{2}\frac{\log\left[\left(\frac{1}{A_s^2}-1\right)\Big/\left(\frac{1}{A_p^2}-1\right)\right]}{\log\left(\frac{\Omega_s}{\Omega_p}\right)}$$

Step 4. Calculate the analog cut-off frequency Ω_c.
For bilinear transformation,

$$\Omega_c = \frac{2}{T}\frac{\tan(\omega_p/2)}{\left[\frac{1}{A_p^2}-1\right]^{1/2N}}$$

For impulse invariant,

$$\Omega_c = \frac{\omega_p/T}{\left[\frac{1}{A_p^2}-1\right]^{1/2N}}$$

Step 5. Determine the analog transfer function of the filter.
When N is even

$$H_a(s) = \prod_{k=1}^{N/2}\frac{\Omega_c^2}{s^2 + b_k\Omega_c s + \Omega_c^2}$$

When N is odd

$$H_a(s) = \frac{\Omega_c}{s+\Omega_c}\prod_{k=1}^{(N-1)/2}\frac{\Omega_c^2}{s^2 + b_k\Omega_c s + \Omega_c^2}$$

where

$$b_k = 2\sin\left(\frac{(2k-1)\pi}{2N}\right)$$

Note: For normalized case take $\Omega_c = 1$ rad/s.

Step 6. Using the chosen transformation determine $H(z)$.

Step 7. Realize the digital filter transfer function $H(z)$ by a suitable structure.

Example 4.9 The specification of the desired lowpass filter is

$$\begin{aligned} 0.8 \le \ |H(\omega)| &\le 1.0; \quad 0 \le \omega \le 0.2\pi \\ |H(\omega)| &\le 0.2; \quad 0.32\pi \le \omega \le \pi \end{aligned}$$

Design Butterworth digital filter using impulse invariant transformation.

Solution Given $A_p = 0.8$ at $\omega_p = 0.2\pi$ rad/s and $A_s = 0.2$ at $\omega_s = 0.32\pi$

$$\frac{\Omega_s}{\Omega_p} = \frac{\omega_s}{\omega_p} = \frac{0.32\pi}{0.2\pi} = 1.6$$

$$N \ge \frac{1}{2}\frac{\log\left(\frac{1}{A_s^2} - 1\right) \Big/ \left(\frac{1}{A_p^2} - 1\right)}{\log\left(\frac{\Omega_s}{\Omega_p}\right)}$$

$$\ge \frac{1}{2}\frac{\log\left[\left(\frac{1}{(0.2)^2} - 1\right) \Big/ \left(\frac{1}{(0.8)^2} - 1\right)\right]}{\log 1.6}$$

$$N \ge 3.9931$$

Choose the order of the filter $N = 4$. The analog cut-off frequency is,

$$\Omega_c = \frac{\omega_p/T}{\left(\frac{1}{A_p^2} - 1\right)^{1/2N}}$$

Let $T = 1$ s.

$$\Omega_c = \frac{\omega_p}{\left(\frac{1}{A_p^2} - 1\right)^{1/2N}} = \frac{0.2\pi}{\left(\frac{1}{(0.8)^2} - 1\right)^{1/8}} = 0.675 \text{ rad/s.}$$

For N is even, the analog transfer function is

$$H_a(s) = \prod_{k=1}^{N/2} \frac{\Omega_c^2}{s^2 + b_k\Omega_c s + \Omega_c^2}$$

where

$$b_k = 2\sin\left(\frac{(2k-1)\pi}{2N}\right)$$

$b_1 = 2\sin\left(\frac{\pi}{8}\right) = 0.765$ and $b_2 = 2\sin\left(\frac{3\pi}{8}\right) = 1.848$

$$
\begin{aligned}
H_a(s) &= \prod_{k=1}^{2} \frac{\Omega_c^2}{s^2 + b_k\Omega_c s + \Omega_c^2} \\
&= \frac{\Omega_c^2}{s^2 + b_1\Omega_c s + \Omega_c^2} \times \frac{\Omega_c^2}{s^2 + b_2\Omega_c s + \Omega_c^2} \\
&= \frac{(0.675)^2}{s^2 + (0.765 \times 0.675)s + (0.675)^2} \times \frac{(0.675)^2}{s^2 + (1.848)(.675)s + (0.675)^2}
\end{aligned}
$$

$$
\begin{aligned}
H_a(s) &= \frac{0.2076}{s^2 + 0.516s + 0.456)(s^2 + 1.247s + 0.456)} \\
&= 0.2076 \Big/ [s - (-0.26 + j0.62)][s - (-0.26 - j0.62)] \\
&\quad \times [s - (-0.62 + j0.26)][s - (-0.6 - j0.26)]
\end{aligned}
$$

Putting into partial fraction we get,

$$
\begin{aligned}
H_a(s) &= \frac{A_1}{s - (-0.26 + j0.62)} + \frac{A_1^*}{s - (-0.26 - j0.62)} \\
&\quad + \frac{A_2}{s - (-0.62 + j0.26)} + \frac{A_2^*}{s - (-0.62 - j0.26)} \\
A_1 &= \left.\frac{0.2076}{[s-(-0.26-j0.62)][s-(-0.62+j0.26)][s-(-0.62-j0.26)]}\right|_{s=-0.26+j0.62} \\
A_1 &= -0.32 + j0.13 \\
A_1^* &= -0.32 - j0.13
\end{aligned}
$$

Similarly

$$
A_2 = 0.32 + j0.76 \quad \text{and} \quad A_2^* = 0.32 - j0.76
$$

The residues A_1, A_1^*, A_2 and A_2^* may be calculated by graphical method. The poles are located in the s-plane as shown in Fig. 4.16.

$$
\begin{aligned}
A_1 &= \frac{0.2076}{1.24\angle 90^\circ \times 0.51\angle 45^\circ 0.95\angle 68^\circ} = 0.3455\angle - 203 \\
&= 0.3455(\cos 203^\circ - j\sin 203^\circ) = -0.32 + j0.13 \\
A_1^* &= -0.32 - j0.13
\end{aligned}
$$

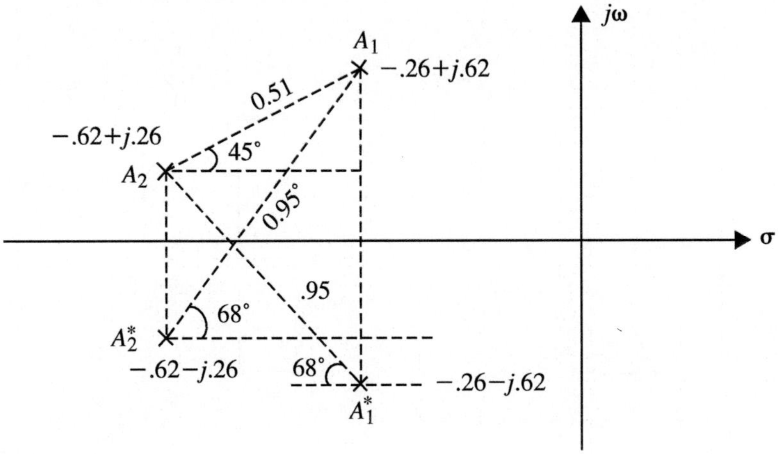

Fig. 4.16 Determination of residues by graphical method

$$A_2 = \frac{0.2076}{0.52\angle 90^\circ 0.51\angle 225^\circ 0.95\angle 112^\circ} = 0.824\angle - 67^\circ$$
$$= 0.824(\cos 67^\circ - j\sin 67^\circ)$$
$$A_2 = 0.32 + j0.76$$
$$A_2^* = 0.32 - j0.76$$

$$H_a(s) = \frac{-0.32 + j0.13}{s - (-0.26 + j0.62)} + \frac{-0.32 - j0.13}{s - (-0.26 - j0.62)} + \frac{0.32 + j0.76}{s - (-0.62 + j0.26)} + \frac{0.32 - j0.76}{s - (-0.62 - j0.26)}$$

$$H(z) = H_a(s)\Big|_{\frac{A_i}{s-p_i} \longrightarrow \frac{A_i}{1-e^{p_i T} z^{-1}}}$$

Let $T = 1$ s.

$$H(z) = \frac{-0.32 + j0.13}{1 - e^{-0.26+j0.62} z^{-1}} + \frac{-0.32 - j0.13}{1 - e^{-0.26-j0.62} z^{-1}} + \frac{0.32 + j0.76}{1 - e^{-0.62+j0.26} z^{-1}} + \frac{0.32 - j0.76}{1 - e^{-0.62-j0.26} z^{-1}}$$

The above equation is simplified as

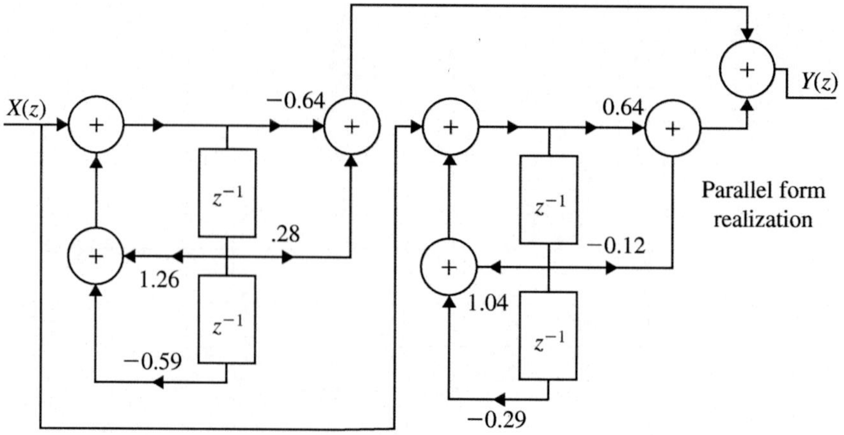

Fig. 4.17 Parallel form realization of $H(z)$ for Example 4.9

$$H(z) = \frac{-0.64 + 0.28z^{-1}}{1 - 1.26z^{-1} + 0.59z^{-2}} + \frac{0.64 - 0.12z^{-1}}{1 - 1.04z^{-1} + 0.29z^{-2}}$$

The above system function is realized in direct form-II and is shown in Fig. 4.17.

Note: For structure realization refer Sect. 4.4.

$$\boxed{H(z) = \frac{-0.64 + 0.28z^{-1}}{1 - 1.26z^{-1} + 0.59z^{-2}} + \frac{0.64 - 0.12z^{-1}}{1 - 1.04z^{-1} + 0.29z^{-2}}}$$

Example 4.10 Design a digital Butterworth filter satisfying the following constraints:

$$0.707 \le |H(e^{j\omega})| \le 1 \qquad \text{for} \quad 0 \le \omega \le \frac{\pi}{2}$$

$$|H(e^{j\omega})| \le 0.2 \qquad \text{for} \quad \frac{3\pi}{4} \le \omega \le \pi$$

with $T = 1$ s using (a) the bilinear transformation; (b) impulse invariant method. Realize the filter in each case using the most convenient realization form.

Solution

(a) **Bilinear Transformation**

Given $A_p = 0.707$ at $\omega_p = (\pi/2)$ and $A_s = 0.2$ at $\omega_s = (3\pi/4)$

$$\frac{\Omega_p}{\Omega_s} = \frac{\tan\frac{\omega_s}{2}}{\tan\frac{\omega_p}{2}} = \frac{\tan\frac{3\pi}{8}}{\tan\frac{\pi}{4}} = 2.414$$

The order of the filter

$$N \geq \frac{1}{2}\frac{\log\left[\left(\frac{1}{A_s^2}-1\right)\Big/\left(\frac{1}{A_p^2}-1\right)\right]}{\log\left(\frac{\Omega_s}{\Omega_p}\right)}$$

$$= \frac{1}{2}\frac{\log\left[\left(\frac{1}{0.2^2}-1\right)\Big/\left(\frac{1}{0.707^2}-1\right)\right]}{\log 2.414}$$

$$N \geq 1.803$$

Let $N = 2$. The analog cut-off frequency

$$\Omega_c = \frac{2}{T}\frac{\tan(\omega_p/2)}{\left(\frac{1}{A_p^2}-1\right)^{1/2N}}$$

$$= 2\frac{\tan(\pi/4)}{\left(\frac{1}{(0.707)^2}-1\right)^{1/4}} = 2 \text{ rad/s.}$$

The analog transfer function when N = even is

$$H_a(s) = \frac{\Omega_c^2}{s^2 + b_k\Omega_c s + \Omega_c^2} \qquad b_k = 2\sin\left[\frac{(2k-1)\pi}{2N}\right] = 2\sin\left(\frac{\pi}{4}\right) = 1.414$$

$$= \frac{4}{s^2 + 2.828s + 4}$$

By using bilinear transformation $H(z)$ is obtained as

$$H(z) = H_a(s)\Big|_{s=\frac{2}{T}\left(\frac{1-z^{-1}}{1+z^{-1}}\right)}$$

$$= \frac{4}{\left[2\left(\frac{1-z^{-1}}{1+z^{-1}}\right)\right]^2 + 2.828 \times 2\left(\frac{1-z^{-1}}{1+z^{-1}}\right) + 4}$$

$$= \frac{4(1+z^{-1})^2}{4(1-z^{-1})^2 + 5.656(1-z^{-2}) + 4(1+z^{-1})^2}$$

$$= \frac{0.2929(1+z^{-1})^2}{1+0.1716z^{-2}}$$

$$\boxed{H(z) = \frac{0.2929(1+2z^{-1}+z^{-2})}{1+0.1716z^{-2}}}$$

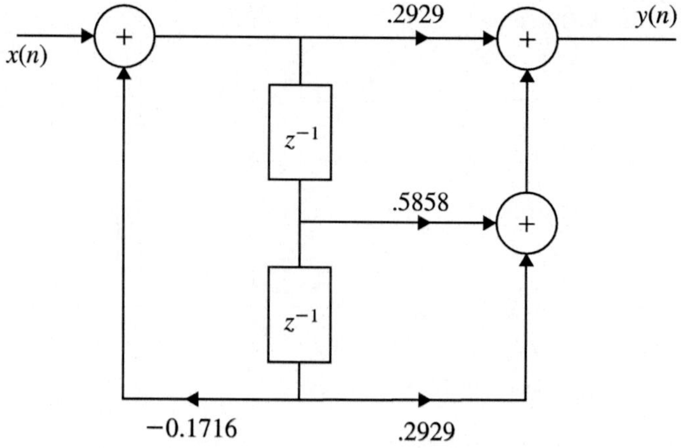

Fig. 4.18 Direct form-II realization of $H(z)$ for the Example 4.10(a)

The above system is realized in direct form-II and is shown in Fig. 4.18a.

(*b*) **Impulse Invariant Method**

Given $A_p = 0.707$ at $\omega_p = (\pi/2)$ and $A_s = 0.2$ at $\omega_s = (3\pi/4)$

$$\frac{\omega_s}{\omega_p} = \frac{3\pi/4}{\pi/2} = \frac{3}{2}$$

$$N \geq \frac{1}{2}\frac{\log\left[\left(\frac{1}{(0.2)^2}-1\right)\Big/\left(\frac{1}{(0.707)^2}-1\right)\right]}{\log\left(\frac{3}{2}\right)}$$

$$N \geq 3.924$$

$$N = 4$$

$$\Omega_c = \left(\frac{1}{A_p^2 - 1}\right)^{\frac{1}{2N}} \cdot \Omega_p$$

$$= \left(\frac{1}{0.707^2 - 1}\right)^{\frac{1}{8}} \cdot \frac{\pi}{2}$$

$$= 1.57$$

The transfer function of analog filter for N being even is,

$$H_a(s) = \prod_{k=1}^{2} \frac{\Omega_c^2}{s^2 + b_k\Omega_c s + \Omega_c^2}$$

where $b_1 = 0.7654$ and $b_2 = 1.8478$

$$H_a(s) = \frac{(1.57)^4}{(s^2 + 1.202s + 2.465)(s^2 + 2.902s + 2.465)}$$
$$= \frac{A}{s+1.45+j0.6} + \frac{A^*}{s+1.45-j0.6} + \frac{B}{s+0.6+j1.45} + \frac{B^*}{s+0.6-j1.45}$$
$$A = \left.\frac{(1.57)^4}{(s + 1.45 - j0.6)(s + 0.6 + j1.45)(s + 0.6 - j1.45)}\right|_{s=-1.45-j0.6}$$
$$= 0.7253 + j1.754$$
$$B = \left.\frac{(1.57)^4}{(s + 1.45 + j0.6)(s + 1.45 - j0.6)(s + 0.6 - j1.45)}\right|_{s=-0.6-j1.45}$$
$$= -0.7253 - j0.3$$

$$H_a(s) = \frac{0.7253 + j1.754}{s + 1.45 + j0.6} + \frac{0.7253 - j1.754}{s + 1.45 - j0.6} + \frac{-0.7253 - 0.3j}{s + 0.6 + j1.45} + \frac{-0.7253 + 0.3j}{s + 0.6 - j1.45}$$

Impulse invariant transformation is

$$\frac{1}{s - p_i} \Longrightarrow \frac{1}{1 - e^{p_i T} z^{-1}}$$

let $T = 1$ s.

$$H(z) = \frac{0.7253 + j1.754}{1 - e^{-1.45} \cdot e^{-j0.6} z^{-1}} + \frac{0.7253 - j1.754}{1 - e^{-1.45} \cdot e^{+j0.6} z^{-1}} + \frac{-0.7253 - 0.3j}{1 - e^{-0.6} \cdot e^{-j1.45} z^{-1}} + \frac{-0.7253 + 0.3j}{1 - e^{-0.6} \cdot e^{j1.45} z^{-1}}$$

$$\boxed{H(z) = \frac{1.454 + 0.1839z^{-1}}{1 - 0.387z^{-1} + 0.055z^{-2}} + \frac{-1.454 + 0.2307z^{-1}}{1 - 0.1322z^{-1} + 0.301z^{-2}}}$$

This can be realized using parallel form as shown in Fig. 4.19.

Example 4.11 For the given specification design an analog Butterworth filter

$$0.9 \leq |H(j\Omega)| \leq 1 \quad \text{for} \quad 0 \leq \Omega \leq 0.2\pi$$
$$|H(j\Omega)| \leq 0.2 \quad \text{for} \quad 0.4\pi \leq \Omega < \pi$$

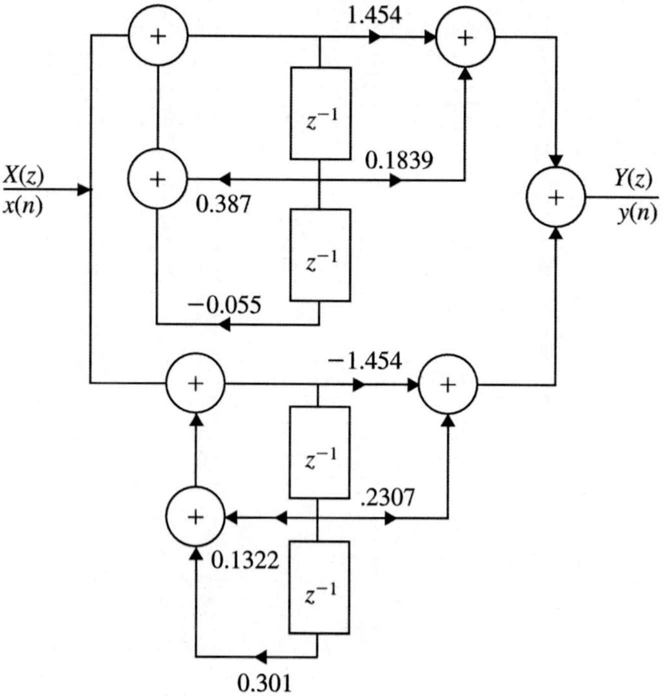

Fig. 4.19 Parallel form realization of $H(z)$ for Example 4.10

Solution Given $\Omega_p = 0.2\pi$, $\Omega_s = 0.4\pi$, $A_p = 0.9$, $A_s = 0.2$.

$$N \geq \frac{1}{2}\frac{\log\left[\left(\frac{1}{A_s^2} - 1\right) \Big/ \left(\frac{1}{A_p^2} - 1\right)\right]}{\log\left(\frac{\Omega_s}{\Omega_p}\right)}$$

$$N \geq \frac{1}{2}\frac{\log\left[\left(\frac{1}{(0.2)^2} - 1\right) \Big/ \left(\frac{1}{(0.9)^2} - 1\right)\right]}{\log\left(\frac{0.4\pi}{0.2\pi}\right)}$$

$$N \geq 3.34$$

$$N = 4$$

$$\Omega_c = \frac{\Omega_p}{(10^{0.1} - 1)^{1/2N}} \quad \text{or} \quad \Omega_c = \frac{\Omega_p}{\left(\frac{1}{A_p^2} - 1\right)^{1/2N}}$$

$$= \frac{0.2\pi}{(0.484)^{1/4}} = 0.24\pi$$

The analog transfer function

$$H_a(s) = \prod_{k=1}^{2} \frac{\Omega_c^2}{s^2 + b_k\Omega_c s + \Omega_c^2}$$
$$= \frac{\Omega_c^2}{s^2 + b_1\Omega_c s + \Omega_c^2} \cdot \frac{\Omega_c^2}{s^2 + b_2\Omega_c s + \Omega_c^2}$$
$$b_k = 2\sin\left(\frac{(2k-1)\pi}{2N}\right), \qquad k = 1, 2,$$

where $b_1 = 0.7653$ and $b_2 = 1.8477$

$$\boxed{H_a(s) = \frac{0.323}{(s^2 + 0.577s + 0.0576\pi^2)(s^2 + 1.393s + 0.0576\pi^2)}}$$

Example 4.12 Determine the order and the poles of lowpass Butterworth analog filter that has a 3 dB attenuation at 500 Hz and an attenuation of 40 dB at 1000 Hz.

Solution Given $-20\log A_p = 3$ dB, $\Omega_p = 2\pi \times 500 = 1000\pi$ rad/s and $-20\log A_s = 40$ dB $\Omega_s = 2\pi \times 1000 = 2000\pi$ rad/s.

$$20\log A_p = -3$$
$$A_p = 0.707$$
$$20\log A_s = -40$$
$$A_s = 0.01$$

$$N \geq \frac{1}{2}\frac{\log\left[\left(\frac{1}{A_s^2} - 1\right) \Big/ \left(\frac{1}{A_p^2} - 1\right)\right]}{\log\left(\frac{\Omega_s}{\Omega_p}\right)}$$
$$N \geq \frac{1}{2}\frac{\log\left[\left(\frac{1}{(0.01)^2} - 1\right) \Big/ \left(\frac{1}{(0.707)^2} - 1\right)\right]}{\log\left(\frac{2000\pi}{1000\pi}\right)}$$
$$N \geq 6.64319$$
$$N = 7$$

The poles of Butterworth filter is given by $s_k = \Omega_c e^{j\Phi_k}$

$$\Omega_c = \frac{\Omega_p}{\left(\frac{1}{A_1^2} - 1\right)^{1/2N}} = \frac{1000\pi}{\left(\frac{1}{(0.707)^2} - 1\right)^{1/14}} = 1000\pi$$

$$s_k = 1000\pi e^{j\Phi_k}, \quad k = 1, \ldots, 7$$

where

$$\phi_k = \frac{\pi}{2} + \frac{(2k-1)\pi}{2N}, \quad k = 1, \ldots, 7$$

The normalized poles of analog Butterworth filter are obtained as follows. (This can be taken from Table 4.1 also).

$$\begin{aligned}
\text{For } k = 1, \quad & \phi_1 = \frac{\pi}{2} + \frac{\pi}{14} = 102.865^\circ \\
\text{For } k = 2, \quad & \phi_2 = 128.57^\circ \\
\text{For } k = 3, \quad & \phi_3 = 154^\circ \\
\text{For } k = 4, \quad & \phi_4 = 180^\circ
\end{aligned}$$

The normalized LHP poles are,

$$\begin{aligned}
s_1 &= 1e^{j102.865^\circ} = -0.2265 + j0.9749 \\
s_2 &= e^{j128.57^\circ} = -0.6235 + j0.7818 \\
s_3 &= e^{j154^\circ} = -0.8987 + j0.4384
\end{aligned}$$

$$\begin{aligned}
s_4 &= e^{j180^\circ} = -1 \\
s_5 &= s_3^* = -0.8987 - j0.4384 \\
s_6 &= s_2^* = -0.6235 - j0.7818 \\
s_7 &= s_1^* = -0.2265 - j0.9749
\end{aligned}$$

These poles in LHP of the s-plane are shown in Fig. 4.20.

The transfer function of analog Butterworth normalized filter is given as

$$\begin{aligned}
H_a(s) &= \frac{1}{(s+s_4)(s+s_1)(s+s_7)(s+s_2)(s+s_6)(s+s_3)(s+s_5)} \\
&= \frac{1}{[(s+1)(s^2+1.8019s+1)(s^2+1.247s+1)(s^2+0.44s+1)]}
\end{aligned}$$

The transfer function of the un-normalized filter is obtained by replaced s by

$$\frac{s}{\Omega_c} = \frac{s}{1000\pi}.$$

Therefore

Fig. 4.20 Location of normalized poles

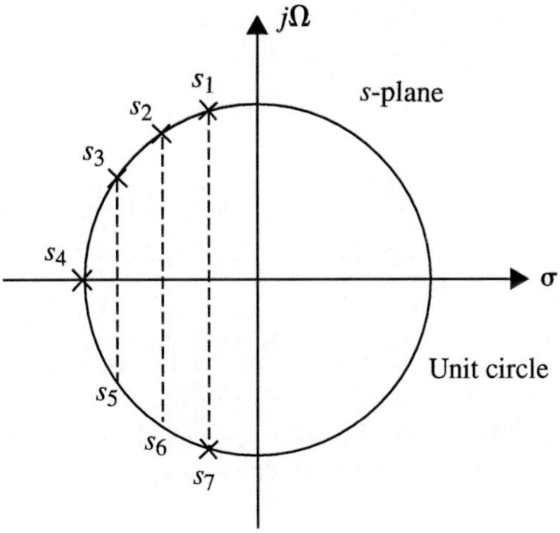

$$H_a(s) = \frac{(\Omega_c)^7}{[(s+\Omega_c)(s^2+1.8019\Omega_c s+\Omega_c^2)(s^2+1.247\Omega_c s+\Omega_c^2)(s^2+0.44\Omega_c s+\Omega_c^2)]}$$

$$H_a(s) = 3\times 10^{24} \Big/ \Big[(s+3140)(s^2+5658s+9.87\times 10^6)(s^2+3893.6s+9.87\times 10^6)(s^2+1381.6s+9.87\times 10^6)\Big]$$

Example 4.13 If $H(s) = \frac{1}{(s+1)(s+2)}$, find $H(z)$ using impulse invariance method for sampling frequency of 5 samples/s.

Solution

$$H(s) = \frac{1}{(s+1)(s+2)}$$

The above function can be put into partial fraction as,

$$H(s) = \frac{1}{(s+1)} - \frac{1}{(s+2)}$$

The sampling frequency is given as $F_s = 5$

$$T = \frac{1}{F_s} = \frac{1}{5} = 0.2\ \text{s}.$$

Using impulse invariance we get,

$$H(z) = \frac{1}{1-e^{-T}z^{-1}} - \frac{1}{(1-e^{2T}z^{-1})}$$
$$= \frac{1}{1-e^{-0.2}z^{-1}} - \frac{1}{(1-e^{-0.4}z^{-1})}$$
$$= \frac{1}{1-0.8187z^{-1}} - \frac{1}{(1-0.6703z^{-1})}$$

$$\boxed{H(z) = \frac{0.1484z^{-1}}{(1-1.489z^{-1}+0.5488z^{-2})}}$$

Example 4.14 Convert the analog filter with system function $H_a(s)$ into digital filter using bilinear transformation.

$$H_a(s) = \frac{(s+0.3)}{(s+0.3)^2+16}$$

Solution

$$H_a(s) = \frac{(s+0.3)}{(s+0.3)^2+16} = \frac{(s+0.3)}{(s+0.3)^2+16}$$

The bilinear transformation is given by

$$s = \frac{2}{T}\frac{(1-z^{-1})}{(1+z^{-1})} = \frac{2}{T}\frac{(z-1)}{(z+1)}$$

Assuming $T = 1$ and substituting for s in $H_a(s)$ we get

$$H(z) = \frac{2\frac{(z-1)}{(z+1)}+0.3}{\left(2\frac{(z-1)}{(z+1)}\right)^2+0.6+\frac{(z-1)}{(z+1)}+16.09}$$
$$= \frac{[2(z-1)+0.3(z+1)](z+1)}{4(z-1)^2+0.6(z-1)(z+1)+16.09(z+1)}$$
$$= \frac{(2.3z^2+0.6z-1.7)}{(4.6z^2+8.09z+19.49)}$$

$$\boxed{H(z) = \frac{(-1.7z^{-2}+0.6z^{-1}+2.3)}{(19.4z^{-2}+8.09z^{-1}+4.6)}}$$

Example 4.15 Find the analog T.F. for a digital Butterworth filter satisfying the following specifications:

$$0.7 \le |H(e^{j\omega})| \le 1, \quad 0 \le \omega \le 0.2\pi$$
$$|H(e^{j\omega})| \le 0.004, \quad 0.6\pi \le \omega \le \pi$$

Assume $T = 1$ s. Apply impulse invariant transformation.

Solution Given $A_p = 0.7$; $\omega_p = 0.2\pi$ and $A_s = 0.004$; $\omega_s = 0.6\pi$

$$\frac{\Omega_s}{\Omega_p} = \frac{0.6\pi}{0.2\pi} = 3$$

$$N \ge \frac{1}{2}\frac{\log\left[\left(\frac{1}{A_s^2}-1\right)/\left(\frac{1}{A_p^2}-1\right)\right]}{\log\frac{\Omega_s}{\Omega_p}}$$

$$\le \frac{\frac{1}{2}\log\left[\left(\frac{1}{(0.004)^2}-1\right)/\left(\frac{1}{(0.7)^2}-1\right)\right]}{\log 3} = 5$$

$$\Omega_c = \frac{1}{T}\frac{\omega_p}{\left(\frac{1}{A_p^2}-1\right)^{1/2N}}$$

$$= \frac{0.2\pi}{\left(\frac{1}{0.49}-1\right)^{0.1}} = 0.2\pi \text{ rad/s}$$

$$\Omega_c = 0.6283$$

For $N = 5$ which is odd, the analog Butterworth filter is expressed by the following system function

$$H_a(s) = \frac{(\Omega_c)^5}{(s+\Omega_c)(s^2+b_1\Omega_c s+\Omega_c^2)(s^2+b_2\Omega_c s+\Omega_c^2)}$$

Since $b_1 = 0.618$ and $b_2 = 1.618$

$$\boxed{H_a(s) = \frac{(0.6283)^5}{(s+0.6283)(s^2+0.3883s+0.395)(s^2+s+0.395)}}$$

Example 4.16 Design a third order Butterworth digital filter using impulse invariant technique. Assume the sampling period $T = 1$ s.

Solution Given $N = 3$. If N is odd, the analog transfer function is

$$H_a(s) = \frac{\Omega_c}{s+\Omega_c} \prod_{k=1}^{(N-1)/2} \frac{\Omega_c^2}{s^2 + b_k\Omega_c s + \Omega_c^2}$$

$$= \frac{\Omega_c}{s+\Omega_c} \cdot \frac{\Omega_c^2}{s^2 + b_k\Omega_c s + \Omega_c^2} \quad \text{(Let } \Omega_c = 1 \text{ for normalized filter).}$$

$$= \frac{1}{(s+1)(s^2 + b_1 s + 1)} \quad \left(b_k = 2\sin\left[\frac{(2k-1)\pi}{2N}\right]\right)$$

$$H_a(s) = \frac{1}{(s+1)(s^2+s+1)} \quad (b_1 = 1)$$

$$= \frac{1}{(s+1)(s+0.5+j0.866)(s+0.5-j0.866)}$$

$$= \frac{A}{s+1} + \frac{B}{s+0.5+j0.866} + \frac{C}{s+0.5-j0.866}$$

$$A = \left.\frac{1}{s^2+s+1}\right|_{s=-1} = 1$$

$$B = \left.\frac{1}{(s+1)(s+0.5-j0.866)}\right|_{s=0.5-j0.866} = -0.5 + j0.288$$

$$C = B^* = -0.5 - j0.288$$

$$H(z) = \frac{1}{s+1} + \frac{-0.5+0.288j}{s+0.5+j0.866} + \frac{-0.5-0.288j}{s+0.5-j0.866}$$

By impulse invariant transformation, the transfer function of the digital filter is,

$$H(z) = \frac{1}{1-e^{-1}z^{-1}} + \frac{-0.5+0.288j}{1-e^{-0.5}\cdot e^{-j0.866}z^{-1}} + \frac{-0.5-0.288j}{1-e^{-0.5}e^{j0.866}z^{-1}}$$

$$\boxed{H(z) = \frac{1}{1-0.368z^{-1}} + \frac{-1+0.66z^{-1}}{1-0.786z^{-1}+0.368z^{-2}}}$$

Example 4.17 Apply impulse invariant method and find $H(z)$ for

$$H(s) = \frac{s+a}{(s+a)^2+b^2}$$

Solution The inverse Laplace transform of given function is

$$h(t) = \begin{cases} e^{-at}\cos bt, & t \geq 0 \\ 0, & \text{otherwise} \end{cases}$$

$$h(n) = h(t)\Big|_{t=nT}$$

$$h(n) = \begin{cases} e^{-anT}\cos(bnT), & t \geq 0 \\ 0, & \text{otherwise} \end{cases}$$

The digital transfer function is

$$\begin{aligned} H(z) &= \sum_{n=0}^{\infty} e^{-anT}\cos bnT z^{-n} \\ &= \sum_{n=0}^{\infty} e^{-anT}\left(\frac{e^{jbnT}+e^{-jbnT}}{2}\right) z^{-n} \end{aligned}$$

$$\begin{aligned} H(z) &= \frac{1}{2}\sum_{n=0}^{\infty}[(e^{-(a-jb)T}z^{-1})^n + (e^{-(a+jb)T}z^{-1})^n] \\ &= \frac{1}{2}\left[\frac{1}{1-e^{-(a-jb)T}z^{-1}} + \frac{1}{1-e^{-(a+jb)T}z^{-1}}\right] \\ &= \frac{1}{2}\left[\frac{1-e^{-(a+jb)T}z^{-1}+1-e^{-(a-jb)T}z^{-1}}{(1-e^{-(a-jb)T}z^{-1})(1-e^{-(a+jb)T}z^{-1})}\right] \end{aligned}$$

$$\boxed{H(z) = \frac{1-e^{-aT}\cos(bT)z^{-1}}{1-2e^{-aT}\cos(bT)z^{-1}+e^{-2aT}z^{-2}}}$$

Example 4.18 An analog filter has a transfer function

$$H(s) = \frac{10}{s^2+7s+10}$$

Design a digital filter equivalent to this using impulse invariant method for $T = 0.2$ s.

Solution Given

$$H(s) = \frac{10}{s^2+7s+10} \quad \text{and} \quad T = 0.2\ \text{s}$$

$$H(s) = \frac{10}{(s+5)(s+2)} = \frac{A}{(s+5)} + \frac{B}{(s+2)}$$

$$A = \left.\frac{10}{(s+2)}\right|_{s=-5} = -3.33$$

$$B = \left.\frac{10}{(s+5)}\right|_{s=-2} = 3.33$$

$$H(s) = \frac{-3.33}{(s+5)} + \frac{3.33}{(s+2)}$$

$$H(z) = \frac{-3.33}{1-e^{-5T}z^{-1}} + \frac{3.33}{1-e^{-2T}z^{-1}}$$

$$= \frac{-3.33}{1-e^{-1}z^{-1}} + \frac{3.33}{1-e^{-0.4}z^{-1}}$$

$$= \frac{-3.33}{1-0.3678z^{-1}} + \frac{3.33}{1-0.67z^{-1}}$$

$$\boxed{H(z) = \frac{1.006z^{-1}}{1-1.0378z^{-1}+0.247z^{-2}}}$$

Example 4.19 Use bilinear transform to design a first order Butterworth LPF with 3 dB cut-off frequency of 0.2π.

Solution Given 3 dB cut-off frequency $\omega_c = 0.2\pi = \omega_p$.

The analog transfer function for first order Butterworth filter is

$$H_a(s) = \frac{\Omega_c}{s+\Omega_c}$$

$$\Omega_c = \frac{2}{T}\tan\left(\frac{\omega_p}{2}\right) = 2\tan\left(\frac{0.2\pi}{2}\right) = 0.6498$$

$$H_a(s) = \frac{0.6498}{s+0.6498}$$

Using bilinear transformation,

$$H(z) = H_a(s)\Big|_{s=\frac{2}{T}\frac{1-z^{-1}}{1+z^{-1}}}$$

Assume $T = 1$ s.

$$
\begin{aligned}
H(z) &= \left.\frac{0.6498}{s + 0.6498}\right|_{s=2\left(\frac{1-z^{-1}}{1+z^{-1}}\right)} \quad [\because T = 1 \text{ s}] \\
&= \frac{0.6498}{\frac{2(1-z^{-1})}{(1+z^{-1})} + 0.6498} \\
&= \frac{0.6498(1 + z^{-1})}{2(1 - z^{-1}) + 0.6498(1 + z^{-1})} \\
&= \frac{0.6498(1 + z^{-1})}{2 - 2z^{-1} + 0.6498 + 0.6498z^{-1}} \\
&= \frac{0.6498(1 + z^{-1})}{2.6498 - 1.3502z^{-1}} \\
&= \frac{0.6498(1 + z^{-1})}{2.6498(1 - 0.5095z^{-1})}
\end{aligned}
$$

$$
\boxed{H(z) = \frac{0.2452(1 + z^{-1})}{1 - 0.5095z^{-1}}}
$$

Example 4.20 Using bilinear transformation design a digital bandpass Butterworth filter with the following specifications:

Sampling frequency = 8 kHz

$\alpha_p = 2$ dB in the passband $800\,\text{Hz} \leq f \leq 1000$ Hz

$\alpha_s = 20$ dB in the stopband $0 \leq f \leq 400\,\text{Hz}$ and $2000 \leq f \leq \infty$

Solution

$$
\begin{aligned}
\frac{\omega_1 T}{2} &= \frac{2 \times \pi \times 400}{2 \times 8000} = \frac{\pi}{20} \qquad \left[T = \frac{1}{f_s} = \frac{1}{8000}\right] \\
\frac{\omega_1 T}{2} &= \frac{2 \times \pi \times 800}{2 \times 8000} = \frac{\pi}{10} \\
\frac{\omega_u T}{2} &= \frac{2\pi \times 1000}{2 \times 8000} = \frac{\pi}{8} \\
\frac{\omega_2 T}{2} &= \frac{2\pi \times 2000}{2 \times 8000} = \frac{\pi}{4}
\end{aligned}
$$

Prewarped analog frequencies are given by

$$\frac{\Omega_1 T}{2} = \tan\frac{\omega_1 T}{2} = \tan\frac{\pi}{20} = 0.1584$$
$$\frac{\Omega_l T}{2} = \tan\frac{\omega_l T}{2} = \tan\frac{\pi}{10} = 0.325$$
$$\frac{\Omega_u T}{2} = \tan\frac{\omega_u T}{2} = \tan\frac{\pi}{8} = 0.4142$$
$$\frac{\Omega_2 T}{2} = \tan\frac{\omega_2 T}{2} = \tan\frac{\pi}{4} = 1$$

First we design a prototype normalized compass filter and then use suitable transformation to obtain the transfer function of bandpass filter. To reduce computational complexity we use above values to find Ω_r and substitute $s = \frac{1-z^{-1}}{1+z^{-1}}$ for bilinear transformation ($\because$ all the above frequencies contains the term $T/2$)

$$A = \frac{-\Omega_1^2 + \Omega_l\Omega_u}{\Omega_1(\Omega_u - \Omega_l)}$$
$$= \frac{-(0.1584)^2 + (0.325)(0.4142)}{0.1584(0.4142 - 0.325)}$$
$$A = 11.303$$

$$B = \frac{\Omega_2^2 - \Omega_l\Omega_u}{\Omega_2(\Omega_u - \Omega_l)} = \frac{1 - (0.4142)(0.325)}{1(0.4142 - 0.325)}$$
$$= \frac{1 - 0.1346}{0.0892} = \frac{0.865385}{0.0892}$$
$$= 9.7016$$
$$\Omega_r = \min\{|A|, |B|\} = 9.7016$$
$$N = \frac{\log_{10}\sqrt{\frac{10^2-1}{10^{0.2}-1}}}{\log_{10}(9.7016)} = \frac{\log_{10}(13.01)}{\log_{10}(9.7016)} = \frac{1.1142}{0.9868}$$
$$= 1.1290$$
$$N = 2$$

The second order normalized Butterworth lowpass filter transfer function given by

$$H(s) = \frac{1}{s^2 + 1.4142s + 1}$$

The transformation for the bandpass filter is

$$s \to \frac{s^2 + \Omega_l \Omega_u}{s(\Omega_u - \Omega_l)} = \frac{s^2 + 0.1346}{s(0.0892)}$$

$$H(s) = \left.\frac{1}{s^2 + 1.4142s + 1}\right|_{s=\frac{s^2+0.1346}{(0.0892)s}}$$

$$= \frac{1}{\left(\frac{s^2+0.1346}{(0.0892)s}\right)^2 + 1.4142\left(\frac{s^2+0.1346}{(0.0892)s}\right) + 1}$$

$$= \frac{0.0079s^2}{s^4 + 0.2692s^2 + 0.0181 + 1.4142(0.0892)s(s^2 + 0.1346) + 0.0079s^2}$$

$$= \frac{0.0079s^2}{s^4 + 0.2692s^2 + 0.0181 + 0.126s(s^2 + 0.1346) + 0.0079s^2}$$

$$= \frac{0.0079s^2}{s^4 + 0.2692s^2 + 0.0181 + 0.126s^3 + 0.0169s + 0.0079s^2}$$

$$H(s) = \frac{0.0079s^4}{s^4 + 0.126s^3 + 0.2771s^2 + 0.0169s + 0.0181}$$

$$H(z) = \left. H(s)\right|_{s=\frac{1-z^{-1}}{1+z^{-1}}}$$

$$= \frac{0.0079\left(\frac{1-z^{-1}}{1+z^{-1}}\right)^2}{\left(\frac{1-z^{-1}}{1+z^{-1}}\right)^4 + 0.126\left(\frac{1-z^{-1}}{1+z^{-1}}\right)^3 + 0.277\left(\frac{1-z^{-1}}{1+z^{-1}}\right)^2 + 0.169\left(\frac{1-z^{-1}}{1+z^{-1}}\right) + 0.0181}$$

$$= 0.0079(1 - z^{-1})^2 \Big/ \Big[(1 - z^{-1})^4 + 0.126(1 - z^{-1})^3(1 + z^{-1}) + 0.2692(1 - z^{-1})^2(1 + z^{-1})^2 + 0.169(1 - z^{-1})(1 + z^{-1})^3 + 0.0181(1 + z^{-1})^4\Big]$$

$$= 0.0079(1 - z^{-2})^2 \Big/ \Big[1 - 4z^{-1} + 6z^{-2} - 4z^{-3} + z^{-4} + 0.126(1 - 2z^{-1} + 2z^{-3} - z^{-4}) + 0.2677(1 - 2z^{-2} + z^{-4}) + 0.169(1 + 2z^{-1} - 2z^{-3} - z^{-4}) + 0.0181(1 + 4z^{-1} + 6z^{-2} + 4z^{-3} + z^{-4})\Big]$$

$$= \frac{0.0079(1 - z^{-2})^2}{7.1871 - 3.8416z^{-1} + 0.644z^{-2} - 4.086z^{-3} + 0.9908z^{-4}}.$$

4.9 Design of Lowpass Digital Chebyshev Filter

4.9.1 Analog Chebyshev Filter

It is designed by approximating the ideal frequency response using error function.

The Chebyshev characteristic provides an alternative method of getting a suitable analog transfer function $H_a(s)$ from which lowpass digital filters are designed using either bilinear transformation or impulse invariance. There are two types of Chebyshev filters available as type I and type II filters. The type I filter characteristic for odd value of N is shown in Fig. 4.21a and type II characteristic for even value of N is shown in Fig. 4.21b. The characteristic of type I filters have equal ripple in the passband and monotonic in the stopband. They are all-pole filters. Type II Chebyshev filter characteristics are shown in Fig. 4.22a and b for odd and even values of N. These filters have equal ripples in the stopband and monotonic in the passband. Further, they have both poles and zeros in the transfer function.

4.9.2 Determination of the Order of the Chebyshev Filter

The type I of Nth order Chebyshev filter can be expressed as,

$$|H(j\Omega)|^2 = \frac{1}{\left[1+\epsilon^2 C_N^2\left(\frac{\Omega}{\Omega_p}\right)\right]}; \quad N = 1, 2, \ldots \tag{5.53a}$$

where ϵ is a parameter which is the function of the ripple in the passband. $C_N(x)$ is the Chebyshev polynomial of order N. In the passband and stopband it is defined as,

$$C_N(x) = \cos(N\cos^{-1} x), \quad |x| \leq 1$$
$$C_N(x) = \cosh(N\cosh^{-1} x), \quad |x| > 1$$

The Chebyshev polynomial is also expressed by the following recursive formula.

$$C_N(x) = 2xC_{N-1}(x) - C_{N-2}(x), \quad N > 1$$

where $C_0(x) = 1$ and $C_1(x) = x$.

Taking logarithm for Eq. (4.53a) we get

$$20\log|H(\Omega)| = -10\log\left[1+\epsilon^2 C_N^2\left(\frac{\Omega}{\Omega_p}\right)\right]$$

At passband $\Omega = \Omega_p$

$$20\log|H(\Omega)| = -\alpha_p$$
$$\alpha_p = 10\log\left[1+\epsilon^2 C_N^2\left(\frac{\Omega_p}{\Omega_p}\right)\right]$$

Using the property of $C_N(1) = 1$, we get

$$\alpha_p = 10\log(1+\epsilon^2)$$
$$\epsilon = \sqrt{10^{0.1\alpha_p} - 1}$$

At stopband $\Omega = \Omega_s$.

$$20\log|H(j\Omega)| = -\alpha_s \log\left[1+\epsilon^2 C_N^2\left(\frac{\Omega_s}{\Omega_p}\right)\right]$$

But $C_N = \left(\frac{\Omega_s}{\Omega_p}\right) = \cosh\left[N\cosh^{-1}\frac{\Omega_s}{\Omega_p}\right]$

$$\alpha_s = 10\log\left[1+\epsilon^2\left\{\cosh\left(N\cosh^{-1}\left(\frac{\Omega_s}{\Omega_p}\right)\right)\right\}^2\right]$$

But

$$\cosh\left(N\cosh^{-1}\left(\frac{\Omega_s}{\Omega_p}\right)\right]^2 = \frac{10^{0.1\alpha_s}-1}{\epsilon^2} = \frac{10^{0.1\alpha_s}-1}{(10^{0.1\alpha_p}-1)}$$
$$N\cosh^{-1}\left(\frac{\Omega_s}{\Omega_p}\right) = \cosh^{-1}\left[\frac{(10^{0.1\alpha_s})-1}{10^{0.1\alpha_p}-1}\right]^{1/2}$$

$$\boxed{N \geq \frac{\cosh^{-1}\left[\frac{10^{0.1\alpha_s}-1}{10^{0.1\alpha_p}-1}\right]^{1/2}}{\cosh^{-1}\left(\frac{\Omega_s}{\Omega_p}\right)}}$$

or

$$\boxed{N \geq \frac{\cosh^{-1}\left[\frac{1}{\epsilon}\left(\frac{1}{A_s^2}-1\right)^{1/2}\right]}{\cosh^{-1}\left(\frac{\Omega_s}{\Omega_p}\right)}}$$

Note: $\cosh^{-1}x$ can be evaluated using the following identity

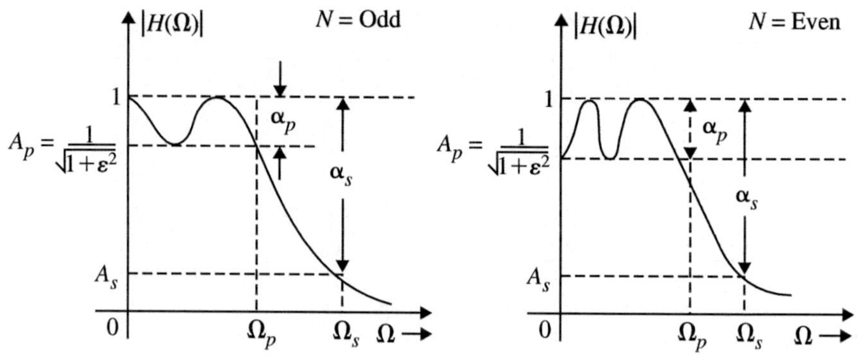

Fig. 4.21 Frequency response of type I Chebyshev filter

$$\boxed{\cosh^{-1} x = \ln[x + \sqrt{(x^2 - 1)}]}$$

They are two types:

I. **Type I**: The magnitude response is equiripple in the passband and monotonic in the stopband as shown in Fig. 4.21.

II. **Type II**: The magnitude response is monotonic in passband and equiripple in stopband as shown in Fig. 4.22.

The magnitude response of type I lowpass filter is given by

$$|H_a(\Omega)|^2 = \frac{1}{1 + \epsilon^2 C_N^2\left(\frac{\Omega}{\Omega_c}\right)} \tag{4.54}$$

where ϵ is attenuation constant and C_N is Chebyshev polynomial

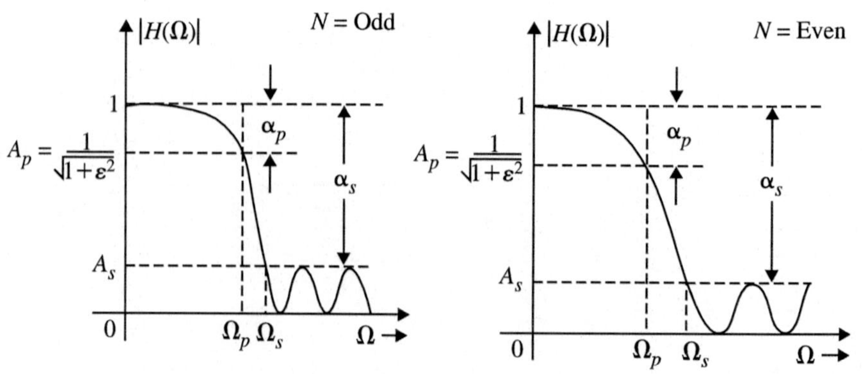

Fig. 4.22 Frequency response of type II Chebyshev filter. **a** N is odd and **b** N is even

$$\epsilon = \left[\frac{1}{A_p^2} - 1\right]^{1/2} \tag{4.55}$$

where $s = j\Omega$ gives $\Omega = s/j$. Substitute $\Omega = s/j$ in above equation

$$H_a(s)H_a(-s) = \frac{1}{1 + \epsilon^2 C_N^2\left(\frac{s/j}{\Omega_c}\right)}$$

$$H_a(s)H_a(-s) = \frac{1}{1 + \epsilon^2 C_N^2(-js_n)} \tag{4.56}$$

If N is even all poles are complex. When N is odd one of the poles is real and all other poles are complex.

4.9.3 Un-normalized Chebyshev Lowpass Filter Transfer Function

When N is even

$$H_a(s) = \prod_{k=1}^{N/2} \frac{B_k\Omega_c^2}{s^2 + b_k\Omega_c^2 s + C_k\Omega_c^2} \tag{4.57}$$

When N is odd

$$H_a(s) = \frac{B_0\Omega_c}{s + C_0\Omega_c} \prod_{k=1}^{N-1/2} \frac{B_k\Omega_c^2}{s^2 + b_k\Omega_c^2 s + C_k\Omega_c^2} \tag{4.58}$$

where

$$b_k = 2Y_N \sin\left(\frac{(2k-1)\pi}{2N}\right) \tag{4.59}$$

$$c_k = Y_N^2 + \cos^2\left(\frac{(2k-1)\pi}{2N}\right) \tag{4.60}$$

$$c_0 = Y_N \tag{4.61}$$

$$Y_N = \frac{1}{2}\left\{\left[\left(\frac{1}{\epsilon^2} - 1\right)^{1/2} + \frac{1}{\epsilon}\right]^{1/N} - \left[\left(\frac{1}{\epsilon^2} + 1\right)^{1/2} + \frac{1}{\epsilon}\right]^{-(1/N)}\right\} \tag{4.62}$$

For N is even, parameter B_k is evaluated by

$$H_a(s)\bigg|_{s=0} = \frac{1}{(1+\epsilon^2)^{1/2}} \tag{4.63}$$

for N is odd, B_k is calculated by

$$H_a(s)\Big|_{s=0} = 1 \quad \text{and} \quad B_0 = B_1 = B_2 = \cdots = B_k. \tag{4.64}$$

4.9.4 Frequency Response of Chebyshev Filter

The frequency response of Chebyshev filter is shown in Fig. 4.23.

$$N \geq \frac{\cos h^{-1}\left\{\frac{1}{\epsilon}\left(\frac{1}{A_s^2}-1\right)^{1/2}\right\}}{\cos h^{-1}\left(\frac{\Omega_s}{\Omega_p}\right)} \tag{4.65}$$

4.9.5 Properties of Chebyshev Filter (Type I)

1. The magnitude oscillates between 1 and $\frac{1}{\sqrt{1+\epsilon^2}}$ with in passband (equiripple in the passband).
2. At cut-off frequency Ω_c, the magnitude is $\frac{1}{\sqrt{1+\epsilon^2}}$.
3. Magnitude is monotonic outside the passband.
4. Chebyshev filters are all-pole design.
5. As N increases, it is approaches to ideal response.

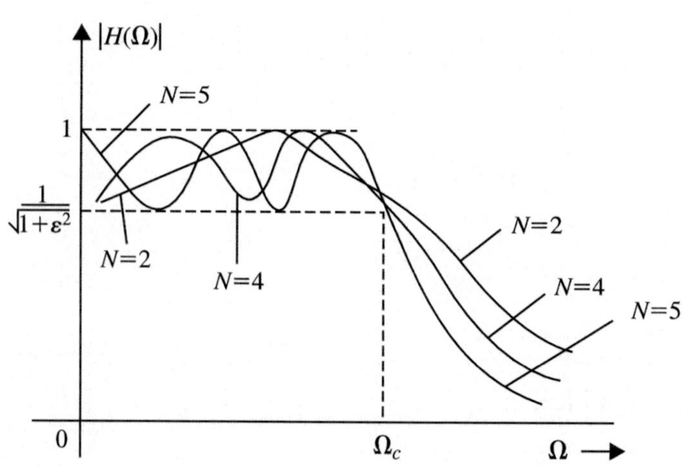

Fig. 4.23 Frequency response of Chebyshev filter type I filter

4.9.6 Design Procedures for Lowpass Digital Chebyshev IIR Filter

1. Choose either bilinear or impulse invariant transformation.
2. Calculate the attenuation constant ϵ

$$\epsilon = \left[\frac{1}{A_p^2} - 1\right]^{1/2}$$

3. Calculate the ratio Ω_s/Ω_p using the following equation which is appropriate.

$$\text{For bilinear transformation, } \frac{\Omega_s}{\Omega_p} = \frac{\tan(\omega_s/2)}{\tan(\omega_p/2)}$$

$$\text{For impulse invariant, } \frac{\Omega_s}{\Omega_p} = \frac{\omega_s}{\omega_p}$$

4. Decide the order N of the filter from

$$N \leq \frac{\cos h^{-1}\left\{\frac{1}{\epsilon}\left[\frac{1}{A_2^s} - 1\right]^{1/2}\right\}}{\cos h^{-1}\left(\frac{\Omega_s}{\Omega_p}\right)}$$

5. Calculate the analog cut-off frequency Ω_c from the equations given below:

$$\text{For bilinear transformation, } \Omega_c = \frac{2}{T}\frac{\tan(\omega_p/2)}{\left[\frac{1}{A_p^2} - 1\right]^{1/2N}}$$

$$\text{For impulse invariant, } \Omega_c = \frac{(\omega_p/T)}{\left[\frac{1}{A_p^2} - 1\right]^{1/2N}}$$

6. Determine analog transfer function, when N is even

$$H_a(s) = \prod_{k=1}^{N/2} \frac{B_k\Omega_c^2}{s^2 + b_k\Omega_c s + c_k\Omega_c^2}$$

When N is odd

$$H_a(s) = \frac{B_0\Omega_c}{s + c_0\Omega_c} \prod_{k=1}^{N-1/2} \frac{B_k\Omega_c^2}{s^2 + b_k\Omega_c s + c_k\Omega_c^2}$$

where

$$b_k = 2Y_N \sin\left(\frac{(2k-1)\pi}{2N}\right)$$
$$c_k = Y_N^2 + \cos^2\left(\frac{(2k-1)\pi}{2N}\right)$$
$$c_0 = Y_N$$

$$Y_N = \frac{1}{2}\left\{\left[\left(\frac{1}{\epsilon^2}+1\right)^{1/2}+\frac{1}{\epsilon}\right]^{1/N} - \left[\left(\frac{1}{\epsilon^2}+1\right)^{1/2}+\frac{1}{\epsilon}\right]^{-(1/N)}\right\}$$

B_k can be calculated from

$$N = \text{even}; \quad H_a(s)\Big|_{s=0} = \frac{1}{(1+\epsilon^2)^{1/2}}$$

$$N = \text{odd}; \quad H_a(s)\Big|_{s=0} = 1$$

(The normal practice is to take $B_0 = B_1 = B_2, \ldots, B_k$).

7. Convert $H_a(s)$ into $H(z)$ using the chosen transform (Bilinear or impulse invariance).
8. Realize the digital filter transfer function by a suitable structure.

4.10 Frequency Transformation

4.10.1 Analog Frequency Transformation

The IIR lowpass analog filters are designed for the given specifications. By applying appropriate approximations, and transformations lowpass digital filters are designed. By using frequency transformations it is possible to design lowpass filters with different passband frequencies, highpass filters, bandpass filters and bandstop filters. In the frequency transformation, the variable s of normalized ($\Omega_c = 1$) lowpass filter is replaced by appropriate variable. From the normalized lowpass filter, the following filters are designed:

1. Low pass to Low pass.
2. Low pass to Highpass filter.
3. Low pass to Bandpass filter.
4. Low pass to Band elimination filter (Band stop filter).

Let Ω_p be the passband frequency of lowpass analog filter. Let Ω'_p be the edge frequency of the lowpass filter which we wish to convert. The frequency transformation to achieve this is given by

$$\boxed{s \rightarrow \frac{\Omega_p}{\Omega'_p}} \tag{4.66}$$

4.10.1.1 Low Pass to High Pass Filter

The transformation that is made to convert the lowpass to highpass filter is given by

$$\boxed{s \rightarrow \frac{\Omega_p \Omega'_p}{s}} \tag{4.67}$$

where Ω_p = Edge frequency of the prototype lowpass filter and Ω'_p = Edge frequency of the highpass filter.

4.10.1.2 Low Pass to Bandpass Filter

Let Ω_l be the lower band edge frequency of bandpass filter and Ω_u be the upper band edge frequency of the same filter. The transfer function from lowpass filter to bandpass filter is achieved by the following transformation.

$$\boxed{s \rightarrow \Omega_p \frac{(s^2 + \Omega_l \Omega_u)}{s(\Omega_u - \Omega_l)}} \tag{4.68}$$

4.10.1.3 Low Pass to Bandstop Filter

Let

Ω_p = Passband edge frequency of the given lowpass filter.
Ω_u = Upper band edge frequency of the bandstop filter.
Ω_l = Lower band edge frequency of the bandstop filter.

The transfer function from lowpass prototype to bandstop filter transfer function is achieved by the following transformation:

$$s \rightarrow \Omega_p \frac{s(\Omega_u - \Omega_l)}{(s^2 + \Omega_u \Omega_l)} \tag{4.69}$$

4.10.2 Digital Frequency Transformation

If we design a lowpass digital filter and wish to convert to desired lowpass filter, high-pass filter, Bandpass filter and bandstop filter, the following frequency transformation may be used.

4.10.2.1 Lowpass to Lowpass Filter

$$\boxed{z^{-1} \rightarrow \frac{z^{-1} - \alpha}{1 - \alpha z^{-1}}} \tag{4.70}$$

where

$\alpha = \frac{\sin[(\omega_p - \omega'_p/2)]}{\sin[(\omega_p + \omega'_p)/2]}.$

$\omega_p =$ Band edge frequency of the lowpass original filter.

$\omega'_p =$ Band edge frequency of the lowpass desired filter.

4.10.2.2 Lowpass to Highpass Filter

$$\boxed{z^{-1} \rightarrow \frac{z^{-1} + \alpha}{1 + \alpha z^{-1}}} \tag{4.71}$$

$$\alpha = \frac{\cos[(\omega_p + \omega'_p)/2]}{\cos[(\omega_p - \omega'_p)/2]}$$

4.10.2.3 Lowpass to Bandpass Filter

$$\boxed{z^{-1} \rightarrow \frac{(z^{-2} - \alpha_1 z^{-1} + \alpha_2)}{(\alpha_2 z^{-2} - \alpha_1 z^{-1} + 1)}} \tag{4.72}$$

$$\alpha_1 = \frac{2\alpha k}{k+1}$$
$$\alpha_2 = \frac{(k-1)}{(k+1)}$$
$$\alpha = \frac{\cos[(\omega_u + \omega_l)/2]}{\cos[(\omega_u - \omega_l)/2]}$$
$$k = \cot\frac{(\omega_u - \omega_l)}{2}\tan\frac{\omega_p}{2}$$

4.10.2.4 Lowpass to Bandstop Filter

$$\boxed{z^{-1} \rightarrow \frac{(z^{-2} - \alpha_1 z^{-1} + \alpha_2)}{(\alpha_2 z^{-1} - \alpha_1 z^{-1} + 1)}} \tag{4.73}$$

$$\alpha_1 = \frac{2\alpha}{(k+1)}$$
$$\alpha_2 = \frac{(1-k)}{(1+k)}$$
$$\alpha = \frac{\cos[(\omega_u + \omega_l)/2]}{\cos[(\omega_u - \omega_l)/2]}$$
$$k = \tan\frac{(\omega_u - \omega_l)}{2}\tan\frac{\omega_p}{2}$$

	Filter type	Transformation
1.	Lowpass	$s \rightarrow \frac{s}{\Omega_c}$
2.	Highpass	$s \rightarrow \frac{\Omega_c}{s}$
3.	Bandpass	$s \rightarrow \frac{Q(s^2+\Omega_0^2)}{\Omega_0 s}$
4.	Bandstop	$s \rightarrow \frac{\Omega_0 s}{Q(s^2+\Omega_0^2)}$

where Ω_0 is centre frequency, $\Omega_0 = \sqrt{\Omega_p \Omega_s}$ where Q is quality factor, $Q = \frac{\Omega_0}{\Omega_s - \Omega_p}$.

Example 4.21 The specification of the desired lowpass filter is

$$0.8 \leq |H(\omega)| \leq 1.0; \qquad 0 \leq \omega < 0.2\pi$$
$$|H(\omega)| \leq 0.2; \qquad 0.32\pi \leq \omega \leq \pi$$

Design Chebyshev digital filter using bilinear transformation.

Solution Given $A_p = 0.8$ at $\omega_p = 0.2\pi$ rad/s and $A_s = 0.2$ at $\omega_s = 0.32\pi$ rad/s. For bilinear transformation we get,

$$\frac{\Omega_s}{\Omega_p} = \frac{\tan(\omega_s/2)}{\tan(\omega_p/2)} = \frac{\tan(0.32\pi/2)}{\tan(0.2\pi/2)} = 1.692$$

The attenuation constant $\epsilon = \left(\frac{1}{A_p^2} - 1\right)^{1/2} = 0.75$

$$N \geq \frac{\cosh^{-1}\left\{\frac{1}{\epsilon}\left[\frac{1}{A_s^2} - 1\right]^{1/2}\right\}}{\cosh^{-1}\frac{\Omega_s}{\Omega_p}} = \frac{\cosh^{-1}\left\{\frac{1}{0.75}\left[\frac{1}{0.2^2} - 1\right]^{1/2}\right\}}{\cosh^{-1} 1.692}$$

$$= \frac{\cosh^{-1} 6.53}{\cosh^{-1} 1.692}$$

$$\cosh^{-1} 6.53 = \ln(6.53 + \sqrt{6.53^2 - 1})$$
$$= 2.56$$
$$\cosh^{-1} 1.692 = \ln(1.692 + \sqrt{1.692^2 - 1})$$
$$= 1.1173$$
$$N = \frac{2.56}{1.1173}$$
$$N \geq 2.295$$
$$N = 3$$

Let $T = 1$ s. The analog cut-off frequency

$$\Omega_c = \frac{2}{T}\frac{\tan(\omega_p/2)}{\left[\frac{1}{A_p^2} - 1\right]^{1/2N}}$$

$$\Omega_c = 2\frac{\tan(0.2\pi/2)}{\left[\frac{1}{(0.8)^2} - 1\right]^{1/6}}$$

$$\Omega_c = 0.715 \text{ rad/s.}$$

For N is odd, the analog transfer function

$$H_a(s) = \frac{B_0\Omega_c}{s + c_0\Omega_c}\prod_{k=1}^{(N-1)/2}\frac{B_k\Omega_c^2}{s^2 + b_k\Omega_c s + c_k\Omega_c^2}$$

$$= \frac{B_0\Omega_c}{s + c_0\Omega_c}\frac{B_1\Omega_c^2}{s^2 + b_1\Omega_c s + c_1\Omega_c^2}$$

$$
\begin{aligned}
Y_N &= \frac{1}{2}\left\{\left[\left(\frac{1}{\epsilon^2}+1\right)^{1/2}+\frac{1}{\epsilon}\right]^{1/N}-\left[\left(\frac{1}{\epsilon^2}+1\right)^{1/2}+\frac{1}{\epsilon}\right]^{-(1/N)}\right\} \\
&= \frac{1}{2}\left\{\left[\left(\frac{1}{0.75^2}+1\right)^{1/2}+\frac{1}{0.75}\right]^{1/3}-\left[\left(\frac{1}{0.75^2}+1\right)^{1/2}+\frac{1}{0.75}\right]^{-(1/3)}\right\} \\
&= 0.7211 - 0.3467 = 0.3744 \\
c_0 &= Y_N = 0.3744 \\
c_k &= Y_N^2 + \cos^2\frac{(2k-1)\pi}{2N} \\
c_1 &= Y_N^2 + \cos^2\left(\frac{\pi}{6}\right) = 0.3744^2 + \cos^2\left(\frac{\pi}{6}\right) = 0.8902 \\
b_k &= 2Y_N \sin\left[\frac{(2k-1)\pi}{2N}\right] \\
b_1 &= 2\times 0.3744 \sin\left[\frac{\pi}{6}\right] = 0.3744
\end{aligned}
$$

$$
\begin{aligned}
H_a(s) &= \frac{0.715B_0}{(s+0.3744\times 0.715)} \times \frac{(0.715)^2 B_1}{s^2+0.3744\times 0.715s + 0.8902\times 0.715^2} \\
&= \frac{0.3655B_0B_1}{(s+0.2677)(s^2+0.2677s+0.4551)}
\end{aligned}
$$

$$
\begin{aligned}
H_a(s)\Big|_{s=0} &= 1 \\
\frac{0.3655B_0B_1}{0.2677\times 0.4551} &= 1 \\
3B_0B_1 &= 1 \\
B_0B_1 &= \frac{1}{3} \\
B_0^2 &= \frac{1}{3} \\
B_0 &= 0.577 \\
\therefore\ B_0 &= B_1 = 0.577
\end{aligned}
$$

$$
\begin{aligned}
H_a(s) &= \frac{0.3655\times 0.577^2}{(s+0.2677)(s^2+0.2677s+0.4551)} \\
&= \frac{0.1217}{(s+0.2677)(s^2+0.2677s+0.4551)} \\
H(z) &= H_a(s)\Big|_{s=\frac{2(1-z^{-1})}{1+z^{-1}}} \qquad \text{(Let } T = 1\text{ s)}
\end{aligned}
$$

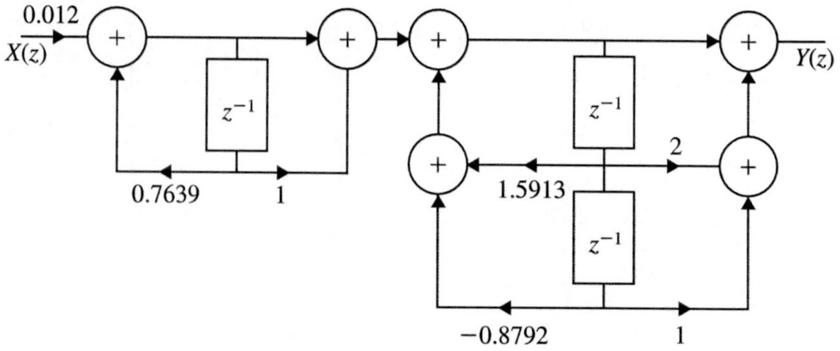

Fig. 4.24 Cascade form realization of $H(z)$ for the Example 4.22

$$H(z) = \frac{0.1217}{\left(\frac{2(1-z^{-1})}{1+z^{-1}} + 0.2677\right)\left(\left(\frac{2(1-z^{-1})}{1+z^{-1}}\right)^2 + 0.2677 \times \frac{2(1-z^{-1})}{1+z^{-1}} + 0.4551\right)}$$

$$= 0.1217(1+z^{-1})^3 \Big/ (2.2677 - 1.7323z^{-1})(4 + 4z^{-2} - 8z^{-1} + 0.5354$$
$$-0.5354z^{-2} + 0.4551 + 0.4551z^{-2} + 0.9102z^{-1})$$

$$= \frac{0.1217(1+z^{-1})^3}{(2.2677 - 1.7323z^{-1})(4.4551 - 7.089z^{-1} + 3.9197z^{-2})}$$

$$= \frac{0.012(1+z^{-1})^3}{(1 - 0.7639z^{-1})(1 - 1.5913z^{-1} + 0.8798z^{-2})}$$

$$\boxed{H(z) = \frac{0.012(1+z^{-1})(1+2z^{-1}+z^{-2})}{(1-0.7639z^{-1})(1-1.5913z^{-1}+0.8792z^{-2})}}$$

The cascade form realization is shown in Fig. 4.24.

Example 4.22 The specifications of the desired lowpass digital filter are

$$0.9 \le |H(\omega)| \le 1; \qquad 0 \le \omega < 0.25\pi$$
$$|H(\omega)| \le 0.24; \qquad 0.5\pi \le \omega \le \pi$$

Design Chebyshev digital filter using impulse invariant transformation.

Solution Given $A_p = 0.9$ at $\omega_p = 0.25\pi$ rad/s and $A_s = 0.24$ at $\omega_s = 0.5\pi$ rad/s.

$$\frac{\Omega_s}{\Omega_p} = \frac{\omega_s}{\omega_p} = \frac{0.5\pi}{0.25\pi} = 2$$

$$\epsilon = \left(\frac{1}{A_p^2} - 1\right)^{1/2} = \left(\frac{1}{(0.9)^2} - 1\right)^{1/2} = 0.484$$

$$N \geq \frac{\cos h^{-1}\left\{\frac{1}{\epsilon}\left[\frac{1}{A_s^2} - 1\right]^{1/2}\right\}}{\cos h^{-1}\frac{\Omega_s}{\Omega_p}}$$

$$\geq \frac{\cos h^{-1}\left\{\frac{1}{0.484}\left[\frac{1}{0.24^2} - 1\right]^{1/2}\right\}}{\cos h^{-1} 2}$$

$$\geq \frac{\cos h^{-1} 2.8126}{\cos h^{-1} 2}$$

$$N \geq 2.136$$

$$N = 3$$

Let $T = 1$ s.

$$\Omega_c = \frac{\omega_p / T}{\left[\frac{1}{A_p^2} - 1\right]^{1/2N}}$$

$$\Omega_c = \frac{0.25\pi}{\left[\frac{1}{(0.9)^2} - 1\right]^{1/6}}$$

$$\Omega_c = 1 \text{ rad/s.}$$

The analog transfer function for N is odd is given by

$$H_a(s) = \frac{B_0 \Omega_c}{s + c_0 \Omega_c} \prod_{k=1}^{(N-1)/2} \frac{B_k \Omega_c^2}{s^2 + b_k \Omega_c s + c_k \Omega_c^2}$$

$$= \frac{B_0 \Omega_c}{s + c_0 \Omega_c} \cdot \frac{B_1 \Omega_c^2}{s^2 + b_1 \Omega_c s + c_1 \Omega_c^2}$$

$$Y_N = \frac{1}{2}\left\{\left[\left(\frac{1}{\epsilon^2} + 1\right)^{1/2} + \frac{1}{\epsilon}\right]^{1/N} - \left[\left(\frac{1}{\epsilon^2} + 1\right)^{1/2} + \frac{1}{\epsilon}\right]^{-(1/N)}\right\}$$

$$= \frac{1}{2}\left\{\left[\left(\frac{1}{0.484^2} + 1\right)^{1/2} + \frac{1}{0.484}\right]^{1/3}\right.$$

$$-\left[\left(\frac{1}{0.484^2}+1\right)^{1/2}+\frac{1}{0.484}\right]^{-(1/3)}\Bigg\}$$

$$
\begin{aligned}
&= \frac{1}{2}\left\{1.6338 - \frac{1}{1.6338}\right\}\\
Y_N &= 0.511\\
c_0 &= Y_N = 0.511\\
c_k &= Y_N^2 + \cos^2\frac{(2k-1)\pi}{2N}\\
c_1 &= 0.511^2 + \cos^2\left(\frac{\pi}{6}\right) = 1.011\\
b_k &= 2Y_N \sin\left[\frac{(2k-1)\pi}{2N}\right]\\
b_1 &= 2\times 0.511 \sin\left[\frac{\pi}{6}\right] = 0.511\\
\therefore\ H_a(s) &= \frac{B_0B_1}{(s+0.511)(s^2+0.511s+1.011)}\\
H_a(s)\Big|_{s=0} &= 1\\
\frac{B_0B_1}{0.511\times 1.011} &= 1 \qquad (\text{Let } B_0 = B_1)\\
B_0^2 &= \frac{1}{1.939}\\
B_0 &= 0.719\\
\therefore\ B_0 &= B_1 = 0.719\\
H_a(s) &= \frac{(0.719)^2}{(s+0.511)(s^2+0.511s+1.011)}\\
&= \frac{0.517}{(s+0.511)(s^2+0.511s+1.011)}\\
&= \frac{A}{(s+0.511)} + \frac{Bs+C}{(s^2+0.511s+1.011)}
\end{aligned}
$$

$$A(s^2+0.511s+1.011)+(Bs+C)(s+0.511) = 0.517$$

Put $s = -0.511$

$$A(1.011) = 0.5166; \quad A = 0.511$$

Put $s = 0$

$$1.01A + 0.51C = 0.517; \quad C = 0$$

Equating the coefficients of s^2 we get

$$A + B = 0$$
$$B = -0.511$$

$A = 0.511$, $B = -0.511$ and $C = 0$

$$H_a(s) = \frac{0.511}{(s+0.511)} - \frac{0.511s}{(s^2+0.511s+1.011)}$$
$$= \frac{0.511}{(s+0.511)} - 0.511\frac{s+0.256-0.256}{(s+0.256)^2+0.972^2}$$
$$= \frac{0.511}{(s+0.511)} - 0.511\frac{s+0.256}{(s+0.256)^2+0.972^2}$$
$$+0.0511 \cdot \frac{0.256}{(s+0.256)^2+0.972^2}$$
$$= \frac{0.511}{(s+0.511)} - 0.511\frac{s+0.256}{(s+0.256)^2+0.972^2} + \frac{0.0511}{0.972}$$
$$\times 0.256 \cdot \frac{0.972}{(s+0.256)^2+0.972^2}$$
$$H(s) = \frac{0.511}{(s+0.511)} - 0.511\frac{s+0.256}{(s+0.256)^2+0.972^2} + 0.135$$
$$\times \frac{0.972}{(s+0.256)^2+0.972}$$
$$H(z) = \frac{0.511}{1-e^{-0.511}z^{-1}} - 0.511\frac{1-e^{-0.256}\cos(0.972)z^{-1}}{1-2e^{-0.256}(\cos 0.972)z^{-1}+e^{-2\times 0.256}z^{-2}}$$
$$+0.135\frac{e^{-0.256}\sin(0.972)z^{-1}}{1-2e^{-0.256}(\cos 0.972)z^{-1}+e^{-2\times 0.256}z^{-2}}$$
$$= \frac{0.511}{1-0.6z^{-1}} + \frac{-0.511+0.223z^{-1}}{1-0.873z^{-1}+0.6z^{-2}} + \frac{0.086z^{-1}}{1-0.873z^{-1}+0.6z^{-2}}$$
$$= \frac{0.511}{1-0.6z^{-1}} + \frac{-0.511+0.223z^{-1}+0.086z^{-1}}{1-0.873z^{-1}+0.6z^{-2}}$$

$$\boxed{H(z) = \frac{0.511}{1-0.6z^{-1}} + \frac{-0.511+0.309z^{-1}}{1-0.873z^{-1}+0.6z^{-2}}}$$

Example 4.23 Determine the system lowest order Chebyshev digital filter that meets the following specification:

(*a*) 1 dB ripple in the passband $0 \leq |\omega| \leq 0.3\pi$.
(*b*) Atleast 60 dB attenuation in the stopband $0.35\pi \leq |\omega| \leq \pi$. Use the bilinear transformation.

Solution

$$-20\log A_p = 1\text{ dB},\quad \omega_p = 0.3\pi$$
$$-20\log A_s = 60\text{ dB},\quad \omega_s = 0.35\pi$$

$$\begin{aligned}20\log A_p &= -1,\\ A_p &= 0.89125,\\ 20\log A_s &= -60,\\ A_s &= 0.001\end{aligned}$$

$$\text{The attenuation constant } \epsilon = \left(\frac{1}{A_p^2} - 1\right)^{\frac{1}{2}}$$
$$= \left(\frac{1}{(0.89)^2} - 1\right)^{\frac{1}{2}} = 0.5123$$

$$\frac{\Omega_s}{\Omega_p} = \frac{\tan\left(\frac{\omega_s}{2}\right)}{\tan\left(\frac{\omega_p}{2}\right)} = \frac{\tan\left(0.35\frac{\pi}{2}\right)}{\tan\left(0.3\frac{\pi}{2}\right)} = \frac{0.6128}{0.5045}$$
$$= 1.203$$

$$N \geq \frac{\cosh^{-1}\left\{\frac{1}{\epsilon}\left[\frac{1}{A_s^2} - 1\right]^{\frac{1}{2}}\right\}}{\cosh^{-1}\frac{\Omega_s}{\Omega_p}}$$
$$\geq \frac{\cosh^{-1}\left\{\frac{1}{0.5123}\left[\frac{1}{(0.001)^2} - 1\right]^{\frac{1}{2}}\right\}}{\cosh^{-1}(1.203)}$$
$$\geq \frac{\cosh^{-1}(1951.98)}{\cosh^{-1}(1.203)}$$
$$\geq \frac{8.2697}{0.6265}$$
$$N \geq 13.1935$$

$$\boxed{N = 14}$$

Example 4.24 Design a Chebyshev filter with a maximum passband attenuation of 2.5 dB at $\Omega_p = 20$ rad/s and the stopband attenuation of $\Omega_s = 50$ rad/s.

Solution Given

$$-20\log A_p = 2.5, \quad \Omega_p = 20 \text{ rad/s}$$
$$A_p = 0.749,$$
$$A_p \approx 0.75$$
$$-20\log A_s = 30, \quad \Omega_s = 50 \text{ rad/s}$$
$$A_s = 0.0316$$

$$\text{The attenuation constant } \epsilon = \left(\frac{1}{A_p^2} - 1\right)^{\frac{1}{2}}$$
$$= \left(\frac{1}{(0.75)^2} - 1\right)^{\frac{1}{2}} = 0.8819$$
$$\frac{\Omega_s}{\Omega_p} = \frac{50}{20} = 2.5$$
$$N \geq \frac{\cosh^{-1}\left\{\frac{1}{\epsilon}\left[\frac{1}{A_s^2} - 1\right]^{\frac{1}{2}}\right\}}{\cosh^{-1}\left(\frac{\Omega_s}{\Omega_p}\right)}$$
$$\geq \frac{\cosh^{-1}\left\{\frac{1}{0.8819}\left[\frac{1}{(0.0316)^2} - 1\right]^{\frac{1}{2}}\right\}}{\cosh^{-1}(2.5)}$$
$$\geq \frac{\cosh^{-1}(35.8654)}{\cosh^{-1}(2.5)}$$
$$\geq \frac{4.2727}{1.5667}$$
$$\geq 2.72$$
$$N = 3$$

The analog cut-off frequency

$$\Omega_c \approx \Omega_p = 20 \text{ rad/s}$$

For N is odd

$$H_a(s) = \frac{B_0\Omega_c}{s + c_0\Omega_c} \prod_{k=1}^{\frac{N-1}{2}} \frac{B_k\Omega_c^2}{s^2 + b_k\Omega_c s + c_k\Omega_c^2}$$
$$= \frac{B_0 \times 20}{s + (c_0 \times 20)} \prod_{k=1}^{\frac{3-1}{2}} \frac{B_k(20)^2}{s^2 + b_k 20s + c_k(20)^2}$$

$$= \frac{20B_0}{s+20c_0} \times \frac{20B_1}{s^2+b_1 20s+(20)^2c_1}$$

$$b_k = 2Y_N \sin\left(\frac{(2k-1)\pi}{2N}\right)$$

$$Y_N = \frac{1}{2}\left\{\left[\left(\frac{1}{\epsilon^2}+1\right)^{\frac{1}{2}}+\frac{1}{\epsilon}\right]^{\frac{1}{N}} - \left[\left(\frac{1}{\epsilon^2}+1\right)^{\frac{1}{2}}+\frac{1}{\epsilon}\right]^{-\left(\frac{1}{N}\right)}\right\}$$

$$= \frac{1}{2}\left\{\left[\left(\frac{1}{(0.8819)^2}+1\right)^{\frac{1}{2}}+\frac{1}{0.8819}\right]^{\frac{1}{3}} - \left[\left(\frac{1}{(0.8819)^2}+1\right)^{\frac{1}{2}}+\frac{1}{0.8819}\right]^{-\frac{1}{3}}\right\}$$

$$= \frac{1}{2}\left\{[1.51187+1.134]^{\frac{1}{3}} - [1.51187+1.134]^{-\frac{1}{3}}\right\}$$

$$= \frac{1}{2}\{1.3831-0.723\}$$

$$Y_N = 0.33$$

$$b_1 = 2Y_N \sin\left(\frac{(2-1)\pi}{2\times 3}\right)$$

$$= 2\times 0.33 \sin\left[\frac{\pi}{6}\right]$$

$$b_1 = 0.33$$

$$c_0 = Y_N = 0.33$$

$$c_k = Y_n^2 + \cos^2\left(\frac{(2k-1)\pi}{2N}\right)$$

$$c_1 = (0.33)^2 + \cos^2\left[\frac{(2-1)\pi}{2\times 3}\right]$$

$$= (0.33)^2 + \cos^2\left[\frac{\pi}{6}\right]$$

$$= (0.33)^2 + 0.75$$

$$c_1 = 0.8589$$

$$H_a(s) = \frac{20B_0}{s+(20\times 0.33)} \times \frac{20^2 B_1}{s^2+(0.33)(20)s+((20)^2\times 0.8589)}$$

$$H_a(s) = \frac{20B_0}{s+6.6} \times \frac{20^2 B_1}{s^2+6.6s+343.56}$$

For $N =$ odd, B_k can be calculated from

$$H_a(s)\Big|_{s=0} = 1, \quad [B_0 = B_1 = B_2 \cdots B_N]$$

$$\frac{20B_0}{6.6} \times \frac{20^2 B_1}{343.56} = 1$$

$$\frac{8000 B_0^2}{2267.496} = 1$$
$$B_0^2 = 0.2834$$
$$B_0 = 0.532$$
$$B_0 = B_1 = 0.532$$
$$H_a(s) = \frac{20 \times 0.532}{s + 6.6} \times \frac{20^2 \times 0.532}{s^2 + 6.6s + 343.56}$$
$$= \frac{10.64}{s + 6.6} \cdot \frac{212.8}{s^2 + 6.6s + 343.56}$$
$$H_a(s) = \frac{2265.76}{(s + 6.6)(s^2 + 6.6s + 343.56)}$$

$$H(z) = H_a(s)\Big|_{s=\frac{2}{T}\frac{(z-1)}{(z+1)}}, \quad \text{Let } T = 1 \text{ s}$$
$$= \frac{2265.76}{\left(\frac{2(z-1)}{z+1} + 6.6\right)\left[\left(\frac{2(z-1)}{(z+1)}\right)^2 + 6.6 \times \frac{2(z-1)}{(z+1)} + 343.56\right]}$$
$$= \frac{2265.76}{\left(\frac{2(z-1)+6.6(z+1)}{(z+1)}\right)\left(\frac{4(z-1)^2+13.2(z-1)(z+1)+343.56(z+1)^2}{(z+1)^2}\right)}$$
$$= \frac{2265.76(z+1)^3}{[2(z-1) + 6.6(z+1)][4(z^2 - 2z + 1) + 13.2(z^2 - 1) + 343.56(z^2 + 2z + 1)]}$$
$$= \frac{2265.76(z+1)^2}{[8.6z + 4.6][4z^2 - 8z + 2 + 13.2z^2 - 13.2 + 343.56z^2 + 687.12z + 343.56]}$$

$$\boxed{H(z) = \frac{2265.76(z+1)^2}{(8.6z + 4.6)[360.76z^2 + 687.72z + 332.36]}}$$

Example 4.25 Design a digital Chebyshev filter to meet the constraints

$$\frac{1}{\sqrt{2}} \le |H(\omega)| \le 1; \quad 0 \le \omega \le 0.2\pi$$
$$0 \le |H(\omega)| \le 0.1; \quad 0.5\pi \le \omega \le \pi$$

by using bilinear transformation. Assume $T = 1$ s.

Solution Given

$$A_p = 0.707 \text{ at } \omega_p = 0.2\pi \text{ rad/s}$$
$$A_s = 0.2 \text{ at } \omega_s = 0.5\pi \text{ rad/s}$$

$$\text{The attenuation constant } \epsilon = \left(\frac{1}{A_p^2} - 1\right)^{\frac{1}{2}} = (2-1)^{\frac{1}{2}}$$

$$\epsilon = 1$$

$$\frac{\Omega_s}{\Omega_p} = \frac{\tan\left(\frac{\omega_s}{2}\right)}{\tan\left(\frac{\omega_p}{2}\right)} = \frac{\tan\left(\frac{0.5\pi}{2}\right)}{\tan\frac{0.2\pi}{2}} = \frac{1}{0.3249}$$

$$\frac{\Omega_s}{\Omega_p} = 3.0378$$

$$n \geq \frac{\cosh^{-1}\left\{\frac{1}{\epsilon}\left[\frac{1}{A_s^2} - 1\right]^{\frac{1}{2}}\right\}}{\cosh^{-1}\frac{\Omega_s}{\Omega_p}}$$

$$\geq \frac{\cosh^{-1}\left\{\frac{1}{1}\left[\frac{1}{0.2^2} - 1\right]^{\frac{1}{2}}\right\}}{\cosh^{-1}(3.078)}$$

$$\geq \frac{\cosh^{-1}(4.899)}{\cosh^{-1}(3.078)}$$

$$\geq \frac{2.2716}{1.7899} \geq 1.269$$

$$N = 2$$

Given T = 1 s. The analog cut-off frequency

$$\Omega_c = \frac{2}{T}\frac{\tan\left(\frac{\omega_p}{2}\right)}{\left[\frac{1}{A_p^2} - 1\right]^{\frac{1}{2}N}}$$

$$\Omega_c = 2\frac{\tan\left(\frac{0.2\pi}{2}\right)}{\left[\frac{1}{(0.707)^2} - 1\right]^{\frac{1}{4}}} = \frac{0.6498}{1} = 0.6498 \text{ rad/s}$$

For N is even

$$H_a(s) = \prod_{k=1}^{\frac{N}{2}} \frac{B_k\Omega_c^2}{s^2 + b_k\Omega_c s + c_k\Omega_c^2}$$

$$= \frac{B_1\Omega_c^2}{s^2 + b_1\Omega_c s + c_1\Omega_c^2}$$

$$
\begin{aligned}
Y_N &= \frac{1}{2}\left\{\left[\left(\frac{1}{\epsilon^2}+1\right)^{\frac{1}{2}}+\frac{1}{\epsilon}\right]^{\frac{1}{N}}-\left[\left(\frac{1}{\epsilon^2}+1\right)^{\frac{1}{2}}+\frac{1}{\epsilon}\right]^{-\left(\frac{1}{N}\right)}\right\}\\
&= \frac{1}{2}\left\{\left[\left(\frac{1}{1^2}+1\right)^{\frac{1}{2}}+\frac{1}{1}\right]^{\frac{1}{2}}-\left[\left(\frac{1}{1^2}+1\right)^{\frac{1}{2}}+\frac{1}{1}\right]^{-\frac{1}{2}}\right\}\\
&= \frac{1}{2}\left\{[4.4142+1]^{\frac{1}{2}}-[4.4142+1]^{-\frac{1}{2}}\right\}\\
&= \frac{1}{2}\{1.554-0.6435\}\\
Y_N &= 0.45525\\
b_k &= 2Y_n \sin\left(\frac{(2k-1)\pi}{2N}\right)\\
b_1 &= 2Y_N \sin\left(\frac{(2-1)\pi}{4}\right)\\
&= 2Y_N \sin\left(\frac{\pi}{4}\right)\\
&= 2\times 0.45525 \sin\left(\frac{\pi}{4}\right)\\
b_1 &= 0.6438
\end{aligned}
$$

$$
\begin{aligned}
c_k &= Y_N^2+\cos^2\left(\frac{(2k-1)\pi}{2N}\right)\\
c_1 &= (0.45525)^2+\cos^2\left(\frac{\pi}{4}\right)\\
&= (0.45525)^2+0.5\\
c_1 &= 0.707\\
H_a(s) &= \frac{B_1\Omega_c^2}{s^2+0.6438\Omega_c s+0.707\Omega_c^2}\\
&= \frac{B_1(0.6498)^2}{s^2+0.6438\times 0.6498s+0.707(0.6498)^2}\\
H_a(s) & \frac{B_1 0.4222}{s^2+0.4183s+0.2985}
\end{aligned}
$$

For N is even.

$$
H_a(s)\Big|_{s=0} = \frac{1}{(1+\epsilon^2)^{\frac{1}{2}}}
$$

$$\frac{B_1 \times 0.4222}{0.2985} = \frac{1}{(1+1^2)^{\frac{1}{2}}}$$

$$\frac{0.4222B_1}{0.2985} = 1$$

$$B_1 = \frac{0.2985}{0.4222}$$

$$B_1 = 0.707$$

$$H_a(s) = \frac{0.707 \times 0.4222}{s^2 + 0.4183s + 0.2985}$$

$$H_a(s) = \frac{0.2985}{s^2 + 0.4183s + 0.2985}$$

Using bilinear transformation we get

$$H(z) = H_a(s)\Big|_{s=\frac{2}{T}\frac{z-1}{z+1}}, \quad T = 1\text{ s}$$

$$= \frac{0.2985}{\left(2\frac{Z-1}{Z+1}\right)^2 + 0.4183\left(s\frac{(Z-1)}{(Z+1)}\right)2 + 0.2985}$$

$$= \frac{0.2985(z+1)^2}{4(z-1)^2 + 0.8366(z-1)(z+1) + 0.2985(z+1)^2}$$

$$= \frac{0.2985(z^2+2z+1)}{4(z^2-2z+1) + 0.8366(z^2-1) + 0.2985(z^2+2z+1)}$$

$$= \frac{0.2985(z^2+2z+1)}{4z^2 - 8z + 4 + 0.8366z^2 - 0.8366 + 0.2985z^2 + 0.597z + 0.2985}$$

$$= \frac{0.2985(z^2+2z+1)}{5.1351z^2 - 7.403z + 3.4619}$$

$$H(z) = \frac{0.2985(z^{-2}+2z^{-1}+1)}{3.4619z^{-2} - 7.403z^{-1} + 5.1351}$$

$$= \frac{0.2985(z^{-2}+2z^{-1}+1)}{5.1351(0.674z^{-2} - 1.4416z^{-1} + 1)}$$

$$\boxed{H(z) = \frac{0.0581(z^{-2}+2z^{-1}+1)}{0.674z^{-2} - 1.4416z^{-1} + 1}}$$

Example 4.26 The transfer function of an analog LPF is

$$H(z) = \frac{1}{s+1}$$

with a bandwidth of 1 rad/s. Use bilinear transform to design a digital filter with a bandwidth of 20 Hz at a sampling frequency of 600 Hz.

Solution Given

$$\begin{aligned} H(z) &= \frac{1}{s+1}, \quad \Omega_p = 1 \text{ rad/s} \\ f'_p &= 20\text{Hz} \\ \Omega'_p &= 2\pi f'_p \\ &= 2\pi \times 20 \\ &= 125.66 \text{ rad/s.} \end{aligned}$$

Analog lowpass filter to lowpass filter transformation from Eq. (4.66) obtain by replacing $s \rightarrow \frac{\Omega_p}{\Omega'_p} \cdot s$

$$\begin{aligned} H(s) &= \frac{1}{\frac{s}{125.66} + 1} \\ &= \frac{125.66}{s + 125.66} \end{aligned}$$

The sampling frequency is given as $f_s = 60Hz$

$$\begin{aligned} T &= \frac{1}{f_s} \\ &= \frac{1}{60} \\ &= 0.0167 \text{ s.} \end{aligned}$$

Using bilinear transformation we get,

$$\begin{aligned} H(z) &= H(s)\Big|_{s=\frac{2}{T}\frac{(1-z^{-1})}{1+z^{-1}}} \\ H(z) &= \frac{125.66}{s+125.66}\Big|_s = \frac{2}{0.0167}\frac{1-z^{-1}}{1+z^{-1}} \\ &= \frac{125.66}{120\frac{(1-z^{-1})}{(1+z^{-1})} + 125.66} \end{aligned}$$

$$= \frac{125.66(1 + z^{-1})}{120(1 - z^{-1}) + 125.66(1 + z^{-1})}$$
$$= \frac{125.66(1 + z^{-1})}{120 - 120z^{-1} + 125.66 + 125.66z^{-1}}$$
$$= \frac{125.66(1 + z^{-1})}{245.66 + 5.66z^{-1}}$$
$$= \frac{125.66(1 + z^{-1})}{245.66(1 + 0.023z^{-1})}$$

$$\boxed{H(z) = \frac{0.512(1 + z^{-1})}{1 + 0.023z^{-1}}}$$

4.11 IIR Filter Design by Approximation of Derivatives

One way of converting an analog filter into a digital filter is to approximate the differential equation governing the analog filter by an equivalent difference equation. Thus, the first order differential equation can be approximated and written as follows.

$$\left.\frac{dy(t)}{dt}\right|_{t=nT} = \frac{y(nT) - y(nT - T)}{T}$$
$$= \frac{y(n) - y(n - 1)}{T}$$

where T is the sampling interval. The above equation can be written with analog system function as $H(s) = s$ and digital system function $H(z) = \frac{(1-z^{-1})}{T}$. The frequency domain equivalence is therefore written as

$$\boxed{s = \frac{(1 - z^{-1})}{T}} \tag{4.74}$$

Thus, the system function for the digital IIR filter is written as

$$\boxed{H(z) = H_a(s)\Big|_{s=\left(\frac{1-z^{-1}}{T}\right)}.} \tag{4.75}$$

The stable analog filter when converted into digital filter should also be stable. The mapping of the LHP in the s-plane to z-plane has to be investigated. Equation (4.74) can be written as

$$z = \frac{1}{(1 - sT)}$$

Substituting $s = j\Omega$ in the above equation we have

$$\begin{aligned} z &= \frac{1}{1 + j\Omega T} \\ &= \frac{1 + j\Omega T}{(1 + j\Omega T)(1 + j\Omega T)} \\ &= \frac{1}{(1 + \Omega^2 T^2)} + \frac{j\Omega T}{(1 + \Omega^2 T^2)} \end{aligned}$$

Let

$$z = x + jy$$

$$x = \frac{1}{(1 + \Omega^2 T^2)} \quad \text{and} \quad y = \frac{\Omega T}{(1 + \Omega^2 T^2)}$$

Consider the following equation

$$\begin{aligned} \left(x - \frac{1}{2}\right)^2 + y^2 &= \left[\frac{1}{(1 + \Omega^2 T^2)} - \frac{1}{2}\right]^2 + \frac{\Omega^2 T^2}{(1 + \Omega^2 T^2)} \\ &= \left(\frac{1}{2}\right)^2. \end{aligned}$$

This is the equation of a circle in the z-plane with centre at $z = \frac{1}{2}$ and radius $\frac{1}{2}$. Thus the mapping takes the points in LHP of the s-plane (for stable analog filter) into corresponding points inside this circle in the z-plane as shown in Fig. 4.25.

From Fig. 4.25 it is evident that the mapping $s = \frac{(1-z^{-1})}{T}$ takes LHP in the s-plane into points inside the circle of radius $\frac{1}{2}$ and at centre $z = \frac{1}{2}$ in the z-plane. Thus, the location of the poles of the digital filter so designed are confined to a small frequency range which means it is suitable for the design of lowpass and bandpass filters of small resonant frequencies and not suitable for highpass digital filter design. The method is illustrated by the following numerical example.

Example 4.27 Convert the analog bandpass filter with the system function

$$H_a(s) = \frac{1}{(s + 0.1)^2 + 5}; \qquad T = 0.1 \text{ s}$$

into a digital IIR filter using the backward difference for the derivative.

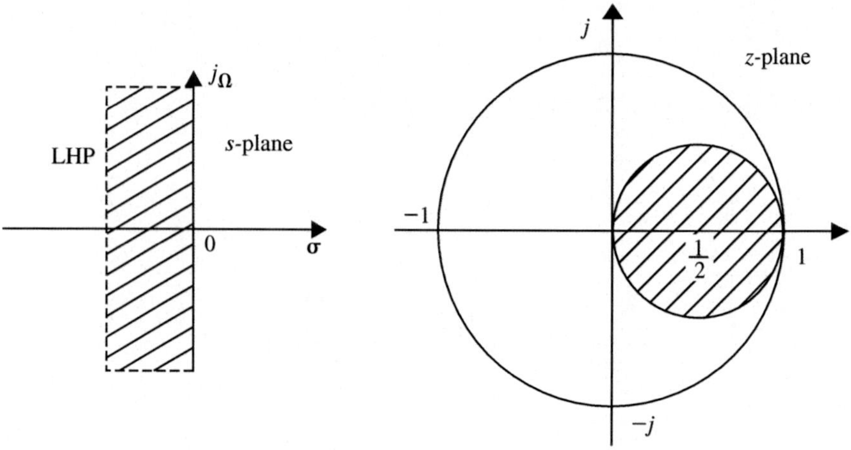

Fig. 4.25 The mapping from s-plane to z-plane by approximation of derivatives

Solution

$$H_a(s) = \frac{1}{(s + 0.1)^2 + 5}$$

Substituting

$$s = \frac{(1 + z^{-1})}{T} = (10 + 10z^{-1})$$

$$H(z) = \frac{1}{(10.10 + 10z^{-1})^2 + 5}$$

$$\boxed{H(z) = \frac{9.34 \times 10^{-3}}{1 - 1.8877z^{-1} + 0.934z^{-2}}}$$

The poles are at

$$z = 0.9438 \pm j0.209$$
$$= 0.9667\angle \pm 12.5°$$

The poles are within the circle at centre $z = \frac{1}{2}$ and radius $\frac{1}{2}$. Hence, the digital filter so designed is stable.

4.12 Frequency Response from Transfer Function $H(z)$

The frequency response of a discrete time system can be determined from the transfer function $H(z)$. Several methods are available to achieve this. The geometric method of evaluation of frequency response is described here.

Consider the transfer function $H(z)$ expressed in terms of poles and zeros are given below:

$$H(z) = \frac{K(z - z_1)(z - z_2)\cdots(z - z_n)}{(z - p_1)(z - p_2)\cdots(z - p_n)} = \prod_{i=1}^{N} \frac{k(z - z_i)}{(z - p_i)}$$

The frequency response is obtained by substituting $z = e^{j\omega T}$ where $0 \leq \omega \leq \omega_s/2$; ω_s is sampling frequency; and T is sampling period.

The frequency response of $H(z)$ written as $H(e^{j\omega T})$ is given below:

$$H(e^{j\omega T}) = \frac{\prod_{i=1}^{N} k(e^{j\omega T} - z_i)}{\prod_{i=1}^{N} (e^{j\omega T} - p_i)}$$

For simplicity consider a system function $H(z)$ having two poles and two zeros. The pole zero diagram is shown in Fig. 4.26. The zeros are located at z_1 and $-z_2$ and the poles are located at p_1 and p_2. Consider any point P on the unit circle. Draw a radial line from the origin to the point P. The line OP makes angle θ from the reference axis. θ and ωT are related as follows.

$$\theta = \omega_1 T$$

Knowing θ and the sampling period T, the frequency ω_1 at point P can be calculated. Draw radial lines from poles and zeros of $H(z)$ to the point P. Then

$$H(e^{j\omega T}) = K \frac{A\angle\theta_{z_1} B\angle\theta_{z_2}}{C\angle\theta_{p_1} D\angle\theta_{p_2}}$$

Knowing K, A, B, C and D, $|H(e^{j\omega T})|$ can be estimated. Similarly, knowing θ_{z_1}, θ_{z_2}, θ_{p_1} and θ_{p_2}, $\angle H(e^{j\omega T})$ can be estimated at $\omega = \omega_1$. By moving the point P on the unit circle, at $\omega = \omega_2$, the magnitude and phase of $H(e^{j\omega T})$ can be estimated. Since the frequency response gets repeated for every $\omega = \frac{\omega_s}{2}$, it is necessary to estimate the magnitude and phase of $H(e^{j\omega T})$ for the frequency interval $0 \leq \omega \leq \frac{\omega_s}{2}$. The following example illustrates the above procedure.

Example 4.28 Consider the following system function of a certain discrete time system

$$H(z) = \frac{0.632z}{(z - 1)(z - 0.368)}$$

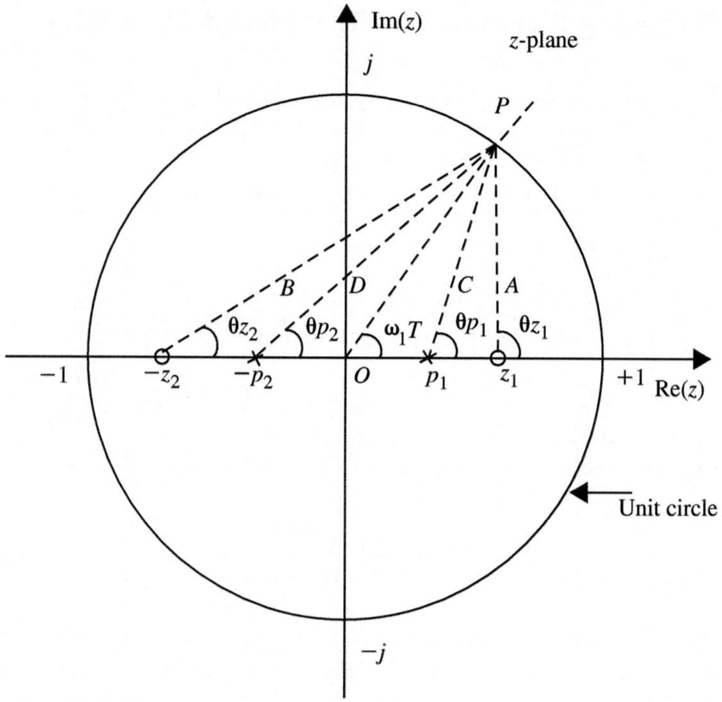

Fig. 4.26 Pole–zero diagram of $H(z)$

Draw the frequency response plot. Assume $T = 1$ s.

Solution The poles and zeros of $H(z)$ are located in the z-plane as shown in Fig. 4.27 and a unit circle is drawn. Any point P is chosen on the unit circle and radial line from the origin is drawn. From Fig. 4.27 ωT is found and knowing $T = 1$, ω is estimated $|H(e^{j\omega T})|$ and $\angle H(e^{j\omega T})$ are also estimated at point P. This is repeated upto $\omega = \frac{\pi}{T} = \pi$. The values are tabulated as shown below.

ω	0	$\frac{\pi}{2}$	π	$\frac{3\pi}{2}$	2π
$\|H(e^{j\omega})\|$	∞	0.42	0.231	0.42	π
$\|H(e^{j\omega})\|$	−90°	−155°	−180°	−205°	−270°

The frequency response polar plot of the discrete time system is shown in Fig. 4.28a. The magnitude and phase responses are shown in Fig. 4.28b and c respectively.

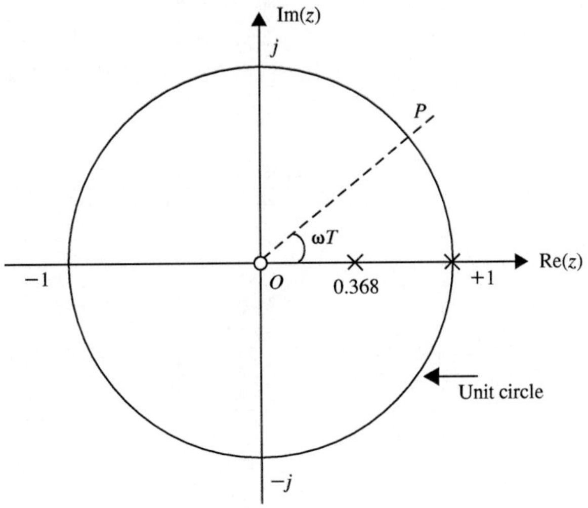

Fig. 4.27 Pole–zero diagram of Example 4.28

4.13 Structure Realization of IIR System

The IIR system can be described by the following difference equation.

$$y(n) = -\sum_{k=1}^{N} a_k y(n-k) + \sum_{k=0}^{M} b_k x(n-k) \tag{4.76}$$

Taking z-transform on both sides we get,

$$Y(z) = -\sum_{k=1}^{N} a_k z^{-k} Y(z) + \sum_{k=0}^{M} b_k z^{-k} X(z) \tag{4.77}$$

$$Y(z) + \sum_{k=1}^{N} a_k z^{-k} Y(z) = \sum_{k=0}^{M} b_k z^{-k} X(z)$$

$$Y(z)\left[1 + \sum_{k=1}^{N} a_k z^{-k}\right] = X(z) \sum_{k=0}^{M} b_k z^{-k}$$

$$\frac{Y(z)}{X(z)} = H(z) = \frac{\sum_{k=0}^{M} b_k z^{-k}}{1 + \sum_{k=1}^{N} a_k z^{-k}} \tag{4.78}$$

$$H(z) = \frac{b_0 + b_1 z^{-1} + b_2 z^{-2} + \cdots + b_M z^{-M}}{1 + a_1 z^{-1} + a_2 z^{-2} + \cdots + a_N z^{-N}} \tag{4.79}$$

where $H(z) \rightarrow$ transfer function of IIR system.

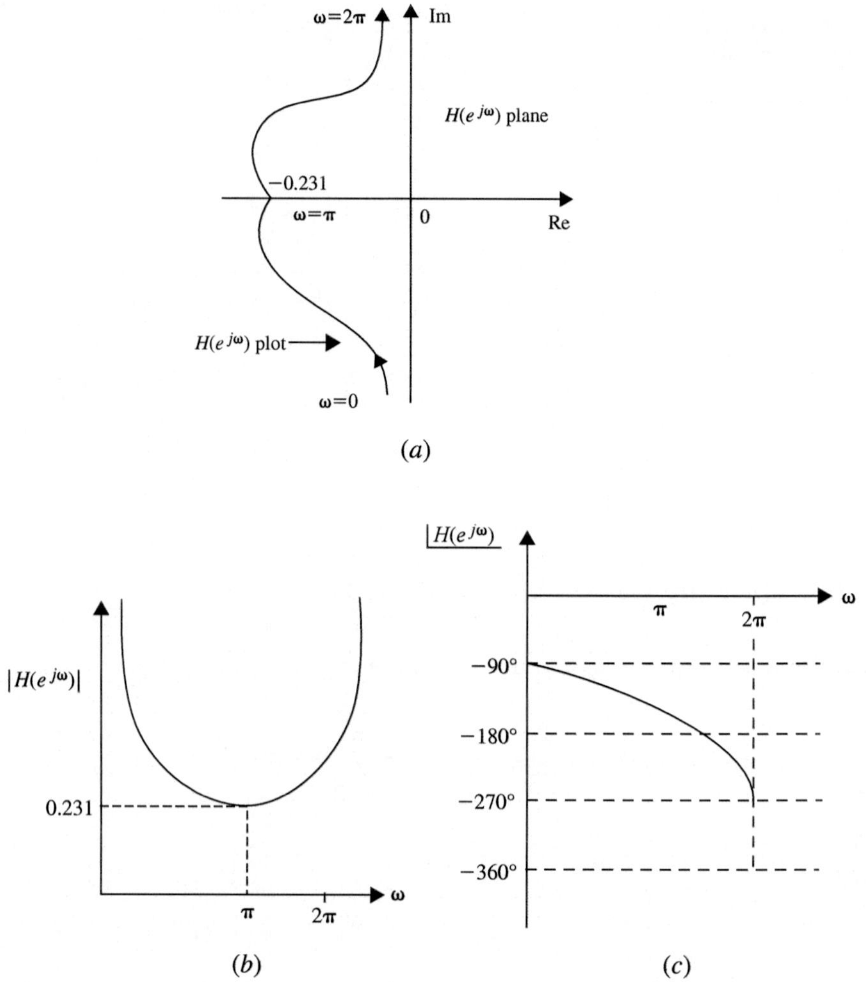

Fig. 4.28 **a** Frequency response polar plot of discrete system. **b** Magnitude response. **c** Phase response

The different types of structures for realizing IIR Systems are:

(*i*) Direct form-I structure
(*ii*) Direct form-II structure
(*iii*) Cascade form structure
(*iv*) Parallel form structure
(*v*) Transposed direct form realization
(*vi*) Lattice structure form

4.13.1 Direct Form-I Structure

Consider the following difference equation of an IIR system.

$$y(n) = -\sum_{k=1}^{N} a_k y(n-k) + \sum_{k=0}^{M} b_k x(n-k)$$

$$\begin{aligned} y(n) = &-a_1 y(n-1) - a_2 y(n-2) - \cdots - a_N y(n-N) \\ &+ b_0 x(n) + b_1 x(n-1) + b_2 x(n-2) + \cdots + +b_M x(n-M) \end{aligned} \tag{4.80}$$

Taking z-transform, we get

$$\begin{aligned} Y(z) = &-a_1 z^{-1} Y(z) - a_2 z^{-2} Y(z) - \cdots - a_N z^{-N} Y(z) \\ &+ b_0 X(z) + b_1 z^{-1} X(z) + b_2 z^{-2} X(z) + \cdots + +b_M z^{-M} X(z) \end{aligned} \tag{4.81}$$

Equation (4.81) is represented in Fig. 4.29.

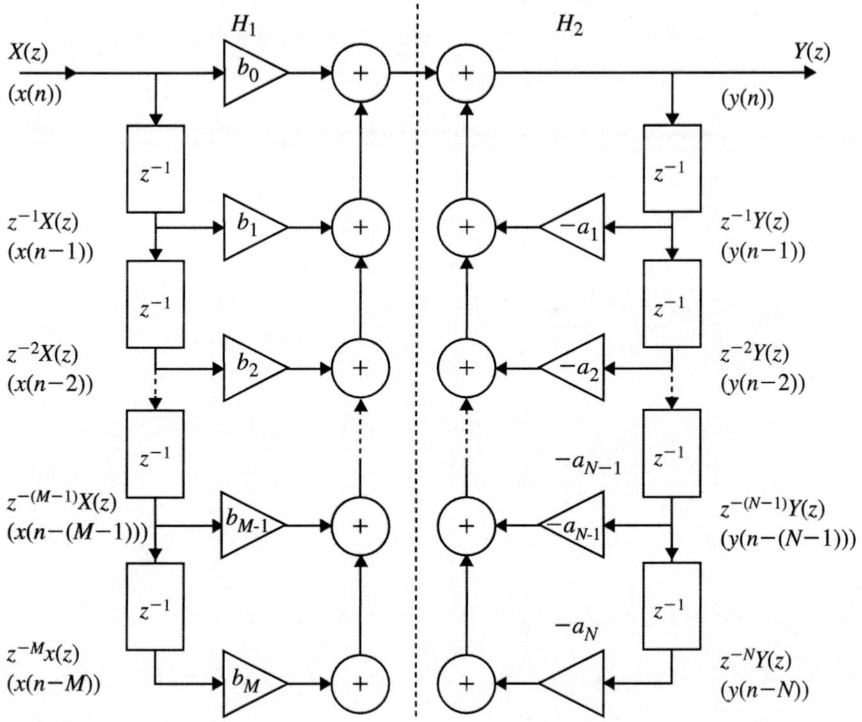

Fig. 4.29 Direct form-I realization

This structure uses different delays (z^{-1}) for input and output. So more memory is required for realizing.

4.13.2 Direct Form-II Structure

Direct form-II structure uses less number of delays compared to Direct form-I. Consider the IIR difference Eq. (4.80)

$$y(n) = -\sum_{k=1}^{N} a_k y(n-k) + \sum_{k=0}^{M} b_k x(n-k)$$

$$\begin{aligned} y(n) &= -a_1 y(n-1) - a_2 y(n-2) - \cdots - a_N y(n-N) \\ &\quad + b_0 x(n) + b_1 x(n-1) + b_2 x(n-2) + \cdots + b_M x(n-M) \end{aligned}$$

Taking z-transform on both sides we get,

$$\begin{aligned} Y(z) &= -a_1 z^{-1} Y(z) - a_2 z^{-2} Y(z) - \cdots - a_N z^{-N} Y(z) \\ &\quad + b_0 X(z) + b_1 z^{-1} X(z) + b_2 z^{-2} X(z) + \cdots + b_M z^{-M} X(z) \end{aligned}$$

$$\begin{aligned} &Y(z) + a_1 z^{-1} Y(z) + a_2 z^{-2} Y(z) + \cdots + a_N z^{-N} Y(z) \\ &\quad = b_0 X(z) + b_1 X(z) z^{-1} + b_2 z^{-2} X(z) + \cdots + b_M z^{-M} X(z) \\ &Y(z)[1 + a_1 z^{-1} + a_2 z^{-2} + \cdots + a_N z^{-N}] \\ &\quad = X(z)[b_0 + b_1 z^{-1} + b_2 z^{-2} + \cdots + b_M z^{-M}] \end{aligned}$$

$$\frac{Y(z)}{X(z)} = \frac{b_0 + b_1 z^{-1} + b_2 z^{-2} + \cdots + b_M z^{-M}}{1 + a_1 z^{-1} + a_2 z^{-2} + \cdots + a_N z^{-N}}$$

Let

$$\frac{Y(z)}{X(z)} = \frac{W(z)}{X(z)} \frac{Y(z)}{W(z)} \text{ where } \frac{W(z)}{X(z)} = \frac{1}{1 + a_1 z^{-1} + \cdots + +a_N z^{-N}} \tag{4.82}$$

$$\text{and } \frac{Y(z)}{W(z)} = b_0 + b_1 z^{-1} + b_2 z^{-2} + \cdots + b_M z^{-M} \tag{4.83}$$

Cross multiplying Eq. (4.82) we get,

$$W(z) + a_1 z^{-1} W(z) + a_2 z^{-2} W(z) + \cdots + a_N z^{-N} W(z) = X(z)$$

$$\therefore W(z) = X(z) - a_1 z^{-1} W(z) - a_2 z^{-2} W(z) - \cdots - a_N z^{-N} W(z) \tag{4.84}$$

Cross multiplying Eq. (4.84)

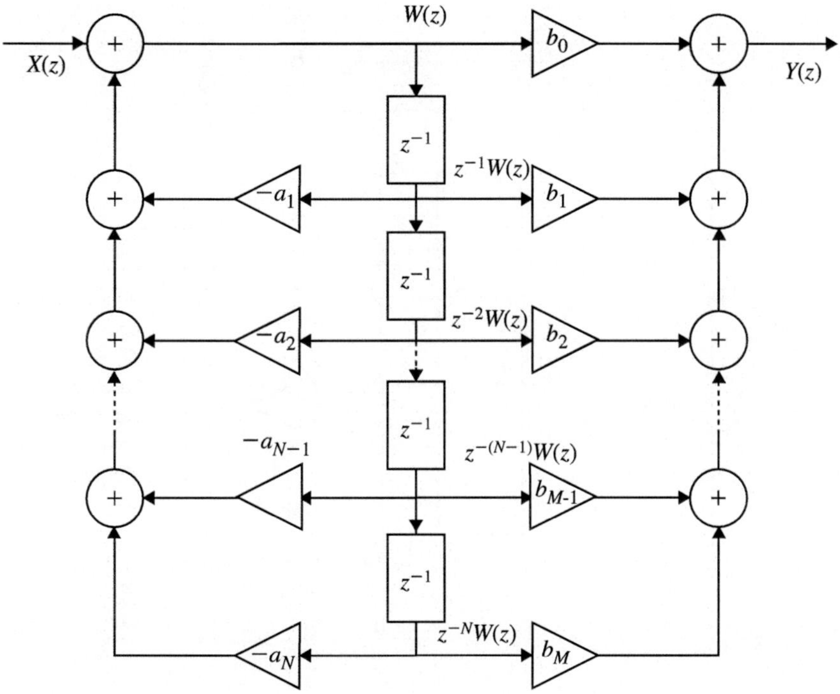

Fig. 4.30 Direct form-II structure of IIR system for $N = M$

$$Y(z) = b_0 W(z) + b_1 z^{-1} W(z) + b_2 z^{-2} W(z) + \cdots + b_M z^{-M} W(z) \tag{4.85}$$

Equations (4.84) and (4.85) represent IIR system in z domain. It can be realized by direct form-II structure as shown in Fig. 4.30.

4.13.2.1 Conversion of DF-I Structure to DF-II Structure

DF-I can be converted to DF-II by interchanging the order of cascading of two systems H_1 and H_2. It can be interchanged by linearity property.

Now we combine the delay elements to get a single system and the resultant structure will be DF-II structure as shown in Fig. 4.31.

4.13.3 Cascade Form Realization

The transfer function $H(z)$ can be expressed as a product of a number of second order or first order sections.

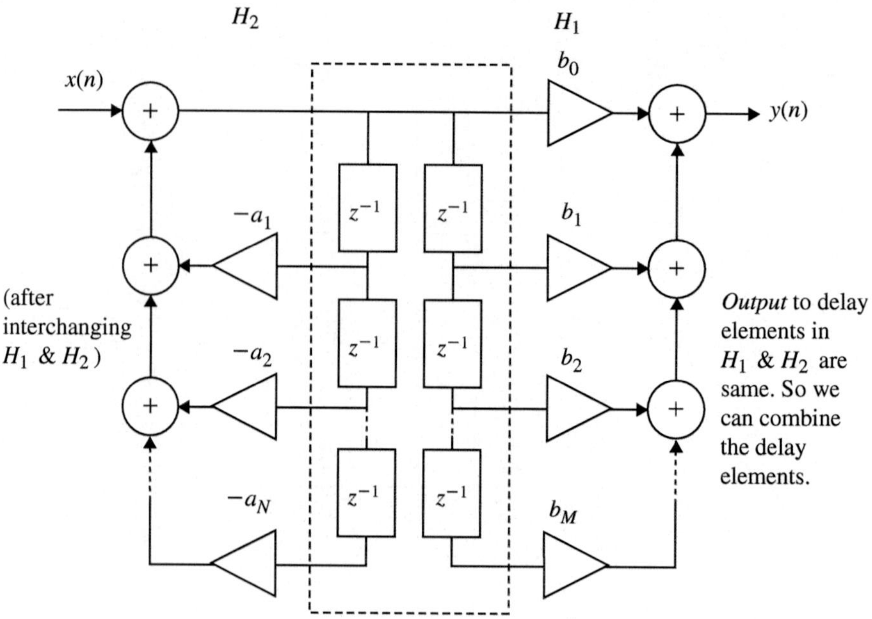

Fig. 4.31 Conversion of Direct form-I to Direct form-II structure

$$H(z) = \frac{Y(z)}{X(z)} = \prod_{i=1}^{k} H_i(z) \tag{4.86}$$

where

$$H_i(z) = \frac{C_{0i} + C_{1i}z^{-1} + C_{2i}z^{-2}}{d_{0i} + d_{1i}z^{-1} + d_{2i}z^{-2}} \quad \text{(2nd order section)} \tag{4.87}$$

or

$$H_i(z) = \frac{C_{0i} + C_{1i}z^{-1}}{d_{0i} + d_{1i}z^{-1}} \quad \text{(1st order section)} \tag{4.88}$$

The individual second order or first order sections can be realized either in direct form-I or DF-II structures. The overall system is obtained by cascading the individual sections.

The cascade form is shown in Fig. 4.32. The difficulty in realizing the system in cascade form are:

(*i*) Decision of pairing poles and zeros.
(*ii*) Deciding the order of cascading the 1st and 2nd order sections.
(*iii*) Scaring multipliers should be provided between individual sections to prevent the filter variable from becoming too large or too small.

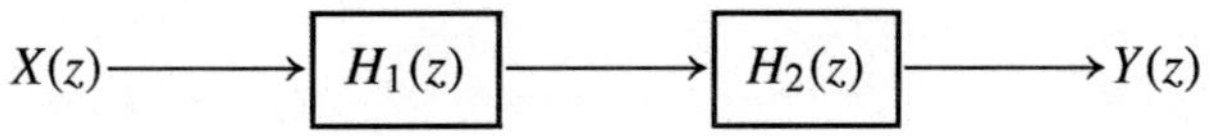

Fig. 4.32 Cascade realization of IIR system

4.13.4 Parallel Form Realization

By applying partial fraction to $H(z)$, the system can be expressed as sum of 1st and 2nd order sections as given below:

$$H(z) = \frac{Y(z)}{X(z)} = C + \sum_{i=1}^{k} H_i(z) \tag{4.89}$$

where,

$$H_i(z) = \frac{C_{0i} + C_{1i}z^{-1}}{d_{0i} + d_{1i}z^{-1} + d_{2i}z^{-2}} \quad \text{(2nd order section)} \tag{4.90}$$

or

$$H_i(z) = \frac{C_{0i}}{d_{0i} + d_{1i}z^{-1}} \quad \text{(1st order section)} \tag{4.91}$$

The individual 1st and 2nd order section can be realized either in DF-I or in DF-II structure. Then the overall system is obtained by connecting the individual sections in parallel as shown in Fig. 4.33.

4.13.5 Transposed Direct Form Realization

4.13.5.1 Signal Flow Graph

It is a graphical representation of the relationship between the variables of a set of linear difference equations. The basic elements of signal flow graph are branches and nodes.

Nodes: Represent system variables, which is equal to the sum of incoming signals from all branches connecting to the node. Nodes having two types:

Source Nodes → have only outgoing branches.
Sink Nodes → have only incoming branches.

Arrow head shows the direction of branch. The branch gain is indicated next to arrow head. The delay is indicated by the branch transmittance z^{-1}.

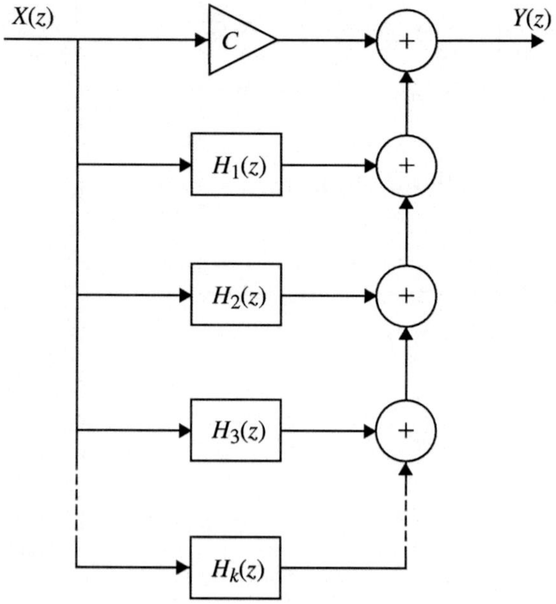

Fig. 4.33 Parallel form realization of IIR filter

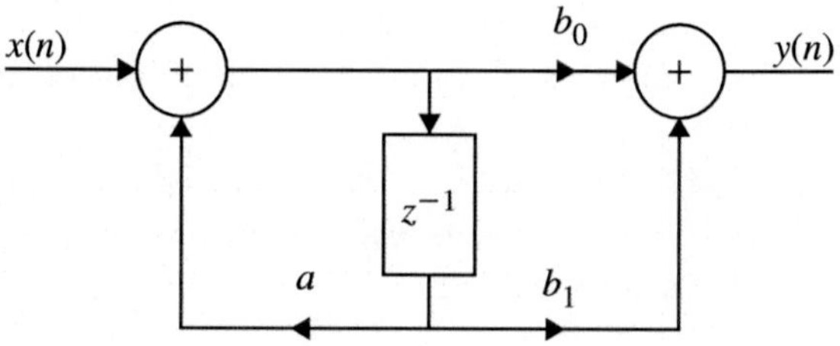

Fig. 4.34 First order system

When branch transmittance is unity it is unlabelled. Consider the block diagram of 1st order digital filter as shown in Fig. 4.34.

The signal flow graph is shown in Fig. 4.35.

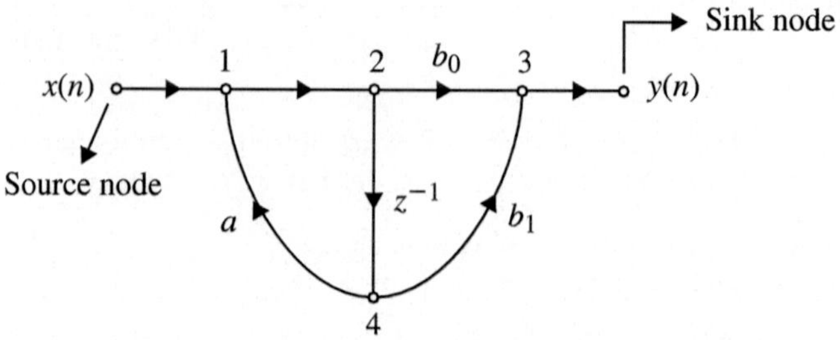

Fig. 4.35 Signal flow graph of first order system

Here we have 4 nodes out of this 2 nodes are running nodes, while other 2 are branching points. The delay is indicated by branch transmittance z^{-1}. Branch transmittance are indicated next to arrow head. The signal flow graph of first order filter is shown in Fig. 4.35.

4.13.6 *Transposition Theorem and Transposed Structure*

The transpose of a structure is defined by the following operations:

(*i*) Reverse the direction of all branches in the signal flow graph.
(*ii*) Interchange the inputs and outputs.
(*iii*) Reverse the roles of all nodes in the flow graph.
(*iv*) Summing points become branching points.
(*v*) Branching points become running points.

According to transportation theorem, the system transfer function remains unchanged by transposition.

Example 4.29 Realize the 2nd order digital filter

$$y(n) = 2n\cos\omega_0 y(n-1) - r^2 y(n-2) + x(n) - r\cos\omega_0 x(n-1).$$

Solution Let

$$x(n) - r\cos(\omega_0)x(n-1) = \omega(n)$$
$$y(n) = 2r\cos\omega_0 y(n-1) - r^2 y(n-2) + \omega(n)$$

The realization is shown in Fig. 4.36.

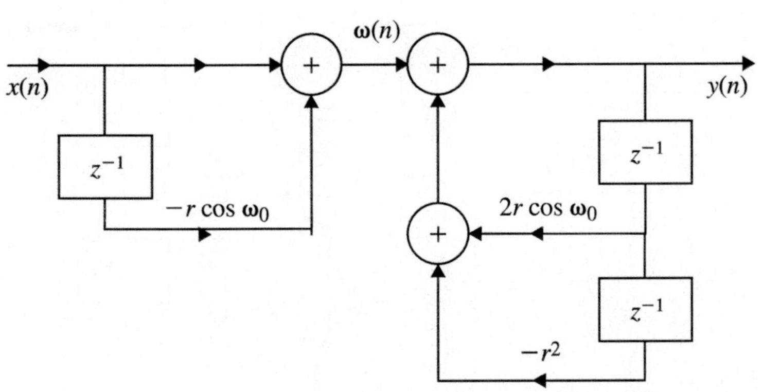

Fig. 4.36 Direct form-I realization for Example 4.29

Example 4.30 Obtain the DF-I realization for the system described by difference equation

$$y(n) = 0.5y(n-1) - 0.25y(n-2) + x(n) + 0.4x(n-1)$$

Solution Let

$$x(n) + 0.4x(n-1) = \omega(n)$$
$$y(n) = 0.5y(n-1) - 0.25y(n-2) + \omega(n)$$

The direct form-I is shown in Fig. 4.37.

Example 4.31 Realize the 2nd order system $y(n) = 2r\cos\omega_0 y(n-1) - r^2 y(n-2) + x(n) - r\cos(\omega_0)x(n-1)$ in DF-II.

Solution

$$y(n) = 2r\cos\omega_0 y(n-1) - r^2 y(n-2) + x(n) - r\cos(\omega_0)x(n-1)$$
$$Y(z)(1 - 2r\cos\omega_0 z^{-1} + r^2 z^{-2}) = X(z)(1 - r\cos\omega_0 z^{-1})$$

$$\frac{Y(z)}{X(z)} = \frac{1 - r\cos\omega_0 z^{-1}}{1 - 2r\cos\omega_0 z^{-1} + r^2 z^{-2}}$$

The direct form-II realization is shown in Fig. 4.38.

Example 4.32 Determine the DF-II realization for the following system $y(n) = -0.1y(n-1) + 0.72y(n-2) + 0.7x(n) - 0.252x(n-2)$.

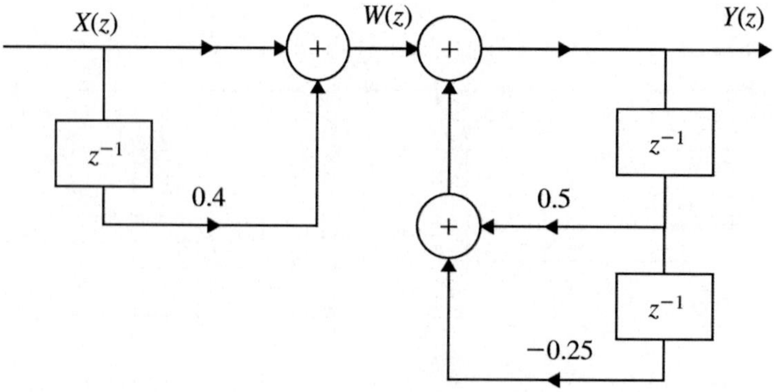

Fig. 4.37 Direct form-I realization for Example 4.30

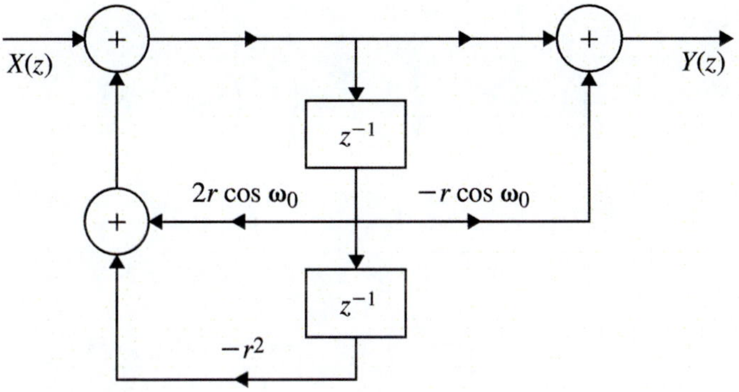

Fig. 4.38 Direct form-II realization for Example 4.31

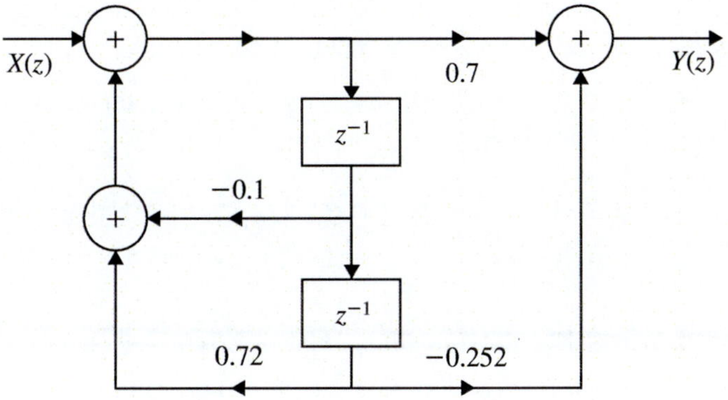

Fig. 4.39 Direct form-II realization for Example 4.32

Solution

$$Y(z)(1 + 0.1z^{-1} - 0.72z^{-2}) = X(z)(0.7 - 0.252z^{-2})$$

$$\frac{Y(z)}{X(z)} = \frac{0.7 - 0.252z^{-2}}{1 + 0.1z^{-1} - 0.72z^{-2}}$$

Direct form-II realization is shown in Fig. 4.39.

Example 4.33 Realize the system with difference equation

$$y(n) = \frac{3}{4}y(n-1) - \frac{1}{8}y(n-2) + \frac{1}{3}x(n-1) + x(n)$$

in cascade form.

Solution

$$Y(z)\left(1-\frac{3}{4}z^{-1}+\frac{1}{8}z^{-2}\right)=X(z)\left(1+\frac{1}{3}z^{-1}\right)$$

$$H(z)=\frac{Y(z)}{X(z)}=\frac{1+\frac{1}{3}z^{-1}}{1-\frac{3}{4}z^{-1}+\frac{1}{8}z^{-2}}$$

$$=\frac{1+\frac{1}{3}z^{-1}}{\left(1-\frac{1}{2}z^{-1}\right)\left(1-\frac{1}{4}z^{-1}\right)}=H_1(z)\cdot H_2(z)$$

$$H_1(z)=\frac{1+\frac{1}{3}z^{-1}}{1-\frac{1}{2}z^{-2}},\quad H_2(z)=\frac{1}{1-\frac{1}{4}z^{-1}}$$

The cascade form realization is shown in Fig. 4.40.

Example 4.34 Obtain the DF-I, DF-II, cascade and parallel form realization for the system (Figs. 4.41, 4.42 and 4.43)

$$y(n)=-0.1y(n-1)+0.2y(n-2)+3x(n)+3.6x(n-1)+0.6x(n-2)$$

Solution

$$y(n)=-0.1y(n-1)+0.2y(n-2)+3x(n)+3.6x(n-1)+0.6x(n-2)$$
$$Y(z)[1+0.1z^{-1}-0.2z^{-2}]=X(z)(3+3.6z^{-1}+0.6z^{-2})$$
$$H(z)=\frac{Y(z)}{X(z)}=\frac{3+3.6z^{-1}+0.6z^{-2}}{1+0.1z^{-1}-0.2z^{-2}}=\left(\frac{3+3.6z^{-1}+0.6z^{-2}}{1+0.1z^{-1}-0.2z^{-2}}\right)$$

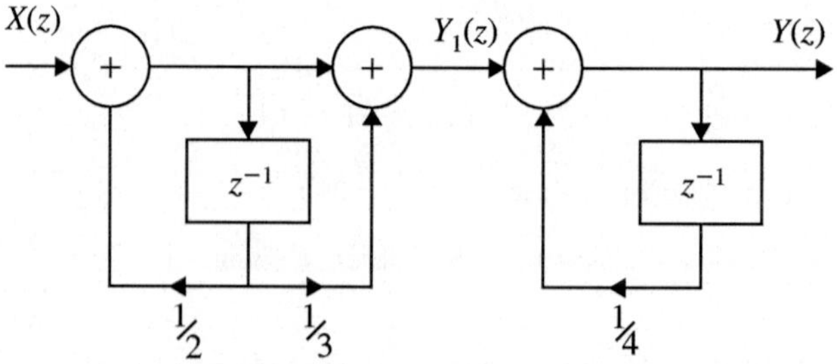

Fig. 4.40 Cascade form realization for Example 4.33

Direct form I and Direct form II realization for Example 4.34 are given in Figs. 4.41a and 4.41b respectively.

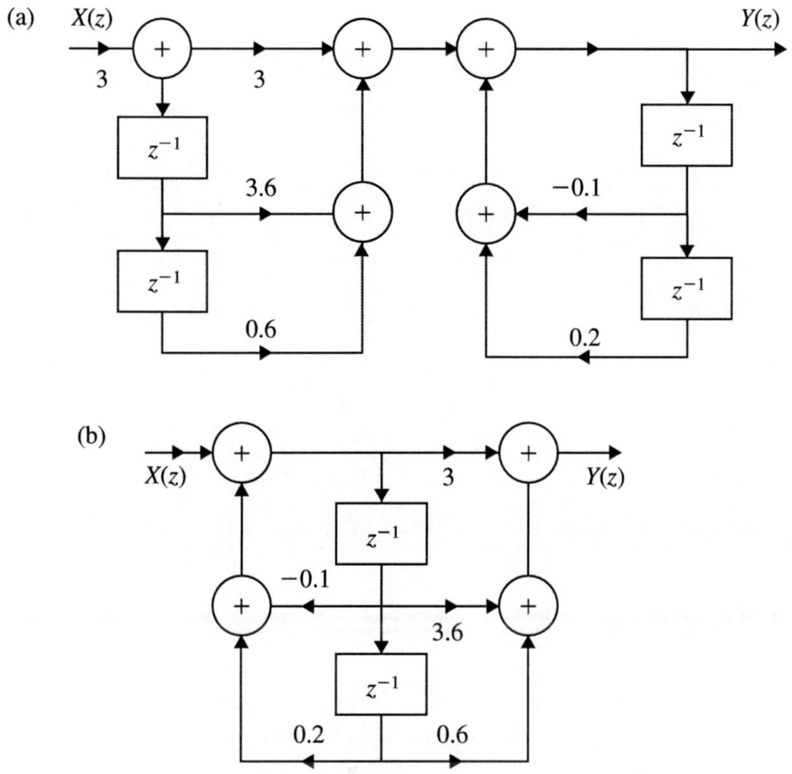

Fig. 4.41 **a**, **b** Direct form-I and form-II realization for Example 4.34

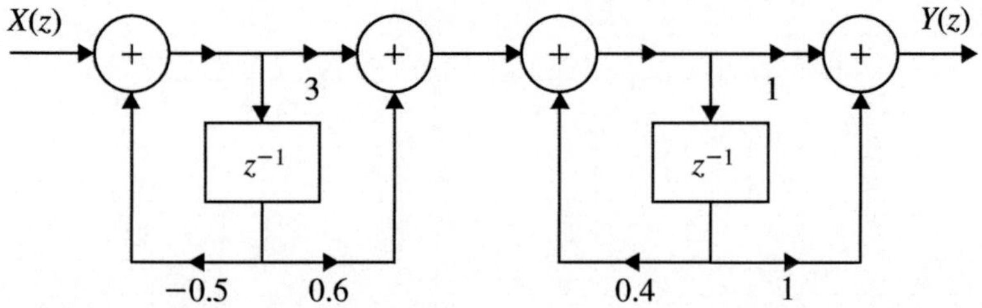

Fig. 4.42 Cascade form realization for Example 4.34

Cascade form realization for Example 4.34 is shown in Fig. 4.42.

Cascade Form

$$\frac{Y(z)}{X(z)} = \frac{3 + 3.6z^{-1} + 0.6z^{-2}}{1 + 0.1z^{-1} - 0.2z^{-2}}$$
$$= \frac{(3 + 0.6z^{-1})(1 + z^{-1})}{(1 + 0.5z^{-1})(1 - 0.4z^{-1})}$$

Let

$$H_1(z) = \frac{3 + 0.6z^{-1}}{1 + 0.5z^{-1}}, \quad H_2(z) = \frac{1 + z^{-1}}{1 - 0.4z^{-1}}$$

Parallel Form

$$H(z) = \frac{3 + 3.6z^{-1} + 0.6z^{-2}}{1 + 0.1z^{-1} - 0.2z^{-2}}$$
$$= -3 + \frac{3.9z^{-1} + 6}{1 + 0.1z^{-1} - 0.2z^{-2}}$$

Parallel form realization for Example 4.34 is shown in Fig. 4.43

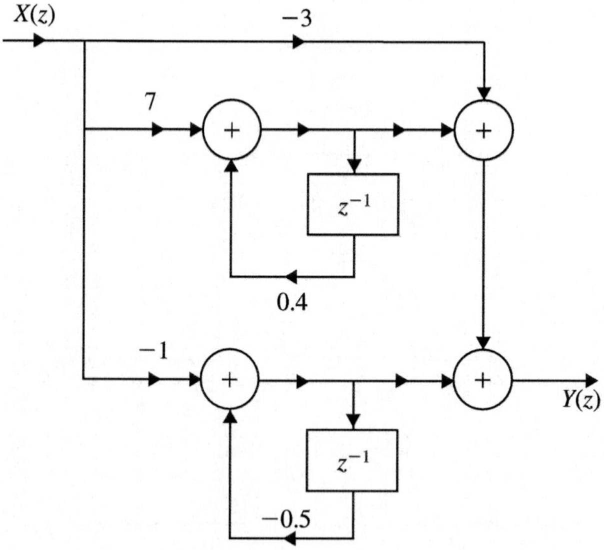

Fig. 4.43 Parallel form realization for Example 4.34

Consider

$$\frac{3.9z^{-1}+6}{1+0.1z^{-1}-0.2z^{-2}} = \frac{3.9z^{-1}+6}{(1-0.4z^{-1})(1+0.5z^{-1})}$$
$$= \frac{A}{1-0.4z^{-1}} + \frac{B}{1+0.5z^{-1}}$$
$$\Rightarrow A = (1-0.4z^{-1})\cdot\left.\frac{3.9z^{-1}+6}{(1-0.4z^{-1})(1+0.5z^{-1})}\right|_{z^{-1}=\frac{1}{0.4}}$$
$$= 7$$
$$\Rightarrow B = (1-0.5z^{-1})\cdot\left.\frac{3.9z^{-1}+6}{(1-0.4z^{-1})(1+0.5z^{-1})}\right|_{z^{-1}=\frac{-1}{0.5}}$$
$$= -1$$
$$\therefore H(z) = -3 + \frac{7}{1-0.4z^{-1}} - \frac{1}{1+0.5z^{-1}}$$
$$= C + H_1(z) + H_2(z)$$

Example 4.35 Obtain the cascade realization for the following system (Fig. 4.44).

$$H(z) = \frac{(1+\frac{3}{2}z^{-1}+\frac{1}{2}z^{-2})(1-\frac{3}{2}z^{-1}+z^{-2})}{(1+z^{-1}+\frac{1}{4}z^{-2})(1+\frac{1}{4}z^{-1}+\frac{1}{2}z^{-2})}$$

Cascade form realization for Example 4.35 is shown in Fig. 4.44

Solution Let $H(z) = H_1(z)H_2(z)$

$$H_1(z) = \frac{1+\frac{3}{2}z^{-1}+\frac{1}{2}z^{-2}}{1+z^{-1}+\frac{1}{4}z^{-2}}, \quad H_2(z) = \frac{1-\frac{3}{2}z^{-1}+z^{-2}}{1+\frac{1}{4}z^{-1}+\frac{1}{2}z^{-2}},$$

Example 4.36 Obtain the cascade realization of the system (Fig. 4.45)

$$H(z) = \frac{(1-\frac{1}{2}z^{-1})(1-\frac{1}{2}z^{-1}+\frac{1}{4}z^{-2})}{(1+\frac{1}{4}z^{-1})(1+z^{-1}+\frac{1}{2}z^{-2})(1-\frac{1}{4}z^{-1}+\frac{1}{2}z^{-2})}$$

Cascade form realization for Example 4.36 is shown in Fig. 4.45

Solution Let

$$H(z) = H_1(z)H_2(z)H_3(z)$$

where

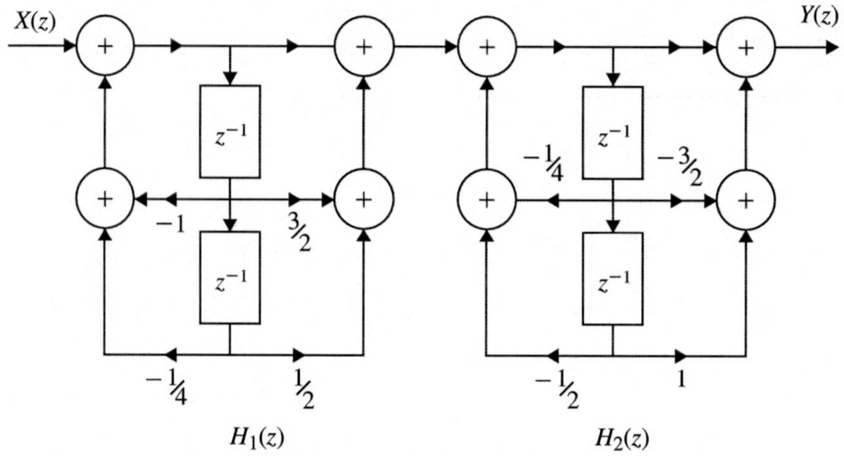

Fig. 4.44 Cascade form realization for Example 4.35

$$H_1(z)=\frac{1-\frac{1}{2}z^{-1}}{1+\frac{1}{4}z^{-1}};\ H_2(z)=\frac{1-\frac{1}{2}z^{-1}+\frac{1}{4}z^{-2}}{1+z^{-1}+\frac{1}{2}z^{-2}};\ H_3(z)=\frac{1}{1-\frac{1}{4}z^{-1}+\frac{1}{2}z^{-2}}$$

Example 4.37 Obtain the direct form-I, direct form-II, cascade and parallel form realization of LTI system governed by the equation (Figs. 4.46, 4.47, 4.48 and 4.49)

$$y(n)=-\frac{3}{8}y(n-1)+\frac{3}{32}y(n-2)+\frac{1}{64}y(n-3)+x(n)+3x(n-1)+2x(n-2)$$

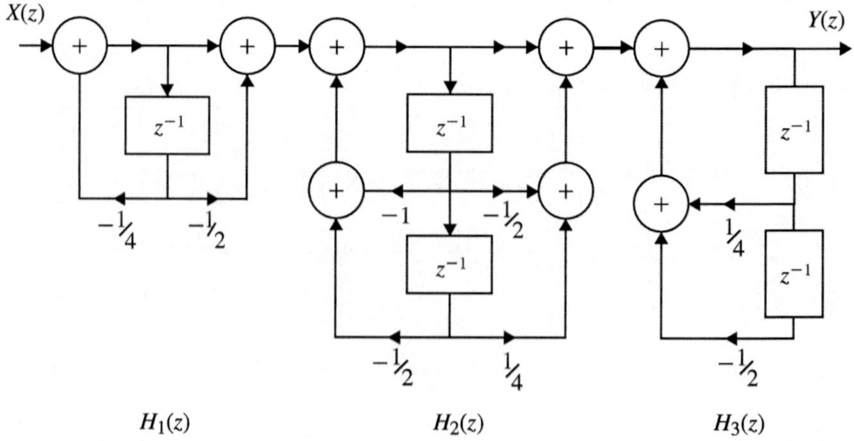

Fig. 4.45 Cascade form realization for Example 4.36

Direct Form-I realization of $H(z)$ for Example 4.37 is shown in Fig. 4.46.

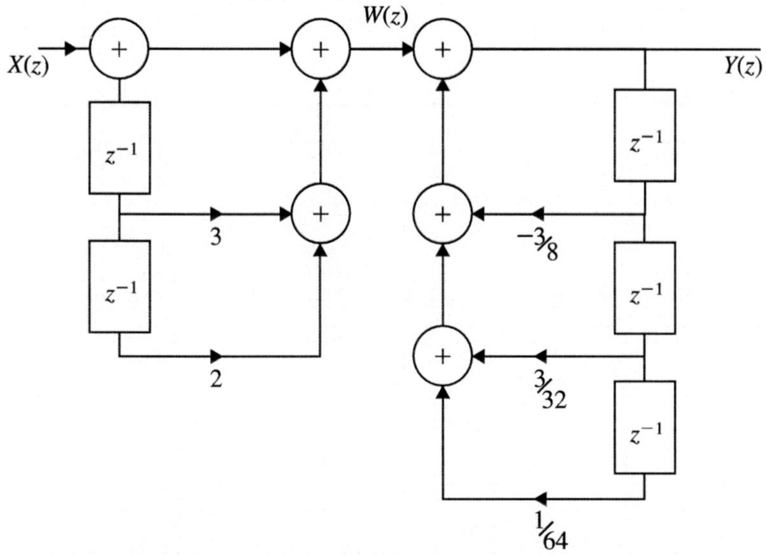

Fig. 4.46 Direct form-I realization of $H(z)$ for the Example 4.37

Fig. 4.47 Direct form-II realization of $H(z)$ for the Example 4.37

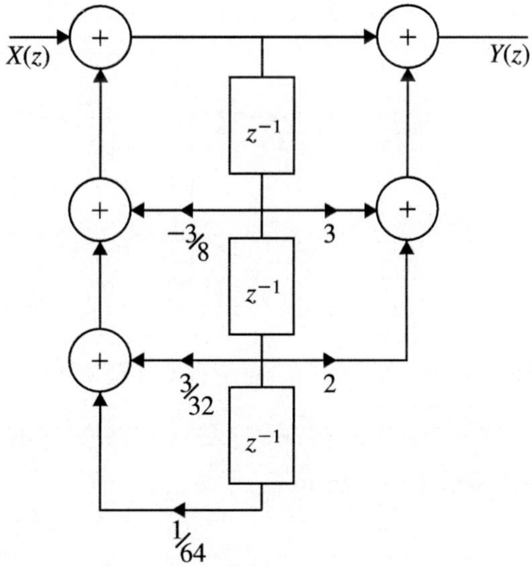

Direct Form-II realization of $H(z)$ for Example 4.37 is shown in Fig. 4.47

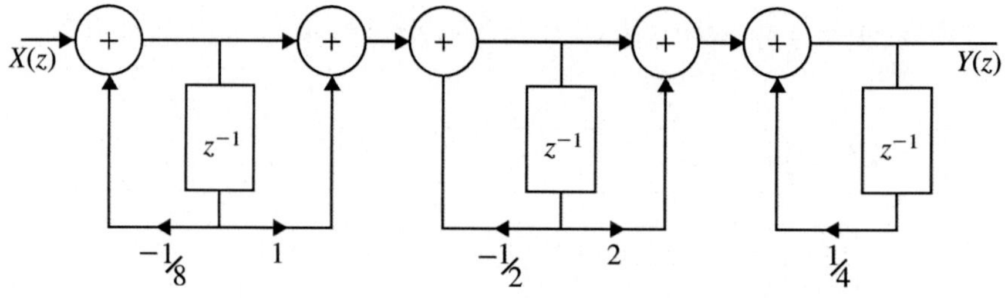

Fig. 4.48 Cascade form realization of $H(z)$ for the Example 4.37

Cascade form realization of $H(z)$ for the Example 4.37 is shown in Fig. 4.48.

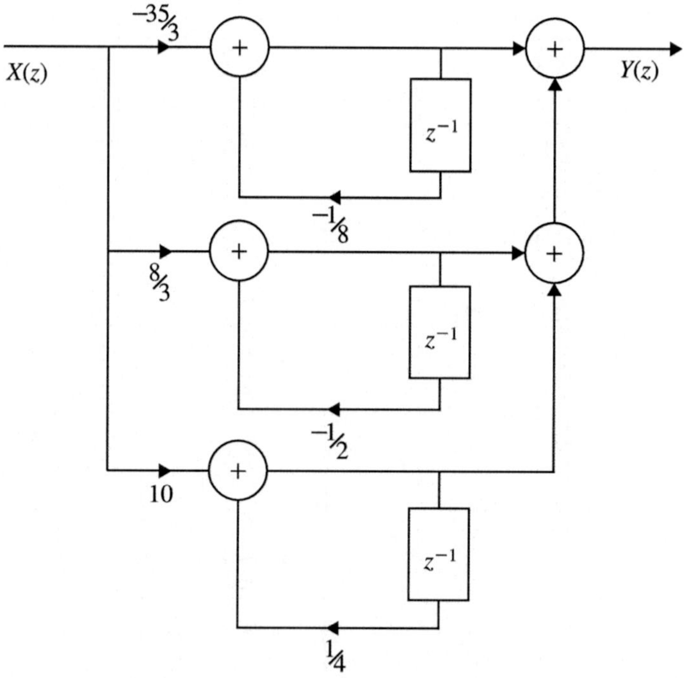

Fig. 4.49 Parallel form realization of $H(z)$ for the Example 4.37

Parallel form realization of $H(z)$ for the Example 4.37 is shown in Fig. 4.49.

Solution Direct form-I

$$y(n) = -\frac{3}{8}y(n-1) + \frac{3}{32}y(n-2) + \frac{1}{64}y(n-3) + x(n) + 3x(n-1) + 2x(n-2)$$

$$Y(z) = -\frac{3}{8}z^{-1}Y(z) + \frac{3}{32}z^{-2}Y(z) + \frac{1}{64}z^{-3}Y(z) + X(z) + 3z^{-1}X(z) + 2z^{-2}X(z)$$

Direct form-II

$$H(z) = \frac{Y(z)}{X(z)} = \frac{Y(z)}{W(z)} \times \frac{W(z)}{X(z)} = \frac{1 + 3z^{-1} + 2z^{-2}}{1 + \frac{3}{8}z^{-1} - \frac{3}{32}z^{-2} - \frac{1}{64}z^{-3}}$$

where

$$\frac{W(z)}{X(z)} = \frac{1}{1 + \frac{3}{8}z^{-1} - \frac{3}{32}z^{-2} - \frac{1}{64}z^{-3}}$$

$$\frac{Y(z)}{W(z)} = 1 + 3z^{-1} + 2z^{-2}$$

$$X(z) = W(z)\left(1 + \frac{3}{8}z^{-1} - \frac{3}{32}z^{-2} - \frac{1}{64}z^{-3}\right)$$

$$W(z) = X(z) - \frac{3}{8}z^{-1}W(z) + \frac{3}{32}z^{-2}W(z) + \frac{1}{64}z^{-3}W(z)$$

$$Y(z) = W(z) + 3z^{-1}W(z) + 2z^{-2}W(z)$$

Cascade Form Realization

$$H(z) = \frac{1 + 3z^{-1} + 2z^{-2}}{1 + \frac{3}{8}z^{-1} - \frac{3}{32}z^{-2} - \frac{1}{64}z^{-3}}$$

$$= \frac{(1 + z^{-1})(1 + 2z^{-1})}{\left(1 + \frac{1}{8}z^{-1}\right)\left(1 + \frac{1}{2}z^{-1}\right)\left(1 - \frac{1}{4}z^{-1}\right)}$$

$$H(z) = H_1(z)H_2(z)H_3(z)$$

where

$$H_1(z) = \frac{1 + z^{-1}}{1 + \frac{1}{8}z^{-1}}; \; H_2(z) = \frac{1 + 2z^{-1}}{1 + \frac{1}{2}z^{-1}}; \; H_3(z) = \frac{1}{1 - \frac{1}{4}z^{-1}}$$

Parallel Form

$$H(z) = \frac{(1 + z^{-1})(1 + 2z^{-1})}{\left(1 + \frac{1}{8}z^{-1}\right)\left(1 + \frac{1}{2}z^{-1}\right)\left(1 - \frac{1}{4}z^{-1}\right)}$$

$$= \frac{A}{1 + \frac{1}{8}z^{-1}} + \frac{B}{1 + \frac{1}{2}z^{-1}} + \frac{C}{1 - \frac{1}{4}z^{-1}}$$

$$A = \left.\frac{(1 + z^{-1})(1 + 2z^{-1})}{\left(1 + \frac{1}{2}z^{-1}\right)\left(1 - \frac{1}{4}z^{-1}\right)}\right|_{z^{-1}=-8} = -\frac{35}{3}$$

$$B = \left.\frac{(1+z^{-1})(1+2z^{-1})}{\left(1+\frac{1}{8}z^{-1}\right)\left(1-\frac{1}{4}z^{-1}\right)}\right|_{z^{-1}=-2} = \frac{8}{3}$$

$$C = \left.\frac{(1+z^{-1})(1+2z^{-1})}{\left(1+\frac{1}{8}z^{-1}\right)\left(1+\frac{1}{2}z^{-1}\right)}\right|_{z^{-1}=4} = 10$$

$$H(z) = -\frac{35}{3}\left(\frac{1}{1+\frac{1}{8}z^{-1}}\right) + \frac{8}{3}\left(\frac{1}{1+\frac{1}{2}z^{-1}}\right) + 10\left(\frac{1}{1-\frac{1}{4}z^{-1}}\right)$$

$$H(z) = H_1(z) + H_2(z) + H_3(z)$$

Example 4.38 Realize the given system in cascade and parallel form (Figs. 4.50 and 4.51)

$$H(z) = \frac{1+\frac{1}{2}z^{-1}}{\left(1-z^{-1}+\frac{1}{4}z^{-2}\right)\left(1-z^{-1}+\frac{1}{2}z^{-2}\right)}$$

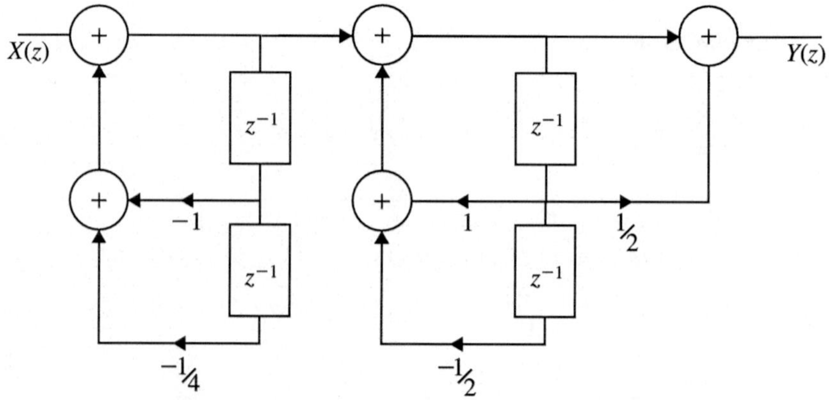

Fig. 4.50 Cascade form realization of $H(z)$ for the Example 4.38

Cascade Form realization of $H(z)$ for the Example 4.38 is shown in Fig. 4.50

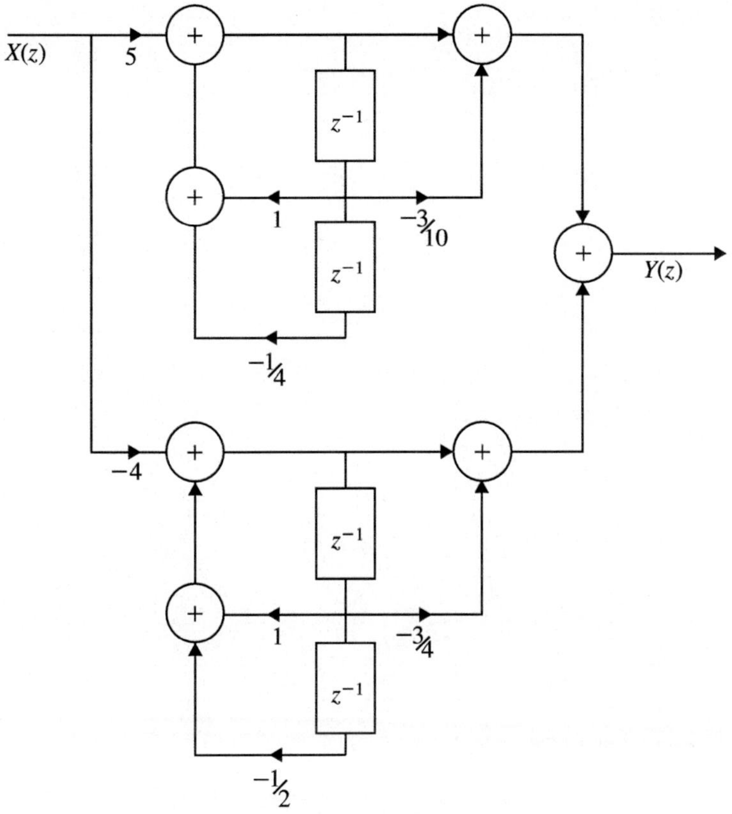

Fig. 4.51 Parallel form realization of $H(z)$ for the Example 4.38

Parallel form realization of $H(z)$ for the Example 4.38 is shown in Fig. 4.51.

Solution Cascade Form Realization

$$\begin{aligned}H(z) &= \frac{1+\frac{1}{2}z^{-1}}{\left(1-z^{-1}+\frac{1}{4}z^{-2}\right)\left(1-z^{-1}+\frac{1}{2}z^{-2}\right)}\\ &= \frac{1}{1-z^{-1}+\frac{1}{4}z^{-2}}\cdot\frac{1+\frac{1}{2}z^{-1}}{1-z^{-1}+\frac{1}{2}z^{-2}}\\ H(z) &= H_1(z)H_2(z)\end{aligned}$$

where

$$H_1(z) = \frac{1}{1+z^{-1}+\frac{1}{4}z^{-2}};\quad H_2(z) = \frac{1+\frac{1}{2}z^{-1}}{1-z^{-1}+\frac{1}{2}z^{-2}}$$

Parallel Form Realization

$$H(z) = \frac{1+\frac{1}{2}z^{-1}}{\left(1-z^{-1}+\frac{1}{4}z^{-2}\right)\left(1-z^{-1}+\frac{1}{2}z^{-2}\right)}$$

$$= \frac{z\left(z^3+\frac{1}{2}z^2\right)}{\left(z^2-z+\frac{1}{4}\right)\left(z^2-z+\frac{1}{2}\right)}$$

$$\frac{H(z)}{z} = \frac{Az+B}{z^2-z+\frac{1}{4}} + \frac{Cz+D}{z^2-z+\frac{1}{2}}$$

$$= \frac{5z-\frac{3}{2}}{z^2-z+\frac{1}{4}} + \frac{-4z+3}{z^2-z+\frac{1}{2}}$$

$$H(z) = \frac{5-\frac{3}{2}z^{-1}}{1-z^{-1}+\frac{1}{4}z^{-2}} + \frac{-4+3z^{-1}}{1-z^{-1}+\frac{1}{4}z^{-2}}$$

$$= H_1(z)H_2(z)$$

$$H(z) = \frac{5-\frac{3}{2}z^{-1}}{1-z^2+\frac{1}{4}z^{-2}} + \frac{-4+3z^{-1}}{1-z^{-1}+\frac{1}{2}z^{-2}}$$

$$= \frac{5\left(1-\frac{3}{10}z^{-1}\right)}{1-z^{-1}+\frac{1}{4}z^{-2}} + \frac{-4\left(1-\frac{3}{4}z^{-1}\right)}{1-z^{-1}+\frac{1}{2}z^{-2}}$$

Example 4.39 The transfer function of a system is given by (Figs. 4.52 and 4.53)

$$H(z) = \frac{(1+z^{-1})^3}{\left(1-\frac{1}{4}z^{-1}\right)\left(1-z^{-1}+\frac{1}{2}z^{-2}\right)}$$

Realize the system in cascade and parallel structures.

Cascade form realization of $H(z)$ for the Example 4.39 is shown in Fig. 4.52.

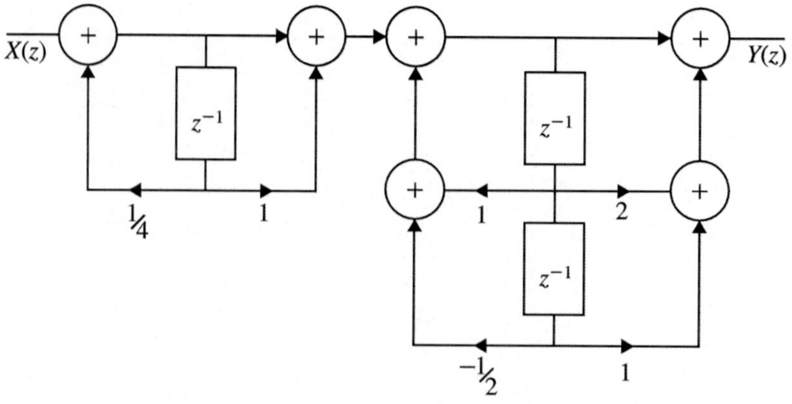

Fig. 4.52 Cascade form realization of $H(z)$ for the Example 4.39

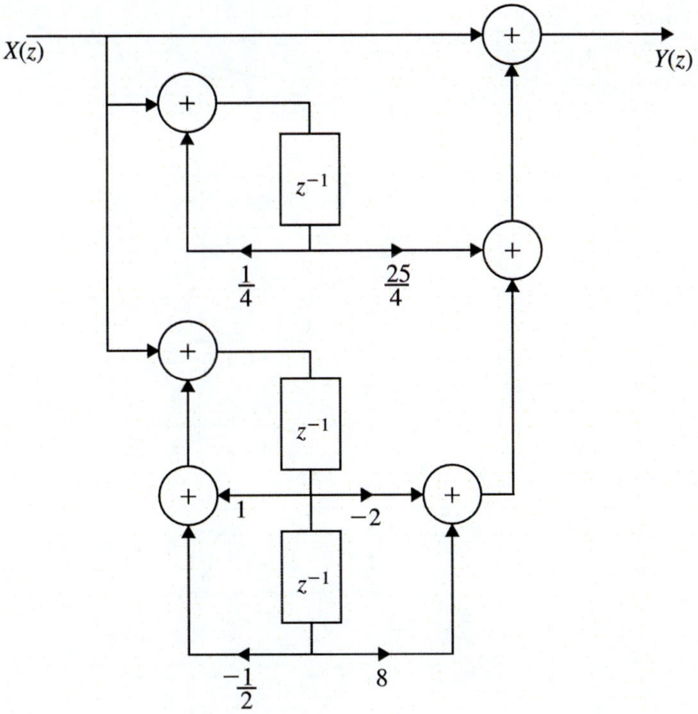

Fig. 4.53 Parallel form realization of $H(z)$ for the Example 4.39

Solution Cascade Form Realization

$$H(z) = \frac{1+z^{-1}}{1-\frac{1}{4}z^{-1}} \cdot \frac{\left(1+z^{-1}\right)^2}{1-z^{-1}+\frac{1}{2}z^{-2}}$$

$$= \frac{1+z^{-1}}{1-\frac{1}{4}z^{-1}} \cdot \frac{1+2z^{-1}+z^{-2}}{1-z^{-1}+\frac{1}{2}z^{-2}}$$

$$H(z) = H_1(z)H_2(z)$$

where

$$H_1(z) = \frac{1+z^{-1}}{1-\frac{1}{4}z^{-1}} \quad \text{and} \quad H_2(z) = \frac{1+2z^{-1}+z^{-2}}{1-z^{-1}+\frac{1}{2}z^{-2}}$$

Parallel form realization of $H(z)$ for the Example 4.39 is shown in Fig. 4.53.

Parallel Form

$$H(z) = \frac{(1+z^{-1})^3}{\left(1-\frac{1}{4}z^{-1}\right)\left(1-z^{-1}+\frac{1}{2}z^{-2}\right)}$$

$$H(z) = \frac{1+3z^{-1}+3z^{-2}+z^{-3}}{1-\frac{5}{4}z^{-1}+\frac{3}{4}z^{-2}-\frac{1}{8}z^{-3}}$$

$$= \frac{z^3+3z^2+3z+1}{z^3-\frac{5}{4}z^{-2}+\frac{3}{4}z-\frac{1}{8}}$$

$$= 1+\frac{\frac{17}{4}z^2+\frac{9}{4}z+\frac{9}{8}}{z^3-\frac{5}{4}z^{-2}+\frac{3}{4}z-\frac{1}{8}}$$

$$= 1+\frac{\frac{17}{4}z^2+\frac{9}{4}z+\frac{9}{8}}{\left(z-\frac{1}{4}\right)\left(z^2-z+\frac{1}{2}\right)}$$

$$\frac{\frac{17}{4}z^2+\frac{9}{4}z+\frac{9}{8}}{\left(z-\frac{1}{4}\right)\left(z^2-z+\frac{1}{2}\right)} = \frac{A}{z-\frac{1}{4}}+\frac{Bz+C}{z^2-z+\frac{1}{2}}$$

$$= \frac{\frac{25}{4}}{z-\frac{1}{4}}+\frac{-2z+8}{z^2-z+\frac{1}{2}}$$

$$H(z) = 1+\frac{\frac{25}{4}z^{-1}}{1-\frac{1}{4}z^{-1}}+\frac{-2z^{-1}+8z^{-2}}{1-z^{-1}+\frac{1}{2}z^{-2}}$$

Example 4.40 Determine the direct form-II and transposed direct form-II for the given system (Fig. 4.54)

$$y(n) = \frac{1}{2}y(n-1)-\frac{1}{4}y(n-2)+x(n)+x(n-1)$$

Transposed form of realization from Direct Form II of Example 4.40 is shown in Fig. 4.54e. Figure 4.54a, 4.54b, 4.54c and 4.54d represent various steps involved to get Fig. 4.54e.

Solution Given

$$Y(z) = \frac{1}{2}z^{-1}Y(z)-\frac{1}{4}z^{-2}Y(z)+X(z)+z^{-1}X(z)$$

$$H(z) = \frac{Y(z)}{X(z)} = \frac{1+z^{-1}}{1-\frac{1}{2}z^{-1}+\frac{1}{4}z^{-2}}$$

Direct Form-II Realization

To get transposed direct form-II do the following operation:

1. Change the direction of all branches.
2. Interchange the input and output.

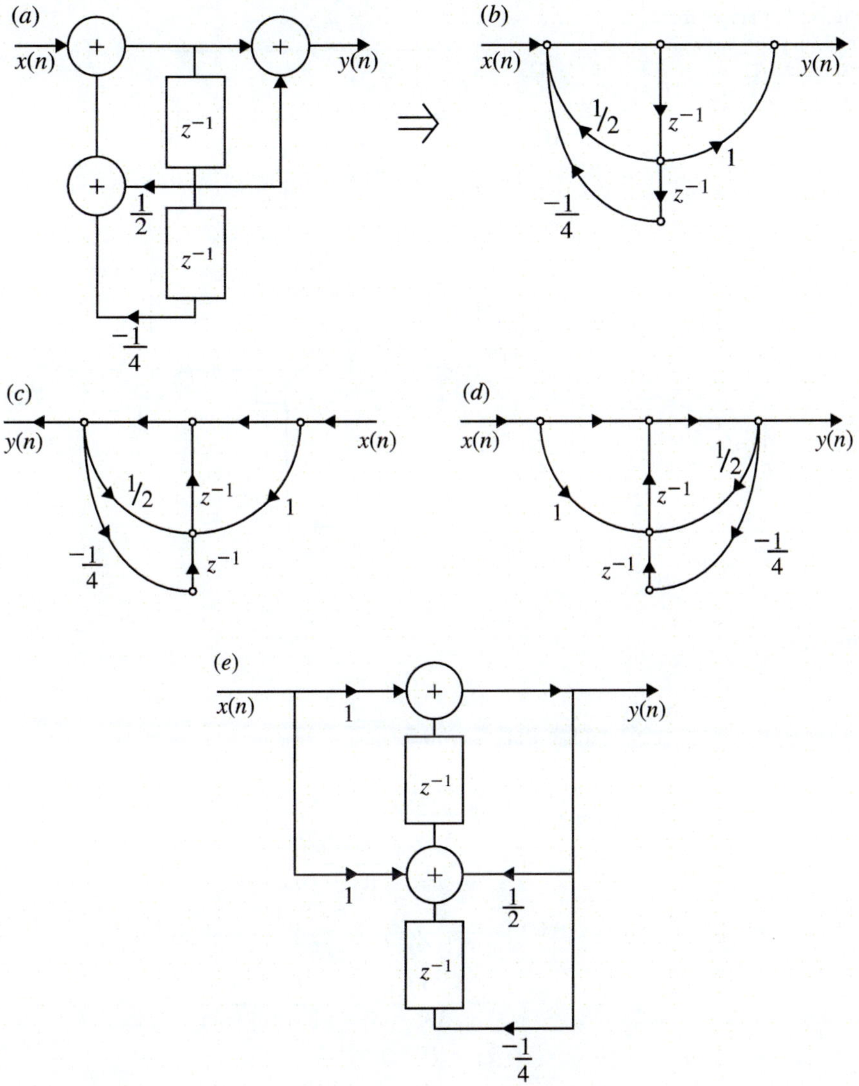

Fig. 4.54 Transposed direct form-II realization for the Example 4.40

3. Change the summing point to branching point and *vice versa*.

Example 4.41 Realize the system given by difference equation (Fig. 4.55)

$$y(n) = -0.1y(n-1) + 0.72y(n-2) + 0.7x(n) - 0.252x(n-2)$$

in parallel form.

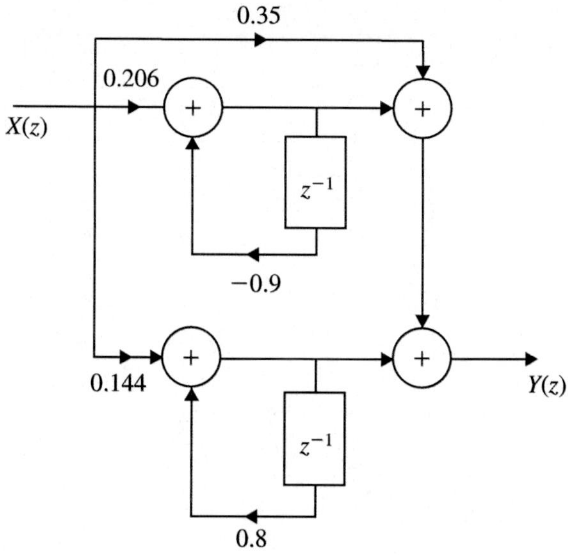

Fig. 4.55 Parallel form realization of $H(z)$ for Example 4.41

Solution

$$
\begin{aligned}
H(z) &= \frac{0.7 - 0.252z^{-2}}{1 + 0.1z^{-1} - 0.72z^{-2}} \\
&= 0.35 + \frac{0.35 - 0.035z^{-1}}{1 + 0.1z^{-1} - 0.72z^{-2}} \\
&= 0.35 + \frac{0.35 - 0.035z^{-1}}{(1 + 0.9z^{-1})(1 - 0.8z^{-1})} \\
&= 0.35 + \frac{A}{(1 + 0.9z^{-1})} + \frac{B}{(1 - 0.8z^{-1})} \\
&= 0.35 + \frac{0.206}{(1 + 0.9z^{-1})} + \frac{0.1444}{(1 - 0.8z^{-1})} \\
&= C + H_1(z) + H_2(z).
\end{aligned}
$$

Parallel Form of realization of $H(z)$ for Example 4.41 is shown in Fig. 4.55.

4.13.7 Lattice Structure of IIR System

Let us consider an all-pole system with system function

$$
H(z) = \frac{1}{1 + \sum_{k=1}^{N} a_N z^{-k}} = \frac{1}{A_N(z)} \tag{4.92}
$$

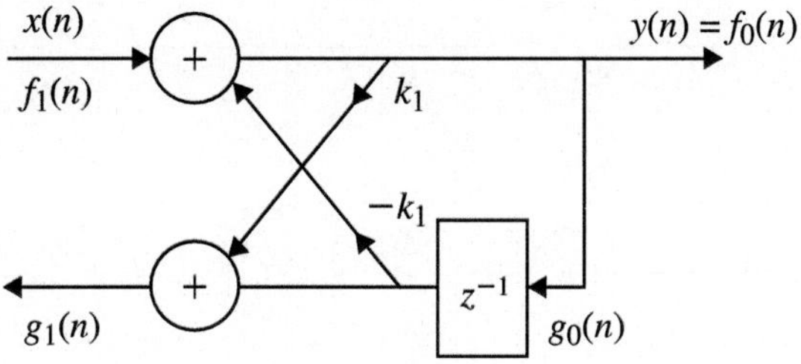

Fig. 4.56 Single stage all-pole lattice structure

The difference equation for this IIR system is derived as follows.

$$\frac{Y(z)}{X(z)} = \frac{1}{1 + \sum_{k=1}^{N} a_N z^{-k}}$$

$$Y(z) + \sum_{k=1}^{N} a_N z^{-k} Y(z) = X(z)$$

$$Y(z) = -\sum_{k=1}^{N} a_N z^{-k} Y(z) + X(z)$$

$$y(n) = -\sum_{k=1}^{N} a_N(k) y(n-k) + x(n) \tag{4.93}$$

or

$$x(n) = y(n) + \sum_{k=1}^{N} a_N(k) y(n-k) \tag{4.94}$$

For $N = 1$

$$x(n) = y(n) + a_1(1) y(n-1) \tag{4.95}$$

The Eq. (4.95) can be realized in lattice structure as shown in Fig. 4.56.

From Fig. 4.56, we can obtain

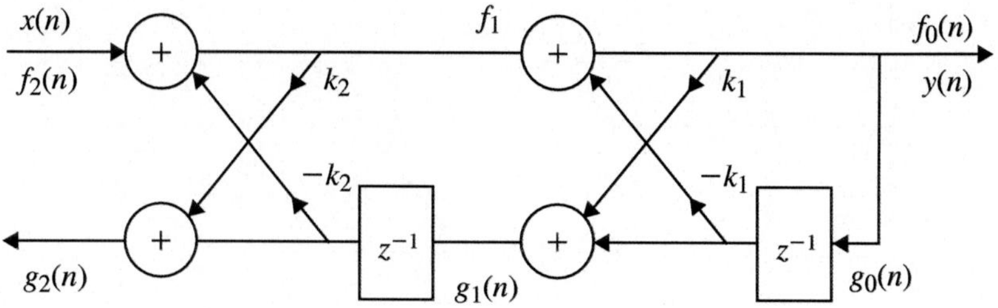

Fig. 4.57 Two stage all-pole lattice filter

$$x(n) = f_1(n) \tag{4.96}$$

$$\begin{aligned} y(n) &= f_0(n) = f_1(n) - k_1 g_0(n-1) \\ &= x(n) - k_1 y(n-1) \end{aligned} \tag{4.97}$$

$$\begin{aligned} g_1(n) &= k_1 f_0(n) + g_0(n-1) \\ &= k_1 y(n) + y(n-1) \end{aligned} \tag{4.98}$$

$$k_1 = a_1(1) \tag{4.99}$$

Now, let us consider for the case $N = 2$, then

$$x(n) = f_2(n) \tag{4.100}$$

$$y(n) = x(n) - a_1(1)y(n-1) - a_2(2)y(n-2) \tag{4.101}$$

This output can also be obtained from a two-stage lattice structure as shown in Fig. 4.57 from which we have

$$f_2(n) = x(n) \tag{4.102}$$

$$f_1(n) = f_2(n) - k_2 g_1(n-1) \tag{4.103}$$

$$g_2(n) = k_2 f_1(n) + g_1(n-1) \tag{4.104}$$

$$f_0(n) = f_1(n) - k_1 g_0(n-1) \tag{4.105}$$

$$g_1(n) = k_1 f_0(n) + g_0(n-1) \tag{4.106}$$

$$y(n) = f_0(n) = g_0(n) \tag{4.107}$$

$$\begin{aligned} &= f_1(n) - k_1 g_0(n-1) \\ &= f_2(n) - k_2 g_1(n-1) - k_1 g_0(n-1) \\ &= f_2(n) - k_2[k_1 f_0(n-1) + g_0(n-2)] - k_1 g_0(n-1) \\ &= f_2(n) - k_1(1+k_2)y(n-1) - k_2 y(n-2) \end{aligned} \tag{4.108}$$

Similarly

$$g_2(n) = k_2 y(n) + k_1(1+k_2)y(n-1) + y(n-2) \tag{4.109}$$

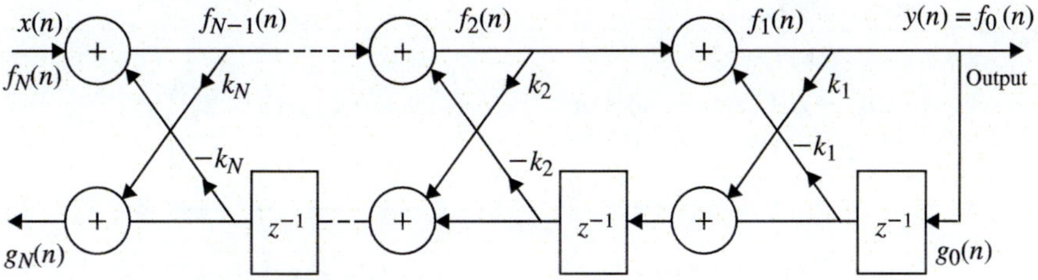

Fig. 4.58 All-pole lattice filter

Comparing Eqs. (4.101) and (4.108) we get

$$a_2(0) = 1, \quad a_2(1) = k_1(1 + k_2), \quad a_2(2) = k_2 \tag{4.110}$$

For a N-stage IIR filter realized in lattice structure as shown in Fig. 4.58, we get

$$f_N(n) = x(n) \tag{4.111}$$

$$f_{m-1}(n) = f_m(n) - k_m g_{m-1}(n-1), \quad m = N, N-1, \ldots, 1 \tag{4.112}$$

$$g_m(n) = k_m f_{m-1}(n) + g_{m-1}(n-1), \quad m = N, N-1, \ldots, 1 \tag{4.113}$$

$$y(n) = f_0(n) = g_0(n) \tag{4.114}$$

4.13.8 Conversion from Direct Form to Lattice Structure

For a 3-stage IIR system

$$x(n) = f_3(n) = f_2(n) + k_3 g_2(n-1) \tag{4.115}$$

$$g_3(n) = k_3 f_2(n) + g_2(n-1) \tag{4.116}$$

from which

$$g_2(n-1) = g_3(n) - k_3 f_2(n) \tag{4.117}$$

Substituting Eq. (4.117) in Eq. (4.115) we get,

$$f_3(n) = f_2(n) + k_3[g_3(n) - k_3 f_2(n)]$$

and

$$
\begin{aligned}
f_2(n) &= \frac{f_3(n) - k_3 g_3(n)}{1 - k_3^2} \\
&= \frac{\begin{bmatrix} y(n) + a_3(1)y(n-1) + a_3(2)y(n-2) + a_3(3)y(n-3) \\ +k_3 a_3(3)y(n) - k_3 a_3(2)y(n-2) \\ -k_3 a_3(n)y(n-1) - a_3 y(n-3) \end{bmatrix}}{1 - k_3^2} \\
&= y(n) + \frac{a_3(1) - a_3(3)a_3(2)}{1 - a_3^2(3)} y(n-1) + \frac{a_3(2) - a_3(2)a_3(3)}{1 - a_3^2(3)} y(n-2) + \cdots
\end{aligned} \tag{4.118}
$$

Comparing Eqs. (4.118) and (4.101) we have

$$
a_2(0) = 1, \quad a_2(1) = \frac{a_3(1) - a_3(3)a_3(2)}{a - a_3^2(3)}, \quad a_2(2) = \frac{a_3(2) - a_3(2)a_3(3)}{1 - a_3^2(3)} \tag{4.119}
$$

In general

$$
\begin{aligned}
a_{m-1}(0) &= 1 \\
k_m &= a_m(m)
\end{aligned}
$$

$$
\boxed{a_{m-1}(k) = \frac{a_m(k) - a_m(m)a_m(m-k)}{1 - a_m^2(m)}} \tag{4.120}
$$

Equation (4.120) can be used to convert direct form to lattice structure.

4.13.9 *Lattice–Ladder Structure*

A general IIR filter containing both poles and zeros can be realized using an all-pole lattice as the basic building block.

Consider an IIR filter with system function

$$
H(z) = \frac{B_M(z)}{A_N(z)} = \frac{\sum_{k=0}^{N} b_M(k) z^{-k}}{1 + \sum_{k=1}^{N} a_N(k) z^{-k}} \tag{4.121}
$$

where $N \geq M$. A lattice structure for Eq. (4.121) can be constructed by filter realizing an all-pole lattice coefficients, k_m, $1 \leq M \leq N$ for the denominator $A_N(z)$, and then adding a ladder part as shown in Fig. 4.59 for $M = N$. The output of the ladder part can be expressed as a weighted linear combination of $\{g_m(n)\}$.

Now the output is given by

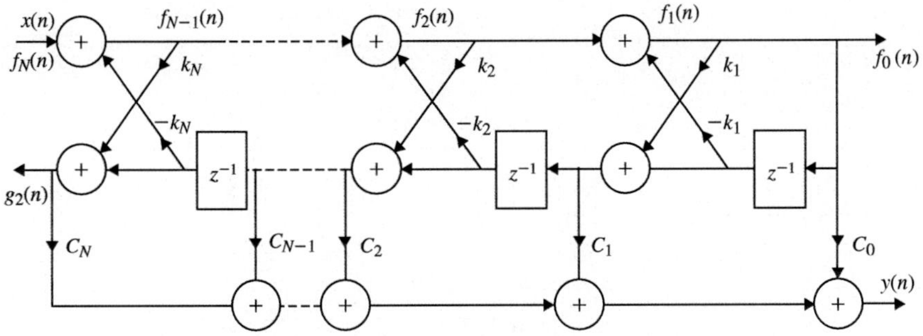

Fig. 4.59 Lattice–ladder structure for realizing a pole–zero IIR filter

$$y(n) = \sum_{m=0}^{M} C_m g_m(n) \tag{4.122}$$

where $\{c_m\}$ are called the ladder coefficients and can be obtained using the recursive relation.

$$\boxed{C_m = b_m - \sum_{i=m+1}^{M} C_i a_i(i-m); \quad m = M, M-1, \ldots, 0} \tag{4.123}$$

Example 4.42 Realize the following system function in lattice ladder structure

$$H(z) = \frac{1 + \frac{3}{2}z^{-1} + \frac{1}{2}z^{-2}}{1 + z^{-1} + \frac{1}{4}z^{-2}}$$

Solution

$$\text{Given} \quad B_M(z) = 1 + \frac{3}{2}z^{-1} + \frac{1}{2}z^{-2}$$

$$A_N(z) = 1 + z^{-1} + \frac{1}{4}z^{-2}$$

$$a_2(0) = 1, \quad a_2(1) = 1, \quad a_2(2) = \frac{1}{4}$$

$$k_2 = a_2(2) = \frac{1}{4}$$

From Eq. (4.121) we have

$$a_{m-1}(k) = \frac{a_m(k) - a_m(m)a_m(m-k)}{1 - a_m^2(m)}$$

For $m = 2$ and $k = 1$

$$\begin{aligned} k_1 = a_1(1) &= \frac{a_2(1) - a_2(2)a_2(1)}{a - a_2^2(2)} \\ &= \frac{1 - \frac{1}{4} \cdot 1}{1 - \left(\frac{1}{4}\right)^2} = \frac{0.75}{0.9375} = 0.8 \end{aligned}$$

For $m = 2$ and $k = 0$

$$a_1(0) = \frac{a_2(0) - a_2(2)a_2(2)}{1 - a_2^2(2)} = \frac{1 - \left(\frac{1}{4}\right)^2}{1 - \left(\frac{1}{4}\right)^2} = 1$$

Therefore, for lattice structure

$$k_1 = 0.8, \quad k_2 = 0.25$$

For ladder structure

$$\begin{aligned} C_m &= b_m - \sum_{i=m+1}^{M} C_i a_i(i-m), \quad m = M, M-1, \ldots, 0 \\ C_2 &= b_2 = \frac{1}{2} \\ C_1 &= b_1 - c_2 a_2(0) \\ &= \frac{3}{2} - \frac{1}{2}(1) = 1 \end{aligned}$$

$$\begin{aligned} C_0 &= b_0 - [C_1 a_1(0) + C_2 a_2(1)] \\ &= 1 - \left[1 \cdot 1 + \frac{1}{2} \cdot 1\right] \\ &= 1 - \left[1 + \frac{1}{2}\right] = -0.5 \end{aligned}$$

The lattice–ladder structure for the given pole–zero is shown in Fig. 4.60.

Example 4.43 Convert the following all-pole IIR filter into a lattice structure.

$$H(z) = \frac{1}{1 + \frac{13}{24}z^{-1} + \frac{5}{8}z^{-2} + \frac{1}{3}z^{-3}}$$

Solution Given

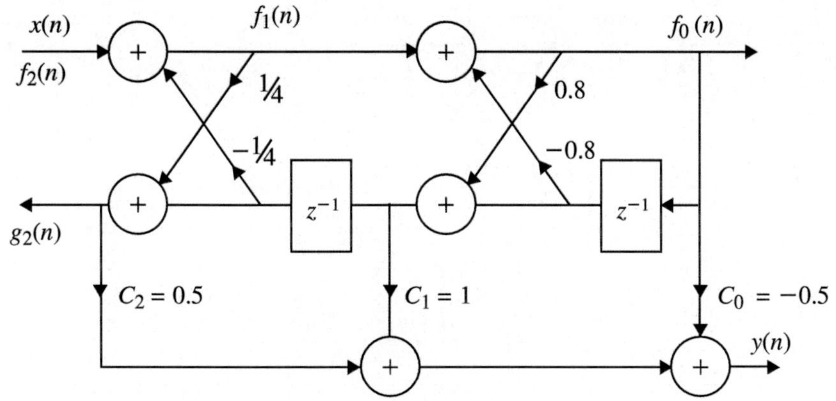

Fig. 4.60 Lattice–Ladder form for the Example 4.42

$$A_N(z) = 1 + \frac{13}{24}z^{-1} + \frac{5}{8}z^{-2} + \frac{1}{3}z^{-3}$$

$$a_3(0) = 1, \quad a_3(1) = \frac{13}{24}, \quad a_3(2) = \frac{5}{8}; \quad a_3(3) = \frac{1}{3}$$

$$K_3 = a_3(3) = \frac{1}{3}$$

From Eq. (4.121) we have

$$a_{m-1}(k) = \frac{a_m(k) - a_m(m)a_m(m-k)}{1 - a_m^2(m)}$$

For $m = 2$, and $k = 1$

$$a_2(1) = \frac{a_3(1) - a_3(3)a_3(2)}{1 - a_3^2(3)}$$

$$= \frac{\frac{13}{24} - \frac{1}{3}\left(\frac{5}{8}\right)}{1 - \left(\frac{2}{3}\right)^2} = \frac{3}{8}$$

For $m = 3$ and $k = 2$

$$k_2 = a_2(2) = \frac{a_3(2) - a_3(3)a_3(1)}{1 - a_3^2(3)}$$

$$= \frac{\frac{5}{8} - \frac{1}{3}\left(\frac{13}{24}\right)}{1 - \left(\frac{1}{3}\right)^2} = \frac{1}{2}$$

For $m = 2$ and $k = 1$

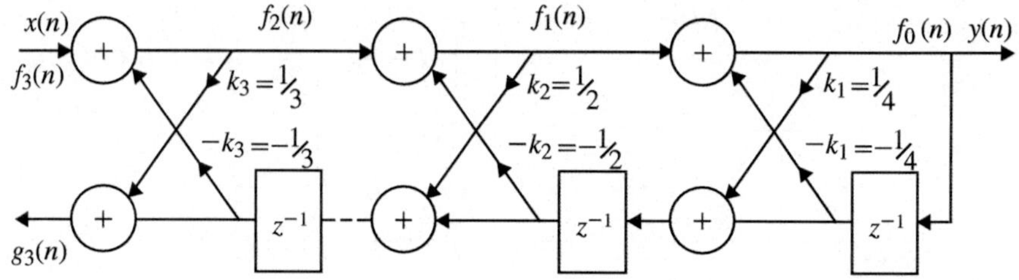

Fig. 4.61 Lattice structure for the Example 4.43

$$k_1 = a_1(1) = \frac{a_2(1) - a_2(2)a_2(1)}{1 - a_2^2(2)}$$
$$= \frac{\frac{3}{8} - \frac{1}{2}\left(\frac{3}{8}\right)}{1 - \left(\frac{1}{2}\right)^2} = \frac{1}{4}$$

Therefore, for lattice structure

$$k_1 = \frac{1}{4}, \quad k_2 = \frac{1}{2}, \quad k_3 = \frac{1}{3}$$

The lattice structure for the given all-pole filter is shown in Fig. 4.61.

Example 4.44 Design an analog Bulterworth filter that has a −2 dB passband altenuation at a frequency of 20 rad/s and atleast −10 dB stopband altenuation at 30 rad/s.

Solution Given

$$K_1 = -2\,\text{dB}$$
$$K_2 = -10\,\text{dB}$$
$$\Omega_1 = 20\,\text{rad/s}$$
$$\Omega_2 = 30\,\text{rad/s}$$
$$A_1 = 10^{K_1/20} = 10^{-2/20} = 0.794$$
$$A_2 = 10^{K_2/20} = 10^{-10/20} = 0.316$$

$$N \geq \frac{1}{2}\left\{\frac{\log\left[\left(\frac{1}{A_2^2} - 1\right) / \left(\frac{1}{A_1^2} - 1\right)\right]}{\log\left(\frac{\Omega_2}{\Omega_1}\right)}\right\}$$

$$N \geq \frac{1}{2}\left\{\frac{\log\left[\left(\frac{1}{(0.316)^2} - 1\right) / \left(\frac{1}{(0.794)^2} - 1\right)\right]}{\log\left(\frac{30}{20}\right)}\right\}$$

$$N \geq 3.37$$

$$\boxed{N = 4}$$

$$\Omega_C = \frac{\Omega_1}{\left(\frac{1}{A_1^2} - 1\right)^{1/2N}}$$
$$= \frac{20}{\left(\frac{1}{(0.794)^2} - 1\right)^{1/8}}$$
$$= 21.38 \text{ rad/s}$$

For N is even

$$H_a(s) = \prod_{k=1}^{N/2} \frac{\Omega_c^2}{(s^2 + b_k \Omega_c s + \Omega_c^2)}$$
$$= \prod_{k=1}^{2} \frac{\Omega_c^2}{(s^2 + b_k \Omega_c s + \Omega_c^2)}$$
$$= \left[\frac{\Omega_c^2}{(s^2 + b_1 \Omega_c s + \Omega_c^2)}\right]\left[\frac{\Omega_c^2}{(s^2 + b_2 \Omega_c s + \Omega_c^2)}\right]$$

$$b_k = 2 \sin \frac{(2K-1)\pi}{2N}$$

$$b_1 = 2 \sin \frac{\pi}{8} = 0.765$$
$$b_2 = 2 \sin \frac{3\pi}{8} = 1.848$$

Substituting the numerical values for Ω_c, b_1 and b_2 in $H_a(s)$ we get

$$H_a(s) = \frac{(21.38)^4}{(s^2 + 0.765 \times 21.38s + 21.38^2)(s^2 + 1.484 \times 21.38s + 21.38^2)}$$

$$\boxed{H_a(s) = \frac{(21.38)^4}{(s^2 + 16.356s + 457.1)(s^2 + 39.51s + 457.1)}}$$

Example 4.45 Design a Bulterworth filter for the following specifications using bilinear transformation.

$$0.8 \le \begin{cases} |H(e^{j\omega})| \le 1 & 0 \le \omega \le 0.2\pi \\ |H(e^{j\omega})| \le 0.2 & 0.6\pi \le \omega \le \pi \end{cases}$$

Solution Given

$$\begin{aligned} A_1 &= 0.8 \\ A_2 &= 0.2 \\ \omega_1 &= 0.2\pi \text{ rad/s} \\ \omega_2 &= 0.6\pi \text{ rad/s} \end{aligned}$$

$$\begin{aligned} \frac{\Omega_2}{\Omega_1} &= \frac{\tan(\omega_2/2)}{\tan(\omega_1/2)} \\ &= \frac{\tan(0.3\pi)}{\tan(0.1\pi)} \\ &= 4.236 \end{aligned}$$

$$N \geq \frac{1}{2}\left\{\frac{\log\left[\left(\frac{1}{A_2^2}-1\right)\Big/\left(\frac{1}{A_1^2}-1\right)\right]}{\log\left(\frac{\Omega_2}{\Omega_1}\right)}\right\}$$

$$\begin{aligned} N &\geq \frac{1}{2}\left\{\frac{\log\left[\left(\frac{1}{(0.2)^2}-1\right)\Big/\left(\frac{1}{(0.8)^2}-1\right)\right]}{\log 4.236}\right\} \\ N &\geq 1.3 \end{aligned}$$

$$\boxed{N = 2}$$

$$\begin{aligned} \Omega_c &= \frac{2}{T}\frac{\tan\left(\frac{\omega_1}{2}\right)}{\left(\frac{1}{A_1^2}-1\right)^{1/2N}} \\ &= \frac{2\tan 0.1\pi}{\left(\frac{1}{(0.8)^2}-1\right)^{1/4}} \\ &= 0.75 \text{ rad/s} \end{aligned}$$

For N is even

$$H_a(s) = \prod_{k=1}^{N/2} \frac{\Omega_c^2}{(s^2 + b_k \Omega_c s + \Omega_c^2)}$$
$$= \frac{\Omega_c^2}{(s^2 + b_1 \Omega_c s + \Omega_c^2)}$$

$$b_k = 2 \sin \left[\frac{(2k-1)\pi}{2N} \right]$$
$$= 2 \sin \frac{\pi}{4}$$
$$= 1.414$$

Substituting b_1 and Ω_c in $H_a(s)$ we get

$$H_a(s) = \frac{0.5625}{s^2 + 1.0605s + 0.5625}$$

Applying bilinear transformation we get

$$H(z) = H_a(s)|_{s=(2/T)\frac{(1-z^{-1})}{(1+z^{-1})}}$$

where $T = 1$ s.

$$H(z) = \frac{0.5625}{\frac{4(1-z^{-1})^2}{(1+z^{-1})^2} + \left(1.0605 \times \frac{2(1-z^{-1})}{(1+z^{-1})}\right) + 0.5625}$$
$$= \frac{0.5625(1 + 2z^{-1} + z^{-2})}{6.6835 - 6.875z^{-1} + 2.4415z^{-2}}$$

$$\boxed{H(z) = \frac{0.0841(1 + 2z^{-1} + z^{-2})}{(1 + 1.026z^{-1} + 0.3613z^{-2})}}$$

Example 4.46 A digital LPF requires to meet the following specifications.

(a) Monotonic passband and stopband.
(b) -3.01 dB cut-off frequency at 0.5π rad/s and stopband attenuation of atleast 15 dB at 0.75π rad/s. Find the system function using bilinear transformation.

Solution From the given specification, the LPF has a monotonic passband and stopband and hence it is a Bulterworth filter. Given

$$\begin{aligned}\omega_1 &= 0.5\pi \text{ rad/s}\\ \omega_2 &= 0.75\pi \text{ rad/s}\\ K_1 &= -3.01\,\text{dB}\\ A_1 &= 10^{K_1/20} = 0.707\\ \alpha_2 &= 15\,\text{dB}\\ A_2 &= 10^{-\alpha_2/20} = 0.1778\end{aligned}$$

$$\begin{aligned}\frac{\Omega_2}{\Omega_1} &= \frac{\tan(\omega_2/2)}{\tan(\omega_1/2)}\\ &= \frac{\tan(0.375\pi)}{\tan(0.25\pi)}\\ &= 2.414\end{aligned}$$

$$N \geq \frac{1}{2}\left\{\frac{\log\left[\left(\frac{1}{A_2^2}-1\right)\Big/\left(\frac{1}{A_1^2}-1\right)\right]}{\log\left(\frac{\Omega_2}{\Omega_1}\right)}\right\}$$

$$N \geq \frac{1}{2}\left\{\frac{\log\left[\left(\frac{1}{(0.1778)^2}-1\right)\Big/\left(\frac{1}{(0.707)^2}-1\right)\right]}{\log 2.414}\right\}$$

$$N \geq 1.94$$

$$\boxed{N = 2}$$

$$\begin{aligned}\Omega_c &= \frac{2}{T}\frac{\tan\left(\frac{\omega_2}{2}\right)}{\left(\frac{1}{A_1^2}-1\right)^{1/2N}}\\ &= \frac{2\tan 0.25\pi}{\left(\frac{1}{(0.707)^2}-1\right)^{1/4}} \qquad [T = 1\text{ s}]\\ &= 2\text{ rad/s}\end{aligned}$$

For N is even

$$H_a(s) = \prod_{k=1}^{N/2} \frac{\Omega_c^2}{(s^2 + b_k \Omega_c s + \Omega_c^2)}$$
$$= \frac{\Omega_c^2}{(s^2 + b_1 \Omega_c s + \Omega_c^2)}$$

$$b_k = 2 \sin \left[\frac{(2k-1)\pi}{2N} \right]$$
$$= 2 \sin \frac{\pi}{4}$$
$$= 1.414$$
$$H_a(s) = \frac{4}{(s^2 + 2.828s + 4)}$$

Applying bilinear transformation we get

$$H(z) = H_a(s)|_{s=(2/T)\frac{(1-z^{-1})}{(1+z^{-1})}}$$

where $T = 1$ s

$$H(z) = \frac{4}{\frac{4(1-z^{-1})^2}{(1+z^{-1})^2} + \left(2.828 \times \frac{2(1-z^{-1})}{(1+z^{-1})}\right) + 4}$$

On simplifying we get

$$\boxed{H(z) = \frac{0.2931(1 + 2z^{-1} + z^{-2})}{(1 + 0.1716z^{-2})}}$$

Example 4.47 Design a digital second order lowpass Bulterworth filter whose analog filter's cut-off frequency is 2200 Hz and sampling frequency is 8000 Hz. Use bilinear transformation.

Solution Given

$$N = 2$$
$$f_c = 2200\,\text{Hz}$$
$$f_{\text{samp}} = 8000\,\text{Hz}$$
$$\Omega_c = 2\pi \frac{f_c}{f_{\text{samp}}}$$
$$= \frac{2\pi \times 2200}{8000}$$

$$= 1.727\ \text{rad/s}$$

$$T = \frac{1}{f_{\text{samp}}}$$

$$= \frac{1}{8000}\ \text{s}$$

For $N = 2$,

$$H_a(s) = \frac{\Omega_c^2}{(s^2 + b_1\Omega_c s + \Omega_c^2)}$$

$$b_k = 2\sin\left[\frac{(2k-1)\pi}{2N}\right]$$

$$b_1 = 2\sin\frac{\pi}{4}$$

$$= 1.414$$

$$H_a(s) = \frac{\Omega_c^2}{(s^2 + b_1\Omega_c s + \Omega_c^2)}$$

$$= \frac{(1.724)^2}{s^2 + 2.441s + (1.724)^2}$$

$$H(z) = H_a(s)|_{s=(2/T)\frac{(1-z^{-1})}{(1+z^{-1})}}$$

where $T = 1/8000$ s.

$$\boxed{H(z) = \frac{2.9825}{\left[\frac{1600(1-z^{-1})^2}{1+z^{-1}}\right]^2 + 38624s + 2.9825}}$$

Example 4.48 For the following specifications design a high pass Bulterworth filter.

$$\alpha_p = 3\ \text{dB}$$

$$\alpha_s = 15\ \text{dB}$$

$$\Omega_p = 1000\ \text{rad/s}$$

$$\Omega_s = 500\ \text{rad/s}$$

Solution Given

$$\begin{aligned}\alpha_1 &= \alpha_p = 3\,\text{dB}\\ \alpha_2 &= \alpha_s = 15\,\text{dB}\\ \Omega_1 &= \Omega_p = 1000\,\text{rad/s}\\ \Omega_2 &= \Omega_s = 500\,\text{rad/s}\\ A_1 &= 10^{-\alpha_1/20} = 0.707\\ A_2 &= 10^{-\alpha_2/20} = 0.177\end{aligned}$$

To design an analog LPF,

$$\begin{aligned}\frac{\Omega_2}{\Omega_1} &= \frac{1000}{500}\\ &= 2\end{aligned}$$

$$\begin{aligned}N &\geq \frac{1}{2}\left\{\frac{\log\left[\left(\frac{1}{A_2^2}-1\right)/\left(\frac{1}{A_1^2}-1\right)\right]}{\log\left(\frac{\Omega_2}{\Omega_1}\right)}\right\}\\ N &\geq \frac{1}{2}\left\{\frac{\log\left[\left(\frac{1}{(0.177)^2}-1\right)/\left(\frac{1}{(0.707)^2}-1\right)\right]}{\log(2)}\right\}\\ N &\geq 2.47\end{aligned}$$

$$\boxed{N = 3}$$

For N is odd,

$$H_a(s) = \frac{\Omega_c}{1+\Omega_c}\prod_{k=1}^{(N-1)/2}\frac{\Omega_c^2}{(s^2+b_k\Omega_c s+\Omega_c^2)}$$

For a normalized LPF, $\Omega_c = 1$ rad/s. Hence

$$\begin{aligned}H_a(s) &= \frac{1}{(s+1)(s^2+b_1 s+1)}\\ b_k &= 2\sin\left[\frac{(2k-1)\pi}{2N}\right]\\ b_1 &= 2\sin\frac{\pi}{6}\\ &= 1\end{aligned}$$

$$H_a(s) = \frac{1}{(s+1)(s^2+s+1)}$$

Converting analog LPF to analog HPF we get

$$H_a(s) = H_a(s)|_{s=\frac{\Omega_c}{s}}$$

where Ω_c is the cut-off frequency of HPF.

$$H_a(s) = \left.\frac{1}{(s+1)(s^2+s+1)}\right|_{s=\frac{1000}{s}}$$

$$\boxed{H_a(s) = \frac{s^3}{(s+1000)(s^2+1000s+10^5)}}$$

Example 4.49 Using a bilinear transform design HPF monotonic passband filter with cut-off frequency of 1000 Hz and down 10 dB at 350 Hz. The sampling frequency is 5000 Hz.

Solution Given

$$\begin{aligned} f_1 &= 1000\,\text{Hz} \\ f_2 &= 350\,\text{Hz} \\ f_{\text{samp}} &= 5000\,\text{Hz} \\ \alpha_1 &= 3\,\text{dB} \\ \alpha_2 &= 10\,\text{dB} \\ A_1 &= 10^{-\alpha_1/20} = 0.707 \\ A_2 &= 10^{-\alpha_2/20} = 0.316 \end{aligned}$$

$$\begin{aligned} \omega_1 &= 2\pi \frac{f_1}{f_{\text{samp}}} \\ &= \frac{2\pi \times 1000}{5000} \\ &= 1.25 \\ \omega_2 &= 2\pi \frac{f_2}{f_{\text{samp}}} \\ &= \frac{2\pi \times 350}{5000} \\ &= 0.43 \end{aligned}$$

$$T = \frac{1}{f_{\text{samp}}}$$
$$= 0.0002 \text{ s}$$

Specifications of analog HPF are,

$$\Omega_1 = \frac{2}{T} \tan \frac{\omega_1}{2}$$
$$= \frac{2}{0.0002} \tan \frac{1.25}{2}$$
$$= 7215 \text{ rad/s}$$
$$\Omega_2 = \frac{2}{T} \tan \frac{\omega_2}{2}$$
$$= \frac{2}{0.0002} \tan \frac{0.43}{2}$$
$$= 2184 \text{ rad/s}$$

Design of analog LPF

$$\frac{\Omega_2}{\Omega_1} = \frac{7215}{2184}$$
$$= 3.303$$

$$N \geq \frac{1}{2} \left\{ \frac{\log \left[\left(\frac{1}{A_2^2} - 1 \right) / \left(\frac{1}{A_1^2} - 1 \right) \right]}{\log \left(\frac{\Omega_2}{\Omega_1} \right)} \right\}$$

$$N \geq \frac{1}{2} \left\{ \frac{\log \left[\left(\frac{1}{(0.316)^2} - 1 \right) / \left(\frac{1}{(0.707)^2} - 1 \right) \right]}{\log(3.303)} \right\}$$
$$N \geq 0.919$$

$$\boxed{N = 1}$$

For N is odd,

$$H_a(s) = \left. \frac{\Omega_c}{s + \Omega_c} \right|_{\Omega_c = 1}$$

$$H_a(s) = \frac{1}{(s+1)}$$

Conversion of analog LPF to analog HPF

$$H_a(s) = H_a(s)\Big|_{s=\frac{\Omega_c}{s}}$$

where $\Omega_c = \Omega_1 = 7215$ rad/s.

$$H_a(s) = \frac{1}{\frac{7215}{s}+1} = \frac{s}{s+7215}$$

Conversion of analog HPF to analog HPF

Applying bilinear transformation, we get

$$H(z) = H_a(s)\Big|_{s=\frac{2}{T}\frac{(1-z^{-1})}{(1+z^{-1})}} = \frac{s}{s+7215}\Big|_{s=\frac{2}{0.0002}\frac{(1-z^{-1})}{(1+z^{-1})}}$$

$$\boxed{H(z) = \frac{0.58(1-z^{-1})}{(1-0.161z^{-1})}}$$

Summary

- Digital filters are broadly classified as Infinite Impulse Response (IIR) and Finite Impulse Response (FIR) filters. The design of IIR filter is presented in this chapter.
- The design methodology for analog filters is a well-developed one. By proper transformations, these analog filters can be converted to digital filters. Impulse invariant transformation and bilinear transformation are used to convert analog filters to digital filters.
- When stable analog filters are converted to digital filters, it is necessary that the poles of these digital filters lie within the unit circle in the s-plane, so that the filter is causal and stable.

- In the design of lowpass filters, when the specifications are given in the frequency domain it may not be always possible to get physically realizable stable filters. The magnitude response needs to be approximated.
- Several approximations to design lowpass filters are available in the literature. However, in this chapter, Butterworth and Chebyshev approximations and approximation of derivatives are used to design lowpass filters.
- By analog to digital frequency transformation, it is possible to convert lowpass prototype analog filter into either a bandpass, highpass or stop band filter.
- Several methods such as direct form-I, direct form-II parallel form, cascade form are available to realize the structure of the digital filter. They are described in this chapter. Finally the method of getting frequency response plot for the system function $H(z)$ is described.

Short Questions and Answers

1. **What are the requirements for an analog filter to be stable and causal?**
   - The analog transfer function $H(s)$ should be a rational function of s and the coefficients of s should be real, so that the system is causal.
   - The poles should lie in the left half of s-plane, so that the system is stable.
   - The number of zeros should be less than or equal to number of poles.
2. **What are the requirements for a digital filter to be stable and causal?**
   - The digital transfer function $H(z)$ should be a rational function of z and the coefficients of z should be real, so that the system is causal.
   - The poles should lie inside of the unit circle in z-plane, so that the system is stable.
3. **Define ripples in a filter.**
   The limits of tolerance in the magnitude of passband and stopband are called ripples. The tolerance in passband is denoted by δ_p and that in stopband is denoted by δ_s.
4. **Mention any two techniques for digitizing the transfer function of an analog filter.**
   The bilinear transformation and the impulse invariant transformations are the two techniques available for digitizing the analog filter transfer function.
5. **Compare the digital and analog filters.**

Digital filter	Analog filter
1. Operates on digital samples of the signal	Operates on analog samples of the signal
2. It is governed by linear difference equation	It is governed by linear differential equation
3. It consists of adders, multipliers and delays implemented digital logic	It consists of electrical components like resistors, capacitors and inductors
4. In digital filters the filter coefficients are designed to satisfy the desired frequency response	In analog filters the approximation problem is solved to satisfy the desired frequency response

6. **What are the advantages and disadvantages of digital filters?** *Advantages*:
   • High thermal stability due to absence of electrical components.
   • The performance characteristics like accuracy, dynamic range, stability and tolerance can be enhanced by increasing the length of the registers.
   • The digital filters are programmable.
   • Multiplexing and adaptive filtering are possible.
   Disadvantages:
   • The bandwidth of the discrete signal is limited by the sampling frequency.
   • The performance of digital filter depends on the hardware used to implement the filter.
7. **Mention the important features of IIR filters.**
   • The physically realizable IIR filter does not have linear phase.
   • The IIR specifications include the desired characteristics for the magnitude response only.
8. **What is impulse invariant transformation?**
   The transformation of analog filter to digital filter without modifying the impulse response of the filter is called impulse invariant transformation i.e., in this transformation the impulse response of the digital filter will be sampled version of the impulse response of analog filters.
9. **What is the main objective of impulse invariant transformation?**
   The main objective is to develop an IIR filter transfer function whose impulse response is the sampled version of impulse response of the analog filter. Therefore, the frequency response characteristics of analog filter is preserved.
10. **How analog poles are mapped to digital poles in impulse invariant/bilinear transformation?**
    In impulse invariant/linear transformation the mapping of analog to digital poles are as follows:
    • The analog poles on the left half of s-plane are mapped into the interior of unit circle in z-plane.
    • The analog poles on the imaginary axis of s-plane are mapped into the unit circle in z-plane.
    • The analog poles on the right half of s-plane are mapped into the exterior of unit circle in z-plane.
11. **What is the importance of poles in filter design?**

The stability of a filter is related to the location of the poles. For a stable analog filter the poles should lie on the left half of the s-plane. For a stable digital filter the poles should lie inside of the unit circle of z-plane.

12. Write the impulse invariant transformation used to transform real poles with and without multiplicity.

The impulse invariant transformation used to transform real pole (at $s = -p_i$) without multiplicity is

$$\frac{1}{(s+p_i)} \xrightarrow{\text{is transformed to}} \frac{1}{1-e^{-p_i T}z^{-1}}$$

13. Write the impulse invariant transform used to transform complex conjugate poles.

$$\frac{(s+a)}{(s+a)^2+b^2} \xrightarrow{\text{is transformed to}} \frac{1-e^{-aT}(\cos bT)z^{-1}}{1-2e^{-aT}(\cos bT)z^{-1}+e^{-2aT}z^{-2}}$$

$$\frac{b}{(s+a)^2+b^2} \xrightarrow{\text{is transformed to}} \frac{1-e^{-aT}(\sin bT)z^{-1}}{1-2e^{-aT}(\cos bT)z^{-1}+e^{-2aT}z^{-2}}$$

14. What is the relation between analog and digital frequency in impulse invariant transformation?

The relation between analog and digital frequency in impulse invariant transformation is given by,

$$\text{Digital frequency, } \omega = \Omega T$$

where, Ω = analog frequency and T = sampling time period.

15. What is aliasing?

The phenomena of high frequency sinusoidal components acquiring the identity of low frequency sinusoidal components after sampling is called aliasing (i.e., aliasing is higher frequencies impersonating lower frequencies). The aliasing, problem will arise if the sampling rate does not satisfy the Nyquist sampling criteria.

16. Why an impulse invariant transformation is not considered to be one-to-one?

In impulse invariant transformation any strip of width $(2\pi)/T$ in the s-plane for values of s in the range $(2k-1)\pi/T \le \Omega \le (2k+1)\pi/T$ (where k is an integer) is mapped into the entire z-plane. The left half portion of each strip in s-plane maps into the interior of the unit circle in z-plane, right half portion of each strip in s-plane maps into the exterior of the unit circle in z-plane and the imaginary axis of each strip in s-plane maps into the unit circle in z-plane. Hence, the impulse invariant transformation is many-to-one. Mapping of s-plane contour to z-plane is shown in Fig. 4.62.

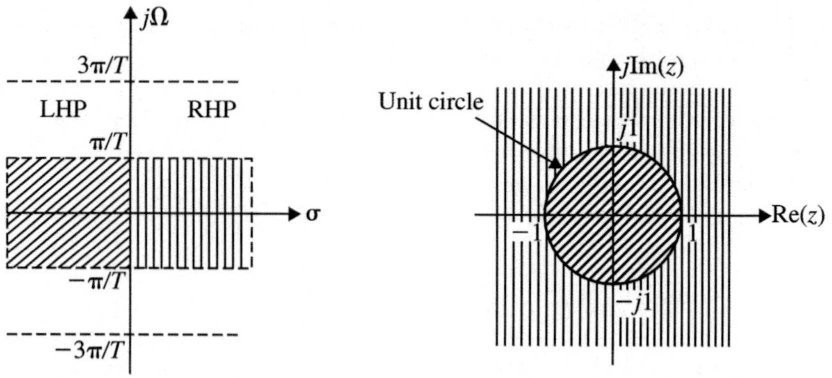

Fig. 4.62 Mapping of s-plane into z-plane in impulse invariant transformation

17. What is aliasing problem in impulse invariant method of designing digital filter? Why it is absent in bilinear transformation?

In impulse invariant mapping, the analog frequencies in the interval $(2k - 1)\pi/T \le \Omega \le (2k+1)\pi/T$ (where k is an integer) maps into corresponding values of digital frequencies in the interval $-\pi \le \omega \le \pi$. Hence the mapping of Ω to ω is many-to-one.

This will result in high frequency components acquiring the identity of the low frequency components if the analog filter is not bandlimited. This effect is called aliasing. The aliasing can be avoided in bandlimited filters by choosing very small values of sampling time. The bilinear mapping is one-to-one mapping and so there is no effect of aliasing.

18. What is bilinear transformation?

The bilinear transformation is a conformal mapping that transforms the s-plane to z-plane. In this mapping the imaginary axis of s-plane is mapped into the unit circle in z-plane, the left half of s-plane is mapped into interior of unit circle in z-plane and right half of s-plane is mapped into exterior of unit circle in z-plane. The bilinear mapping is a one-to-one mapping and it is accomplished when

$$s = \frac{2}{T}\frac{1 - z^{-1}}{1 + z^{-1}}$$

19. What is the relation between analog and digital frequency in bilinear transformation?

The relation between analog and digital frequency in bilinear transformation is given by,

$$\text{Digital frequency, } \omega = 2\tan^{-1}\left(\frac{\Omega T}{2}\right)$$

where, Ω = analog frequency and T = sampling time period.

20. What is frequency warping?
In bilinear transformation the relation between analog and digital frequencies is non-linear. When the s-plane is mapped into z-plane using bilinear transformation, this non-linear relationship introduces distortion in frequency axis, which is called frequency warping.

21. What is pre-warping? Why it is employed?
In IIR filter design using bilinear transformation, the conversion of the specified digital frequencies to analog frequencies is called pre-warping. The pre-warping is necessary to eliminate the effect of warping on amplitude response.

22. Explain the technique of pre-warping.
In IIR filter design using bilinear transformation the specified digital frequencies are converted to analog frequencies which are called pre-warp frequencies. Using the pre-warped frequencies, the analog filter transfer function is designed and then it is transferred to digital filter transfer function.

23. Compare the impulse invariant and bilinear transformation.

Impulse invariant	Bilinear iransformation
1. It is many-to-one mapping.	It is one-to-one mapping
2. The relation between analog and digital frequency is linear.	The relation between analog and digital frequency is non-linear
3. To prevent the problem of aliasing the analog filters should be bandlimits.	There is no problem of aliasing and the filters need not be bandlimits

24. Obtain the impulse response of digital filter to correspond to an analog filter with impulse response $h_a(t) = 0.3e^{-2t}$ and with a sampling rate of l.0 kHz using impulse invariant method.
Given that, $h_a(t) = 0.3e^{-2t}$ and sampling frequency, $F = 1\text{kHz} = 1 \times 10^3$ Hz.

$$\therefore \text{ Sampling time, } T = \frac{1}{F} = \frac{1}{1 \times 10^3} = 10^{-3} \text{ s}$$

$$\text{Impulse response of digital fitter } h(n) = h_e(t)|_{t=nT} = 0.3e^{-2t}|_{t=nT} = 0.3e^{-2nT}$$
$$= 0.3(e^{-2T})^n = 0.3(e^{-2\times 10^{-3}})^n$$
$$= 0.3(0.998)^n; \quad \text{for } n \geq 0.$$

25. Given that, $H_a(s) = 1/(s + 1)$. By impulse invariant method, obtain the digital filter transfer function and the difference equation of digital filter.
Given that, $H_a(s) = 1/(s + 1)$. In impulse invariant transformation,

$$\frac{1}{s + p_i} \xrightarrow{\text{is transformed to}} \frac{1}{1 - e^{-p_i T} z^{-1}}$$

Let $T = 1$ s. Transfer function of digital filter

$$H(z) = \frac{1}{1 - e^{-T} z^{-1}} = \frac{1}{1 - e^{-1} z^{-1}} = \frac{1}{1 - 0.368 z^{-1}}$$

We know that,

$$H(z) = \frac{Y(z)}{X(z)}$$

$$\frac{Y(z)}{X(z)} = \frac{1}{1 - 0.368 z^{-1}}$$

On cross multiplying we get

$$Y(z) - 0.368 z^{-1} Y(z) = X(z)$$
$$\therefore\ Y(z) = X(z) + 0.368 z^{-1} Y(z)$$

On taking inverse z-transform we get,

$$y(n) = x(n) + 0.368 y(n-1)$$

26. **Sketch the mapping of s-plane to z-plane in bilinear transformation.**
See (Fig. 4.63).
Mapping of s-plane contour to z-plane contour for question 26 is shown in Fig. 4.63.
27. **How does the order of the filter affect the frequency response of Butterworth filter?**
The magnitude response of Butterworth filter is shown in Fig. 4.64, from which it can be observed that the magnitude response approaches the ideal response as the order of the filter is increased.

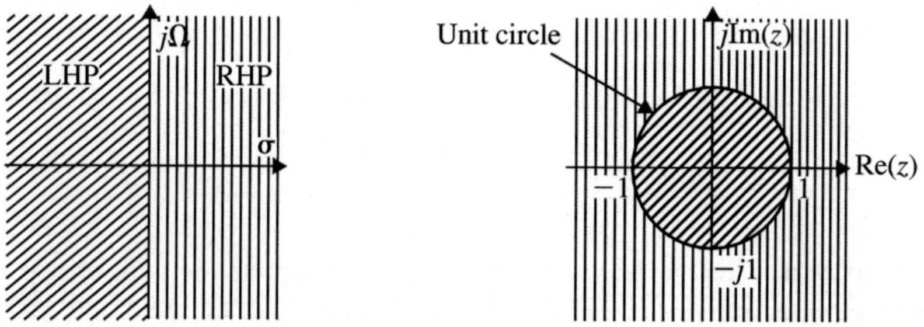

Fig. 4.63 Mapping of s-plane to z-plane

Magnitude response of Butterworth lowpass filter for various values of N for question 27 is shown in Fig. 4.64

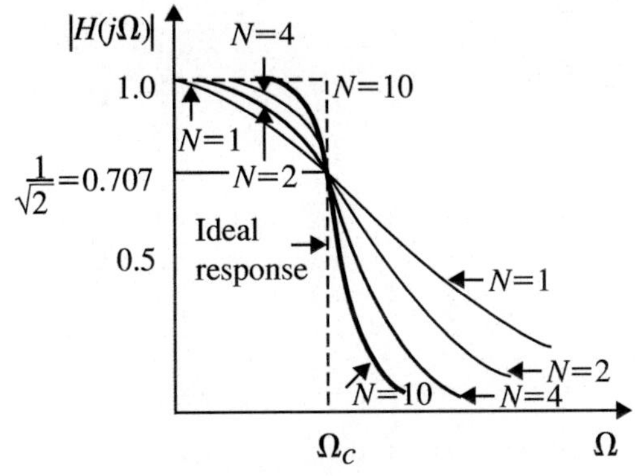

Fig. 4.64 Magnitude response of Butterworth lowpass filter for various values of N

28. **Write the transfer function of un-normalized Butterworth lowpass filter.**
When N is even, transfer function of analog lowpass Butterworth filter is,

$$H_a(s) = \sum_{k=1}^{N/2} \frac{\Omega_c^2}{s^2 + b_k \Omega_c s + \Omega_c^2}$$

When N is odd, transfer function of analog lowpass Butterworth filter is,

$$H_a(s) = \frac{\Omega_c}{s + \Omega_c} \sum_{k=1}^{(N-1)/2} \frac{\Omega_c^2}{s^2 + b_k \Omega_c s + \Omega_c^2}$$

where

$$b_k = 2 \sin[\frac{(2k-1)\pi}{2N}]$$

N = Order of the fitter and Ω_c = Analog cut-off frequency.

29. **How will you choose the order N for a Butterworth filter?**
Calculate a parameter N_1 using the following equation and correct, it to nearest integer.

$$N_1 = \frac{1}{2} \frac{\log \left\{ \left[\frac{1}{A_p^2} - 1 \right] \Big/ \left[\frac{1}{A_s^2 - 1} \right] \right\}}{\log \left(\frac{\Omega_s}{\Omega_p} \right)}$$

Choose the order N of the filter such that $N \geq N_1$.

30. **Write the properties of Butterworth filter.**

(*i*) The Butterworth filters have all-pole designs.

(*ii*) At the cut-off frequency Ω_c, the magnitude of normalized Butterworth filter is $1/\sqrt{2}$.

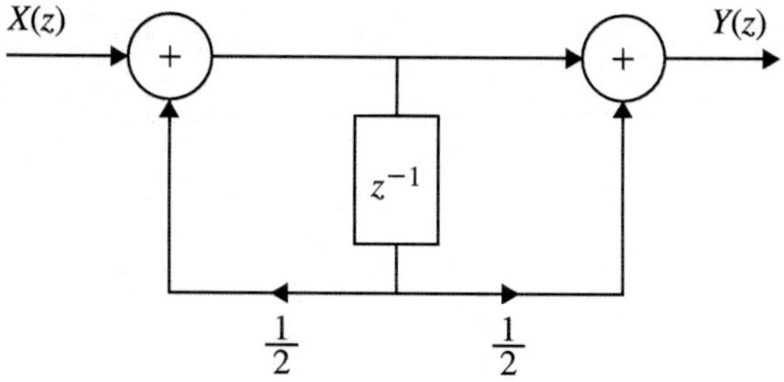

Fig. 4.65 Realization of $H(z) = \frac{(1+\frac{1}{2}z^{-1})}{(1-\frac{1}{2}z^{-1})}$

(*iii*) The filter order N, completely specifies the filter and as the value of N increases the magnitude response approaches the ideal response.

31. Find the digital transfer function $H(z)$ by using impulse invariant method for the analog T.F. $H(s) = \frac{1}{(s+2)}$. Assume $T = 0.1$ s and 0.5 s.

$$H(z) = \frac{1}{(1 - 0.8187z^{-1})} \quad \text{for } T = 0.1 \text{ s}$$

$$H(z) = \frac{1}{(1 - 0.3679z^{-1})} \quad \text{for } T = 0.5 \text{ s}$$

32. Realize the following system using Direct Form-II.

$$y(n) - \frac{1}{2}y(n-1) = x(n) + \frac{1}{2}x(n-1)$$

$$H(z) = \frac{(1 + \frac{1}{2}z^{-1})}{(1 - \frac{1}{2}z^{-1})}$$

See (Fig. 4.65). Realization of $H(z)$ of question 32 is shown in Fig. 4.65

33. Transform the single pole lowpass Butterworth filter with system function $H(s) = \frac{\Omega_p}{(\frac{\Omega_p}{s}+\Omega_p)}$ into a bandpass filter with upper and lower band edge frequencies Ω_u and Ω_l respectively.

$$H(s) = \frac{\Omega_p(s^2 + \Omega_l\Omega_u)}{\Omega_p(s^2 + \Omega_l + \Omega_u) + s(\Omega_u - \Omega_l)}$$

34. Convert the analog bandpass filter.

$$H_a(s) = \frac{1}{(s + 0.1)^2 + 9}$$

into a digital IIR filter by use of the mapping $s = \frac{1}{T}\frac{(z-1)}{(z+1)}$.

$$H(z) = \frac{T^2(1 + 2z^{-1} + z^{-2})}{[9.01T^2 + (18.02T^2 - 2)z^{-1} + (9.01T^2 - 2T + 1)z^{-2}]}$$

35. Write the equation for frequency transformation from lowpass to bandpass filter.

$$s \rightarrow \Omega_p \frac{(s^2 + \Omega_l \Omega_u)}{s(\Omega_u - \Omega_l)}$$

36. Find the digital filter equivalent for $H(s) = \frac{1}{(s+8)}$

$$H(z) = \frac{1}{(1 - e^{-8T} z^{-1})}$$

37. What are the parameters (specifications) of a Chebyshev filter?
The parameters of a Chebyshev filter are (*a*) passband ripple, (*b*) passband cut-off frequency, (*c*) stopband frequency, (*d*) attenuation beyond stopband frequency.

38. Give the location of poles of normalized Butterworth filter.
The poles of the normalized Butterworth filter occur on a unit circle at equally spaced points in the complex s-plane. The pole locations are identified as

$$s_k = e^{j\pi/2} e^{j(2k+1)\frac{\pi}{N}}, \quad k = 0, 1, 2 \ldots (N - 1)$$

N is order of the filter. For stable filter only LHP poles are taken.

39. State two advantages of bilinear transformation
The two advantages of bilinear transformation are:

(*a*) It avoids aliasing in frequency components.
(*b*) The transformation of stable analog filter results in a stable digital filter.

40. What is Chebyshev approximation?
In Chebyshev approximation, the error is defined as the difference between the ideal brickwall characteristic and the actual response and this is minimized over a prescribed band of frequencies. There are two types of Chebshev transfer functions. In the type 1, approximation, the magnitude characteristic is equiripple in the passband and monotonic in the stopband. In type 2 approximation, the magnitude response is monotonic in the passband and equiripple in the stopband.

41. What is Butterworth approximation?
The frequency response characteristic of the lowpass Butterworth filter is monotonic in both the passband and stopband. The response approximates to the ideal response as the order N of the filter increases. (Flat Characteristics).

42. What are the parameters that can be obtained from the Chebyshev filter specifications?
For a given set of specifications, the order N of the filter can be determined. Knowing N, the poles and zeros can be located.

43. Write down the transfer functions of a first order Butterworth normalized lowpass filter and highpass filter.

The normalized first order Butterworth normalized lowpass filter is

$$h_a(s) = \frac{1}{(s+1)}$$

The first order highpass filter normalized transfer function is

$$h_a(s) = \frac{1}{\frac{1}{s}+1} = \frac{s}{s+1}$$

44. Compare the lowpass Butterworth filter with lowpass Chebyshev filter.

(*a*) In the Butterworth filter, the frequency response decreases monotonically while the magnitude response of the Chebyshev filter contains ripples in the passband and monotically decreases in the stopband.
(*b*) The poles of the Butterworth filter lie on unit circle whereas the poles of Chebyshev filter lie on ellipse.
(*c*) The transition band is more in Butterworth filter compared to Chebyshev filter.
(*d*) For the same specifications, the order of the Butterworth filter is higher than that of Chebyshev filter.

45. Why IIR filters do not have linear phase?

A linear phase filter must have a system function

$$H(z) = \pm z^{-N} H(z^{-1})$$

where z^{-N} represents a delay of N units. For this it is necessary to have a mirror image pole outside the unit circle for every pole inside the unit circle in the z-plane. This means the filter will be unstable. Hence a causal and stable IIR filter cannot have linear phase if it is to be physically realizable.

46. What are the disadvantages of Impulse invariance method of designing filters?

If the sampling interval T is large, the IIR filter designed using impulse invariance method results in aliasing due to sampling. This method is not suitable to design highpass filter due to the spectrum aliasing which results from the sampling process.

Long Answer Type Questions

1. Explain the impulse invariance method of IIR filter design.
2. Explain bilinear transformation method of IIR filter design.
3. Draw the direct form-I and direct form-II structure of IIR system.
4. Convert the analog transfer function of the second order Butterworth filter into digital transfer function using bilinear transformation.

5. What is meant by aliasing? Explain with example.
6. Describe the procedure for the design of digital filters from analog filters and advantages and disadvantages of digital filters.
7. State and prove the conditions satisfied by a stable and causal discrete time filter in the z-transform domain.
8. Design a Butterworth filter using impulse invariant method for the following specifications:

$$0.8 \le |H(e^{j\omega})| \le 1; \quad 0 \le \omega \le 0.2\pi$$
$$|H(e^{j\omega})| \le 0.2; \quad 0.6 \le \omega \le \pi$$

9. Mention the advantages and disadvantages of FIR and IIR filters.
10. Design a Butterworth digital filter to meet the following constraints.

$$0.9 \le |H(\omega)| \le 1; \quad 0 \le \omega \le \frac{\pi}{2}$$
$$|H(\omega)| \le 0.2; \quad \frac{3\pi}{4} \le \omega \le \pi$$

Use bilinear transformation mapping technique. Assume $T = 1$ s.
11. Develop impulse invariant mapping technique for designing IIR filter.
12. Realize the given transfer function using direct form-I and parallel methods

$$H(z) = \frac{(4z^2 + 11z - 2)}{(z+1)(z-3)}.$$

13. Derive the equation for calculating the order of the Butterworth filter.
14. Design and realize a digital filter using bilinear transformation for the following specifications:

    (*a*) Monotonic passband and stopband.
    (*b*) -3 dB cutoff at 0.5π rad.
    (*c*) magnitude down at least 15 dB at $\omega = 0.75\pi$ rad.

15. Using impulse invariant method find $H(z)$ at $T = 1$ s.

$$H(s) = \frac{2}{s^2 + 8s + 15}$$

16. Enumerate the various steps involved in the design of LF digital Butterworth IIR filter.
17. Obtain direct form-II, cascade and parallel realizations of a discrete time system described by the difference equation

$$y(n) - \frac{5}{8}y(n-1) - \frac{1}{16}y(n-2) = x(n) - 3x(n-1) + 3x(n-2) - x(n-3)$$

Ans: Direct Form-II

$$\frac{Y(z)}{X(z)} = \frac{(1 - 3z^{-1} + 3z^{-2} - z^{-3})}{\left(1 - \frac{5}{8}z^{-1} - \frac{1}{16}z^{-2}\right)}$$

Cascade Form

$$H_1(z) = \frac{(1 - z^{-1})}{(1 - 0.7125z^{-1})}$$

$$H_2(z) = \frac{(1 - 2z^{-1} + z^{-2})}{(1 + 0.086z^{-1})}$$

Parallel Form

$$H(z) = -208 + 16z^{-1} + \frac{209 - 149z^{-1}}{\left(1 - \frac{5}{8}z^{-1} - \frac{1}{16}z^{-2}\right)}$$

18. Determine the system function $H(z)$ of the lowest order Chebyshev digital filter that meets the following specifications:

    (a) 1 dB ripple in the passband $0 \leq \omega \leq 0.3\pi$.
    (b) At least 60 dB attenuation in the stopband $0.35 \leq |\omega| \leq \pi$. Use the bilinear transformation.

19. Design a Chebyshev filter with a maximum passband attenuation of 2.5 dB at $\Omega_p = 20$ rad/s and the stopband attenuation of 30 dB at $\Omega_s = 50$ rad/s.
20. Find the direct form-I and direct form-II realization of the filter

$$H(z) = \frac{1 - 2z^{-2} + 3z^{-3}}{1 - 0.2z^{-3}}$$

Chapter 5
Design of Finite Impulse Response (FIR) Digital Filters

Learning Objectives

After completing this chapter, you should be able to

- study the characteristic of practical frequency selective digital filters;
- realize the Finite Impulse Response (FIR) filter structure;
- study the characteristics of FIR filters with linear phase;
- study different types of symmetry of the impulse response of linear phase FIR filters;
- design FIR filter by frequency sampling method;
- design linear phase FIR filters using windows;
- design FIR differentiators.

5.1 Introduction

In Chap. 3, the design of IIR filter was considered. These filters are designed by transformation from analog domain to digital domain. They are called recursive filters since feedbacks are used. Hence, IIR filters should be carefully designed to ensure that the system function $H(z)$ is stable. In the case of FIR filter design which is non-recursive in nature, the stability problem does not arise. Since the system function $H(z)$ is a polynomial of z^{-1} it always guarantees the stability of the filter. Further, unlike IIR filter design which has non-linear phase characteristic, the FIR digital filter has linear phase. A filter with a non-linear phase characteristic causes phase distortion in the signal which is undesirable in much practical applications such as data transmission, music and video. However, to meet the same frequency response specifications, the order of FIR filter transfer function is much higher than

S. Palani, *Discrete Time Systems and Signal Processing*,
https://doi.org/10.1007/978-3-031-32421-5_5

that of an IIR digital filter. Hence, FIR filters require more parameters and more memory and have more computational complexity.

The basic approach to the design of FIR digital filter is based on direct approximation of the specified magnitude response in addition to the requirement of linear phase response. Unlike IIR filter, it does not have any connection with the analog filter. Linear phase FIR filters are designed based on whether the impulse response function $h(n)$ is positive or negative and the order of the filter is odd or even. Thus there are four types of linear phase FIR filters. FIR filters are also designed based on Fourier coefficients which have infinite length. By truncating these coefficients of infinite Fourier series, FIR filters are designed. This is called windowing. Several windowing techniques are available. FIR filter design using the above techniques is discussed in this chapter.

5.1.1 LTI System as Frequency Selective Filters

An LTI system performs a type of filtering among the various frequency components at its input. The nature of this filtering is determined by the frequency response characteristics $H(\omega)$, which in turn depend on the choice of the system parameters [a_k and b_k coefficient]. Thus, by proper selection of the coefficients, we can design frequency selective filters that pass signals with frequency components in some bands while they attenuate signals that contain frequency components in other frequency bands.

In general, an LTI system modifies the input spectrum $X(\omega)$ according to its frequency response $H(\omega)$ to yield an output signal with spectrum $Y(\omega) = H(\omega)X(\omega)$. $H(\omega)$ acts as a weighting function or a spectral shaping function to the different frequency components in the input signal.

Let us consider a signal $x(n)$ with frequency content in a band of frequencies $\omega_1 < \omega < \omega_2$ (i.e. band limited signal)

$$X(\omega) = 0 \quad \text{for } \omega \geq \omega_2 \quad \text{and} \quad \omega \leq \omega_1 \tag{5.1}$$

If it is passed through a filter with frequency response,

$$H(\omega) = \begin{cases} Ce^{-j\omega\alpha}, & \omega_1 < \omega < \omega_2 \\ 0, & \text{otherwise, where } C \text{ and } \alpha = +\text{ve constants} \end{cases} \tag{5.2}$$

The signal at the output of the filter has the spectrum which is expressed as

$$\begin{aligned} Y(\omega) &= X(\omega)H(\omega) \\ &= X(\omega)Ce^{-j\omega\alpha} \qquad (5.3) \\ &= CX(\omega)e^{-j\omega\alpha} \qquad (5.4) \end{aligned}$$

Using the following identity, we get

$$F[x(n-\alpha)] = X(\omega)e^{-j\omega\alpha}$$
$$\therefore\ Y(\omega) = CF[x(n-\alpha)]$$

Taking inverse Fourier transform on both sides, we get

$$y(n) = Cx(n-\alpha)$$

i.e. the filter output is simply a delayed and amplitude scaled version of the input signal. A pure delay is usually tolerable and is not considered a distortion of the signal. Likewise the amplitude scaling. The filter defined by

$$H(\omega) = \begin{cases} Ce^{-j\omega\alpha}, & \omega_1 < \omega < \omega_2 \\ 0, & \text{otherwise} \end{cases}$$

is called an ideal filter.

$$\begin{aligned} H(\omega) &= |H(\omega)|\angle H(\omega) \\ |H(\omega)| &= C, \quad \angle H(\omega) = \theta(\omega) = -\alpha\omega \\ |H(\omega)|\ \text{(amplitude)} &= \text{constant} \\ |H(\omega)|\ \text{(phase)} &= \text{linear function of frequency } \omega \end{aligned} \tag{5.5}$$

In general, any deviation of the frequency response characteristics of a linear filter from the ideal response results in signal distortion.

(*i*) *Amplitude distortion:* If the filter has a variable magnitude frequency response characteristic in the passband, then the filter has amplitude distortion.
(*ii*) *Phase distortion:* If the phase characteristic is not linear within the desired frequency band, the signal undergoes phase distortion.
(*iii*) *Signal delay:* The derivative of the phase with respect to frequency has the units of delay.

$$\tau(\omega) = -\frac{d\theta(\omega)}{d\omega} = -\frac{d}{d\omega}(-\alpha\omega) = \alpha \tag{5.6}$$

Filters are classified according to their frequency response characteristics. The ideal frequency responses $H(\omega)$ of four major types of filters are as follows:

(*i*) Ideal frequency response of LPF

$$H_d(\omega) = \begin{cases} Ce^{-j\alpha\omega}, & -\omega_c \le \omega \le \omega_c \\ 0, & -\pi < \omega < -\omega_c \ \text{ and } \ \omega_c < \omega < \pi \end{cases} \tag{5.7}$$

(*ii*) Ideal frequency response of HPF

$$H_d(\omega) = \begin{cases} Ce^{-j\alpha\omega}, & -\pi \le \omega \le -\omega_c,\ \omega_c \le \omega \le \pi, \\ 0, & -\omega_c < \omega < \omega_c \end{cases} \tag{5.8}$$

(*iii*) Ideal frequency response of BPF

$$H_d(\omega) = \begin{cases} Ce^{-j\alpha\omega}, & -\omega_{c_2} \le \omega \le -\omega_{c_1} \text{ and } \omega_{c_1} \le \omega \le \omega_{c_2} \\ 0, & -\pi \le \omega \le -\omega_{c_2}\ \&\ -\omega_{c_1} \le \omega \le \omega_{c_1}\ \&\ \omega_{c_2} \le \omega \le \pi \end{cases} \tag{5.9}$$

(iv) Ideal frequency response of BSF

$$H_d(\omega) = \begin{cases} Ce^{-j\alpha\omega}, & -\pi \le \omega \le -\omega_{c_2}\ \&\ -\omega_{c_1} \le \omega \le \omega_{c_1}\ \&\ \omega_{c_2} \le \omega \le \pi \\ 0, & -\omega_{c_2} < \omega < -\omega_{c_1} \text{ and } \omega_{c_1} < \omega < \omega_{c_2} \end{cases} \tag{5.10}$$

The magnitude responses of ideal lowpass, highpass, bandpass and bandstop filters are shown in Fig. 5.1a–d, respectively.

5.2 Characteristic of Practical Frequency Selective Filters

Causality has very important implications is the design of frequency selective filters. Causality implies that the frequency response characteristic $H(\omega)$ of the filter cannot be zero, except at a finite set of points in the frequency range. In addition, $H(\omega)$ cannot have an infinitely sharp cutoff from passband to stopband. That is $H(\omega)$ cannot drop from unity to zero abruptly. Therefore, the ideal filters are non-causal and hence physically un-realizable for the real-time signal processing applications.

The input $x(n)$ and output $y(n)$ of a LTI system is governed by the following N^{th} order difference equation.

$$y(n) = -\sum_{k=1}^{N} a_k y(n-k) + \sum_{k=0}^{M} b_k x(n-k)$$

On taking Fourier transform on both sides, we get

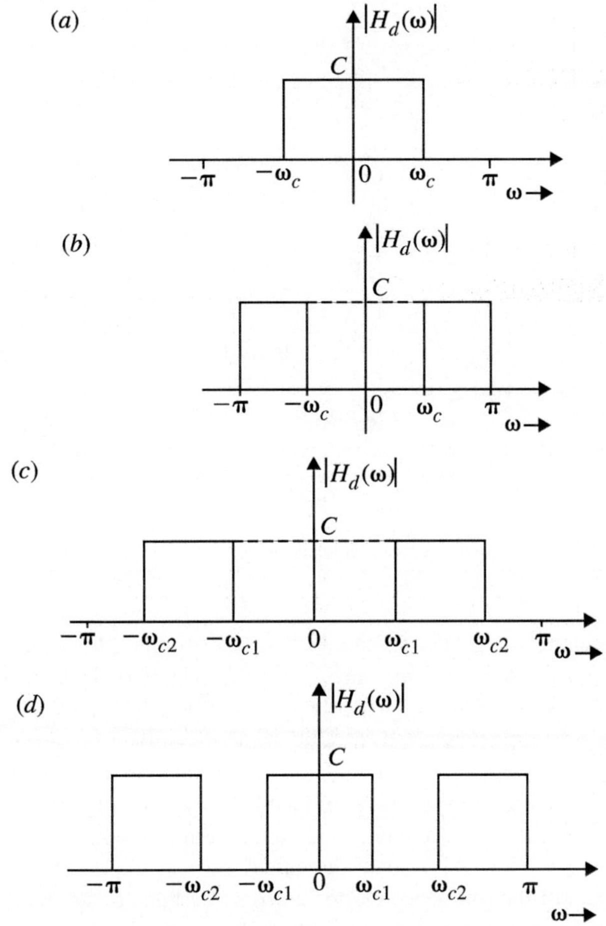

Fig. 5.1 Magnitude response of ideal lowpass, highpass, bandpass and bandstop filter

$$Y(\omega) = -\sum_{k=1}^{N} a_k e^{-j\omega k} Y(\omega) + \sum_{k=0}^{M} b_k e^{-j\omega k} X(\omega)$$

$$Y(\omega)\left[1 + \sum_{k=1}^{N} a_k e^{-j\omega k}\right] = \sum_{k=0}^{M} b_k e^{-j\omega k} X(\omega)$$

$$H(\omega) = \frac{Y(\omega)}{X(\omega)}$$

$$= \frac{\sum_{k=0}^{M} b_k e^{-j\omega k}}{1 + \sum_{k=1}^{N} a_k e^{-j\omega k}}; \quad N \geq M \tag{5.11}$$

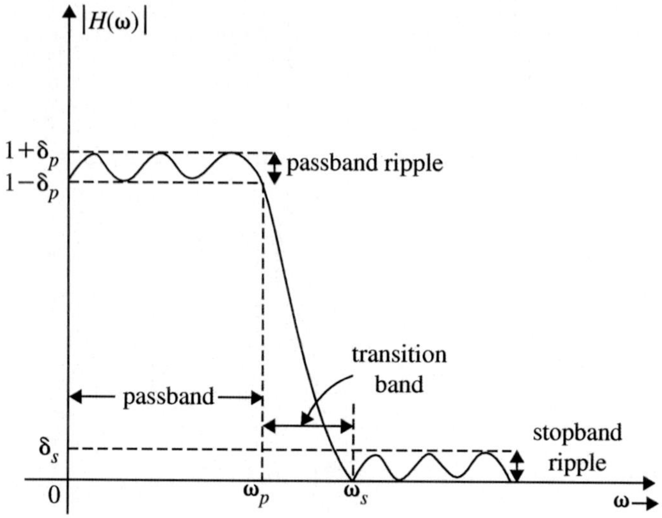

Fig. 5.2 Magnitude response of a practical lowpass filter

which are causal and physically realizable. The basic digital filter design problem is to approximate any of the ideal frequency response characteristics with a system that has frequency response $H(\omega)$ by properly selecting the coefficients $\{a_k\}$ and $\{b_k\}$.

In practice, it is not necessary to insist that the magnitude $|H(\omega)|$ is constant in the entire passband of the filters. A small amount of ripple in the passband is usually tolerable. Similarly, it is not necessary for the filter response $|H(\omega)|$ to be zero in the stopband. A small non-zero value or a small amount of ripple in the stopband is also tolerable. The magnitude response of a practical lowpass filter is shown in Fig. 5.2.

The transition of the frequency response from passband to stopband defines the transition band or transition region of the filter. The band-edge frequency ω_p defines the edge of the passband, while the frequency ω_s denotes the beginning of the stopband. Thus the width of the transition band is $\omega_s - \omega_p$. The width of the passband is usually called the bandwidth of the filter.

In any filter design problem, one may specify

(*i*) the maximum passband ripple δ_p;
(*ii*) the maximum stopband ripple δ_s;
(*iii*) passband edge frequency ω_p;
(*iv*) stopband edge frequency ω_s.

Based on these specifications, we may select the parameters a_k and b_k in the frequency response characteristic which best approximates the desired specifications.

5.3 Structures for Realization of the FIR Filter

In general, an FIR system is described by the following difference equation:

$$y(n) = \sum_{k=0}^{N-1} b_k x(n-k) \tag{5.12}$$

$$Y(z) = \sum_{k=0}^{N-1} b_k z^{-k} X(z)$$

$$H(z) = \frac{Y(z)}{X(z)} = \sum_{k=0}^{N-1} b_k z^{-k} = b_0 + b_1 z^{-1} + b_2 z^{-2} + \cdots + b_{N-1} z^{-(N-1)} \tag{5.13}$$

where $H(z)$ is transfer function.

$$\begin{aligned} H(z) &= Z[h(n)] \\ &= \sum_{k=0}^{N-1} h(k) z^{-k} = h(0) + h(1)z^{-1} + h(2)z^{-2} + \cdots + h(N-1)z^{-(N-1)} \end{aligned} \tag{5.14}$$

On comparing Eqs. (5.13) and (5.14), we get

$$b_k = h(k) \quad \text{for} \quad k = 0, 1, 2, 3, \ldots, N-1 \tag{5.15}$$

The different types of structures for realizing FIR systems are as follows:

1. Direct form realization.
2. Cascade form realization.
3. Linear phase realization.

5.3.1 *Direct Form Realization*

The direct form structure that can be obtained from the z-domains equation governing the FIR system is shown in Fig. 5.3.

$$\begin{aligned} Y(z) &= \sum_{k=0}^{N-1} b_k z^{-k} X(z) \\ &= b_0 X(z) + b_1 z^{-1} X(z) + b_2 z^{-2} X(z) + \cdots + b_{N-1} z^{-(N-1)} X(z) \end{aligned} \tag{5.16}$$

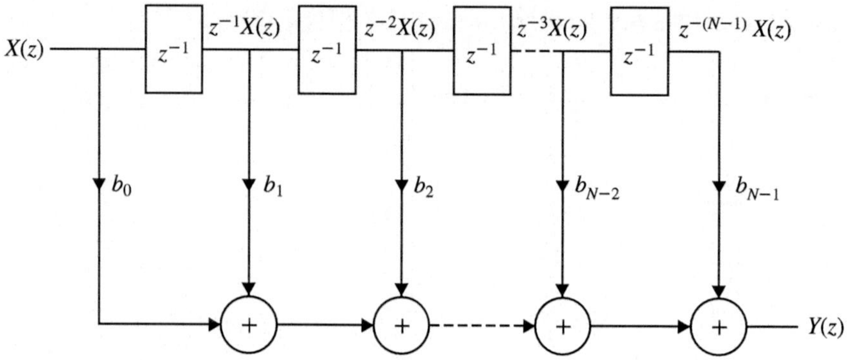

Fig. 5.3 Direct form realization

5.3.2 *Cascade Form Realization*

The transfer function of FIR system is $(N-1)$th order polynomial in z. This polynomial can be factorized into second-order factors (when N is odd) and the transfer function can be expressed as a product of second-order factors.

$$\text{When } N \text{ is odd, } H(z) = \prod_{i=1}^{(N-1)/2} (c_{0i} + c_{1i}z^{-1} + c_{2i}z^{-2}) \tag{5.17}$$
$$= H_1 H_2 \ldots H_{(N-1)/2}$$

When N is even, $H(z)$ will have one first-order and $(N-2)/2$ second-order section.

$$\text{When } N \text{ is even, } H(z) = (c_{01} + c_{11}z^{-1}) \prod_{i=2}^{N/2} (c_{0i} + c_{1i}z^{-1} + c_{2i}z^{-2}) \tag{5.18}$$
$$= H_1 H_2 \ldots H_{N/2}$$

5.3.3 *Linear Phase Realization*

In FIR systems for linear phase response, the impulse response should be symmetrical. For this,

$$h(k) = h(N-1-k) \tag{5.19}$$

When N is Even

$$H(z) = \sum_{k=0}^{N-1} h(k)z^{-k} = \sum_{k=0}^{(N/2)-1} h(k)z^{-k} + \sum_{k=(N/2)}^{N-1} h(k)z^{-k} \tag{5.20}$$

Let $m = N - 1 - k \Longrightarrow k = N - 1 - m$
When $k = \frac{N}{2}; \Longrightarrow m = N - 1 - \frac{N}{2} = \frac{N}{2} - 1$
When $k = N - 1; \Longrightarrow m = N - 1 - (N - 1) = 0$

$$\therefore\ H(z) = \sum_{k=0}^{(N/2)-1} h(k)z^{-k} + \sum_{m=0}^{(N/2)-1} h(N-1-m)z^{-(N-1-m)}$$

On replacing m by k, we get

$$\begin{aligned}
H(z) &= \sum_{k=0}^{(N/2)-1} h(k)z^{-k} + \sum_{k=0}^{(N/2)-1} h(N-1-k)z^{-(N-1-k)} \\
&= \sum_{k=0}^{(N/2)-1} h(k)z^{-k} + \sum_{k=0}^{(N/2)-1} h(k)z^{-(N-1-k)} \\
H(z) &= \sum_{k=0}^{(N/2)-1} h(k)[z^{-k} + z^{-(N-1-k)}]
\end{aligned}$$

Therefore,

$$\boxed{\begin{aligned}
Y(z) &= \sum_{k=0}^{(N/2)-1} h(k)[z^{-k}X(z) + z^{-(N-1-k)}X(z)] \quad \text{where } h(k) = b_k \\
&= b_0[X(z) + z^{-(N-1)}X(z)] + b_1[z^{-1}X(z) + z^{-(N-2)}X(z)] \\
&\quad + \cdots + b_{(N/2)-1}[z^{-[(N/2)-1]}X(z) + z^{-(N/2)}X(z)]
\end{aligned}} \tag{5.21}$$

Equation (5.21) is realized in Fig. 5.4.

When N is Odd

$$H(z) = \sum_{k=0}^{N-1} h(k)z^{-k} = \sum_{k=0}^{\frac{N-3}{2}} h(k)z^{-k} + h\left(\frac{N-1}{2}\right)z^{-\left(\frac{N-1}{2}\right)} + \sum_{k=\frac{N+1}{2}}^{N-1} z^{-1} \tag{5.22}$$

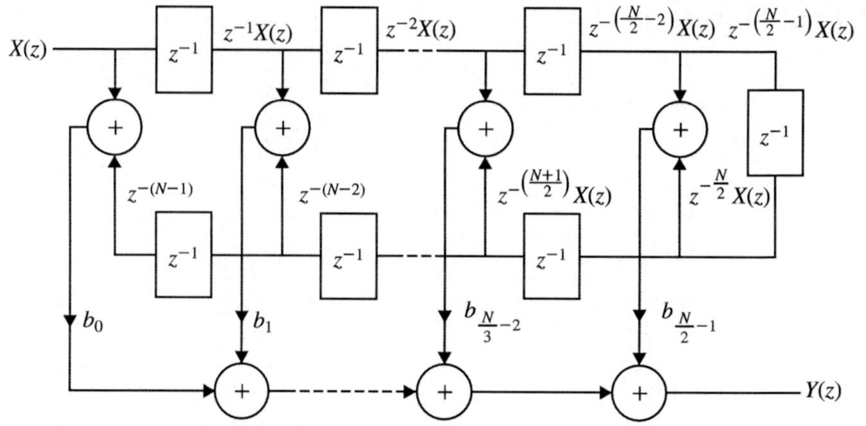

Fig. 5.4 Linear phase realization when N is even

$$\text{Let } m = N-1-k \Longrightarrow k = N-1-m$$
$$\text{When } k = \frac{N+1}{2}; \Longrightarrow m = \frac{N-3}{2}$$
$$\text{When } k = N-1; \Longrightarrow m = 0$$

$$\therefore\ H(z) = \sum_{k=0}^{\frac{N-3}{2}} h(k)z^{-k} + h\left(\frac{N-1}{2}\right) z^{-\left(\frac{N-1}{2}\right)} \sum_{m=0}^{\frac{N-3}{2}} h(N-1-m)z^{-(N-1-m)}$$

On replacing m by k, we get

$$H(z) = \sum_{k=0}^{\frac{N-3}{2}} h(k)z^{-k} + h\left(\frac{N-1}{2}\right) z^{-\left(\frac{N-1}{2}\right)} \sum_{k=0}^{\frac{N-3}{2}} h(N-1-k)z^{-(N-1-k)}$$
$$= \sum_{k=0}^{\frac{N-3}{2}} h(k)[z^{-k} + z^{-(N-1-k)}] + h\left(\frac{N-1}{2}\right) z^{-\left(\frac{N-1}{2}\right)} \quad \because (b_k = h(k))$$

$$Y(z) = b_{\frac{N-1}{2}} z^{-\left(\frac{N-1}{2}\right)} X(z) + \sum_{k=0}^{\frac{N-3}{2}} b_k (z^{-k} X(z)) + z^{-(N-1-k)} X(z)$$

$$\boxed{\begin{aligned} Y(z) = b_{\frac{N-1}{2}} z^{-\left(\frac{N-1}{2}\right)} X(z) + b_0[X(z) + z^{-(N-1)}X(z)] + \cdots \\ + b_{\frac{N-3}{2}}\left[z^{-\left(\frac{N-3}{2}\right)} X(z) + z^{-(N+2)}X(z)\right] \end{aligned}} \tag{5.23}$$

Equation (5.23) is realized in Fig. 5.5.

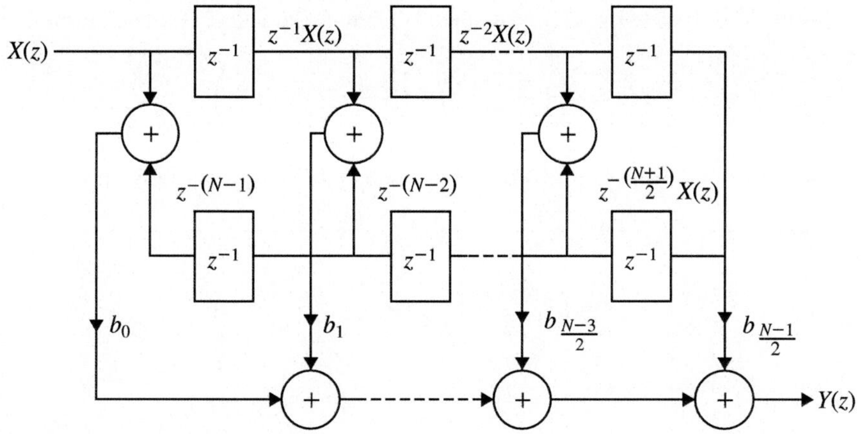

Fig. 5.5 Linear phase realization when N is odd

Linear phase realization when N is odd is shown in Fig. 5.5

Example 5.1 Draw the direct form structure of the FIR system described by the following transfer function (Fig. 5.6):

$$H(z) = 1 + \frac{1}{2}z^{-1} + \frac{3}{4}z^{-2} + \frac{1}{4}z^{-3} + \frac{1}{2}z^{-4} + \frac{1}{5}z^{-5}$$

Direct form realization of $H(z)$ for Example 5.1 is shown in Fig. 5.6

Solution

$$Y(z) = X(z) + \frac{1}{2}z^{-1}X(z) + \frac{3}{4}z^{-2}X(z) + \frac{1}{4}z^{-3}X(z) + \frac{1}{2}z^{-4}X(z) + \frac{1}{5}z^{-5}X(z)$$

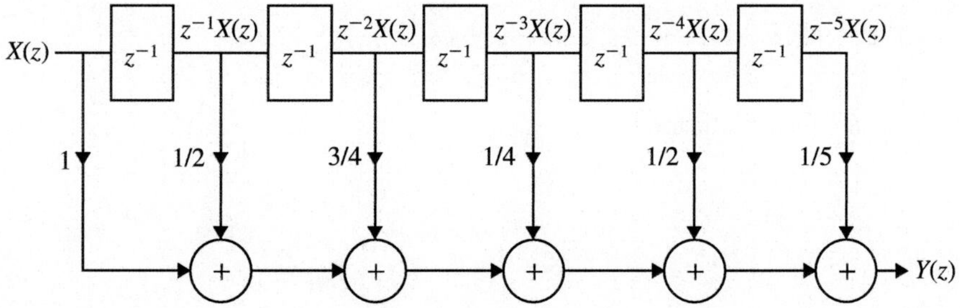

Fig. 5.6 Direct form (or) transversal realization of $H(z)$ for Example 5.1

Example 5.2 Realize the FIR system in (1) Direct form and (2) Cascade form.

$$H(z) = \left(1 + \frac{1}{2}z^{-1}\right)\left(1 + \frac{1}{2}z^{-1} + \frac{1}{4}z^{-2}\right)$$

Direct form realization of FIR filter for Example 5.2 is shown in Fig. 5.7

Solution Given (Fig. 5.7)

$$H(z) = \left(1 + \frac{1}{2}z^{-1}\right)\left(1 + \frac{1}{2}z^{-1} + \frac{1}{4}z^{-2}\right)$$

(1) Direct form realization

$$\begin{aligned} H(z) &= \frac{Y(z)}{X(z)} \\ &= 1 + \frac{1}{2}z^{-1} + \frac{1}{4}z^{-2} + \frac{1}{2}z^{-1} + \frac{1}{4}z^{-2} + \frac{1}{8}z^{-3} \\ &= 1 + z^{-1} + \frac{1}{2}z^{-2} + \frac{1}{8}z^{-3} \end{aligned}$$

$$Y(z) = X(z) + z^{-1}X(z) + \frac{1}{2}z^{-2}X(z) + \frac{1}{8}z^{-3}X(z)$$

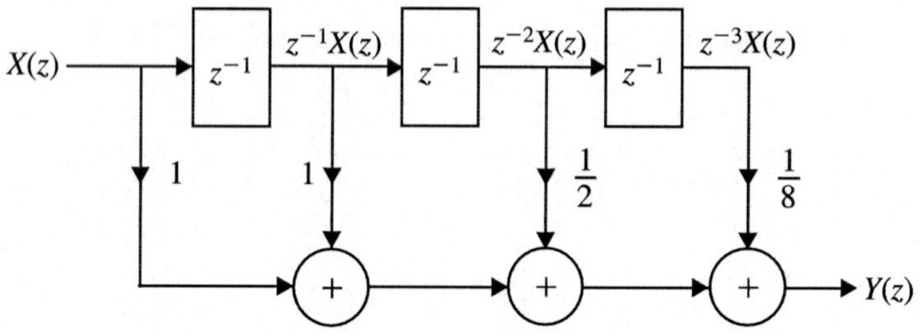

Fig. 5.7 Direct form realization of FIR filter for Example 5.2

(2) Cascade form realization

$$\begin{aligned} H(z) &= \left(1 + \frac{1}{2}z^{-1}\right)\left(1 + \frac{1}{2}z^{-1} + \frac{1}{4}z^{-2}\right) \\ &= H_1(z) \cdot H_2(z) \end{aligned}$$

where

$$H_1(z) = 1 + \frac{1}{2}z^{-1}$$
$$H_2(z) = 1 + \frac{1}{2}z^{-1} + \frac{1}{4}z^{-2}$$
$$h_1(n) = \left\{1, \frac{1}{2}\right\}$$
$$h_2(n) = \left\{1, \frac{1}{2}, \frac{1}{4}\right\}$$

Cascade form realization of FIR filter for Example 5.2 is shown in Fig. 5.8

Here $h_1(n)$ and $h_2(n)$ do not satisfy the condition $h(n) \neq h(\infty - 1 - \infty)$. Therefore, both systems are realized in direct form (Fig. 5.8).

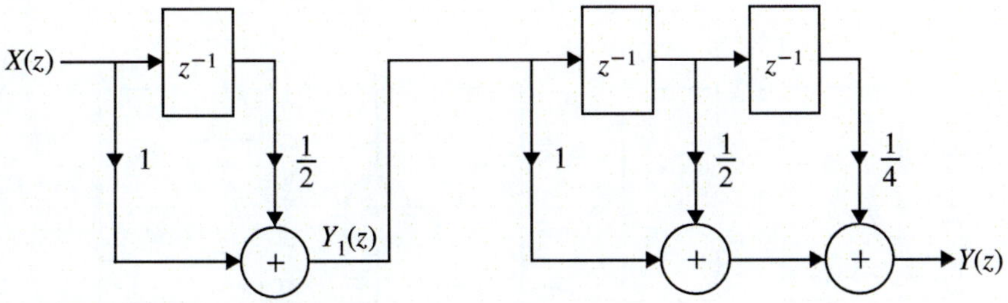

Fig. 5.8 Cascade form realization of FIR filter for Example 5.2

Example 5.3 Obtain cascade realization of linear phase FIR filter having system function

$$H(z) = \left(1 + \frac{1}{2}z^{-1} + z^{-2}\right)\left(2 + \frac{1}{4}z^{-1} + 2z^{-2}\right)$$

using minimum number of multipliers.

Solution Given

$$\begin{aligned} H(z) &= \left(1 + \frac{1}{2}z^{-1} + z^{-2}\right)\left(2 + \frac{1}{4}z^{-1} + z^{-2}\right) \\ &= H_1(z)H_2(z) \end{aligned}$$

where

$$H_1(z) = 1 + \frac{1}{2}z^{-1} + z^{-2},$$

$$H_2(z) = 2 + \frac{1}{4}z^{-1} - 2z^{-2}$$

$$h_1(n) = \left\{1 + \frac{1}{2}, 1\right\} \text{ and } N = 3,$$

$$h_2(n) = \left\{2 + \frac{1}{4}, 2\right\} \text{ and } N = 3$$

Cascade connection of linear phase FIR filter of Example 5.3 is shown in Fig. 5.9

Here $h_1(n)$ and $h_2(n)$ satisfy the condition $h(n) = h(N - 1 - n)$ (Fig. 5.9).

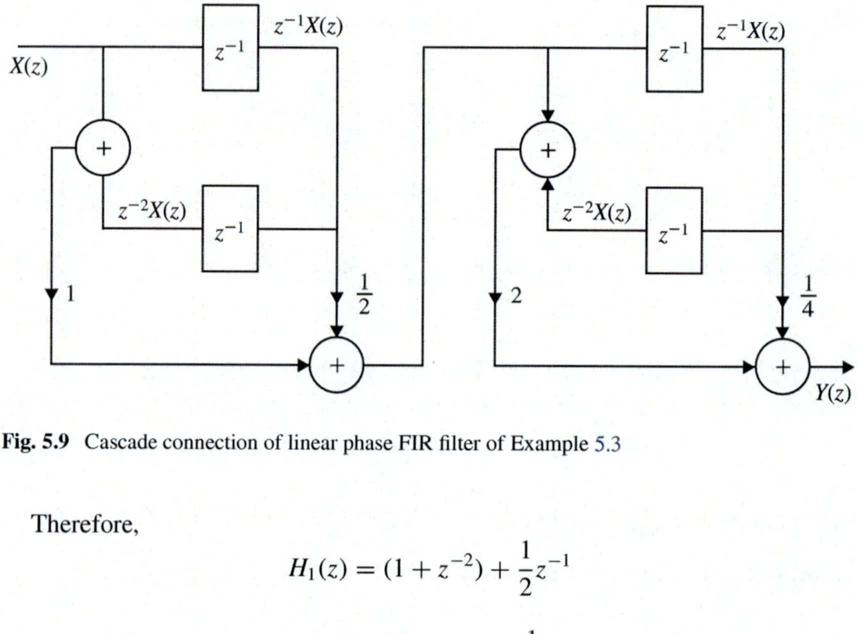

Fig. 5.9 Cascade connection of linear phase FIR filter of Example 5.3

Therefore,

$$H_1(z) = (1 + z^{-2}) + \frac{1}{2}z^{-1}$$

$$H_2(z) = 2(1 + z^{-2}) + \frac{1}{4}z^{-1}$$

Example 5.4 Realize the following system with minimum number of multipliers:

$$(i) \qquad H(z) = \frac{1}{4} + \frac{1}{2}z^{-1} + \frac{3}{4}z^{-2} + \frac{1}{2}z^{-3} + \frac{1}{4}z^{-4}$$

$$(ii) \qquad H(z) = 1 + \frac{1}{2}z^{-1} + \frac{1}{2}z^{-2} + z^{-3}$$

$$(iii) \qquad H(z) = \left(1 + \frac{1}{2}z^{-1} + z^{-2}\right)\left(1 + \frac{1}{4}z^{-1} + z^{-2}\right)$$

Solution (*i*) Given

$$H(z) = \frac{1}{4} + \frac{1}{2}z^{-1} + \frac{3}{4}z^{-2} + \frac{1}{2}z^{-3} + \frac{1}{4}z^{-4}$$

$$H(z) = \sum_{n=0}^{\infty} h(n)z^{-n} = h(0) + h(1)z^{-1} + h(2)z^{-2} + h(3)z^{-3} + \cdots$$

Comparing these two equations, we get the following impulse response.

$$h(n) = \left\{\frac{1}{4}, \frac{1}{2}, \frac{3}{4}, \frac{1}{2}, \frac{1}{4}\right\} \quad \text{and} \quad N = 5$$

Here $h(n)$ satisfies the condition $h(n) = h(N - 1 - n)$

$$\therefore\ H(z) = \frac{Y(z)}{X(z)} = \frac{1}{4} + \frac{1}{2}z^{-1} + \frac{3}{4}z^{-2} + \frac{1}{2}z^{-3} + \frac{1}{4}z^{-4}$$

$$= \frac{Y(z)}{X(z)} = \frac{1}{4}\left(1 + z^{-4}\right) + \frac{1}{2}\left(z^{-1} + z^{-3}\right) + \frac{3}{4}z^{-2}$$

$$Y(z) = \frac{1}{4}\left(X(z) + z^{-4}X(z)\right) + \frac{1}{2}\left(z^{-1}X(z) + z^{-3}X(z)\right) + \frac{3}{4}z^{-2}X(z)$$

$Y(z)$ is realized as shown in Fig. 5.10.

(*ii*) Given

$$H(z) = 1 + \frac{1}{2}z^{-1} + \frac{1}{2}z^{-2} + z^{-3}$$

Impulse response

$$h(n) = \left\{1, \frac{1}{2}, \frac{1}{2}, 1\right\} \quad \text{and} \quad N = 4$$

Here $h(n)$ satisfies the condition

$$h(n) = h(N-1-n)$$
$$\therefore \quad \frac{Y(z)}{X(z)} = [1 + z^{-3}] + \frac{1}{2}[z^{-1} + z^{-2}]$$
$$Y(z) = [X(z) + z^{-3}X(z)] + \frac{1}{2}[z^{-1}X(z) + z^{-2}X(z)]$$

$Y(z)$ is realized as shown in Fig. 5.11.

(*iii*) Given

$$H(z) = \left(1 + \frac{1}{2}z^{-1} + z^{-2}\right)\left(1 + \frac{1}{4}z^{-1} + z^{-2}\right)$$
$$= H_1(z)H_2(z)$$

where

$$H_1(z) = 1 + \frac{1}{2}z^{-1} + z^{-2} = (1 + z^{-2}) + \frac{1}{2}z^{-1}$$
$$H_2(z) = 1 + \frac{1}{4}z^{-1} + z^{-2} = (1 + z^{-2}) + \frac{1}{4}z^{-1}$$

Consider

$$H_1(z) = \frac{Y_1(z)}{X(z)} = (1 + z^{-2}) + \left(\frac{1}{2}z^{-1}\right)$$
$$Y_1(z) = [X(z) + z^{-2}X(z)] + \frac{1}{2}z^{-1}X(z)$$

Consider

$$H_2(z) = \frac{Y(z)}{Y_1(z)} = (1 + z^{-2}) + \left(\frac{1}{4}z^{-1}\right)$$
$$Y(z) = [Y_1(z) + z^{-2}Y_1(z)] + \frac{1}{4}z^{-1}Y_1(z)$$

$Y(z)$ is realized as shown in Fig. 5.12.

5.3.4 *Lattice Structure of an FIR Filter*

Let us consider an FIR filter with system function

Fig. 5.10 Linear phase realization of $H(z)$ for Example 5.4(i)

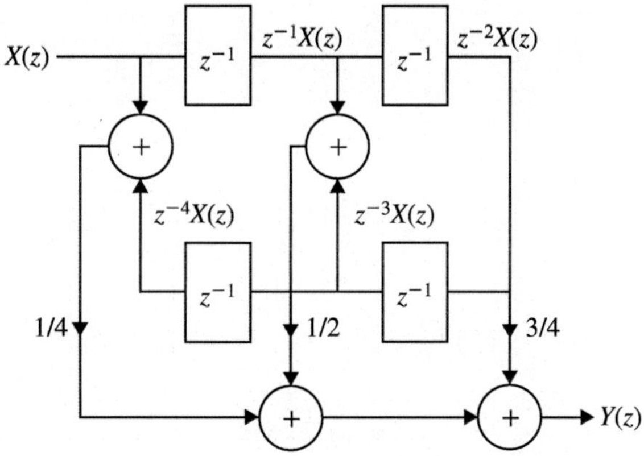

Fig. 5.11 Linear phase realization of $H(z)$ for Example 5.4(ii)

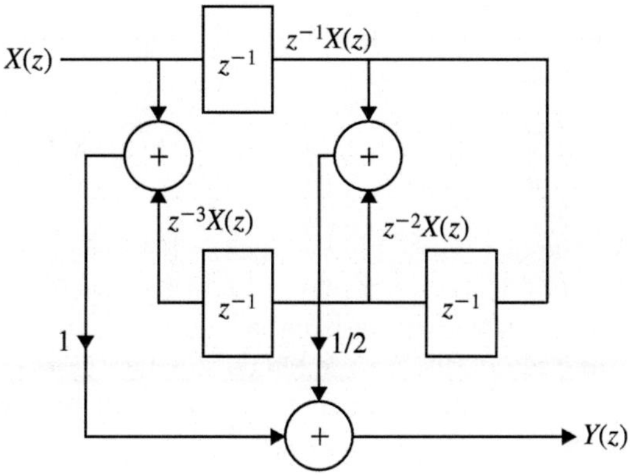

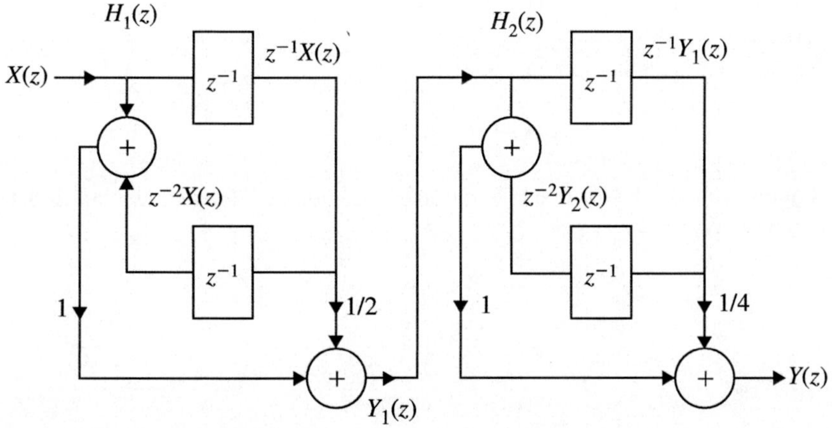

Fig. 5.12 Linear phase realization of $H(z)$ for Example 5.4(iii)

$$H(z) = A_m(z) = 1 + \sum_{k=1}^{m} \alpha_m(k) z^{-k}, \quad m \geq 1$$

$$\frac{Y(z)}{X(z)} = 1 + \sum_{k=1}^{m} \alpha_m(k) z^{-k}$$

$$Y(z) = X(z) + \sum_{k=1}^{m} \alpha_m(k) z^{-k} X(z)$$

Taking inverse z-transform on both sides, we get

$$y(n) = x(n) + \sum_{k=1}^{m} \alpha_m(k) x(n-k) \tag{5.24}$$

Interchanging the role of input and output in Eq. (5.24), we get

$$x(n) = y(n) + \sum_{k=1}^{m} \alpha_m(k) y(n-k) \tag{5.25}$$

We find that Eq. (5.25) describes an IIR system having the system function $H(z) = \frac{1}{A_m(z)}$, while the system described by the difference equation in Eq. (5.24) represents an FIR system with system function $H(z) = A_m(z)$.

Based on this, we use lattice structure system described in Chap. 3 to obtain a lattice structure for an all-zero FIR system by interchanging the role of input and output.

For an all-pole filter, the input $x(n) = f_N(n)$ and the output $y(n) = f_0(n)$. For an all-zero FIR filter of order $M - 1$, the input $x(n) = f_0(n)$ and the output $y(n) = f_{m-1}(n)$.

For $m = 1$, Eq. (5.24) reduces to

$$y(n) = x(n) + \alpha_1 x(n-1) \tag{5.26}$$

This output can also be obtained from a single-stage lattice filter shown in Fig. 5.13 from which we have

$$\begin{aligned} x(n) &= f_0(n) = g_0(n) \\ y(n) &= f_1(n) = f_0(n) + k_1 g_0(n-1) \\ &= x(n) + k_1 x(n-1) \end{aligned} \tag{5.27}$$

and

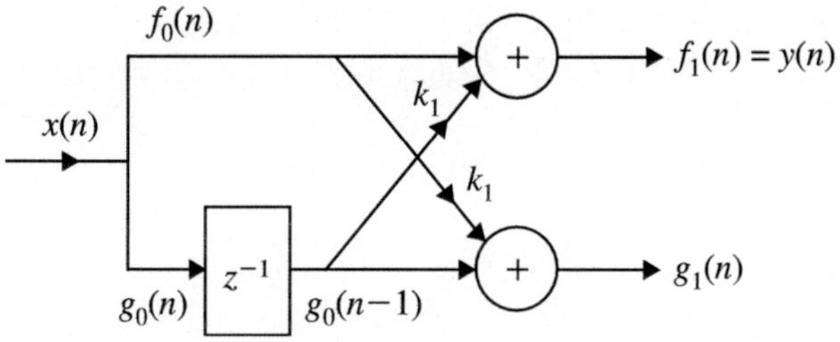

Fig. 5.13 Single-stage all-zero lattice structure

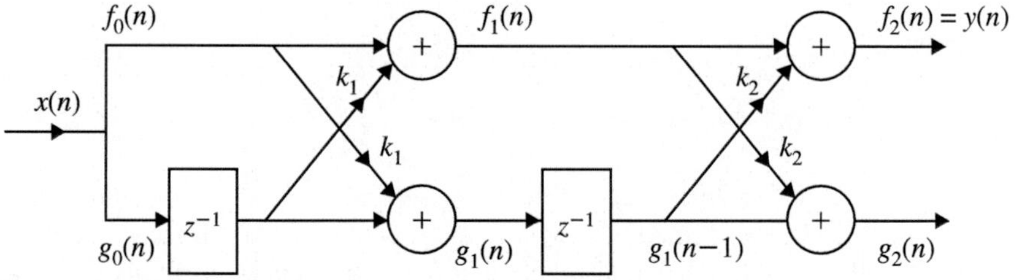

Fig. 5.14 Two-stage all-zero lattice structure

$$\begin{aligned} g_1(n) &= k_1 f_0(n) + g_0(n-1) \\ &= k_1 x(n) + x(n-1) \end{aligned} \tag{5.28}$$

Comparing Eq. (5.26) with Eq. (5.27), we get

$$\begin{aligned} \alpha_1(0) &= 1 \\ \alpha_1(1) &= k_1 \end{aligned}$$

Now let us consider an FIR filter for which $m = 2$.

$$y(n) = x(n) + \alpha_2(1)x(n-1) + \alpha_2(2)x(n-2) \tag{5.29}$$

By cascading two lattice stages as shown in Fig. 5.14, it is possible to obtain the output $y(n)$.

From Fig. 5.14, the output from second stage is obtained as

$$\begin{aligned} y(n) = f_2(n) &= f_1(n) + k_2 g_1(n-1) \\ g_2(n) &= k_2 f_1(n) + g_1(n-1) \end{aligned} \tag{5.30}$$

Substituting for $f_1(n)$ and $g_1(n-1)$ from Eq. (5.28) in Eq. (5.30), we get

$$y(n) = f_2(n) = x(n) + k_1x(n-1) + k_2[k_1x(n-1) + x(n-2)]$$
$$= x(n) + k_1(1+k_2)x(n-1) + k_2x(n-2) \quad (5.31)$$

Equation (5.31) is identical to Eq. (5.29) from which we have

$$\begin{aligned}
\alpha_2(0) &= 1, \\
\alpha_2(2) &= k_2, \\
\alpha_2(1) &= k_1(1+k_2) \\
&= \alpha_1(1)[1+\alpha_2(2)]
\end{aligned} \quad (5.32)$$

Similarly

$$g_2(n) = \alpha_2x(n) + k_1(1+k_2)x(n-1) + x(n-2) \quad (5.33)$$

From Eqs. (5.31) and (5.33), we observe that the filter coefficients for the lattice filter that produces $f_2(n)$ are $\{1, \alpha_2(1), \alpha_2(2)\}$ while the coefficients for filter with output $g_2(n)$ are $\{\alpha_2(2),\ \alpha_2(1), 1\}$. We also note that these two sets of coefficients are in reverse order.

For a $m-1$ stage filter

$$\begin{aligned}
f_0(n) &= g_0(n) \\
f_m(n) &= f_{m-1}(n) + k_m g_{m-1}(n-1) \\
g_m(n) &= k_m f_{m-1}(n) + g_{m-1}(n-1), \quad m = 1, 2, ...m-1
\end{aligned} \quad (5.34)$$

The output of $m-1$ stage filter

$$y(n) = f_{m-1}(n).$$

5.3.4.1 Conversion of Lattice Coefficients to Direct Form Filter Coefficients

For $m = 3$, Eq. (5.24) can be written as

$$y(n) = x(n) + \alpha_3(1)x(n-1) + \alpha_3(2)x(n-2) + \alpha_3(3)x(n-3) \quad (5.35)$$

We have

$$\begin{aligned}y(n) = f_3(n) &= f_2(n) + k_3 g_2(n-1)\\ &= x(n) + \alpha_2(1)x(n-1) + \alpha_2(2)x(n-2) + k_3\alpha_2(2)x(n-1)\\ &\quad + k_3\alpha_2(1)x(n-2) + k_3 x(n-3)\\ &= x(n) + [\alpha_2(1) + k_3\alpha_2(1)]x(n-1) + [\alpha_2(2) + k_3\alpha_2(1)]x(n-2)\\ &\quad + k_3 x(n-3)\end{aligned} \tag{5.36}$$

Comparing Eqs. (5.35) and (5.36), we get

$$\begin{aligned}\alpha_3(0) &= 1\\ \alpha_3(1) &= \alpha_2(1) + k_3\alpha_2(1)\\ &= \alpha_2(1) + \alpha_3(3)\alpha_2(1)\\ \alpha_3(2) &= \alpha_2(2) + k_3(3)\alpha_2(1)\\ &= \alpha_2(2) + \alpha_3(3)\alpha_2(1)\\ \alpha_3(3) &= k_3\end{aligned} \tag{5.37}$$

From Eqs. (5.32) and (5.37) for a general case, we find that

$$\begin{aligned}\alpha_m(0) &= 1\\ \alpha_m(m) &= k_m\\ \alpha_m(k) &= \alpha_{m-1}(k) + k_m(m)\alpha_{m-1}(m-k)\end{aligned} \tag{5.38}$$

Equation (5.38) can be used to convert the lattice filter coefficients to direct form FIR filter coefficients.

Example 5.5 Consider an FIR lattice filter coefficients

$$k_1 = \frac{1}{3}, \quad k_2 = \frac{1}{4}, \quad k_3 = \frac{1}{2}$$

Determine the FIR filter coefficients for the direct form structure.

Solution From the given data and from Eq. (5.38), we find that

$$\begin{aligned}\alpha_3(0) &= 1\\ \alpha_3(3) &= k_3 = \frac{1}{2}\\ \alpha_1(1) &= k_1 = \frac{1}{3}\\ \alpha_2(2) &= k_2 = \frac{1}{4}\end{aligned}$$

From Eq. (5.38), we write

$$\alpha_m(k) = \alpha_{m-1}(k) + k_m\alpha_{m-1}(m-k)$$

For $m = 2$ and $k = 1$

$$\begin{aligned}\alpha_2(1) &= \alpha_1(1) + k_2\alpha_1(1)\\ &= \frac{1}{3} + \frac{1}{4}\cdot\frac{1}{3} = \frac{1}{3} + \frac{1}{12} = \frac{4+1}{12} = \frac{5}{12}\end{aligned}$$

For $m = 3$ and $k = 1$

$$\begin{aligned}\alpha_3(1) &= \alpha_2(1) + \alpha_3(3)\alpha_2(2)\\ &= \frac{5}{12} + \frac{1}{2}\cdot\frac{1}{4}\\ &= \frac{5}{12} + \frac{1}{8}\\ &= \frac{40+12}{96}\\ &= \frac{52}{96} = \frac{13}{24}\end{aligned}$$

For $m = 2$ and $k = 2$

$$\begin{aligned}\alpha_3(2) &= \alpha_3(2) + \alpha_3(3)\alpha_2(1)\\ &= \frac{1}{4} + \frac{1}{2}\cdot\frac{5}{12}\\ &= \frac{1}{4} + \frac{5}{24}\\ &= \frac{6+5}{24}\\ &= \frac{11}{24}\end{aligned}$$

$$\boxed{\alpha_3(0) = 1,\quad \alpha_3(1) = \frac{13}{24},\quad \alpha_3(2) = \frac{11}{2},\quad \alpha_3(3) = \frac{1}{2}}$$

5.3.4.2 Conversion of Direct Form FIR Filter Coefficients to Lattice Coefficients

For a three-stage direct form FIR filter

$$y(n) = x(n) + \alpha_3(1)x(n-1) + \alpha_3(2)x(n-2) + \alpha_3(3)x(n-3) \tag{5.39}$$

For a three-stage lattice structure

$$\begin{aligned}y(n) = f_3(n) &= f_2(n) + k_3 g_2(n-1)\\ g_3(n) &= k_3 f_2(n) + g_2(n-1)\\ y(n) = f_3(n) &= f_2(n) + k_3[g_3(n) - k_3 f_2(n)]\\ y(n) = f_3(n) &= f_2(n)[1-k_3^2] + k_3 g_3(n)\\ f_2(n)[1-k_3^2] &= f_3(n) - k_3 g_3(n)\end{aligned}$$

$$\begin{aligned}f_2(n) &= \frac{f_3(n) - k_3 g_3(n)}{1-k_3^2}\\ &= \Big[x(n) + \alpha_3(1)x(n-1) + \alpha_3(2)x(n-2) + \alpha_3(3)x(n-3)\\ &\quad -k_3^2 x(n) - \alpha_3(3)\alpha_3(2)x(n-1) - \alpha_3(3)\alpha_3(1)x(n-2)\\ &\quad -\alpha_3(3)x(n-3)\Big] \Big/ 1-\alpha_3^2(3)\\ &= x(n) + \frac{\alpha_3(1) - \alpha_3(2)\alpha_3(2)}{1-\alpha_3^2(2)}x(n-1) + \frac{\alpha_3(2) - \alpha_3(3)\alpha_3(1)}{1-\alpha_3^2(3)}x(n-2)\end{aligned} \tag{5.40}$$

Comparing Eqs. (5.39) and (5.40), we get

$$\begin{aligned}\alpha_2(0) &= 1\\ \alpha_2(1) &= \frac{\alpha_3(1) - \alpha_3(3)\alpha_3(2)}{1-\alpha_3^2(3)}\\ \alpha_2(2) &= \frac{\alpha_3(2) - \alpha_3(3)\alpha_3(1)}{1-\alpha_3^2(3)}\end{aligned}$$

In general, for a m-stage filter

$$\boxed{\begin{aligned}\alpha_{m-1}(0) &= 1\\ k_m &= \alpha_m(m)\\ \alpha_{m-1}(k) &= \frac{\alpha_m(k) - \alpha_m(m)\alpha_m(m-k)}{1-\alpha_m^2(m)}, \qquad 1 \le k \le m-1\end{aligned}} \tag{5.41}$$

Example 5.6 An FIR filter is given by the difference equation

$$y(n) = x(n) + \frac{4}{3}x(n-1) + \frac{1}{2}x(n-2) + \frac{2}{3}x(n-3)$$

Determine its lattice form.

Solution Given

$$y(n) = x(n) + \frac{4}{3}x(n-1) + \frac{1}{2}x(n-2) + \frac{2}{3}x(n-3)$$

$$\alpha_3(0) = 1, \quad \alpha_3(1) = \frac{4}{3}, \quad \alpha_3(2) = \frac{1}{2}, \quad \alpha_3(3) = \frac{2}{3}$$

From Eq. (5.41), we get

$$\begin{aligned}
\alpha_2(0) &= 1 \\
k_3 = \alpha_3(3) &= \frac{2}{3} \\
\alpha_{m-1}(k) &= \frac{\alpha_m(k) - \alpha_m(m)\alpha_m(m-k)}{1 - \alpha_m^2(m)}, \quad 1 \le k \le 2
\end{aligned}$$

For $m = 3$ and $k = 1$

$$\begin{aligned}
\alpha_2(1) &= \frac{\alpha_3(1) - \alpha_3(3)\alpha_3(2)}{1 - \alpha_3^2(3)} \\
&= \frac{\frac{4}{3} - \frac{2}{3} \cdot \frac{1}{2}}{1 - (\frac{2}{3})^2} \\
&= \frac{\frac{4}{3} - \frac{2}{6}}{1 - \frac{4}{9}} \\
&= 1.8
\end{aligned}$$

For $m = 3$ and $k = 2$

$$\begin{aligned}
k_2 = \alpha_2(2) &= \frac{\alpha_3(2) - \alpha_3(3)\alpha_3(1)}{1 - \alpha_3^2(3)} \\
&= \frac{\frac{1}{2} - \frac{2}{3} \cdot \frac{4}{3}}{1 - (\frac{2}{3})^2} \\
&= \frac{\frac{1}{2} - \frac{8}{9}}{1 - \frac{4}{9}} \\
&= -0.7
\end{aligned}$$

For $m = 2$ and $k = 1$

$$
\begin{aligned}
k_1 = \alpha_1(1) &= \frac{\alpha_2(1) - \alpha_2(2)\alpha_2(1)}{1 - \alpha_2^2(2)} \\
&= \frac{1.8 - (-0.7)(1.8)}{1 - (-0.7)^2} \\
&= \frac{3.06}{0.51} \\
&= 6
\end{aligned}
$$

$$
\boxed{k_1 = 6, \quad k_2 = -0.7, \quad k_3 = 0.6667}
$$

The lattice structure is shown in Fig. 5.15.

Example 5.7 **(a)** An FIR filter is given by the difference equation

$$
y(n) = 2x(n) + \frac{4}{5}x(n-1) + \frac{3}{2}x(n-2) + \frac{2}{3}x(n-3)
$$

Determine the lattice form.

Lattice form representation for Example 5.7(a) is shown in Fig. 5.16.

Solution Given (Fig. 5.16)

$$
\begin{aligned}
y(n) &= 2x(n) + \frac{4}{5}x(n-1) + \frac{3}{2}x(n-2) + \frac{2}{3}x(n-3) \\
&= 2\left[x(n) + \frac{2}{5}x(n-1) + \frac{3}{4}x(n-2) + \frac{1}{3}x(n-3)\right] \\
&= k_0\left[1 + \sum_{k=1}^{3} \alpha_m(k)x(n-k)\right]
\end{aligned}
$$

where $k_0 = 2$

$$
\alpha_3(0) = 1,\ \alpha_3(1) = \frac{2}{5},\ \alpha_3(2) = \frac{3}{4},\ \alpha_3(3) = \frac{1}{3}
$$

From Eq. (5.24), we get

$$
\begin{aligned}
\alpha_2(0) &= 1 \\
k_3 = \alpha_3(3) &= \frac{1}{3} \\
\alpha_{m-1}(k) &= \frac{\alpha_m(k) - \alpha_m(m)\alpha_m(m-k)}{1 - \alpha_m^2(m)}, \quad 1 \le k \le 2
\end{aligned}
$$

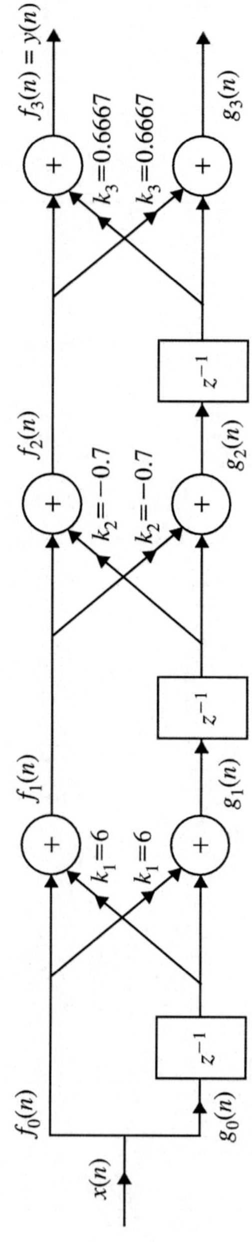

Fig. 5.15 Lattice structure for Example 5.6

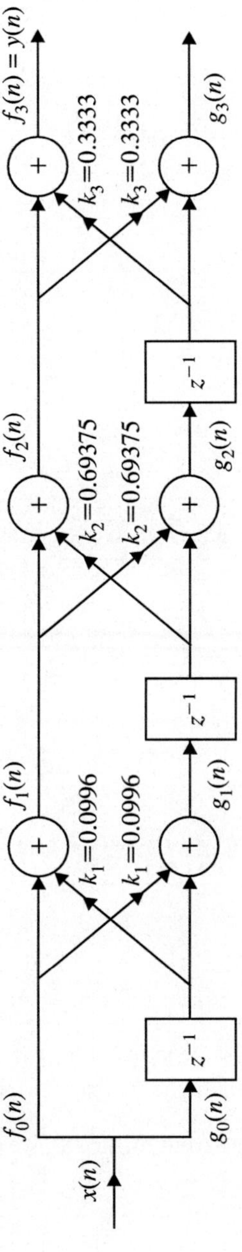

Fig. 5.16 Lattice Form for Example 5.7(*a*)

For $m = 3$ and $k = 1$

$$\alpha_2(1) = \frac{\alpha_3(1) - \alpha_3(3)\alpha_3(2)}{1 - \alpha_3^2(3)}$$
$$= \frac{\frac{2}{5} - \frac{1}{3} \cdot \frac{3}{4}}{1 - (\frac{1}{3})^2}$$
$$\alpha_2(1) = 0.16875$$

For $m = 3$ and $k = 2$

$$k_2 = \alpha_2(2) = \frac{\alpha_3(2) - \alpha_3(3)\alpha_3(1)}{1 - \alpha_3^2(3)}$$
$$= \frac{\frac{3}{4} - \frac{1}{3} \cdot \frac{2}{5}}{1 - (\frac{1}{3})^2}$$
$$= \frac{111}{160}$$

$$\boxed{k_2 = 0.69375}$$

For $m = 2$ and $k = 1$

$$k_1 = \alpha_1(1) = \frac{\alpha_2(1) - \alpha_2(2)\alpha_2(1)}{1 - \alpha_2^2(2)}$$
$$= \frac{0.16875 - (0.69375)(0.16875)}{1 - (0.69375)^2}$$

$$\boxed{k_1 = 0.0996}$$

Lattice form coefficients are

$$k_1 = 0.0996, \quad k_2 = 0.69375, \quad k_3 = 0.3333.$$

Example 5.7 **(b)** Given a three-stage lattice filter with coefficients $k_1 = \frac{1}{4}$, $k_2 = \frac{1}{4}$ and $k_3 = \frac{1}{3}$. Determine the FIR filter coefficients for the direct form structure.

Solution From the given data and from Eq. (5.38), we can find that

$$\alpha_3(0) = 1, \quad \alpha_3(3) = k_3 = \frac{1}{3}$$

$$\alpha_1(1) = k_1 = \frac{1}{4}, \quad \alpha_2(2) = k_2 = \frac{1}{4}$$

We know that

$$\alpha_m(k) = \alpha_{m-1}(k) + k_m\alpha_{m-1}(m-k)$$

For $m = 2$ and $k = 1$

$$\alpha_2(1) = \alpha_1(1) + k_2\alpha_1(1)$$
$$\alpha_2(1) = \frac{1}{4} + \frac{1}{4}.\frac{1}{4} = \frac{1}{4} + \frac{1}{16}$$
$$= \frac{4+1}{16} = \frac{5}{16}$$

For $m = 3$ and $k = 1$

$$\alpha_3(1) = \alpha_2(1) + \alpha_3(3)\alpha_2(2)$$
$$= \frac{5}{16} + \frac{1}{3}.\frac{1}{4} = \frac{5}{16} + \frac{1}{12}$$
$$= \frac{60+16}{192} = \frac{76}{192}$$
$$= 0.3958$$
$$= \frac{19}{48}$$

For $m = 3$ and $k = 2$

$$\alpha_3(2) = \alpha_2(2) + k_3\alpha_2(1)$$
$$= \frac{1}{4} + \frac{1}{3}.\frac{5}{16}$$
$$\alpha_3(2) = \frac{1}{4} + \frac{5}{48} = \frac{12+5}{48} = \frac{17}{48}$$

Direct form coefficients are

$$\boxed{\alpha_3(0) = 1, \quad \alpha_3(1) = \frac{19}{48}, \quad \alpha_3(2) = \frac{17}{48}, \quad \alpha_3(3) = \frac{1}{3}}$$

5.4 FIR Filters

The filters designed by using finite number of samples of impulse response are called FIR filters. These finite number of samples are obtained from the infinite duration desired impulse response $h_d(n)$. Here $h_d(n)$ is the inverse Fourier transform of $H_d(\omega)$, where $H_d(\omega)$ is the ideal (desired) frequency response. The various methods of designing FIR filters differ only in the method of determining the samples of $h(n)$ from the samples of $h_d(n)$. They are discussed below.

5.4.1 Characteristics of FIR Filters with Linear Phase

Let $h(n)$ be a causal finite duration sequence defined over the interval $0 \leq n \leq N-1$ and the samples of $h(n)$ be real. The Fourier transform of $h(n)$ is given by

$$H(\omega) = \sum_{n=0}^{N-1} h(n)e^{-j\omega n} \tag{5.42}$$

which is periodic with period 2π. Therefore,

$$H(\omega) = H(\omega + 2\pi m), \quad m = 0, \pm 1, \pm 2, \ldots \tag{5.43}$$

with the constraint that $h(n)$ is real and $H(\omega)$ is complex.

$$H(\omega) = \pm|H(\omega)|e^{j\theta(\omega)}$$

The operator '$\pm$' represents real part of $H(\omega)$ taking on both +ve and −ve values. When $h(n)$ is real then $|H(\omega)|$ is symmetric function and $\angle H(\omega)$ or $\angle\theta(\omega)$ is anti-symmetric function, i.e.

$$\begin{aligned} |H(\omega)| &= |H(-\omega)| \\ \angle\theta(\omega) &= -\angle\theta(-\omega) \end{aligned} \tag{5.44}$$

For many practical FIR filters, exact linearity of phase is a desired goal.

Let us assume that the phase of $H(\omega)$ is a linear function of ω, i.e.

$$\begin{aligned} &\theta(\omega) \propto \omega \\ &\theta(\omega) \propto -\alpha\omega, \qquad -\pi \leq \omega \leq \pi \end{aligned} \tag{5.45}$$

where α is a constant phase delay in samples. From Eqs. (5.43) and (5.45), we get

$$H(\omega) = \pm|H(\omega)|e^{-j\alpha\omega}$$

From Eq. (5.24), we get

$$H(\omega) = \sum_{n=0}^{N-1} h(n)e^{-j\omega n}$$

$$\sum_{n=0}^{N-1} h(n)e^{-j\omega n} = \pm|H(\omega)|e^{-j\alpha\omega} \tag{5.46}$$

$$\sum_{n=0}^{N-1} h(n)[\cos\omega n - j\sin\omega n] = \pm|H(\omega)|[\cos\alpha\omega - j\sin\alpha\omega]$$

On equating the real and imaginary parts, we get

$$\pm\,|H(\omega)|\cos\alpha\omega = \sum_{n=0}^{N-1} h(n)\cos\omega n \tag{5.47}$$

$$\pm|H(\omega)|\sin\alpha\omega = \sum_{n=0}^{N-1} h(n)\sin\omega n \tag{5.48}$$

Dividing Eq. (5.20) by Eq. (5.48), we get

$$\frac{\sin\alpha\omega}{\cos\alpha\omega} = \frac{\sum_{n=0}^{N-1} h(n)\sin\omega n}{\sum_{n=0}^{N-1} h(n)\cos\omega n} \tag{5.49}$$

$$\sin\alpha\omega\sum_{n=0}^{N-1} h(n)\cos\omega n = \cos\alpha\omega\sum_{n=0}^{N-1} h(n)\sin\omega n$$

$$\sum_{n=0}^{N-1} h(n)[\sin\alpha\omega\cos\omega n - \cos\alpha\omega\sin\omega n] = 0 \tag{5.50}$$

$$\sum_{n=0}^{N-1} h(n)\sin(\alpha - n)\omega = 0 \tag{5.51}$$

The solution of Eq. (5.51) exits when

$$\alpha = \frac{N-1}{2} \quad \text{and} \quad h(n) = h(N-1-n), \quad 0 \le n \le N-1 \tag{5.52}$$

From the condition $\alpha = \frac{N-1}{2}$, we can say that for every value of N there is only one value of phase delay α for which linear phase can be obtained easily.

From the condition $h(n) = h(N-1-n)$, we can say that for this value of α, the $h(n)$ has a special kind of symmetry. The symmetry impulse response is shown in Fig. 5.17.

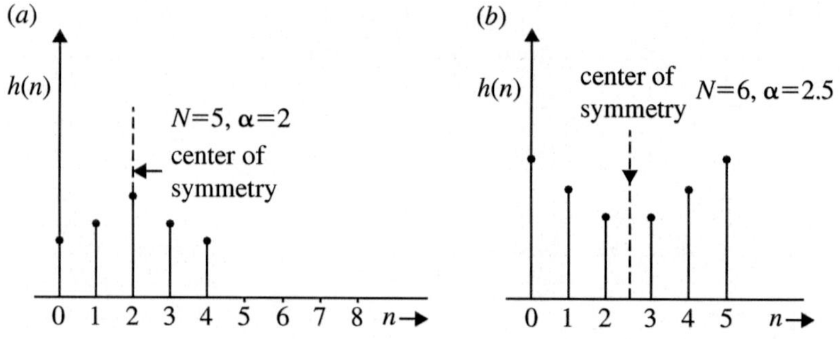

Fig. 5.17 Symmetry impulse response for N = odd and N = even

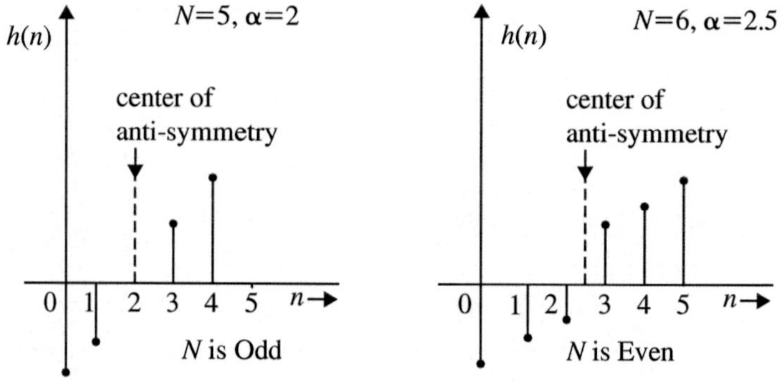

Fig. 5.18 Anti-symmetry impulse response for N = odd and N = even

The definition of linear phase filter $\theta(\omega) = -\omega\alpha$ requires to have both constant group delay and constant phase delay.

If only constant group delay is required, an another type of linear phase filter exists, in which the phase of $H(\omega)$ is a piece-wise linear function of ω. For this case, $H(\omega)$ can be expressed as

$$H(\omega) = \pm|H(\omega)|e^{+j(\beta-\omega\alpha)} \tag{5.53}$$

Solution of Eq. (5.53) exists only if $\alpha = \frac{N-1}{\alpha}$, $\beta = \pm\frac{\pi}{2}$ and

$$h(n) = -h(N-1-n) \quad \text{for} \quad 0 \leq n \leq -1 \tag{5.54}$$

The anti-symmetric impulse response is shown in Fig. 5.18.

5.4.1.1 Phase Delay and Group Delay

The phase delay and group delay are the two important parameters that characterize the frequency response characteristics of a digital filter. Suppose the system is excited by the following input which is sinusoidal of frequency ω_0 and amplitude A.

$$x(n) = A\cos(\omega_0 n + \phi)$$

For the LTI system, the output may be expressed by the following equation:

$$\begin{aligned} y(n) &= A|H(e^{j\omega_0})|\cos\left(\omega_0\left(n + \frac{\theta(\omega_0)}{\omega_0}\right) + \Phi\right) \\ &= A|H(e^{j\omega_0})|\cos(\omega_0(n - \tau_p(\omega_0)) + \Phi) \end{aligned}$$

where

$$\boxed{\tau_p(\omega_0) = \frac{\theta(\omega_0)}{\omega_0}}$$

τ_p is called the phase delay. Now the output is a time delayed version of the input $x(n)$.

When the input signal contains many sinusoidal components with different frequencies which are different from harmonics, each component will go through different phase delays when they are passed through a LTI discrete system and the signal delay now is determined by what is named as group delay denoted by the letter $\tau_g(\omega)$. The group delay is defined as

$$\boxed{\tau_g(\omega) = \frac{d\theta(\omega)}{d\omega}}$$

Example 5.8 Determine the frequency response of FIR filter defined by $y(n) = 0.25x(n) + x(n-1) + 0.25x(n-2)$. Calculate the phase delay and group delay.

Solution Given

$$y(n) = 0.25x(n) + x(n-1) + 0.25x(n-2)$$

Taking Fourier transform on both sides

$$Y(e^{j\omega}) = 0.25X(e^{j\omega}) + e^{-j\omega}X(e^{j\omega}) + 0.25e^{-j2\omega}X(e^{j\omega})$$

$$H(e^{j\omega}) = \frac{Y(e^{j\omega})}{X(e^{j\omega})} = 0.25 + e^{-j\omega} + 0.25e^{-2j\omega}$$
$$= e^{-j\omega}(0.25e^{j\omega} + 1 + 0.25e^{-j\omega})$$
$$= e^{-j\omega}(1 + 0.5\cos\omega)$$
$$H(e^{j\omega}) = e^{-j\omega}\bar{H}(e^{j\omega})$$

We know that

$$H(e^{j\omega}) = e^{j\theta(\omega)}\bar{H}(e^{j\omega})$$

Comparing these two equations, we get

$$\theta(\omega) = -\omega$$

The phase delay

$$\boxed{\tau_p = \frac{-\theta(\omega)}{\omega} = \frac{\omega}{\omega} = 1}$$

The group delay

$$\boxed{-\frac{d\theta(\omega)}{d\omega} = -\frac{d}{d\omega}(-\omega) = 1}$$

5.4.2 *Frequency Response of Linear Phase FIR Filter*

Depending on the value of N and the type of symmetry of the filter impulse response, there are four possible types of linear phase. The following are the four cases of impulse response for the linear phase FIR filters.

Case I. Symmetric impulse response when N is odd.
Case II. Symmetric impulse response when N is even.
Case III. Anti-symmetric impulse response when N is odd.
Case IV. Anti-symmetric impulse response when N is even.

5.4.2.1 Symmetric Impulse Response of the Linear Phase FIR Filters When *N* Is Odd

The Fourier transform of $h(n)$ is

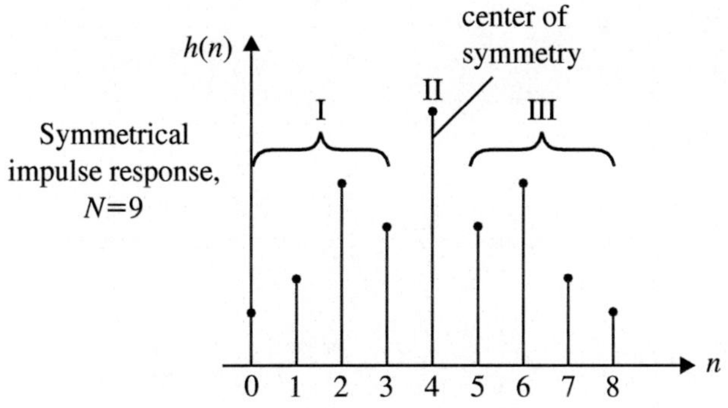

Fig. 5.19 Symmetry impulse response when N = odd

$$H(\omega) = H(e^{j\omega}) = \sum_{n=-\infty}^{\infty} h(n)e^{-j\omega n}$$

Since the impulse response of FIR filter has only N samples, the limits of summation can be changed to $n = 0$ to $N - 1$

$$H(\omega) = \sum_{n=-\infty}^{N-1} h(n)e^{-j\omega n}$$

Let impulse response $h(n)$ be symmetric and it has odd number of samples. Let $N = 9$. The symmetrical impulse response is shown in Fig. 5.19.

When N is an odd number, the symmetrical impulse response will have the centre of symmetry at $n = (N - 1)/2$. Hence, $H(e^{j\omega})$ can be expressed as

$$H(\omega) = \sum_{n=0}^{\frac{N-3}{2}} h(n)e^{-j\omega n} + h\left[\frac{N-1}{2}\right] e^{-j\omega[\frac{N-1}{2}]} + \sum_{n=(N+1)/2}^{N-1} h(n)e^{-j\omega n} \tag{5.55}$$

Let $m = N - 1 - n$. Therefore, $n = N - 1 - m$.

$$\text{When } n = \frac{N+1}{2}; \quad m = N - 1 - \frac{N+1}{2} = \frac{N-3}{2}$$
$$\text{When } n = N - 1; \quad m = N - 1 - (N - 1) = 0$$

$$H(\omega) = \sum_{n=0}^{\frac{N-3}{2}} h(n)e^{-j\omega n} + h\left[\frac{N-1}{2}\right] e^{-j\omega[\frac{N-1}{2}]} + \sum_{m=0}^{\frac{N-3}{2}} h(N-1-m)e^{-j\omega(N-1-m)}$$

Replacing m by n we get

$$H(\omega) = \sum_{n=0}^{\frac{N-3}{2}} h(n)e^{-j\omega n} + h\left[\frac{N-1}{2}\right]e^{-j\omega[\frac{N-1}{2}]} + \sum_{n=0}^{\frac{N-3}{2}} h(N-1-n)e^{-j\omega(N-1-n)}$$

For symmetrical impulse response $h(N-1-n) = h(n)$. Therefore,

$$H(\omega) = \sum_{n=0}^{\frac{N-3}{2}} h(n)e^{-j\omega n} + h\left[\frac{N-1}{2}\right]e^{-j\omega[\frac{N-1}{2}]} + \sum_{n=0}^{\frac{N-3}{2}} h(n)e^{(-j\omega(N-1)+j\omega n)}$$

$$H(\omega) = e^{-j\omega[\frac{N-1}{2}]}\left[h\left[\frac{N-1}{2}\right] + \sum_{n=0}^{\frac{N-3}{2}} h(n)\left[e^{j\omega[\frac{N-1}{2}]-j\omega n} + e^{-j\omega[\frac{N-1}{2}]+j\omega n}\right]\right]$$

$$= e^{-j\omega[\frac{N-1}{2}]}\left[h\left[\frac{N-1}{2}\right] + \sum_{n=0}^{\frac{N-3}{2}} h(n)\left[e^{j\omega[\frac{N-1}{2}-n]} + e^{-j\omega[\frac{N-1}{2}-n]}\right]\right]$$

$$= e^{-j\omega[\frac{N-1}{2}]}\left[h\left[\frac{N-1}{2}\right] + \sum_{n=0}^{\frac{N-3}{2}} 2h(n)\cos\omega\left(\frac{N-1}{2}-n\right)\right]$$

Let $k = \dfrac{N-1}{2} - n; \quad n = \dfrac{N-1}{2} - k$

When $n = 0, k = \dfrac{N-1}{2}$ and

When $n = \dfrac{N-3}{2}, k = 1$

$$= e^{-j\omega[\frac{N-1}{2}]}\left[h\left[\frac{N-1}{2}\right] + \sum_{k=1}^{\frac{N-1}{2}} 2\left(\frac{N-1}{2}-k\right)\cos\omega k\right]$$

Replacing k by n, we get

$$H(\omega) = e^{-j\omega[\frac{N-1}{2}]}\left[h\left[\frac{N-1}{2}\right] + 2\sum_{n=1}^{\frac{N-1}{2}} h\left(\frac{N-1}{2}-n\right)\cos\omega n\right] \quad (5.56)$$

Equation (5.56) is the frequency response of linear phase FIR filter when impulse response is symmetric and N is odd.

Magnitude function of $H(\omega)$ is given by

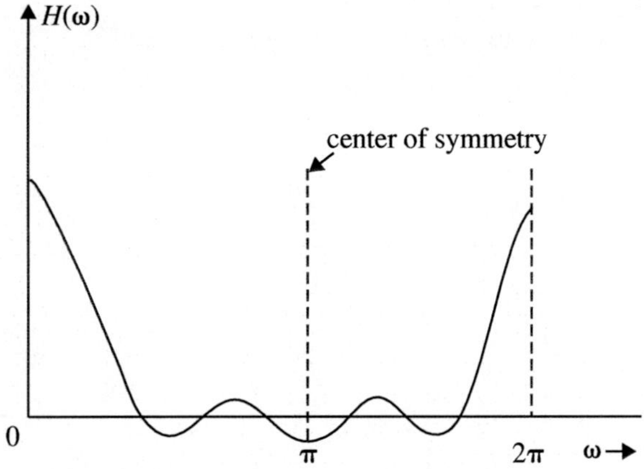

Fig. 5.20 Magnitude response of $H(\omega)$

$$|H(\omega)| = h\left[\frac{N-1}{2}\right] + 2\sum_{n=1}^{\frac{N-1}{2}} h\left(\frac{N-1}{2} - n\right)\cos\omega n \tag{5.57}$$

Phase function of $H(\omega)$ is given by

$$\angle H(\omega) = -\omega\left[\frac{N-1}{2}\right] = -\omega\alpha \tag{5.58}$$

where $\alpha = (N-1)/2$. The magnitude response of $H(\omega)$ is shown in Fig. 5.20.

The magnitude response of $H(\omega)$ is symmetric with $\omega = \pi$ when the impulse response is symmetric and N is odd.

5.4.2.2 Symmetric Impulse Response of the Linear Phase FIR Filters When *N* Is Even

Let $h(n)$ be symmetric and impulse response, for $N = 10$, which is shown in Fig. 5.21.

The Fourier transform of $h(n)$ is

$$H(\omega) = \sum_{n=0}^{N-1} h(n)e^{-j\omega n}$$

$$= \sum_{n=0}^{\frac{N}{2}-1} h(n)e^{-j\omega n} + \sum_{n=\frac{N}{2}}^{N-1} h(n)e^{-j\omega n} \tag{5.59}$$

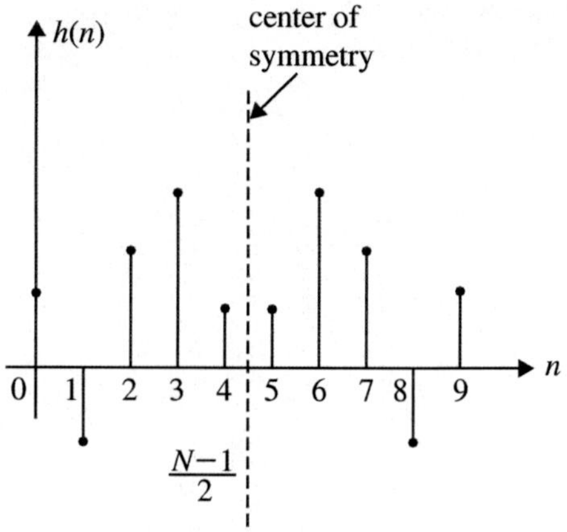

Fig. 5.21 Symmetry impulse response when N = even

Let $m = N - 1 - n \Longrightarrow n = N - 1 - m$

When $n = \frac{n}{2}; \Longrightarrow m = N - 1 - \frac{N}{2} = \frac{N}{2} - 1$

When $n = N - 1; \Longrightarrow m = N - 1 - N + 1 = 0$

$$H(\omega) = \sum_{n=0}^{\frac{N}{2}-1} h(n)e^{-j\omega n} + \sum_{m=0}^{\frac{N}{2}-1} h(N-1-m)e^{-j\omega(N-1-m)}$$

replace m by n and $h(n) = h(N-1-m)$

$$H(\omega) = \sum_{n=0}^{\frac{N}{2}-1} h(n)e^{-j\omega n} + \sum_{n=0}^{\frac{N}{2}-1} h(n)e^{-j\omega(N-1)+j\omega n}$$

$$= e^{-j\omega(\frac{N-1}{2})}\left[\sum_{n=0}^{\frac{N}{2}-1} h(n)\left(e^{j\omega(\frac{N-1}{2})-j\omega n} + e^{-j\omega(\frac{N-1}{2})+j\omega n}\right)\right]$$

$$= e^{-j\omega(\frac{N-1}{2})}\left[\sum_{n=0}^{\frac{N}{2}-1} h(n)\left(e^{j\omega(\frac{N-1}{2}-n)} + e^{-j\omega(\frac{N-1}{2}-n)}\right)\right]$$

$$= e^{-j\omega(\frac{N-1}{2})}\left[\sum_{n=0}^{\frac{N}{2}-1} 2h(n)\cos\left(\frac{N-1}{2} - n\right)\omega\right] \quad \left(\because \cos\theta = \frac{e^{i\theta} + e^{-i\theta}}{2}\right)$$

$$= e^{-j\omega(\frac{N-1}{2})}\left[\sum_{n=0}^{\frac{N}{2}-1} 2h(n)\cos\left(\frac{N}{2} - n - \frac{1}{2}\right)\omega\right]$$

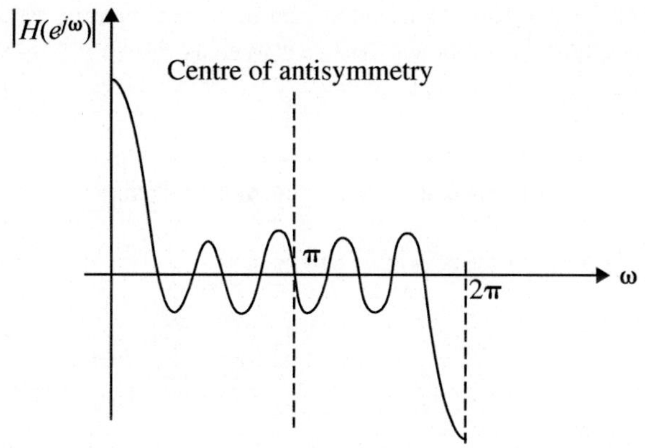

Fig. 5.22 Magnitude function of $H(\omega)$

$$\text{Let } k = \frac{N}{2} - n \Longrightarrow n = \frac{N}{2} - k$$

$$\text{When } n = 0; \Longrightarrow k = \frac{N}{2}$$

$$\text{When } n = \frac{N}{2} - 1; \Longrightarrow k = 1$$

$$H(\omega) = e^{-j\omega\left(\frac{N-1}{2}\right)}\left[\sum_{k=1}^{\frac{N}{2}} 2h\left(\frac{N}{2} - k\right)\cos\omega\left(k - \frac{1}{2}\right)\right]$$

Replacing k by n, we get

$$H(\omega) = e^{-j\omega\left(\frac{N-1}{2}\right)}\left[\sum_{n=1}^{\frac{N}{2}} 2h\left(\frac{N}{2} - n\right)\cos\omega\left(n - \frac{1}{2}\right)\right] \tag{5.60}$$

Equation (5.60) is the frequency response of linear phase FIR filter when impulse response is symmetric and N is even.

Magnitude function of $H(\omega)$ is given by

$$\boxed{|H(\omega)| = \sum_{n=1}^{N/2} 2h\left(\frac{N}{2} - n\right)\cos\left(n - \frac{1}{2}\right)\omega} \tag{5.61}$$

Phase function of $H(\omega)$ is given by

$$\boxed{\angle|H(e^{j\omega}) = -\omega\left[\frac{N-1}{2}\right] = -\omega\alpha \quad \text{where } \alpha = \frac{N-1}{2}}$$

The magnitude function of $H(\omega)$ is anti-symmetric with $\omega = \pi$, when impulse response is symmetric and N is even as shown in Fig. 5.22.

5.4.2.3 Frequency Response of Linear Phase FIR Filters When Impulse Response Is Anti-symmetric When *N* Is Odd

The Fourier transform of $h(n)$ is

$$H(\omega) = \sum_{n=-\infty}^{\infty} h(n)e^{-j\omega n}$$

Since the impulse response has only N samples, the limit of summation can be changed from $N = 0$ to $N - 1$.

$$\therefore\ H(\omega) = \sum_{n=0}^{N-1} h(n)e^{-j\omega n}$$

The impulse response is anti-symmetric at $n = \left(\frac{N-1}{2}\right)$, and $h\left(\frac{N-1}{2}\right) = 0$. Hence, $H(\omega)$ can be expressed as

$$\begin{aligned} H(\omega) &= \sum_{n=0}^{\frac{N-3}{2}} h(n)e^{-j\omega n} + h\left(\frac{N-1}{2}\right)e^{-j\omega\left(\frac{N-1}{2}\right)} + \sum_{n=\frac{N+1}{2}}^{N-1} h(n)e^{-j\omega n} \\ &= \sum_{n=0}^{\frac{N-3}{2}} h(n)e^{-j\omega n} + \sum_{n=\frac{N+1}{2}}^{N-1} h(n)e^{-j\omega n} \end{aligned} \tag{5.62}$$

$$\begin{aligned} &\text{Let } m = N - 1 - n \Longrightarrow n = N - 1 - m \\ &\text{When } n = \frac{N+1}{2}; \Longrightarrow m = N - 1 - \left(\frac{N+1}{2}\right) = \frac{N-3}{2} \\ &\text{When } n = N - 1; \Longrightarrow m = N - 1 - (N - 1) = 0 \end{aligned}$$

Using the above relations, Eq. (5.62) can be written as

$$H(\omega) = \sum_{n=0}^{\frac{N-3}{2}} h(n)e^{-j\omega n} + \sum_{m=0}^{\frac{N-3}{2}} h(N - 1 - m)e^{-j\omega(N-1-m)}$$

On replacing m by n, we get

$$H(\omega) = \sum_{n=0}^{\frac{N-3}{2}} h(n)e^{-j\omega n} + \sum_{n=0}^{\frac{N-3}{2}} h(N-1-m)e^{-j\omega(N-1-n)}$$

For anti-symmetric impulse response, $h(N-1-n) = -h(n)$. Therefore,

$$H(\omega) = \sum_{n=0}^{\frac{N-3}{2}} h(n)e^{-j\omega n} + \sum_{n=0}^{\frac{N-3}{2}} (-h(n))e^{-j\omega(-n)-j\omega(N-1)}$$

$$H(\omega) = \left[\sum_{n=0}^{\frac{N-3}{2}} h(n)\left[e^{-j\omega n+j\omega(\frac{N-1}{2})} - e^{-j\omega(-n)-j\omega(N-1)+j\omega(\frac{N-1}{2})}\right]\right] e^{-j\omega(\frac{N-1}{2})}$$

$$= \left[\sum_{n=0}^{\frac{N-3}{2}} h(n)\left[e^{j\omega(\frac{N-1}{2}-n)} - e^{-j\omega(\frac{N-1}{2}-n)}\right]\right] e^{-j\omega(\frac{N-1}{2})}$$

because

$$\sin\theta = \frac{e^{j\theta - e^{-j\theta}}}{2j}$$

$$H(\omega) = \left[\sum_{n=0}^{\frac{N-3}{2}} h(n)2j\sin\left[\omega\left(\frac{N-1}{2}-n\right)\right]\right] e^{-j\omega(\frac{N-1}{2})}$$

The operator j can be written as $e^{(j\pi)/2}$

$$\therefore\ H(\omega) = \left[\sum_{n=0}^{\frac{N-3}{2}} 2h(n)e^{\frac{j\pi}{2}}\sin\left[\omega\left(\frac{N-1}{2}-n\right)\right]\right] e^{-j\omega(\frac{N-1}{2})}$$

$$= \left[\sum_{n=0}^{\frac{N-3}{2}} 2h(n)\sin\left[\omega\left(\frac{N-1}{2}-n\right)\right]\right] e^{j(\frac{\pi}{2}-\omega(\frac{N-1}{2}))}$$

$$\text{Let } k = \frac{N-1}{2} - n \Longrightarrow n = \frac{N-1}{2} - k$$

$$\text{When } n = 0; \Longrightarrow k = \frac{N-1}{2}$$

$$\text{When } n = \frac{N-3}{2}; \Longrightarrow k = \frac{N-1}{2} - \frac{N-3}{2} = 1$$

$$H(\omega) = \left[\sum_{k=0}^{\frac{N-1}{2}} 2h\left(\frac{N-1}{2} - k\right) \sin \omega k \right] e^{j\left(\frac{\pi}{2} - \omega\left(\frac{N-1}{2}\right)\right)}$$

On replacing k by n, we get

$$H(\omega) = \left[\sum_{k=0}^{\frac{N-1}{2}} 2h\left(\frac{N-1}{2}\right) \sin \omega n \right] e^{j\left(\frac{\pi}{2} - \omega\left(\frac{N-1}{2}\right)\right)} \tag{5.63}$$

Equation (5.63) is the frequency response of linear phase FIR filter when impulse response in anti-symmetric and N is odd. The magnitude function is given by Eq. (5.64). Magnitude function

$$|H(\omega)| = \sum_{n=1}^{\frac{N-1}{2}} 2h\left(\frac{N-1}{2}\right) \sin \omega n \tag{5.64}$$

The phase function is given by Eq. (5.65)

$$\angle H(\omega) = \frac{\pi}{2} - \omega\left(\frac{N-1}{2}\right) = \beta - \alpha\omega \tag{5.65}$$

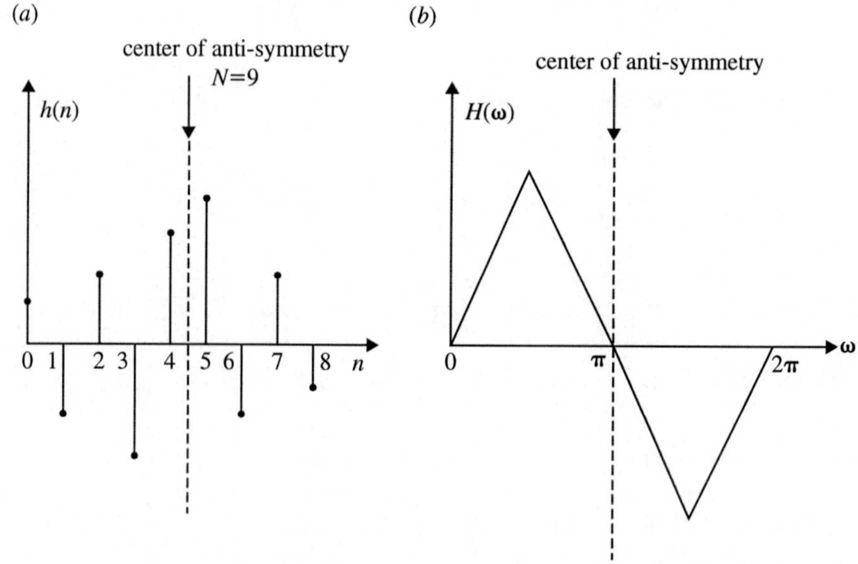

Fig. 5.23 **a** Anti-symmetric impulse response for $N = 9$. **b** Magnitude function of $H(\omega)$

where $\beta = \frac{\pi}{2}$ and $\alpha = \frac{N-1}{2}$.

The sketch of symmetrical impulse response when $N = 8$ and its corresponding magnitude response is shown in Fig. 5.23(a) and (b), respectively.

5.4.2.4 Frequency Response of Linear Phase FIR Filters When Impulse Response Is Anti-symmetric When N Is Even

Let the anti-symmetric impulse response with N as even be as shown in Fig. 5.24.

The Fourier transform of $h(n)$ for impulse response of N samples is given as,

$$H(\omega) = \sum_{n=0}^{N-1} h(n)e^{-j\omega n}$$

The impulse response is asymmetric in between $n = \frac{N}{2} - 1$ and $n = \frac{N}{2}$

$$H(\omega) = \sum_{n=0}^{\frac{N}{2}-1} h(n)e^{-j\omega n} + \sum_{n=\frac{N}{2}}^{N-1} h(n)e^{-j\omega n} \tag{5.66}$$

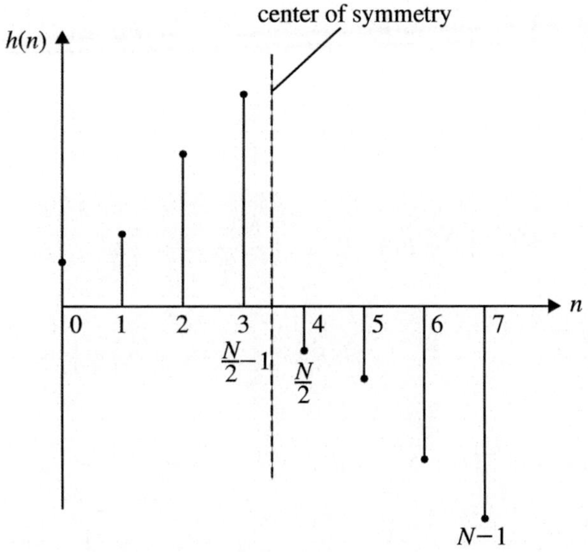

Fig. 5.24 Anti-symmetric impulse response for N = even

Let $m = N - 1 - n \Longrightarrow n = N - 1 - m$
When $n = \frac{N}{2}; \Longrightarrow m = N - 1 - \frac{N}{2} = \frac{N}{2} - 1$
When $n = N - 1; \Longrightarrow m = N - 1 - (N - 1) = 0$

$$H(\omega) = \sum_{n=0}^{\frac{N}{2}-1} h(n)e^{-j\omega n} + \sum_{m=0}^{\frac{N}{2}-1} h(N-1-m)e^{-j\omega(N-1-m)}$$

On replacing m by n, we get

$$H(\omega) = \sum_{n=0}^{\frac{N}{2}-1} h(n)e^{-j\omega n} + \sum_{n=0}^{\frac{N}{2}-1} h(N-1-n)e^{-j\omega(N-1-n)}$$

For anti-symmetric $h(N-1-n) = -h(n)$

$$\begin{aligned}
H(\omega) &= \sum_{n=0}^{\frac{N}{2}-1} h(n)e^{-j\omega n} - \sum_{n=0}^{\frac{N}{2}-1} h(n)e^{-j\omega(N-1)+j\omega n} \\
&= e^{-j\omega\left(\frac{N-1}{2}\right)}\left[\sum_{n=0}^{\frac{N}{2}-1} h(n)e^{-j\omega\left(\frac{N-1}{2}\right)-j\omega n} - \sum_{n=0}^{\frac{N}{2}-1} h(n)e^{-j\omega\left(\frac{N-1}{2}\right)+j\omega n}\right] \\
&= e^{-j\omega\left(\frac{N-1}{2}\right)}\left[\sum_{n=0}^{\frac{N}{2}-1} h(n)\left[e^{j\omega\left(\frac{N-1}{2}-n\right)} - e^{-j\omega\left(\frac{N-1}{2}-n\right)}\right]\right] \\
&= e^{-j\omega\left(\frac{N-1}{2}\right)}\left[\sum_{n=0}^{\frac{N}{2}-1} h(n)\left[2j\sin\left(\frac{N-1}{2}-n\right)\omega\right]\right] \\
&= e^{-j\omega\left(\frac{N-1}{2}\right)}e^{j\pi/2}\left[\sum_{n=0}^{\frac{N}{2}-1} 2h(n)\sin\left(\frac{N-1}{2}-n\right)\omega\right] \\
&= e^{j\left(\frac{\pi}{2}-\omega\left(\frac{N-1}{2}\right)\right)}\left[\sum_{n=0}^{\frac{N}{2}-1} 2h(n)\sin\left(\frac{N-1}{2}-n\right)\omega\right] \\
&= e^{j\left(\frac{\pi}{2}-\omega\left(\frac{N-1}{2}\right)\right)}\left[\sum_{n=0}^{\frac{N}{2}-1} 2h(n)\sin\left(\frac{N}{2}-n-\frac{1}{2}\right)\omega\right]
\end{aligned}$$

Substitute $k = \frac{N}{2} - n \Longrightarrow n = \frac{N}{2} - k$

$$\text{When } n = 0; \Longrightarrow k = \frac{N}{2}$$

$$\text{When } n = \frac{N}{2} - 1; \Longrightarrow k = \frac{N}{2} - \frac{N}{2} + 1 = 1$$

$$H(\omega) = e^{j\left(\frac{\pi}{2} - \omega\left(\frac{N-1}{2}\right)\right)} \left[\sum_{k=1}^{\frac{N}{2}} 2h\left(\frac{N}{2} - k\right) \sin\left(k - \frac{1}{2}\right)\omega \right]$$

Replacing k by n, we get

$$H(\omega) = e^{j\left(\frac{\pi}{2} - \omega\left(\frac{N-1}{2}\right)\right)} \left[\sum_{n=1}^{\frac{N}{2}} 2h\left(\frac{N}{2} - n\right) \sin\left(n - \frac{1}{2}\right)\omega \right] \tag{5.67}$$

Equation (5.67) represents frequency response of linear phase FIR filter with impulse response as anti-symmetric when N is even. The magnitude response is given by Eq. (5.68) and is shown in Fig. 5.25.

$$\boxed{|H(e^{j\omega})| = \sum_{n=1}^{\frac{N}{2}} 2h\left(\frac{N}{2} - n\right) \sin\left(n - \frac{1}{2}\right)\omega} \tag{5.68}$$

The phase response is given by

$$\boxed{\begin{aligned} \angle H(e^{j\omega}) &= \tfrac{\pi}{2} - \omega\left(\tfrac{N-1}{2}\right) \\ \angle H(\omega) &= \beta - \omega\alpha \end{aligned}} \tag{5.69}$$

where $\alpha = \frac{N-1}{2}$ and $\beta = \frac{\pi}{2}$.

5.5 Design Techniques for Linear Phase FIR Filters

Some of the well-known method of design techniques for linear phase FIR filters are as follows:

1. Fourier serious method.
2. Window method.
3. Frequency sampling method.

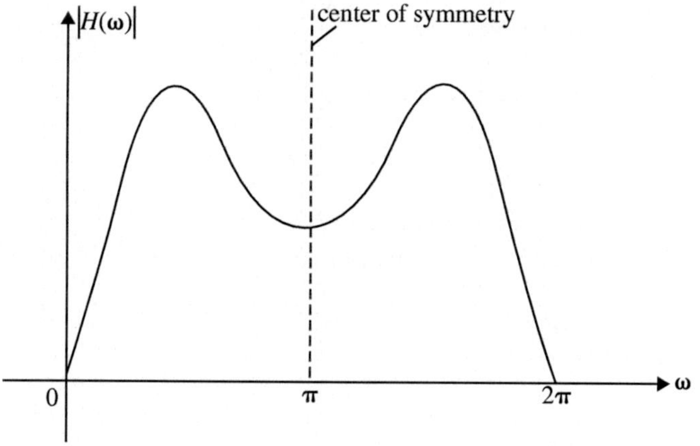

Fig. 5.25 Magnitude function of $H(e^{j\omega})$

5.5.1 Fourier Series Method of FIR Filter Design

Fourier series analysis exists only for periodic function. That is, any periodic function can be expressed as a linear combination of complex exponentials. The frequency response of a digital filter is periodic, with period equal to the sampling frequency. Therefore, the desired frequency response of an FIR digital filter can be represented by Fourier series as

$$H_d(\omega)|_{\omega=\omega T} = H_d(\omega T) = \sum_{n=-\infty}^{\infty} h_d(n)e^{-j\omega nT} \tag{5.70}$$

where the Fourier series coefficients $h_d(n)$ are the desired impulse response and are given by

$$h_d(n) = \frac{1}{\omega_s}\int_{-\omega_s/2}^{\omega_s/2} H_d(\omega T)e^{j\omega nT}\,d\omega \tag{5.71}$$

where

$$f_s = \text{sampling frequency in Hz.}$$
$$\omega_s = 2\pi f_s = \text{sampling frequency in rad/sec.}$$
$$T = \frac{1}{f_s} = \text{sampling period in sec.}$$

The impulse response $h_d(n)$ has infinite number of samples. For FIR filters, we truncate this infinite impulse response to a finite duration sequence of length N, where N is odd.

$$\therefore\ h(n) = h_d(n); \quad |n| \le \left(\frac{N-1}{2}\right) \tag{5.72}$$

On taking z-transform of Eq. (5.72), we get

$$H(z) = \sum_{n=-\frac{(N-1)}{2}}^{\frac{N-1}{2}} h(n)z^{-n} \tag{5.73}$$

The transfer function of Eq. (5.73) represents non-causal filter. Hence, the transfer function is multiplied by $z^{-\frac{(N-1)}{2}}$. This modification does not affect the amplitude response of the filter.

$$H(z) = z^{-\left(\frac{N-1}{2}\right)} \sum_{n=-\frac{(N-1)}{2}}^{\frac{N-1}{2}} h(n)z^{-n} \tag{5.74}$$

$$= z^{-\left(\frac{N-1}{2}\right)} \left[\sum_{n=-\frac{(N-1)}{2}}^{-1} h(n)z^{-n} + h(0) + \sum_{n=1}^{\frac{N-1}{2}} h(n)z^{-n} \right]$$

$$H(z) = z^{-\left(\frac{N-1}{2}\right)} \left[\sum_{n=1}^{\frac{N-1}{2}} h(-n)z^{n} + h(0) + \sum_{n=1}^{\frac{N-1}{2}} h(n)z^{-n} \right]$$

$$H(z) = z^{-\left(\frac{N-1}{2}\right)} \left[h(0) + \sum_{n=1}^{\frac{N-1}{2}} h(n)[z^{n} + z^{-n}] \right] \quad \because h(n) = h(-n) \tag{5.75}$$

The abrupt truncation of the Fourier series results in oscillations in the passband and stopband. These oscillations are due to the slow convergence of the Fourier series at the points of discontinuity. This effect is known as 'Gibbs phenomenon'. This oscillation can be reduced by multiplying the desired impulse response coefficients by an appropriate window function.

The specifications of lowpass, highpass, bandpass and bandstop filters design by Fourier series are given below:

$$\text{Lowpass, } H_d(\omega) = \begin{cases} 1, & -\omega_c \le \omega \le \omega_c \\ 0, & -\frac{\omega_s}{2} \le \omega \le -\omega_c \text{ and } \omega_c < \omega \le \frac{\omega_s}{2} \end{cases} \tag{5.76}$$

$$\text{Highpass, } H_d(\omega) = \begin{cases} 1, & -\frac{\omega_s}{2} \le \omega \le -\omega_c \text{ and } \omega_c \le \omega \le \frac{\omega_s}{2} \\ 0, & -\omega_c \le \omega \le \omega_c \end{cases} \tag{5.77}$$

Bandpass, $H_d(\omega)$

$$= \begin{cases} 1, & -\omega_{c_2} \le \omega \le -\omega_{c_1} \text{ and } \omega_{c_1} \le \omega \le \omega_{c_2} \\ 0, & -\frac{\omega_s}{2} \le \omega_c < -\omega_{c_2},\ -\omega_{c_1} \le \omega \le \omega_{c_1}, \text{ and } \omega_{c_2} < \omega \le \frac{\omega_s}{2} \end{cases} \tag{5.78}$$

Bandstop, $H_d(\omega)$

$$= \begin{cases} 1, & -\frac{\omega_s}{2} \le \omega_c < -\omega_{c_2},\ -\omega_{c_1} \le \omega \le \omega_{c_1}, \text{ and } \omega_{c_2} \le \omega \le \frac{\omega_s}{2} \\ 0, & -\omega_{c_2} < \omega < -\omega_{c_1} \text{ and } \omega_{c_1} < \omega < \omega_{c_2} \end{cases} \tag{5.79}$$

Example 5.9 Design an ideal lowpass filter with a frequency response

$$H_d(e^{j\omega}) = \begin{cases} 1, & \text{for } -\frac{\pi}{2} \le |\omega| \le \frac{\pi}{2} \\ 0, & \text{for } \frac{\pi}{2} \le |\omega| \le \pi \end{cases}$$

Find the values of $h(n)$ for $N = 11$. Find $H(z)$.

Solution $h_d(n)$ is shown in Fig. 5.26a.

$$\begin{aligned} h_d(n) &= \frac{1}{2\pi}\int_{-\pi}^{\pi} H_d(e^{j\omega})e^{j\omega n}d\omega \\ &= \frac{1}{2\pi}\int_{-\pi/2}^{\pi/2} e^{j\omega n}d\omega \\ &= \frac{1}{2\pi jn}[e^{j\omega n}]_{-\pi/2}^{\pi/2} \\ &= \frac{1}{\pi n}\left[\frac{e^{j(\pi/2)n} - e^{-j(\pi/2)n}}{2j}\right] \\ &= \frac{\sin(\pi/2)n}{\pi n}, \qquad -\infty \le n \le 11 \end{aligned}$$

Here $n = 11$

$$\therefore\ h(n) = \begin{cases} \frac{\sin(\pi/2)n}{\pi n}, & \text{for } |n| \le 5 \\ 0, & \text{otherwise} \end{cases}$$

For $n = 0$

$$\begin{aligned} h_d(n) &= \lim_{n\to 0} \frac{\sin(\pi/2)n}{\pi n} \\ &= \frac{1}{2}\lim_{n\to 0}\frac{\sin(\pi/2)n}{(\pi/2)n} \qquad \left[\because \lim_{\theta\to 0}\frac{\sin\theta}{\theta} = 1\right] \\ &= \frac{1}{2} \end{aligned}$$

From the frequency response, $H_d(e^{j\omega})$ we find that $\alpha = 0$, and therefore, filter coefficients are symmetrical about $n = 0$, i.e. $h(n) = h(-n)$.

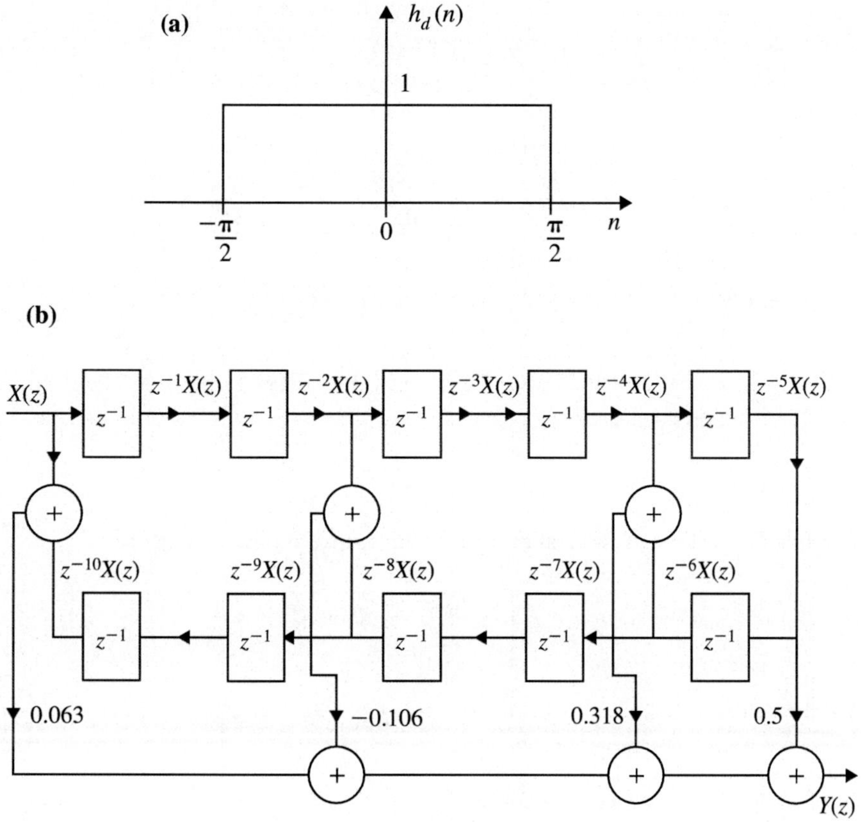

Fig. 5.26 **a** Impulse response of LPF for Example 5.9. **b** Structure realization for Example 5.9

$$\text{For } n = 1: \quad h(1) = h(-1) = \frac{\sin(\pi/2)}{\pi} = \frac{1}{\pi} = 0.3183$$

$$\text{For } n = 2: \quad h(2) = h(-2) = \frac{\sin 2(\pi/2)}{2\pi} = 0$$

$$\text{For } n = 3: \quad h(3) = h(-3) = \frac{\sin 3(\pi/2)}{3\pi} = -0.106$$

$$\text{For } n = 4: \quad h(4) = h(-4) = \frac{\sin 2\pi}{4\pi} = 0$$

$$\text{For } n = 5: \quad h(5) = h(-5) = \frac{\sin 5(\pi/2)}{5\pi}$$
$$= 0.06366$$

$$
\begin{aligned}
H(z) &= z^{-\left(\frac{N-1}{2}\right)}\left[\sum_{n=1}^{\frac{N-1}{2}} h(n)[z^n + z^{-n}] + h(0)\right] \\
&= z^{-5}\left[h(0) + \sum_{n=1}^{5} h(n)[z^n + z^{-n}]\right] \\
&= z^{-5}\left[0.5 + h(1)(z + z^{-1}) + h(2)(z^2 + z^{-2}) + h(3)(z^3 + z^{-3})\right. \\
&\quad \left. + h(4)(z^4 + z^{-4}) + h(5)(z^5 + z^{-5})\right] \\
H(z) &= 0.53z^{-5} + 0.3183(z^{-4} + z^{-6}) + 0 - 0.106(z^{-2} + z^{-8}) \\
&\quad + 0.063(z^{-10} + 1) \\
Y(z) &= 0.5z^{-5}X(z) + 0.3183(z^{-4} + z^{-6})X(z) - 0.106[X(z)z^{-2} + X(z)z^{-8}] \\
&\quad + 0.063[X(z)z^{-10} + X(z)]
\end{aligned}
$$

The structure realization is shown in Fig. 5.26b.

Example 5.10 Design an ideal highpass filter with a frequency response

$$
H_d(e^{j\omega}) = \begin{cases} 1, & \text{for } -\frac{\pi}{4} \le |\omega| \le \pi \\ 0, & \text{for } |\omega| < \frac{\pi}{4} \end{cases}
$$

Find the values of $h(n)$ for $N = 11$. Find $H(z)$. Plot the magnitude response.

Solution $h_d(n)$ is shown in Fig. 5.27a.

$$
\begin{aligned}
h_d(n) &= \frac{1}{2\pi}\int_{-\pi}^{\pi} H_d(e^{j\omega})e^{j\omega n}\,d\omega \\
&= \frac{1}{2\pi}\left[\int_{-\pi}^{-\pi/4} e^{j\omega n}\,d\omega + \int_{\pi/4}^{\pi} e^{j\omega n}\,d\omega\right] \\
&= \frac{1}{2\pi jn}\left[e^{j\omega n}\Big|_{-\pi}^{-\pi/4} + e^{j\omega n}\Big|_{\pi/4}^{\pi}\right] \\
&= \frac{1}{\pi n(2j)}\left[e^{-j(\pi/4)n} - e^{-j\pi n} + e^{j\pi n} - e^{j(\pi/4)n}\right]
\end{aligned}
$$

$$
h_d(n) = \begin{cases} \frac{1}{\pi n}[\sin \pi n - \sin(\pi/4)n], & \text{for } |n| \le 5 \\ 0, & \text{otherwise} \end{cases}
$$

For $n = 0$

$$h(n) = \lim_{n\to 0} \frac{\sin \pi n}{\pi n} - \lim_{n\to 0} \frac{\sin(\pi/4)n}{4(\pi n/4)}$$
$$= 1 - \frac{1}{4} = 0.75$$

$$\text{For } n = 1: \quad h(1) = h(-1) = \frac{\sin \pi - \sin(\pi/4)}{\pi}$$
$$= -0.225$$
$$\text{For } n = 2: \quad h(2) = h(-2) = \frac{\sin 2\pi - \sin(2\pi/4)}{2\pi}$$
$$= -0.159$$
$$\text{For } n = 3: \quad h(3) = h(-3) = \frac{\sin 3\pi - \sin(3\pi/4)}{3\pi}$$
$$= -0.075$$
$$\text{For } n = 4: \quad h(4) = h(-4) = \frac{\sin 4\pi - \sin \pi}{4\pi}$$
$$= 0$$
$$\text{For } n = 5: \quad h(5) = h(-5) = \frac{\sin 5\pi - \sin(5\pi/4)}{5\pi}$$
$$= 0.045$$

$$H(z) = z^{-5}\left[\sum_{n-1}^{N} h(n)(z^n + z^{-n}) + h(0)\right]$$
$$= z^{-5}\left[h(0) + h(1)(z + z^{-1}) + h(2)(z^2 + z^{-2}) + h(3)(z^3 + z^{-3}) + h(4)(z^4 + z^{-4}) + h(5)(z^5 + z^{-5})\right]$$
$$H(z) = 0.75z^{-5} - 0.225(z^{-4} + z^{-6}) - 0.159(z^{-3} + z^{-7}) - 0.075(z^{-2} + z^{-8}) + 0.045(1 + z^{-10})$$
$$Y(z) = 0.75z^{-5}X(z) - 0.225[X(z)z^{-4} + X(z)z^{-6}] - 0.159[X(z)z^{-3} + X(z)z^{-7}] - 0.075[X(z)z^{-2} + X(z)z^{-8}] + 0.045[X(z) + X(z)z^{-10}]$$

The structure realization of FIR highpass filter is shown in Fig. 5.27b.

Magnitude Response

The frequency response of FIR filter when impulse response is symmetric and N is odd.

$$H(e^{j\omega}) = \sum_{n=0}^{\frac{N-1}{2}} a(n) \cos \omega n$$

where

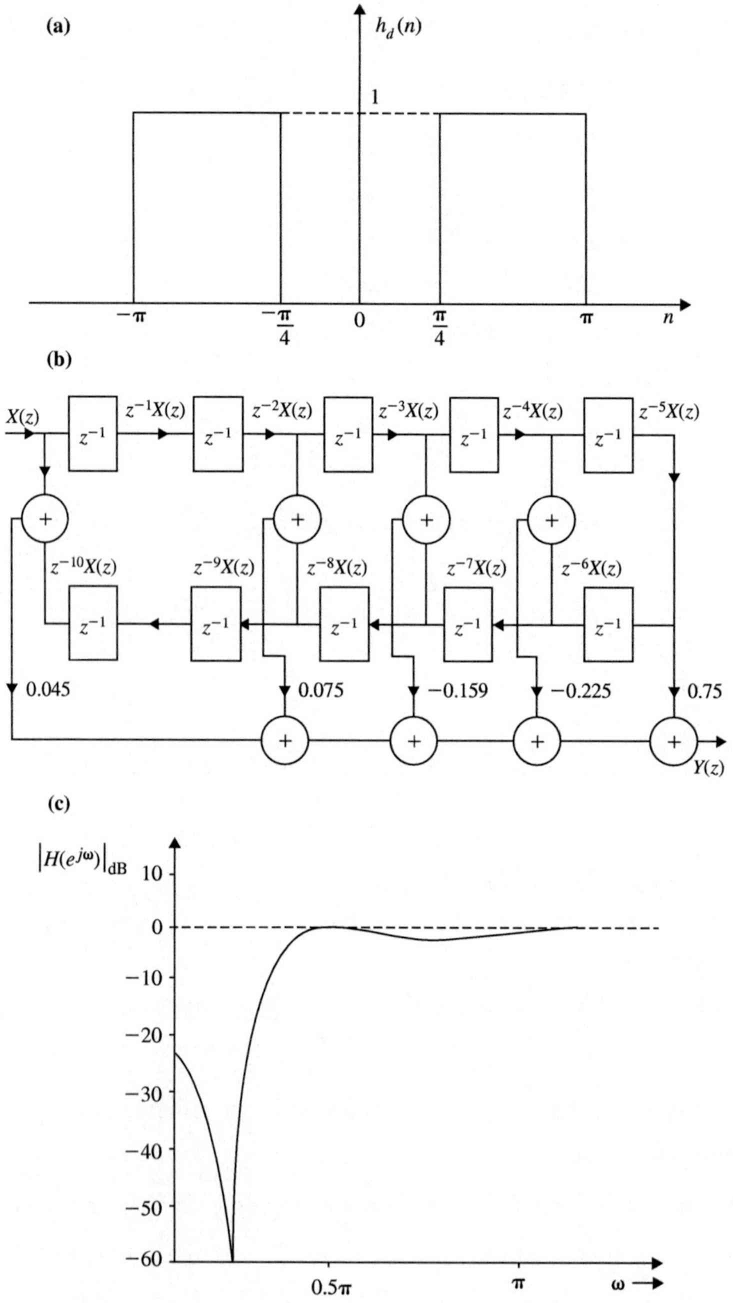

Fig. 5.27 **a** Impulse response for Example 5.10. **b** Structure realization for Example 5.10. **c** Magnitude response for Example 5.10

$$a(0) = h\left(\frac{N-1}{2}\right) = h(5) = 0.75$$
$$a(n) = 2h\left(\frac{N-1}{2} - n\right)$$
$$a(1) = 2h(5-1) = 2h(4) = -0.45$$
$$a(2) = 2h(5-2) = 2h(3) = -0.318$$
$$a(3) = 2h(5-3) = 2h(2) = -0.15$$
$$a(4) = 2h(5-4) = 2h(1) = 0$$
$$a(5) = 2h(5-5) = 2h(0) = 0.09$$

$$\begin{aligned} H(e^{j\omega}) &= a(0) + a(1)\cos\omega + a(2)\cos 2\omega + a(3)\cos 3\omega + a(4)\cos 4\omega \\ &\quad + a(5)\cos 5\omega \\ &= 0.75 - 0.45\cos\omega - 0.318\cos 2\omega - 0.15\cos 3\omega + 0.09\cos 5\omega \end{aligned}$$

The magnitude response is shown in Fig. 5.27c.

ω (deg/sec)	0	10	30	50	80	100	120	140	160	170
$H(e^{j\omega})$	−0.08	−0.066	0.122	0.61	1.11	0.98	0.94	1.26	1.01	0.96
$\lvert H(e^{j\omega})\rvert_{\text{dB}}$	−22	−23.62	−18.2	−4.2	0.95	−0.132	−0.537	2	0.16	−0.31

Example 5.11 Design an ideal bandpass filter with a frequency response

$$H_d(e^{j\omega}) = \begin{cases} 1, & \text{for } \frac{\pi}{4} \le |\omega| \le \frac{3\pi}{4} \\ 0, & \text{otherwise} \end{cases}$$

Find the values of $h(n)$ for $N = 11$. Find $H(z)$ and also plot the frequency response.

Solution The impulse response $h_d(n)$ is shown in Fig. 5.28a.

$$\begin{aligned} h_d(n) &= \frac{1}{2\pi}\int_{-\pi}^{\pi} H_d(e^{j\omega})e^{j\omega n} \\ &= \frac{1}{2\pi}\left[\int_{-(3\pi/4)}^{-\pi/4} e^{j\omega n}d\omega + \int_{\pi/4}^{(3\pi)/4} e^{j\omega n}d\omega\right] \\ &= \frac{1}{2\pi jn}\left[e^{(-j\pi n)/4} - e^{(j3\pi n)/4} + e^{(j3\pi n)/4} - e^{(j\pi n)/4}\right] \end{aligned}$$

$$h_d(n) = \begin{cases} \frac{1}{\pi n}[\sin[(3\pi)/4]n - \sin(\pi/4)n], & \text{for } |n| \le 5 \\ 0, & \text{otherwise} \end{cases}$$

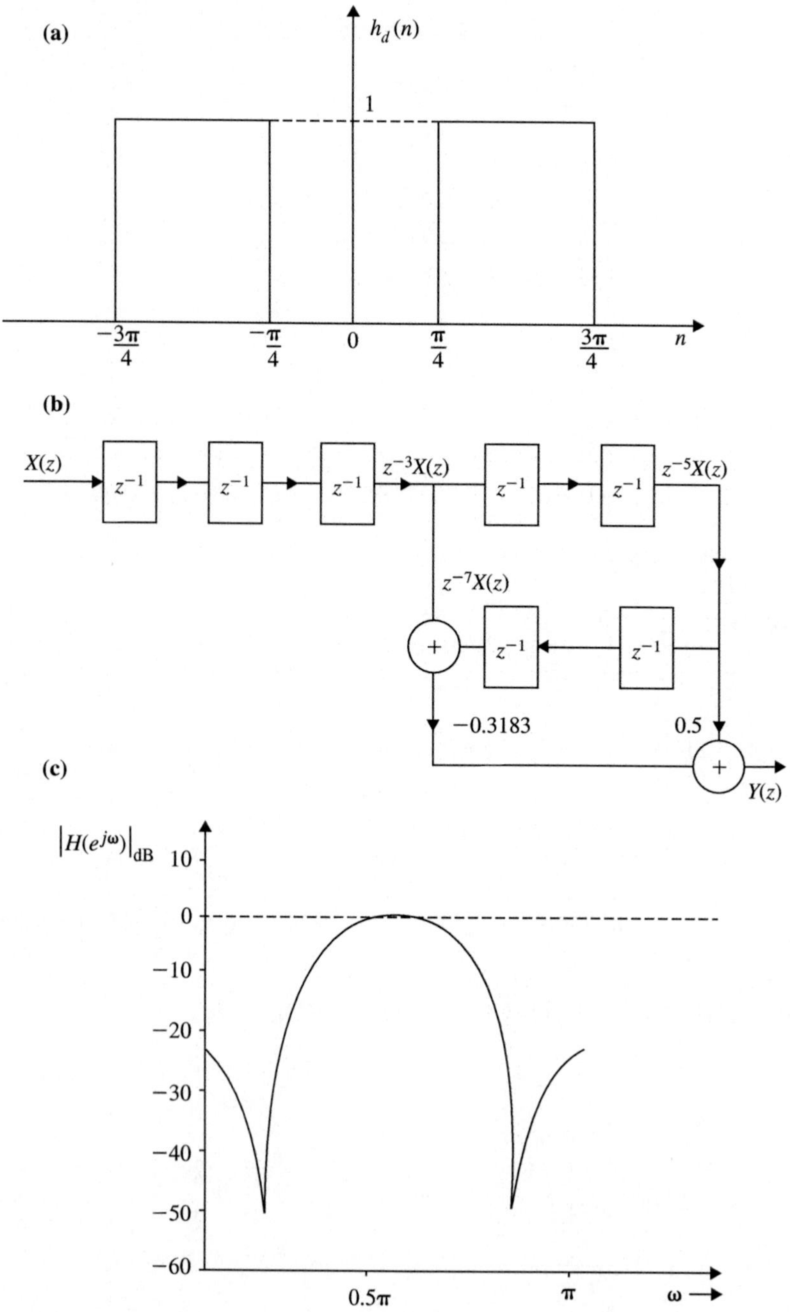

Fig. 5.28 **a** Impulse response for Example 5.11. **b** Structure realization for Example 5.11. **c** Frequency response of bandpass filter of Example 5.11

For $n = 0$

$$h_d(0) = \lim_{n\to 0} \frac{\sin(3\pi/4)n}{\frac{4}{3}\left(\frac{\pi n 3}{4}\right)} - \lim_{n\to 0} \frac{\sin(\pi/4)n}{4\left(\frac{\pi n}{4}\right)}$$
$$= \frac{3}{4} - \frac{1}{4} = 0.5$$

$$\text{For } n = 1: \quad h(1) = h(-1) = \frac{\sin\frac{3\pi}{4} - \sin\frac{\pi}{4}}{\pi} = 0$$
$$\text{For } n = 2: \quad h(2) = h(-2) = \frac{\sin\frac{3\pi}{2} - \sin\frac{\pi}{2}}{2\pi} = -\frac{2}{2\pi} = -0.3183$$
$$\text{For } n = 3: \quad h(3) = h(-3) = \frac{\sin\frac{9\pi}{4} - \sin\frac{3\pi}{4}}{3\pi} = 0$$
$$\text{For } n = 4: \quad h(4) = h(-4) = \frac{\sin 3\pi - \sin\pi}{4\pi} = 0$$
$$\text{For } n = 5: \quad h(5) = h(-5) = \frac{\sin\frac{15\pi}{4} - \sin\frac{5\pi}{4}}{5\pi} = 0$$

$$H(z) = z^{-5}\left[\sum_{n=1}^{5} h(n)(z^n + z^{-n}) + h(0)\right]$$
$$= 0.5z^{-5} - 0.3183(z^{-3} + z^{-7})$$

The filter coefficients of the causal filters are

$$h(0) = h(10) = h(1) = h(9) = h(2) = h(8) = h(4) = h(6) = 0$$
$$h(3) = h(7) = -0.3183$$
$$h(5) = 0.5$$

The filter structure is shown in Fig. 5.28b.

$$|H(e^{j\omega})| = \sum_{n=1}^{\frac{N-1}{2}} 2h\left(\frac{N-1}{2} - n\right)\cos\omega n + h\left(\frac{N-1}{2}\right)$$
$$= \sum_{n=1}^{\frac{N-1}{2}} a(n)\cos\omega n + a(0)$$

where

$$a(0) = h\left(\frac{N-1}{2}\right) = h(5) = 0.5$$

$$a(n) = 2h\left(\frac{N-1}{2} - n\right)$$

$$a(1) = 2h(5-1) = 2h(4) = 0$$

$$a(2) = 2h(5-2) = 2h(3) = -0.6366$$

$$a(3) = 2h(5-3) = 2h(2) = 0$$

$$a(4) = 2h(5-4) = 2h(1) = 0$$

$$a(5) = 2h(5-5) = 2h(0) = 0$$

$$|H(e^{j\omega})| = 0.5 - 0.6366\cos 2\omega$$

ω (deg/sec)	0	20	45	60	90	120	135	180
$H(e^{j\omega})$	−0.1366	0.012	0.5	0.818	1.1366	0.818	0.5	−0.1366
$\lvert H(e^{j\omega})\rvert_{\text{dB}}$	−17.3	−38.17	−6.02	−1.74	1.11	−1.74	−6.02	−17.32

The magnitude response is shown in Fig. 5.28c.

Example 5.12 Design an ideal band reject filter with a desired frequency response

$$H_d(e^{j\omega}) = 1 \quad \text{for } |\omega| \le \frac{\pi}{3} \text{ and } |\omega| \ge \frac{2\pi}{3}$$

Find the values of $h(n)$ for $N = 11$. Find $H(z)$.

Solution The impulse response $h_d(n)$ is shown in Fig. 5.29a.

$$\begin{aligned} h_d(n) &= \frac{1}{2\pi}\int_{-\pi}^{\pi} H_d(e^{j\omega})e^{j\omega n}\,d\omega \\ &= \frac{1}{2\pi}\left[\int_{-\pi}^{-2\pi/3} e^{j\omega n}d\omega + \int_{-(\pi/3)}^{\pi/3} e^{j\omega n}d\omega + \int_{2\pi/3}^{\pi} e^{j\omega n}d\omega\right] \\ &= \frac{1}{2\pi jn}\left[e^{-j2\pi n/3} - e^{-j\pi n} + e^{j\pi n/3} - e^{-j\pi n/3} + e^{j\pi n} - e^{j2\pi n/3}\right] \end{aligned}$$

$$h_d(n) = \begin{cases} \frac{1}{\pi n}\left[\sin \pi n + \sin \frac{\pi}{3}n - \sin \frac{2\pi}{3}n\right], & |n| \le 5 \\ 0, & \text{otherwise} \end{cases}$$

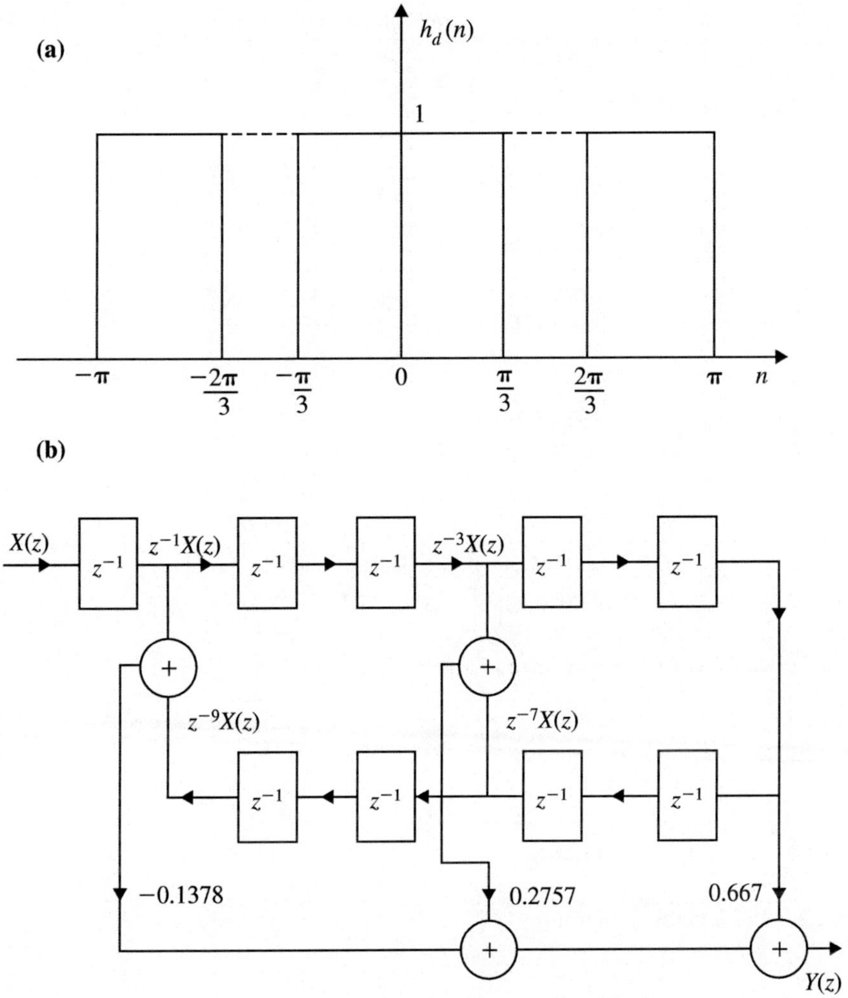

Fig. 5.29 **a** Impulse response for 5.12. **b** Structure realization for Example 5.12

The filter coefficients are symmetrical about $n = 0$ satisfying the condition $h(n) = h(-n)$. For $n = 0$

$$h_d(0) = \lim_{n\to 0}\left[\frac{\sin \pi n}{\pi n} + \frac{\sin(\pi/3)n}{\pi n} - \frac{\sin(2\pi/3)n}{\pi n}\right]$$

$$= \left[1 + \frac{1}{3} - \frac{2}{3}\right] = 0.667$$

$$\text{For } n = 1: \quad h(1) = h(-1) = \frac{\sin \pi + \sin \frac{\pi}{3} - \sin \frac{2\pi}{3}}{\pi} = 0$$

$$\text{For } n = 2: \quad h(2) = h(-2) = \frac{\sin 2\pi + \sin \frac{2\pi}{3} - \sin \frac{4\pi}{3}}{2\pi} = 0.2757$$

$$\text{For } n = 3: \quad h(3) = h(-3) = \frac{\sin 3\pi + \sin \pi - \sin 2\pi}{3\pi} = 0$$

$$\text{For } n = 4: \quad h(4) = h(-4) = \frac{\sin 4\pi + \sin \frac{4\pi}{3} - \sin \frac{8\pi}{3}}{4\pi} = -0.1378$$

$$\text{For } n = 5: \quad h(5) = h(-5) = \frac{\sin 5\pi + \sin \frac{5\pi}{3} - \sin \frac{10\pi}{3}}{5\pi} = 0$$

$$\begin{aligned} H(z) &= z^{-5}\left[\sum_{n=1}^{5} h(n)(z^{n} + z^{-n}) + h(0)\right] \\ &= 0.667z^{-5} + 0.2757(z^{-3} + z^{-7}) - 0.1378(z^{-1} + z^{-9}) \\ Y(z) &= 0.667z^{-5}X(z) + 0.2757[X(z)z^{-3} + X(z)z^{-7}] \\ &\quad -0.1378[X(z)z^{-1} + X(z)z^{-9}] \end{aligned}$$

The structure realization is shown in Fig. 5.29b.

5.5.2 *Window Method*

5.5.2.1 Rectangular Window

Rectangular window function,

$$W_R(n) = \begin{cases} 1; & 0 \le n \le N-1 \\ 0; & \text{otherwise} \end{cases} \tag{5.80}$$

(or)

$$W_R(n) = \begin{cases} 1; & -\left(\frac{N-1}{2}\right) \le n \le \frac{N-1}{2} \\ 0; & \text{otherwise} \end{cases} \tag{5.81}$$

The spectrum of the Rectangular window is given by

$$
\begin{aligned}
W_R(\omega) &= \sum_{n=-((N-1)/2)}^{\frac{N-1}{2}} e^{-j\omega n} = e^{j\omega\frac{(N-1)}{2}} + \cdots + e^{j\omega} + 1 + e^{-j\omega} \\
&+ e^{-2j\omega} + \cdots + e^{-j\omega\left(\frac{N-1}{2}\right)} \qquad (5.82)\\
&= e^{j\omega\frac{(N-1)}{2}}\left[1 + e^{-j\omega} + \cdots + e^{-j\omega(N-1)}\right] \\
&= e^{j\omega\frac{(N-1)}{2}}\left[\frac{1-e^{-j\omega N}}{1-e^{-j\omega}}\right] \quad \left(\because\ 1 + x + x^2 + \cdots + x^{N-1} = \frac{1-x^N}{1-x}\right)
\end{aligned}
$$

$$
\begin{aligned}
&= \frac{e^{\frac{j\omega N}{2}}\left(1-e^{-j\omega N}\right)}{e^{\frac{j\omega}{2}}\left(1-e^{-j\omega}\right)} = \frac{e^{\frac{j\omega N}{2}} - e^{-\frac{j\omega N}{2}}}{e^{\frac{j\omega}{2}} - e^{-\frac{j\omega}{2}}} \\
&= \frac{\sin(\omega N/2)}{\sin(\omega/2)} \qquad (5.83)
\end{aligned}
$$

The window spectrum for $N = 31$ is shown in Fig. 5.30. The spectrum of $W_R(\omega)$ has two important features and they are as follows:

1. Width of mainlobe $= \frac{4\pi}{N}$
2. Peak sidelobe magnitude (dB) $= -13$ dB

The magnitude response $|H(\omega)|$ of the lowpass filter designed using Rectangular window is shown in Fig. 5.30. The approximated filter response differs from the ideal desired response, i.e. in the passband a series of overshoots and undershoots occur. In the stopband, the FIR filter has a non-zero response. This can be explained in terms of the features of the window spectrum.

The width of the transition region is related to the width of the mainlobe of $W_R(\omega)$. Since the mainlobe width of $W_R(\omega)$ is equal to $\frac{4\pi}{N}$, the size of this transition region can be reduced to any desired size by increasing the size N of the window sequence. The increase in N also increases the number of computations necessary to implement the FIR filter.

In the passband, the sidelobe effect of the $W_R(\omega)$ appears as both overshoots and undershoots in the desired response. In the stopband, these effects appear as a non-zero response. These sidelobe effects do not diminish significantly, but remain constant as the duration of Rectangular window is increased.

It is observed that whatever be the number of elements of $h_d(n)$ included in the $h(n)$, the magnitudes of the overshoot and leakage will not change significantly. This result is known as Gibb's phenomenon. To reduce these sidelobe effects, we must consider alternate window sequences having spectrum exhibiting smaller sidelobes. We can observe that the sidelobes of the window spectrum represent the contribution of the high frequency components. For the Rectangular window, these high frequency components are due to the sharp transitions from 0 to 1 and 1 to 0 at the edges of window sequence. Hence, the amplitude of these high frequency components, i.e. sidelobe levels can be reduced by replacing these sharp transitions by more gradually ones. So we go for Raised Cosine window sequences.

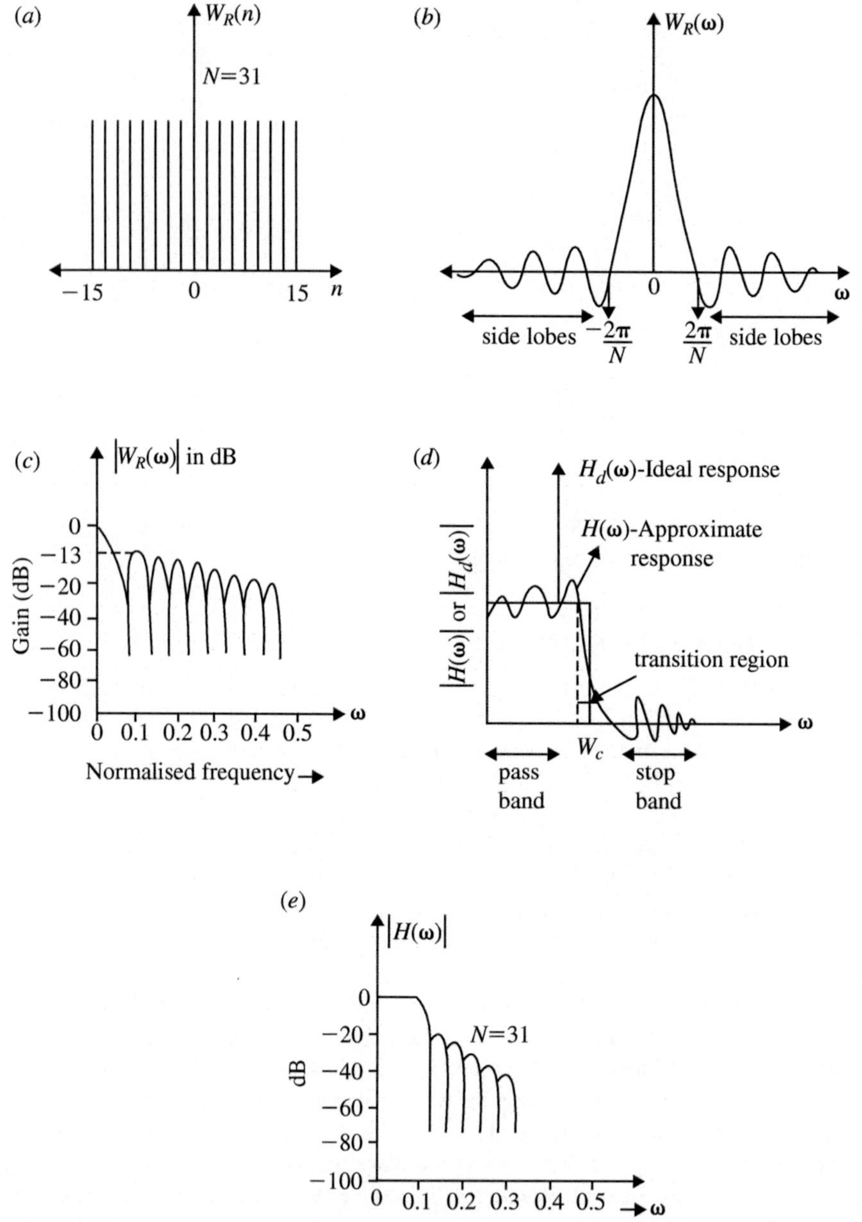

Fig. 5.30 Rectangular window sequence and its frequency response. **a** Rectangular window sequence; **b** Magnitude response of Rectangular window; **c** Log-magnitude response of Rectangular window; **d** Magnitude response of LPF approximated using Rectangular window and **e** Log-magnitude response of FIR LPF designed using Rectangular window

Design Procedure

Step 1: Choose the desired frequency response of the filter, $H_d(\omega)$.

Step 2: Take inverse Fourier transform of $H_d(\omega)$ to obtain the desired impulse response, $h_d(n)$.

Step 3: Choose the window sequence $W(n)$ and multiply $h_d(n)$ by $W(n)$ [i.e. $h(n) = W(n) \times h_d(n)$] to convert the infinite impulse response to finite impulse response.

$$W(n) = \begin{cases} 1, & 0 \leq n \leq N-1 \\ 0, & \text{otherwise} \end{cases}$$

Step 4: The transfer function of the filter $H(z)$ is obtained by taking z-transform of $h(n)$.

Step 5: Realize the filter using suitable realization method (either direct form (or) linear phase realization).

Example 5.13 Design a LPF using Rectangular window by taking 9 samples of $W(n)$ and with cut-off frequency of 1.2 rad/s.?

Solution Step 1: Choose the Desired Frequency Response

$$H_d(\omega) = \begin{cases} e^{-j\omega\alpha}, & -\omega_c \leq \omega \leq \omega_c \quad \text{[for LPF]} \\ 0, & -\pi \leq \omega \leq -\omega_c \quad \text{and} \quad \omega_c \leq \omega \leq \pi \end{cases}$$

Step 2: To Find $H_d(n)$

Take inverse Fourier transform of $H_d(\omega)$. That is,

$$\begin{aligned} h_d(n) &= \frac{1}{2\pi}\int_{-\pi}^{\pi} H_d(\omega)e^{j\omega n}d\omega \\ &= \frac{1}{2\pi}\int_{-\omega_c}^{\omega_c} e^{-j\omega\alpha}e^{j\omega n}d\omega \\ &= \frac{1}{2\pi}\int_{-\omega_c}^{\omega_c} e^{j\omega(n-\alpha)}d\omega \\ &= \frac{1}{2\pi}\left[\frac{e^{j\omega(n-\alpha)}}{j(n-\alpha)}\right]_{-\omega_c}^{\omega_c} \\ &= \frac{1}{2j\pi(n-\alpha)}\left[e^{j\omega_c(n-\alpha)} - e^{-j\omega_c(n-\alpha)}\right] \\ &= \frac{1}{\pi(n-\alpha)}[\sin\omega_c(n-\alpha)] \\ \therefore\ h_d(n) &= \frac{1}{\pi(n-\alpha)}[\sin\omega_c(n-\alpha)] \end{aligned}$$

Step 3: Conversion of Infinite to Finite Sequence

$$h(n) = W(n) \times h_d(n)$$

Here, for Rectangular window,

$$W(n) = \begin{cases} 1, & \text{for } 0 \le n \le (N-1) \\ 0, & \text{for otherwise} \end{cases}$$

$$\therefore h(n) = \begin{cases} \frac{\sin \omega_c(n-\alpha)}{\pi(n-\alpha)}; & 0 \le n \le (N-1) \\ 0, & \text{otherwise} \end{cases}$$

where

$$\omega_c = 1.2 \text{ rad/sec.},$$
$$n = 0 \text{ to } 8 \quad [\because 9 \text{ samples}]$$
$$\alpha = \frac{N+1}{2} = \frac{9-1}{2} = 4$$

Therefore,

$$h(0) = \frac{\sin 1.2(0-4)}{\pi(0-4)} = -0.079$$
$$h(1) = \frac{\sin 1.2(1-4)}{\pi(1-4)} = -0.0469$$
$$h(2) = \frac{\sin 1.2(2-4)}{\pi(2-4)} = 0.107$$
$$h(3) = \frac{\sin 1.2(3-4)}{\pi(3-4)} = 0.296$$
$$h(4) = \frac{\sin 1.2(4-4)}{\pi(4-4)} = 0$$

Apply L' Hospital's rule $\lim\limits_{\theta \to 0} \frac{\sin A\theta}{\theta} = \lim\limits_{\theta \to 0} \frac{\cos A\theta \cdot A}{A\theta} = A$

$$\therefore h(4) = \frac{\sin 1.2(n-4)}{\pi(n-4)} = \frac{1.2}{\pi} = 0.3819$$
$$h(5) = \frac{\sin 1.2(5-4)}{\pi(5-4)} = 0.296$$
$$h(6) = \frac{\sin 1.2(6-4)}{\pi(6-4)} = 0.107$$
$$h(7) = \frac{\sin 1.2(7-4)}{\pi(7-4)} = -0.0469$$
$$h(8) = \frac{\sin 1.2(8-4)}{\pi(8-4)} = -0.079.$$

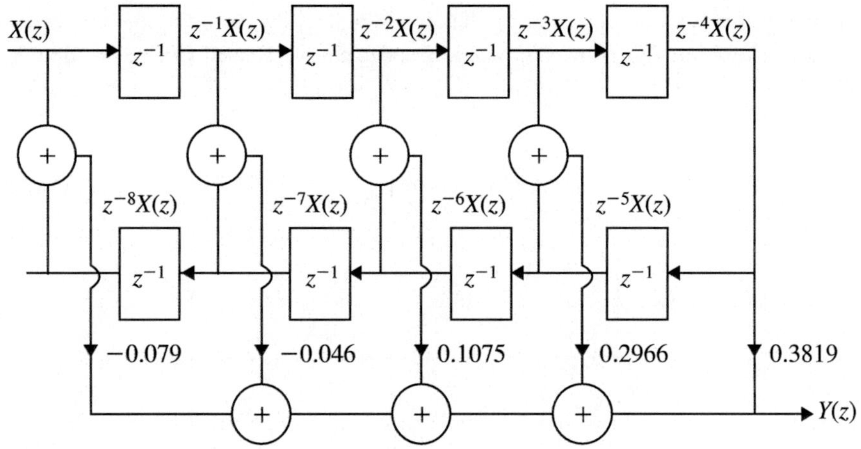

Fig. 5.31 Linear phase realization of $H(z)$ for Example 5.13

Step 4: To Find $H(z)$

$$
\begin{aligned}
H(z) &= \sum_{n=0}^{N-1} h(n)z^{-n} \\
&= h(0) + h(1)z^{-1} + h(2)z^{-2} + h(3)z^{-3} + h(4)z^{-4} + h(5)z^{-5} \\
&\quad + h(6)z^{-6} + h(7)z^{-7} + h(8)z^{-8} \\
&= -0.079 - 0.0469z^{-1} + 0.107z^{-2} + 0.296z^{-3} + 0.3819z^{-4} \\
&\quad + 0.296z^{-5} + 0.107z^{-6} - 0.0469z^{-7} - 0.079z^{-8} \\
&= -0.079(1 + z^{-8}) - 0.0469(z^{-1} + z^{-7}) \\
&\quad + 0.107(z^{-2} + z^{-6}) + 0.296(z^{-3} + z^{-5}) + 0.3819z^{-4}
\end{aligned}
$$

Step 5: Linear Phase Realization

The structure realization is shown in Fig. 5.31.

5.5.2.2 Raised Cosine Windows

The raised cosine windows are smoother at the ends, but closer to one at the middle. The smoother ends and the broader middle section produce less distortion of $h_d(n)$ around $n = 0$.

The window function is of the form

$$
W_\alpha(n) = \begin{cases} \alpha + (1-\alpha)\cos\left(\frac{2\pi n}{N-1}\right), & -\left(\frac{N-1}{2}\right) \le n \le \frac{N-1}{2} \\ 0, & \text{otherwise} \end{cases} \tag{5.84}
$$

If $\alpha = 0.5$, the window is called as Hanning Window. If $\alpha = 0.54$, the window is called as Hamming Window. The window spectrum for Raised Cosine window is given by

$$W_\alpha(\omega) = \sum_{n=-\left(\frac{N-1}{2}\right)}^{\frac{N-1}{2}} \left[\alpha + (1-\alpha)\cos\left(\frac{2\pi n}{N-1}\right)\right] e^{-j\omega n} \tag{5.85}$$

$$= \underbrace{\alpha \sum_{n=-\left(\frac{N-1}{2}\right)}^{\frac{N-1}{2}} e^{-j\omega n}}_{X} + \underbrace{\left(\frac{1-\alpha}{2}\right) \sum_{n=-\left(\frac{N-1}{2}\right)}^{\frac{N-1}{2}} e^{-j\left(\omega-\frac{2\pi}{N-1}\right)n}}_{Y}$$

$$+ \underbrace{\left(\frac{1-\alpha}{2}\right) \sum_{n=-\left(\frac{N-1}{2}\right)}^{\frac{N-1}{2}} e^{-j\left(\omega+\frac{2\pi}{N-1}\right)n}}_{Z} \tag{5.86}$$

$$X = \alpha\left[e^{j\omega\left(\frac{N-1}{2}\right)} + \cdots + e^{j\omega} + 1 + e^{-j\omega} + \cdots + e^{-j\omega\left(\frac{N-1}{2}\right)}\right]$$

$$= \alpha\left[\frac{1-e^{-j\omega N}}{1-e^{-j\omega}}\right] e^{j\omega\left(\frac{N-1}{2}\right)}$$

$$= \alpha\left[\frac{e^{\frac{j\omega N}{2}} - e^{-\frac{j\omega N}{2}}}{e^{\frac{j\omega}{2}} - e^{-\frac{j\omega}{2}}}\right]$$

$$X = \alpha\left[\frac{\sin\frac{\omega N}{2}}{\sin\frac{\omega}{2}}\right] \tag{5.87}$$

$$Y = \frac{1-\alpha}{2}\left[e^{j\left(\omega-\frac{2\pi}{N-1}\right)}\left(\frac{N-1}{2}\right) + \cdots + e^{j\left(\omega-\frac{2\pi}{N-1}\right)}\right.$$

$$\left. +1 + e^{-j\left(\omega-\frac{2\pi}{N-1}\right)} + \cdots + e^{-j\left(\omega-\frac{2\pi}{N-1}\right)}\left(\frac{N-1}{2}\right)\right]$$

$$= \frac{1-\alpha}{2}\left[\frac{e^{j\left(\omega-\frac{2\pi}{N-1}\right)}\left(\frac{N-1}{2}\right)\left[1-e^{-j\left(\omega-\frac{2\pi}{N-1}\right)N}\right]}{e^{j\left(\frac{\omega}{2}-\frac{\pi}{N-1}\right)}\left[1-e^{-j\left(\omega-\frac{2\pi}{N-1}\right)}\right]}\right]$$

$$= \frac{1-\alpha}{2}\left[\frac{e^{j\left(\frac{\omega N}{2}-\frac{\pi N}{N-1}\right)} - e^{-j\left(\frac{\omega N}{2}-\frac{\pi N}{N-1}\right)}}{e^{j\left(\frac{\omega N}{2}-\frac{\pi}{N-1}\right)} - e^{-j\left(\frac{\omega}{2}-\frac{\pi}{N-1}\right)}}\right]$$

$$= \frac{1-\alpha}{2}\left[\frac{\sin\left(\frac{\omega N}{2}-\frac{\pi N}{N-1}\right)}{\sin\left(\frac{\omega}{2}-\frac{\pi}{N-1}\right)}\right] \tag{5.88}$$

Similarly

$$z = \frac{1-\alpha}{2}\left[\frac{\sin\left(\frac{\omega N}{2}+\frac{\pi N}{N-1}\right)}{\sin\left(\frac{\omega}{2}+\frac{\pi}{N-1}\right)}\right] \tag{5.89}$$

Substituting Eqs. (5.87)–(5.89) in Eq. (5.85), we get

$$\boxed{\begin{aligned} W_\alpha(\omega) = \alpha\frac{\sin\frac{\omega N}{2}}{\sin\frac{\omega}{2}} + \frac{1-\alpha}{2}\left[\frac{\sin\left(\frac{\omega N}{2}-\frac{\pi N}{N-1}\right)}{\sin\left(\frac{\omega}{2}-\frac{\pi}{N-1}\right)}\right] \\ +\frac{1-\alpha}{2}\left[\frac{\sin\left(\frac{\omega N}{2}+\frac{\pi N}{N-1}\right)}{\sin\left(\frac{\omega}{2}+\frac{\pi}{N-1}\right)}\right] \end{aligned}} \tag{5.90}$$

5.5.2.3 Hanning Window

The Hanning window sequence can be obtained by substituting $\alpha = 0.5$ in Eq. (5.84)

$$W_C(n) = \begin{cases} 0.5 + 0.5\cos\left(\frac{2\pi n}{N-1}\right); & -\left(\frac{N-1}{2}\right) \le n \le \frac{N-1}{2} \\ 0; & \text{otherwise} \end{cases} \tag{5.91}$$

(or)

$$W_C(n) = \begin{cases} 0.5 - 0.5\cos\frac{2n\pi}{N-1}; & 0 \le n \le N-1 \\ 0; & \text{otherwise} \end{cases} \tag{5.92}$$

Frequency response of the Hanning window is given by

$$\boxed{W_C(\omega) = 0.5\frac{\sin\frac{\omega N}{2}}{\sin(\omega/2)} + 0.25\frac{\sin\left(\frac{\omega N}{2}-\frac{\pi N}{N-1}\right)}{\sin\left(\frac{\omega}{2}-\frac{\pi}{N-1}\right)} + 0.25\frac{\sin\left(\frac{\omega N}{2}+\frac{\pi N}{N-1}\right)}{\sin\left(\frac{\omega}{2}+\frac{\pi}{N-1}\right)}}$$

$$\begin{aligned} \text{Width of mainlobe} &= \frac{8\pi}{N} \\ \text{Peak sidelobe magnitude (dB)} &= -31\text{ dB} \end{aligned} \tag{5.93}$$

In the log-magnitude response of $W_C(\omega)$, the magnitude of the first sidelobe is -31dB, an improvement of 6dB over triangular window. Mainlobe width is same as in triangular window. But the magnitude of sidelobes is reduced, as shown in Fig. 5.32. Hence, the Hanning window is preferable to triangular window since the magnitude responses of lowpass filter have improved stopband characteristics.

Design Procedure

Step 1: Choose the desired frequency response of the filter $[H_d(\omega)]$.
Step 2: Take inverse Fourier transform of $H_d(\omega)$ to obtain the desired impulse response $h_d(n)$.
Step 3: Choose a window sequence $W(n)$.

$$W(n) = \begin{cases} 0.5 - 0.5\cos\left(\frac{2\pi n}{N-1}\right), & 0 \le n \le N-1 \\ 0, & \text{otherwise} \end{cases}$$

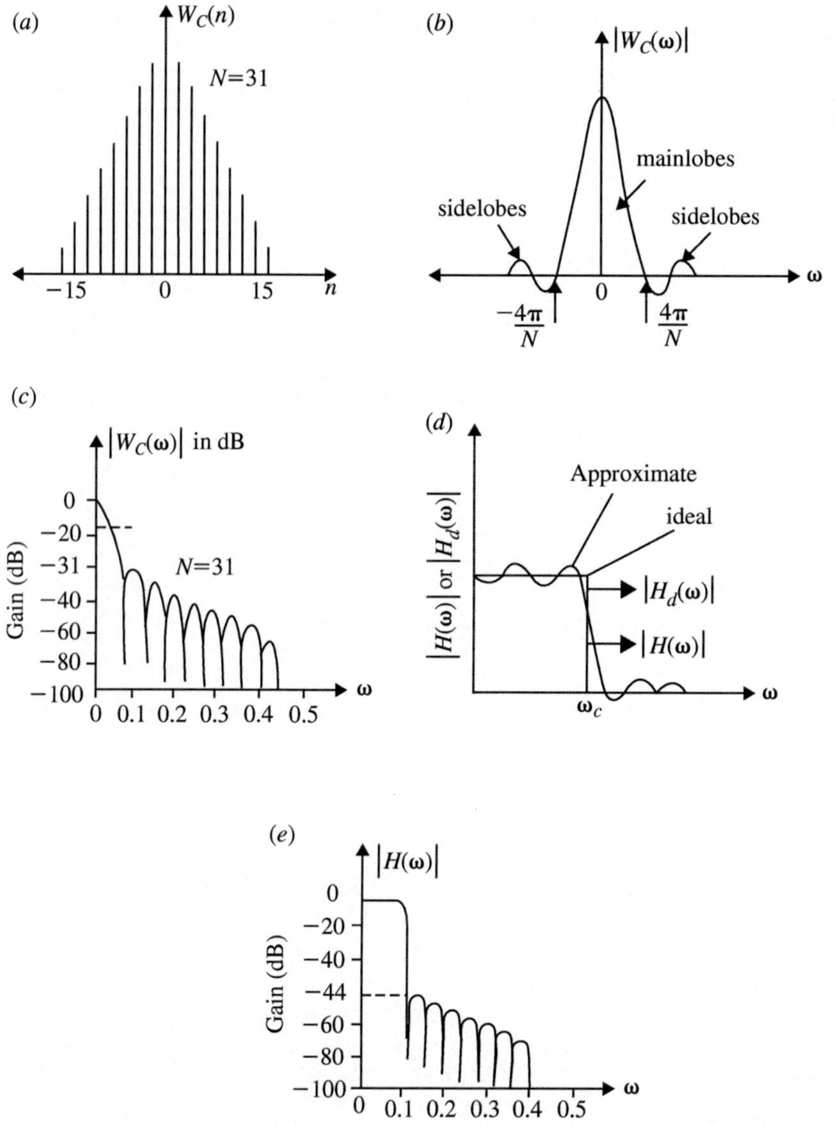

Fig. 5.32 Hanning window sequence and its frequency response. **a** Rectangular window sequence; **b** Magnitude response of Hanning window; **c** Log-magnitude response of Hanning window; **d** Magnitude response of LPF approximated using Hanning window and **e** Log-magnitude response of FIR filter using Hanning window

Multiply $h_d(n)$ by $W(n)$ to convert the infinite impulse response to a finite impulse response $h(n)$. That is

$$h(n) = h_d(n) \times W(n)$$

Step 4: The transfer function of the filter, i.e. $H(z)$ is obtained by taking z-transform of $h(n)$.

Step 5: Realize the filter using suitable realization method (use Direct Form or Linear Phase).

Example 5.14 Design a BPF to pass frequencies in the range 1–2 rad/s. using Hanning window, with $N = 5$.

Solution Step 1: Choose $H_d(\omega)$ such that

$$H_d(\omega) = \begin{cases} e^{-j\omega\alpha}, & -\omega_{c_2} \le \omega \le -\omega_{c_1} \text{ and } \omega_{c_1} \le \omega \le \omega_{c_2} \\ 0, & \text{otherwise} \end{cases}$$

Step 2: To Find $h_d(n)$

$$\begin{aligned}
h_d(n) &= \frac{1}{2\pi}\left[\int_{-\omega_{c_2}}^{-\omega_{c_1}} e^{-j\omega\alpha} e^{j\omega n} d\omega + \int_{\omega_{c_1}}^{\omega_{c_2}} e^{-j\omega\alpha} e^{j\omega n} d\omega\right] \\
&= \frac{1}{2\pi}\left[\int_{-\omega_{c_2}}^{-\omega_{c_1}} e^{j\omega(n-\alpha)} d\omega + \int_{\omega_{c_1}}^{\omega_{c_2}} e^{j\omega(n-\alpha)} d\omega\right] \\
&= \frac{1}{2\pi}\left[\frac{e^{j\omega(n-\alpha)}}{j(n-\alpha)}\right]_{-\omega_{c_2}}^{-\omega_{c_1}} + \frac{1}{2\pi}\left[\frac{e^{j\omega(n-\alpha)}}{j(n-\alpha)}\right]_{\omega_{c_1}}^{\omega_{c_2}} \\
&= \frac{1}{2\pi j(n-\alpha)}\left[e^{-j\omega_{c_1}(n-\alpha)} - e^{-j\omega_{c_2}(n-\alpha)} + e^{j\omega_{c_2}(n-\alpha)} - e^{j\omega_{c_1}(n-\alpha)}\right] \\
&= \frac{1}{2\pi j(n-\alpha)}\left[\left(e^{j\omega_{c_2}(n-\alpha)} - e^{-j\omega_{c_2}(n-\alpha)}\right) - \left(e^{j\omega_{c_1}(n-\alpha)} - e^{-j\omega_{c_1}(n-\alpha)}\right)\right] \\
&= \frac{1}{2\pi j(n-\alpha)} \times 2j\left[\sin\omega_{c_2}(n-\alpha) - \sin\omega_{c_1}(n-\alpha)\right] \\
&= \frac{1}{\pi(n-\alpha)}\left[\sin\omega_{c_2}(n-\alpha) - \sin\omega_{c_1}(n-\alpha)\right]
\end{aligned}$$

Given that $\omega_{c_1} = 1$ rad/sec. and $\omega_{c_2} = 2$ rad/sec., $\alpha = \frac{N-1}{2} \longrightarrow \alpha = 2$.

Step 3: To Find $h(n)$

$$
\begin{aligned}
W(n) &= 0.5 - 0.5\cos\left(\frac{2\pi n}{N-1}\right)\\
h(n) &= h_d(n) \times W(n)\\
&= \frac{1}{\pi(n-2)}[\sin 2(n-2) - \sin(n-2)]\left[0.5 - 0.5\cos\left(\frac{2\pi n}{4}\right)\right]
\end{aligned}
$$

$$
\begin{aligned}
h(0) &= 0\\
h(1) &= 0.0108\\
h(2) &= 0.318 \quad \text{[Applying L'Hospital's rule]}\\
h(3) &= 0.0108\\
h(4) &= 0
\end{aligned}
$$

Step 4: Take z-transform

$$H(z) = \sum_{n=0}^{4} h(n)z^{-n} = h(0) + h(1)z^{-1} + h(2)z^{-2} + h(3)z^{-3} + h(4)z^{-4}$$

$$\boxed{H(z) = 0.0108[z^{-1} + z^{-3}] + 0.318z^{-2}}$$

Step 5: Draw the realization structure. This is shown in Fig. 5.33.

5.5.2.4 Hamming Window

The Hamming window sequence can be obtained by substituting $\alpha = 0.54$ in Eq. (5.84)

$$W_H(n) = \begin{cases} 0.54 + 0.46\cos\left(\frac{2\pi n}{N-1}\right); & -\left(\frac{N-1}{2}\right) \le n \le \frac{N-1}{2} \\ 0; & \text{otherwise} \end{cases} \tag{5.94}$$

(or)

$$W_H(n) = \begin{cases} 0.54 - 0.46\cos\frac{2n\pi}{N-1}; & 0 \le n \le N-1 \\ 0; & \text{otherwise} \end{cases} \tag{5.95}$$

Frequency response,

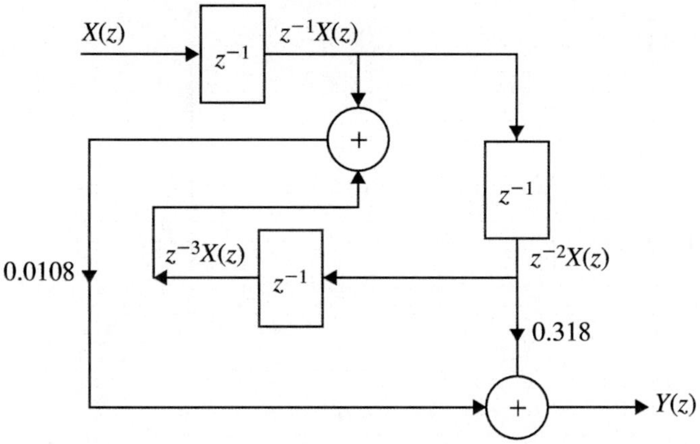

Fig. 5.33 Linear phase realization of $H(z)$ for Example 5.14

$$W_H(\omega) = 0.54\frac{\sin\frac{\omega N}{2}}{\sin(\omega/2)} + 0.23\frac{\sin\left(\frac{\omega N}{2} - \frac{\pi N}{N-1}\right)}{\sin\left(\frac{\omega}{2} - \frac{\pi}{N-1}\right)} + 0.23\frac{\sin\left(\frac{\omega N}{2} + \frac{\pi N}{N-1}\right)}{\sin\left(\frac{\omega}{2} + \frac{\pi}{N-1}\right)} \quad (5.96)$$

$$\text{Width of mainlobe} = \frac{8\pi}{N}$$

$$\text{Peak sidelobe magnitude (dB)} = -41 \text{ dB}$$

The Hamming window and its frequency response are shown in Fig. 5.34. The magnitude of the first sidelobe has been reduced to −41dB, an improvement of 10dB compared to the Hanning window. But sidelobe magnitude at high frequencies is almost constant.

Design Procedure

Step 1: Choose the desired frequency response of the filter, i.e. $H_d(\omega)$.

Step 2: Take the inverse Fourier transform of $H_d(\omega)$ to obtain the desired impulse response $h_d(n)$.

Step 3: Choose a window sequence (here) Hamming window sequence with

$$W(n) = \begin{cases} 0.54 - 0.46\cos\left(\frac{2\pi n}{N-1}\right); & 0 \le n \le N-1 \\ 0; & \text{otherwise} \end{cases}$$

Now multiply $h_d(n)$ and $W(n)$ to get $h(n)$

$$h(n) = h_d(n) \times W(n)$$

This step converts the infinite impulse response to finite impulse response $h(n)$.

Step 4: The transfer function of the filter, $H(z)$ is obtained by taking the z-transform of $h(n)$.

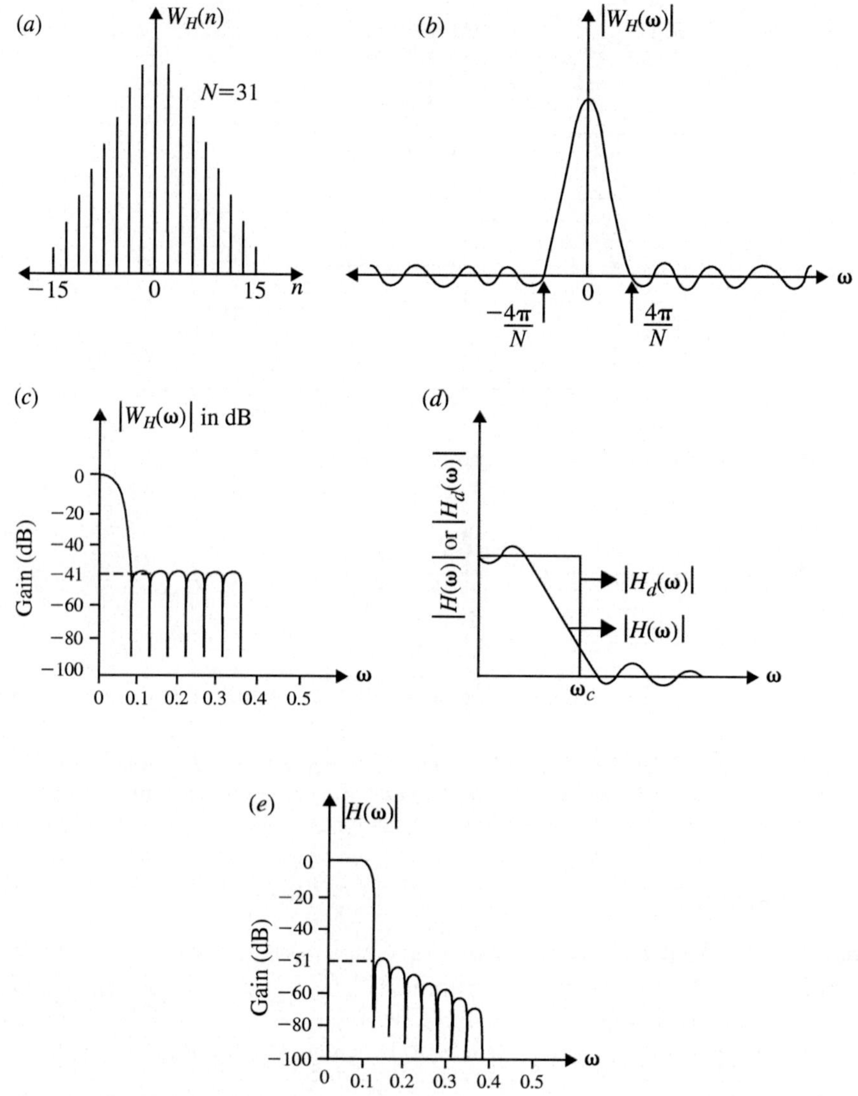

Fig. 5.34 Hamming window sequence and its frequency response. **a** Hamming window sequence; **b** Magnitude response of Hamming window; **c** Log-magnitude response of Hamming window; **d** Magnitude response of LPF approximated using Hamming window and **e** Log-magnitude response of FIR filter using Hamming window

Step 5: Realize the filter using suitable realization method (use Direct Form or Linear Phase).

Example 5.15 Design a HPF with Hamming window. The cut-off frequency is 1.2 rad s^{-1} $N = 9$.

Solution Step 1: Choose $H_d(\omega)$: The Hamming window characteristic is shown in Fig. 5.35a.

$$H(e^{j\omega}) = H_d(\omega) = \begin{cases} e^{-j\omega\alpha}; & -\pi \leq \omega \leq -\omega_c \text{ and } \omega_c \leq \omega \leq \pi \\ 0; & \text{otherwise} \end{cases}$$

Step 2: To Find $h_d(n)$

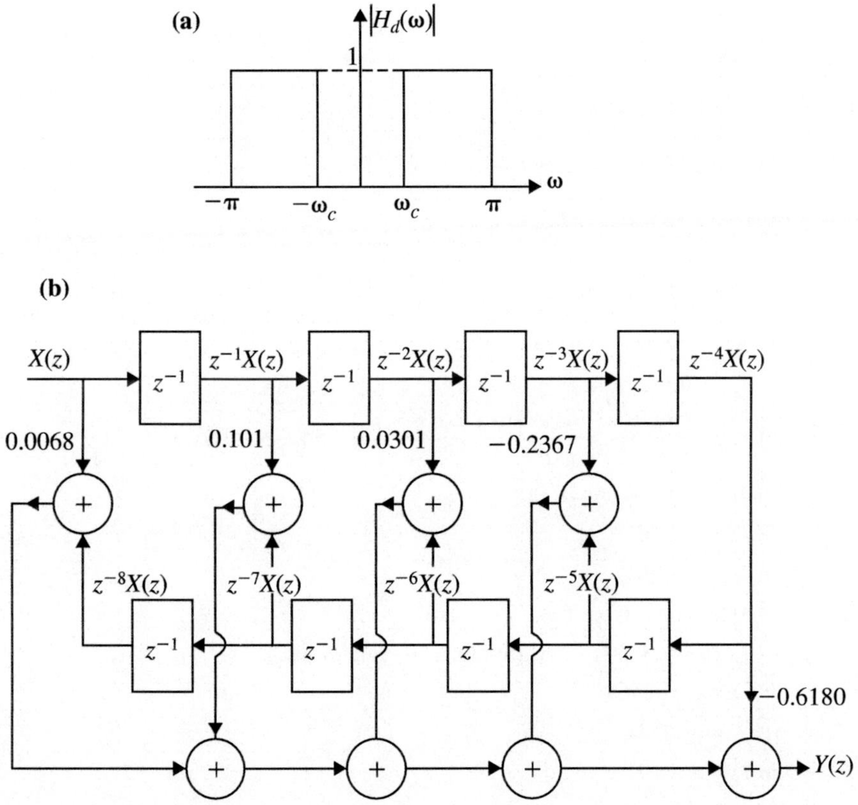

Fig. 5.35 **a** Hamming window characteristic. **b** Linear phase realization of $H(z)$ for Example 5.15

$$
\begin{aligned}
h_d(n) &= \frac{1}{2\pi}\int_{-\pi}^{\pi} H_d(\omega)e^{j\omega n}d\omega \\
&= \frac{1}{2\pi}\int_{-\pi}^{-\omega_c} e^{-j\omega\alpha}e^{j\omega n}d\omega + \frac{1}{2\pi}\int_{\omega_c}^{\pi} e^{-j\omega\alpha}e^{j\omega n}d\omega \\
&= \frac{1}{2\pi}\left[\int_{-\pi}^{-\omega_c} e^{j\omega(n-\alpha)}d\omega + \int_{\omega_c}^{\pi} e^{j\omega(n-\alpha)}d\omega\right] \\
&= \frac{1}{2\pi}\left[\left[\frac{e^{j\omega(n-\alpha)}}{j(n-\alpha)}\right]_{-\pi}^{-\omega_c} + \left[\frac{e^{j\omega(n-\alpha)}}{j(n-\alpha)}\right]_{\omega_c}^{\pi}\right] \\
&= \frac{1}{j2\pi(n-\alpha)}\left[e^{-j\omega_c(n-\alpha)} - e^{-j\pi(n-\alpha)} + e^{j\pi(n-\alpha)} - e^{j\omega_c(n-\alpha)}\right] \\
&= \frac{1}{j2\pi(n-\alpha)}(2j)\left[\sin(n-\alpha)\pi - \sin\omega_c(n-\alpha)\right] \\
&= \frac{1}{\pi(n-\alpha)}\left[\sin\pi(n-\alpha) - \sin\omega_c(n-\alpha)\right]
\end{aligned}
$$

Step 3: To Find $h(n)$

$$
\begin{aligned}
W(n) &= \left[0.54 - 0.46\cos\left(\frac{2\pi n}{N-1}\right)\right] \\
h(n) &= h_d(n) \times W(n) \\
&= \frac{[\sin\pi(n-\alpha) - \sin\omega_c(n-\alpha)]}{\pi(n-\alpha)}\left[0.54 - 0.46\cos\left(\frac{2\pi n}{N-1}\right)\right]
\end{aligned}
$$

where $\alpha = \frac{N-1}{2} = 4$

$$
\begin{aligned}
h(n) &= \frac{[\sin\pi(n-4) - \sin 1.2(n-4)]}{\pi(n-4)}\left[0.54 - 0.46\cos\left(\frac{n\pi}{4}\right)\right] \\
h(0) &= 6\times10^{-3} = 0.0063 \\
h(1) &= 0.0101 \\
h(2) &= 0.0581 \\
h(3) &= -0.2367 \\
h(4) &= -0.6180 \quad \text{(Applying L'Hospital's rule)} \\
h(5) &= -0.2567 \\
h(6) &= 0.0581 \\
h(7) &= 0.0101 \\
h(8) &= 6.34\times10^{-3}
\end{aligned}
$$

Step 4: Take z-transform:

$$H(z) = \sum_{n=0}^{N-1} h(n) z^{-n}$$

$$H(z) = 0.0063[1 + z^{-8}] + 0.101[z^{-1} + z^{-7}] + 0.0581[z^{-2} + z^{-6}]$$
$$-0.2367[z^{-3} + z^{-5}] + (-0.6180) z^{-4}$$

Step 5: Draw the realization structure. The structure is realized as shown in Fig. 5.35b.

5.5.2.5 Blackman Window

The Blackman window function is another type of cosine window and is given by the following equation:

$$W_B(n) = \begin{cases} 0.42 + 0.5 \cos \frac{2n\pi}{N-1} + 0.08 \cos \frac{4n\pi}{N-1}; & -\left(\frac{N-1}{2}\right) \le n \le \frac{N-1}{2} \\ 0; & \text{otherwise} \end{cases} \tag{5.97}$$

(or)

$$W_B(n) = \begin{cases} 0.42 - 0.5 \cos \frac{2n\pi}{N-1} + 0.08 \cos \frac{4n\pi}{N-1}; & 0 \le n \le N-1 \\ 0; & \text{otherwise} \end{cases} \tag{5.98}$$

$$\text{Width of mainlobe} = \frac{12\pi}{N}$$
$$\text{Peak sidelobe magnitude (dB)} = -58 \text{ dB}$$

The frequency response and impulse response characteristics of Blackman window are shown in Fig. 5.36.

The frequency response of Blackman window is shown in Fig. 5.36. The magnitude of the first sidelobe is −58dB and the sidelobe magnitude decreases with frequency. This desirable feature is achieved at the expense of increased mainlobe width.

Design Procedure

1. Choose the desired frequency response $H_d(\omega)$. Take inverse Fourier transform of $H_d(\omega)$ to obtain the desired impulse response $h_d(n)$.
2. Choose a window sequence $W(n)$ and multiply $h_d(n)$ and $W(n)$ to convert the infinite impulse response to finite impulse response $h(n)$.
3. The transfer function $H(z)$ is obtained by taking z-transform of $h(n)$.
4. Realize the filter using suitable realization method.

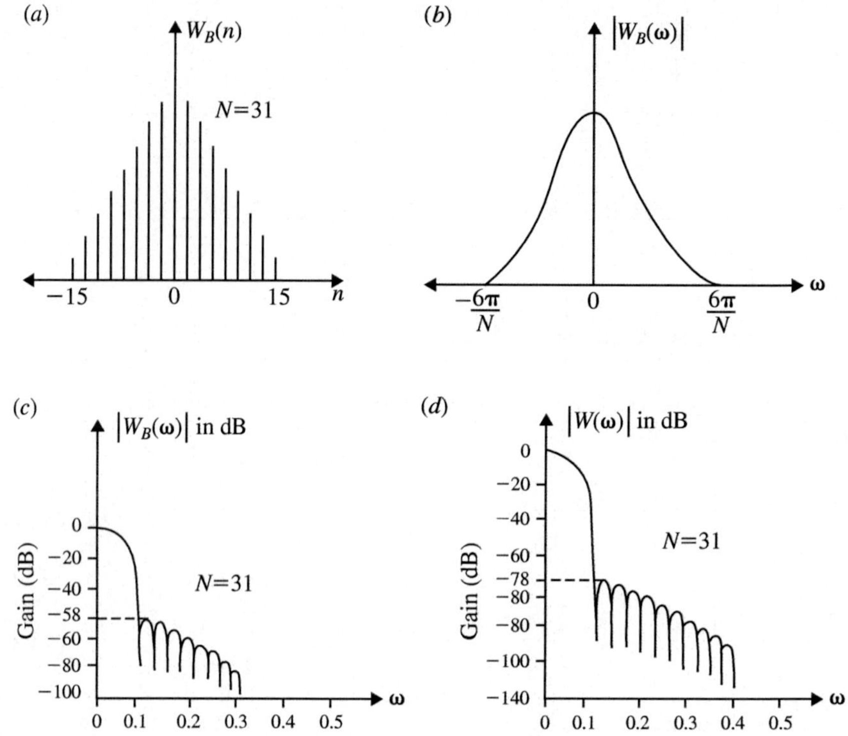

Fig. 5.36 Blackman window sequence and its frequency response. **a** Blackman window sequence; **b** Magnitude response of Blackman window; **c** Log-magnitude response of Blackman window and **d** Log-magnitude response of FIR LPF approximated using Blackman window

Example 5.16 Design an ideal bandpass filter using Blackman window to pass frequencies in the range 1.2 to 1.7 rad/s with $N = 7$.

Solution

Step 1. Choose desired frequency response

$$H_d(e^{j\omega}) = \begin{cases} e^{-j\omega\alpha}, & -\omega_{c_2} \le \omega \le \omega_{c_1} \text{ and } \omega_{c_1} \le \omega \le \omega_{c_2} \\ 0, & \text{else} \end{cases}$$

$$\therefore\ H_d(e^{j\omega}) = \begin{cases} e^{-j\omega\alpha}, & -1.7 \le \omega \le -1.2 \text{ and } 1.2 \le \omega \le 1.7 \\ 0, & \text{else} \end{cases}$$

$$h_d(n) = \frac{1}{2\pi}\int_{-\pi}^{\pi} H_d(e^{j\omega})e^{j\omega n}d\omega$$

$$h_d(n) = \frac{1}{2\pi}\int_{-1.7}^{-1.2} e^{-j\omega\alpha}\cdot e^{j\omega n}d\omega + \frac{1}{2\pi}\int_{1.2}^{1.7} e^{-j\omega\alpha}\cdot e^{j\omega n}d\omega$$

$$= \frac{1}{2\pi}\left[\frac{e^{j\omega(n-\alpha)}}{j(n-\alpha)}\right]_{-1.7}^{-1.2} + \frac{1}{2\pi}\left[\frac{e^{j\omega(n-\alpha)}}{j(n-\alpha)}\right]_{1.2}^{1.7}$$

$$= \frac{1}{2\pi}\left[\frac{e^{-j1.7(n-\alpha)}}{j(n-\alpha)} - \frac{e^{-1.2j(n-\alpha)}}{j(n-\alpha)} + \frac{e^{1.7j(n-\alpha)}}{j(n-\alpha)} - \frac{e^{j1.2(n-\alpha)}}{j(n-\alpha)}\right]$$

$$= \frac{1}{2\pi}\left[\frac{2j\sin 1.7(n-\alpha) - 2j\sin 1.2(n-\alpha)}{j(n-\alpha)}\right]$$

$$= \frac{1}{\pi(n-\alpha)}[\sin 1.7(n-\alpha) - \sin 1.2(n-\alpha)]$$

where $\alpha = \frac{N-1}{2} = 3$

$$\therefore\ h_d(n) = \frac{1}{\pi(n-3)}[\sin 1.7(n-3) - \sin 1.2(n-3)]$$

$$W(n) = \begin{cases} 0.42 - 0.5\cos\left(\frac{2\pi n}{6}\right) + 0.08\cos\left(\frac{4\pi n}{6}\right); & 0 \le n \le 6 \\ 0; & \text{otherwise} \end{cases}$$

Step 2:

$$h(n) = W(n)h_d(n) \qquad 0 \le n \le N-1$$

$$= \left[0.42 - 0.5\cos\left(\frac{2\pi n}{6}\right) + 0.08\cos\left(\frac{4\pi n}{6}\right)\right]$$

$$\times \frac{1}{\pi(n-3)}[\sin 1.7(n-3) - \sin 1.2(n-3)]; \quad 0 \le n \le 6$$

$$h(0) = 0$$

$$h(1) = -0.019$$

$$h(2) = 0.0119$$

$$h(3) = \text{Apply } L'\text{ Hospital's Rule}$$

$$h(4) = 0.0119$$

$$h(5) = -0.019$$

$$h(6) = 0$$

$$h(3) = \left(\left[0.42 - 0.5\sin\left(\frac{2\pi\times 3}{6}\right) - 0.08\sin\left(\frac{4\pi\times 3}{6}\right)\right]\right.$$

$$\times [\sin 1.7(n-3) - \sin 1.2(n-3)] \Big/ \pi(n-3)$$

$$= \left[0.42 - 0.5\cos\left(\frac{2\pi\times 3}{6}\right) + 0.08\cos\left(\frac{4\pi\times 3}{6}\right)\right]$$

$$\left.\times\left[\frac{\sin 1.7(n-3)}{1.7(n-3)}\cdot 1\cdot 7 - \frac{\sin 1.2(n-3)}{1.2(n-3)}\cdot 1\cdot 2\right]\right) \Big/ \pi$$

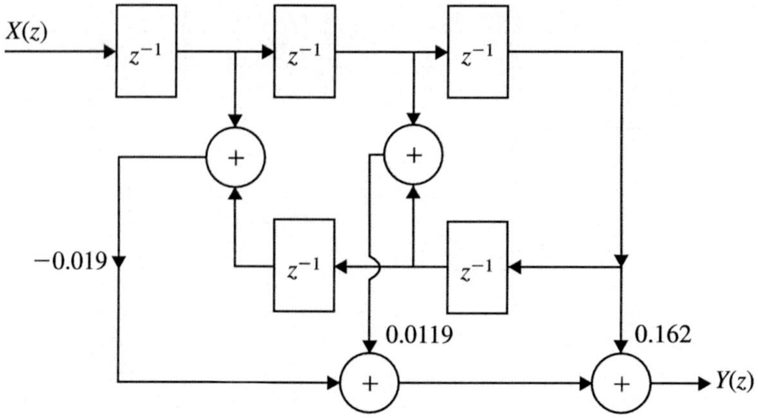

Fig. 5.37 Linear phase realization of $H(z)$ for Example 5.16

$$= \frac{1 \times (0.5)}{\pi} = 0.162$$

Step 3:

$$\begin{aligned} H(z) &= \sum_{n=0}^{N-1} h(n)z^{-n} = h(0) + h(1)z^{-1} + h(2)z^{-2} + h(3)z^{-3} + h(4)z^{-4} \\ &\quad + h(5)z^{-5} + h(6)z^{-6} \\ H(z) &= -0.019z^{-1} + 0.0119z^{-2} + 0.162z^{-3} + 0.0119z^{-4} - 0.019z^{-5} \\ &= -0.019(z^{-1} + z^{-5}) + 0.0119(z^{-2} + z^{-4}) + 0.162z^{-3} \end{aligned}$$

Step 4: **Realization Structure** The structure realization is shown in Fig. 5.37.

5.5.2.6 Kaiser Window

Kaiser window function is given in the following form:

$$W_K(n) = \begin{cases} \dfrac{I_0\left(\alpha\sqrt{1-\left(\frac{2n}{N-1}\right)^2}\right)}{I_0(\alpha)}; & -\left(\frac{N-1}{2}\right) \le n \le \frac{N-1}{2} \\ 0; & \text{otherwise} \end{cases} \tag{5.99}$$

or

$$W_K(n) = \begin{cases} \dfrac{I_0\left(\alpha\sqrt{\left(\frac{N-1}{2}\right)^2-\left(n-\frac{N-1}{2}\right)^2}\right)}{I_0\left(\alpha\frac{N-1}{2}\right)}; & 0 \le n \le \frac{N-1}{2} \\ 0; & \text{otherwise} \end{cases} \tag{5.100}$$

The parameter α is an independent variable that can be varied to control the sidelobe levels with respect to the mainlobe peak. The modified Bessel function of the first kind $I_0(x)$ is given by

$$I_0(x) = 1 + \sum_{k=1}^{\infty}\left[\frac{1}{k!}\left(\frac{x}{2}\right)^k\right]^2 = 1 + \sum_{k=1}^{\infty}\left(\frac{(0.5)^2}{(k!)^2}\right)^k \tag{5.101}$$

$$= 1 + \frac{(0.25x^2)^2}{(k!)^2} = 1 + \frac{0.25x^2}{(1!)^2} + \frac{(0.25x^2)^2}{(2!)^2} + \cdots \tag{5.102}$$

This equation can be used to compute $I_0(\alpha)$ and $I_0\left(\alpha\sqrt{\left(\frac{2n}{N-1}\right)^2}\right)$. Figure 5.38 shows the Kaiser window sequence and its frequency response. The width of the mainlobe can be adjusted by varying the length N of the window sequence (Table 5.1).

Design Procedure

(*i*) Determine 'δ' using the formula

$$\delta = \min(\delta_p, \delta_s)$$

where

$$\delta_p = \frac{10^{0.05A_p} - 1}{10^{0.05A_p} + 1}$$
$$\delta_s = 10^{-0.05A_s}$$

(*ii*) Find the actual stop band attenuation using 'δ'

$$A_s' = -20\log_{10}\delta$$

(*iii*) Determine Kaiser window design parameter using the following equation:

$$\alpha_k = \begin{cases} 0, & A_s' \le 21 \\ 0.5842(A_s' - 21)^{0.4} + 0.07886(A_s' - 21), & 21 \le A_s' \le 50 \\ 0.1102(A_s - 8.7), & A_s' > 50 \end{cases}$$

(*iv*) Determine the design parameter 'D' using the following equation:

$$D = \begin{cases} 0.9222, & A_s' \le 21 \\ \frac{A_s' - 7.95}{14.36}, & A_s' > 21 \end{cases}$$

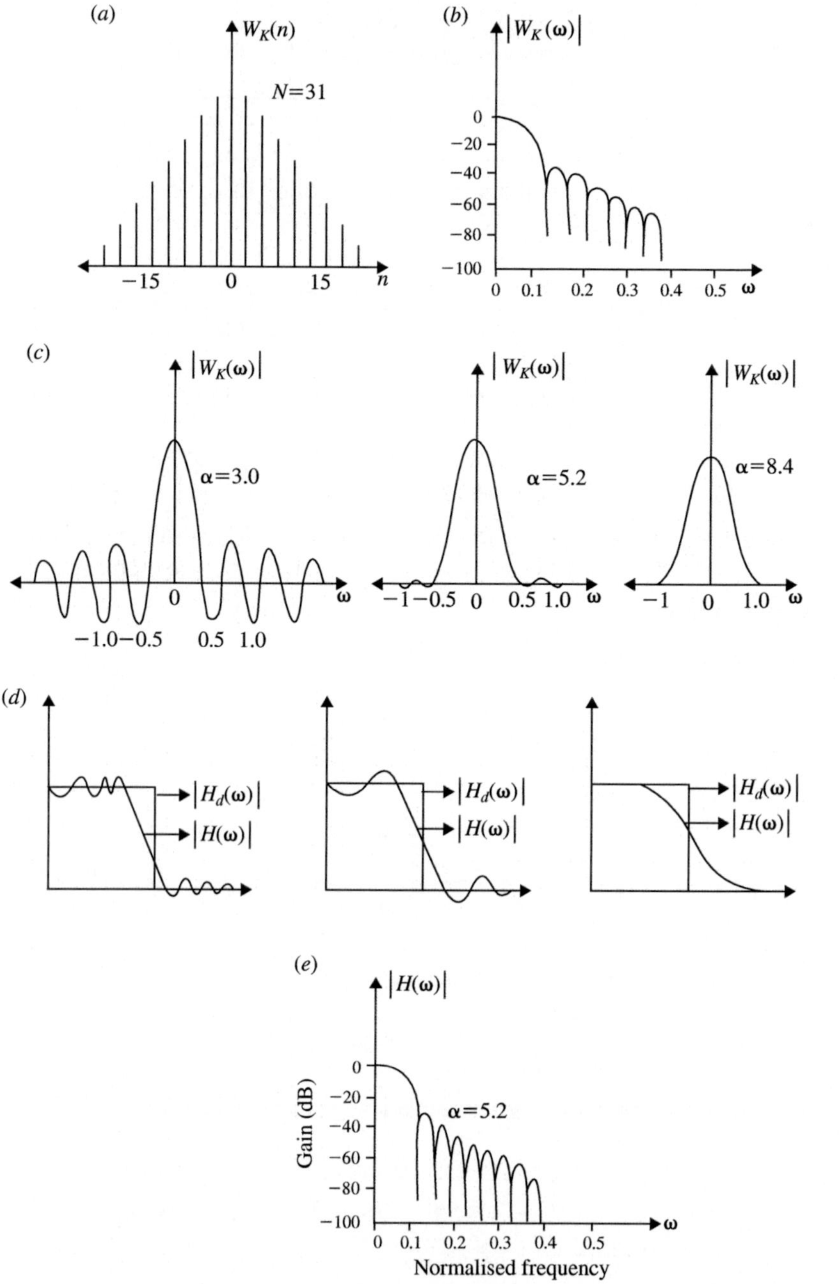

Fig. 5.38 Kaiser window sequence and its frequency response. **a** Kaiser window sequence; **b** Log-magnitude response of Kaiser window; **c** Magnitude frequency response of Kaiser window for different values of α (when $N = 31$); **d** Magnitude response of Kaiser window for different values of α (when $N = 31$) and **e** Log-magnitude response of FIR LPF designed using Kaiser window

(*v*) Determine the order of filter using the following equation:

$$N \geq \frac{FD}{\Delta_f} + 1$$
$$\Delta_f = f_s - f_p$$

where D = design parameter and F = sampling frequency.

(*vi*) Select the desired frequency response $H_d(\omega)$

$$\omega_c = \frac{2\pi f_c}{F} = \frac{2\pi(1/2)(f_p + f_s)}{F}$$

where $f_c = (1/2)(f_p + f_s)$.

Table 5.1 Comparison of window function in frequency domain characteristics

Type of window	Approximate width of Mainlobe	Peak sidelobe Magnitude (dB)
Rectangular	$4\pi/N$	−13
Hanning	$8\pi/N$	−31
Hamming	$8\pi/N$	−41
Blackman	$12\pi/N$	−58

Comparison of window function with frequency domain characteristic is shown in Table 5.1

(*vii*) Find $h_d(n)$ using inverse Fourier transform

$$h_d(n) = \frac{1}{2\pi}\int_{-\pi}^{\pi} H_d(\omega)e^{j\omega n}d\omega$$

(*viii*) Find the window function using $\omega(n)$ equation. Bessel function

$$I_0(x) = 1 + \frac{0.25x^2}{(1!)^2} + \frac{(0.25x^2)^2}{(2!)^2} + \frac{(0.25x^2)^3}{(3!)^2} + \frac{(0.25x^2)^4}{(4!)^2} + \frac{(0.25x^2)^5}{(5!)^2} + \frac{(0.25x^2)^6}{(6!)^2}$$

$$\omega(n) = \frac{I_0\left[\alpha_k\sqrt{\left(\frac{N-1}{2}\right)^2\left(n - \frac{N-1}{2}\right)^2}\right]}{I_0\left[\alpha_k\left(\frac{N-1}{2}\right)\right]}; \quad 0 \leq n \leq N-1$$

$$\omega(n) = \frac{I_0\left[\alpha\sqrt{1 - \left(\frac{2n}{N-1}\right)^2}\right]}{I_0[\alpha]}; \; -\left(\frac{N-1}{2}\right) \leq n \leq \frac{N-1}{2}; \text{ and } |n| \leq \frac{N-1}{2}$$

(*ix*) Find $h(n)$ using $h(n) = h_d(n)\omega(n)$.
(*x*) Find $H(z)$ using z-transform

$$H(z) = \sum_{n=0}^{N-1} h(n)z^{-n}$$

(*xi*) Draw suitable realization.

Example 5.17 Design a FIR lowpass filter using Kaiser window filter with the following specifications.

$$\begin{aligned}
\text{Passband cut-off frequency} &= 150 \text{ Hz} \\
\text{Stopband cut-off frequency} &= 250 \text{ Hz} \\
\text{Passband ripple} &= 0.1 \text{ dB} \\
\text{Stopband attenuation} &= 40 \text{ dB} \\
\text{Sampling frequency} &= 1000 \text{ Hz}
\end{aligned}$$

Solution Given

$$\begin{aligned}
f_p &= 150 \text{ Hz} \\
f_s &= 250 \text{ Hz} \\
A_p &= 0.1 \text{ dB} \\
A_s &= 40 \text{ dB} \\
F &= 1000 \text{ Hz}
\end{aligned}$$

Step 1: Find δ using the following equation:

$$\begin{aligned}
\delta_p &= \frac{10^{0.05A_p} - 1}{10^{0.05A_p} + 1} = \frac{10^{0.05\times0.1} - 1}{10^{0.05\times0.1} + 1} = 5.76 \times 10^{-3} \\
&= 0.00576 \\
\delta_s &= 10^{-0.05A_s} = 10^{-0.05\times40} = 0.01 \\
\delta &= \min(\delta_p, \delta_s) \\
\therefore\ \delta &= 0.00576
\end{aligned}$$

Step 2: Find actual stopband attenuation, A_s', using the following equation:

$$\begin{aligned}
A_S' &= -20\log_{10}\delta \\
&= -20\log_{10}(0.00576) \\
\therefore\ A_s' &= 44.8 \text{ dB}
\end{aligned}$$

Step 3: Find the Kaiser window parameter, α_k, using the following equation:

$$\alpha_k = 0.5842(A'_s - 21)^{0.4} + 0.07886(A'_s - 21)$$
$$\alpha_k = 0.5842(44.8 - 21)^{0.4} + 0.07886(44.8 - 21)$$
$$\therefore\ \alpha_k = 3.953$$

Step 4: Find the design parameter 'D' (because $A'_s > 21$) using the following equation:

$$D = \frac{A'_s - 7.95}{14.36} = \frac{44.8 - 7.95}{14.36}$$
$$\therefore\ D = 2.566$$

Step 5: Find the order of the filter, N, using the following equation:

$$\begin{aligned} N \geq \frac{F \cdot D}{\Delta f} + 1 &= \frac{F \cdot D}{f_s - f_p} + 1 \\ &= \frac{1000 \times 2.566}{(250 - 150)} + 1 \\ &= 26.66 \\ \therefore\ N &= 27 \end{aligned}$$

Step 6: To find cut-off frequency, ω_c, the following equation is used:

$$\begin{aligned} f_c &= \frac{1}{2}(f_s + f_p) \\ &= \frac{1}{2}(250 + 150) = 200\ \text{Hz} \\ \omega_c &= \frac{2\pi f_c}{F} \quad \text{(normalized by } F) \\ \omega_c &= \frac{2\pi \times 200}{1000} = 1.2566 \\ \therefore\ \omega_c &= 1.2566 \end{aligned}$$

Step 7: To find $h_d(n)$, the following procedure is followed.

For a lowpass filter, the frequency response is given by

$$H_d(e^{j\omega}) = \begin{cases} e^{-j\omega\alpha}, & -\omega_c \leq \omega \leq \omega_c \\ 0, & \text{else} \end{cases}$$

$$h_d(n) = \frac{1}{2\pi}\int_{-\pi}^{\pi} H_d(e^{j\omega})e^{j\omega n}\,d\omega$$

$$h_d(n) = \frac{1}{2\pi}\int_{-\omega_c}^{\omega_c} e^{-j\omega\alpha}e^{j\omega n}\,d\omega$$

as $\omega_c = 1.2566$ (from Step 6)

$$h_d(n) = \frac{1}{2\pi}\int_{-1.2566}^{1.2566} e^{j\omega(n-\alpha)} d\omega$$

$$h_d(n) = \frac{1}{2\pi}\frac{e^{j\omega(n-\alpha)}}{j(n-\alpha)}\Bigg|_{-1.2566}^{1.2566}$$

$$h_d(n) = \frac{1}{2\pi}\left[\frac{e^{j1.2566(n-\alpha)} - e^{-j1.2566(n-\alpha)}}{j(n-\alpha)}\right]$$

$$h_d(n) = \frac{\sin 1.2566(n-\alpha)}{\pi(n-\alpha)}$$

because

$$\sin\theta = \frac{e^{i\theta} - e^{-i\theta}}{2i}$$

As $\alpha = \frac{N-1}{2} = \frac{27-1}{2} = 13$

$$h_d(n) = \frac{\sin 1.2566(n-13)}{\pi(n-13)} \qquad 0 \le n \le N-1 \text{ and } 0 \le n \le 26$$

$$h_d(0) = -0.01438 \qquad h_d(7) = 0.050$$

$$h_d(1) = 0.0156 \qquad h_d(8) = -1.1797 \times 10^{-5}$$

$$h_d(2) = 0.0275 \qquad h_d(9) = -0.075$$

$$h_d(3) = -1.1797 \times 10^{-5} \qquad h_d(10) = -0.0623$$

$$h_d(4) = -0.033 \qquad h_d(11) = 0.09355$$

$$h_d(5) = -0.0233 \qquad h_d(12) = 0.3027$$

$$h_d(6) = 0.0267 \qquad h_d(13) = \frac{1.2566}{\pi} = 0.3999$$

Step 8: To find the window function $W(n)$. As N is very large

$$W(n) = \frac{I_0\left[\alpha_k\sqrt{1-\left(\frac{2n}{N-1}\right)^2}\right]}{I_0[\alpha_k]} \qquad 0 \le n \le N-1;\ 0 \le n \le 26$$

$$W(n) = \frac{I_0\left[3.953\sqrt{1-(\frac{2n}{26})^2}\right]}{I_0[3.953]} \qquad [\because N = 27]$$

$$I_0[x] = 1 + \frac{0.25x^2}{(1!)^2} + \frac{(0.25x^2)^2}{(2!)^2} + \frac{(0.25x^2)^3}{(3!)^2} + \cdots + \frac{(0.25x^2)^6}{(6!)^2}$$

$$W(0) = \frac{I_0[3.953]}{I_0[3.953]} = \frac{10.824}{10.824} = 1$$
$$W(1) = \frac{I_0[3.938]}{10.824} = 0.99$$
$$W(2) = \frac{I_0[3.9]}{10.824} = 0.9582$$
$$W(3) = \frac{I_0[3.843]}{10.824} = 0.9125$$
$$W(4) = \frac{I_0[3.758]}{10.824} = 0.8485$$
$$W(5) = \frac{I_0[3.646]}{10.824} = 0.77$$
$$W(6) = \frac{I_0[3.504]}{10.824} = 0.684$$
$$W(7) = \frac{I_0[3.328]}{10.824} = 0.5905$$
$$W(8) = \frac{I_0[3.113]}{10.824} = 0.4955$$

$$W(9) = \frac{I_0[2.85]}{10.824} = 0.3998$$
$$W(10) = \frac{I_0[2.5239]}{10.824} = 0.30967$$
$$W(11) = \frac{I_0[2.105]}{10.824} = 0.22689$$
$$W(12) = \frac{I_0[1.519]}{10.824} = 0.1539$$
$$W(13) = \frac{I_0[0]}{10.824} = 0.09242$$

Step 9: To Find $h(n)$

$$h(n) = h_d(n)W(n) \qquad 0 \le n \le 26$$

$h(0) = -0.01438$ $\qquad$ $h(7) = 0.029526$

$h(1) = 0.015444$ $\qquad$ $h(8) = -5.8419 \times 10^{-6}$

$h(2) = 0.02635$ $\qquad$ $h(9) = -0.02998$

$h(3) = -1.0758 \times 10^{-5}$ $\qquad$ $h(10) = -0.01929$

$h(4) = -0.028$ $\qquad$ $h(11) = 0.02122$

$h(5) = -0.0179$ $\qquad$ $h(12) = 0.04658$

$h(6) = 0.01826$ $\qquad$ $h(13) = 0.036958$

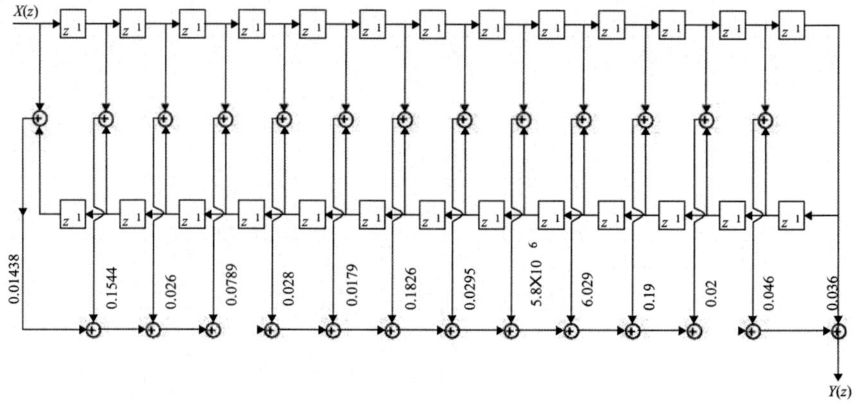

Fig. 5.39 Linear phase realization of $H(z)$ for Example 5.17

Applying the linear phase condition of a FIR filter, that is, $h(n) = h(N-1-n)$, we get

$$
\begin{aligned}
h(0) &= h(26) & h(7) &= h(19)\\
h(1) &= h(25) & h(8) &= h(18)\\
h(2) &= h(24) & h(9) &= h(17)\\
h(3) &= h(23) & h(10) &= h(16)\\
h(4) &= h(22) & h(11) &= h(15)\\
h(5) &= h(21) & h(12) &= h(14)\\
h(6) &= h(20) & h(13) &= h(13)
\end{aligned}
$$

Step 10: To Find $H(z)$

$$
\begin{aligned}
H(z) = {} & -0.01438(1+z^{-26})+0.01544(z^{-1}+z^{-25})+0.02635(z^{-2}+z^{-24})\\
& -1.075 \times 10^{-5}(z^{-3}+z^{-23}) - 0.0028(z^{-4}+z^{-22})\\
& -0.0179(z^{-5}+z^{-21}) + 0.01826(z^{-6}+z^{-20})\\
& +0.029525(z^{-7}+z^{-19}) - 5.8419 \times 10^{-6}(z^{-8}+z^{-18})
\end{aligned}
$$

$$
\begin{aligned}
& -0.02998(z^{-9}+z^{-17}) - 0.01929(z^{-10}+z^{-16})\\
& +0.02122(z^{-11}+z^{-15})+0.04658(z^{-12}+z^{-14})+0.036958(z^{-13})
\end{aligned}
$$

Step 11: Draw the realization structure. This is shown in Fig. 5.39.

5.5.3 Frequency Sampling Method

In this method, the ideal frequency response is sampled at a sufficient number of points.

Let $h(n)$ be the filter coefficients of an FIR filter and $H(k)$ be the DFT of $h(n)$

$$H(k) = \sum_{k=0}^{N-1} h(n)e^{-j\frac{2\pi kn}{N}}, \qquad k = 0, 1, \ldots, N-1$$

$$h(n) = \frac{1}{N}\sum_{k=0}^{N-1} H(k)e^{j\frac{2\pi kn}{N}}, \qquad n = 0, 1, \ldots, N-1$$

The DFT samples $H(k)$ for an FIR sequence can be regarded as samples of the filter z-transform evaluated at N-points equally spaced around the unit circle.

$$H(k) = H(z)\Big|_{z=e^{j\frac{2\pi k}{N}}}$$

The transfer function $H(z)$ of an FIR filter is given by

$$H(z) = \sum_{n=0}^{N-1} h(n)z^{-n}$$

$$\begin{aligned} H(z) &= \sum_{n=0}^{N-1}\left[\frac{1}{N}\sum_{k=0}^{N-1} H(k)e^{j\frac{2\pi kn}{N}}\right]z^{-n} \\ &= \sum_{k=0}^{N-1}\frac{H(k)}{N}\sum_{k=0}^{N-1} H(k)\left(e^{j\frac{2\pi k}{N}}z^{-1}\right)^{n} \\ &= \sum_{k=0}^{N-1}\frac{H(k)}{N}\,\frac{1-\left(e^{j\frac{2\pi k}{N}}z^{-1}\right)^{N}}{1-e^{\frac{j2\pi k}{N}}z^{-1}} \\ &= \frac{1-z^{-N}}{N}\sum_{k=0}^{N-1}\frac{H(k)}{1-e^{\frac{j2\pi k}{N}}z^{-1}} \end{aligned}$$

Using the following identity, we get

$$H(e^{j\omega})\Big|_{\omega=\frac{2\pi k}{N}} = H(e^{j\frac{2\pi kn}{N}}) = H(k)$$

i.e. $H(k)$ is the kth DFT component obtained by sampling the frequency response $H(e^{j\omega})$. As such this approach for designing FIR filter is called the frequency sampling. There are two design procedures.

1. Type-1 design procedure.
2. Type-2 design procedure.

Type-1 Design Procedure

Step 1: Choose the desired frequency response $H_d(\omega)$.
Step 2: Sample $H_d(\omega)$ at 'N' points by taking $\omega = \frac{2\pi k}{N}$ and generate the sequence

$$\tilde{H}(k) = H_d(\omega)|_{\omega=\frac{2\pi k}{N}}, \qquad k = 0 \text{ to } N-1 \tag{5.103}$$

Step 3: Compute $h(n)$ using the following equations:

$$\text{When } N \text{ is odd } h(n) = \frac{1}{N}\left\{\tilde{H}(0) + 2\sum_{k=1}^{\frac{N-1}{2}} \text{Re}\left[\tilde{H}(k)e^{\frac{j2\pi kn}{N}}\right]\right\} \tag{5.104}$$

$$\text{When } N \text{ is even } h(n) = \frac{1}{N}\left\{\tilde{H}(0) + 2\sum_{k=1}^{(N/2)-1} \text{Re}\left[\tilde{H}(k)e^{\frac{j2\pi kn}{N}}\right]\right\} \tag{5.105}$$

Step 4: Take z-transform of $h(n)$

$$H(z) = \sum_{n=0}^{N-1} h(n)z^{-n} \tag{5.106}$$

Step 5: Draw the realization structure.

Type-2 Design Procedure

Step 1: Choose the desired frequency response $H_d(\omega)$.
Step 2: Sample $H_d(\omega)$ at 'N' points by taking $\omega = \frac{\pi(2k+1)}{N}$ and generate the sequence, i.e.

$$\tilde{H}(k) = H_d(\omega)\Big|_{\omega=\frac{\pi(2k+1)}{N}}, \qquad k = 0, 1, \ldots, N-1 \tag{5.107}$$

Step 3: Compute $h(n)$ using the following equations:

$$\text{When } N \text{ is odd, } h(n) = \frac{2}{N}\left\{\sum_{k=0}^{\frac{N-3}{2}} \text{Re}\left[\tilde{H}(k)e^{\frac{jn\pi(2k+1)}{N}}\right]\right\} \tag{5.108}$$

$$\text{When } N \text{ is even, } h(n) = \frac{2}{N}\left\{\sum_{k=0}^{(N/2)-1} \text{Re}\left[\tilde{H}(k)e^{\frac{jn\pi(2k+1}{N}}\right]\right\} \tag{5.109}$$

Step 4: Take z-transform of $h(n)$

$$H(z) = \sum_{n=0}^{N-1} h(n) z^{-n} \tag{5.110}$$

Step 5: Draw the realization structure.

Example 5.18 Determine the filters coefficient of a linear phase FIR filter of length $N = 15$, which has a symmetric unit sample response and a frequency response that satisfies the following conditions.

$$H_r\left(\frac{2\pi k}{15}\right) = \begin{cases} 1; & k = 0, 1, 2, 3 \\ 0.4; & k = 4 \\ 0; & k = 5, 6, 7 \end{cases}$$

Solution Step 1:

$$\tilde{H}(k) = H_d(\omega) \Big/ \omega = \frac{2\pi k}{15} = \begin{cases} 1e^{-j\omega\alpha}; & k = 0, 1, 2, 3 \\ 0.4e^{-j\omega\alpha}; & k = 4 \\ 0; & k = 5, 6, 7 \end{cases}$$

$$\alpha = \frac{N-1}{2} = 7$$

$$\tilde{H}(k) = \begin{cases} e^{-j\frac{2\pi k}{15}\cdot\alpha}; & k = 0, 1, 2, 3 \\ 0.4e^{-j\frac{2\pi k}{15}\alpha}; & k = 4 \\ 0; & k = 5, 6, 7 \end{cases}$$

Step 2: When

$$\begin{aligned}
k &= 0, & \tilde{H}(0) &= 1 \\
k &= 1, & \tilde{H}(1) &= e^{-j\frac{2\pi}{15}(1)\times 7} = e^{-j\frac{14\pi}{15}} \\
k &= 2, & \tilde{H}(2) &= e^{-j\frac{2\pi}{15}(2)\times 7} = e^{-j\frac{28\pi}{15}} \\
k &= 3, & \tilde{H}(3) &= e^{-j\frac{2\pi}{15}(3)\times 7} = e^{-j\frac{42\pi}{15}} \\
k &= 4, & \tilde{H}(4) &= 0.4e^{-j\frac{2\pi}{15}(4)\times 7} = 0.4e^{-j\frac{56\pi}{15}} \\
k &= 5, & \tilde{H}(5) &= 0 \\
k &= 6, & \tilde{H}(6) &= 0 \\
k &= 7, & \tilde{H}(7) &= 0
\end{aligned}$$

Step 3: To Find $h(n)$. Here $N = 15$ (odd)

$$\therefore\ h(n) = \frac{1}{N}\left[\tilde{H}(0) + 2\sum_{k=1}^{7}\text{Re}\left[\tilde{H}(k)e^{j\frac{2\pi kn}{N}}\right]\right]$$

$$h(n) = \frac{1}{15}\left[1 + 2\sum_{k=1}^{3}\text{Re}\left[e^{-j7\times\frac{2\pi k}{15}} \times e^{j\frac{2\pi nk}{15}}\right] + 2\text{Re}\left[0.4 \times e^{-j\frac{56\pi}{15}} \times e^{j\frac{2\pi nk}{15}}\right]\right]$$

$$= \frac{1}{15}\left[1 + 2\sum_{k=1}^{3}\cos\frac{2\pi k}{15}(n-7) + 0.8\cos\frac{8\pi}{15}(n-7)\right]$$

$$= \frac{1}{15} + \frac{2}{15}\cos\frac{2\pi}{15}(n-7) + \frac{2}{15}\cos\frac{4\pi}{15}(n-7) + \frac{2}{15}\cos\frac{6\pi}{15}(n-7)$$
$$+\frac{0.8}{15}\cos\frac{8\pi}{15}(n-7)$$

$$h(0) = -0.71155 \qquad h(8) = 4.778$$
$$h(1) = 0.575 \qquad h(9) = 0.459$$
$$h(2) = 0.973 \qquad h(10) = -1.6015$$
$$h(3) = -0.499 \qquad h(11) = -0.499$$
$$h(4) = -1.6015 \qquad h(12) = 0.973$$
$$h(5) = 0.459 \qquad h(13) = 0.575$$
$$h(6) = 4.778 \qquad h(14) = -0.71155$$
$$h(7) = 7.053$$

Example 5.19 A lowpass filter is to be designed with the following desired frequency response

$$H_d(e^{j\omega}) = \begin{cases} e^{-j2\omega}, & -\pi/4 \le \omega \le \pi/4 \\ 0; & \pi/4 \le |\omega| \le \pi \end{cases}$$

Determine the filter coefficients $h_d(n)$ if window function

$$\omega(n) = \begin{cases} 1, & 0 \le n \le 4 \\ 0, & \text{otherwise} \end{cases}$$

Solution Given

$$H_d(e^{j\omega}) = \begin{cases} e^{-j2\omega}, & -\pi/4 \le \omega \le \pi/4 \\ 0; & \pi/4 \le |\omega| \le \pi \end{cases}$$

$$\begin{aligned} h_d(n) &= \frac{1}{2\pi}\int_{-\pi}^{\pi} H_d(e^{j\omega})e^{j\omega n}d\omega \\ &= \frac{1}{2\pi}\int_{-\pi/4}^{\pi/4} e^{-j2\omega}.e^{j\omega n}d\omega \\ &= \frac{1}{2\pi}\int_{-\pi/4}^{\pi/4} e^{j\omega(n-2)}d\omega \\ &= \frac{1}{2\pi}\left[\frac{e^{j\omega(n-2)}}{j(n-2)}\right]_{-\pi/4}^{\pi/4} \\ &= \frac{1}{2\pi}\left[\frac{e^{j(\pi/4)(n-2)} - e^{-j(\pi/4)(n-2)}}{j(n-2)}\right] \\ &= \frac{1}{2\pi}\left[\frac{\sin(\pi/4)(n-2)}{(n-2)}\right] \qquad n \ne 2 \end{aligned}$$

From the given frequency response, it is clear that

$$N - 1 = 4 \Rightarrow N = 5 \Rightarrow \alpha = \frac{N-1}{2} = 2$$

if $n = 2$. Apply L'Hospital's rule.

$$h_d(n) = \lim_{n\to 2} \frac{\cos(\pi/4)(n-2) - (\pi/4)}{\pi(-1)}$$

$$h_d(2) = 1 \cdot \frac{\pi/4}{\pi} = \frac{1}{4}$$

$$\therefore\ h_d(n) = \begin{cases} \frac{\sin(\pi/4)(n-2)}{\pi(n-2)}, & n \ne 2 \\ \frac{1}{4}, & n = 2 \end{cases}$$

To determine finite impulse response $h(n)$

$$h(n) = h_d(n)\omega(n) \quad \left[\therefore\ \omega(n) = \begin{cases} 1, & 0 \le n \le 4 \\ 0, & \text{else} \end{cases}\right]$$

$$h(n) = \begin{cases} h_d(n), & 0 \le n \le 4 \\ 0, & \text{else} \end{cases}$$

Therefore, filter coefficients are

$$h(0) = \frac{\sin(\pi/4)(-2)}{\pi(n-2)} = 0.159$$

$$h(1) = \frac{\sin(\pi/4)(-1)}{\pi(-1)} = 0.225$$

$$h(2) = \frac{1}{4}$$

$$h(3) = \frac{\sin(\pi/4)(1)}{\pi(1)} = 0.225$$

$$h(4) = \frac{\sin(\pi/4)(2)}{\pi(2)} = 0.159$$

Example 5.20 A filter is to be designed with the following desired frequency response:

$$H_d(e^{j\omega}) = \begin{cases} 0, & -\pi/4 \le \omega \le \pi/4 \\ e^{-j2\omega}; & \pi/4 \le |\omega| \le \pi \end{cases}$$

Determine the filter coefficients $h_d(n)$ using Hanning window with $N = 5$.

Solution Given

$$H_d(e^{j\omega}) = \begin{cases} 0, & -\pi/4 \le \omega \le \pi/4 \\ e^{-j2\omega}, & \pi/4 \le |\omega| \le \pi \end{cases}$$

$$h_d(n) = \frac{1}{2\pi}\int_{-\pi}^{\pi} H_d(e^{j\omega})e^{j\omega n}d\omega$$

$$= \frac{1}{2\pi}\left[\int_{-\pi}^{-\pi/4} e^{-j2\omega}\cdot e^{j\omega n}d\omega + \int_{\pi/4}^{\pi} e^{-j2\omega}\cdot e^{j\omega n}d\omega\right]$$

$$= \frac{1}{2\pi}\left[\int_{-\pi}^{-\pi/4} e^{j\omega(n-2)}d\omega + \int_{\pi/4}^{\pi} e^{j\omega(n-2)}d\omega\right]$$

$$= \frac{1}{2\pi}\left[\left[\frac{e^{j\omega(n-2)}}{j(n-2)}\right]_{-\pi}^{-(\pi/4)} + \left[\frac{e^{j\omega(n-2)}}{j(n-2)}\right]_{\pi/4}^{\pi}\right]$$

$$= \frac{1}{2\pi}\left[\frac{e^{-j(\pi/4)(n-2)} - e^{-j\pi(n-2)} + e^{j\pi(n-2)} - e^{j(\pi/4)(n-2)}}{j(n-2)}\right]$$

$$h_d(n) = \left[\frac{\sin\pi(n-2) - \sin(\pi/4)(n-2)}{\pi(n-2)}\right] \quad n \ne 2$$

if $n = 2$. Apply L'Hospital's rule.

$$h_d(n) = \lim_{n\to 2}\frac{\cos\pi(n-2)(-\pi) - \cos(\pi/4)(n-2)[-(\pi/4)]}{\pi(-1)}$$

$$h_d(2) = \frac{\pi - (\pi/4)}{\pi} = \frac{(3\pi/4)}{4} = \frac{3}{4}$$

$$\therefore\ h_d(n) = \begin{cases} \frac{\sin\pi(n-2)-\sin(\pi/4)(n-2)}{\pi(n-2)}, & n \neq 2 \\ \frac{3}{4}, & n = 2 \end{cases}$$

To determine $h(n)$,

$$h(n) = h_d(n)\omega(n)$$

where

$$\omega_{\text{Han}}(n) = \begin{cases} 0.5 - 0.5\cos\frac{2\pi n}{N-1}, & 0 < n < N-1 \\ 0, & \text{otherwise} \end{cases}$$

Therefore,

$$\omega_{\text{Han}}(n) = \begin{cases} 0.5 - 0.5\cos\frac{2\pi n}{4}, & 0 < n < 4 \\ 0, & \text{otherwise} \end{cases}$$

Therefore, the filter coefficients are

$$\begin{array}{lll} h_d(0) = -0.159 & \omega(0) = 0 & h(0) = 0 \\ h_d(1) = -0.225 & \omega(1) = 0.5 & h(1) = -0.1125 \\ h_d(2) = \dfrac{3}{4} & \omega(2) = 1 & h(2) = \dfrac{3}{4} \\ h_d(3) = -0.225 & \omega(3) = 0.5 & h(3) = -0.1125 \\ h_d(4) = -0.159 & \omega(4) = 0 & h(4) = 0 \end{array}$$

Example 5.21 The desired response of a LPF is

$$H_d(e^{j\omega}) = \begin{cases} e^{-j3\omega}, & -3\pi/4 \le \omega \le 3\pi/4 \\ 0; & 3\pi/4 \le |\omega| \le \pi \end{cases}$$

Determine the frequency response of the filter for $N = 7$ using Hamming window.

Solution Given

$$H_d(e^{j\omega}) = \begin{cases} e^{-j3\omega} & -3\pi/4 \le \omega \le 3\pi/4 \\ 0, & 3\pi/4 \le |\omega| \le \pi \end{cases}$$

$$h_d(n) = \frac{1}{2\pi}\int_{-3\pi/4}^{3\pi/4} e^{-j3\omega}e^{j\omega n}\,d\omega$$

$$= \frac{1}{2\pi}\left[\int_{-3\pi/4}^{3\pi/4} e^{j\omega(n-3)}\,d\omega\right]$$

$$= \frac{1}{2\pi}\left[\frac{e^{j\omega(n-3)}}{j(n-3)}\right]_{-3\pi/4}^{3\pi/4}$$

$$= \frac{1}{2\pi}\left[\frac{e^{j(3\pi/4)(n-3)} - e^{-j(3\pi/4)(n-3)}}{j(n-3)}\right]$$

$$h_d(n) = \left[\frac{\sin(3\pi/4)(n-3)}{\pi(n-3)}\right] \qquad n \neq 3$$

For $n = 3$, apply L'Hospital's rule.

$$h_d(n) = \operatorname*{Lim}_{n\to 3} \frac{\cos(3\pi/4)(n-3)[(3\pi/4)-1]}{\pi(-1)}$$

$$h_d(3) = 1 \cdot \frac{(3\pi/4)}{\pi} = \frac{3}{4}$$

$$\therefore\ h_d(n) = \begin{cases} \frac{\sin(3\pi/4)(n-3)}{\pi(n-3)}, & n \neq 3 \\ \frac{3}{4}, & n = 3 \end{cases}$$

To determine $h(n)$,

$$h(n) = h_d(n)\omega(n) \quad \text{where } \omega_{\text{Ham}}(n) = \begin{cases} 0.54 - 0.46\cos\frac{2\pi n}{6}, & 0 < n < 6 \\ 0, & \text{otherwise} \end{cases}$$

Therefore, filter coefficients are

$h_d(0) = 0.0750$	$\omega(0) = 0.08$	$h(0) = 0.006$
$h_d(1) = -0.1592$	$\omega(1) = 0.31$	$h(1) = -0.0494$
$h_d(2) = 0.2251$	$\omega(2) = 0.77$	$h(2) = 0.1733$
$h_d(3) = 0.75$	$\omega(3) = 1$	$h(3) = 0.75$
$h_d(4) = 0.2251$	$\omega(4) = 0.77$	$h(4) = 0.1733$
$h_d(5) = -0.1592$	$\omega(5) = 0.31$	$h(5) = -0.0494$
$h_d(6) = 0.0750$	$\omega(6) = 0.08$	$h(6) = 0.006$

The transfer function of the system is

$$\begin{aligned} H(z) &= \sum_{n=0}^{N-1} h(n)z^{-n} \\ &= h(0)z^{-0} + h(1)z^{-1} + h(2)z^{-2} + h(3)z^{-3} + h(4)z^{-4} + h(5)z^{-5} \\ &\quad + h(6)z^{-6} \\ &= h(0)[1 + z^{-6}] + h(1)[z^{-1} + z^{-5}] + h(2)[z^{-2} + z^{-4}] + h(3)z^{-3} \\ &= 0.006[1 + z^{-6}] - 0.0494[z^{-1} + z^{-5}] + 0.1733[z^{-2} + z^{-4}] + 0.75z^{-3} \end{aligned}$$

The frequency response of the filter is

$$\begin{aligned}
H(e^{j\omega}) &= \left[h\left[\frac{N-1}{2}\right] + 2\sum_{n=1}^{\frac{N-1}{2}} h\left(\frac{N-1}{2} - n\right)\cos\omega n \right] e^{-j\omega\left(\frac{N-1}{2}\right)} \\
&= \left[h(3) + 2\sum_{n=1}^{3} h\,(3-n)\cos\omega n \right] e^{-j3\omega} \\
&= (h(3) + 2h(2)\cos\omega + 2h(1)\cos 2\omega + 2h(0)\cos 3\omega)e^{-j3\omega} \\
H(e^{j\omega}) &= (0.75 + 0.3466\cos\omega - 0.0988\cos 2\omega + 0.012\cos 3\omega)e^{-j3\omega}
\end{aligned}$$

Magnitude response

$$H(e^{j\omega}) = 0.75 + 0.3466\cos\omega - 0.0988\cos 2\omega + 0.012\cos 3\omega$$

Phase response

$$\angle H(e^{j\omega}) = -3\omega$$

Example 5.22 A lowpass filter should have the frequency response given below. Find filter coefficients $h_d(n)$. Also determine τ if $h_d(n) = h_d(-n)$

$$H_d(e^{j\omega}) = \begin{cases} e^{-j\omega\tau}, & -\omega_c \le \omega \le \omega_c \\ 0; & \omega_c \le |\omega| \le \pi \end{cases}$$

Solution Given

$$H_d(e^{j\omega}) = \begin{cases} e^{-j\omega\tau} & -\omega_c \le \omega \le \omega_c \\ 0; & \omega_c \le |\omega| \le \pi \end{cases}$$

$$\begin{aligned}
h_d(n) &= \frac{1}{2\pi}\int_{-\omega_c}^{\omega_c} e^{-j\omega\tau} e^{j\omega n} d\omega \\
&= \frac{1}{2\pi}\int_{-\omega_c}^{\omega_c} e^{j\omega(n-\tau)} d\omega \\
&= \frac{1}{2\pi}\left[\frac{e^{j\omega(n-\tau)}}{j(n-\tau)}\right]_{-\omega_c}^{\omega_c} \\
&= \frac{1}{2\pi}\left[\frac{e^{j\omega_c(n-\tau)} - e^{-j\omega_c(n-\tau)}}{j(n-\tau)}\right] \\
h_d(n) &= \left[\frac{\sin\omega_c(n-\tau)}{\pi(n-\tau)}\right] \qquad n \ne \tau
\end{aligned}$$

if $n = \tau$; apply L'Hospital's rule.

$$h_d(n) = \operatorname*{Lim}_{n \to \tau} \frac{\cos \omega_c(n-\tau)(-\omega_c)}{\pi(-1)}$$
$$h_d(\tau) = \frac{\omega_c}{\pi}$$
$$\therefore\ h_d(n) = \begin{cases} \frac{\sin \omega_c(n-\tau)}{\pi(n-\tau)}, & n \neq \tau \\ \frac{\omega_c}{\pi}, & n = \tau \end{cases}$$

To determine τ, given

$$h(n) = h_d(-n)$$
$$h_d(-n) = \frac{\sin \omega_c(n-\tau)}{\pi(n-\tau)} \Rightarrow \frac{-\sin \omega_c(n+\tau)}{-\pi(n+\tau)} = \frac{\sin \omega_c(n+\tau)}{\pi(n+\tau)}$$
$$= \frac{\sin \omega_c(n-\tau)}{\pi(n-\tau)} = \frac{\sin \omega_c(n+\tau)}{\pi(n+\tau)}$$

This exists only if

$$(n-\tau) = n + \tau$$

To achieve this τ should be 0.

$$\tau = 0$$

Example 5.23 A LPF has the desired response as given below

$$H_d(e^{j\omega}) = \begin{cases} e^{-j3\omega}, & 0 \le \omega \le \pi/2 \\ 0; & \pi/2 \le \omega \le \pi \end{cases}$$

Determine filter coefficients $h(n)$ for $N = 7$ using type-1 frequency sampling technique.

Solution Given

$$H_d(e^{j\omega}) = \begin{cases} e^{-j3\omega}, & 0 \le \omega \le \pi/2 \\ 0, & \pi/2 \le \omega \le \pi \end{cases}$$
$$\tilde{H}(k) = H_d(\omega)\Big|_{\omega=\omega_k=\frac{2\pi k}{N}} \qquad k = 0, 1, \ldots, N-1$$
$$= H_d(\omega)\Big|_{\omega_k=\frac{2\pi k}{N}} \qquad k = 0, 1, \ldots, 6$$

The magnitude response is shown in Fig. 5.40.

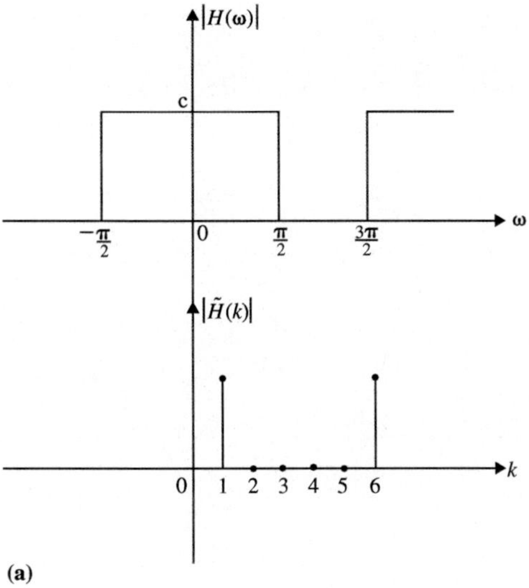

(a)

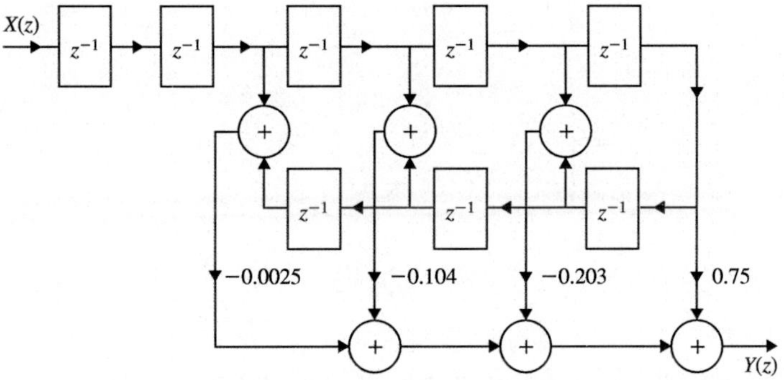

(b)

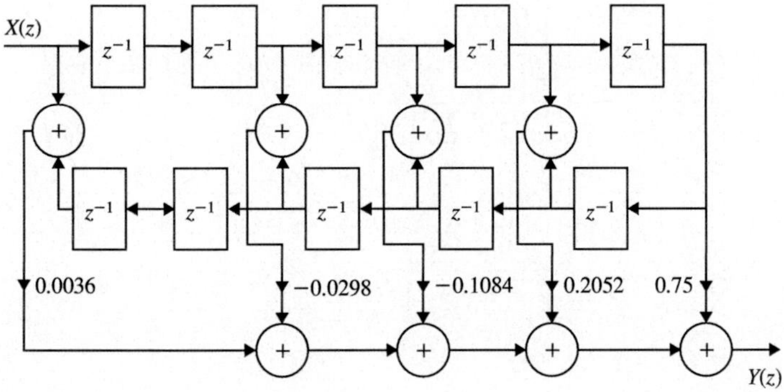

Fig. 5.40 Magnitude response for Example 5.23. **a** Structure realization for Example 5.24. **b** Structure realization for Example 5.24

$$\omega_k = \frac{2\pi k}{7}$$

$$\begin{aligned}
\text{where} \quad & k = 0 \quad \omega_0 = 0 \\
& k = 1 \quad \omega_1 = \frac{2\pi}{7} \\
& k = 2 \quad \omega_2 = \frac{4\pi}{7} \\
& k = 3 \quad \omega_3 = \frac{6\pi}{7} \\
& k = 4 \quad \omega_4 = \frac{8\pi}{7}
\end{aligned}$$

$$k = 5 \quad \omega_5 = \frac{10\pi}{7}$$

$$k = 6 \quad \omega_6 = \frac{12\pi}{7} \rightarrow \omega_k > \frac{3\pi}{2}$$

$$\tilde{H}(k) = \begin{cases} e^{-j3\omega_k}, & k = 0, 1 \\ 0, & k = 2, 3, 4, 5 \\ e^{-j3\omega_k}, & k = 6 \end{cases}$$

$$= \begin{cases} e^{-j6\pi k/7}, & k = 0, 1 \\ 0, & k = 2, 3, 4, 5 \quad \therefore N \text{ is odd} \\ e^{-j6\pi k/7}, & k = 6 \end{cases}$$

$$h(n) = \frac{1}{N}\left[\tilde{H}(0) + 2\sum_{k=1}^{\frac{N-1}{2}} \text{Re}\left[\tilde{H}(k)e^{j\frac{2\pi kn}{N}}\right]\right]$$

$$= \frac{1}{7}\left[\tilde{H}(0) + 2\sum_{k=1}^{3} \text{Re}\left[\tilde{H}(k)e^{j\frac{2\pi kn}{N}}\right]\right]$$

$$= \frac{1}{7}\left[\tilde{H}(0) + 2\sum_{k=1}^{3} \text{Re}\left[e^{j\frac{-j6\pi k}{7}} \cdot e^{j\frac{2\pi kn}{7}}\right]\right]$$

$$= \frac{1}{7}\left[1 + 2\sum_{k=1} \text{Re}\left[e^{j\frac{2\pi k}{7}(n-3)}\right]\right]$$

$$= \frac{1}{7}\left[1 + 2\sum_{k=1} \cos\frac{2\pi k}{7}(n-3)\right]$$

$$h(n) = \frac{1}{7}\left[1 + 2\cos\frac{2\pi}{7}(n-3)\right]$$

The filter coefficients are

$$
\begin{aligned}
h(0) &= -0.11456 \\
h(1) &= 0.07927 \\
h(2) &= 0.321 \\
h(3) &= 0.4285 \\
h(4) &= h(2) = 0.321 \\
h(5) &= h(1) = 0.07927 \\
h(6) &= h(0) = -0.1145
\end{aligned}
$$

Example 5.24 Design a highpass filter with frequency response

$$H_d(e^{j\omega}) = \begin{cases} e^{-j\omega\alpha}, & -\pi/4 \le \omega \le \pi \\ 0; & |\omega| \le \pi/4 \end{cases}$$

using (*i*) Hanning and (*ii*) Hamming window with $N = 11$.

Solution Hanning Window

Step 1:

$$H_d(e^{j\omega}) = \begin{cases} e^{-j\omega\alpha}, & -\pi/4 \le \omega \le -\pi,\ \pi/4 \le \omega \le \pi \\ 0, & \text{otherwise} \end{cases}$$

Step 2:

$$
\begin{aligned}
h_d(n) &= \frac{1}{2\pi}\left[\int_{-\pi}^{-\pi/4} e^{-j\omega\alpha}e^{j\omega n}d\omega + \int_{\pi/4}^{\pi} e^{-j\omega\alpha}e^{j\omega n}d\omega\right] \\
&= \frac{1}{2\pi}\left[\int_{-\pi}^{-\pi/4} e^{j\omega(n-\alpha)}d\omega + \int_{\pi/4}^{\pi} e^{j\omega(n-\alpha)}d\omega\right] \\
&= \frac{1}{2\pi}\left[\frac{e^{j\omega(n-\alpha)}}{j(n-\alpha)}\right]_{-\pi}^{-\pi/4} + \left[\frac{e^{j\omega(n-\alpha)}}{j(n-\alpha)}\right]_{\pi/4}^{\pi} \\
&= \frac{1}{2\pi}\left[\frac{e^{-j(\pi/4)(n-\alpha)} - e^{-j\pi(n-\alpha)} + e^{j\pi(n-\alpha)} - e^{j(\pi/4)(n-\alpha)}}{j(n-\alpha)}\right] \\
&= \frac{1}{2\pi j(n-\alpha)}2j\left[\sin\pi(n-\alpha) - \sin\frac{\pi}{4}(n-\alpha)\right] \\
h_d(n) &= \frac{\sin\pi(n-\alpha) - \sin\frac{\pi}{4}(n-\alpha)}{\pi(n-\alpha)}, \qquad n \ne \alpha
\end{aligned}
$$

When $n = \alpha$; apply L'Hospital's rule.

$$h_d(n) = \lim_{n \to \alpha} \frac{\cos \pi(n-\alpha)\pi - \cos(\pi/4)(n-\alpha)\pi/4}{\pi}$$

$$= \frac{\pi - (\pi/4)}{\pi} = 0.75$$

$$h_d(n) = \begin{cases} \frac{\sin \pi(n-\alpha) - \sin(\pi/4)(n-\alpha)}{\pi(n-\alpha)}, & n \neq \alpha \\ 0.75, & n = \alpha \end{cases}$$

Step 3:

$$h(d) = h_d(n) + \omega_H(n)$$

$$\omega_H(n) = \begin{cases} 0.5 - 0.5\cos\left(\frac{2\pi n}{N-1}\right), & 0 \le n \le N-1 \\ 0, & \text{otherwise} \end{cases}$$

$$\omega_H(n) = \begin{cases} 0.5 - 0.5\cos\left(\frac{\pi n}{5}\right), & 0 \le n \le 10 \\ 0, & \text{otherwise} \end{cases}$$

$$\begin{aligned} \omega_H(0) &= 0 & \omega_H(6) &= 0.9045 \\ \omega_H(1) &= 0.0954 & \omega_H(7) &= 0.6545 \\ \omega_H(2) &= 0.3454 & \omega_H(8) &= 0.3454 \\ \omega_H(3) &= 0.6545 & \omega_H(9) &= 0.0954 \\ \omega_H(4) &= 0.9045 & \omega_H(10) &= 0 \\ \omega_H(5) &= 1 \end{aligned}$$

$$\alpha = \frac{N-1}{2} = 5$$

$$h_d(n) = \begin{cases} \frac{\sin \pi(n-5) - \sin(\pi/4)(n-5)}{\pi(n-5)}, & n \neq 5 \\ 0.75, & n = 5 \end{cases}$$

$$h_d(0) = \frac{\sin \pi(-5) - \sin(\pi/4)(-5)}{\pi(-5)} = 0.0450$$

$$\begin{aligned} h_d(1) &= 0 & h_d(6) &= -0.225 \\ h_d(2) &= -0.075 & h_d(6) &= -0.159 \\ h_d(3) &= -0.159 & h_d(6) &= -0.075 \\ h_d(4) &= -0.225 & h_d(6) &= 0 \\ h_d(5) &= 0.75 & h_d(6) &= 0.0450 \end{aligned}$$

$$h(n) = h_d(n)\omega_H(n)$$

$$\begin{aligned} h(0) &= 0.045 \times 0 = 0 \\ h(1) &= 0 \times 0.0954 = 0 \\ h(2) &= -0.075 \times 0.3454 = -0.0259 \\ h(3) &= -0.159 \times 0.6545 = -0.104 \\ h(4) &= -0.225 \times 0.9045 = -0.203 \\ h(5) &= 0.75 \times 1 = 0.75 \end{aligned}$$

$$\begin{aligned} h(6) &= -0.225 \times 0.9045 = -0.203 \\ h(7) &= -0.159 \times 0.6545 = -0.104 \\ h(8) &= -0.075 \times 0.3454 = -0.0259 \\ h(9) &= 0 \times 0.954 = 0 \\ h(10) &= 0.045 \times 0 = 0 \end{aligned}$$

Step 4:

$$\begin{aligned} H(z) &= \sum_{n=0}^{10} h(n) z^{-n} \\ H(z) &= -0.0259[z^{-2} + z^{-8}] - 0.104[z^{-3} + z^{-7}] - 0.203[z^{-4} + z^{-6}] \\ &\quad +0.75z^{-5} \end{aligned}$$

Step 5: The structure realization is shown in Fig. 5.40a.

Hamming Window

Step 1:

$$H_d(e^{j\omega}) = \begin{cases} e^{-j\omega\alpha}, & -\pi/4 \le \omega \le -\pi,\ \pi/4 \le \omega \le \pi \\ 0, & \text{otherwise} \end{cases}$$

Step 2:

$$h_d(n) = \begin{cases} \frac{\sin(n-\alpha) - \sin(\pi/4)(n-\alpha)}{\pi(n-\alpha)}, & n \ne \alpha \\ 0.75, & n = \alpha \end{cases}$$

Step 3:

$$h(d) = h_d(n)\omega_H(n)$$

$$\omega_H(n) = \begin{cases} 0.54 - 0.46\cos\left(\frac{2\pi n}{N-1}\right), & 0 \le n \le N-1 \\ 0.54 - 0.46\cos\left(\frac{\pi n}{5}\right), & 0 \le n \le 10 \end{cases}$$

$$\omega_H(0) = 0.08 \qquad \omega_H(6) = 0.9121$$
$$\omega_H(1) = 0.1678 \qquad \omega_H(7) = 0.6821$$
$$\omega_H(2) = 0.3978 \qquad \omega_H(8) = 0.3978$$
$$\omega_H(3) = 0.6821 \qquad \omega_H(9) = 0.1678$$
$$\omega_H(4) = 0.9121 \qquad \omega_H(10) = 0.08$$
$$\omega_H(5) = 1$$

$$h(n) = h_d(n)\omega_H(n) \quad [\because\ h_d(n) \text{ same as above problem}]$$

$$\begin{aligned}
h(0) &= 0.045 \times 0.08 = 0.0036\\
h(1) &= 0 \times 0.1678 = 0\\
h(2) &= -0.075 \times 0.3978 = -0.0298\\
h(3) &= -0.159 \times 0.6821 = -0.1084\\
h(4) &= -0.225 \times 0.9121 = -0.2052\\
h(5) &= 0.75 \times 1 = 0.75\\
h(6) &= -0.255 \times 0.9121 = -0.2052\\
h(7) &= -0.159 \times 0.6821 = -0.1084\\
h(8) &= -0.075 \times 0.3978 = -0.0298\\
h(9) &= 0 \times 0.1678 = 0\\
h(10) &= 0.045 \times 0.08 = 0.0036
\end{aligned}$$

Step 4:

$$H(z) = \sum_{n=0}^{N-1} h(n)z^{-n} = \sum_{n=0}^{10} h(n)z^{-n}$$
$$H(z) = 0.0036[1 + z^{-10}] - 0.0298[z^{-2} + z^{-8}] - 0.1084[z^{-3} + z^{-7}]$$
$$-0.2052[z^{-4} + z^{-6}] + 0.75z^{-5}$$

Step 5: The structure realization is shown in Fig. 5.40b.

Example 5.25 Design an ideal differentiator with frequency response

$$H(e^{j\omega}) = j\omega, \qquad -\pi \le \omega \le \pi$$

using (*i*) Rectangular and (*ii*) Hamming window with $N = 8$. For both cases, plot magnitude response.

Solution Rectangular Window

Step 1:

$$H(e^{j\omega}) = j\omega = e^{j(\pi/2)}\omega$$
$$H(e^{j\omega}) = e^{j(\pi/2)}e^{-j\alpha\omega}H(e^{j\omega}), \quad \alpha = 0$$
$$H_d(e^{j\omega}) = j\omega e^{-j\omega\alpha} \quad \text{for } -\pi \le \omega \le \pi, \; N = 8$$

Step 2:

$$h_d(n) = \frac{1}{2\pi}\int_{-\pi}^{\pi} j\omega e^{-j\omega\alpha}e^{j\omega n}d\omega$$
$$= \frac{1}{2\pi}j\int_{-\pi}^{\pi} \omega e^{j\omega(n-\alpha)}d\omega$$
$$= \frac{1}{2\pi}\left[\frac{\omega e^{j\omega(n-\alpha)}}{j(n-\alpha)}\right]_{-\pi}^{\pi} - \left[\frac{e^{j\omega(n-\alpha)}}{[j(n-\alpha)]^2}\right]_{-\pi}^{\pi}$$

$$= \frac{1}{2\pi}\left[\frac{\omega e^{j\omega(n-\alpha)}}{j(n-\alpha)} - \frac{e^{j\omega(n-\alpha)}}{[j(n-\alpha)]^2}\right]_{-\pi}^{\pi}$$
$$= \frac{1}{2\pi}\left[\frac{\pi e^{j\pi(n-\alpha)}}{j(n-\alpha)} - \frac{e^{j\pi(n-\alpha)}}{j^2(n-\alpha)^2} - \frac{(-\pi)e^{-j\pi(n-\alpha)}}{j(n-\alpha)} + \frac{e^{-j\pi(n-\alpha)}}{j(n-\alpha)^2}\right]$$
$$= \frac{1}{2\pi}\left[\pi\frac{2\cos\pi(n-\alpha)}{j(n-\alpha)} + \frac{2j\sin\pi(n-\alpha)}{(n-\alpha)^2}\right]$$
$$h_d(n) = \frac{\cos\pi(n-\alpha)}{n-\alpha} - \frac{\sin\pi(n-\alpha)}{\pi(n-\alpha)^2} \quad \text{where } N = \text{even}$$
$$h_d(n) = \frac{-\sin\pi(n-\alpha)}{\pi(n-\alpha)^2}$$

Step 3:

$$h(n) = h_d(n)\omega_H(n)$$
$$\omega_H(n) = 1$$
$$h(n) = h_d(n)$$

$h_d(n)$ is anti-symmetry. So

$$h_d(n) = -h_d(N-1-n)$$

where

$$\alpha = (N-1)/2 = \frac{7}{2}.$$

$$h_d(0) = -h_d(7) = \frac{-\sin(-7/2)\pi}{\pi(-7/2)} = -0.026$$
$$h_d(1) = -h_d(6) = 0.0509$$
$$h_d(2) = -h_d(5) = -0.1415$$
$$h_d(3) = -h_d(4) = 1.27$$
$$h(n) = h_d(n)$$
$$h_d(0) = -h_d(7) = -0.026$$
$$h_d(1) = -h_d(6) = 0.0509$$
$$h_d(2) = -h_d(5) = -0.1415$$
$$h_d(3) = -h_d(4) = 1.27$$

Step 4:

$$\begin{aligned} H(z) &= \sum_{n=0}^{7} h(n)z^{-n} \\ &= -0.026[1 - z^{-7}] + 0.0509[z^{-1} - z^{-6}] - 0.1415[z^{-2} - z^{-5}] \\ &\quad +1.27[z^{-3} - z^{-4}] \end{aligned}$$

Step 5: The structure realized as shown in Fig. 5.41a.

Frequency response of ideal differentiator for anti-symmetric impulse response and N = even is

$$H(e^{j\omega}) = \sum_{n=1}^{\frac{N}{2}} d(n) \sin \omega \left(n - \frac{1}{2}\right)$$

where

$$d(n) = \sum_{n=1}^{\frac{N}{2}} 2h\left(\frac{N}{2} - n\right)$$

$$d(1) = 2h\left(\frac{N}{2} - 1\right) = 2h(4 - 1) = 2h(3) = 2.54$$
$$d(2) = 2h(4 - 2) = 2h(2) = -0.283$$
$$d(3) = 2h(4 - 3) = 2h(1) = 0.1010$$
$$d(4) = 2h(0) = -0.52$$

$$H(e^{j\omega}) = d(1) \sin \frac{\omega}{2} + d(2) \sin \frac{3\omega}{2} + d(3) \sin \frac{5\omega}{2} + d(4) \sin \frac{7\omega}{2}$$

(a)

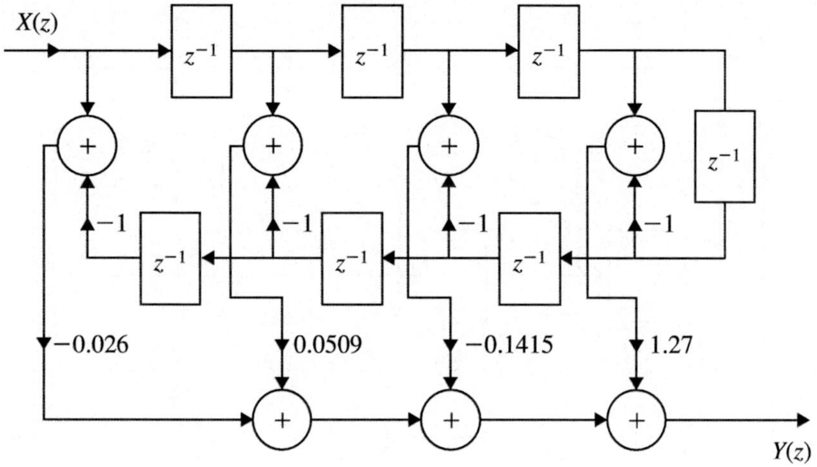

(b)

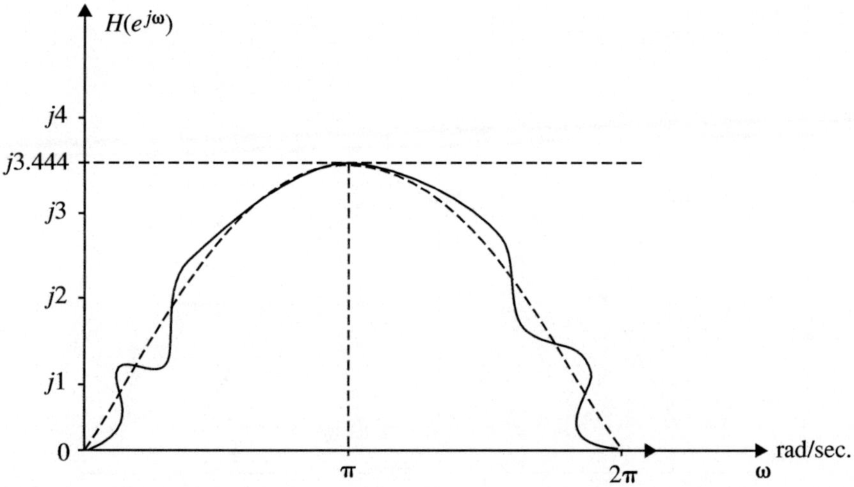

Fig. 5.41 **a** Structure realization (rectangular window) for Example 5.25. **b** Frequency response plot. **c** Structure realization (Hamming window) for Example 5.25. **d**

$$= 2.54 \sin \frac{\omega}{2} - 0.283 \sin \frac{3\omega}{2} + 0.1010 \sin \frac{5\omega}{2} - 0.52 \sin \frac{7\omega}{2}$$

$$H(e^{j\omega}) = j\left(2.54 \sin \frac{\omega}{2} - 0.283 \sin \frac{3\omega}{2} + 0.1010 \sin \frac{5\omega}{2} - 0.52 \sin \frac{7\omega}{2}\right)$$

(c)

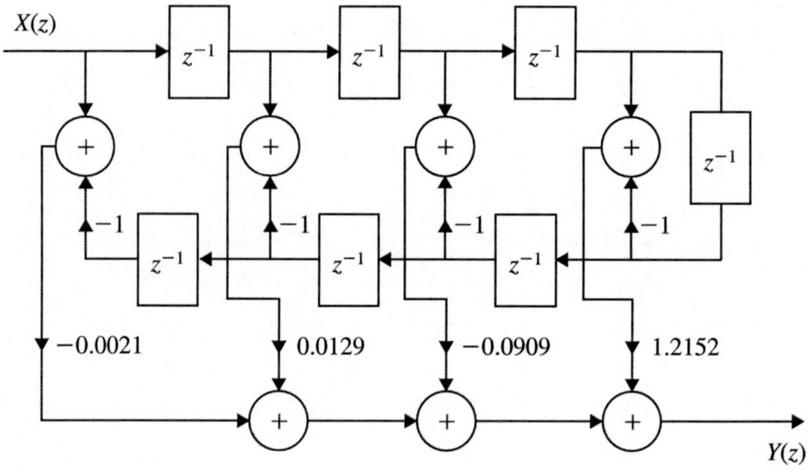

(d)

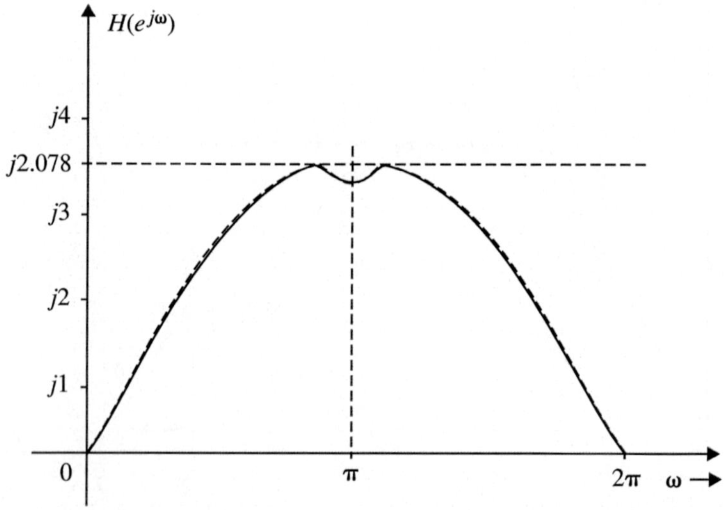

Fig. 5.41 (continued)

ω deg/sec	0	20	40	60	80	100	180	240	270	360
$H(e^{j\omega})$	0	$j0.5839$	$j0.3889$	$j1.2975$	$j0.5795$	$j1.9945$	$j3.444$	$j1.662$	$j1.89$	0

The frequency response plot is shown in Fig. 5.41b.

Hamming Window

Step 1 and 2: **Step 1 and 2:** Same as Rectangular window.
Step 3:

$$\omega_H(n) = \begin{cases} 0.54 - 0.46\cos\left(\frac{2\pi n}{N-1}\right), & 0 \le n \le N-1 \\ 0.54 - 0.46\left(\cos\frac{\pi n}{5}\right), & 0 \le n \le N \end{cases}$$

$$\begin{aligned}
\omega_H(0) &= \omega_H(7) = 0.08 \\
\omega_H(1) &= \omega_H(6) = 0.2532 \\
\omega_H(2) &= \omega_H(5) = 0.6424 \\
\omega_H(3) &= \omega_H(4) = 0.954
\end{aligned}$$

$h_d(n)$ same as above.

$$\begin{aligned}
h(0) &= -h(7) = h_d(0)\omega_H(0) = -0.0021 \\
h(1) &= -h(6) = h_d(1)\omega_H(1) = 0.0129 \\
h(2) &= -h(5) = h_d(2)\omega_H(2) = -0.0909 \\
h(3) &= -h(4) = h_d(3)\omega_H(3) = 1.2152
\end{aligned}$$

Step 4:

$$\begin{aligned}
H(z) = &-0.0021[1 - z^{-7}] + 0.0129[z^{-1} - z^{-6}] - 0.0909[z^{-2} - z^{-5}] \\
&+1.2152[z^{-3} - z^{-4}]
\end{aligned}$$

Step 5: The structure realization is shown in Fig. 5.41c.

Frequency response

$$\bar{H}(e^{j\omega}) = \sum_{n=1}^{\frac{N}{2}} d(n)\sin\left(n - \frac{1}{2}\right)\omega$$

where

$$\begin{aligned}
d(n) &= \sum_{n=1}^{\frac{N}{2}} 2h\left(\frac{N}{2} - n\right) \\
&= \sum_{n=1}^{4} 2h\,(4-n) \\
d(1) &= 2h(4-1) = 2h(3) = 2.4304 \\
d(2) &= 2h(4-2) = 2h(2) = -0.1818 \\
d(3) &= 2h(4-3) = 2h(1) = 0.0258 \\
d(4) &= 2h(0) = -0.0042
\end{aligned}$$

$$\begin{aligned}
\bar{H}(e^{j\omega}) &= d(1)\sin\frac{\omega}{2} + d(2)\sin\frac{3\omega}{2} + d(3)\sin\frac{5\omega}{2} + d(4)\sin\frac{7\omega}{2} \\
&= 2.4304\sin\frac{\omega}{2} - 0.1818\sin\frac{3\omega}{2} + 0.0258\sin\frac{5\omega}{2} - 0.0042\sin\frac{7\omega}{2} \\
H(e^{j\omega}) &= j\bar{H}(e^{j\omega}) \\
&= j\left(2.4304\sin\frac{\omega}{2} - 0.1818\sin\frac{3\omega}{2} + 0.0258\sin\frac{5\omega}{2} - 0.0042\sin\frac{7\omega}{2}\right)
\end{aligned}$$

ω deg/sec	0	45	90	120	180	240	270	315	360
$\lvert H(e^{j\omega})\rvert$	0	$j0.7875$	$j1.568$	$j2.078$	$j2.6422$	$j2.078$	$j1.568$	$j0.7875$	0

Example 5.26 Design an ideal Hilbert transformer having frequency response

$$H(e^{j\omega}) = \begin{cases} j, & -\pi \le \omega \le 0 \\ -j, & 0 \le \omega \le \pi \end{cases}$$

using (*i*) Rectangular window and (*ii*) Blackman window with $N = 11$.

Solution Step 1:

$$H(e^{j\omega}) = \begin{cases} j, & -\pi \le \omega \le 0 \\ -j, & 0 \le \omega \le \pi \end{cases}$$

Step 2:

$$
\begin{aligned}
h_d(n) &= \frac{1}{2\pi}\left[\int_{-\pi}^{0} je^{j\omega n}d\omega + \int_{0}^{\pi} -je^{j\omega n}d\omega\right]\\
&= \frac{j}{2\pi}\left[\left[\frac{e^{j\omega n}}{jn}\right]_{-\pi}^{0} - \left[\frac{e^{j\omega n}}{jn}\right]_{0}^{\pi}\right]\\
&= \frac{j}{2\pi}\frac{1}{jn}\left[\left[e^{0} - e^{-j\pi n} - e^{j\pi n} + e^{0}\right]\right]\\
&= \frac{1}{2\pi n}\left[2 - e^{j\pi n} + e^{-j\pi n}\right]\\
&= \frac{1}{2\pi n}\left[2 - 2\cos\pi n\right]\\
&= \frac{1-\cos\pi n}{\pi n} = \begin{cases}\frac{2\sin^2(\frac{\pi n}{2})}{\pi n}, & n \neq 0\\ 0, & n = 0\end{cases}\\
h_d(n) &= \frac{1-\cos\pi n}{\pi n} \qquad \therefore\ h_d(n)\ \text{is anti-symmetry}\\
h_d(0) &= \frac{1-\cos 0}{\pi 0} = 0\\
h_d(1) &= \frac{1-\cos\pi}{\pi} = \frac{2}{\pi} = -h_d(-1)\\
h_d(2) &= 0 = -h_d(-2)\\
h_d(3) &= \frac{2}{3\pi} = -h_d(-3)\\
h_d(4) &= 0 = -h_d(-4)\\
h_d(5) &= \frac{2}{5\pi} = -h_d(-5)
\end{aligned}
$$

Step 3:

$$
\begin{aligned}
h(n) &= h_d(n)\omega_H(n) \quad (\text{where } \omega_h(n) = 1)\\
h(n) &= h_d(n)\\
h(0) &= 0\\
h(1) &= -h(-1) = \frac{2}{\pi}\\
h(2) &= -h(-2) = 0
\end{aligned}
$$

$$
\begin{aligned}
h(3) &= -h(-3) = \frac{2}{3\pi}\\
h(4) &= -h(-4) = 0\\
h(5) &= -h(-5) = \frac{2}{5\pi}
\end{aligned}
$$

Step 4:

$$
\begin{aligned}
H(Z) &= z^{-5}\sum_{n=-5}^{5} h(n)z^{-n} \\
&= z^{-5}\left[\frac{2}{\pi}[z - z^{-1}] + \frac{2}{3\pi}[z^3 - z^{-3}] + \frac{2}{5\pi}[z^5 - z^{-5}]\right] \\
&= \frac{2}{\pi}[z^{-4} - z^{-6}] + \frac{2}{3\pi}[z^{-2} - z^{-8}] + \frac{2}{5\pi}[1 - z^{-10}]
\end{aligned}
$$

Step 5: The structure realization is shown in Fig. 5.42a.

Blackman Window

Step 1 and 2: **Step 1 and 2:** Same as Rectangular window.
Step 3:

$$
\begin{aligned}
&\omega_B(n) = 0.42 + 0.5\cos\frac{\pi n}{5} + 0.08\cos\frac{2\pi n}{5} \qquad \text{for} - 5 \le n \le 5 \\
&\omega_B(0) = 1 \\
&\omega_B(1) = \omega_B(-1) = 0.849
\end{aligned}
$$

$$
\begin{aligned}
\omega_B(2) &= \omega_B(-2) = 0.509 \\
\omega_B(3) &= \omega_B(-3) = 0.2 \\
\omega_B(4) &= \omega_B(-4) = 0.4 \\
\omega_B(5) &= \omega_B(-5) = 0
\end{aligned}
$$

$$
\begin{aligned}
h(n) &= h_d(n)\omega_B(n) \\
h(0) &= 0 \\
h(1) &= -h(-1) = \left(\frac{2}{\pi}\right)0.849 = 0.5405 \\
h(2) &= -h(-2) = 0 = 0 \\
h(3) &= -h(-3) = \left(\frac{2}{3\pi}\right)0.2 = 0.0423 \\
h(4) &= -h(-4) = 0 = 0 \\
h(5) &= -h(-5) = \left(\frac{2}{5\pi}\right)0 = 0
\end{aligned}
$$

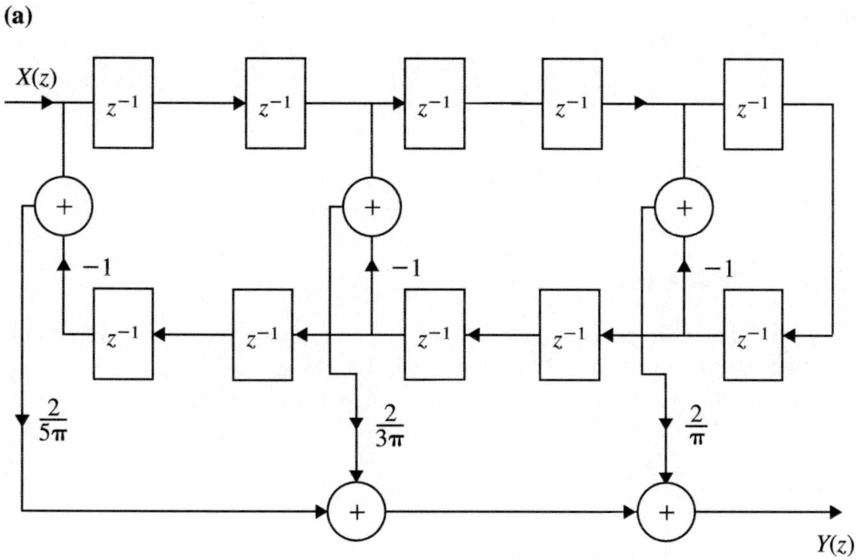

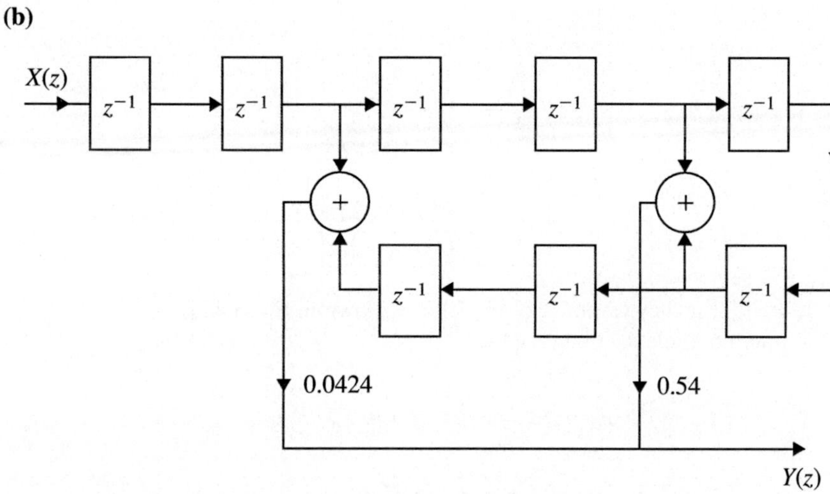

Fig. 5.42 **a** Structure realization (rectangular window) for Example 5.26. **b** Structure realization (Blackman window) for Example 5.26

Step 4:

$$H(z) = z^{-5}\sum_{n=-5}^{5} h(n)z^{-n}$$
$$= z^{-5}[0.54(z - z^{-1}) + 0.0424(z^3 - z^{-3})]$$
$$= 0.54[z^{-4} - z^{-6}] + 0.0424[z^{-2} - z^{-8}]$$

Step 5: The structure realization is shown in Fig. 5.42b.

Example 5.27 Using a rectangular window technique design a lowpass filter with passband gain of unity, cut-off frequency of 100 Hz and working at a sampling frequency of 5 kHz. The length of the impulse response should be 7.

Solution Given $f_c = 1000\,\text{Hz}$, $F = 5000\,\text{Hz}$

$$H_d(e^{i\omega}) = \begin{cases} 1, & -\omega_c \leq \omega \leq \omega_c \\ 0, & -\pi \leq -\omega \leq -\omega_c \text{ and } \omega_c \leq \omega \leq \pi \end{cases}$$

where

$$\omega_c = 2\pi f_c$$
$$= 2\pi \times 1000.T$$

$$\boxed{\omega_c = \frac{2000\pi}{5000} = \frac{2\pi}{5}}$$

The desired frequency response of the LPF is shown in Fig. 5.43a.

The filter coefficients are given by

$$h_d(n) = \frac{1}{2\pi}\int_{-\pi}^{\pi} H_d(e^{j\omega}).e^{j\omega n}d\omega$$
$$= \frac{1}{2\pi}\int_{-\omega c}^{\omega c} e^{j\omega n}d\omega$$
$$= \frac{1}{2\pi}\int_{\frac{-2\pi}{5}}^{\frac{2\pi}{5}} e^{j\omega n}d\omega$$

$$= \frac{1}{2\pi}\left[\frac{e^{j\omega n}}{jn}\right]_{\frac{-2\pi}{5}}^{\frac{2\pi}{5}}$$
$$= \frac{1}{n\pi}\left[\frac{e^{\frac{j2\pi n}{5}} - e^{\frac{-j2\pi n}{5}}}{2j}\right]$$

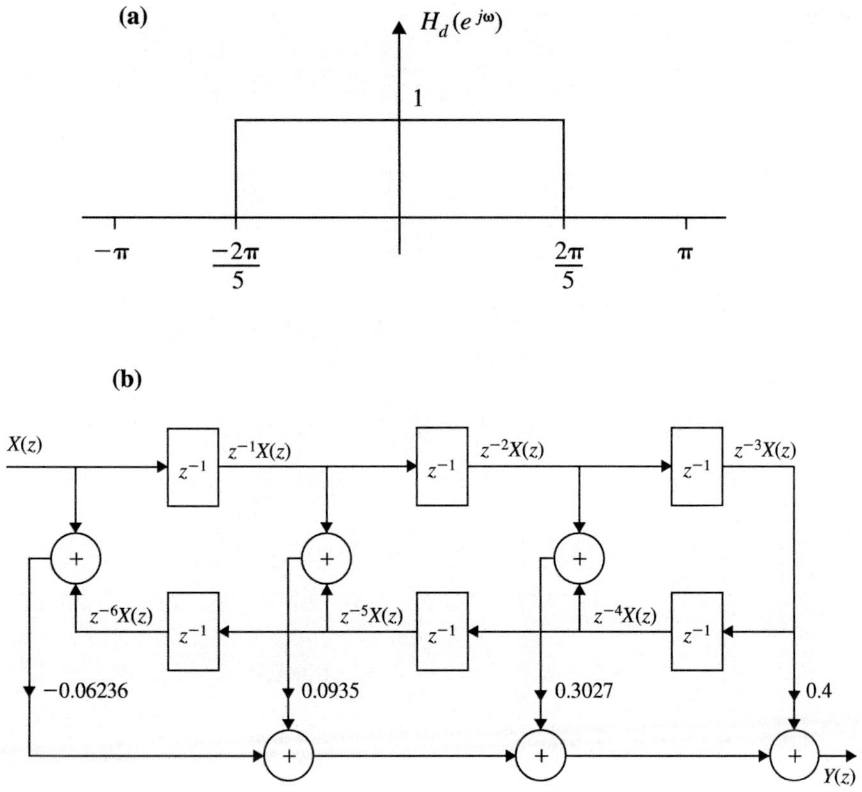

Fig. 5.43 **a** Frequency Response of for Example 5.27. **b** Linear phase realization of FIR filter for Example 5.27

$$h_d(n) = \frac{\sin \frac{2\pi n}{5}}{n\pi}, -\infty \leq n \leq \infty$$

The rectangular window for $N = 7$ is given by

$$\omega_R(n) = \begin{cases} 1, & -3 \leq n \leq 3 \\ 0, & \text{otherwise} \end{cases}$$

The filter coefficient

$$h(n) = h_d(n) \times \omega_R(n)$$

$$h_(n) = \begin{cases} h_d(n), & -3 \leq n \leq 3 \\ 0, & \text{otherwise} \end{cases}$$

For $n = 0$,

$$h(0) = h_d(0) = \lim_{n\to 0} \frac{\sin\frac{2\pi n}{5}}{n\pi}$$

$$= \lim_{n\to 0} \frac{2}{5}\frac{\sin\frac{2\pi n}{5}}{\frac{2\pi n}{5}}$$

$$= \lim_{n\to 0} \frac{2}{5}$$

$$h(0) = \frac{2}{5}$$

For $n = 1$;

$$h(1) = h(-1) = \frac{\sin\frac{2\pi}{5}}{\pi} = 0.3027$$

For $n = 2$;

$$h(2) = h(-2) = \frac{\sin\frac{4\pi}{5}}{2\pi} = 0.0935$$

For $n = 3$;

$$h(3) = h(-3) = \frac{\sin\frac{6\pi}{5}}{3\pi} = -0.06236$$

The filter coefficients of realizable filter are

$$h(0) = h(6) = -0.06236$$
$$h(1) = h(5) = 0.0935$$
$$h(2) = h(4) = 0.3027$$
$$h(3) = 0.4$$

It is symmetric about $\alpha = \frac{N-1}{2} = \frac{7-1}{2} = 3$. The transfer function of realizable filter is

$$H(z) = \sum_{n=0}^{N-1} h(n)z^{-n}$$

$$= \sum_{n=0}^{6} h(n)z^{-n}$$

$$= h(0) + h(1)z^{-1} + h(2)z^{-2} + h(3)z^{-3}$$
$$+h(4)z^{-4} + h(5)z^{-5} + h(6)z^{-6}$$

$$H(z) = \frac{Y(z)}{X(z)}$$

$$= h(0)[1 + z^{-6}] + h(1)[z^{-1} + z^{-5}] + h(2)[z^{-2}$$
$$+ z^{-4}] + h(3)z^{-3}$$
$$Y(z) = -0.06236[1 + z^{-6}]X(z) + 0.0935[z^{-1} + z^{-5}]X(z)$$
$$+ 0.3027[z^{-2} + z^{-4}]X(z) + 0.4z^{-3}X(z)$$

Linear phase realization of the filter is shown in Fig. 5.43b.

Example 5.28 Design a bandpass filter which approximates the ideal filter with cut-off frequencies at 0.2 rad/s and 0.3 rad/s. The filter order is $N = 7$. Use Hamming window.

Solution Given $\omega c_1 = 0.2$ rad/s, $\omega c_2 = 0.3$ rad/s, $N = 7$.

Step 1: Choose $H_d(\omega)$ ideal filter such that

$$H_d(\omega) = \begin{cases} 1, & -\omega c_2 \le \omega \le -\omega c_1 \text{ and } \omega c_1 \le \omega \le -\omega c_2 \\ 0, & \text{otherwise} \end{cases}$$

Step 2: To Find $h_d(n)$

$$h_d(n) = \frac{1}{2\pi}\left[\int_{-\omega c_2}^{-\omega c_1} H_d(\omega)e^{j\omega n}d\omega + \int_{\omega c_1}^{\omega c_2} H_d(\omega)e^{j\omega n}d\omega\right]$$
$$= \frac{1}{2\pi}\left[\int_{-0.3}^{-0.2} e^{j\omega n}d\omega + \int_{0.2}^{0.3} ej\omega nd\omega\right]$$
$$= \frac{1}{2\pi}\left[\left[\frac{e^{j\omega n}}{jn}\right]_{-0.3}^{-0.2} + \left[\frac{e^{j\omega n}}{jn}\right]_{0.2}^{0.3}\right]$$
$$= \frac{1}{2n\pi j}\left[e^{-j0.2n} - e^{-j0.3n} + e^{j0.3n} - e^{j0.2n}\right]$$
$$= \frac{1}{n\pi}\left[\frac{e^{j0.3n} - e^{-j0.3n}}{2j} - \frac{e^{j0.2n} - e^{-j0.2n}}{2j}\right]$$
$$= \frac{1}{n\pi}[\sin 0.3n - \sin 0.2n]$$
$$h_d(n) = \frac{1}{n\pi}[\sin 0.3n - \sin 0.2n]$$

Step 3: To Find $h(n)$

$$h(n) = h_d(n) \times \omega_H(n)$$

Hamming window function is

$$\omega_H(n) = \begin{cases} 0.54 \;\; +0.46\cos(\frac{2\pi n}{N-1}) & -\frac{(N-1)}{2} \le n \le \frac{(N-1)}{2} \\ 0, & \text{otherwise} \end{cases}$$

$$\omega_H(0) = 0.54 + 0.46 = 1$$
$$\omega_H(1) = \omega_H(-1) = 0.54 + 0.46\cos\frac{2\pi}{6} = 0.77$$
$$\omega_H(2) = \omega_H(-2) = 0.54 + 0.46\cos\frac{4\pi}{6} = 0.31$$
$$\omega_H(3) = \omega_H(-3) = 0.54 + 0.46\cos\frac{6\pi}{6} = 0.08$$

$$h_d(n) = \frac{1}{n\pi}[\sin 0.3n - \sin 0.2n], \qquad -3 \le n \le 3$$
$$h_d(0) = \lim_{n\to 0}\frac{1}{\pi}\left[0.3\frac{\sin 0.3n}{0.3n} - 0.2\frac{\sin 0.2n}{0.2n}\right]$$
$$= \frac{0.3 - 0.2}{\pi} = 0.03183$$
$$h_d(1) = h_d(-1) = \frac{1}{\pi}[\sin 0.3 - \sin 0.2] = 0.0308$$
$$h_d(2) = h_d(-2) = \frac{1}{2\pi}[\sin 0.6 - \sin 0.4] = 0.0278$$
$$h_d(3) = h_d(-3) = \frac{1}{3\pi}[\sin 0.9 - \sin 0.6] = 0.0232$$
$$h(n) = h_d(n) \times \omega_H(n)$$
$$h(0) = h_d(0) \times \omega_H(0) = 0.03183$$
$$h(1) = h(-1) = h_d(+1) \times \omega_H(+1) = 0.02926$$
$$h(2) = h(-2) = h_d(2) \times \omega_H(2) = 0.0086$$
$$h(3) = h(-3) = h_d(3) \times \omega_H(3) = 0.01856$$

Step 4: To Find $H(z)$

$$H(z) = z^{-(\frac{N-1}{2})}\left[\sum_{n=1}^{\frac{N-1}{2}} h(n)[z^n + z^{-n}] + h(0)\right]$$
$$= z^{-3}\left[h(0) + \sum_{n=1}^{3} h(n)[z^n + z^{-n}]\right]$$
$$= z^{-3}\left[h(0) + h(1)[z^1 + z^{-1}] + h(2)[z^2 + z^{-2}] + h(3)[z^3 + z^{-3}]\right]$$

$$\boxed{H(z) = \left[h(0)z^{-3} + h(1)[z^{-2} + z^{-4}] + h(2)[z^{-1} + z^{-5}] + h(3)[1 + z^{-6}]\right]}$$

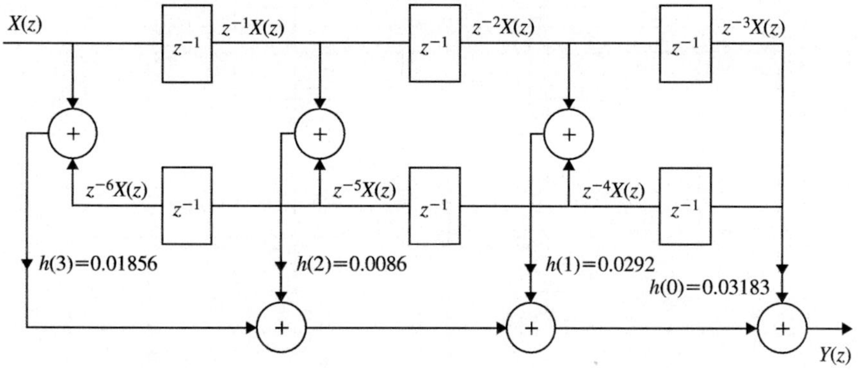

Fig. 5.44 Realization of FIR filter for Example 5.28

$$\boxed{\begin{array}{l} H(z) = 0.03183z^{-3} + 0.02926[z^{-2} + z^{-4}] + 0.0086[z^{-1} + z^{-5}] \\ +0.01856[1 + z^{-6}] \end{array}}$$

Step 5: Realization of the filter

$$\begin{aligned} Y(z) = h(0)z^{-3}X(z) &+ h(1)[z^{-2} + z^{-4}]X(z) \\ &+h(2)[z^{-1} + z^{-5}]X(z) \\ &+h(3)[1 + z^{-6}]X(z) \end{aligned}$$

FIR filter is realized as shown in Fig. 5.44.

Example 5.29 Obtain linear phase structure with minimum member of multipliers for the system described by the equation.

$$y(n) = x(n) + \frac{1}{2}x(n-1) - \frac{1}{4}x(n-2) + \frac{1}{2}x(n-3) + x(n-4)$$

Solution Given

$$y(n) = x(n) + \frac{1}{2}x(n-1) - \frac{1}{4}x(n-2) + \frac{1}{2}x(n-3) + x(n-4)$$

Taking z-transform on both sides, we get

$$\begin{aligned} Y(z) &= X(z) + \frac{1}{2}z^{-1}X(z) - \frac{1}{4}z^{-2}X(z) + \frac{1}{2}z^{-3}X(z) + z^{-4}X(z) \\ &= X(z)\left[1 + \frac{1}{2}z^{-1} - \frac{1}{4}z^{-2} + \frac{1}{2}z^{-3} + z^{-4}\right] \end{aligned}$$

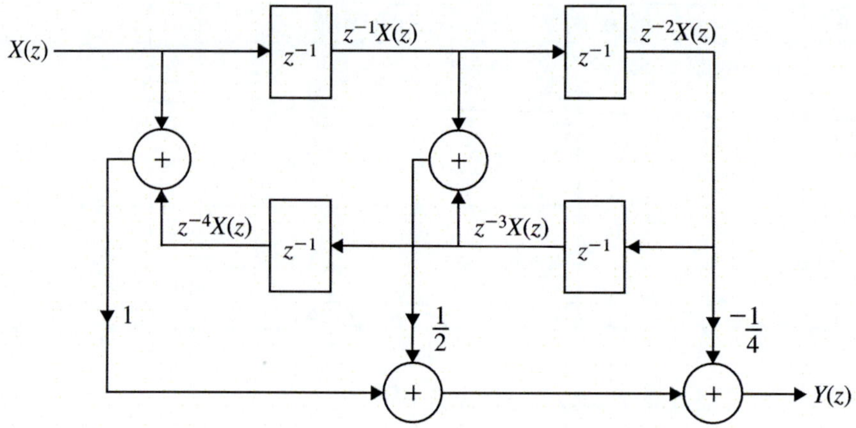

Fig. 5.45 Linear Phase Realization for Example 5.29

$$\frac{Y(z)}{X(z)} = 1 + \frac{1}{2}z^{-1} - \frac{1}{4}z^{-2} + \frac{1}{2}z^{-3} + z^{-4}$$

$$H(z) = 1 + \frac{1}{2}z^{-1} - \frac{1}{4}z^{-2} + \frac{1}{2}z^{-3} + z^{-4}$$

$$H(z) = \sum_{n=0}^{\infty} h(n)z^{-n}$$
$$= h(0) + h(1)z^{-1} + h(2)z^{-2} + h(3)z^{-3} + h(4)z^{-4} + \cdots$$

Comparing these equations, we get impulse response.

$$h(0) = 1, \quad h(1) = \frac{1}{2}, \quad h(2) = \frac{-1}{4}, \quad h(3) = \frac{1}{2}, \quad h(4) = 1$$

$$h(n) = \{1, \frac{1}{2}, \frac{-1}{4}, \frac{1}{2}, 1\} \text{ and } N = 5$$

Here $h(n)$ satisfies the condition $h(n) = h(N - 1 - n)$

$$H(z) = \frac{Y(z)}{X(z)} = (1 + z^{-4}) + \frac{1}{2}[z^{-1} + z^{-3}] - \frac{1}{4}z^{-2}$$
$$Y(z) = [X(z) + z^{-4}X(z)] + \frac{1}{2}[z^{-1}X(z) + z^{-3}X(z)] - \frac{1}{4}z^{-2}X(z)$$

$Y(z)$ is realized as shown in Fig. 5.45.

Example 5.30 Design a linear phase FIR digital filter for the given specifications using Hamming window of length $M = 7$.

$$H_d(\omega) = \begin{cases} e^{-j3\omega}, & \text{for } |\omega| \le \frac{\pi}{6} \\ 0, & \text{for } \frac{\pi}{6} \le |\omega| \le \pi \end{cases}$$

Solution Given

$$H_d(\omega) = \begin{cases} e^{-j3\omega}, & |\omega| \le \frac{\pi}{6} \\ 0, & \frac{\pi}{6} \le |\omega| \le \pi \end{cases}$$

The frequency response having a term $e^{-j\omega(\frac{N-1}{2})}$ gives $h(n)$ which is symmetrical about $\alpha = \frac{N-1}{2} = 3$, i.e. we get a causal sequence.

$$\begin{aligned} h_d(n) &= \frac{1}{2\pi}\int_{\frac{-\pi}{6}}^{\frac{\pi}{6}} e^{-j\omega n} d\omega \\ &= \frac{1}{2\pi}\int_{\frac{-\pi}{6}}^{\frac{\pi}{6}} e^{j(n-3)\omega} d\omega \\ &= \frac{1}{2\pi}\left[\frac{e^{j(n-3)\omega}}{j(n-3)}\right]_{\frac{-\pi}{6}}^{\frac{\pi}{6}} \\ &= \frac{1}{\pi(n-3)}\left[\frac{e^{j\frac{\pi}{6}(n-3)} - e^{-j\frac{\pi}{6}(n-3)}}{2j}\right] \\ h_d(n) &= \frac{\sin\frac{\pi}{6}(n-3)}{\pi(n-3)}, \qquad 0 \le n \le N-1, \quad 0 \le n \le 6 \\ h_d(0) = h_d(6) &= \frac{\sin(\frac{\pi}{2})}{3\pi} = 0.1061 \\ h_d(1) = h_d(5) &= \frac{\sin(\frac{\pi}{3})}{2\pi} = 0.1378 \\ h_d(2) = h_d(4) &= \frac{\sin(\frac{\pi}{6})}{\pi} = 0.1591 \\ h_d(3) &= \lim_{n\to 3}\frac{\sin\frac{\pi}{6}(n-3)}{6.\frac{\pi}{6}(n-3)} = \frac{1}{6} = 0.1666 \end{aligned}$$

Hamming window function is

$$\omega_H(n) = \begin{cases} 0.54 - 0.46\cos(\frac{2\pi n}{N-1}), & 0 \le n \le N-1 \\ 0, & \text{otherwise} \end{cases}$$

$$\omega_H(n) = \begin{cases} 0.54 - 0.46\cos(\frac{2\pi n}{6}), & 0 \le n \le 6 \\ 0, & \text{otherwise} \end{cases}$$

$$\omega_H(0) = 0.54 - 0.46 = 0.08$$
$$\omega_H(1) = 0.54 - 0.46\cos\left(\frac{2\pi}{6}\right) = 0.31$$
$$\omega_H(2) = 0.54 - 0.46\cos\left(\frac{4\pi}{6}\right) = 0.77$$
$$\omega_H(3) = 0.54 - 0.46\cos\left(\frac{6\pi}{6}\right) = 1$$
$$\omega_H(4) = \omega_H(2) = 0.77$$
$$\omega_H(5) = \omega_H(1) = 0.31$$
$$\omega_H(6) = \omega_H(0) = 0.08$$

$$h(n) = h_d(n) \times \omega_H(n)$$
$$h(0) = h_d(0) \times \omega_H(0)$$
$$h(0) = 0.0048$$
$$h(6) = h(0) = 0.0048$$
$$h(1) = h(5) = h_d(1) \times \omega_H(1) = 0.04271$$
$$h(2) = h(4) = h_d(2) \times \omega_H(2) = 0.1225$$
$$h(3) = h_d(3) \times \omega_H(3) = 0.16667$$

$$\begin{aligned} H(z) &= \sum_{n=0}^{N-1} h(n)z^{-n} \\ &= h(0) + h(1)z^{-1} + h(2)z^{-2} + h(3)z^{-3} + h(4)z^{-4} + h(5)z^{-5} + h(6)z^{-6} \\ &= h(0)[1 + z^{-6}] + h(1)[z^{-1} + z^{-5}] + h(2)[z^{-2} + z^{-4}] + h(3)z^{-3} \end{aligned}$$

$$\boxed{\begin{aligned} &H(z) = 0.0048[1 + z^{-6}] + 0.04271[z^{-1} + z^{-5}] \\ &+0.1225[z^{-2} + z^{-4}] + 0.16667z^{-3} \end{aligned}}$$

Realization of the filter is

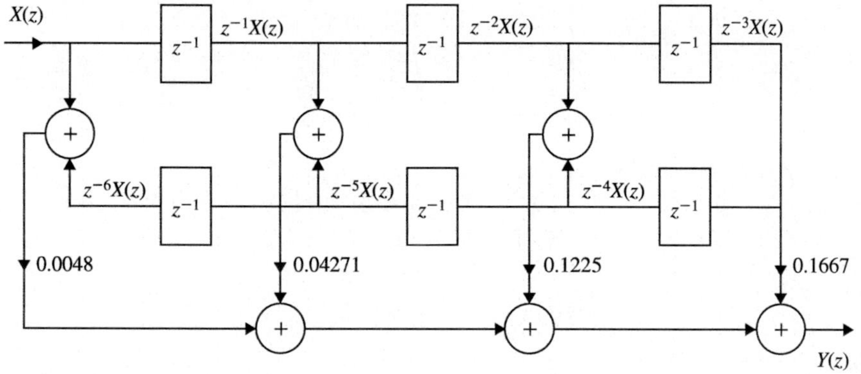

Fig. 5.46 FIR filter structure for Example 5.30

$$
\begin{aligned}
H(z) &= \frac{Y(z)}{X(z)} \\
&= 0.0048[1 + z^{-6}] + 0.04271[z^{-1} + z^{-5}] + 0.1225[z^{-2} + z^{-4}] \\
&\quad +0.16667z^{-3} \\
Y(z) &= 0.0048[X(z) + z^{-6}X(z)] + 0.04271[z^{-1}X(z) + z^{-5}X(z)] \\
&\quad +0.1225[z^{-2}X(z) + z^{-4}X(z)] + 0.16667z^{-3}X(z)
\end{aligned}
$$

The structure of the filter is shown in Fig. 5.46.

Example 5.31 Design and implement linear phase FIR filter of length $N = 15$ which has the following unit sample sequence.

$$
H(k) = \begin{cases} 1, & k = 0, 1, 2, 3 \\ 0, & k = 4, 5, 6, 7 \end{cases}
$$

Solution Linear phase FIR filter is symmetric about

$$
\alpha = \frac{N-1}{2} = \frac{15-1}{2} = 7
$$

$$
|H(k)| = \begin{cases} 1, & \text{for } 0 \le k \le 3 \text{ and } 11 \le k \le 14 \\ 0, & \text{for } 4 \le k \le 10 \end{cases}
$$

$$
H(k) = \begin{cases} e^{-j\omega\alpha}, & k = 0, 1, 2, 3 \\ 0, & k = 4, 5, 6, 7 \end{cases}
$$

$$H(k) = \begin{cases} e^{\frac{-j2\pi k\alpha}{15}}, & k = 0, 1, 2, 3 \\ 0, & k = 4, 5, 6, 7 \end{cases}$$

$$H(k) = \begin{cases} e^{\frac{-j14\pi k}{15}}, & k = 0, 1, 2, 3 \\ 0, & 4 \le k \le 10 \\ e^{\frac{-j14\pi(k-15)}{15}}, & 11 \le k \le 14 \end{cases}$$

$$H(k) = \begin{cases} e^{\frac{j14\pi k}{15}}, & k = 0, 1, 2, 3 \\ 0, & k = 4, 5, 6, 7 \end{cases}$$

$$h(n) = \frac{1}{N}\left[H(0) + 2\sum_{k=1}^{\frac{N-1}{2}} Re(H(k)e^{\frac{j2\pi nk}{15}})\right]$$

$$= \frac{1}{15}\left[1 + 2\sum_{k=1}^{7} Re[e^{\frac{j14\pi k}{15}}.e^{\frac{j2\pi nk}{15}}]\right]$$

$$= \frac{1}{15}\left[1 + 2\sum_{k=1}^{3} \cos\frac{2\pi k(7-n)}{15}\right]$$

$$h(n) = \frac{1}{15}\left[1 + 2\cos\frac{2\pi(7-n)}{15} + 2\cos\frac{4\pi(7-n)}{15} + 2\cos\frac{6\pi(7-n)}{15}\right]$$

$$h(0) = h(14) = \frac{1}{15}\left[1 + 2\cos\frac{14\pi}{15} + 2\cos\frac{28\pi}{15} + 2\cos\frac{42\pi}{15}\right] = -0.05$$

$$h(1) = h(13) = \frac{1}{15}\left[1 + 2\cos\frac{12\pi}{15} + 2\cos\frac{24\pi}{15} + 2\cos\frac{36\pi}{15}\right] = 0.041$$

$$h(2) = h(12) = \frac{1}{15}\left[1 + 2\cos\frac{10\pi}{15} + 2\cos\frac{20\pi}{15} + 2\cos\frac{30\pi}{15}\right] = 0.0666$$

$$h(3) = h(11) = -0.0365$$

$$h(4) = h(10) = -0.1078$$

$$h(5) = h(9) = 0.034$$

$$h(6) = h(8) = 0.3188$$

$$h(7) == 0.466$$

$$H(z) = \sum_{n=0}^{N-1} h(n)z^{-n}$$

$$= h(0)z^{-1} + h(1)z^{-2} + \cdots$$

$$= h(0)[1 + z^{-14}] + h(1)[z^{-1} + z^{-13}] + h(2)[z^{-2} + z^{-12}]$$

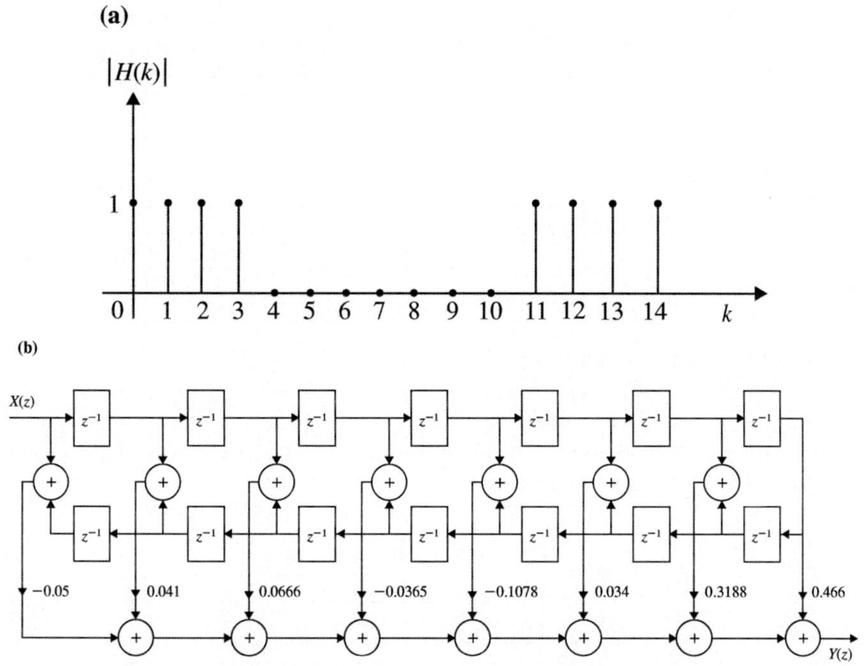

Fig. 5.47 **a** Ideal magnitude response for Example 5.31. **b** Structure realization for Example 5.31

$$
\begin{aligned}
&+h(3)[z^{-3}+z^{-11}]+h(4)[z^{-4}+z^{-10}]+h(5)[z^{-5}+z^{-9}]\\
&+h(6)[z^{-6}+z^{-8}]+h(7)z^{-7}\\
H(z) = &-0.05[1+z^{-14}]+0.041[z^{-1}+z^{-13}]+0.0666[z^{-2}+z^{-12}]\\
&-0.0365[z^{-3}+z^{-11}]-0.1078[z^{-4}+z^{-10}]+0.034[z^{-5}+z^{-9}]\\
&+0.3188[z^{-6}+z^{-8}]+0.466z^{-7}
\end{aligned}
$$

The structure is shown in Fig. 5.47.

Example 5.32 Design a FIR filter approximating the ideal frequency response.

$$
H_d(e^{j\omega}) = \begin{cases} e^{-j\alpha\omega}, & \text{for } |\omega| \le \frac{\pi}{6} \\ 0, & \text{for } \frac{\pi}{6} \le |\omega| \le \pi \end{cases}
$$

Determine the filter coefficients for $N = 13$.

Solution We know that

$$
h_d(n) = \frac{1}{2\pi}\int_{-\pi}^{\pi} H_d(e^{j\omega})e^{j\omega n}d\omega
$$

$$= \frac{1}{2\pi}\int_{\frac{-\pi}{6}}^{\frac{\pi}{6}} e^{-j\alpha\omega n} e^{j\omega n} d\omega$$

$$\begin{aligned}
&= \frac{1}{2\pi}\int_{\frac{-\pi}{6}}^{\frac{\pi}{6}} e^{j(n-\alpha)\omega} d\omega \\
&= \frac{1}{2}\left[\frac{e^{j(n-\alpha)\omega}}{j(n-\alpha)}\right]_{\frac{-\pi}{6}}^{\frac{\pi}{6}} \\
&= \frac{1}{\pi(n-\alpha)}\left[\frac{e^{j(n-\alpha)\frac{\pi}{6}}}{2j} - \frac{e^{-j(n-\alpha)\frac{\pi}{6}}}{2j}\right] \\
&= \frac{\sin(n-\alpha)\frac{\pi}{6}}{\pi(n-\alpha)}
\end{aligned}$$

From the frequency response, we can find that the filter coefficients are symmetrical about $\alpha = \frac{N-1}{2} = 6$, satisfying $h(n) = h(N-1-n)$

$$h_n = \begin{cases} h_d(n), & \text{for } 0 \le n \le 12 \\ 0, & \text{otherwise} \end{cases}$$

$$h_n = \begin{cases} \frac{\sin\frac{\pi}{6}(n-\alpha)}{\pi(n-\alpha)}, & 0 \le n \le 12 \\ 0, & \text{otherwise} \end{cases}$$

$$\begin{aligned}
h(n) &= h(N-1-n) \\
h(n) &= h(13-1-0) = h(12) = 0 \\
h(1) &= h(11) = 0.0318 \\
h(2) &= h(10) = 0.0689 \\
h(3) &= h(9) = 0.106 \\
h(4) &= h(8) = 0.1379 \\
h(5) &= h(7) = 0.159 \\
h(6) &= \operatorname*{Lim}_{n\to 6} \frac{\sin\frac{\pi}{6}(n-6)}{6\frac{\pi}{6}(n-6)} = \frac{1}{6} = 0.1667
\end{aligned}$$

$$\begin{aligned}
H(z) &= \sum_{n=0}^{N-1} h(n)z^{-n} \\
&= \sum_{n=0}^{12} h(n)z^{-n} \\
&= h(0) + h(1)z^{-1} + h(2)z^{-2} + h(3)z^{-3} + \cdots + h(12)z^{-12}
\end{aligned}$$

$$= h(0)[1+z^{-12}] + h(1)[z^{-1}+z^{-11}] + h(2)[z^{-2}+z^{-10}]$$
$$+h(3)[z^{-3}+z^{-9}] + h(4)[z^{-4}+z^{-8}] + h(5)[z^{-5}+z^{-7}] + h(6)z^{-6}$$

$$\boxed{\begin{aligned}&H(z) = 0.0318[z^{-1}+z^{-11}] + 0.0689[z^{-2}+z^{-10}]\\&+0.106[z^{-3}+z^{-9}] + 0.1379[z^{-4}+z^{-8}]\\&+0.159[z^{-5}+z^{-7}] + 0.1667z^{-6}\end{aligned}}$$

The structure realization is shown in Fig. 5.48.

Example 5.33 Design a filter with

$$H_d(e^{-j\omega}) = \begin{cases} e^{-j3\omega}, & \frac{-\pi}{4} \le \omega \le \frac{\pi}{4} \\ 0, & \frac{\pi}{4} < |\omega| \le \pi \end{cases}$$

Solution Given

$$H_d(e^{-j\omega}) = e^{-j3\omega}$$

$$\alpha = 3 = \frac{N-1}{2}$$

$$N = 7$$

The frequency response is having a term $e^{-j\omega(\frac{N-1}{2})}$ which gives $h(n)$ symmetrical about $n = \frac{N-1}{2} = 3$. Therefore, we get casual sequence.

$$h_d(n) = \int_{\frac{-\pi}{4}}^{\frac{\pi}{4}} e^{-j3\omega} e^{j\omega n} d\omega$$

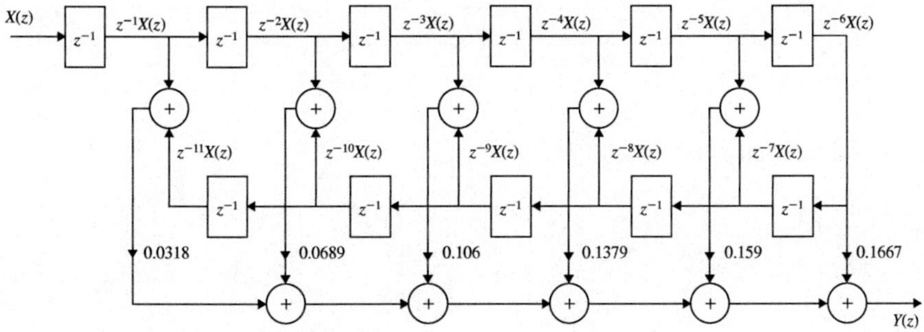

Fig. 5.48 Structure realization for Example 5.32

$$= \int_{\frac{-\pi}{4}}^{\frac{\pi}{4}} e^{j(n-3)\omega} d\omega$$

$$= \int_{\frac{-\pi}{4}}^{\frac{\pi}{4}} \left[\frac{e^{j(n-3)\omega}}{j(n-3)} \right]_{\frac{-\pi}{4}}^{\frac{\pi}{4}}$$

$$h_d(n) = \frac{\sin \frac{\pi}{4}(n-3)}{\pi(n-3)}$$

For $N = 7$, the filter coefficients are

$$h(n) = h_d(n)$$
$$h(0) = h(6) = 0.075$$
$$h(1) = h(5) = 0.159$$
$$h(2) = h(4) = 0.22$$
$$h(3) = 0.25$$

$$H(z) = \sum_{n=0}^{N-1} h(n)z^{-n} = \sum_{n=0}^{6} h(n)z^{-n}$$
$$= h(0) + h(1)z^{-1} + h(2)z^{-2} + h(3)z^{-3} + h(4)z^{-4} + h(5)z^{-5} + h(6)z^{-6}$$
$$= h(0)[1 + z^{-6}] + h(1)[z^{-1} + z^{-5}] + h(2)[z^{-2} + z^{-4}] + h(3)z^{-3}$$

$$\boxed{H(z) = 0.075[1 + z^{-6}] + 0.159[z^{-1} + z^{-5}] + 0.22[z^{-2} + z^{-4}] + 0.25z^{-3}}$$

Example 5.34 Design a filter with

$$H_d(e^{-j\omega}) = \begin{cases} e^{-j3\omega}, & \frac{-\pi}{4} \le \omega \le \frac{\pi}{4} \\ 0, & \frac{\pi}{4} \le |\omega| \le \pi \end{cases}$$

using a Hamming window with $N = 7$.

Solution Given

$$H_d(e^{-j\omega}) = \begin{cases} e^{-j3\omega}, & \frac{-\pi}{4} \le \omega \le \frac{\pi}{4} \\ 0, & \frac{\pi}{4} \le -|\omega| \le \pi \end{cases}$$

$\alpha = 3$; $\frac{N-1}{2} = 3$; $N = 7$. Impulse response $h(n)$ is symmetric about $\alpha = 3$. Since the frequency response is having a term $e^{-j\omega\alpha}$

$$h_d(n) = \frac{1}{2\pi}\int_{\frac{-\pi}{4}}^{\frac{\pi}{4}} e^{-j3\omega}e^{j\omega n}d\omega$$

$$= \frac{1}{2\pi}\int_{\frac{-\pi}{4}}^{\frac{\pi}{4}} e^{j(n-3)\omega}d\omega$$

$$h_d(n) = \frac{\sin\frac{\pi}{4}(n-3)}{\pi(n-3)}$$

$$h_d(0) = h_d(6) = 0.075$$
$$h_d(1) = h_d(5) = 0.159$$
$$h_d(2) = h_d(4) = 0.22$$
$$h_d(3) = \underset{n\to 3}{\text{Lim}}\frac{\sin\frac{\pi}{4}(n-3)}{4.\frac{\pi}{4}(n-3)}$$
$$h_d(3) = \frac{1}{4} = 0.25$$

Hamming window function is

$$\omega_H(n) = \begin{cases} 0.54 - 0.46\cos(\frac{2\pi n}{N-1}), & 0 \le n \le N-1 \\ 0, & \text{otherwise} \end{cases}$$

$$\omega_H(n) = \begin{cases} 0.54 - 0.46\cos(\frac{2\pi n}{6}), & 0 \le n \le 6 \\ 0, & \text{otherwise} \end{cases}$$

$$\omega_H(0) = 0.08$$
$$\omega_H(1) = 0.31$$
$$\omega_H(2) = 0.77$$
$$\omega_H(3) = 1$$
$$\omega_H(4) = 0.77$$
$$\omega_H(5) = 0.31$$
$$\omega_H(6) = 0.08$$

The filter coefficients are

$$h(n) = h_d(n) \times \omega_H(n)$$
$$h(0) = h(6)$$
$$= h_d(0) \times \omega_H(0)$$

$$
\begin{aligned}
&= 0.006 \\
h(1) &= h(5) \\
&= h_d(0) \times \omega_H(1) \\
&= 0.049 \\
h(2) &= h(4) \\
&= h_d(0) \times \omega_H(2) \\
&= 0.1694 \\
h(3) &= h_d(3)\omega_H(3) \\
&= 0.25
\end{aligned}
$$

The transfer function of the filter is

$$
\begin{aligned}
H(z) &= \sum_{n=0}^{N-1} h(n)z^{-n} \\
&= \sum_{n=0}^{6} h(n)z^{-n}
\end{aligned}
$$

$$
\begin{aligned}
&= h(0) + h(1)z^{-1} + h(2)z^{-2} + h(3)z^{-3} + h(4)z^{-4} + h(5)z^{-5} \\
&\quad + h(6)z^{-6} \\
&= h(0)[1 + z^{-6}] + h(1)[z^{-1} + z^{-5}] + h(2)[z^{-2} + z^{-4}] + h(3)z^{-3}
\end{aligned}
$$

$$
\boxed{\begin{aligned} &H(z) = 0.006[1 + z^{-6}] + 0.049[z^{-1} + z^{-5}] \\ &+0.1694[z^{-2} + z^{-4}] + 0.25z^{-3} \end{aligned}}
$$

The structure realized as shown in Fig. 5.49.

Example 5.35 A lowpass filter is required to be designed with the desired frequency response.

$$
H_d(\omega) = \begin{cases} e^{-j2\omega}, & -0.25\pi \le \omega \le 0.25\pi \\ 0, & 0.25\pi \le \omega \le \pi \end{cases}
$$

Obtain the filter coefficients, $h(n)$, using Hamming window function. Also fluid the frequency response $H(\omega)$ of the designed filter.

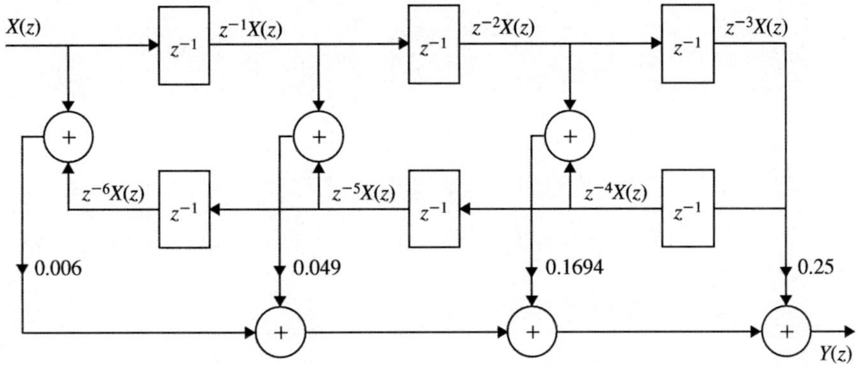

Fig. 5.49 Structure realization for Example 5.34

Solution Given

$$H_d(\omega) = \begin{cases} e^{-j2\omega}, & -0.25\pi \le \omega \le 0.25\pi \\ 0, & 0.25\pi \le \omega \le \pi \end{cases}$$

$$\begin{aligned} h_d(n) &= \frac{1}{2\pi}\int_{-\pi}^{\pi} H_d(\omega)e^{j\omega n}d\omega \\ &= \frac{1}{2\pi}\int_{-0.25\pi}^{0.25\pi} e^{-j2\omega}e^{j\omega n}d\omega \\ &= \frac{1}{2\pi}\int_{-0.25\pi}^{0.25\pi} e^{j(n-2)\omega}d\omega \\ &= \frac{1}{2\pi}\left[\frac{e^{j(n-2)\omega}}{j(n-2)}\right]_{-0.25\pi}^{0.25\pi} \\ &= \frac{1}{\pi(n-2)}\left[\frac{e^{j0.25\pi(n-2)} - e-j0.25\pi(n-2)}{2j}\right] \\ h_d(n) &= \frac{1}{\pi(n-2)}.\sin 0.25\pi(n-2) \end{aligned}$$

From the given frequency, it is clear that

$$\frac{N-1}{2} = 2 \to N-1 = 4 \to N = 5$$

$$\alpha = \frac{N-1}{2} = 2$$

If $n = 2$,

$$\begin{aligned} h_d(n) &= \lim_{n\to 2} \frac{\sin(\frac{\pi}{4})(n-2)}{4.\frac{\pi}{4}(n-2)} \\ &= \frac{1}{4} \end{aligned}$$

$$H_d = \begin{cases} \frac{\sin 0.25(n-2)}{\pi(n-2)} & , n \neq 2 \\ \frac{1}{4} & , n = 2 \end{cases}$$

$$\begin{aligned} h_d(0) = h_d(4) &= \frac{\sin 0.25(-2)}{\pi(-2)} = 0.159 \\ h_d(1) = h_d(3) &= \frac{\sin 0.25(-1)}{\pi(-1)} = 0.225 \\ h_d(2) &= = \frac{1}{4} = 0.25 \end{aligned}$$

Hamming window

$$\begin{aligned} \omega_H(n) &= \begin{cases} 0.54 - 0.46\cos(\frac{2\pi n}{N-1}), & 0 \leq n \leq N \leq 1 \\ 0, & \text{otherwise} \end{cases} \\ &= \begin{cases} 0.54 - 0.46\cos(\frac{2\pi n}{4}), & 0 \leq n \leq N \leq 4 \\ 0, & \text{otherwise} \end{cases} \end{aligned}$$

$$\begin{aligned} \omega_H(0) &= 0.54 - 0.46 = 0.08 \\ \omega_H(1) &= 0.54 \\ \omega_H(2) &= 1 \\ \omega_H(3) &= 0.54 \\ \omega_H(4) &= 0.54 - 0.46 = 0.08 \\ h(n) &= h_d(n) \times \omega_H(n) \\ h(0) &= h(4) = h_d(0) \times \omega_H(0) = 0.0127 \\ h(1) &= h(3) = h_d(1) \times \omega_H(1) = 0.1215 \\ h(2) &= h_d(2) \times \omega_H(2) = 0.25 \end{aligned}$$

Filter coefficients are

$$\boxed{\begin{aligned} &h(0) = h(4) = 0.0127 \\ &h(1) = h(3) = 0.1215 \\ &h(2) = 0.25 \end{aligned}}$$

Summary

- Digital filters are classified as FIR and IIR filters. They have their own merits and demerits. The choice of the filter is made depending upon the application.
- The difference equation which describes FIR filter is non-recursive; the output response is a function only of past and present values of the input. When FIR is implemented in this form it is always stable.
- FIR filters have exactly linear phase characteristic which has unique properties and helps in the design and applications.
- The design of FIR filters is divided into five independent stages. They are as follows:

  (*a*) Filter specifications.
  (*b*) Coefficient calculation.
  (*c*) Realization.
  (*d*) Analysis of errors.
  (*e*) Implementation.

- FIR filters are designed usually by the following methods among others.

  (*a*) Frequency sampling method.
  (*b*) Fourier series method.
  (*c*) Design using windows.

- The following windowing techniques are generally used to design FIR filters.

  (*a*) Rectangular window.
  (*b*) Raised cosine window.
  (*c*) Hanning window.
  (*d*) Hamming window.
  (*e*) Blackman window.
  (*f*) Kaiser window.

- There are different types of structures for realizing FIR filter system. The following structures are described in this chapter:

(a) Direct from realization.
(b) Cascade form realization.
(c) Linear phase realization.
(d) Lattice structure realization.

Short Questions and Answers

1. **What are FIR Filters?**
The specifications of the desired filter will be given in terms of ideal frequency response $H_d(\omega)$. The impulse response $h_d(n)$ of desired filter can be obtained by inverse Fourier transform of $H_d(\omega)$, which consists of infinite number of samples. The filters designed by selecting finite number of samples of impulse response are called FIR filters.
2. **What are advantages of FIR filters?**
The advantages of FIR filters are as follows:

   - Linear phase FIR filters can be easily designed.
   - Efficient realizations of FIR filter exist as both recursive and non-recursive structures.
   - FIR filters realized non-recursively are always stable.
   - The roundoff noise can be made small in non-recursive realization of FIR fitter.

3. **What are disadvantages of FIR filters?**
The disadvantages of FIR filters are as follows:

   - The duration of impulse response should be large to realize sharp cut-off filters.
   - The non-integral delay can lead to problems in some signal processing applications.

4. **What are the conditions to be satisfied for constant phase delay in linear phase FIR filters?**
The conditions for constant phase delay are as follows:

$$\text{Phase delay, } \alpha = \frac{N-1}{2} \; (i.e.\text{phase delay is constant})$$
$$\text{Impulse response, } h(n) = h(N-1-n) \; (i.e.\text{impulse response is symmetric})$$

5. **What are the possible types of impulse response for linear phase FIR filter?**
There are four types of impulse response for linear phase FIR filter.

   - Symmetric impulse response for N is odd.
   - Symmetric impulse response for N is even.
   - Anti-symmetric impulse response for N is odd.
   - Anti-symmetric impulse response for N is even.

6. **Write the magnitude and phase function of FIR filter when impulse response is symmetric and N is odd.**
Magnitude function, $|H(\omega)| = h\left(\frac{N-1}{2}\right) + \sum_{n=1}^{N-1} 2h\left(\frac{N-1}{2} - n\right) \cos \omega n$

Phase function, $\angle H(\omega) = -\alpha\omega$

7. **Write the magnitude and phase function of FIR filter when impulse response is symmetric and N is even.**

$$\text{Magnitude function, } |H(\omega)| = \sum_{n=1}^{N/2} 2h\left(\frac{N-1}{N} - n\right) \cos\left(\omega\left(n - \frac{1}{2}\right)\right)$$

Phase function, $\angle H(\omega) = -\alpha\omega$

8. **Write the magnitude and phase function of FIR filter when impulse response is anti-symmetric and N is odd.**

$$\text{Magnitude function, } |H(\omega)| = \sum_{n=1}^{\frac{N-1}{2}} 2h\left(\frac{N-1}{N} - n\right) \sin \omega n$$

Phase function, $\angle H(\omega) = \beta - \alpha\omega$

9. **Write the magnitude and phase function of FIR filter when impulse response is anti-symmetric and N is even**

$$\text{Magnitude function, } |H(\omega)| = \sum_{n=1}^{N/2} 2h\left(\frac{N}{2} - n\right) \sin\left(\omega\left(n - \frac{1}{2}\right)\right)$$

Phase function, $\angle H(\omega) = \beta - \alpha\omega$

10. **List the well-known design techniques for linear phase FIR filter.**
There are three well-known methods of design techniques for linear phase FIR filters. They are as follows:

- Fourier series method and window method.
- Frequency sampling method.
- Optimal filter design methods.

11. **What is Gibb's phenomenon (or Gibb's oscillation)?**
In FIR filter design by Fourier series method (or Rectangular window method) the finite duration impulse response is truncated to finite duration impulse response. The abrupt truncation of impulse response introduces oscillations in the passband and stopband. This effect is known as Gibb's phenomenon (or Gibb's oscillation).
12. **Write the procedure for designing FIR filter using windows.**

- Choose the desired frequency response of the filter $H_d(\omega)$.

- Take inverse Fourier transform of $H_d(\omega)$ to obtain the desired impulse response $h_d(n)$.
- Choose a window sequence $\omega(n)$ and multiply $h_d(n)$ by $\omega(n)$ to convert the infinite duration impulse response $h(n)$.
- The transfer function $H(z)$ of the filter is obtained by taking z-transform of $h(n)$.

13. **Write the procedure for FIR filters designed by frequency sampling method.**
    - Choose the desired frequency response $H_d(\omega)$.
    - Take N-samples of $H_d(\omega)$ to generate the sequence $H(k)$.
    - Take inverse DFT of $H(k)$ to get the impulse response $h(n)$.
    - The transfer function $H(z)$ of the filter is obtained by taking z-transform of impulse response.
14. **What is the drawback in FIR filter design using windows and frequency sampling method? How is it overcome?**
    The FIR filter designs by window and frequency sampling method do not have precise control over the critical frequencies such as ω_p and ω_s.
    This drawback can be overcome by designing FIR filter using the Chebyshev approximation technique. In this technique, an error function is used to approximate the ideal frequency response, in order to satisfy the desired specifications.
15. **Write the characteristics features of Rectangular window.**
    - The mainlobe width is equal to $4\pi/N$.
    - The maximum sidelobe magnitude is -13 dB.
    - The sidelobe magnitude does not decrease significantly with increasing ω.
16. **List the features of FIR filter designed using Rectangular window.**
    - The width of the transition region is related to the width of the mainlobe of window spectrum.
    - Gibb's oscillations are noticed in the passband and stopband.
    - The attenuation in the stopband is constant and cannot be varied.
17. **How the transition width of the FIR filter can be reduced in design using windows?**
    In FIR filter design using windows, the width of the transition region is related to the width of the mainlobe in window spectrum. If the mainlobe width is narrow, then the transition region in FIR filter will be small. In general, the width of mainlobe is $x\pi/N$, where $x = 4$ or 8 or 12 and N is the length of the window sequence used for designing the filter. Hence, the width of mainlobe can be reduced by increasing the value of N, which in turn reduces the width of the transition region in the FIR filter.
18. **Why Gibb's oscillations are developed in Rectangular window and how can it be eliminated or reduced?**
    Gibb's oscillations in Rectangular window are due to sharp transition from 1 to 0

at the edges of window sequence. These oscillations can be eliminated or reduced by replacing the sharp transition by gradual transition. This is the motivation for development of triangular and cosine windows.

19. **List the characteristics of FIR filters designed using windows.**
    - The width of transition band depends on the type of window.
    - The width of transition band can be made narrow by increasing the value of N where N is the length of window sequence.
    - The attenuation in the stopband is fixed for a given window, except in case of Kaiser window where it is variable.

20. **Write the frequency response of Hanning window.**
    Frequency response of Hanning window is

$$W_C(\omega) = 0.5\frac{\sin\omega(N/2)}{\sin\omega/2} + 0.25\frac{\sin(\omega N/2 - \pi N/(N-1))}{\sin(\omega/2 - \pi/(N-1))} + 0.25\frac{\sin(\omega N/2 + \pi N/(N-1))}{\sin(\omega/2 + \pi/(N-1))}$$

21. **Write the frequency response of Hamming window.**
    Frequency response of Hamming window is

$$W_H(\omega) = 0.54\frac{\sin\omega(N/2)}{\sin\omega/2} + 0.23\frac{\sin(\omega N/2 - \pi N/(N-1))}{\sin(\omega/2 - \pi/(N-1))} + 0.23\frac{\sin(\omega N/2 + \pi N/(N-1))}{\sin(\omega/2 + \pi/(N-1))}$$

22. **Give the equation for Hanning window function.**
    Hanning window function is

$$W_C(n) = \begin{cases} 0.5 - 0.5\cos(2\pi n/(N-1)); & \text{for } 0 \le n \le (N-1) \\ 0; & \text{else} \end{cases}$$

23. **List the features of Hanning window spectrum.**
    - The mainlobe width is equal to $8\pi/N$.
    - The maximum sidelobe magnitude is -31 dB.
    - The sidelobe magnitude decreases with increasing ω.

24. **Compare the Rectangular window and Hanning window.**

	Rectangular Window	Hanning Window
1.	The mainlobe width is equal to $4\pi/N$.	The mainlobe width is equal to $8\pi/N$.
2.	The maximum sidelobe magnitude in window spectrum is −13dB.	The maximum sidelobe magnitude in window spectrum is −31dB.
3.	In window spectrum, the sidelobe magnitude slightly decreases with increasing ω.	In window spectrum, the sidelobe magnitude decreases with increasing ω.
4	In FIR filter designed using Rectangular window, the minimum stopband attenuation is 22dB.	In FIR filter designed using Hanning window, the minimum stopband attenuation is 44dB.

25. **Write the equation for Hamming window function.**
Hanning window function is

$$W_H(n) = \begin{cases} 0.54 - 0.46\cos(2\pi n/(N-1)); & \text{for } 0 \leq n \leq (N-1) \\ 0; & \text{else} \end{cases}$$

26. **Sketch the log-magnitude spectrum of Hamming window.**
27. **Compare the Rectangular window and Hamming window.**

	Rectangular Window	Hamming Window
1.	The mainlobe width is equal to $4\pi/N$.	The mainlobe width is equal to $8\pi/N$.
2.	The maximum sidelobe magnitude in window spectrum is −13dB.	The maximum sidelobe magnitude in window spectrum is −41dB.
3.	In window spectrum, the sidelobe magnitude slightly decreases with increasing ω.	In window spectrum, the sidelobe magnitude remains constant.
4	In FIR filter designed using Rectangular window, the minimum stopband attenuation is 22dB.	In FIR filter designed using Hamming window, the minimum stopband attenuation is 51dB.

28. **List the features of Hamming window spectrum.**

- The mainlobe width is equal to $8\pi/N$.
- The maximum sidelobe magnitude is −41 dB.
- The sidelobe magnitude remains constant for increasing ω.

29. **Compare the Hanning window and Hamming window.**

	Hanning Window	Hamming Window
1.	The mainlobe width in window spectrum is equal to $8\pi/N$.	The mainlobe width in window spectrum is equal to $8\pi/N$.
2.	The maximum sidelobe magnitude in window spectrum is −31 dB.	The maximum sidelobe magnitude in window spectrum is −41 dB.
3.	In window spectrum, the sidelobe magnitude decreases with increasing ω.	In window spectrum, the sidelobe magnitude remains constant. Here, the increased sidelobe attenuation is achieved at the expense of constant attenuation at high frequencies.
4	In FIR filter designed using Hanning window, the minimum stopband attenuation is 44 dB.	In FIR filter designed using Hamming window, the minimum stopband attenuation is 51 dB.

30. **Compare Hamming window and Blackman window.**

	Hamming Window	Blackman Window
1.	The mainlobe width in window spectrum is equal to $8\pi/N$.	The mainlobe width in window spectrum is equal to $12\pi/N$.
2.	The maximum sidelobe magnitude in window spectrum is −41 dB.	The maximum sidelobe magnitude in window spectrum is −58 dB.
3.	The higher value of sidelobe attenuation is achieved at the expense of constant attenuation at high frequencies.	The higher value of sidelobe attenuation is achieved at the expense of increased mainlobe width.
4.	In widow spectrum, the sidelobe magnitude remains constant with increasing ω.	In window spectrum, the sidelobe magnitude decreases rapidly with increasing ω.
5	In FIR filter designed using Hamming window, the minimum stopband attenuation is 51 dB.	In FIR filter designed using Blackman window, the minimum stopband attenuation is 78 dB.

31. **List the features of Blackman window spectrum.**

- The mainlobe width is $12\pi/N$.
- The maximum sidelobe magnitude is −58dB.
- The sidelobe magnitude decreases with increasing ω.
- The sidelobe attenuation in Blackman window is the highest among windows, which is achieved at the expense of increased mainlobe width. However, the mainlobe width can be reduced by increasing the value of N.

32. **Write the expression for Kaiser window function.**
Kaiser window function is

$$W_k(n) = \begin{cases} \frac{I_0(\beta)}{I_0(\alpha)}; & \text{for } -(N-1)/2 \le n \le (N-1)/2 \\ 0; & \text{else.} \end{cases}$$

$$\text{where } \beta = \alpha\left(1 - \left(\frac{2n}{(N-1)}\right)^2\right)^{0.5}$$

$$I_0(x) = 1 + \frac{0.25x^2}{(1!)^2} + \frac{(0.25x^2)^2}{(2!)^2} + \frac{(0.25x^2)^3}{(3!)^2} + \cdots$$

The series of $I_0(x)$ is used to compute $I_0(\beta)$ and $I_0(\alpha)$ and are computed for any desired accuracy. Usually 25 terms of the series are sufficient for most practical purposes.

33. **List the desirable features of Kaiser window spectrum.**
    - The widths of mainlobe and the peak sidelobe are variable.
    - The parameter α in the Kaiser window function is an independent variable that can be varied to control the sidelobe levels with respect to mainlobe peak.
    - The width of the mainlobe in window spectrum can be varied by varying the length N of the window sequence.

34. **Compare the Hamming window and Kaiser window.**

	Hamming Window	**Kaiser Window**
1.	The width of mainlobe in window spectrum is $8\pi/N$.	The width of mainlobe in window spectrum depends on the values of α and N.
2.	The maximum sidelobe magnitude in windows spectrum is fixed at -41dB.	The maximum sidelobe magnitude with respect to peak of mainlobe is variable using the parameter α.
3.	In window spectrum, the sidelobe magnitude remains constant with increasing ω.	In window spectrum, the sidelobe magnitude decreases with increasing ω.
4	In FIR filter designed using Hamming window, the minimum stopband attenuation is 51 dB.	In FIR filter designed using Kaiser window, the minimum stopband attenuation is variable and depends on the value of α.

35. **Define an IIR filter.**
    The filter designed by considering all the infinite samples of impulse response is called IIR filter. The impulse response is obtained by taking inverse Fourier transform of ideal frequency response.

36. **Compare IIR and FIR filter.**

	IIR Filter	FIR Filter
1.	All infinite samples of impulse response are considered.	Only N samples of impulse response are considered.
2.	The impulse response cannot be directly converted to digital filter transfer function.	The impulse response can be directly converted to digital filter transfer function.
3.	The design involves design of analog filter and then transforming analog filter to digital filter.	The digital filter can be directly designed to achieve the desired specification.
4.	The specifications include the desired characteristics for magnitude response only.	The specifications include the desired characteristics for both magnitude and phase response.
5	Linear phase characteristics cannot be achieved.	Linear phase filters can be easily designed.

37. **What are the properties that are maintained same in the transformation of analog to digital filters? (Or mention two properties that an analog filter should have for effective transformation.)**
The analog filters should be stable and causal for effective transformation to digital filters. While transforming the analog filters to digital filters, these two properties (i.e. stability and causality) are maintained same, which means that the transformed digital filter should also be stable and physically realizable.
38. **What is the condition to be satisfied by linear phase FIR filter?**

For an FIR filter to have linear phase, its phase response should satisfy the following equations:

$$\theta(\omega) = -\alpha\omega$$
$$\theta(\omega) = \beta - \alpha\omega$$

If the first equation is satisfied, the filter will have constant group and constant phase delay responses. For this, the impulse response of the filter must have positive symmetry. If the second equation is satisfied, the filter will have constant group delay. In this case, the impulse response of the filter will have negative symmetry.
39. **In the design of FIR filter, how is Kaiser window different from other windows?**
In other windows, the width of the mainlobe and the attenuation of the sidelobes depend only upon the length N of the filter and they cannot be controlled independently whereas in Kaiser window it is possible to control the length of the filter and the transition width of the mainlobe by introducing additional parameter.

40. **Show that the filter with $h(n) = \{-1, 0, 1\}$ is a linear phase filter.**

$$H(e^{j\omega}) = -e^{-j\omega} 2 \sin \omega$$
$$\theta(\omega) = \pi - \omega$$

Hence, it has linear phase.

41. **What are the desirable characteristics of the frequency response of window function?**
The desirable characteristics of the frequency response of window function are as follows:

(*a*) The central lobe of the frequency response of the window should be narrow and contain maximum energy.
(*b*) The sidelobes should decrease in energy as ω tends to π.
(*c*) The highest sidelobe level should be small.

42. **Why is linear phase important in digital signal processing applications?**
The linear phase does not alter the shape of the signal and hence the output response is not distorted. Hence, digital filters with linear phase are designed.

43. **List out the well-known design techniques for linear phase FIR filter.**
The following techniques are used to design linear phase FIR filter.

(*a*) Frequency sampling method.
(*b*) Windows method.
(*c*) Optimal or minimax design.

44. **What are the disadvantages of FIR filter?**
The disadvantages of FIR filter are as follows:

(*a*) Memory requirement and execution time are very high.
(*b*) FIR filters require narrow transition band which increases the cost as it requires large arithmetic operations and hardware components such as multipliers, adders and time delay elements.

45. **What is frequency sampling?**
The filter coefficients $h(n)$ can be obtained as the inverse DFT of the frequency samples,

$$h(n) = \frac{1}{N} \sum_{k=0}^{N-1} H(k) e^{j\left(\frac{2\pi}{N}\right)nk}$$

where $H(k) = 0, 1, \ldots, (N-1)$ are samples of the ideal frequency response. For linear phase, with positive symmetrical impulse response, the above equation is written as

$$h(n) = \frac{1}{N} \left[\sum_{k=1}^{\frac{N}{2}-1} 2|H(k)| \cos \frac{2\pi k(n-\alpha)}{N} + H(0) \right]$$

where $\alpha = (N-1)/2$. The resulting filter will have a frequency response which is exactly same as the original response at the sampling instants. This is called frequency sampling.

46. **For what type of filters frequency sampling method is suitable?**
The frequency sampling method is very suitable for the design of non-recursive FIR frequency selective filters such as lowpass, highpass and bandpass filters and also for filters with arbitrary frequency response.

47. **What is meant by FIR filter and why is it stable?**
Consider the following difference equation and the transfer function of a digital filter which are described as given below.

$$y(n) = \sum_{n=0}^{N-1} h(k)x(n-k)$$

$$H(z) = \sum_{n=0}^{N-1} h(k)z^{-k}$$

where $h(n)$ are the coefficients of the filter. When the digital filter is implemented in this form, it is called FIR filter. The poles of $H(z)$ all fall at the origin of the z-plane and therefore FIR filters are always stable.

48. **Find the number of delays required in a direct form-I structure that realizes a second-order transfer function.**
Four delay elements are required to realize a second-order system in direct form-I structure (Figs. 5.50 and 5.51).

49. **The unit sample response of an FIR filter is**

$$h(n) = a^n\{u(n) - u(n-2)\}$$

Draw the direct form realization of this system.

$$H(z) = \frac{1}{(1 + az^{-1})}$$

50. **A first-order filter structure is shown in Fig.** 5.52. **Find the transfer function.**

$$H(z) = \frac{(1 + z^{-1})}{(1 - bz^{-1})}$$

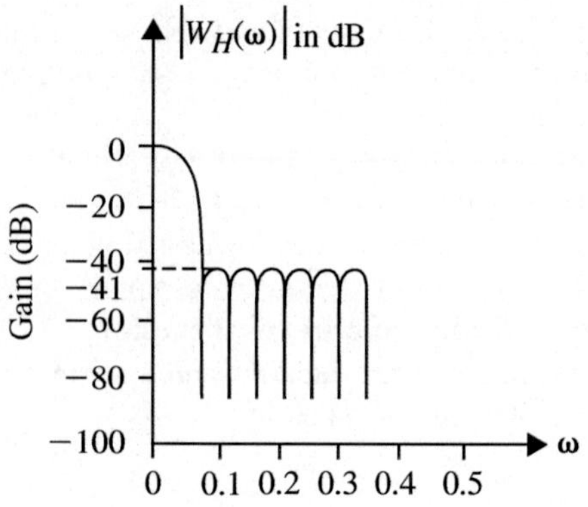

Fig. 5.50 Log-magnitude spectrum of Hamming window

Log magnitude of Hamming window of question 26 is shown in Fig. 5.50

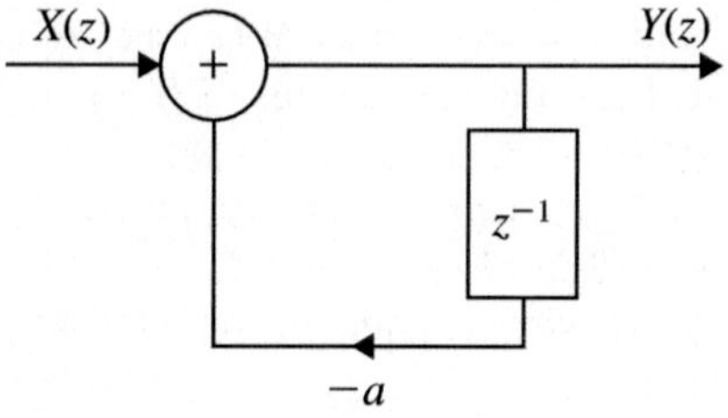

Fig. 5.51 Direct form realization for Problem 49

Direct form realization for question 49 is shown in Fig. 5.51

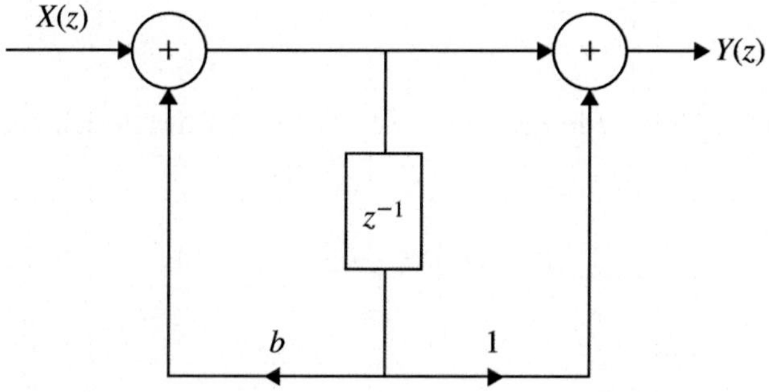

Fig. 5.52 A structure for Problem 50

First order filter structure for question 50 is shown in Fig. 5.52.

Long Answer Type Questions

1. Describe the design of FIR filter using frequency sampling technique.
2. A bandpass FIR filter of length 7 is required. It is to have lower and upper cut-off frequencies of 3 kHz and 5 kHz, respectively, and is intended to be used with a sampling frequency of 24 kHz. Determine the filter coefficients using Hanning Window. Consider the filter to be causal.
3. Compare the Hamming window and Blackman window.
4. Explain the polyphase decomposition for FIR filter structure.
5. What is the principle of designing FIR filter using frequency sampling method?
6. Design a second-order band reject filter with ω_1 and ω_2 as cut-off frequency and sampling interval as T.
7. List the three well-known methods of design techniques for FIR filters and explain any one.
8. Design an ideal lowpass filter for a cut-off frequency $\omega_c = 0.4$ rad using (8-tap) window design method.
9. Determine the filter coefficients $h(n)$ of length $M = 15$ obtained by sampling its frequency response as

$$H\left[\left(\frac{2\pi}{15}\right)k\right] = \begin{cases} 1, & k = 0, 1, 2, 3, 4 \\ 0.4, & k = 5 \\ 0, & k = 6, 7 \end{cases}$$

using rectangular window.
10. Design a digital filter with

$$H_d(e^{j\omega}) = \begin{cases} 1, & 2 \leq |\omega| \leq \pi \\ 0, & \text{otherwise} \end{cases}$$

using Hamming window with $N = 7$. Draw the frequency response.

11. List the various steps in designing FIR filters.
12. The desired frequency response of a desired filter is

$$H_d(\omega) = \begin{cases} e^{-j3w}, & \frac{-\pi}{4} \leq \omega \leq \frac{\pi}{4} \\ 0, & \frac{\pi}{4} \leq |\omega| \leq \pi \end{cases}$$

Determine the filter coefficients if the window function is defined as

$$\omega(x) = \begin{cases} 1, & 0 \leq x \leq 5 \\ 0, & \text{otherwise} \end{cases}$$

13. Illustrate the steps involved in the design of linear phase FIR filter by the frequency sampling method.

14. An FIR filter is given by

$$y(n) = 2x(n) + \frac{4}{5}x(n-1) + \frac{3}{2}x(n-2) + \frac{2}{3}x(n-3)$$

Find the lattice structure coefficients.
15. What are the issues in designing FIR filter using window method?
16. Mention the advantages and disadvantages of FIR and IIR filters.
17. Design a digital lowpass FIR filter of length II with cut-off frequency of 0.5 kHz and sampling rate 2 kHz using Hamming window.
18. Derive the frequency response of a linear phase FIR filter with symmetric impulse response.
19. Explain the design procedure for designing FIR filter using window function.
20. Explain the concept of optimum equiripple filter.
21. Design a Nine tap linear phase filter having the ideal response.

$$H_d(e^{j\omega}) = \begin{cases} 1, & |\omega| \le \frac{\pi}{6} \\ 0, & \frac{\pi}{6} \le |\omega| \le \frac{\pi}{3} \\ 1, & \frac{\pi}{3} \le |\omega| \le \pi \end{cases}$$

using Hamming window and draw the realization for the same.
22. Design an FIR digital filter to approximate an ideal lowpass filter with passband gain of unity, cut-off frequency of 850 kHz and working at a sampling frequency of $f_s = 5$ kHz. The length of the impulse response should be 5. Use a rectangular window.
23. The desired response of a lowpass filter is

$$H_d(e^{j\omega}) = \begin{cases} e^{-j3\omega}; & \frac{-3\pi}{4} \le \omega \le \frac{3\pi}{4} \\ 0, & \frac{3\pi}{4} \le |\omega| \le \pi \end{cases}$$

Determine the frequency response of the filter for $M = 7$ using a Hamming window.

Chapter 6
Digital Signal Processor

Learning Objectives

After completing this chapter, you should be able to

- ✠ understand the key issues in general- and special-purpose processors for digital signal processing.
- ✠ understand the working of different computer architectures for signal processing.
- ✠ choose the right digital signal processor for the given application.
- ✠ implement DSP algorithms on general-purpose DSPs.

6.1 Introduction

The DSP processors were introduced in the early 1980s to handle information which was available in digital form. Since then it has made very rapid progress to handle very complex problems. The DSP processors are divided into two broad categories as general- purpose and special-purpose processors. The special-purpose programmable digital signal processors are designed with features that are specifically required for digital signal processing applications. Some of the features include multiplier and multiplier accumulator (MAC), modified bus structures, multiple access memory, multi-port memory and pipelining. But the conventional microprocessor which is a general-purpose microprocessor does not have these features. An advanced microprocessor or a RISC processor may use some of the techniques of programmable digital signal processors or may even have instructions that are specifically required for DSP

S. Palani, *Discrete Time Systems and Signal Processing*,
https://doi.org/10.1007/978-3-031-32421-5_6

applications, or they may have performances close to that of programmable Digital Signal Processors for certain operations. However, in terms of low power requirement, cost, real-time input/output capability and availability of on-chip memories, the programmable DSPs have an advantage over advanced microprocessors. Some of the P-DSPs include fixed point devices such as Texas Instruments TMS320C5x, TMS320C54x and Motorola DSP563x and floating point processors such as Texas Instruments TMS320C4x and TMS320C67xx. Some of the features specifically required for performing digital signal processing operations are described ahead.

6.2 Multiplier Accumulator (MAC) Unit

Most of the digital signal processing computations such as convolution and correlation require array multiplication. Array multiplication can be done using a single multiplier and adder that is shown in Fig. 6.1. One of the important requirements of these array multipliers is that they have to process the signals in real time, i.e. before the next sample of the input and output signals arrive at the input of the array, the array multiplication should be completed. There are two approaches to solve this problem. One of the approaches is that a dedicated MAC unit may be implemented in hardware, which integrates multiplier and accumulator in a single hardware unit. It is adopted by the Motorola DSP processor DSP5600x. The other approach is to have separate multiplier and accumulator. This approach is adopted in Texas Instruments DSP processor, TMS320C5x. The output of the multiplier is stored in the product register. The content of this product register can be added to accumulator register ACC in the central ALU. In both approaches, the MAC operation can be completed in one clock cycle.

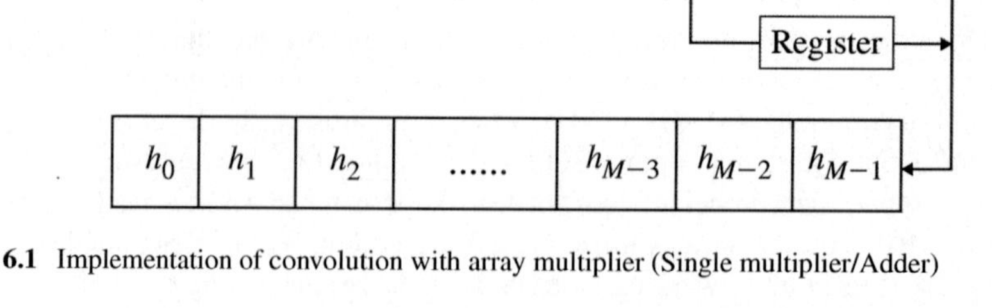

Fig. 6.1 Implementation of convolution with array multiplier (Single multiplier/Adder)

- y_n, the output at the nth sampling instant is obtained by multiplying the array $x_n = [x_n, x_{n-1}, x_{n-2}, \ldots, x_{n-M+3}, x_{n-M+2}, x_{n-M+1}]$ corresponding to the present and the past $M-1$ samples of input with array $h = [h_0, h_1, h_2, \ldots, h_{M-3}, h_{M-2}, h_{M-1}]$ which correspond to the impulse response sequence.
- y_{n+1}, the output at the $(n+1)$th sampling instant is obtained by multiplying the input signal array x_{n+1} with array h.
- The input signal array x_{n+1} is obtained by shifting array x_n towards the right so that the $(n+1)$th sample of the input data x_{n+1} becomes the first element and all the elements of x_n are shifted towards the right by 1 position. Instead of shifting the elements of x_n towards the right all at a time after finishing the vector multiplication, each of the elements may be shifted separately soon after the MAC operation that uses these elements is over. This is achieved in P-DSP by using a special instruction called Multiply Accumulate with Data Shift (MACD).

For example:

MACD pgm, dma multiplies the content of the program memory (pgm) with the content of the data memory with address (dma) and stores the result in the product register. The content of the product register is added to the accumulator before the new product is stored. Further, the content of dma is copied to the next location whose address is dma$+1$.

The efficient operation of a digital signal processor depends on its architectural features, execution speed, type of arithmetic and word length. There are two types of special-purpose hardware used in DSP. To execute DSP algorithms such as digital filters and FFT, an algorithm-specific DSP hardware is designed. For DSP applications in telecommunication digital audio and control applications, application-specific DSP hardware is designed. The general-purpose processor used nowadays in general requires operations performed sequentially and faster. Further, many DSP algorithms require repetitive arithmetic operations. The standard microprocessors available do not meet these requirements. However, in DSP hardware design, the hardware architecture and the instruction set for DSP operations are given utmost priority. This is achieved using the concepts of parallelism which include the following:

1. Hardware architecture.
2. VLIW architecture.
3. Pipelining.
4. On-chip memory/cache.
5. Instruction sets dedicated to DSP.

The architectural features of DSP are described below.

6.3 Bus Structures and Memory Access Schemes

The MACD (MAC operation with data move) instruction requires four memory accesses per instruction cycle. Instruction cycle is the time that elapses since an instruction is fetched till the particular instruction completes execution including the time taken for writing the result into a register or memory. The four memory access/clock periods required for the MACD instructions are as follows:

(i) Fetch the MACD instruction from the program memory.
(ii) Fetch one of the operands from the program memory.
(iii) Fetch the second operands from the data memory.
(iv) Write the content of the data memory with address dma into the location with the address dma + 1.

The impulse response coefficients are stored in the program memory, and the sample of input data are stored in the data memory.

6.3.1 Von Neumann Architecture

Von Neumann concepts where operations are performed sequentially were introduced in the year 1946 in the computer architecture design.

During the first clock period, the instruction code can be fed from the program memory to the control unit. In the second clock period, one of the operands is fed to the processing unit from the program memory. In the third clock period, the second operand is fed to the control unit from the data memory. In the fourth clock period, write the content of the data memory with address dma with location with the address dma + 1. The MACD instruction to be executed in a machine with 'Von Neumann Architecture' is shown in Fig. 6.2. It requires four clock cycles because it has a single address bus and a single data bus for accessing the program as well as data memory area. In a computer with Von Neumann architecture, the CPU can be either reading an instruction or reading/writing data from/to the memory. Both cannot occur at the same time since the instruction and data use the same signal pathways and memory. The Von Neumann architecture consists of three buses, namely the data bus, the address bus and control bus.

- **Data bus**: It transfers the data between the CPU and its peripherals. It is bidirectional. The CPU can read or write data in the peripheral.
- **Address bus**: CPU uses the address bus to indicate which peripherals it wants to access and within each peripheral with a specific register. It is unidirectional. The CPU always writes the address which is read by the peripherals.
- **Control bus**: It carries signals that are used to manage and synchronize the exchange between the CPU and its peripherals, as well as that indicates if the CPU wants to read or write the peripheral.

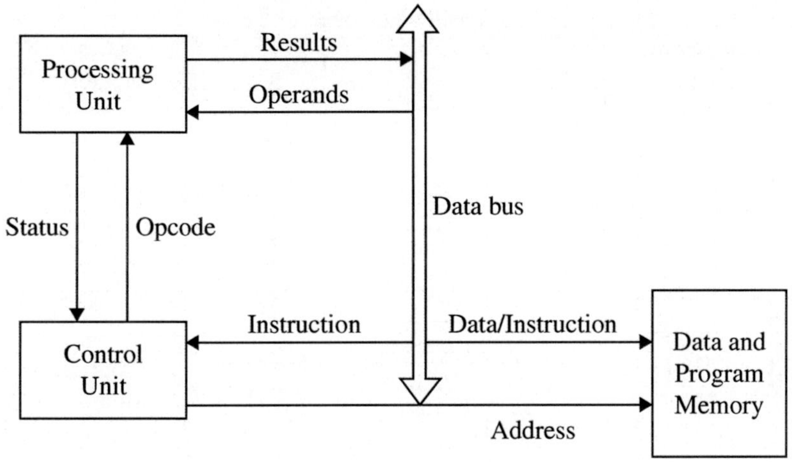

Fig. 6.2 Von Neumann architecture

One of the ways by which the number of clock cycles required for the memory access can be reduced is to use more than one bus for both address and data. That is implemented in 'Harvard Architecture'.

6.3.2 Harvard Architecture

The Harvard architecture which has two separate buses for the program and data memory is shown in Fig. 6.3. Hence, the content of program memory and data memory can be accessed in parallel. The instruction code can be fed from the program memory to the control unit while the operand is fed to the processing unit from the data memory. The processing unit consists of the registers and processing elements such as MAC units, multiplier, ALU and shifter. With the Harvard architecture, the number of memory accesses/clock cycles is two. This can be increased further by using more buses. Several P-DSPs follow the 'Modified Harvard architecture' which is shown in Fig. 6.4.

One set of bus is used to access a memory that has both program and data while another has data alone. Data can also be transferred from one memory to another. This type of architecture is used in P-DSPs from Texas Instruments and Analog devices.

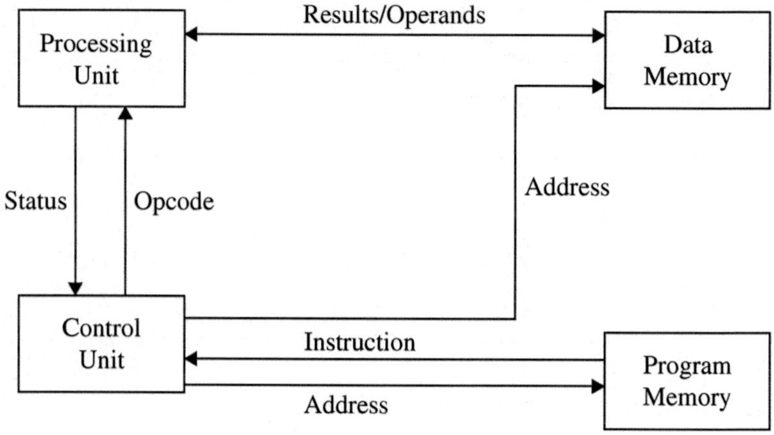

Fig. 6.3 Harvard architecture

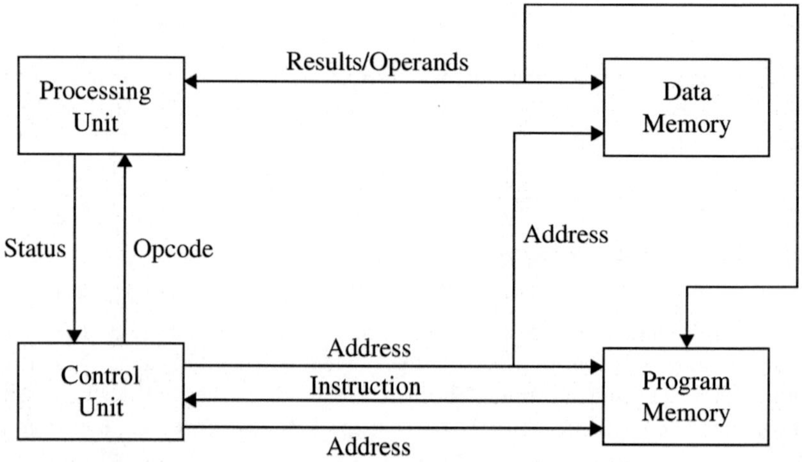

Fig. 6.4 Modified Harvard architecture

6.3.3 Multiple Access Memory

The number of memory accesses/clock periods can also be increased by using high-speed memory that permits more than one memory access/clock period. For example: **DARAM**—Dual-Access RAM permits two memory accesses/clock periods.

If DARAM is connected to a processing unit of the P-DSP (by using Harvard architecture) with two independent data and address buses, four memory accesses/clock periods can be achieved.

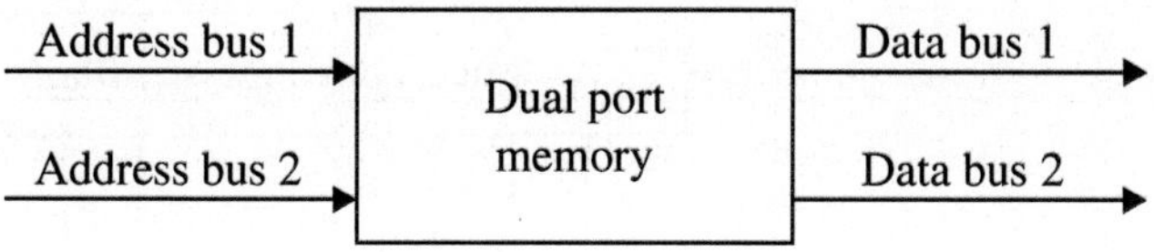

Fig. 6.5 Dual-ported memory

6.3.4 Multi-ported Memory

Another technique that is used for increasing the number of accesses/clock periods is to use multi-ported memory. It has two independent data and address buses as shown in Fig. 6.5. Using this, memory accesses can be achieved in a clock period. It is mostly needed for storing the program and data in two different memory chips in order to permit simultaneous access to both program and data memory.

Limitation

It increases the cost compared to two single-port memory since multi-ported memory requires a larger number of pins and a larger chip area. Larger number of pins require more expensive package and a larger dye size.

6.4 VLIW Architecture

Another architecture used for P-DSPs is the Very Long Instruction Word (VLIW) architecture. In this architecture, P-DSPs have a number of processing units such as ALUs, MAC units and shifters. The VLIW is accessed from memory and is used to specify the operands and operations to be performed by each of the data paths. The block diagram of VLIW architecture is shown in Fig. 6.6.

- Multiple functional units share a common multi-ported register file for fetching the operands and storing the results.
- Parallel random access by the functional units to the register file is facilitated by the read/write cross bar.
- Execution of the operations in the functional units is carried out with load/store operation of data between a RAM and the register file.

Advantages

1. The increased performance can be achieved with VLIW architecture which depends on the degree of parallelism in the algorithm and the number of functional units.
2. The throughput is higher since the algorithm involves execution of independent operations.
3. It is potentially easier to program.

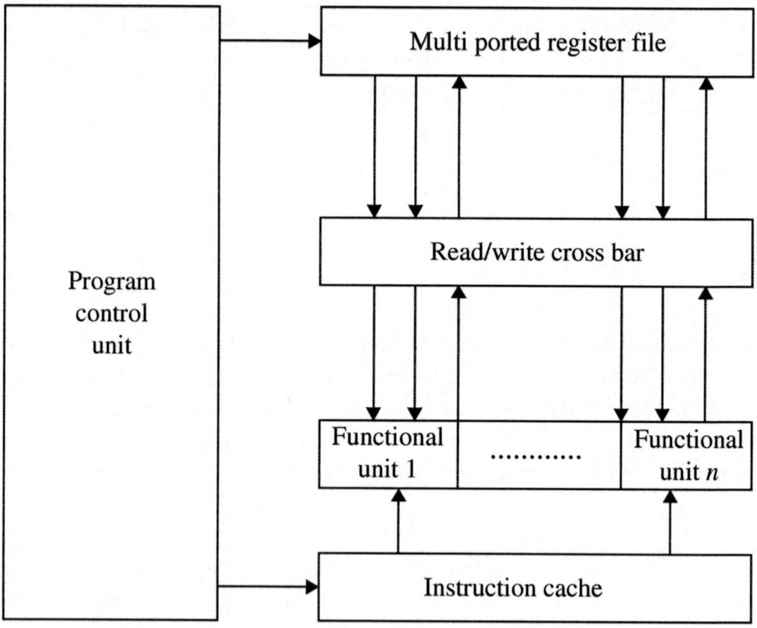

Fig. 6.6 Block diagram of the VLIW architecture

4. It is potentially scalable.
5. It has better programming efficiency.

Disadvantages

1. It may not always be possible to have independent stream of data for processing.
2. The number of functional units is also limited by the hardware cost for the multi-ported register file and cross bar switch.
3. High power consumption, and high program memory bandwidth.
4. Misleading MIP ratings.

6.5 Pipelining

One of the approaches adopted for increasing the efficiency of the advanced microprocessors as well as P-DSPs is by using instruction pipelining. An instruction cycle starting with the fetching of an instruction and ending with the execution of the instruction including the time storage of the results can be split into a number of microinstructions. Execution of each of the microinstructions is also referred to as one phase. For example, an instruction cycle requiring four microinstructions can be said to be in four phases as follows:

Value of T	Fetch Phase	Decode Phase	Read Phase	Execute Phase
1	I_1			
2		I_1		
3			I_1	
4				I_1
5	I_2			
6		I_2		
7			I_2	
8				I_2
9	I_3			
10		I_3		
11			I_3	
12				I_3

Fig. 6.7 Instruction cycles of processor with no pipelining

1. **Fetch phase** in which the instruction is fetched from the program memory.
2. **Decode phase** in which the instruction is decoded in the instruction register.
3. **Memory read phase** in which the operand required for the execution of the instruction may be read from the data memory.
4. **Execution phase** in which the execution as well as the storage of the results in either one of the registers or memory is carried out.

Each phase may be carried out separately by four functional units. Let us assume that each of the four phases takes equal time for completion.

Processor with No Pipelining

The instruction cycles of the processor with no pipelining are shown in Fig. 6.7. In a conventional microprocessor with no pipelining, each of the functional units is busy only 25% of the time since only one instruction is processed at the CPU at a time. Figure 6.7 shows that each functional unit is busy when a program containing three instructions I_1, I_2 and I_3 is executed.

Processor with Pipelining

In a processor with pipelining, the functional units can be kept busy almost all the time by processing a number of instructions simultaneously in the CPU. Consider the processor with four functional units. Four instructions I_1, I_2, I_3 and I_4 can be processed simultaneously as shown in Fig. 6.8. When I_1 enters the decode phase, I_2 can enter the opcode fetch phase. When I_1 enters the operand read phase, I_2 enters the decode phase and I_3 enters the opcode fetch phase. When I_1 enters the execute phase, I_2 enters the operand read phase, I_3 enters the decode phase and I_4 enters the opcode fetch phase. The pipelining is fully loaded now, and all the functional units have useful work to do.

Value of T	Fetch Phase	Decode Phase	Read Phase	Execute Phase
1	I_1			
2	I_2	I_1		
3	I_3	I_2	I_1	
4	I_4	I_3	I_2	I_1
5	I_5	I_4	I_3	I_2
6	I_6	I_5	I_4	I_3
7	I_7	I_6	I_5	I_4
8	I_8	I_7	I_6	I_5
9	I_9	I_8	I_7	I_6
10		I_9	I_8	I_7
11			I_9	I_8
12				I_9

Fig. 6.8 Instruction cycle of a processor with pipelining

Table 6.1 Pipeline depth of some P-DSPs

P-DSPs name/family	Pipeline depth
Analog devices	2
Motorola DSP5600x	3
TMS320C5x	4
TMS320C54x	5

Let T denote the time required for each phase of the instruction. One clock cycle of the processor corresponds to T. In a period of $12T$, only three instructions can be executed in a machine without pipelining, whereas in a machine with pipelining nine instructions can be executed in the same period. Hence, the throughput is increased by a factor of 3 in this case. Therefore, for executing a program with N instructions, the time required for execution is $(N + 4)T$ in a machine with pipelining, whereas in a machine without pipelining, the time required for executing N instructions is $4NT$.

The number of instructions that are processed simultaneously in the CPU, also referred to as depth of pipelining, differs in different families of P-DSPs. The pipeline depths of some of the P-DSPs are given in Table 6.1.

6.6 Architecture of TMS320C5x

TMS320C5x is a 16-bit fixed point processor. The DSP chips have IC number with prefix TMS320 (Texas Instruments). The next letter indicates that CMOS technology is used for the IC and the on-chip non-volatile memory is a ROM. If the next letter is E (e.g. TMS320E5x), it indicates that the technology used is CMOS and the on-chip non-volatile memory is an EPROM. If it is neither of these (e.g. TMS3205x), it indicates that NMOS technology is used for the IC and on-chip non-volatile memory is a ROM. Under C5x itself there are three processors, 'C50, 'C51 and 'C54, that have identical instruction set but have differences in the capacity of on-chip ROM and RAM. The instruction set of TMS320C5x and other DSP chips is superior to the instruction set of conventional microprocessors such as 8085 and Z80, as most of the instructions require only a single cycle for execution. The block diagram of the internal architecture of TMS320C5x is shown in Fig. 6.9. It has advanced Harvard architecture because they have separate memory bus for program and data and has instructions that enable data transfer between the program and data memory area.

6.6.1 Bus Structure

Separate program and data buses allow simultaneous access to program instructions and data, which provide a high degree of parallelism. For example, in multiply accumulate operation, while data is multiplied, a previous product can be loaded into, added to or subtracted from the accumulator and at the same time, a new address can be generated. Thus, everything is executed in one cycle. Such a parallelism supports a powerful set of arithmetic, logic and bit manipulation operations that all can be performed in a single machine cycle.

The TMS320C5x architecture has four buses, and their functions are as follows:

(i) **Program Bus (PB)**: It carries the instruction code and immediate operands from program memory to the CPU.
(ii) **Program Address Bus (PAB)**: It provides addresses to program memory for both reads and writes.
(iii) **Data read Bus (DB)**: It interconnects various elements of the CPU to data memory space.
(iv) **Data read Address Bus (DAB)**: It provides the address to access the data memory space.

The program and data buses can work together to transfer data from on-chip data memory and internal or external program memory to the multiplier for single-cycle multiply/accumulate operations.

6.6.2 Central Arithmetic Logic Unit (CALU)

It consists of the following elements:

(i) **(16 x 16) bit parallel multiplier**: It performs 16×16 multiplication of numbers represented in 2's complement form. The 16-bit temporary register O (TREGO) holds the multiplicand. The other operand for the multiplication can

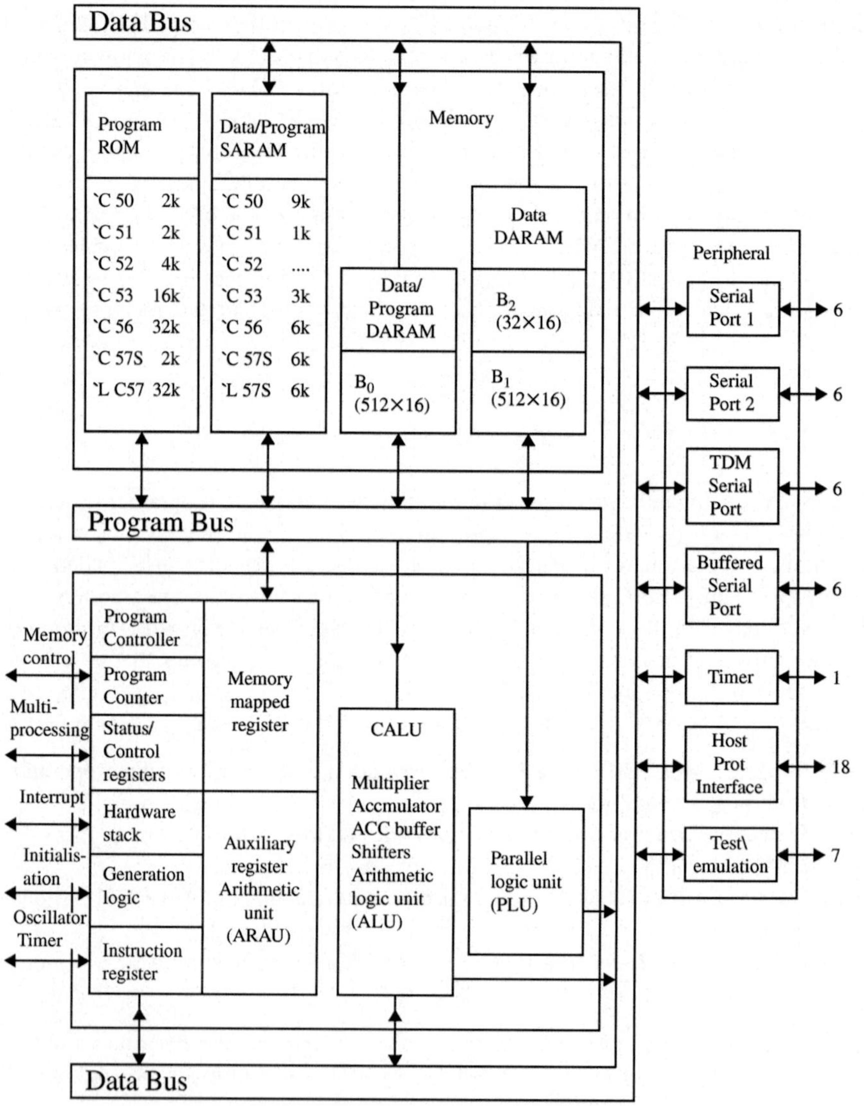

Fig. 6.9 Internal architecture of TMS320C5x

be specified using one of the addressing modes. The 32-bit PREG (Product register) holds the result of multiplication.

(ii) **Accumulator (ACC)**: It is a 32-bit register used for arithmetic and logical operations. The higher order word and lower order words of ACC are represented as ACCH and ACCL, respectively. It can be loaded either directly or from memory.

(iii) **Arithmetic Logic Unit (ALU)**: One of the operands for the ALU operation comes from ACC. Either the higher order word or lower order word of ACC can be loaded from memory. The results of the operation performed in central ALU are stored in ACC.

(iv) **Accumulator Buffer (ACCB)**: It is a 32-bit register used for temporary storage of ACC.

(v) **0–16-bit left barrel shifter and right barrel shifter**: It permits the contents of memory to be left-shifted by 0 to 16 before they are either fed to ALU or stored from ALU to memory.

6.6.3 Auxiliary Register ALU (ARAU)

It consists of following elements:

(i) Eight 16-bit auxiliary registers (AR_s) AR0-AR7.
(ii) 3-bit auxiliary register pointer (ARP).
(iii) Unsigned 16-bit ALU.

ARAU calculates indirect addresses by using inputs coming from ARS, 16-bit index register (INDX) and auxiliary register compare register (ARCR). ARAU can auto index the current AR while the data memory location is being addressed and can index either by ± 1 or by the contents of the INDX. As a result, accessing data does not require the CALU for address manipulation. Therefore, the CALU is free for other operations in parallel. This makes the instructions to be executed faster compared to the conventional microprocessor. For example:

In8085 (MOV A, M; INX H)
These instructions enable the accumulator to be loaded using indirect addressing mode, and HL register used as the address pointer is incremented. These two instructions can be replaced by a single instruction in C5x.

LACC *+, O—any one of the auxiliary registers can be used as the address pointer and incremented. The register that will be used is specified by the content of the ARP. Some of the other registers of ARAU and their functions are as follows.

INDEX Register (INDX)
The 16-bit INDX is used by the ARAU as a step value (addition or subtraction by more than 1) to modify the address in the ARS during indirect addressing. For example, when the ARAU steps across a row of a matrix, the indirect address is

incremented by 1. However, when the ARAU steps down a column, the address is incremented by the dimension of the matrix.

Auxiliary Register Compare Register (ARCR)
It is a 16-bit register used for address boundary comparison. For example:

CMPR Instruction
It compares the ARCR to the selected AR and places the result of the comparison in the TC bit of ST1 (Status Register 1).

Block Move Address Register (BMAR)
The 16-bit BMAR holds an address value to be used with block moves and multiply/accumulate operation. This register provides the 16-bit address for an indirect addressed second operand.

Block Repeat Registers (RPTC, BRCR, PASR, PAER)
All these registers are 16-bit wide.

1. **Repeat counter register (RPTC)** holds the repeat count in a repeat single instruction operation and is loaded by the RPT and RPTZ instructions.
2. **Block repeat counter register (BRCR)** holds the count value for the block repeat feature. This value is loaded before a block repeat operation is initiated.
3. **Block repeat program address start register (PASR)** indicates the 16-bit address where the repeated block of code starts.
4. **Block repeat program address end register (PAER)** indicates the 16-bit address where the repeated block of code ends. The PASR and PAER are loaded by the RPTB instructions.

6.6.4 Parallel Logic Unit (PLU)

It performs boolean operations or the bit manipulation required of high-speed controller. The PLU can set, clear, test or toggle the bits in a status register control register or any data memory location. The PLU allows logic operations to be performed on data memory values directly without affecting the contents of the ACC or PREG. Results of a PLU function are written back to the original data memory location.

6.6.5 Memory-Mapped Registers

The 'C5x has 96 memory-mapped registers mapped into page 0 of the data memory space. All 'C5x DSPs have 28 CPU registers and 16 input/output (I/O) port registers but have different peripheral and reserved registers. Since the memory-mapped registers are a component of the data memory space, they can be written to and read from in the same way as any other data memory location. The memory-mapped registers are used for indirect data address points, temporary storage, CPU status and control, or integer arithmetic processing through the ARAU.

6.6.6 *Program Controller*

The program controller contains logic circuits that perform the following operations:

1. Decodes the instructions.
2. Manages the CPU pipeline.
3. Stores the status of CPU operation.
4. Decodes the conditional operations.

Parallelism of the architecture lets the processor perform three concurrent memory operations: fetch an instruction, read an operand and write an operand in any given machine cycle. It consists of the following elements:

1. 16-bit Program Counter (PC).
2. 16-bit status registers ST0, ST1, Processor Mode Status Register (PMST) and Circular Buffer Control Register (CBCR).
3. (16 × 16) bit hardware stack.
4. Address generation logic.
5. Instruction register.
6. Interrupt flag register and interrupt mask register.

6.6.7 *Status Registers*

The status registers can be stored in data memory and loaded from data memory, thereby allowing the processor status to be saved and restored for subroutines. The bit assignment details for ST0 and ST1 are given in Fig. 6.10.

(*a*)

15–13	12	11	10	9	8–0
ARP	OV	OVM	I	INTM	DP

(*b*)

15–13	12	11	10	9	8–7	6	5	4	3–2	1–0
ARB	CNF	TC	SXM	C	II	HM	I	XF	II	PM

Fig. 6.10 **a** Status register (ST0) bit assignment and **b** status register (ST1) bit assignment

Significance of the various bits of ST0 and ST1 are as follows:

Auxiliary Register Pointer (ARP): It selects the AR to be used in indirect addressing. When the ARP is loaded, the previous ARP value is copied to the auxiliary register buffer (ARB) in ST1.
Overflow Flag Bit (OV): This bit indicates that an arithmetic operation overflows in the ALU.
Overflow Mode Bit (OVM): This bit enables/disables the accumulator overflow saturation mode in the ALU.
Interrupt Mode Bit (INTM): This bit globally masks or enables all interrupts.
Data Memory Page Pointer Bits (DP): These bits specify the address of the current data memory page.
Auxiliary Register Buffer (ARB): It holds the previous value contained in the ARP in ST0. Whenever the ARP is loaded, the previous ARP value is copied to the ARB, except when using the LST #0 instruction. When the ARB is loaded using the LST #1 instruction, the same value is also copied to the ARP.
On-chip RAM Configuration Control Bit (CNF): It enables the on-chip dual-access RAM block 0 (DARAM B0) to be addressable in data memory space or program memory space.

(i) If CNF = 0, the on-chip DARAM B0 is mapped into data memory space. The CNF bit can be cleared by CLRC CNF instruction.
(ii) If CNF = 1, the on-chip DARAM B0 is mapped into program memory space. The CNF bit can be set by SETC CNF instruction.

Test/Control Flag Bit (TC): It stores the results of the ALU or parallel logic unit (PLU) test bit operation. The status of the TC bit determines if the conditional branch, call and return instructions are to be executed.
Sign Extension Mode Bit (SXM): It enables/disables sign extension of an arithmetic operation. The SXM bit does not affect the operation of certain arithmetic or logic instruction: ADDC, ADDS, SUBB or SUBS instruction.
Carry Bit (C): This 1-bit field indicates an arithmetic operation carries or borrows in the ALU. The single-bit shift and rotate instructions affect the C bit.
Hold Mode Bit (HM): This 1-bit field determines whether the central processing unit (CPU) stops or continues execution when acknowledging an active HOLD signal.
Pin Status Bit (XF): It determines the level of the external flag (XF) output pin.
Product Shift Mode Bits (PM): These bits determine the product shifter (P-SCALER) mode and shift value for the PREG output into the ALU.

Table 6.2 gives the functions of PM bits.

Table 6.2 PM bits and their functions

PM bits		Functions
b_1	b_0	P-scalar mode for PREG output
0	0	No shift
0	1	Left-shifted by 1 bit, LSB zero-field
1	0	Left-shifted by 4 bits, LSB zero-field
1	1	Right-shifted by 6 bits, 6 LSB lost

6.6.8 On-Chip Memory

The 'C5x has a total address range of 224K words × 16 bits. The memory space is divided into four individually selectable memory segments:

(i) 64K words → Program memory space.
(ii) 64K words → Local data memory space.
(iii) 64K words → Input/Output ports.
(iv) 32K words → Global data memory space.

The large on-chip memory of 'C5x includes

(i) Program Read-Only Memory (ROM).
(ii) Data/Program Dual-Access RAM (DARAM).
(iii) Data/Program Single-Access RAM (SARAM).

Program ROM

The 'C5x DSP carries a 16-bit on-chip maskable programmable ROM. The program memory can reside both on- and off chip. The device configuration can be changed by setting and clearing the MP/MC bit in the processor mode status register (PMST). If MP/MC pin is low, then 'C5x selects an on-chip ROM. If MP/MC pin is high, then 'C5x devices start execution from off-chip memory.

Data/Program Dual-Access RAM

All 'C5x DSPs carry a 1056-word × 16-bit on-chip dual-access RAM (DARAM). The DARAM is divided into three memory blocks:

(i) 512-word data or program DARAM block B_0.
(ii) 512-word data DARAM block B_1.
(iii) 32-word data DARAM block B_2.

DARAM blocks B_1 and B_2 are always configured as data memory. DARAM block B_0 can be configured as data or program memory.

Data/Program Single-Access RAM

All 'C5x DSPs except 'C52 carry a 16-bit on-chip single-access RAM of various sizes which is divided into 2K word and 1K word blocks that continue in program or data memory space. The SARAM can be configured by software in one of following three ways:

(i) All SARAM are configured as data memory.
(ii) All SARAM are configured as program memory.
(iii) SARAM are configured as both data memory and program memory.

All 'C5x CPUs support parallel accesses to these SARAM blocks. However, one SARAM block can be accessed only once per machine cycle. In other words, the CPU can read from or write to one SARAM block while accessing another SARAM block.

On-Chip Memory Protection

The 'C5x DSPs have a maskable option that protects the contents of on-chip memories. When the related bit is set, no externally originating instruction can access the on-chip memory spaces.

6.6.9 On-Chip Peripherals

The on-chip peripherals available in 'C5x are as follows:

(i) Clock generator.
(ii) Hardware Timer.
(iii) Software-Programmable Wait State Generators.
(iv) Parallel Input/Output ports.
(v) Host Port Interface (HPI).
(vi) Serial Port.
(vii) Buffered Serial Port (BSP).
(viii) Time Division Multiplexed (TDM) Serial Port.
(ix) User-Maskable Interrupts.

Clock Generator

The clock generator consists of an internal oscillator and a Phase Locked Loop (PLL) circuit. The generator can be driven internally by a crystal resonator or driven externally by a clock source. The PLL circuit can generate an internal CPU clock by multiplying the clock source by a specific factor and so a clock source with frequency lower than that of the CPU can be used.

Hardware Timer

A 16-bit hardware timer with a 4-bit pre-scalar is available. The timer can be stopped, restarted, reset or disabled by specific status bits. The time counter register (TIM) gives the current count of the timer. The timer Period Register (PRD) defines the period for the timer. Timer Control Register (TCR) controls the operations of the timer.

Software-Programmable Wait State Generators

Software-Programmable Wait State logic is incorporated in 'C5x allowing wait state generation without any external hardware for interfacing with slower off-chip memory and I/O devices. It consists of multiple wait state generating circuits. Each circuit is user-programmable to operate in different wait states for off-chip memory accesses.

Parallel Input/Output Ports

The 'C5x has 64K input/output ports. Sixteen of these ports are memory-mapped in data memory space. Each of the input/output ports can be addressed by IN or OUT instruction or any instruction that reads from or writes to data memory. The IS signal indicates a read or write operation through an input/output port.

Serial Port

Three different kinds of serial ports are available:

(i) General-purpose serial port.
(ii) Time division multiplexed serial port.
(iii) Buffered serial port.

Each 'C5x contains at least one general-purpose, high-speed synchronous, full-duplex serial port interface that provides direct communication with serial devices such as codes, serial analog to digital converters and other serial system. It is capable of operating at up to one-fourth of the machine cycle.

Five 16-bit registers that control and operate the serial port interface are as follows:

(i) Serial Port Control (SPC): It contains the mode control and status bit of the serial port.
(ii) Data Receive Register (DRR): It holds the incoming serial data.
(iii) Data Transmit Register (DXR): It holds the outgoing serial data.
(iv) Data Transmit Shift Register (XSR): It controls the shifting of the data from the DXR to the output pin.
(v) Data Receive Shift Register (RSR): It controls the storing of the data from the input pin to the DRR.

Host Port Interface

It is available on the 'C57s and 'LC57. It is an 8-bit parallel input/output port that provides an interface to a host processor. Information is exchanged between the DSP and the host processor through on-chip memory that is accessible to both the host processor and the 'C57.

Buffered Serial Port

It is available on the 'C56 and 'C57. It is a full-duplex, double-buffered serial port and an auto buffering unit. It provides flexibility on the data stream length. It supports high-speed data transfer and reduces interrupt latencies.

TDM Serial Port

It is available on the 'C50, 'C51 and 'C53. It is a full-duplex serial port that can be configured by software either for synchronous operations or time division multiplexed operations. It provides simple and efficient interface for multiprocessing applications.

User-Maskable Interrupts

The 'C5x has four external interrupts ($INT1 - INT4$) and five internal interrupts (a timer interrupt and four serial interrupts are user-maskable). When an interrupt service routine (ISR) is executed, the contents of the program counter are saved on an 8-level hardware stack and the contents of 11 specific CPU registers, ACC, ACCB, PREG, ST0, ST1, PMST, TREG0, TREG1, TREG2, INDX and ARCR, are saved in one deep stack (shadow registers). When a return from interrupt instruction is executed, the CPU register contents are restored.

6.7 Addressing Modes

The 'C5x supports the following six addressing modes:

1. Direct addressing.
2. Indirect addressing.
3. Immediate addressing.
4. Dedicated register addressing.
5. Memory-mapped register addressing.
6. Circular addressing.

6.7.1 Direct Addressing

The data memory used in C5x processors is split into 512 pages each of 128 words long. In the direct addressing mode of C5x, only lower order 7 bits of the address are specified in the instruction. The upper 9 bits are taken from the data memory page pointer (DP). The DP in ST0 holds the address of the current data memory page as shown in Fig. 6.11.

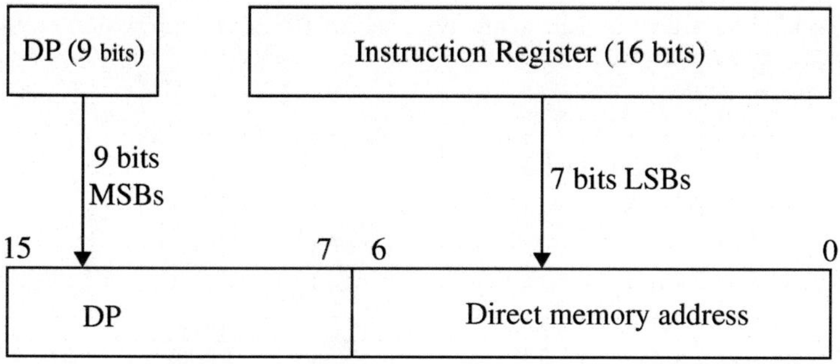

Fig. 6.11 16-bit data memory address bus

Table 6.3 Various options in indirect addressing

Symbol	Value of AR after instruction execution
*	AR unaltered
*+	AR incremented by 1
*−	AR decremented by 1
*0+	AR incremented by the content of INDX
*0−	AR decremented by the content of INDX
*BR0+	AR incremented by the content of INDX with reverse carry propagation
*BR0−	AR decremented by the content of INDX with reverse carry propagation

6.7.2 Indirect Addressing

The Auxiliary registers (AR0 – AR7) are used for accessing data, using the indirect addressing mode. In indirect addressing, out of eight ARs the one which is currently used for accessing data is denoted by the register ARP. The contents of ARP can be temporarily stored in the ARB register. The indirect addressing mode permits the AR used for addressing to be updated automatically either after or before the operand is fetched. Hence, a separate instruction is not required to update the AR. However, it requires the contents of an AR to be incremented or decremented by an 8-bit constant using SBRK and ADRK instructions, i.e.

SBRK #K → subtracts the constant K from AR pointed by ARP.
ADRK #K → adds the constant K to the AR pointed by ARP.

The symbol used to indicate the indirect addressing mode and the action taken after executing the instruction are given in Table 6.3.

For example, let the value of ARP → 2, AR_2 → 1250 h and INDX → 2h and content of data memory location 1240–1260 h be filled with data 2345 h. The contents of ACC and AR_2 after the following sequence of LACC instructions executed are given below

	Instruction executed	Contents after execution ACC	AR_2
(i)	LACC *, 0	2345h	1250h
(ii)	LACC *+, 1	468Ah	1251h
(iii)	LACC *−, 2	9786Ah	1250h
(iv)	LACC *0+, 4	11A28h	1252h
(v)	LACC *0−, 3	8D14h	125h0h

6.7.3 *Immediate Addressing*

The immediate addressing mode can be used to load either a 16-bit constant or a constant of length 13, 9 or 8. Accordingly, it is referred to as long immediate or short immediate addressing mode. This mode is indicated by the symbol #. For example:

ADD #55h → adds 55h to Accumulator (ACC) [short immediate addressing].
ADD #1267h → adds 1267h to ACC [long immediate addressing].

Therefore, in immediate addressing, the instruction word(s) contains the value of the immediate operand. The 'C5x has both 1 word (8-bit, 9-bit and 13-bit constants) short immediate instructions and 2-word (16-bit constant) long immediate instructions. Table 6.4 lists the instructions that support immediate addressing.

Table 6.4 Instructions that support immediate addressing

	8-bit constant		9-bit constant	13-bit constant
Short immediate addressing	ADD ADRK LACL LAR RPT SBRK SUB	LDP		MPY
Long immediate addressing (16-bit constant)	ADD LACC OR SPLK SPLK	AND LAR RPT SUB	SPL MPY RPTZ XOR	CPL OPL XPL

Short Immediate Addressing

In short immediate instructions, the operand is contained within the instruction machine code. For example:

ADD #OFEh—lower 8 bits are the operand and are added to the ACC.
Long Immediate Addressing

In long immediate instructions, the operand is contained in the second word of a 2-word instruction. There are two long immediate addressing modes: one-operand instruction and two-operand instructions. For example:

ADD #1245h—the second word (1245h) of the 2-word instruction is added to ACC.
BLDD #2345h, 012h—the source address (operand 1) is fetched via PAB, the decimation address (operand 2) uses the direct addressing mode. Bits 15–8 of machine code contain the opcode. Bit 7 with a value of 0 defines the direct addressing and bits 6–0 contain the dma.

6.7.4 Memory-Mapped Register Addressing

The registers corresponding to data memory page 0 are referred to as memory-mapped registers (MMRS). With memory-mapped register addressing, the MMRS can be modified without affecting the current data page pointer value. The memory-mapped register addressing mode operates like the direct addressing mode, except that the 9 MSBs of the address are forced to 0 instead of being loaded with the contents of the DP. This allows the memory-mapped registers of data page 0 to be modified directly without the overhead of changing the DP or auxiliary register. The following instructions operate in the memory-mapped register addressing mode.

- LAMM—Load accumulator with memory-mapped register.
- LMMR—Load memory-mapped register.
- SAMM—Store accumulator in memory-mapped register.
- SMMR—Store memory-mapped register.

For example:

(i) LMMR AR0, #1500h

Before execution		**After execution**	
Data memory		*Data memory*	
1500h	6785h	1500h	6785h
AR0	1325	AR0	6785h

(ii) SMMR AR0, #1600h

Before execution		After execution	
Data memory		*Data memory*	
1600h	6785h	1600h	1325h
AR0	1325h	AR0	1325h

(iii) LAMM*

Before execution		After execution	
ARP	1	ARP	1
AR1	325h	AR1	325h
Data memory		*Data memory*	
825h	1234h	825h	1234h
25h	2345h	25h	2345h
ACC	6789h	ACC	2345h

After execution of the LAMM* instruction, the value in data memory location 25h is loaded into ACC. 25h corresponds to the lower order 7 bits of AR1 and the higher bits of PAB are made to be 0 as the MMR corresponds to page 0.

6.7.5 Dedicated Register Addressing

The dedicated register addressing mode operates like the long immediate addressing mode, except that the address comes from one of two special-purpose memory-mapped registers in the CPU:

(i) Block Move Address Register (BMAR).
(ii) Dynamic Bit Manipulation Register (DBMR).

The advantage of this addressing mode is that the address of the block of memory to be acted upon can be changed during execution of the program. For example:

1. **BLDD BMAR, DAT 100; DP = 0** If BMAR contains the value of 300h, then the content of data memory location 300h is copied to data memory location 100 on the current data page.
2. **OPL DAT10; DP = 6**
   If DBMR contains the value of OFFF0h and the address O30Ah contains the value of 01h, then the content of data memory location O30Ah is ORed with the content of DMAR. The resulting value OFFF1h is stored back in memory location O30Ah.

6.7.6 *Circular Addressing*

Circular addressing is the most sophisticated ’C5x addressing mode. Many algorithms such as a convolution, correlation and finite impulse response (FIR) filters can use circular buffer in memory to implement a sliding window, which contains the most recent data to be processed. There are five memory-mapped registers that control the circular buffer operation, and they are

- CBSR1—circular buffer 1 start register.
- CBSR2—circular buffer 2 start register.
- CBER1—circular buffer 1 end register.
- CBER2—circular buffer 2 end register.
- CBCR—circular buffer control register.

The 8-bit CBCR enables and disables the circular buffer operation. First, the start and end addresses are loaded into the corresponding buffer register. Next, a value between the start and end registers for the circular buffer is loaded into an AR. The corresponding circular buffer enable bit in the CBCR should be set.

6.8 Instruction Sets

6.8.1 *Addition/Subtraction Instructions*

In the addition/subtraction instructions of ’C5x, one of the operands is ACC. The other operand can be PREG, ACCB or the content of memory fetched using one of the addressing modes.

6.8.1.1 Addition Instruction

1. **ADD dma, [shift] (direct addressing)**: The contents of the data memory address (dma) or a 16-bit constant are shifted left as defined by the shift code and added to the contents of the accumulator and the result is stored in ACC. For example:
   ADD 55h, 2: ACC is added to the content of a data memory with dma 55h in the current page after shifting it left by two positions.
   ADD {immediate}, [shift], [ARn] (indirect addressing): short immediate—**ADD #K and long immediate** — ADD #2K, [shift].

Examples

(a) **ADD # 23h**: ACC is added to the immediate constant 23h.
(b) **ADD # 2345h, 2**: data 2345h is left-shifted by two positions before it is added to ACC.

Before execution		After execution	
ACC	1234h	ACC	9F48h

(c) **ADD *, 1,** AR_2

Before execution		After execution	
ARP	1	ARP	2
AR1	2100h	AR1	2100h
Data memory		*Data memory*	
2100h	4563h	2100h	4563h
ACC	1234h	ACC	9CFAh

2. **ADRK #K**: The 8-bit immediate constant value is added to the current auxiliary register (AR). The result is stored in the AR. For example:
ADRK #25h

Before execution		After execution	
ARP	2	ARP	2
AR2	4200h	AR2	4225h

3. **ADCB**: The contents of the accumulator buffer (ACCB) and the value of the C bit are added to the contents of the ACC and the result is stored in ACC. The content of ACCB is unaffected. The C bit is set, if the result of the addition generates a carry; otherwise the C bit is cleared. For example:

Before execution			After execution		
ACC	1 (C)	1235h	ACC	0 (C)	1258h
ACCB		2h	ACCB		2h

Other addition instructions: ADDB, ADDC, ADDS and ADDT.

6.8.1.2 Subtraction Instructions

1. **SBB**: The contents of the accumulator buffer (ACCB) are subtracted from the contents of the ACC. The result is stored in the ACC and the contents of the ACCB

are unaffected. The C bit is cleared if the result generates a borrow, otherwise the C bit is set. For example:

Before execution			After execution of SBB instruction		
ACC	X C	2000h	ACC	1 C	1000h
ACCB		1000h	ACCB		1000h

2. **SUB dma, [shift]**: ACC is subtracted from the content of data memory address (dma) in the current page after shifting it left by specified shift position. For example:
   (i) **SUB 25h, 2**: ACC is subtracted from the content of data memory with dma 25h after shifting it left by two positions.
   (ii) **SUB *, 2**: ACC is subtracted from the content of location pointed by the current AR after shifting it left by two positions.

   Other subtraction instructions: SBBB, SUBB, SUBC, SUBS, SUBT and SBRK.

6.8.2 Multiplication Instruction

In multiply instructions, one of the operands is taken from TREG0 and other operand is specified using one of the addressing modes.

1. **MPY**: Multiply numbers in 2's complement form.
   - Direct addressing: MPY dma—The contents of TREG0 are multiplied by the contents of data memory address and the result is stored in the PREG.
   - Indirect addressing: MPY {ind}, [ARn]
   - Short immediate: MPY #K
   - Long immediate: MPY #LK
2. **MAC**: Multiply and accumulate.
   MAC pma, dma (direct addressing): The contents of the product register (PREG) are shifted, as defined by the PM bits and added to the ACC. The result is stored in the ACC. The contents of the data memory address (dma) are loaded into TREG0. The contents of the dma are multiplied by the contents of pma. The result is stored in PREG. The C bit is set, if the result of addition generates a carry, otherwise C bit is cleared.

3. **MACD**: Multiply and accumulate with data move. **MACD pma, dma (direct addressing)**: The contents of dma are multiplied by the contents of pma and the result is stored in PREG. The previous value of the PREG is shifted as specified by the PM bits and added to the ACC before the PREG is loaded with the product. For example:

MACD 0FF0Fh, 15h

Before execution		After execution	
Data memory		*Data memory*	
315h	45h	315h	45h
Data memory		*Data memory*	
316h	28h	316h	45h
Programme memory		*Programme memory*	
FF0Fh	8h	FF0Fh	8h
TREG0	1234h	TREG0	45h
PREG	4563h	PREG	0045h
ACC	1234h	ACC	5797h

Other multiplication instructions:

1. MPYU: multiply unsigned numbers.
2. MPYA: multiply and add the product to ACC.
3. MPYS: multiply and subtract the product from ACC.
4. MADD: multiply and add the product to ACC address of an operand given by BMAR.
5. MPYA: multiply and add the product to ACC and move the on-chip RAM by one word.

6.8.3 Shift/Logical Instructions

6.8.3.1 Logical Instructions

AND Instruction

1. Direct addressing: **AND dma.**
2. Indirect addressing: **AND** {Ind}, [AR_n].
3. Long immediate: **AND # LK, [shift].**

One of the operands for the AND instructions is ACC. The other operand can be the content of a memory location specified using direct and indirect addressing. Alternately, a long constant can be specified using immediate addressing. For example: **AND #2345h, 2**

Before execution		After execution	
ACC	1234h	ACC	0014h

Other AND instruction is ANDB: The contents of the ACC are ANDed with the contents of the ACCB. The result is stored in the ACC and the contents of the ACCB are unaffected.

OR instruction

When OR instruction is given to the ACC with a long constant K, the content of a dma or pma and the result are stored in the ACC.

1. Direct addressing: **OR dma.**
2. Indirect addressing: **OR** {ind} [, AR_n].
3. Long immediate: **OR # lk, [shift].**

For example: **OR * AR1**

Before execution		After execution	
ARP	0	ARP	1
AR0	1100h	AR0	1100h
Data memory		*Data memory*	
1100h	1235h	1100h	1235h
ACC	1111h	ACC	2346h

Other OR instruction is ORB: The contents of the ACC are ORed with the contents of the accumulator buffer.

XOR Instruction

When XOR instruction is given to the ACC with a long constant K, the content of a dma or pma and the result are stored in ACC.

1. Direct addressing: **XOR dma.**
2. Indirect addressing: **XOR {ind} [, shift].**
3. Long immediate: **XOR #lk [, shift].**

For example: **XOR 50h**

Before execution		After execution	
DP	1	DP	2
Data memory		*Data memory*	
FFD0h	4563h	FFD0h	4563h
ACC	1234h	ACC	5757h

Other XOR instruction is XORB: The contents of the ACC are X-ORed with the contents of the accumulator buffer (ACCB).

Shift Instructions

ROL: Rotate ACC left once: rotate the contents of the ACC left by 1 bit. The value of the carry bit is shifted into LSB of the ACC. The MSB bit of the original ACC is shifted into the C bit.

ROLB: Rotate the contents of both ACC and ACCB left once.

ROR: Rotate the contents of the ACC right once.

RORB: Rotate the contents of both ACC and ACCB right once.

Other shift/logical Instruction

BSAR: Rotate ACC right by $n(1-16)$ bits.
EXAR: Exchange the contents of ACC and ACCB.
NEG: Find 2's complement of ACC.
CMPL: Find 1's complement of ACC.
BIT: Copy bit n of a memory onto TC.
BITT: Copy bit n of a memory onto TC bit in ST1. n is given by the 4 LSBs of TREG2.

6.8.4 *Load/Store Instructions*

LACB: Load the contents of the ACC to ACCB.
LACC: Load data memory value, with left shift, to ACC (LACC dma, shift).
: Load long immediate, with left shift to ACC (LACC #lk, shift).
LACL: Load data memory value to ACCL and zeros filled to ACCH (LACL dma).
: Load short immediate to ACCL and zeros filled to ACCH (LACL #K).
LACT: Load data memory value with left shift specified by TREG1 to ACC.
LAMM: Load contents of memory-mapped register to ACCL; zeros filled to ACCH.
SACB: Store the contents of the ACC in ACCB.
SACH: Store ACCH, with left shift in data memory address.
SACL: Store ACCL, with left shift in data memory address.
SAMM: Store ACCL in memory-mapped register.
LAR: Load data memory value to ARX (Auxiliary register).
SAR: Store ARX in data memory location.
LDP: Load data memory value to DP (Data Pointer) bits.
MAR: Modify auxiliary register.
SPLK: Store long immediate in data memory location.
LPH: Load data memory value to PREG higher byte.
LT: Load data memory value to TREG0.
PAC: Load PREG with shift specified by PM bits to ACC.
SPH: Store PREG higher byte, with shift specified by PM bits in data memory location.
LST: Load data memory value to ST0.
: Load data memory value to ST1.
SST: Store ST0 in data memory location.
: Store ST1 in data memory location.

For example:

(i) **LST #0, 00h**

Before execution		After execution	
Data memory		*Data memory*	
300h	1025h	300h	1025h
ST0	1245h	ST0	1025h
ST1	09A0h	ST1	09A0h

(ii) **LST #1, 00h**

Before execution		After execution	
Data memory		*Data memory*	
300h	1025h	300h	1025h
ST0	5E07h	ST0	1E07h
ST1	1234h	ST1	1025h

(iii) **LST** #1, *, **AR_3** (indirect addressing)

Before execution		After execution	
Data memory		*Data memory*	
300h	0587h	300h	0587h
ST0	5E07h	ST0	1E07h
ST1	09A0h	ST1	0587h
ARP	2	ARP	0

(iv) **LT 300h**

Before execution		After execution	
Data memory		*Data memory*	
300h	60h	300h	60h
TREG0	5h	TREG0	60h

(v) **LDP #30h**

Before execution		After execution	
DP	01h	DP	30h

(vi) **LACC *, 4**

Before execution		After execution	
ARP	3	ARP	3
AR3	0100h	AR3	0100h
Data memory		*Data memory*	
100h	0FFh	100h	0FFh
ACC	0EF0h	ACC	0FF0h

(vii) **LACB**

Before execution		After execution	
ACC	0000 1234h	ACC	1234 5678h
ACCB	1234 5678h	ACCB	1234 5678h

(viii) **LAR AR4**

Before execution		After execution	
ARP	4	ARP	4
AR4	300h	AR4	16h
300h	16h	300h	16h

The instructions SAMM, LAMM, SMMR and LMMR have already been explain in Sect. 6.7.4.

6.8.5 Move Instructions

The data move instruction copies the data from one memory location to the next higher location. It can use either direct or indirect addressing mode.

BLDP: Block move data from data memory to program memory.
BLDD: Block move data from one data memory to another.
BLPD: Block move data from program memory to data memory.
TBLR: Block move data from program memory to data memory. The program memory address is contained in ACC lower order word. The dma can be specified using either direct or indirect addressing.
TBLW: Block move data from data memory to program memory. The program memory address is contained in ACCC. The dma can be specified using either direct or indirect addressing.

For example:

(i) **BLDP 00h**

Before execution		**After execution**	
Data memory		*Data memory*	
400h	2523h	400h	2523h
BMAR	1850h	BMAR	1850h
Programme memory		*Programme memory*	
1850h	1254h	1850h	2523h

(ii) **BLPD 00h**

Before execution		**After execution**	
Data memory		*Data memory*	
300h	1245h	300h	1100h
Programme memory		*Programme memory*	
2850h	1100h	2850h	1100h
BMAR	2850h	BMAR	2850h

(iii) **BLPD #250h, 00h**

Before execution		**After execution**	
Data memory		*Data memory*	
300h	1234h	300h	1100h
Programme memory		*Programme memory*	
250h	1100h	250h	1100h

6.8.6 Branch and Call Instructions

The 'C5x has both conditional and unconditional branch and call instructions. The branch and call instructions permit more than one condition to be tested using single instruction. Branching occurs only if all the conditions are satisfied.

B: Branch unconditionally.
BACC: Branch unconditionally to the address given by ACC.
BACCD: Delayed branch to pma specified by ACCL.
BCND: Branch conditionally.
BCNDD: Delayed branch conditionally to pma.
BANZ: Branch conditionally if ARn not zero.
BANZD: Delayed branch to pma if AR not zero.
BD: Delayed branch unconditionally to pma.
CALA: Call a subroutine using indirect addressing.
CALAD: Delayed call to subroutine addressed by ACCL.

CALL: Call a subroutine unconditionally.
CALLD: Delayed call to subroutine unconditionally.
CC: Call a subroutine conditionally.
CCD: Delayed call to subroutine conditionally.

6.8.7 PUSH and POP Instructions

PUSH: Pushes the values down one level in the seven lower locations of the stack. The contents of ACCL are copied to the top of the stack.
PUSHD: Pushes a data memory location to the top of the stack instead of ACC after pushing the contents of the stack one level down.
POP: Pops the top of the stack to ACC.
POPD: Pops the top of the stack to a data memory.

When the stack is popped, the bottom word is left unaffected and hence the bottom two words contain the same values.

6.8.8 RET Instructions

RET: Return from subroutine: The contents of the top of the stack are copied to the program counter. The stack is piped one level after the contents are copied.
RETCD: Delayed return from subroutine conditionally.
RETD: Delayed return from subroutine.

6.8.9 Repeat Instructions

RPT: Repeat next instruction: The iteration count can be specified using direct addressing, indirect addressing, short and long immediate addressing.

1. Direct addressing: **RPT dma.**
2. Indirect addressing: **RPT *.**
3. Short immediate addressing: **RPT #7.**
4. Long immediate addressing: **RPT #2345h.**

RPTB: Block of instructions repeated. The number of repeats specified in block repeat count register (BRCR). **RPTZ**: Repeat preceded by clearing both ACC and PREG. For example:
RPTZ #K: Clears both ACC and PREG and then executes the next instruction K times.

6.8.10 IN and OUT Instructions

IN: Reads a 16-bit number from input port and stores it in the data memory location.
OUT: Reads a 16-bit number from data memory and writes it onto the output port.

6.8.11 NORM Instruction

It is useful for converting a fined point number into a floating point number. The number to be converted is stored in ACC. In a sign magnitude representation the MSB is a sign extended bit, and only the remaining bits denote the magnitude. Every time the NORM instruction is executed, it removes an extra bit in ACC, which denotes the bit corresponding to a sign extension. By repeated use of the NORM instruction, the ACC can be made to contain only the magnitude. The exponent is stored in the current AR.

6.9 Architecture of 54x

The block diagram of TMS320C54x internal hardware is shown in Fig. 6.12. It consists of the CPU containing the various functional units such as ALU, MAC unit, Exp encoder, barrel shifter, memory-mapped registers, system control interface, peripheral interface, memory and external interface, program address generation logic (PAGEN) and data address generation logic (DAGEN) end eight 16-bit buses which interconnect these units. The comparison of the features of 5x and 54x is tabulated in Table 6.5.

6.9.1 Bus Structure

'C54x consists of eight 16-bit buses (i.e. four program/data buses and four address buses). Program bus (PB) carries the instruction code and immediate operands from program memory.

Data buses (CB, DB and EB)

CB and DB carry the operands that are read from data memory. EB carries the data to be written to memory. Address buses (PAB, CAB, DAB and EAB) carry the addresses needed for instruction execution.

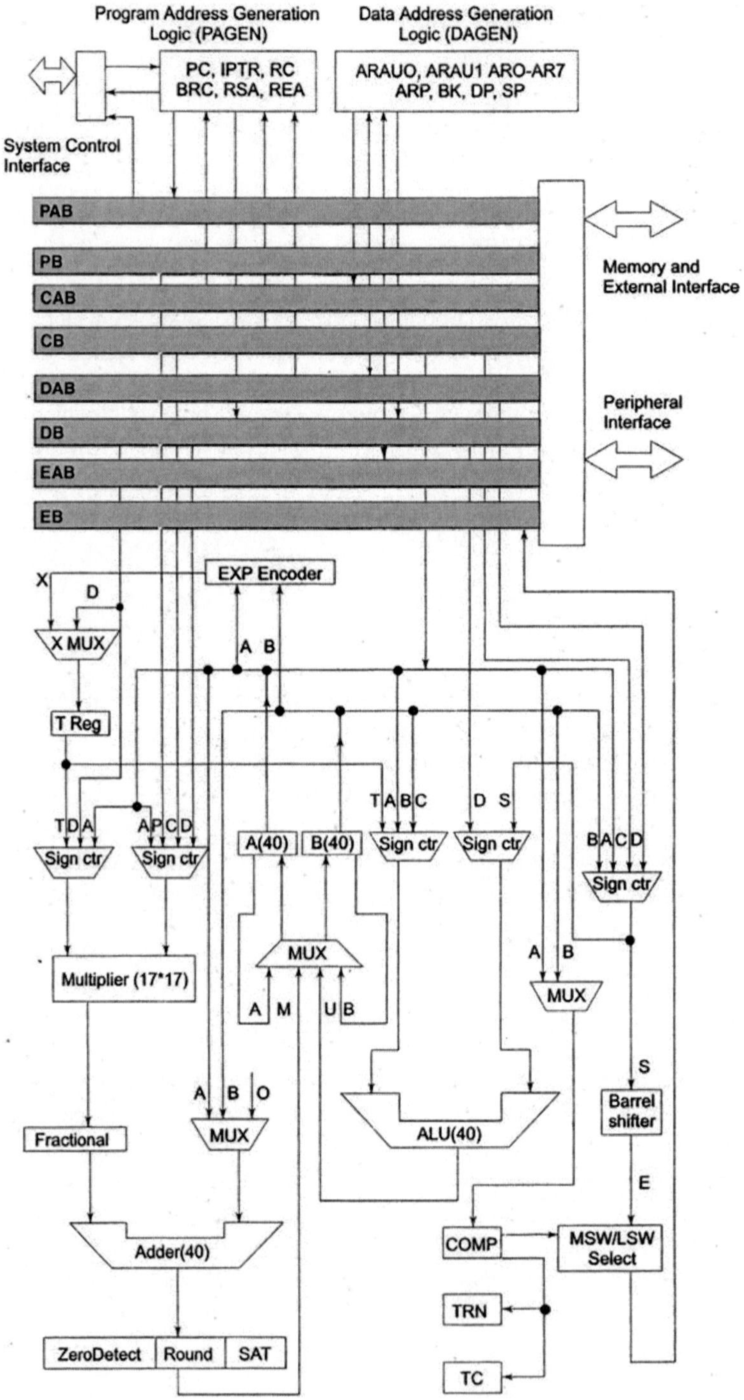

Fig. 6.12 Block diagram of 54x internal hardware

Table 6.5 Comparison of the features of 5x and 54x

Description	5x	54x
Name of program bus	One, PB	One, PB
Name of the data bus	One, DB	DB and CB (for read) EB (for write)
Name of address buses	PAB, DAB	PAB, CAB, DAB, EAB
Main ALU	32-bit ALU	40-bit ALU
Accumulators	32-bit ACC	40-bit ACCA and ACCB
Barrel shutter	0–16-bit left shift 0–16-bit right shift	40 bit: 0–31 left shift 0–15 right shift
Multiplier	16 × 16 bit	17 × 17-bit
Adder	32 bit	40 bit
Auxiliary Register ALU	ARAU	ARAU0 & ARAU1
Auxiliary registers	AR0–AR7	AR0–AR7
Stack pointer (SP)	Not available	16 bit: SP
Circular buffer register	Two 16-bit start & end register	16-bit BK
Status registers	16-bit PMST, ST0, ST1	16-bit PMST, ST0, ST1
Block repeat registers	16-bit BRCR, PASR, PAER	16-bit BRC, RSA, REA
Program counter	16-bit PC	16-bit PC
Extended program memory	Not available	7-bit XPC
Interrupt registers	16-bit IMR and IFR	16-bit IMR and IFR
General-purpose I/O	BIO and XF	Same as that of 5x
Wait state generator	PDWSR	SWWSR
Hardware timer	16-bit timer	Same as that of 5x
Clock generator	PLL based	Same as that of 5x
Synchronous serial port	Full duplex and double buffered	Same as that of 5x
TDM serial ports	Up to 7 devices using TDM can communicate serially	Same as that of 5x
Buffered serial port	Standard 5x serial port with additional auto buffering unit	Same as that of 5x
Host port interface	8-bit standard HPI	8-bit standard HPI or enhanced 8-bit and 16-bit HPI
Multichannel buffered serial port including internal programmable clock and other advanced features	Not available	Available
On-chip ROM for lookup table for A law, μ law commanding, sine wave generation	Not available	Available

6.9.2 Internal Memory Organization

The ’54x memory is organized into three individually selectable spaces: program, data and input/output spaces (DARAM and SARAM). It contains both RAM and ROM.

On-chip ROM: The on-chip ROM is part of program memory space (20K word) and in some cases, part of the data memory. Therefore, the ROM may be mapped into both data and program space (8K words).

On-chip Dual-Access RAM (DARAM): It consists of several blocks and each block can be accessed twice per machine cycle; the CPU can read from and write to a single block of DARAM in the same cycle. It is mapped into program space (5K word).

On-chip Single-Access RAM (SARAM): It consists of several blocks. Each block is accessible once per machine cycle for either a read or a write. It is mapped into program space.

6.9.3 Central Processing Unit (CPU)

The ’C54x CPU contains

- 40-bit Arithmetic Logic Unit (ALU).
- Two 40-bit accumulator registers.
- Barrel shifter.
- Multiply/Accumulate Block.
- 16-bit Temporary Register.
- 16-bit Transition Register (TRN).
- Compare, Select and Store Unit (CSSU).
- Exponent encoder.

Status Register (ST0)

The status register ST0 bits are shown in Fig. 6.13.

ARP (auxiliary register pointer): It selects the auxiliary register to be used in indirect addressing.

TC (test/control flage): TC stores the results of the arithmetic logic unit (ALU) test bit operations. TC is affected by the BIT, BITF, BITT, CMPM, CMPR, CMPS and SFTC instructions. The status (set or cleared) of TC determines if the conditional branch, call, execute and return instructions are executed.

Fig. 6.13 Status register ST0

15–13	12	11	10	9	8–0
ARP	TC	C	OVA	OVB	P

C (carry)—C = 1 if result of addition generates a carry. C = 0 if result of subtraction generates a borrow.
OVA (overflow flag for ACCA)—OVA = 1 when an overflow occurs in either the ALU or multiplier's adder in ACCA.
OVB (overflow flag for ACCB)—OVA = 1 when an overflow occurs in either the ALU or multiplier's adder in ACCB.
DP (data memory page pointer): This 9-bit field is concatenated with the seven LSBs of an instruction word to form direct memory address of 16-bit address for single data memory operand addressing. This operation is done if the compiler mode bit is STI (CPL) = 0.

Status Register ST1

The status register ST1 bits are shown in Fig. 6.14.

BRAF (Block repeat active flag).

–if BRAF = 1, block repeat is active.
–if BRAF = 0, block repeat is de-active and BRC decrements below 0.

CPL (Compiler mode): CPL = 0, the relative direct addressing mode using the data page pointer is selected if CPL $\neq 0$ using stack pointer the relative direct addressing mode is selected.
XF (XF status) External flagepin: The SSBX instruction can set XF, and RSBX instruction can reset XF.
HM (Hold mode): HM indicates whether the processor continues internal execution when acknowledging an active HOLD signal.
INTM (Interrupt mode): If INTM = 0, all unmasked interrupts are enabled. If INTM = 1, all maskable interrupts are disabled.
OVM (overflow mode): OVM determines what is loaded into the designation accumulator when an overflow occurs.
SXM (Sign-extension mode): SXM = 0, sign extension is suppressed. SXM = 1, data is sign extended before being used by the ALU.
C16 (Dual 16-bit/double precision arithmetic mode): It determines the arithmetic mode of the ALU's operation. If C16 = 0 double precision arithmetic mode, and C16 = 1 dual 16-bit arithmetic mode.
CMPT (Compatibility mode): It determines the compatibility mode for the ARP.
ASM (Accumulator shift mode): Specify the shift value within –16 through 15 range is coded as a 2's complement form.
Processor Mode Status Register (PMST): It is loaded with memory-mapped register instruction such as STM.

15	14	13	12	11	10	9	8	7	6	5	4–0
BRAF	CPL	XF	HM	INTM	O	OVM	SXM	C16	FRCT	CMPT	ASM

Fig. 6.14 Status register ST1

Auxiliary Registers

'C54x also includes eight auxiliary registers and a software stack to enable a highly optimized C compiler. The eight 16-bit auxiliary registers (AR0–AR7) can be accessed by the CPU and modified by the auxiliary register arithmetic units (ARAUS). The primary function of the auxiliary registers is to generate 16-bit addresses for data space.

Barrel Shifter

The 40-bit barrel shifter of 'C54x can perform arithmetic and logical shifts by up to 31 bits left or by up to 16 bits rights in a single instruction cycle. Shifter inputs can come directly from data memory or from either of the two accumulators. Shifter outputs can be sent to the ALU or stored in memory.

The barrel shifter is also used for scaling operations such as

- Pre-scaling an input data memory operand or the ACC value before an ALU operation.
- Performing a logical or arithmetic shift of the ACC value.
- Normalizing the ACC.
- Post-scaling the ACC before storing the accumulator value into data memory.

Compare, Select and Store Unit (CSSU)

The compare, select and store unit (CSSU) is an application-specific hardware unit dedicated to add/compare/select (ACS) operations of the viterbi operator. Figure 6.15 shows the CSSU, which is used with the ALU to perform fast ACS operations. The CSSU allows the 'C54x to support various viterbi butterfly algorithms used in equalizers and channel decoders. The add function of the viterbi operator is performed by the ALU. This function consists of a double addition function (Met 1 D1 and Met

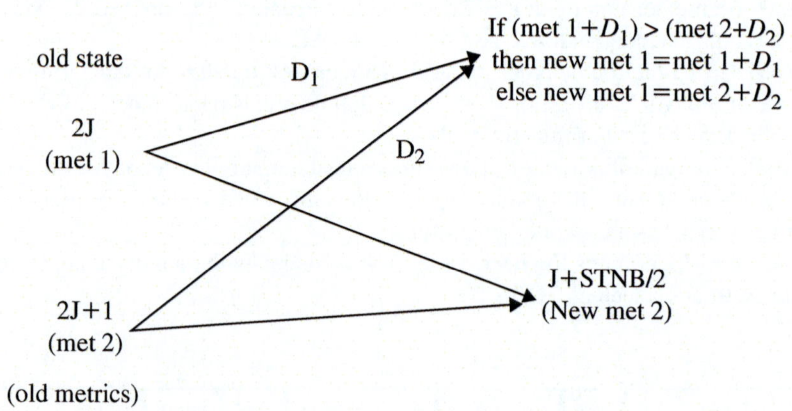

Fig. 6.15 Viterbi operator (STNB = no. of states, met path metrics and D = branch metrics)

2 ID2). Double addition is completed in one machine cycle if the ALU is configured for dual 16-bit mode by setting the C16 bit in ST1.

The CSSU implements the compare and select operation via the CMPS instruction, a comparator and the 16-bit transition (TRN) register. This operation compares two 16-bit parts of the specified accumulator and shifts the decision into bit 0 of TRN. This decision is also stored in the TC bit of ST0. Based on the decision, the corresponding 16-bit part of the accumulator is stored in data memory: TRN register contains information of the path transition decisions to new states. This information can be used for a back-tracking routine that finds the optimal path, which results in decoding the code.

6.9.4 Pipeline

The 'C54x DSP has a six-level deep instruction pipeline. The six stages of the pipeline are independent of each other, which allows overlapping execution of instructions. During any given cycle, one of six different instructions can be active, each at a different stages of completion. The six levels and function of the pipeline structure are

- Program pre-fetch.
- Program fetch.
- Decode.
- Access.
- Read.
- Execute.

6.9.5 On-Chip Peripherals

The 'C54x has the following on-chip peripherals:

- General-purpose input/output pins.
- Software-programmable wait state generator.
- Programmable book-switching logic.
- Host port interface.
- Hardware timer.
- Clock generator.
- Serial ports.
- Synchronous serial ports.
- Buffered serial ports.
- Time Division Multiplexed (TDM) serial ports.

6.9.6 *Data Addressing*

The 'C54x offers seven basic data addressing modes:

- Immediate addressing: uses the instruction to encode a fined value.
- Absolute addressing: uses the instruction to encode a fined address.
- Accumulator addressing: uses accumulator A to access a location in program memory as data.
- Direct addressing: uses 7 bits of the instruction to encode the lower 7 bits of an address. The 7 bits are used with the data page pointer (DP) or the stack pointer (SP) to determine the actual memory address.
- Indirect addressing: uses the ARS to access memory.
- Memory-mapped register addressing: uses the memory-mapped registers without modifying either the current DP value or the current SP value.
- Stack addressing: manages adding and removing in terms of the stack.

During the execution of instructions using direct, indirect or memory-mapped register addressing, the data-address generation logic (DAGEN) computes the addresses of data memory operands.

6.10 Simple Assembly Language Program

Example 6.1 To write an assembly language program for 32-bit addition.

Solution In this example, two 32-bit numbers X and Y will be added as follows:

$$\begin{array}{rl} X & \Longrightarrow X_1\ X_0 \\ Y & \Longrightarrow \underline{Y_1\ Y_0} \\ \{+\} & \overline{Z_1\ Z_0} \end{array}$$

where X_1 and Y_1 are MSB (16 bit) of X and Y, respectively; X_0 and Y_0 are LSB (16 bit) of X and Y, respectively.

Program:

```
LDP    #100H     ;  Load the data page pointer
LACC   0001, 10H ;  Load the higher accumulator with X1
                    ACC = X100
ADDS   0000      ;  Add the ACC with X0
                    ACC = X1X0
ADDS   0002      ;  Add the ACC with Y0
                    ACC = X1X0 + 00Y0
ADD    0003, 10H ;  Add the ACC with Y1 with shift
                    ACC = X1X0 + Y1Y0
SACL   0004      ;  Store the content of lower accumulator in dma 8004;
                      ACCL = Z0
SACH   0005      ;  Store the content of higher accumulator in dma 8005;
                      ACCH = Z1
H:     B
```

For example

Inputs	Outputs
8000–2439 (X_0)	8004–355F (Z_0)
8001–5523 (X_1)	8005–C6A8 (Z_1)
8002–1126 (Y_0)	
8003–7185 (Y_1)	

Example 6.2 To write a program for 64-bit addition.

Solution In this example, two 64-bit numbers X and Y will be added as follows:

$$\begin{array}{ll} X & \Longrightarrow X_3\ X_2\ X_1\ X_0 \\ Y & \Longrightarrow \underline{Y_3\ Y_2\ Y_1\ Y_0} \\ \{+\} & Z_3\ Z_2\ Z_1\ X_0 \end{array}$$

Program:

```
LDP   #100H     ; Load the data page pointer
                ; First 32-bit addition
LACC  0001, 10H ; Load the higher accumulator with X1
                  ACC = X100
ADDS  0000      ; Add the accumulator with X0
                  ACC = X1X0
ADDS  0004      ; Add the lower accumulator with Y0
                  ACC = X1X0 + 00Y0
ADDS  0005, 10H ; Add the higher accumulator with Y1
                  ACC = X1X0 + Y1Y0
SACL  0008      ; Store the content of lower accumulator (ACCL) in
                    dma 8008
                  ACCL = Z0
SACH  0009      ; Store the content of higher accumulator (ACCH) in
                    dma 8009
                  ACCH = Z1
                ; second 32-bit addition
LACC  0003, 10H ; load the higher accumulator with X3
                  ACC = X300
ADDC  0002      ; Add the accumulator with X2 and first 32 bit
                    addition CARRY
                  ACC = X3X2 + CARRY
ADDS  0006      ; ACC = X3X2 + 00Y2 + CARRY
ADD   0007, 10H ; ACC = X3X2 + Y3Y2 + CARRY
SACL  0010      ; Store ACCL in dma 8010;
                  ACCL = Z2
SACH  0011      ; Store ACCH in dma 8011;
                  ACCH = Z3

H:    B
```

For example

Inputs	Outputs
8000–1123 (X_0)	8008–6465 (Z_0)
8001–8279 (X_1)	8009–53CB (Z_1)
8002–A453 (X_2)	8010–2B99 (Z_2)
8003–CB21 (X_3)	8011–EE34 (Z_3)
8004–5342 (Y_0)	
8005–D152 (Y_1)	
8006–8745 (Y_2)	
8007–2312 (Y_3)	

Example 6.3 To write an assembly language program for 32-bit subtraction.

Solution In this example two 32-bit numbers X and Y will be subtracted as follows:

$$\begin{array}{rl} X & \Longrightarrow X_1\ X_0 \\ Y & \Longrightarrow \underline{Y_1\ Y_0} \\ \{-\} & \overline{Z_1\ Z_0} \end{array}$$

Program:

```
LDP    #100H     ; Load data page pointer
LACC   0001, 10H ; Load the higher accumulator with X1
                   ACC = X100
ADDS   0000      ; Add the accumulator with X0
                   ACC = X1X0
SUBS   0002      ; Subtract Y0 from ACC
                   ACC = X1X0 − −00Y0
SUB    0003, 10H ; Subtract Y1 from ACC with shift
                   ACC = X1X0 − −Y1Y0
SACL   0004      ; Store ACCL in dma 8004
                   ACCL = Z0
SACH   0005      ; Store ACCH in dma 8005
                   ACCH = Z1
```

For example

Inputs	Outputs
8000–7725 (X_0)	8004–2311 (Z_0)
8001–894A (X_1)	8005–4422 (Z_1)
8002–5414 (Y_0)	
8003–4528 (Y_1)	

Example 6.4 To write an assembly language program for 64-bit subtraction.

Solution In this example, two 64-bit numbers X and Y will be subtracted as follows:

$$\begin{array}{rl} X & \Longrightarrow X_3\ X_2\ X_1\ X_0 \\ Y & \Longrightarrow \underline{Y_3\ Y_2\ Y_1\ Y_0} \\ \{-\} & Z_3\ Z_2\ Z_1\ Z_0 \end{array}$$

Program:

LDP	#100H	;	Load the data page pointer
		;	First 32-bit subtraction
LACC	0001, 10H	;	Load the higher accumulator with X_1 ACC = $X_1 00$
ADDS	0000	;	Add the accumulator with X_0 ACC = $X_1 X_0$
SUBS	0004	;	Subtract the accumulator with Y_0 ACC = $X_1 X_0 - -00Y_0$
SUB	0005, 10H	;	Subtract Y_1 from accumulator with shift ACC = $X_1 X_0 - -Y_1 Y_0$
SACL	0008	;	Store ACCL in dma 8008 ACCL = Z_0
SACH	0009	;	Store ACCH in dma 8009 ACCH = Z_1
		;	second 32-bit subtraction
LACC	0003, 10H	;	load the higher accumulator with X_3 ACC = $X_3 00$
ADDS	0002	;	Add the accumulator with X_2 ACC = $X_3 X_2$
SUBB	0006	;	Subtract the Y_2 CARRY from ACC ACC = $X_3 X_2 - -00Y_2 - -$CARRY
SUB	0007, 10H	;	ACC = $X_3 X_2 - -Y_3 Y_2 - -$CARRY
SACL	0010	;	Store ACCL in dma 8010 ACCL = Z_2
SACH	0011	;	Store ACCH in dma 8011 ACCH = Z_3
H:	B		

For example

Inputs	Outputs
8000–894A (X_0)	8008–4422 (Z_0)
8001–7725 (X_1)	8009–2311 (Z_1)
8002–6525 (X_2)	8010–1223 (Z_2)
8003–BC13 (X_3)	8011–2411 (Z_3)
8004–4528 (Y_0)	
8005–5414 (Y_1)	
8006–5302 (Y_2)	
8007–9802 (Y_3)	

Example 6.5 To write a program for 32-bit integer multiplication.

Solution
Program:

```
LDP   #100H       ; Load data page pointer
LACC  #037AH, 0   ; Load the ACC with multiplicand
                    i.e. ACC = 0000037A
SACL  0000, 0     ; Store ACC in memory location 8000
                    8000 = 037A
LACC  #012EH, 0   ; Load the ACC with multiplier
                    i.e. ACC = 0000012E
SACL  0001, 0     ; Store ACCL in memory location 8001
                    8001 = 012E
LT    0000        ; Load the product (TREG) register with content of
                      memory location 8000
                    TREG = 037A
MPY   0001        ; Multiply the content of product register (TREG) with
                      content of memory location 8001 and the result is
                      stored in TREG
PAC               ; Move the content from product register to ACC
                    ACC = 000419EC
SACL  0002, 0     ; Store the content in ACCL memory location 8002
SACH  0003, 0     ; Store the content of ACCH in memory location 8003
H:    B  H
```

Outputs
8002 : 19EC
8003 : 0004

Example 6.6 To write a program to calculate the value of the function

$$Y = A * X_1 + B * X_2 + C * X_3$$

Solution
Program: The constants A, B, C, X_1, X_2 and X_3 are to be stored in dma from 8000 to 8005 as shown below

Data memory Address (DMA)	Data
8000	A
8001	B
8002	C
8003	X_1
8004	X_2
8005	X_3

```
LDP    #100H   ;  Load data page pointer
LACL   #0      ;  Clear the accumulator
LT     0000    ;  Load the T register with constant A; T = A
MPY    0003    ;  Multiply the T register with X1; T = AX1
LTA    0001    ;  Loads the T register from data memory and adds the
                    contents of T register with ACC and stores the result in
                    accumulator ACC = AX1; T = B
MPY    0004    ;  Multiply the T register with X2
                  T = BX2; ACC = AX1
LTA    0002    ;  T = C; ACC = AX1 + BX2
MPY    0005    ;  T = CX3; ACC = AX1 + BX2
APAC           ;  Adds the contents of T register with accumulator and stores
                    the result in accumulator ACC = AX1 + BX2 + CX3
SACL   0006, 0 ;  Store the ACCL in dma 8006.
```

Output

If $A = 1$, $B = 2$, $C = 3$, $X_1 = 4$, $X_2 = 5$ and $X_3 = 6$ gives 8006 : 0020

Note:

LTA: Instruction is equivalent to two instructions: LT and APAC.

LT: It loads the T register from data memory.

APAC: It adds the content of T register with the accumulator and stores the result in the accumulator.

Example 6.7 To write a program that explains the usage of RPT instruction.

Solution RPT instruction is an immediate addressing mode instruction. The 8-bit constant N specified by the RPT instruction is loaded into the repeat counter of 'C50. This causes the instruction following RPT instruction to be executed $N + 1$ times.

In this example, the value 4 is added 5 times continuously, and the result is stored in data memory 8000.

Program:

```
LDP    #100H
LACL   #4       ;  Load the lower accumulator with 4
SACL   0000, 0  ;  Store ACCL in dma 8000
LACL   #0       ;  Clear the ACC
RPT    #4       ;  Set repeat count is 4
ADD    0000, 0  ;  4 is added with ACC 5 times
                   ACC = 14
SACL   0000, 0  ;  Store ACCL in dma 8000
H:     B  H
```

Example 6.8 To find the two's complement of a given number.

Solution CMPL instruction replaces the content of the accumulator with its logical inversion i.e. one's complement. By adding one with its complement number, we will get the 2's complement of a number.

Program:

```
LDP     #100H
LACL    #7       ;  Load 7 to ACC
                    ACC = 0007
CMPL             ;  Complements the accumulator contents (1's complement)
                    ACC = FFF8
ADD     #1       ;  Add ACC with 1 (2's complement)
                    ACC = FFF9
SACL    0000, 0  ;  Store ACCL in dma 8000
H:      B  H
```

Example 6.9 To write a program for square waveform and sawtooth waveform generation.

Solution

Square Waveform Generation

```
  . MMREGS
  . TEXT
START:
        LDP     #1204
        LACC    #0H     ;  Load the ACC with lower amplitude
Loop:   SACL    0       ;  Store ACCL in dma
                           ACC = 0000
        RPT     #0FFH   ;  Frequency of the square wave
        OUT     0, 04   ;  Address for DAC
        CMPL            ;  Complements the ACC contents
                           ACC = FFFF
        B       loop    ;  Go to pma loop
.END
```

Sawtooth Waveform Generation

```
  . MMREGS
  . TEXT
START:
        LDP     #120H
        LACC    #0H     ;  Load the accumulator with lower amplitude
        SACL    0       ;  Store ACCL in dma
        OUT     0, 04H  ;  Address for DAC
Loop:   LACC    0       ;  Load the ACC with content in dma
        OUT     0, 04H  ;  Address for DAC
        ADD     #05h    ;  Change this value for frequency
        SACL    0       ;  Store ACCL in dma
        SUB     #0FFFh  ;  Change upper amplitude
        BCND    Loop    ;  LEQ
        B       START
.END
```

Example 6.10 To write a program for linear convolution

$$y(n) = x(n) * h(n).$$

Solution

Program:

```
        .text
        .mmregs
START:
        LDP     #0002H
        LAR     AR3, #0200H  ;  y(n) starting
        LAR     AR4, #0007   ;  N1 + N2 − 1 (length of linear
                                convolution sequence)
        LAR     AR1, #0100H  ;  x(n) data array
Loop:   MAR     *, AR1       ;  modify the auxiliary register
        LACC    *+
        SACL    050H         ;  starting of the scope of multiplication
        LAR     AR2, #0153H  ;  end of the array, to be multiplied
                                with h(n) {150 + N1 − 1}
        MAR     *, AR2
        ZAP
        RPT     #0003        ;  N1 − 1 times so that N1 times
        MACD    0C100H, *−
        APAC                 ;  to accumulate the final product sample
        MAR     *, AR3
        SACL    *+
        MAR     *, AR4
        BANZ    LOP, *−
        H: B    H
```

Input:

$x(n)$ data memory: 8100–0001
8101–0003
8102–0001
8104–0003

$h(n)$ data memory: C100–0001
C101–0002
C102–0001

Output:

Data memory location: 8200–0001
8201–0005
8202–0008
8203–0008
8204–0007
8205–0003

Example 6.11 To write a program for circular convolution

$$X_3(m) = x_1(n) \textcircled{N} x_2(n).$$

Solution

Program:

```
.MMREGS                              ; program initialisation
.Text
        LDP     #100H
        LACC    0000H                ; length of the input is given is 8000
        SUB     #0001H
        SACL    0001H
        LAR     AR0, 1H
        LAR     AR2, #0010H
Loop3:  LAR     AR1 #0060H           ; give the inputs x1(n) in AR1
        LAR     AR3 #0050H           ; give the inputs x2(n) in AR3
        LAR     AR4, 1H
        ZAP
 Loop:  MAR     #, AR3               ; multiply x1(n) and x2(n) and
                                       add the multiplication
        LT      *+, AR1              ; output
        MPY     *+
        SPL     5H
        ADD     5H
        MAR     *, AR4
        BANZ    Loop, *-, AR2        ; outputs of correlation are stored
                                       in AR2
        SACL    *+
        CALL    ROTATE
Loop2:  MAR     *, AR0
        BANZ    Loop3, *-
        H: B  H


ROTATE:
          LDP      #100H             ;      rotate the values of x2(n)
          LACC     0001H
          SUB      #1H
          SACL     0002H
          LACC     0050H
          SACB
          LAR      AR3 #8051H
          LAR      AR5 #8070H
          LAR      AR6 2H
  Loop1:  MAR      *, AR3
          LACC     *+, 0, AR5
          SACL     *+, 0, AR6
          BANZ     Loop1, *-
          LACB
          MAR      *, AR5
          SACL     *+
          LACC     #8070H
          SAMM     BMAR
          LAR      AR3, #0050H
          MAR      *, AR3
          RPT      0H
          BLOD     BMAR, *-
          RET
```

Input:

$x_1(n)$:	8000–0004
	8050–0002
	8051–0001
	8052–0002
	8053–0001
$x_2(n)$:	8060–0001
	8061–0002
	8062–0003
	8063–0004

Output:

$x_3(n)$:	8010–000E
	8011–0010
	8012–000E
	8013–0010

Summary

- DSP algorithms require extensive use of arithmetic operations such as multiplications and additions, and therefore the amount of data flow through the CPU is very high. Standard microprocessors do not possess the required hardware architecture and instruction set for the above purpose.
- To overcome the limitations of the standard microprocessors, digital signal processors are designed which include Harvard architecture, pipelining, dedicated hardware such as fast hardware multiplier–accumulator and shifters, fast internal memories and DSP connected special instructions.
- In the latest generation DSP processors, new architectures such as **Very Long Instruction Word (VLIW) and Static Super Scalar** are used. These architectures have multiple data paths and arithmetic units. Parallelism at the instruction level enhances the performance of the processors.
- Digital signal processors are classified as general-purpose and special-purpose processors. The functioning of general-purpose DSP processors is similar to standard microprocessors except that they have specially designed architecture and instruction sets. On the other hand, special-purpose processors are used to perform certain specific algorithms such as digital FIR filtering and for execution of application-dependent operations. While special-purpose processors are faster in execution, they are not as flexible as general-purpose processors.

Short Questions and Answers

1. **What are the classifications of digital signal processors?**
   The digital signal processors are classified into two categories. They are
   (i) General-purpose digital signal processors.
   (ii) Special-purpose digital signal processors.
2. **What is meant by general-purpose digital signal processors?**
   General-purpose digital signal processors are basically high-speed microprocessors with architecture and instruction sets optimized for DSP operation. Examples are
   (i) Fixed point processors such as TMS320C5x, TMS320C54x, ADSP-219x and ADSP-219xx.
   (ii) Float point processors such as TMS320C3x, TMS320C67x and ADSP-21xxxx.
3. **What is meant by special-purpose digital signal processor?**
   Special-purpose digital signal processors consist of hardware, designed for specific DSP algorithm such as FFT and designed for specific DSP applications such as PCM and filtering. Examples are
   (i) MT93001—Mitel's multichannel telephony voice echo canceller.
   (ii) P-DSP 16515A, TM-44 and TM-66—FFT processor.
   (iii) UPDSP 16256 and model 3092—programmable FIR filter.
4. **What are the factors that influence the selection of DSPs?**
   - Architectural features.
   - Execution speed.
   - Type of arithmetic.
   - Word length.
5. **What are the applications of P-DSPs?**
   The application of P-DSPs are digital cell phones, automated inspection, voice mail, motor control, video conferences, noise cancellation, medical imaging, speech synthesis, satellite communication, etc.
6. **What are the advantages and disadvantages of the VLIW architecture?**
   Advantages:
   - Increased performance.
   - Better compiler targets.
   - Potentially easier to program.
   - Potentially scalable.
   - Can add more execution units, and allow more instructions to be packed into the VLIW instruction.

Disadvantages:

- New kind of programmer/compiler complexity.
- Increased memory use.
- Program must keep track of instruction scheduling.
- High power consumption.
- Misleading MIP ratings.

7. **What is pipelining?**
   Pipelining a processor means breaking down its instructions into a series of discrete pipeline stages which can be completed in sequence by specialized hardware.
8. **What are the different stages in pipelining and explain?**
   (i) Fetch phase: The next instruction is fetched from the address stored in the program counter.
   (ii) Decode phase: Instruction in the instruction register is decoded and the address in the program counter is incremented.
   (iii) Memory read phase: It reads the data from the data buses and also writes data to the data buses.
   (iv) Execute phase: It executes the instruction currently in the instruction register and also completes the write process.
9. **What is pipeline depth?**
   The number of pipeline stages is referred to as the pipeline depth.
10. **What is the pipeline depth of TMS320C50 and TMS320C54x?**
    TMS320C50-4 and TMS320C54x-6.
11. **What are the different buses of TMS320C5x and other functions?**
    The TMS320C5x processor has four buses.
    (i) Program bus (PB): It carries the instruction code and immediately operates from program memory to the CPU.
    (ii) Program address bus (PAB): It provides an address to the program memory space for both read and write.
    (iii) Data read bus (DB): It interconnects various elements of the CPU to the data memory space.
    (iv) Data read address bus (DAB): It provides an address to access the data memory space.
12. **What are the addressing modes of TMS320C5x?**
    The addressing modes in TMS320C5x are
    (i) Immediate addressing.
    (ii) Indirect addressing.
    (iii) Register addressing.
    (iv) Memory-mapped register addressing.
    (v) Direct addressing.
    (vi) Circular addressing mode.

13. **What is the advantage of the Harvard architecture of TMS320 series?**
The Harvard architecture has two separate memories for their instruction and data. It is capable of simultaneously reading an instruction code and reading or writing a memory or peripheral.
14. **Differentiate between Von Neumann and Harvard architectures.**
In Von Neumann architecture, the CPU can be either reading an instruction or reading/writing data from/to memory. Both cannot occur at the same time since the instruction and data use the same signal pathways and memory, whereas the Harvard architecture has two memories for their instruction and data, requiring dedicated buses for each of them.
15. **State the merits and demerits of multi-ported memories?**
Merit: It increases the number of accesses/clock period. For example, in dual-ported memory, program and data memory can be accessed simultaneously.
Demerits: It requires a larger number of pins and larger chip area, and it is more expensive.
16. **List the various registers used with ARAV.**
   - Eight auxiliary registers (AR_0–AR_7).
   - Auxiliary register pointer (ARP).
   - Unsigned 16-bit ALU.
17. **What are the elements of control processing units of TMS320C5x?**
   - Central arithmetic logic unit (CALU).
   - Parallel logic unit (PLU).
   - Auxiliary register arithmetic unit (ARAU).
   - Memory-mapped registers.
   - Program controller.
18. **What is the function of parallel logic unit?**
The parallel logic unit is a second logic unit that executes a logic operation on data without affecting the contents of the accumulator.
19. **What are the arithmetic instructions of 'C5x?**
ADD, ADDB, ADDC, SUB, SUBB, MPY and MPYU.
20. **What are the logical instructions of 'C5x?**
AND, ANDB, OR, ORB, XOR and XORB.
21. **What are load/store instruction?**
LACB, LACC, LACL, LAMM, LAR, SACB, SACH, SACL, SAR and SAMM.
22. **What are the shift instructions?**
ROR, ROL, ROLB, RORB and BSAR.

Long Answer Type Questions

1. Explain how the Harvard architecture as used by the TMS320 family differs from the strict Harvard architecture. Compare this with the architecture of a standard Von Neumann processor.
2. A multiplier–accumulator, with three pipe stages, is required for a digital signal processor. Sketch a block diagram of a suitable configuration for the MAC. With the aid of a timing diagram, explain how the MAC works.
3. In relation to the DSP processor, explain SIMD and VLIW techniques. In each case, clearly point out the advantages and disadvantages of the technique in signal processing.
4. Explain the operation of CSSU and TMS320C54x and explain their use considering the viterbi operator.
5. Explain what is meant by instruction pipelining. Explain with an example, how pipelining increases the throughput efficiency.
6. Explain the operation of TDM serial ports in P-DSPs.
7. With a suitable diagram, describe the functions of the multiplier/adder unit of TMS320C54x.
8. Explain the function of auxiliary registers in the indirect addressing mode to point the data memory location.
9. Write a program to use the auxiliary register in the memory pointing and looping.
10. Write a program to compute the following equation:

$$Y = A * X_1 + B * X_2 + C * X_3$$

11. Write a program to perform addition of two 64-bit numbers.
12. Write short notes on

    (i) 32-bit accumulator.
    (ii) 16×16 bit parallel multiplier.
    (iii) Shifter.

13. Give the key features of the digital signal processor.
14. Explain the memory-mapped addressing mode used in P-DSPs.
15. What are the different ways in which the auxiliary register pointer can be updated in 5x?
16. Explain the immediate addressing mode of 'C5x with examples.
17. Explain the arithmetic instruction of 'C5x.
18. Write short notes on

    (i) Barrel shifter.
    (ii) Exponent encoder.
    (iii) Compare Select and Store Unit (CSSU).

Index

S. Palani, *Discrete Time Systems and Signal Processing*,
https://doi.org/10.1007/978-3-031-32421-5